# PHOTOCHEMISTRY OF ORGANIC COMPOUNDS

# Postgraduate Chemistry Series

A series designed to provide a broad understanding of selected growth areas of chemistry at postgraduate student and research level. Volumes concentrate on material in advance of a normal undergraduate text, although the relevant background to a subject is included. Key discoveries and trends in current research are highlighted and volumes are extensively referenced and cross-referenced. Detailed and effective indexes are an important feature of the series. In some universities, the series will also serve as a valuable reference for final year honours students.

## *Editorial Board*

## *Titles in the Series*

Protecting Groups in Organic Synthesis
James R. Hanson

Organic Synthesis with Carbohydrates
Geert-Jan Boons and Karl J. Hale

Organic Synthesis using Transition Metals
Roderick Bates

Stoichiometric Asymmetric Synthesis
Mark Rizzacasa and Michael Perkins

Catalysis in Asymmetric Synthesis (Second Edition)
Vittorio Caprio and Jonathan M. J. Williams

Reaction Mechanisms in Organic Synthesis
Rakesh Kumar Parashar

Forthcoming
Practical Biotransformations: A Beginner's Guide
Gideon Grogan

# PHOTOCHEMISTRY OF ORGANIC COMPOUNDS
## From Concepts to Practice

**Petr Klán**

*Department of Chemistry, Faculty of Science, Masaryk University*
*Czech Republic, klan@sci.muni.cz*

**Jakob Wirz**

*Department of Chemistry, University of Basel, Switzerland*
*J.Wirz@unibas.ch*

A John Wiley and Sons, Ltd, Publication

*Registered office*
John Wiley & Sons Ltd, The Atrium, Southern Gate, Chichester, West Sussex, PO19 8SQ, United Kingdom

For details of our global editorial offices, for customer services and for information about how to apply for permission to reuse the copyright material in this book please see our website at www.wiley.com.

*Library of Congress Cataloging-in-Publication Data*

Klán, Petr.
Photochemistry of organic compounds : from concepts to practice / Petr Klán, Jakob Wirz.
    p. cm.
Includes bibliographical references and index.
ISBN 978-1-4051-9088-6 (cloth : alk. paper) – ISBN 978-1-4051-6173-2 (pbk.: alk. paper)
1. Organic photochemistry. 2. Organic compounds–Synthesis. I. Wirz, Jakob. II. Title.
QD275.K53 2009
572'.435–dc22
2008044442

A catalogue record for this book is available from the British Library.

ISBN 978-1-405-19088-6 (H/B)
ISBN 978-1-405-16173-2 (P/B)

Typeset in 10/12pt Times by Thomson Digital, Noida, India.

*This book is dedicated to our families, teachers and students*

# Contents

# Special Topics

# Case Studies

# Foreword

This is a timely book, in several respects. First, photochemists and students of photochemistry badly need to replace the aging basic textbooks and reference books that they have been using, or at least to complement them with up-to-date material. Second, as illustrated by nearly any recent conference on photochemistry, most studies involving photochemistry are now performed by scientists whose initial training was not in photochemistry and who often do not think of themselves primarily as photochemists, but rely on light-induced transformations merely as a convenient tool. It is no wonder that these individuals ask for a comprehensive yet understandable and current introductory text. Third, like it or not, mankind is inexorably being dragged into an era in which the ability to generate energy in a sustainable manner acquires paramount importance, and this almost inevitably implies wise exploitation of solar energy. Who, if not chemists, chemical engineers, and material scientists, along with physicists and biologists, will come up with the new materials, catalysts, and processes that are needed for efficient, affordable, and environmentally benign capture of solar energy and its conversion into electricity and fuels? The numbers of researchers working in these areas will soon have to increase rapidly as nations all over the world finally begin to face reality and start to invest in solar energy research in a serious fashion. The demand for a good textbook covering photophysics and photochemistry and suitable for training increasing numbers of students is bound to grow.

The text written by Professors Klán and Wirz is masterful and well balanced. It fills the current need for a modern treatment of the subject admirably. This is not surprising, given the authors' vast experience and stature in the field. I was struck by the broad and comprehensive view of photochemistry and photophysics that they offer. I also very much appreciate their thoughtful ability to explain difficult concepts in a lucid way, accessible even to beginners. I suspect that it must have been tempting to gloss over some of the more difficult theoretical aspects of the subject, but if the authors were tempted, it does not show. They avoid glib statements of generalities and instead provide simplified yet accurate discussions, complemented with a truly impressive number of references to original literature. I like the organization, especially the inclusion of numerous case studies and special topics in a way that does not perturb the flow of the text, and I like the inclusion of problems and their selection. The graphical summary in Chapter 7 is particularly useful.

I congratulate the authors on their accomplishment and expect the book to be a tremendous success.

Josef Michl
*University of Colorado at Boulder*

# Preface

The absorption of light by matter opens a new dimension of physical and chemical processes. Photochemistry is intrinsically an interdisciplinary field pertaining to all natural sciences and many technical disciplines. This text aims to provide a hands-on guide for scientists in all fields, inspiring and aiding them to pursue their own research dealing with desired or unwanted effects associated with light absorption, as well as for undergraduate and graduate students of chemistry. We try to face rather than evade difficult aspects, yet to provide the reader with simple concepts and guidelines, sometimes to the point of terrible simplification.

Coverage of the relevant literature is extensive, with over 1500 references being provided, but it obviously cannot be comprehensive. Both current research (up to mid-2008) and early pioneering contributions are cited and recommended reading draws attention to outstanding, more profound treatments of special topics. We have shamelessly borrowed from excellent presentations in the literature and from our teachers, but have certainly missed some essential work as a result of our predilections. The text is interlaced with 32 *Special Topics* that provide information about some special applications of photochemistry in chemistry, physics, medicine, technology and in practical life, and about important photochemical processes that occur in the human body, green plants, the atmosphere and even deep space. Moreover, 39 *Case Studies* offer a glance at particular examples, often accompanied by detailed descriptions of laboratory procedures. Solved *Problems* at the end of the chapters and in some sections are intended as a practical aid to practice and refresh the readers' preceding studies.

The essentials of a quantum mechanical treatment of the interaction of electromagnetic radiation with molecules are summarized on a descriptive basis in the introductory Chapter 1, along with a few historical remarks. The basic concepts of photophysics and photochemistry and of energy transfer as well as a classification of photoreactions are presented in Chapter 2. Laboratory proceedings and equipment are described in Chapter 3, which includes quantum yield measurements, Stern–Volmer kinetics and the global analysis of spectral data. Chapter 4 provides simple but useful models describing electronic excitation and the associated profound changes of electronic structure, which are designed to help the reader to understand and undertake to predict photochemical reactivity. Chapter 5 describes exemplary cases of mechanistic investigations and time-resolved studies of important reactive intermediates such as carbenes, radicals and enols.

The extensive Chapter 6 is divided into eight sections dealing with the photochemistry of the most typical organic chromophores. The information is organized according to structural categories that are common in organic chemistry, such as alkanes, alkenes, aromatic compounds and oxygen atom-containing compounds, and emphasizes a visual presentation of the material. The sections are introduced by a brief description of the photophysical properties of the corresponding chromophores and their typical photoreactions. A list of recommended review articles and selected theoretical and computational photochemistry references is also provided. The compiled information of each section is then categorized by the mechanisms of photoreactions in individual subsections. The mechanism is first discussed in general terms; a number of examples follow, in which detailed reaction schemes of the mechanism are presented and discussed. Basic information about these particular reactions, such as the multiplicity of the reactive excited species, key reaction intermediates and chemical yields, is often provided and extensively referenced to the primary and secondary literature. The last section (6.8) focuses on the reactions of auxiliary chromophores such as photosensitizers, photocatalysts and photoinitiators.

Chapter 7, on retrosynthetic photochemistry, is a graphical compilation of reaction schemes, listed according to the target structures, which can be synthesized by the photochemical reactions described in Chapter 6.

## Acknowledgements

We are deeply indebted to our many colleagues who read parts of the text and drew our attention to its shortcomings by providing valued criticism: Silvia Braslavsky, Silvio Canonica, Georg Gescheidt, Richard S. Givens, Axel Griesbeck, Dominik Heger, Martin Jungen, Michael Kasha, Jaromír Literák, Ctibor Mazal, Josef Michl, Pavel Müller, Peter Šebej, Jack Saltiel, Vladimír Šindelář, Aneesh Tazhe Veetil and Andreas Zuberbühler.

Petr Klán
Jakob Wirz

# 1

# Introduction

## 1.1 Who's Afraid of Photochemistry?

Photochemistry has become an integral part of all branches of science: chemistry, biochemistry, medicine, biophysics, materials science, analytical chemistry, signal transmission, and so on. In our daily life, we are surrounded by products that are produced with the aid of photochemistry or that exploit photochemistry or photophysics to perform their function. Examples include information technology (computer chips and communication networks, data storage, displays, circuit boards and e-paper, precise time measurement), nanotechnology, sustainable technologies (solar energy storage, waste water cleaning), security and analytical devices (holograms, sensors), cosmetics (skin protection, hair colouring, etc.) and lighting (LEDs).

The highly reactive excited states and intermediates generated by pulsed excitation can be characterized by fast spectroscopic techniques. Novel instrumentation with rising sensitivity and spatial and temporal resolution is becoming available, ultimately allowing one to detect single molecules and to follow their reactions and their motions in space. To exploit this ever-increasing wealth of information, knowledge of photochemistry's arts and pitfalls will be required of researchers in many interdisciplinary fields. The high reactivities and short lifetimes of electronically excited molecules are responsible for their utmost sensitivity to pre-association in ground-state complexes, organized media and supramolecular structures, because their reactions are likely to occur with reaction partners that are in the immediate neighbourhood when light is absorbed.

The interaction of light and matter produces astounding effects and the outcome may be hard to predict. Will light do the desired trick for you? A well-trained chemist will have a reliable notion of the reactions that might result from the addition of, say, sodium borohydride to a solution of testosterone – but would he or she dare to predict or make an educated guess about what will happen when a solution containing testosterone is irradiated (Scheme 1.1)?

*Photochemistry of Organic Compounds: From Concepts to Practice*   Petr Klán and Jakob Wirz
Copyright © 2009 P. Klán and J. Wirz

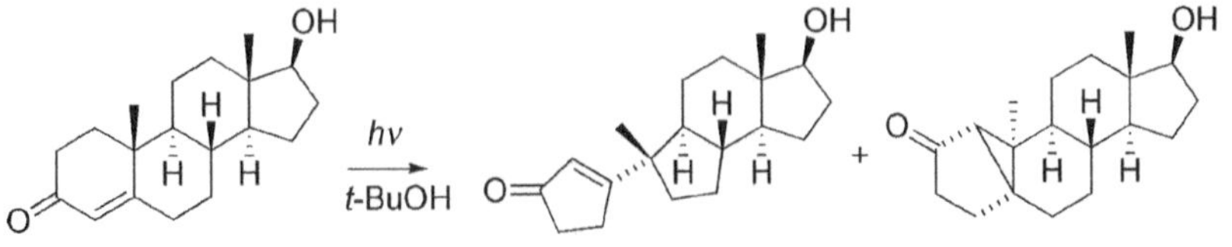

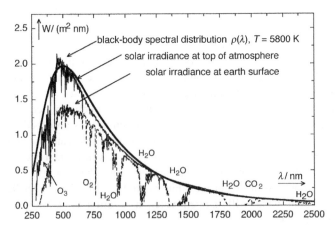

**Scheme 1.1**

Many synthetic chemists shy away from using key photochemical reaction steps[1] that might substantially reduce the number of reaction steps required to synthesize a desired product. This is bound to change with the increasing emphasis on green chemistry. The chemical reactivity of electronically excited molecules differs fundamentally from that in the ground state. In fact, the general rules and guidelines of photoreactivity are often the *opposite* of those in ground-state chemistry (recall the Woodward–Hoffmann rules as an example). Herein lies the great potential of synthetic photochemistry: in many cases, photochemistry achieves what ground-state chemistry cannot. Light adds a new dimension to chemistry. Photochemical synthesis on a gram scale is simple and requires only relatively inexpensive equipment.

Sunlight directly or indirectly (by providing food and fuel) drives most chemical transformations in biota. The spectral distribution of sunlight is close to that of a *black body* at a temperature of 5800 K (Figure 1.1, thick solid line). Outside the Earth's atmosphere, the integrated average power hitting a surface at right-angles to the propagation direction of solar irradiation (*solar constant*) is 1366 W m$^{-2}$. It varies by

**Figure 1.1**   Spectra of the solar radiation outside the Earth's atmosphere (—) and at ground level (---) compared with the spectrum of a black body at $T = 5800$ K (—). The spectra are plotted using the ASTM Standard G173-03e1, Standard Tables for Reference Solar Spectral Irradiances, by permission. The broad minima in the Earth's surface spectrum are due to absorption in the Earth's atmosphere by the substances indicated. The sharp lines in both spectra at lower wavelengths (Fraunhofer lines) are due to absorption by peripheral solar gas

several percent over the year with the Earth's distance from the Sun and with the Sun's activity. The total amount of radiation received by Earth is determined by its cross-section, $\pi R^2$. The average per surface area, $4\pi R^2$, is thus one-quarter of the solar constant, $342\,W\,m^{-2}$. However, the spectrum of the solar radiation measured at the Earth's surface, also shown in Figure 1.1 (dashed line), is modified by absorption and scattering while travelling through the atmosphere. The ozone layer of the stratosphere absorbs strongly below approximately 280 nm (which is a biologically hazardous spectral region denoted UVC radiation), so that only UVB (280–315 nm), UVA (315–400 nm) and visible radiation penetrate. Water, carbon dioxide and dioxygen molecules selectively absorb in the visible and especially in the infrared region. Latitude, time of day, seasonal variations, variations in the ozone layer, clouds, fumes, haze, and so on further reduce the solar radiation received at ground level. Thus, the global average on the Earth's surface comes to about $200\,W\,m^{-2}$. For comparison, the metabolism of an adult human consumes about $9000\,kJ\,d^{-1}$, producing about 100 W as heat.

*Photosynthesis* (Scheme 1.2) is the source of food and fossil fuel and has played an important role in the origin of life.[a] A fully grown beech tree assimilates about $10\,m^3$ of $CO_2$ (contained in $4 \times 10^4\,m^3$ of air) on a sunny day, producing the same amount of $O_2$ and 12 kg of carbohydrates. Globally, photosynthesis stores about $2 \times 10^{12}\,t$ of $CO_2$ in biomass per year, the equivalent of $2 \times 10^{19}\,kJ$, most of which, however, is eventually returned to the atmosphere by reoxidation.

$$H_2O\ (l) + CO_2\ (g) \xrightarrow{h\nu} (CH_2O)\ (s) + O_2\ (g),\ \Delta H^\circ(298\ K) = +467\ kJ\ mol^{-1}$$

**Scheme 1.2** *Photosynthesis*

Solar radiation also plays a major role in abiotic environmental photochemistry, taking place mostly in surface waters. However, the present state of knowledge about these processes is limited, because individual photosensitive substances are present at very high levels of dilution. On a cloudless summer noon, surface waters in central Europe receive approximately $1\,kW\,m^{-2}$ of sunlight or about 2 mol of photons per hour between the wavelengths of 300 and 500 nm, the region of interest for photochemical reactions. About 1300 times this dose is accumulated per year.[2]

Solar radiation is becoming increasingly appreciated for its influence on living matter and as a source of natural energy. The current human energy consumption[3] of $4.6 \times 10^{20}\,J$ per year or $1.4 \times 10^{13}\,W$ amounts to about 0.015% of the solar energy reaching the Earth's surface or 2.4% of natural photosynthetic energy storage. To manage the rising problem of global warming and, at the same time, to address the gap in energy supply created by the expected decrease in carbon fuel consumption is one of the 21st century's main challenges.[4–6] Even under the optimistic assumption that the total consumption will level

---

[a] Although Charles Darwin was coy about the origin of life, he famously speculated in a letter to a friend about a 'warm little pond', in which all manner of chemical substances might accumulate over time. About 100 years later, Miller and Urey shot flashes of lightning through a flask containing an 'ocean' of liquid water and an 'atmosphere' of hydrogen-rich gases, methane, ammonia and hydrogen sulfide. Light was added in later experiments. They identified all sorts of compounds in the resulting 'primordial soup', including various amino acids.

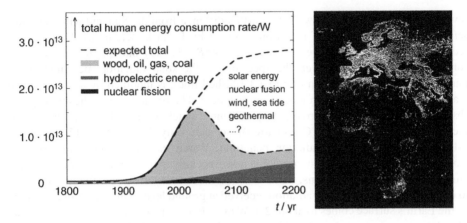

**Figure 1.2** Left: human energy consumption rate. Current per capita rate ≈2300 W (EU 6000 W, USA 12 000 W). Right: night-time satellite picture of Africa and Europe showing the distribution of artificial light sources

off, as shown in Figure 1.2 (left) (notwithstanding the developing nations' expected growing energy demand; Figure 1.2, right), it is far from clear how the gap will be filled when the sources of fossil fuels are exhausted, and whether the demand that their consumption be strongly curtailed to reduce the rate of global warming can be met. Uranium ores worth exploiting are also limited, leaving few conceivable options for substantial contributions through sustainable technology, primarily solar energy storage (photovoltaics, hydrogen production), nuclear fusion, wind energy, sea tide and geothermal sources. Major efforts to increase the efficiency of all processes requiring energy and to find substitutes for energy-demanding behaviour will be needed to balance the worldwide energy demand against our non-renewable resources during this century. Light-driven processes will be a major contributor in the transformation to a low-carbon economy.[7] The production of photovoltaic cells is currently rising at a pace of about 40% per year. More dramatic increases in the exploitation of renewable energy sources will follow as soon as these techniques can compete with the rising cost of fossil fuels.

A remarkably detailed understanding of the properties and chemical behaviour of transient intermediates has been deduced from well-designed conventional methods of physical organic chemistry using standard laboratory techniques such as quenching and sensitization, trapping or radical clocks. However, working backwards from the structure of stable photoproducts and the variation of their yield as a function of various additives requires demanding preparative and analytical efforts; moreover, the lifetimes of intermediates can only be estimated from such data by making some assumptions about the rate constants of quenching or trapping (Section 3.9.6). On the other hand, flash photolysis (Section 3.7) provides absolute rate constants with little effort, but yields precious little hard information allowing the identification of the observed transients. Kinetic data obtained by optical flash photolysis alone are thus prone to false assignments and the combination of both methods is highly recommended. Once a hypothetical reaction mechanism has been advanced, the quantitative comparison of the effect of added reagents on product distributions, quantum yields and transient kinetics provides a stringent test for the assignment of the transients observed by flash photolysis (Section 3.9.8, Equation 3.38). Other time-resolved spectroscopic techniques such as MS, IR, Raman, EPR, CIDNP and X-ray diffraction provide detailed structural information permitting unambiguous assignment of transient intermediates and their chemical and physical properties can now be determined under most conditions.

---

### Special Topic 1.1: Historical remarks

Many of the above considerations regarding the impact of photochemistry on the future of mankind were expressed 100 years ago in prophetic statements by Giacomo Luigi Ciamician (1857–1922).[8,9] The first attempts to relate the colour of organic compounds to their molecular structure date back to the mid-19th century, when synthetic dyes became one of the chemical industry's most important products. In 1876, Witt introduced the terms *chromophore* (a molecular group that carries the potential for generating colour) and *auxochrome* (polar substituents that increase the depth of colour). Dilthey, Witzinger and others further developed the basic model. Such colour theories had to remain empirical and rather mystical until the advent of molecular quantum mechanics after 1930, and it took another 20 years until simple molecular orbital theories such as Platt's free electron model (FEMO), Hückel molecular orbital theory (HMO) and Pariser, Parr and Pople's configuration interaction model (PPP SCF CI) provided lucid and satisfactory models for the interpretation of electronic spectra of conjugated molecules. Such models were also successful in rationalizing trends in series of related molecules and in describing the electronic structure and reactivity of excited states.

More sophisticated *ab initio* methods can and will increasingly provide accurate predictions of excited-state energies, transition moments and excited-state potential energy surfaces of fairly large organic molecules. However, these methods are less amenable to generalization, intuitive insight or prediction of substituent effects. The translation of quantitative *ab initio* results into the language of simple and lucid MO theories will remain a necessity.

General lines of thought on how to understand the reactivity of electronically excited molecules emerged only after 1950. T. Förster, M. Kasha, G. Porter, E. Havinga, G. Hammond, H. Zimmerman, J. Michl, N. Turro and L. Salem were among the intellectual leaders who developed the basic concepts for structure–reactivity correlations in photochemistry (Section 4.1). Spectroscopic techniques together with computational methods began to provide adequate characterization of excited states and their electronic structure. Simple models such as correlation diagrams were used for the qualitative prediction of potential energy surfaces. Matrix isolation at cryogenic temperatures permitted the unambiguous identification of reactive intermediates.

The rapid development of commercially available lasers and electronic equipment allowed for the real-time detection of primary transient intermediates by flash photolysis. Photochemistry thus emerged as the principal science for the study of organic reaction mechanisms in general, because it can be employed to generate the intermediates that are postulated to intervene in chemical reactions of the ground state. Cutting edge research provides unprecedented spatial and temporal resolution to monitor structural dynamics with atomic-scale resolution, to detect single enzyme or DNA molecules at work and to construct light-driven molecular devices.

**Figure 1.3** Some of photochemistry's icons. Top (from left to right): Giacomo Ciamician (1857–1922), Theodor Förster (1910–1974), Michael Kasha (1920–). Bottom: George Hammond (1921–2005), George Porter (1920–2002), Ahmed Zewail (1946–). Photographs reproduced by permission of the Scientists (M. K., A. Z.), or of successors at their former institutions

## Special Topic 1.2: Quantity calculus

By convention, physical quantities are organized in the International System (SI) of quantities and units, which is built upon seven *base quantities* (Table 1.1), each of which is regarded as having its own *dimension*. The current definitions of the corresponding *base units* are given in the IUPAC Green Book, *Quantities, Units and Symbols in Physical Chemistry*.[10] A clear distinction should be drawn between the names of units and their symbols, e.g. mole and mol, respectively.

**Table 1.1**  *The SI base quantities*

| Base quantity | | | SI unit | |
| --- | --- | --- | --- | --- |
| Name | Symbol for quantity | Symbol for dimension | Name | Symbol |
| Length | $l$ | L | metre | m |
| Mass | $m$ | M | kilogram | kg |
| Time | $t$ | T | second | s |
| Electric current | $I$ | I | ampere | A |
| Thermodynamic temperature | $T$ | $\Theta$ | kelvin | K |
| Amount of substance | $n$ | N | mole | mol |
| Luminous intensity | $I_v$ | J | candela | cd |

A *measurement* amounts to the comparison of an object with a reference quantity of the same *dimension* (e.g. by holding a metre stick to an object). The value of a *physical quantity Q* can be expressed as the product of a *numerical value* or *measure*, $\{Q\}$, and the associated *unit* $[Q]$, Equation 1.1.

$$Q = \{Q\}[Q], \; e.g., \; h = 6.626 \times 10^{-34} \, \text{J s}$$

**Equation 1.1**

The symbols used to denote units are printed in roman font; those denoting physical quantities or mathematical variables are printed in italics and should generally be single letters that may be further specified by subscripts and superscripts, if required. The unit of any physical quantity can be expressed as a product of the SI base units, the exponents of which are integer numbers, e.g. $[E] = \text{m}^2 \, \text{kg s}^{-2}$. Dimensionless physical quantities, more properly called quantities of dimension one, are purely numerical physical quantities such as the refractive index $n$ of a solvent. A physical quantity being the product of a number and a unit, the unit of a dimensionless quantity is also one, because the neutral element of multiplication is one, not zero.

*Quantity calculus*, the manipulation of numerical values, physical quantities and units, obeys the ordinary rules of algebra.[11] Combined units are separated by a space, e.g. $\text{J K}^{-1} \text{mol}^{-1}$. The ratio of a physical quantity and its unit, $Q/[Q]$, is a pure number. Functions of physical quantities must be expressed as functions of pure numbers, e.g. $\log(k/\text{s}^{-1})$ or $\sin(\omega t)$. The *scaled quantities* $Q/[Q]$ are particularly useful for headings in tables and axis labels in graphs, where pure numbers appear in the table entries or on the axes of the graph. We will also make extensive use of scaled quantities $Q/[Q]$ in practical 'engineering' equations. Such equations are very convenient for repeated use and it is immediately clear, which units must be used in applying them. For example, a practical form of the ideal gas equation is shown in Equation 1.2.

$$p = 8.3145 \frac{n}{\text{mol}} \frac{T}{K} \left(\frac{V}{\text{m}^3}\right)^{-1} \text{Pa}$$

**Equation 1.2**

If the pressure is required in psi units, the equation can be multiplied by the conversion factor $1.4504 \times 10^{-4}$ psi/Pa $= 1$ to give Equation 1.3.

$$p = 1.206 \times 10^{-3}\, \frac{n}{\text{mol}}\, \frac{T}{\text{K}}\, \left(\frac{V}{\text{m}^3}\right)^{-1}\, \text{psi}$$

**Equation 1.3**

In this text, the symbols $H^+$ and $e^-$ generally used by chemists are adopted as symbols for the *proton* and *electron*, rather than p and e, respectively, as recommended by IUPAC. The symbol $e$ represents the elementary charge; the charge of the electron is $-e$, that of the proton is $e$. In chemical schemes these charges will be represented by $\ominus$ and $\oplus$, respectively. Some fundamental physical quantities and energy conversion factors are given in Tables 8.1 and 8.2.

## 1.2   Electromagnetic Radiation

We are dealing with the interactions of light and matter. The expression 'light' is used here somewhat loosely to include the near-ultraviolet (UV, $\lambda = 200$–$400$ nm) and visible regions (VIS, $\lambda = 400$–$700$ nm) of the entire electromagnetic spectrum (Figure 1.4), which spans over 20 orders of magnitude.

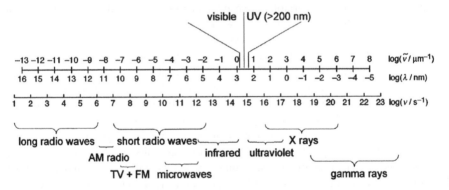

**Figure 1.4**   Electromagnetic spectrum

*The first law of photochemistry* (Grothus, 1817; Draper, 1843) states that only absorbed light is effective in photochemical transformation.

*The second law of photochemistry* (Einstein, 1905) states that light absorption is a quantum process. Usually, one photon is absorbed by a single molecule.

The particle nature of light was postulated in 1905 by Einstein to explain the *photoelectric effect*. When light is incident on a metal surface in an evacuated tube, electrons may be ejected from the metal. This is the operational basis of photomultipliers and image intensifiers, which transform light to an amplified electric signal (see Section 3.1).

The kinetic energy of these electrons is independent of the light intensity. This surprising result was not understood until Einstein proposed that light energy is quantized in small packets called *photons*. The photon is the quantum of electromagnetic energy, the smallest possible amount of light at a given frequency $\nu$. A photon's energy is given by the *Einstein equation* (Equation 1.4), where $h = 6.626 \times 10^{-34}$ J s is Planck's constant.[b]

$$E_p = h\nu$$

**Equation 1.4**

Einstein made the bold prediction that photons with an energy below that needed to remove an electron from a particular metal would not be able to eject an electron, so that light of a frequency below a certain threshold $\nu_{th}$ would not give rise to a photoelectric effect, no matter how high the light intensity. Moreover, he predicted that a plot of the photoelectrons' maximum kinetic energy, $E_{max}$, against the frequency of light $\nu$ should be a straight line with a slope of $h$ (Figure 1.5). R. A. Millikan verified this 10 years later.[12]

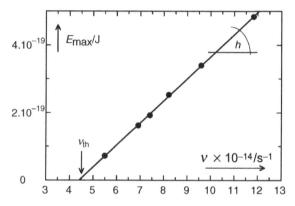

**Figure 1.5** Maximum kinetic photoelectron energy $E_{max}$ ejected from sodium metal as a function of light frequency. Adapted from ref. 12

The frequency $\nu$ of light is inversely proportional to its wavelength $\lambda$, $\nu = c/\lambda$, where $c = 2.998 \times 10^8$ m s$^{-1}$ is the speed of light; it is often replaced by the wavenumber, $\tilde{\nu} = \nu/c = 1/\lambda$, which corresponds to the number of waves per unit length. The SI base units are $[\nu] = s^{-1}$ and $[\tilde{\nu}] = m^{-1}$. We will generally use the derived units cm$^{-1}$ ($= 100$ m$^{-1}$) for wavenumbers of vibrational transitions and $\mu$m$^{-1}$ ($= 10\,000$ cm$^{-1}$) for wavenumbers of electronic transitions.

The energy transferred to a molecule by the absorption of a photon is $\Delta E = h\nu = hc\tilde{\nu}$. The energy of 1 mol of photons (1 einstein), $\lambda \approx 300$ nm, amounts to $N_A E_p = N_A hc\tilde{\nu} \approx 400$ kJ mol$^{-1}$ and is sufficient for the homolytic cleavage of just about any single bond in organic molecules. For example, a similar amount of energy would be taken up by

---

[b] The constant $h$ was introduced in 1899 by Max Planck to derive a formula reproducing the intensity distribution of a *black-body radiator* (Section 2.1.3). To this end, Planck had to assume that a hot body emits light in quanta of energy $h\nu$, but he considered this assumption to be an amazing mathematical trick rather than a fundamental property of nature.

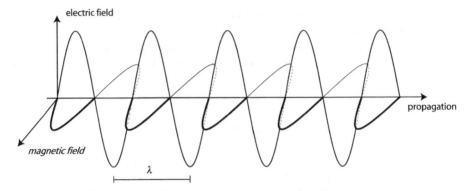

electric field

propagation

*magnetic field*

$\lambda$

**Figure 1.6**  Electromagnetic wave along a propagation axis $x$. Note the scaling parameter $\lambda$. For $\lambda = 300$ nm, a full wave spans a length that is more than two orders of magnitude larger than that of an average molecule. The propagation axis could be replaced by time $t$ to indicate the electromagnetic wave's oscillations at a given point in space. For a light wave of $\lambda = 300$ nm, the scale shown would then represent a time span of $10^{-15}$ s

naphthalene $[C_{p,m}(g) = 136\,\mathrm{J\,K^{-1}\,mol^{-1}}]$ if it were immersed in a heat bath at 3000 K. This might suggest that organic molecules are indiscriminately destroyed by irradiation with UV light. Fortunately, this is not the case and we shall see why. The physical description of light is mind-boggling. Classical optics can be fully 'understood' by mathematically treating light beams as electromagnetic waves (Figure 1.6).

Other phenomena are best described in terms of light's particle nature (Equation 1.4). These seemingly contradictory properties are inseparable parts of the dual nature of light. Both must be taken into account when considering a simple process such as the absorption of light by matter. The above statements will surprise few readers because they have heard them many times before. But consider the experiment depicted in Figure 1.7.

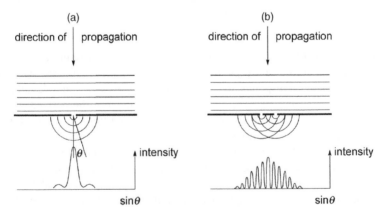

(a)

direction of | propagation

intensity

$\sin\theta$

(b)

direction of | propagation

intensity

$\sin\theta$

**Figure 1.7**  Parallel waves (representing either light or water waves in a ripple tank) encountering a barrier with (a) one small slit producing diffraction and (b) two small slits whose separation is equal to 10 times their width, producing an interference–diffraction pattern

The diffraction patterns generated by parallel waves encountering a barrier with (a) one or (b) two holes (Figure 1.7) are predicted by the wave theory of light. The peaks and troughs appearing in the intensity distribution of the right-hand diagram with two slits are due to constructive and destructive interference of the diffracted light beams emerging from the two slits. What happens when the distant light source's intensity is reduced so much that only one photon arrives at the barrier per second? Now the photons cannot interact with each other while passing the barrier and they will be registered one at a time when they hit the detector screen. However, if we wait long enough to obtain good statistics on the intensity distribution, we will get exactly the same pattern of peaks and troughs as with a strong light source. Thus a single photon approaching barrier (b) in Figure 1.7 near the two holes has wave-like properties and will exhibit interference 'with itself' while passing one of the barrier's slits![c] In Section 1.4, we shall see that electrons, which in classical physics can be treated as particles carrying the elementary charge $-e$, $e = 1.602 \times 10^{-19}$ C, also exhibit wave-like properties.

## 1.3 Perception of Colour

*Recommended review.*[13]

*Vision* is initiated by the absorption of light in the eye's retina, which hosts about 5 million cone cells and 90 million rod cells (Special Topic 6.1). Humans have three kinds of cones with different response curves. The cones are less sensitive to light than the rods (which support vision at low light levels), but they allow the perception of colour. Moreover, their response times to stimuli are faster than those of rods. An attempt to chart colours in a triangle of three primary light sources goes back to Maxwell. The modified diagram shown in Figure 1.8 was defined in 1931 by the International Commission on Illumination. Pure spectral colours, those seen by looking at a monochromatic light source, are located on the curved solid line from violet (400 nm) to red (700 nm). The straight line connecting 380 and 700 nm at the bottom represents the non-spectral colours of purple that can be obtained by mixing violet and red light.

Any colour perceived by direct viewing of a light source corresponds to a point in the area enclosed by the curved line and the tilted bottom line. Following the dotted line in Figure 1.8, we would proceed from spectral (or saturated) blue (480 nm) through pale blue, white and pale yellow to saturated yellow (580 nm). A pair of colours that can produce the sensation of 'white', i.e. which can be joined by a straight line that passes through the white point, is called a *complementary pair*. Thus blue and yellow form a complementary pair, as do orange (600 nm) and blue–green (488 nm), also called 'cyan'. Such colours are called *additive colours*. Physically, spectral yellow (580 nm) is clearly different from a mixture of red (700 nm) and green light (520 nm). However, both stimulate our eyes in a similar manner, so we do not notice that difference. The additive reproduction process usually uses red, green and blue light (RGB) to produce the other colours. Combining one of these additive primary colours with another in equal amounts produces the additive secondary colours cyan, magenta and yellow (CMYK, K for black). Combining all three

---

[c] 'I am afraid I can't put it more clearly', Alice replied very politely, 'for I can't understand it myself to begin with' (Lewis Carroll, *Alice in Wonderland*). 'We all *know* what light is, but it is difficult to *tell* what it is.' (Samuel Johnson).

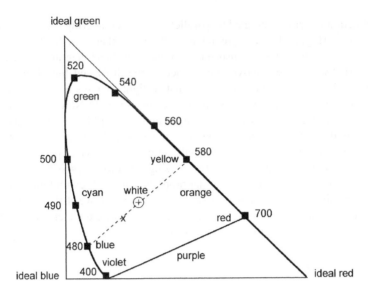

**Figure 1.8** Colour 'triangle'. Any perceived colour is represented by a point in the area enclosed by the curved line. The wavelengths of pure spectral colours are shown in nm

primary colours in equal intensities produces white. Computer monitors and television screens are common applications of additive colour mixing.

The dominant wavelength of a perceived colour (say, of point x in the diagram) is that reached on the curved line by drawing a straight line from the white point through x (480 nm). The colour associated with this dominant wavelength is called the *hue*. The amount of hue that makes up the light's composition is known as *saturation*, i.e. the colours on the curved line are fully saturated. As the dominant wavelength or hue is diluted with white light, saturation decreases. All of the above considerations are independent of the intensity of the light.

Now consider the colours perceived by diffuse reflection from a dyed surface that is irradiated by white light. If the dye absorbs at, say, 480 nm (blue light), the colour perceived is the complementary one, yellow (white light minus blue). The colours produced by paints, dyes, inks and so on are called *subtractive colours*. Because there is no spectral colour 'purple', no dye absorbing at a single wavelength can produce the colour green as its complementary pair. A green dye such as chlorophyll has two absorption bands, one near 450 nm and another near 700 nm.

Examples of colours with low saturation are the blue sky and sunlight's reddish colour at sunset. These colours are due to *Rayleigh scattering* of sunlight (Figure 1.1) in the atmosphere, which arises from fluctuations in the density of air and also from nanometre-sized particles. The scattering probability varies as $1/\lambda^4$. Tangential sunlight at dusk and dawn is thus depleted of short wavelengths (blue) and is reddish. Conversely, sunlight scattered by the atmosphere is enriched in short wavelengths so that the sky appears blue.

Vision is a very complex physiological process. The retina is not just the host of the primary light receptors, the rods and cones. Behind them is a complex system of nerves that perform processing tasks before transmitting the light signals to the brain. For example, the colour perceived by viewing a dyed surface that is illuminated with different light sources remains largely the same, although the spectrum of the reflected light may differ widely, depending on the light source. Because this correction of colour perception is not perfect, you will want to see new clothes outdoors before you buy them.

## 1.4 Electronic States: Elements of Molecular Quantum Mechanics

*Recommended textbooks and review articles.*[14–20]

This section will give a bird's-eye view of the elements of quantum chemistry that are needed to deal with electronic spectroscopy, photophysical processes and primary photoreactions. It can be skipped by readers knowledgeable in quantum mechanics and may be hard going for those who are not. Most of the following chapters will not require a deep understanding of these concepts. Nevertheless, readers without a solid background in quantum mechanics should find it rewarding to grasp the take-home lessons that are given here and will be referred to later. For more detailed treatments of theories and models of computational chemistry,[14] computational photochemistry,[15–17] and quantum mechanics in general,[18] the reader is referred to specialized textbooks. Simple quantum chemical models to describe electronically excited states and their reactivity will be discussed in Chapter 4.

The stationary states of a molecule are limited to a set of discrete energy levels $E$, which are defined by the Schrödinger equation:

$$\hat{H}\Psi = E\Psi$$

**Equation 1.5** *Schrödinger equation*

The *wavefunction* $\Psi = \Psi(q_{el}, q_{nucl})$ depends on the coordinates of the electrons, $q_{el}$, and of the nuclei, $q_{nucl}$, in the molecule (three coordinates per particle). It has the mathematical properties of a wave and exhibits interference when interacting with other waves. The quantum mechanical *Hamiltonian operator* $\hat{H}$ is a differential operator,[d] a mathematical prescription that is to be exerted upon the wavefunction. The dominant terms of $\hat{H}$ determine (*i*) the kinetic energies of the electrons and nuclei by differentiating the wavefunction twice with respect to its coordinates ($-\partial^2/\partial x_i^2$) and (*ii*) the potential energies due to the Coulomb potentials surrounding all charged particles, namely the negative potential energies due to the attractive forces between the electrons and the nuclei ($-e^2 Z_i/(4\pi\varepsilon_0 r_{ij})$) and the positive ones due to the repulsive forces between the electrons ($e^2/(4\pi\varepsilon_0 r_{ij})$) and between the nuclei ($e^2 Z_i Z_j/(4\pi\varepsilon_0 r_{ij})$); $Z_i$ and $Z_j$ are the nuclear charges and $r_{ij}$ the distances between particles $i$ and $j$. These pair-wise attraction and repulsion terms reduce to $-Z_{ij}/r_{ij}$, $1/r_{ij}$ and $Z_i Z_j/r_{ij}$, respectively, in atomic units (Table 8.3). They imply that no particle moves independently of the others.

A wavefunction $\Psi$ is a proper solution (called *eigenfunction*) of Equation 1.5, if the function $\Psi$ is reproduced unchanged by the mathematical operation $\hat{H}\Psi$ except for the constant factor $E$ called the *eigenvalue*, which is the energy of an allowed state. There are many solutions of Equation 1.5 for a given system of electrons and nuclei, each one associated with an eigenvalue $E$. The well-known lowest-energy eigenvalues for the hydrogen atom and their associated eigenfunctions, called *atomic orbitals* (AOs), are

---

[d] The circumflex ('hat', ^) serves to distinguish an operator from an algebraic quantity.

*Introduction*

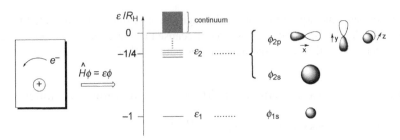

**Figure 1.9** Energies and atomic orbitals of a hydrogen atom. $R_H = 13.6\,eV$

shown in Figure 1.9. The symbol $\phi$ is used instead of $\Psi$ to designate AO wavefunctions and $\varepsilon$ is used instead of $E$ for one-electron orbital energies.

Exact eigenfunctions $\Psi$ are known only for hydrogen-like atoms (atoms or ions with only one electron). For atoms with more than one electron, let alone for molecules of chemical interest, the Schrödinger Equation 1.5 cannot be solved to yield exact wavefunctions. However, powerful approximation methods allow us to determine the energy levels of such systems and the associated wavefunctions to a high degree of accuracy. Electrons being much lighter than nuclei, $m_H/m_e \approx 1800$, they move much more rapidly, much like the fleas near a grazing cow. The *Born–Oppenheimer (BO) approximation* neglects the correlation between nuclear and electronic motion. Electronic distributions and energies are calculated for an arbitrary selection of fixed nuclear positions and potential energy surfaces (PESs) are constructed point by point by repeating this procedure for different molecular structures. As an example, Figure 1.10 shows the potential energy curve for a hypothetical diatomic molecule with a minimum at

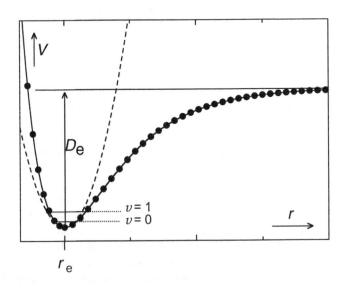

**Figure 1.10** Potential energy diagram for a diatomic molecule. The dashed line is the harmonic approximation to the potential energy that is often used to determine vibrational wavefunctions and energies near the bottom of the potential well

bond length $r_e$ and a well depth $D_e$. The so-called potential energy includes not only the potential energy of the nuclei and electrons, but also the kinetic energy of the latter. The total energy including nuclear kinetic energy is represented by horizontal lines ($v = 0$ for the lowest vibrational level). The PESs of molecules with more than two nuclei, $N > 2$, are difficult to visualize. They are frequently shown as two- or, at best, three-dimensional cuts along one or two 'relevant' internal coordinates through their $(3N - 6)$-dimensional space. This should always be kept in mind, because such pictures can be misleading.

As a result of the BO approximation, wavefunctions can be factored into the product of an electronic wavefunction, $\Psi'_{el}$, and a vibrational wavefunction, $\chi$. Rotational states are not considered here, because they are not resolved in solution spectra.

$$\Psi \approx \Psi'_{el}(q_{el}, q_{nucl})\chi(q_{nucl})$$

**Equation 1.6**   *Separation of electronic and vibrational wavefunctions*

The vibrational wavefunction depends only on the nuclear coordinates $q_{nucl}$. The nuclei move in a fixed electronic potential that is often approximated as a harmonic potential (dashed curve in Figure 1.10) to determine vibrational wavefunctions near the potential's bottom. The electronic wavefunction carries the complete information about the motion and distribution of electrons. It still depends on both sets of coordinates $q_{el}$ and $q_{nucl}$, but the latter are now fixed parameters rather than independent variables; the nuclei are considered to be fixed in space. The BO assumption is a mild approximation that is entirely justified in most cases. Not only does it enormously simplify the mathematical task of solving Equation 1.5, it also has a profound impact from a conceptual point of view: The notions of *stationary electronic states* and of a PES are artefacts of the BO approximation. It does break down, however, when two electronic states come close in energy and it must be abandoned for the treatment of radiationless decay processes (see Section 2.1.5).

Magnetic fields in molecules arise from three sources: the motion of charges in space, electronic spin and nuclear spin. The last two are fundamental properties of electrons and some nuclei, which have no classical analogy. So far, we have ignored the magnetic interactions operating in molecular systems altogether. They were not even mentioned as contributions to the potential energy part of the Hamiltonian operator (Equation 1.5), because they are much weaker than the Coulombic terms dealt with. However, magnetic interactions are the only driving force for photophysical processes that involve a change in *multiplicity*, so they must be considered explicitly to deal with such processes (Section 4.8). Moreover, we shall see below that the symmetry of electronic spin functions must always be treated explicitly, because of its strong impact on the distribution of electrons via the *Pauli principle*.

In classical mechanics, the angular momentum $l$ of a mass point running on a circular orbit is defined by the *vector*[e] $l = r \times p$ (Figure 1.11), which has an orientation in space that

---

[e] Vectors have a magnitude and a direction; in this text, the symbols for vector quantities (and matrices, Section 3.7.5) are printed in bold italics. In a Cartesian coordinate system, a vector $a$ is fully defined by its components $a_x$, $a_y$ and $a_z$, the projections of $a$ onto the Cartesian axes. The length or magnitude of $a$ is given by the Pythagorean expression $a = |a| = (a_x^2 + a_y^2 + a_z^2)^{1/2}$. Two vectors of equal length and direction are equal, independent of their position in space. The sum (or difference) of two vectors is defined by the sum (or difference) of their components, $a \pm b = c$, $c_x = a_x \pm b_x$, etc. The scalar product of two vectors is designated $ab$ and is a scalar; $ab = ab\cos\varphi$, where $\varphi$ is the smaller angle enclosed by $a$ and $b$. The vector product (or cross product), designated $a \times b$, is a vector $c$ of length $c = ab\sin\varphi$, which equals the area of the parallelogram spanned by $a$ and $b$. The vector $c$ is perpendicular to the plane spanned by $a$ and $b$ and points in the direction given by the right-hand rule (the direction of the thumb, if the index of the right hand is rotated from $a$ to $b$ through the angle $\varphi$).

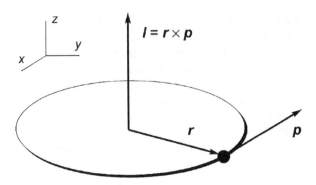

**Figure 1.11**   Angular momentum **l** of a mass point running on a circular orbit

may be defined by its components in a Cartesian coordinate system, $\ell_x$, $\ell_y$ and $\ell_z$, and a length $\ell = mvr = (\ell_x^2 + \ell_y^2 + \ell_z^2)^{1/2}$, the magnitude of the angular momentum.

As the energies of electrons in atomic orbitals are quantized, so are their angular momenta. The quantum of angular momentum is $\hbar = h/2\pi$. Only one component of the vector *l* of angular momentum can be specified precisely. The atomic orbitals (Figure 1.9) obtained by solution of the Schrödinger equation are thus specified by two quantum numbers, the orbital angular momentum quantum number *l* and the angular momentum quantum number $m_l$ of one component, arbitrarily chosen to be the z-component. The quantum number *l* is a non-negative integer and for a given value of *l* there are $2l + 1$ permitted values of $m_l$ (Equation 1.7). The quantum number $l = 0$ is associated with s-orbitals, $l = 1$ with p-orbitals, etc. The magnitude of the orbital magnetic moment $\boldsymbol{\mu}$ of an electron in an orbital with quantum number *l* is $\mu = \mu_B[l(l + 1)]^{1/2}$. The constant $\mu_B = e\hbar/2m_e = 9.274 \times 10^{-2}\,\text{J}\,\text{T}^{-1}$ is called the *Bohr magneton*.

$$l = 0, 1, 2, \ldots \quad \text{and} \quad m_l = l, l-1, \ldots, -l$$

**Equation 1.7**   *Orbital angular momentum quantum numbers*

A classical analogy for visualizing the quantization of $\boldsymbol{\mu}$ is that of a spinning top in the gravitational field. In response to gravity, the top does not fall down, but precesses around the axis of the force, the spin axis sweeping out a conical surface. The response of an electron in an orbital with $l \neq 0$ to an external magnetic field *H* along the z-axis will be similar: the orbital magnetic moment $\boldsymbol{\mu}$ will precess around the z-axis. The fundamental difference is that the projection of the magnetic moment on to the z-axis is allowed to assume only the discrete set of values $m_l$ defined by Equation 1.7.

The *electron spin angular momentum* quantum number is restricted to the value $s = 1/2$, so that analogously to Equation 1.7 (replace *l* by *s* and $m_l$ by $m_s$), electron spin can take only two orientations with respect to the z-axis, specified by the spin magnetic quantum numbers $m_s = +1/2$ (spin 'up', ↑) and $m_s = -1/2$ (spin 'down', ↓). Stern and Gerlach first demonstrated the existence of the electron spin magnetic moment by passing a beam of silver atoms through an inhomogeneous magnetic field. Silver atoms have a single unpaired electron, so that the beam separates into two components, one for atoms with $m_s = +\frac{1}{2}$ and the other for $m_s = -\frac{1}{2}$.

The allowed quantum numbers of total angular momentum $j$ that can arise from a system of two sources of angular momentum with quantum numbers $j_1$ and $j_2$ are defined by the series given in Equation 1.8.

$$j = j_1 + j_2, j_1 + j_2 - 1, \ldots, |j_1 - j_2|$$

**Equation 1.8**  *Combination of angular momentum quantum numbers*

A pictorial representation of Equation 1.8 is shown in Figure 1.12. Given two sticks of length $j_1$ and $j_2$, the task is to join them with a third stick of length $j$ in all possible ways to form triangles. The length $j$ of the third stick has to start at length $j_1 + j_2$ and can be reduced only in unit steps. Note that 'triangles' with lengths $j = 0$ or $j = j_1 + j_2$ are permitted.

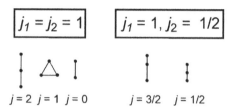

**Figure 1.12**  Vectorial addition of angular momenta

Lower-case letters are used to label the quantum numbers for orbitals ($s$ for electronic spin, $l$ for orbital angular momentum and $j$ for total angular momentum) and capital letters ($S$ for total electronic, $L$ for total orbital and $J$ for total angular momentum) to label the overall state of an atom or molecule. For example, the total angular momentum of a p-electron ($s = \frac{1}{2}$, $l = 1$) can have the values $j = s + l = \frac{3}{2}$ and $j = s + l - 1 = |s - l| = \frac{1}{2}$, so for an atom with a single p-electron $J$ can take the values $\frac{3}{2}$ and $\frac{1}{2}$.

When there are two electrons in a molecule, their total spin angular momentum quantum number $S$ is given by Equation 1.9.

$$S = s_1 + s_2, s_1 - s_2 = 1, 0$$

**Equation 1.9**  *Total spin angular momentum for two electrons*

If there are three electrons, $S$ (a non-negative integer or half integer) is obtained by coupling the third spin to $S = 1$ and $S = 0$ for the first two spins, giving $S = \frac{3}{2}$, $\frac{1}{2}$ and $\frac{1}{2}$, and so on for more electrons. When all electron spins in a molecule are paired ($\downarrow \uparrow$), the total electronic spin quantum number $S$ is 0; there is no net electronic spin. Radicals with a single unpaired electron have $S = \frac{1}{2}$. For a system with two unpaired electron spins, $S = 1$. A state with $S = 0$ can have only one value of $M_S$, $M_S = 0$, and is a singlet. A state with $S = 1$ can have any of the three values $M_S = 1, 0$ or $-1$ (Equation 1.7) and is a triplet. The *multiplicity* $M$ of a state is equal to $2S + 1$. Thus, for $S = 0$ the multiplicity is 1 (*singlet state*), for $S = \frac{1}{2}$ it is 2 (*doublet state*) and for $S = 1$ it is 3 (*triplet state*). The multiplicity indicates the number of distinct magnetic sublevels belonging to an

*Introduction*

electronic state. The multiplicity of molecular species is commonly shown as an upper left index, e.g. $^3O_2$ and $^2NO$ for dioxygen in the triplet ground state and for the NO radical, respectively. In photochemistry, the labels $S_1$ and $T_1$ are also used to designate the lowest excited singlet and triplet state of molecules with an even number of electrons (see Section 2.1.1). The interaction of electron spin with the orbital magnetic moment (*spin–orbit coupling*, SOC) must be considered to calculate the rates of photophysical processes involving a change of multiplicity, which play a dominant role in photochemistry (Section 2.1.1; SOC will be discussed in Sections 2.1.6 and 5.4.4).

Provided that we do not consider magnetic interactions in the Hamiltonian operator explicitly, the electronic wavefunction $\Psi'_{el}$ can be further separated into the product of the so-called space part $\Psi_{el}$ and a spin function $\sigma$ as shown in Equation 1.10 for a two-electron system (linear combinations of spin functions are needed for systems with more electrons). For a single electron, the spin function $\sigma = \alpha$ (or $\uparrow$) represents the state with $m_s = {}^1\!/_2$ and the function $\beta$ (or $\downarrow$) that with $m_s = -{}^1\!/_2$.

$$\Psi \approx \Psi_{el}(q_{el}, q_{nucl})\chi(q_{nucl})\sigma$$

***Equation 1.10***  *Separation of electronic, vibrational and spin wavefunctions*

The spatial part of the wavefunction $\Psi_{el}$ (e.g. $2p_y$, Figure 1.13) holds the information about the spatial distribution of the electrons surrounding fixed nuclear positions. Max Born first proposed that the square of the spatial wavefunction multiplied by a volume element $dv$, $\Psi_{el}^2 dv$, is equal to the probability of finding the electrons in that volume element $dv$. This is easily visualized for atomic orbitals containing a single electron; $\phi^2 dv = \phi^2(x_1,y_1,z_1)dxdydz$ is the probability of finding the electron in the volume element $dv = dxdydz$ surrounding a given position $x_1,y_1,z_1$. For species with many electrons, $\Psi_{el}^2 dv$ is the probability of finding one electron in a volume element around a defined position in space, $x_1,y_1,z_1$, another at $x_2,y_2,z_2$, and so on.

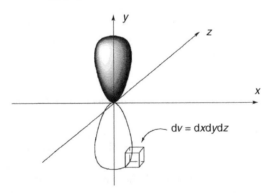

**Figure 1.13**  Spatial wavefunction of a $2p_y$ AO. The shading (filled/empty) denotes the change in sign of the wavefunction. The probability $(2p_y)^2 dv$ is $\geq 0$ throughout

The Born interpretation of $\Psi_{el}^2$ as a probability function requires that the wavefunction $\Psi_{el}$ be *normalized*, namely that integration of $\Psi_{el}^2 dv$ over all space (Equation 1.11), equals unity, because the system of electrons described by $\Psi_{el}$ must be located somewhere in space. Integration over all space requires multiple integration over the three coordinates for each of the $n$ electrons, which is abbreviated as $\langle \Psi_{el} | \Psi_{el} \rangle$.[f] Wavefunctions can always be normalized by dividing $\Psi_{el}$ by $\langle \Psi_{el} | \Psi_{el} \rangle$.

$$\int \Psi_{el}^2 d\nu = \int\int \cdots \int\int \Psi_{el}^2 dx_1 dy_1 dz_1 \ldots dx_n dy_n dz_n = \langle \Psi_{el} | \Psi_{el} \rangle$$

**Equation 1.11**

The sublevels of an electronic state with multiplicity $M > 1$ are nearly *degenerate*, that is, they have nearly the same energy. Any small energy differences between the sublevels are due to the weak magnetic interactions within the molecule. The splitting increases in response to strong external magnetic fields.

The foremost impact of electronic spin is due to the *Pauli principle*, a fundamental postulate of quantum mechanics. It can be stated as follows: *The total wavefunction (Equation 1.10) must be antisymmetric with respect to the interchange of any electron pair $e_i$ and $e_j$*, i.e. of their coordinates $q_{e_i}$ and $q_{e_j}$ (Equation 1.12).

$$\Psi(\ldots q_{e_i}, \ldots q_{e_j}, \ldots; q_n) = -\Psi(\ldots q_{e_j}, \ldots q_{e_i} \ldots; q_n)$$

**Equation 1.12**   *Antisymmetry of the total wavefunction*

Wavefunctions must be either symmetric (delete the minus sign from Equation 1.12) or antisymmetric in order to be consistent with the Born interpretation: electrons being indistinguishable, $\Psi^2$ must be invariant with respect to an interchange of any pair of electrons, because the probability of finding $e_i$ in a volume element around the coordinates $q_{e_i}$ and $e_j$ around $q_{e_j}$ must be the same when the labels $i$ and $j$ are exchanged. Both symmetric and antisymmetric wavefunctions would satisfy this condition, but the Pauli principle allows only antisymmetric wavefunctions.

Electronic spin exerts a strong influence on the energies associated with wavefunctions as a consequence of the Pauli principle. Consider the factored wavefunction, Equation 1.10. The product of two symmetric or two antisymmetric functions is symmetric, that of a symmetric and an antisymmetric function is antisymmetric. The spin functions for two-electron systems representing a singlet state ($S = 0$) are antisymmetric; those representing a triplet state ($S = 1$) are symmetric.[18] Vibrational wavefunctions $\chi(q_{nucl})$ are invariant with respect to electron exchange, i.e. symmetric, as they do not depend on the electronic coordinates. Therefore, *the electronic wavefunction for the triplet state must be antisymmetric* in order to satisfy the Pauli principle.

Now consider an electronic wavefunction $\Psi_{el}$ describing the distribution of two electrons in a triplet state. Because it is antisymmetric, it must change sign, when the

---

[f] Wavefunctions may be complex functions; $\Psi_{el}$ then defines a complex number $a + ib$ ($i = \sqrt{-1}$) rather than a real one for each point in the multidimensional parameter space of the electronic coordinates. In that case $\Psi_{el}^2$ must be replaced by $\Psi_{el}^* \Psi_{el}$, where $\Psi_{el}^*$ is the complex conjugate of $\Psi_{el}$ ($a + ib$ is replaced by $a - ib$). In the shorthand notation $\langle \Psi_{el} | \Psi_{el} \rangle$, the first wavefunction is understood to be the complex conjugate of the second.

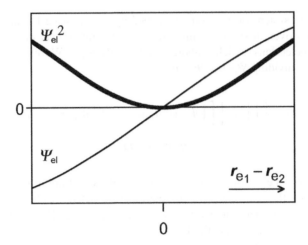

**Figure 1.14**   Fermi hole

distance between the two electrons defined by the vector $r_{e_1}-r_{e_2}$ vanishes (Figure 1.14). The square of the wavefunction, $\Psi_{el}^2$, must then have a broad minimum around that point. Two electrons with parallel spin can never be at the same point in space and, because $\Psi_{el}$ is a continuous function, they have a low probability of being near each other. This dip in the distribution of two electrons is called the *Fermi hole*. It has nothing to do with the Coulombic repulsion between the electrons that also operates to keep electrons apart. Rather, it functions somewhat like traffic lights that (ideally) prevent cars from crashing. Electronic wavefunctions for a singlet state are symmetric under electron exchange, so that the traffic rule does not apply. This is the basis of the first of *Hund's rules* (see Section 4.1).

The Born interpretation also requires that wavefunctions be either symmetric or antisymmetric with respect to all symmetry operations of a molecule, that is, when the coordinates of all the electrons and nuclei are exchanged by symmetry-equivalent coordinates. For example, the electronic distribution around an isolated atom must be spherically symmetric in the absence of external fields.

Once an eigenfunction $\Psi_n$ has been determined, all observable properties of a system in that state are defined. The index $n$ is a label to distinguish different solutions of Equation 1.5, which represent different states. For each observable property of a given state, there is an associated operator $\hat{O}$ and the property is defined by performing the integration $\langle \Psi_n | \hat{O} | \Psi_n \rangle$. Thus the energy associated with $\Psi_n$ is determined by the Hamiltonian operator $\hat{O} = \hat{H}$, $E_n = \langle \Psi_n | \hat{H} | \Psi_n \rangle$. Similarly, the dipole moment vector $\mu_n$ is given by $\mu_n = \langle \Psi_n | \hat{M} | \Psi_n \rangle$, where $\hat{M}$ is the dipole moment operator (Section 2.1.4).

But how can we calculate the energy from approximate wavefunctions $\Phi_n \approx \Psi_n$ that are not eigenfunctions of the operator $\hat{H}$, so that the operation $\hat{H}\Phi_n$ will not reproduce $\Phi_n$ unchanged? We can always multiply both sides of the approximate equation $\hat{H}\Phi_n \approx E_n\Phi_n$ with $\Phi_n$ from the left: $\Phi_n\hat{H}\Phi_n \approx \Phi_n E_n\Phi_n$. On the right-hand side, we can replace $\Phi_n E_n\Phi_n$

by $E_n \Phi_n^2$, because the energy $E$ is a constant, $\Phi_n E_n = E_n \Phi_n$. This is not so on the left-hand side, because $\hat{H}$ is a differential operator: $\Phi_n \hat{H} \neq \hat{H} \Phi_n$. Integration of $\Phi_n \hat{H} \Phi_n \approx E_n \Phi_n^2$ over all space yields Equation 1.13. For normalized wavefunctions $\Phi_n$, $\langle \Phi_n | \Phi_n \rangle = 1$. Most operators $\hat{O}$ for other observables are much simpler than $\hat{H}$, so that the evaluation of Equation 1.13, replacing $\hat{H}$ by $\hat{O}$ and $E_n$ by the observable $o_n$, with a given approximate wavefunction $\Phi_n$ is straightforward.

$$\langle \Phi_n | \hat{H} | \Phi_n \rangle \approx E_n \langle \Phi_n | \Phi_n \rangle, \text{ i.e. } E_n \approx \frac{\langle \Phi_n | \hat{H} | \Phi_n \rangle}{\langle \Phi_n | \Phi_n \rangle}$$

**Equation 1.13**

Observables calculated from approximate wavefunctions as in Equation 1.13 are called *expectation values*, an expression used in probability theory. In practice, we will always have to be satisfied with approximate wavefunctions. How can we choose between different approximations? And if our trial wavefunction has adjustable parameters (such as the coefficients of atomic orbitals in molecular orbitals; see Section 4.1), how can we choose the adjustable parameters' best values? Here, Rayleigh's *variation theorem* is of great value. It tells us that the expectation value for the ground state energy $E_1$, $\langle E_1 \rangle$, calculated from an approximate wavefunction $\Phi_1$ is always larger than the true energy $E_1$ (Equation 1.14). Proof of the variation theorem is given in textbooks on quantum mechanics.[18]

$$\frac{\langle \Phi_1 | \hat{H} | \Phi_1 \rangle}{\langle \Phi_1 | \Phi_1 \rangle} = \langle E_1 \rangle \geq \frac{\langle \Psi_1 | \hat{H} | \Psi_1 \rangle}{\langle \Psi_1 | \Psi_1 \rangle} = E_1$$

**Equation 1.14**   *Variation theorem*

Thus, the wavefunction giving the lowest eigenvalue $E_1$ will be the best. Having defined a trial wavefunction $\Phi$ with adjustable parameters, we want to optimize it by determining those values of the parameters that give the lowest expectation value for the energy. If we use a trial function $\Phi$ that is a *linear combination* (LC) of an *orthonormal*[g] *basis set*, e.g. a set of orthonormal AOs $\phi_i$ (LCAO) (Equation 1.15),

$$\Phi = c_1 \phi_1 + c_2 \phi_2 + \cdots = \sum_i c_i \phi_i$$

**Equation 1.15**

where we require $\Phi$ to be normalized, $<\Phi|\Phi> = 1$, then we must solve a set of simultaneous differential equations $(\partial E / \partial c_i) = 0$, the *secular equations*, in order to find the best values of the coefficients $c_i$. This is a well-known problem of linear algebra; nontrivial solutions (the trivial solution is a vanishing wavefunction; $c_i = 0$ for all $i$) exist

---

[g] The functions $\phi_i$ of the basis set are orthonormal, i.e. orthogonal, $\langle \phi_i | \phi_j \rangle = 0$ for $i \neq j$ and normalized, $\langle \phi_i | \phi_i \rangle = 1$, for all $i$.

only if the *secular determinant*[h] (Equation 1.16) vanishes.

$$||H_{ij}-\delta_{ij}E|| = \begin{Vmatrix} H_{11}-E & H_{12} & \cdots & H_{1n} \\ H_{21} & H_{22}-E & \cdots & H_{2n} \\ \cdot & \cdot & & \cdot \\ \cdot & \cdot & & \cdot \\ \cdot & \cdot & & \cdot \\ H_{n1} & H_{n2} & \cdots & H_{nn}-E \end{Vmatrix} = 0$$

**Equation 1.16**   *Secular determinant*

The elements $H_{ij}$ of Equation 1.16 represent the integrals $\langle \Phi_i|\hat{H}|\Phi_j \rangle$ and the symbol $\delta_{ij}$ used in the shorthand notation for the secular determinant on the left is *Kronecker's delta*, which takes the value of 1 for $i=j$ and 0 otherwise, as can be seen from the extended notation on the right. Solving Equation 1.16 yields a set of energies $E_n$ that are called the *eigenvalues* of the matrix $(H_{ij}-\delta_{ij}E)$. The associated sets of coefficients $c_{in}$, the *eigenvectors*, are then obtained by inserting each solution $E_n$ into the secular equations. The corresponding *eigenfunction* $\Phi_n$ is given by Equation 1.15 with the coefficients defined by the eigenvector associated with $E_n$. An explicit example will be given in Section 4.3.

The same procedure can be used to calculate the interaction energy between two systems. Given the wavefunctions $\Psi_n$ and $\Psi_m$ for the isolated separate systems, we can use them as trial wavefunctions as in Equation 1.15 to determine the energies of the combined system. The probability of radiative or nonradiative transitions between zero-order BO wavefunctions $\Psi_n$ and $\Psi_m$ is proportional to the square of the integrals $\langle \Psi_n|\hat{O}|\Psi_m \rangle$, where the operator $\hat{O}$ depends on the process considered (see Sections 2.1.4–2.1.6).

From a mathematical point of view, the task of finding (approximate) eigenfunctions of Equation 1.5 for a molecule is no more complicated than solving the Newtonian equations for a mechanical system with a similar number of bodies such as the solar system. An important difference is that the interactions between all particles in a molecule are of comparable magnitude, on the order of electronvolts ($1\,eV \times N_A = 96.4\,kJ\,mol^{-1}$). In calculations of satellite trajectories or planetary movements, on the other hand, one can start with a small number of bodies (e.g. Sun, Earth, Moon, satellite) and subsequently add the interactions with other, more distant or lighter bodies as weak perturbations. This greatly simplifies the task of calculating a satellites' trajectory.

Nevertheless, *perturbation theory* is also of great importance in quantum mechanics. The formidable task of solving the Schrödinger Equation 1.5 is simplified by removing some particularly intractable terms from the Hamiltonian operator $\hat{H}$. Using eigenfunctions of the simplified operator $\hat{H}^0$, it is possible to estimate the eigenvectors and

---

[h]The determinant of a *square matrix* $A=(A_{ij})$ of size $n \times n$ (see footnote c in Section 3.7.5), denoted $||A||$, is a number that is calculated from the elements $A_{ij}$ of the matrix by Laplacian expansion by minors (see textbooks of mathematics). The results for $1 \times 1$, $2 \times 2$ and $3 \times 3$ determinants are shown below.

$$||A_{11}|| = A_{11}; \quad \begin{Vmatrix} A_{11} & A_{12} \\ A_{21} & A_{22} \end{Vmatrix} = A_{11}A_{22} - A_{12}A_{21}$$

$$\begin{Vmatrix} A_{11} & A_{12} & A_{13} \\ A_{21} & A_{22} & A_{23} \\ A_{31} & A_{32} & A_{33} \end{Vmatrix} = A_{11}A_{22}A_{33} + A_{12}A_{23}A_{31} + A_{13}A_{21}A_{32} - A_{12}A_{21}A_{33} - A_{11}A_{23}A_{32} - A_{13}A_{22}A_{31}$$

eigenvalues of the more complete operator $\hat{H} = \hat{H}^0 + \hat{h}$, where $\hat{h}$ is called the *perturbation operator*.

Perturbation theory is also invaluable to predict trends in related series of compounds (substituent effects). To calculate the energy of benzene *ab initio* amounts to evaluating the energy released when six carbon nuclei, six protons and 42 electrons combine to form benzene, which is $6 \times 10^5$ kJ mol$^{-1}$. The accuracy needed to predict chemical reactivity is on the order of 1 kJ mol$^{-1}$ (a decrease of 1 kJ mol$^{-1}$ in activation energy amounts to a 50% increase in the corresponding rate constant). This is reminiscent of the attempt to determine the weight of the captain by weighing the ship with and without the captain aboard. Perturbation theory evaluates the small energy *difference* between a pair of closely related compounds directly, making no attempt to calculate the parent system on an absolute scale. We shall make extensive use of perturbation theory in Chapter 4.

## 1.5 Problems

1. Calculate the energy of one mole of photons, $\lambda = 400$ nm. $[E_p = 300$ kJ mol$^{-1}]$

2. Determine the work function $\varphi$ for sodium metal (the energy needed to remove an electron from sodium metal) and the value of Planck's constant $h$ from Figure 1.5. $[\varphi \approx 1.84$ eV, $h \approx 6.6 \times 10^{-34}$ J s. Note: the first ionization potential of sodium, $I_1 = 5.14$ eV, refers to ionization of a sodium atom in the gas phase]

3. Name the colour of light of 400, 500 and 600 nm and of a light source emitting at both 490 and 630 nm with equal intensity. [Figure 1.8]

4. Name the colour of dyes with absorption maxima at 400 nm, at 500 nm and at both 400 and 600 nm (equal intensity). [Figure 1.8]

5. What are the advantages of the Born–Oppenheimer approximation and when is it necessary to go beyond? [Separation of electronic and vibrational wavefunctions, existence of PES; calculation of IC and ISC rate constants, near-degenerate states]

6. How many angular momentum quantum numbers $j$ are possible for an electron in a d-orbital? [6]

7. Given the ground-state energies of atoms A and B, (a) $E_A = -20$ and $E_B = -30$ eV or (b) $E_A = E_B = -25$ eV, and the matrix element for their interaction at a given distance, $\langle \Psi_A | \hat{H} | \Psi_B \rangle = 1$ eV, calculate the lowest eigenvalue (for cases a and b) of the combined system A$\cdots$B. [(a) $-30.1$ eV, (b) $-26$ eV]

8. If plants exist on planets of other stars, their colour would probably be different from plants on Earth. Explain. [Ref. 21]

# 2

# A Crash Course in Photophysics and a Classification of Primary Photoreactions

## 2.1 Photophysical Processes

An authoritative compilation of definitions and concepts is available in the *Glossary of Terms Used in Photochemistry*, 3rd edition,[22] which can be downloaded from the IUPAC webpages or from those of the photochemical societies (http://pages.unibas.ch/epa or http://www.i-aps.org/).

### 2.1.1 State Diagrams

The expression '*Jabłonski diagram*'[a,25] is commonly used to designate state diagrams of the kind depicted in Figure 2.1. The French community prefers to name such diagrams after *J. Perrin*.[26] Actually, several authors had used similar state or term diagrams to describe the long-lived emission observed from irradiated solid solutions, but none prior to Lewis, Lipkin and Magel in 1941[27] had attributed the metastable species to a triplet state of the dissolved organic molecules. Firm evidence that the emitting species were normal organic molecules in a triplet state was soon provided by Terenin in 1943[28] and by Lewis and Kasha in 1944.[23] Irrefutable proof identifying the triplet state of an organic molecule was finally obtained in 1958 by Hutchison and Mangum,[29] who reported ESR spectra of naphthalene held in a small number of known orientations in a durene crystal.

---

[a] Excerpts from a personal communication from Professor M. Kasha, Florida State University, to J. W., 25 July 2008:
'Alexander Jabłonski intended that his diagram applies to dye molecules. In those the lowest $S_1$–$T_1$ separation is relatively small (generally about 1000–2000 cm$^{-1}$). Jabłonski gave a neat kinetic analysis which indicated that thermal excitation from the metastable excited (T) state to the nearby excited (S) state was achievable in general. Note that the T-state designation for the metastable excited state was the result of the 1944 Lewis and Kasha study.[23] Jabłonski did not accept this interpretation. A paper I presented at the memorial symposium in Torun, Poland, in honor of Jabłonski,[24] was commissioned by the Polish Physical Society and was much admired by the Polish physicists. At the meeting I was informed that in his last year or so Jabłonski finally accepted the idea that the "metastable" state was really the lowest excited triplet state of the dye.'

---

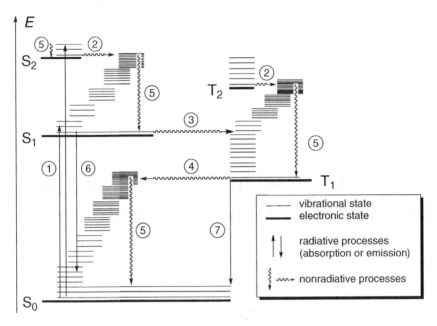

**Figure 2.1** State diagram (commonly called Jabłonski diagram) depicting molecular states and photophysical processes. The vertical position of the thick horizontal lines represents the electronic energy minimum. Vibrational energy levels are shown as thin lines. The width of the horizontal lines and their position along the abscissa are chosen merely to avoid congestion in the graphical diagram and have no physical connotation

Molecular[b] electronic states are represented by thick horizontal lines (—) that are arranged in vertical order to indicate relative energies and are labelled consecutively by increasing energy, beginning with the singlet ground state $S_0$, followed by excited singlet states $S_1$, $S_2$, etc. Vibrational states, represented here by thin lines (—), are usually not shown explicitly. When they are shown as in Figure 2.1, the lowest thin line of each electronic state represents the zero-point energy level ($v = 0$) of that state. States of a given *multiplicity* are collected in separate columns.[c] For molecules with a singlet ground state, the left-hand column collects the electronic singlet states and that to the right the triplet states. The lowest triplet state is labelled $T_1$, followed by $T_2$, $T_3$, and so on.

*Photophysical processes* are *radiative* or *radiationless transitions* by which molecules are promoted from one electronic state to another. No structural change results, although the bond lengths and angles generally differ somewhat in different electronic states. Radiative transitions are associated with the absorption or emission of a photon and are represented as straight arrows (↑ or ↓), while radiationless transitions are not associated

---

[b] We use the term *molecule* to comprise all (meta)stable molecular species including ions and radicals.

[c] Horizontal displacement does *not* indicate a change in structure. The energy levels represent minima on the potential energy surfaces of a given electronic state. The corresponding structures will be somewhat different for each state. Such a representation is not meaningful if an excited state has no energy minimum near the structure of the ground state.

with absorption or emission and are shown as wavy arrows ($\rightsquigarrow$). A *vibronic transition* is a transition between two states that differ in both the electronic and the vibrational quantum numbers of a molecule. The term *vibronic* is a fusion of vibrational and electronic.

*Comments on Figure 2.1*:

① The time scale of electronic absorption is the time during which a molecule interacts strongly with a photon. The wavelengths of near-UV photons are more than two orders of magnitude larger than the size of an average organic molecule. Assume that the 'length' of a photon is on the order of its wavelength, say $\lambda = 300$ nm. Then the time needed for a photon to traverse a molecule at the speed of light, $c = 3 \times 10^8$ m s$^{-1}$, is $t \approx \lambda/c = 10^{-15}$ s. A more tenable derivation of the same is obtained from the *Heisenberg uncertainty principle*,[d] for the variables time $t$ and energy $E$ (Equation 2.1).

$$\delta E \delta t \geq \hbar/2 = h/4\pi$$

**Equation 2.1**   *Heisenberg uncertainty principle*

The 'uncertainties' $\delta$ are defined as the standard deviations of simultaneous measurements of time and energy. During an electronic transition of a molecule in the presence of an electromagnetic field, the indeterminacy $\delta E$ of the molecular energy is on the order of the photon energy, $E_p = h\nu = hc/\lambda$. Hence $\delta t \geq \lambda/(4\pi c) \approx 1 \times 10^{-16}$ s for $\lambda = 300$ nm.

Absorption of electromagnetic radiation in the UV–VIS range always populates an electronically excited state, notwithstanding the high density of vibrationally (and rotationally) excited states in that region. Transitions to high-energy vibrational states by light absorption are 'forbidden' (not observed); only transitions involving electronic excitation carry appreciable intensity, hence the name electronic spectra. Transitions to the first excited vibrational levels occur in the infrared region of the electromagnetic spectrum and very weak transitions to the second level (overtones) or to states in which two vibrational degrees of freedom are excited (combination bands) are observable in the near-infrared ($0.1 < \tilde{\nu}/\mu\text{m}^{-1} < 1$). Following electronic absorption, the isoenergetic vibrational states come to play in the radiationless transitions ②–④, which are considered next.

② *Internal conversion* (IC) is an isoenergetic radiationless transition between two electronic states of the same multiplicity.

③, ④ *Intersystem crossing* (ISC) is an isoenergetic radiationless transition between two electronic states of different multiplicity. In molecules with a singlet ground state, ISC can take place from a singlet state to a triplet state, process ③, or vice versa, ④.

IC and ISC, ②–④, are processes of energy redistribution within the excited molecule by which electronic energy is distributed over many vibrational modes. This takes time, in much the same way as it is difficult to get change for a large banknote denominated in a foreign currency on a flea market (see Section 2.1.5). IC and ISC are essentially irreversible processes, because they are associated with an entropy increase (high density of states in the lower-energy electronic state) and because the following process ⑤ is very fast in solution.

---

[d] Indeterminacy would be a more adequate translation of *Unschärfe*.

⑤ *Vibrational relaxation* (also called vibrational cooling, vibrational deactivation or thermalization) encompasses all processes by which excess vibrational energy that a molecule has acquired by a vibronic transition (absorption, IC or ISC) is transferred to the surrounding medium. It occurs through collisions with other molecules. In solution, the time required to transfer a vibrational quantum to the surrounding medium is on the order of a vibrational period $1/\nu_{vib}$, which corresponds to the time elapsing between collisions with the solvent (100 fs); thermalization of large molecules ($\geq 10$ atoms) usually occurs with a half-life of a few picoseconds in solution, but may take much longer in poorly conducting rare gas matrices at low temperature. Molecules in high vacuum (isolated molecules) cannot undergo vibrational relaxation by collision with other molecules and the loss of excess vibrational energy by infrared radiation is very slow. However, a hot molecule will still undergo intramolecular vibrational redistribution (IVR) by which the energy that is originally localized in the mode populated by a vibronic transition is distributed among the other vibrational modes, that is, the molecule acts as its own heat bath. IVR occurs on a sub-picosecond time scale in 'large' molecules (benzene or larger), but may require much longer in small molecules with low densities of vibrational states.

⑥ *Fluorescence*[e,30] is the spontaneous emission of radiation by an excited molecule, typically in the first excited singlet state $S_1$, with *retention* of spin multiplicity.

⑦ *Phosphorescence*[f,30] is the spontaneous emission of radiation from an excited molecule, typically in the first excited triplet state $T_1$, involving a *change* in spin multiplicity.

*Primary photochemical processes* are processes initiated from an electronically excited state that yield a primary photoproduct that is chemically different from the original reactant. Photochemical processes are always in competition with photophysical processes that eventually restore the reactant in the ground state. Photoreactions leading to new products can be efficient only if they are faster than the competing photophysical processes. Therefore, it is essential to have a feeling for the time scales of the latter (Table 2.1, Figure 2.2). Photophysical processes of molecules in solution usually obey a

**Table 2.1**  *Summary of photophysical processes*

| Process | Name | Time scale $(\tau = 1/k_{process})/s$ | Section |
|---------|------|------------------|---------|
| ① | Absorption | $10^{-15}$ | 2.1.2 |
| ② | Internal conversion | $10^{-12}$–$10^{-6}$ | 2.1.7, 2.1.5 (Kasha rule) |
| ③ | Intersystem crossing (S → T) | $10^{-12}$–$10^{-6}$ | 2.1.6 (El Sayed rule) |
| ④ | Intersystem crossing (T → S) | $10^{-9}$–$10^{1}$ | 2.1.6 |
| ⑤ | Vibrational relaxation | $10^{-13}$–$10^{-12}$ | 2.1.5 |
| ⑥ | Fluorescence | $10^{-9}$–$10^{-7}$ | 2.1.3 |
| ⑦ | Phosphorescence | $10^{-6}$–$10^{-3}$ | 2.1.6 |

---

[e] The word is derived from the mineral fluorospar and was coined in 1852 by G. G. Stokes to describe the emission observed when fluorospar or a solution of quinine sulfate were exposed to UV radiation that was separated from sunlight by means of a prism. Stokes stated that the emitted light is always of longer wavelength than the exciting light. This statement is now known as Stokes' law and the wavelength shift is referred to as the *Stokes shift* (see Section 2.1.9). Similar experiments were in fact reported earlier by D. Brewster (1833) and E. Becquerel (1842).

[f] The term is derived from Greek φως = light and φορειν = to bear and was used in the Middle Ages for materials that glow in the dark after exposure to light.

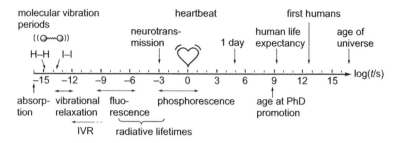

**Figure 2.2** Time scales

first-order rate law. When several processes contribute to the decay of a large number of molecules in an excited state M* that was generated by absorption of a short light pulse at time $t = 0$, the decay of the initial concentration $c_{M^*}(0)$ will be given by Equation 2.2:

$$c_{M^*}(t) = c_{M^*}(0)e^{-(\Sigma k)t}$$

**Equation 2.2**  *First-order rate law*

The *lifetime* of M*, $\tau(M^*)$, is defined as the time in which the initial concentration $c_{M^*}(0)$ is reduced to $c_{M^*}(0)e^{-1} \approx 0.368 c_{M^*}(0)$, that is, $\tau(M^*) = 1/\Sigma k$, where $\Sigma k$ is the sum of the rate constants of all processes contributing to the decay of M*. More detailed rules for estimating the rate constants of photophysical processes for individual molecules will be given in the following sections.

The difference between photophysical and photochemical processes is not always clear cut. For example, the *E–Z* isomerization of alkenes represents a photochemical process by the above definition, yet the same process might be seen as photophysical in terminal alkenes, where geometric isomers are indistinguishable in the absence of isotopic labels.

The term *luminescence* denotes the spontaneous emission of radiation from an electronically excited species and comprises fluorescence and phosphorescence, and also chemiluminescence (Section 5.6).

### 2.1.2  Beer–Lambert Law

The *absorbance* $A(\lambda)$ of a sample is defined by Equation 2.3:[g,10,22]

$$A(\lambda) = \log(P_\lambda^0/P_\lambda^{tr}) = -\log[T(\lambda)]; \quad T(\lambda) = 10^{-A(\lambda)}$$

**Equation 2.3**

where $P_\lambda^0$ and $P_\lambda^{tr}$ are the incident and transmitted *spectral radiant power*, respectively, and $T(\lambda) = P_\lambda^{tr}/P_\lambda^0$ is the internal *transmittance* of the sample at the given wavelength $\lambda$. For

---

[g] *Radiant power*, P, is the radiant energy, Q, emitted, transferred or received at all wavelengths per unit time, $P = dQ/dt$. If the radiant energy Q is constant over the time interval considered, $P = Q/t$. The SI unit of radiant power is watt, $[P] = W$. The *spectral radiant power* is a differential, $P_\lambda = dP(\lambda)/d\lambda$, representing the *radiant power* P per wavelength interval. Spectral radiant power expressed in wavenumbers, $P_{\tilde{\nu}} = dP(\tilde{\nu})/d\tilde{\nu}$, is equal to $P_\lambda/\tilde{\nu}^2$, because $|d\lambda/d\tilde{\nu}| = 1/\tilde{\nu}^2$. The factor $1/\tilde{\nu}^2$ cancels in the ratio $T(\lambda) = P_\lambda^0/P_\lambda^{tr} = P_{\tilde{\nu}}^0/P_{\tilde{\nu}}^{tr}$. The units of spectral radiant power are $[P_\lambda] = W\ m^{-1}$ and $[P_{\tilde{\nu}}] = W\ m$. The *intensity* or *irradiance*, symbol *I*, is the radiant power received at a surface, $[I] = W\ m^{-2}$. The ratios $P^0/P^{tr}$ and $I^0/I^{tr}$ are equal. The expression 'intensity' has traditionally been used indiscriminately for photon flux, irradiance or radiant power. When considering an object exposed to radiation, it should therefore be used only for qualitative descriptions.

absorbance and transmittance, the wavelength $\lambda$ can be replaced by the corresponding wavenumber, $\tilde{\nu} = 1/\lambda$, $A(\lambda) = A(\tilde{\nu})$ and $T(\lambda) = T(\tilde{\nu})$. The terms 'optical density' (OD) for $A(\lambda)$ and 'transmission' for $T(\lambda)$ should no longer be used and the term 'extinction' should be reserved for the total effect of absorption, scattering and luminescence. The accepted term for the quantity $1 - T(\lambda)$ is *absorptance*, symbol $\alpha$.

The *Beer–Lambert law* (Equation 2.4) holds only for absorbing species that do not exhibit any concentration-dependent aggregation (Lambert law).

$$A(\lambda) = \varepsilon(\lambda)cl$$

**Equation 2.4**   *Beer–Lambert law for a single absorbing compound*

Equation 2.5 is the extension of this law for solutions containing a mixture of $n$ absorbing species $B_i$, $i = 1, 2, \dots n$. When there are no specific interactions between these species, the contributions of the individual components to the total absorbance, that is, the partial absorbances $A_i(\lambda) = \varepsilon_i(\lambda)c_il$, are additive. The proportionality constants $\varepsilon_i(\lambda)$ are called *molar (decadic) absorption coefficients* of species $B_i$ at wavelength $\lambda$; again, $\varepsilon_i(\lambda) = \varepsilon_i(\tilde{\nu})$. The pathlength $l$ is commonly measured in centimetres and the concentrations $c_i$ in molarity, $M = mol\,dm^{-3}$, so that the unit of the molar absorption coefficient becomes $[\varepsilon(\lambda)] = dm^3\,mol^{-1}\,cm^{-1} = M^{-1}\,cm^{-1}$.

$$A(\lambda) = \sum_{i=1}^{n} A_i(\lambda) = l \sum_{i=1}^{n} \varepsilon_i(\lambda)c_i$$

**Equation 2.5**   *Beer–Lambert law for a mixture of absorbing compounds*

### 2.1.3   Calculation of Fluorescence Rate Constants from Absorption Spectra

In addition to the spontaneous emission of excited molecules, fluorescence and phosphorescence (Section 2.1.1), the interaction of electromagnetic radiation with excited molecules gives rise to *stimulated emission*, the microscopic counterpart of (stimulated) absorption. Albert Einstein derived the existence of a close relationship between the rates of absorption and emission in 1917, before the advent of quantum mechanics (see Special Topic 2.1).

---

### Special Topic 2.1: Einstein coefficients of absorption and emission

Consider a gas of molecules with two electronic states, an excited state $m$ and the ground state $n$ (Figure 2.3), in equilibrium with an electromagnetic field of radiant energy density $\rho$ emanating from a black-body light source such as the Sun.

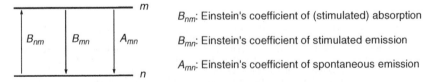

$B_{nm}$: Einstein's coefficient of (stimulated) absorption

$B_{mn}$: Einstein's coefficient of stimulated emission

$A_{mn}$: Einstein's coefficient of spontaneous emission

**Figure 2.3**   Two-level scheme with radiative processes

The frequency distribution $\rho_\nu = d\rho/d\nu$ of a *black-body radiator* is given by Planck's law (Equation 2.6):

$$\rho_\nu = \frac{8\pi h \tilde{\nu}^3}{e^{h\nu/k_B T} - 1}$$

**Equation 2.6**   *Planck distribution of a black-body radiator*

where $T$ is the temperature of the black body.

Once the gas has reached equilibrium with the radiation field, the populations of its energy levels $m$ and $n$, $N_m$ and $N_n$, must be constant, $dN_m/dt = dN_n/dt = 0$. With the Einstein coefficients defined in Figure 2.3 one obtains Equation 2.7:

$$dN_m/dt = B_{nm} N_n \rho_\nu - N_m [A_{mn} + B_{mn}\rho_\nu] = 0$$

**Equation 2.7**

Solving for $\rho_\nu$ yields Equation 2.8:

$$\rho_\nu = \frac{A_{mn}/B_{nm}}{N_n/N_m - B_{mn}/B_{nm}}$$

**Equation 2.8**

The ratio of equilibrium populations $N_n/N_m$ in the denominator of Equation 2.8 must obey Boltzmann's equation (Equation 2.9):

$$N_n/N_m = e^{\Delta E/k_B T} = e^{h\nu/k_B T}$$

**Equation 2.9**   *Boltzmann's law*

where $\Delta E = h\nu$ is the energy difference between two molecular states $m$ and $n$.

Equation 2.8 has the same form as Equation 2.6 if and only if $B_{mn} = B_{nm} = B$, i.e. $B_{mn}/B_{nm} = 1$. Combining Equation 2.8 with Equation 2.6, we obtain the final relation Equation 2.10, where we have dropped the now superfluous indices $m$ and $n$:

$$A = 8\pi h \tilde{\nu}^3 B$$

**Equation 2.10**

The probability of stimulated emission by an excited molecule, $B\rho_\nu$, is the same as that of the reverse process, absorption by the ground-state molecule. This is a consequence of the law of microscopic reversibility: if the number of excited molecules $N_m$ is equal to the number of ground-state molecules $N_n$, then the rates of stimulated absorption and emission must be equal. If by some means a population inversion can be produced, $N_m > N_n$, the net effect of interaction with electromagnetic radiation of frequency $\nu$ will be stimulated emission (Figure 2.4). This is the operating principle of the *laser*. LASER is an acronym for light amplification by stimulated emission of radiation (Section 3.1).

Polyatomic molecules have broad absorption and emission bands due to the contributions of vibronic transitions. Einstein's equation (Equation 2.10) relates the coefficients of absorption and emission in a two-level system. Several extensions of

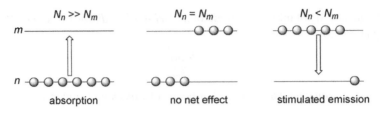

**Figure 2.4**   Net effects of interaction with electromagnetic radiation

Equation 2.10, which permit the determination of fluorescence rate constants $k_f$ from the broad absorption spectra of molecules, have been derived. The best known is that of Strickler and Berg.[31] A simpler version that is easier to use (Equation 2.11)[32] is sufficient for most practical purposes:

$$\frac{k_f}{s^{-1}} = 2900 n^2 \left(\frac{\tilde{\nu}_{max}}{\mu m^{-1}}\right)^2 \int_{\substack{absorption \\ band}} \frac{\varepsilon}{M^{-1}\,cm^{-1}} \frac{d\tilde{\nu}}{\mu m^{-1}}$$

***Equation 2.11***   *Determination of fluorescence rate constants from absorption spectra*

where $n$ is the refractive index of the solvent, $\varepsilon$ is the *molar absorption coefficient*, and $\tilde{\nu}_{max}$ is the wavenumber of the absorption band's maximum. However, overlap of absorption bands frequently hampers integration over a single absorption band. It may be helpful in this respect to assume a mirror-image relationship (Section 2.1.9) between the first absorption band and fluorescence. In general, fluorescence rate constants $k_f$ determined from lifetime measurements (Equation 3.33) are in satisfactory agreement ($\pm 20\%$) with those calculated from the absorption spectrum (Equation 2.11). Even quick-and-dirty estimates will give the right order of magnitude for $k_f$, which may be quite useful. In some cases, however, experimental values of $k_f$ (determined from $k_f = \Phi_f / \tau_f$; see Section 3.9.7, Equation 3.33) are found to be much smaller than those calculated by Equation 2.11. This indicates that the electronic transition to the lowest excited state $S_1$ may be hidden under an absorption band of stronger intensity. Such an observation led to the important identification of a 'forbidden' state in linear polyenes (see Section 4.7).

### 2.1.4   Dipole and Transition Moments, Selection Rules

An electric dipole can be represented by two equal but opposite charges, $+q$ and $-q$, that are held at a distance $r$. The *dipole moment* $\boldsymbol{\mu}$ is a *vector* pointing from the negative to the positive charge.[h] The magnitude of $\boldsymbol{\mu}$, $\mu = qr$, is commonly measured in the non-SI unit debye, $1\,D = 3.336 \times 10^{-30}\,C\,m$. Thus, for an electron held at a distance of 100 pm from a proton, $\mu = 4.8$ D. In quantum mechanics, the dipole moment $\boldsymbol{\mu}$ of a molecule in the ground state is obtained from the ground-state wavefunction $\Psi_0$ as the expectation value of

---

[h] Confusingly, the chemical symbol for a dative bond as in $H_3N \rightarrow BH_3$ indicates the direction of electron donation and, accordingly, the dipole moment vector points from + towards − in many chemistry textbooks. However, IUPAC defines the direction of the dipole moment as pointing from the negative to the positive charge.

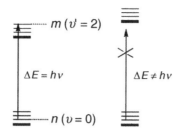

**Figure 2.5** The resonance condition for absorption of light with frequency $\nu$

the dipole moment operator (Equation 2.12). The vectors $r_i = (x_i, y_i, z_i)$ define the positions of the electrons $i$, $r_\mu = (x_\mu, y_\mu, z_\mu)$ those of the nuclei $\mu$, $Z_\mu$ are the charges of the nuclei, and $e$ is the elementary charge. The three components of $\mu$, $\mu_x$, $\mu_y$ and $\mu_z$, are obtained by inserting the respective components of the vectors $r$.

$$\hat{M} = e(\Sigma r_i - \Sigma Z_\mu r_\mu), \quad \mu = \langle \Psi_0 | \hat{M} | \Psi_0 \rangle$$

**Equation 2.12**

For a quantum mechanical treatment of light absorption, the dual nature of light must be taken into account. Light can be absorbed only if the excitation energy matches the photon energy, $\Delta E = h\nu$ that is, if the absorption of a photon by the molecule leads to one of the discrete excited states of the molecule (Figure 2.5).

When the Bohr resonance condition is met, the probability of light absorption may be calculated from the interaction of the electric vector of the light wave of frequency $\nu$ with the electrons and nuclei in the molecule. Molecules being much smaller than the wavelengths of UV–VIS radiation (Figure 1.6), the electromagnetic radiation field acting on a molecule is treated as a linear electric field and magnetic interactions are neglected. We assume that the molecule is initially in the lowest vibrational level ($\nu = 0$) of the electronic ground state $n$ and reaches some vibrational level of the excited state $m$, for example, $\nu' = 2$ as shown in Figure 2.5. Conversely, for emission, one would assume that $\nu' = 0$ and $\nu$ could take any value. The *transition moment* for a single vibronic absorption band $M_{n,0 \to m,\nu'}$, is then given by Equation 2.13.

$$M_{n,0 \to m,\nu'} = e\langle \Psi_{n,0} | \hat{M} | \Psi_{m,\nu'} \rangle$$

**Equation 2.13** *Transition moment*

Like the dipole moment $\mu$, the vector $M_{n,0 \to m,\nu'}$ has a magnitude, measured in debye, and a direction relative to the Cartesian coordinate system of the molecule. Computer programs make use of Equations 2.12 and Equation 2.13 to calculate dipole and transition moments, respectively, from the appropriate wavefunctions $\Psi_{n,0}$ and $\Psi_{m,\nu'}$. The comparison of calculated and experimental (Section 3.6) transition moments is an essential test to validate simplified models of electronic excitation (Section 4.4). Note that the dipole and transition moments of a molecule are not related; in fact, they may well be orthogonal to each other.

To calculate the transition moment $M_{n,0 \to m,\nu'}$, we use the approximate wavefunctions that are factored into the product of an electronic function $\Psi_{el}$, a vibrational wavefunction

$\chi$ and a spin function $\sigma$ (Equation 1.10). The dipole moment operator $\hat{M}$ is a multiplicative operator, $\hat{M}\Psi = \Psi\hat{M}$, and does not affect the spin functions, so that Equation 2.13 can be factored into a product of integrals, Equation 2.14:

$$M_{n,0 \to m,\upsilon'} = e\langle\Psi_{\text{el},n}\chi_0|\hat{M}|\Psi_{\text{el},m}\chi_{\upsilon'}\rangle\langle\sigma_n|\sigma_m\rangle$$

**Equation 2.14**

The spin functions are *orthonormal*: The integral $\langle\sigma_n|\sigma_m\rangle$ equals unity if $\sigma_n = \sigma_m$ and it vanishes otherwise, so that the transition moment vanishes for singlet–triplet and triplet–singlet transitions (or doublet–quartet transitions in radicals). This is our first *selection rule*: *Electronic transitions between states of different multiplicity are forbidden* (the transition moment is zero). Note that selection rules, of which we will encounter several more in Chapter 4, do not hold strictly; they are broken like all rules in real life. They hold only within the limited view of some approximations, in the present case the neglect of magnetic interactions. The multiplicity selection rule is a strong one in the absence of heavy atoms, because magnetic interactions in most organic molecules are very weak. As we have seen in Section 2.1.1, absorption coefficients of singlet–triplet transitions and rate constants of phosphorescence, which are directly related to the absorption coefficients (see Special Topic 2.1 in Section 2.1.3), are orders of magnitude smaller than those for the corresponding spin-allowed transitions, singlet–singlet absorption and fluorescence.

Using the definition of $\hat{M}$ (Equation 2.12), the first integral of Equation 2.14 can be expanded as follows:

$$\langle\Psi_{\text{el},n}\chi_0|\hat{M}|\Psi_{\text{el},m}\chi_{\upsilon'}\rangle = \langle\Psi_{\text{el},n}|\Sigma\mathbf{r}_i|\Psi_{\text{el},m}\rangle\langle\chi_0|\chi_{\upsilon'}\rangle - \langle\Psi_{\text{el},n}|\Psi_{\text{el},m}\rangle\langle\chi_0|\Sigma Z_\mu\mathbf{r}_\mu|\chi_{\upsilon'}\rangle$$

**Equation 2.15**

Note the factorization of the two terms; the vibrational wavefunctions $\chi$ can be separated because they do not depend on the electronic positions $\mathbf{r}_i$ and the electronic wavefunctions can be separated if we disregard their parametric dependence on the precise nuclear positions.[i] The second term then vanishes due to the orthogonality of the electronic wavefunctions $\Psi_{\text{el},n}$ and $\Psi_{\text{el},m}$, so that the transition dipole moment of a spin-allowed transition becomes

$$M_{n,0 \to m,\upsilon'} = e\langle\Psi_{\text{el},n}|\Sigma\mathbf{r}_i|\Psi_{\text{el},m}\rangle\langle\chi_0|\chi_{\upsilon'}\rangle = M_{\text{el},n \to m}\langle\chi_0|\chi_{\upsilon'}\rangle$$

**Equation 2.16**

Equation 2.16 represents the transition moment for a single *vibronic transition* from the zeroth vibrational level of the ground state to a single vibrational level $\upsilon'$ of the electronically excited state. The vibrational overlap integrals $\langle\chi_0|\chi_{\upsilon'}\rangle$ are called *Franck–Condon integrals*; they will generally be different from zero, because the two vibrational wavefunctions refer to different electronic potential energy surfaces (vibrational wavefunctions calculated from a single harmonic potential are orthonormal). The transition moment of a single vibronic band is thus determined by the product of two factors, the *electronic transition moment* (Equation 2.17) and the Franck–Condon integral

---

[i] This approximation must be dropped for forbidden electronic transitions, i.e. when $\langle\Psi_{\text{el},n}|\Sigma\mathbf{r}_i|\Psi_{\text{el},m}\rangle = 0$ (symmetry-forbidden transitions, Section 4.4).

$\langle \chi_0 | \chi_{\upsilon'} \rangle$, the overlap between the lowest vibrational level of the ground state ($\upsilon = 0$) and a given vibrational level $\upsilon'$ of the excited state.

$$M_{\text{el},n \rightarrow m} = e \langle \Psi_{\text{el},n} | \Sigma r_i | \Psi_{\text{el},m} \rangle$$

**Equation 2.17**

The electronic transition moment can be easily calculated on the basis of further simplifications that will be introduced in Section 4.4. For symmetrical molecules, group theory provides a simple tool for predicting which, if any, of the three components ($x$, $y$ or $z$) of the electronic transition vector $M_{\text{el},n \rightarrow m}$ are different from zero (Section 4.4). The effect of vibrational overlap will be considered in Section 2.1.9.

To obtain the total intensity of an absorption band, we have to sum the individual contributions for each vibrational level $\upsilon'$ (Equation 2.18).

$$M_{n \rightarrow m} = e \langle \Psi_{\text{el},n} | \Sigma r_i | \Psi_{\text{el},m} \rangle \sum_{\upsilon'} \langle \chi_0 | \chi_{\upsilon'} \rangle$$

**Equation 2.18** *Overall transition moment of an electronic transition*

If some low-lying vibrational levels $\upsilon > 0$ of the electronic ground state are substantially populated, we must also consider contributions from $M_{n,\upsilon > 0 \rightarrow m,\upsilon'}$ (so-called *hot bands*). Note that transitions from $\upsilon > 0$ to $\upsilon' = 0$ are located to the *red* of the 0–0 transition ($\upsilon = 0$ to $\upsilon' = 0$). Hot bands can be identified on the basis of their temperature dependence given by the Boltzmann population of vibrationally excited states in thermal equilibrium.

The square of the overall transition moment $M_{n \rightarrow m}$ (Equation 2.18) is proportional to the *oscillator strength* $f_{nm}$ (Equation 2.19), where the frequency $\bar{\nu}_{nm}$ and the wavenumber $\bar{\nu}_{nm}$ are average values for the electronic transition and D is the debye unit, 1 D $= 3.336 \times 10^{-30}$ C m. The dimension of $f_{nm}$ is unity.

$$f_{nm} = \frac{8\pi^2 m_e \bar{\nu}_{nm}}{3he^2} |M_{n \rightarrow m}|^2 \approx 4.7 \times 10^{-3} \frac{\bar{\nu}_{nm}}{\mu m^{-1}} \frac{|M_{n \rightarrow m}|^2}{D^2}$$

**Equation 2.19** *Oscillator strength*

The oscillator strength was originally defined as unity for an electron oscillating harmonically in three dimensions, an early model of an atom. It can be determined experimentally by integration of an absorption band (Equation 2.20), where $\varepsilon$ is the *molar absorption coefficient*.

$$f_{nm} = 4.3 \times 10^{-5} \int \frac{\varepsilon(\tilde{\nu})}{M^{-1} \, cm^{-1}} \frac{d\tilde{\nu}}{\mu m^{-1}}$$

**Equation 2.20** *Experimental determination of an oscillator strength*

The width of an absorption band is typically about $0.3 \, \mu m^{-1}$. Thus for a very strong transition, $\varepsilon \approx 10^5 \, M^{-1} \, cm^{-1}$, the oscillator strength $f$ is indeed on the order of 1.

## 2.1.5 Rate Constants of Internal Conversion; the Energy Gap Law

*Radiationless transitions* (IC and ISC) represent a conversion of electronic energy of an initial, excited state to vibrational energy in a lower-energy electronic state (cf. Figure 2.1). Within the Born–Oppenheimer approximation (Section 1.3), radiationless transitions between electronic states are impossible; the potential energy surfaces represent impenetrable walls. To remove that artefact, the correlation of electronic and nuclear motions must be included explicitly in the Hamiltonian operator. Moreover, radiationless transitions involving a multiplicity change (ISC) remain forbidden unless magnetic interactions are included. These relatively small terms are introduced by time-dependent *perturbation theory* (Equation 2.21).

$$k_{i \to f} = \frac{2\pi}{\hbar} V_{if}^2 \rho_f$$

**Equation 2.21**   *Fermi golden rule*

Equation 2.21 was introduced Dirac,[33] and later dubbed *golden rule No. 2* by Fermi. It is now commonly referred to as *Fermi's golden rule*. The rate constant for a radiationless transition between an initial and final state $k_{i \to f}$ is thus proportional to the product of two factors, the density $\rho_f$ of excited vibrational levels of the final state that match the energy of the initial state and the square of the vibronic coupling term between the initial and final BO state, $V_{if} = \langle \Psi_i | \hat{h} | \Psi_f \rangle$, where $\hat{h}$ is a *perturbation operator* that couples nuclear with electronic motion to promote IC and couples electron spin with orbital angular momentum (spin–orbit coupling, SOC) to allow for ISC. As a result, the density of states is weighted by the square of the Franck–Condon integrals $\langle \chi_i | \chi_f \rangle$ (see Section 2.1.9). Once the radiationless transition has occurred, intramolecular vibrational redistribution (IVR) and dissipation of the excess energy to the solvent, vibrational relaxation, are very fast, so that radiationless processes IC and ISC are generally irreversible.

Quantitative calculations of rate constants $k_{i \to f}$ are demanding, but the *energy gap law* for 'large' molecules is well established empirically: *rate constants of IC and ISC decrease exponentially as the energy difference $\Delta E$ between the two electronic states increases in a series of related molecules*. The physical background for this has been analysed:[34–38] There are two opposing effects: the density of vibrational states increases strongly with increasing energy gap, yet the even stronger exponential decrease of the Franck–Condon overlap integrals $\langle \chi_i | \chi_f \rangle$ between the vibrational wavefunction $\chi_i$ of the initial state and those of the final ones $\chi_f$ (Figure 2.6) dominates at large energy

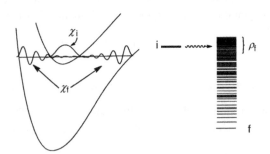

**Figure 2.6**   Left: Franck–Condon integrals $\langle \chi_i | \chi_f \rangle$ for IC are small for large energy gaps, because highly excited vibrational wavefunctions $\chi_f$ oscillate rapidly and their amplitude is small in the region of overlap with $\chi_i$. Right: the density of vibrational levels in the final state increases with the energy gap

gaps [recall the Maxwell–Boltzmann velocity distribution $\rho(v)$ of an ideal gas, where the exponential term $\exp(-mv^2/2kT)$ dominates over the density-of-states term $v^2$ at high velocities $v$].

The dominant accepting modes, those vibrations that contribute most to vibronic coupling between the initial and final states, are those with the highest frequency. Their Franck–Condon integrals are comparatively large, because fewer vibrational quanta are needed to match the energy of the electronically excited state. A beautiful experimental example of both the energy gap law and the influence of accepting modes was provided by Siebrand[39] (Figure 2.7): the rate constants for the radiationless $T_1 \rightarrow S_0$ decay of aromatic hydrocarbons, $k_{TS}$, decrease exponentially with the energy gap and the rate constants for the hydrogenated compounds ($\tilde{\nu}_{CH} \approx 3000\,\text{cm}^{-1}$) are nearly an order of magnitude larger than those of the perdeuterated compounds ($\tilde{\nu}_{CD} \approx 2250\,\text{cm}^{-1}$). To model the observed *isotope effect* (see Special Topic 5.2), constant values of $E_0^H = 0.4\,\mu\text{m}^{-1}$ and $E_0^D = 0.55\,\mu\text{m}^{-1}$ were subtracted from the triplet energies $E_T$ (in $\mu\text{m}^{-1}$) of the hydrogenated and deuterated compounds, respectively, and the difference was divided by the relative number of hydrogen atoms in the molecule, $\eta = N_H/(N_C + N_H)$.

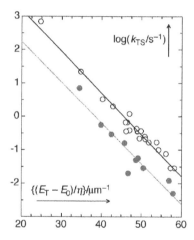

**Figure 2.7** Nonradiative decay rate constants of aromatic hydrocarbon triplets.[39] Filled circles refer to perdeuterated compounds

The roughly exponential dependence of nonradiative transition rate constants on the energy gap $\Delta E$ can be turned into useful empirical rules-of-thumb to estimate rate constants of IC and ISC. As the relevant energy gaps are usually determined from absorption or emission spectra, we replace $\Delta E$ by $\Delta\tilde{\nu}$, $\Delta\tilde{\nu} = \Delta E/hc$. Rate constants of internal conversion can then be roughly estimated from Equation 2.22.

$$\log(k_{IC}/\text{s}^{-1}) \approx 12 - 2\Delta\tilde{\nu}/\mu\text{m}^{-1}$$

**Equation 2.22** *Energy gap law for internal conversion*

### 2.1.6   Rate Constants of Intersystem Crossing, El Sayed Rules

Many photoreactions of organic molecules proceed via the triplet state. Hence intersystem crossing is involved twice, first upon formation of the triplet state from the excited singlet state $S_1$ and second when the primary photoproduct, for example a triplet biradical, returns to the singlet manifold to form a stable product in the ground state. One of these relatively slow steps often determines the overall rate of the reaction. Long-lived excited triplet states are readily quenched (Section 2.2.2), allowing for a deliberate manipulation of the product distribution. ISC requires a spin flip, which can only be induced by magnetic interactions. The magnetic coupling of electron spin to other types of motion is small. Rates of ISC are thus generally smaller than those of IC. Indeed, the electronic coupling term $V_{if}$ of Equation 2.21 vanishes within the approximations used to calculate IC rate constants and spin–orbit coupling (SOC) must be included in the Hamiltonian operator to pick up the dominant terms of these weak magnetic interactions. In biradicals with spatially well-separated radical centres, hyperfine coupling, the interactions of electron spin with the magnetic field of nuclei, may become dominant (Section 5.4.4).[40]

Rates of ISC are around five orders of magnitude slower than those of IC in organic molecules not carrying heavy atoms and also obey the energy gap law (Equation 2.23).

$$\log(k_{ISC}/s^{-1}) \approx 7(\text{or } 10 \text{ for n, } \pi^* \Leftrightarrow \pi, \pi^*) - 2\Delta\tilde{\nu}/\mu m^{-1}$$

**Equation 2.23**   *Energy gap law for intersystem crossing*

However, when ISC involves a change in orbital type,[j] such as in the $^1n,\pi^* \rightarrow {}^3\pi,\pi^*$ transitions of $\alpha,\beta$-unsaturated ketones (Figure 2.8), the reduction amounts to only about two

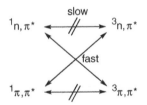

**Figure 2.8**   El Sayed rules for ISC

orders of magnitude. These qualitative rules for ISC are known as *El Sayed's rules*.[41] Accurate calculations of the SOC term $V_{if}$ appearing in the Fermi golden rule (Equation 2.21) are feasible with modern quantum mechanical software packages and the results are consistent with El Sayed's rules. Nevertheless, it is clearly desirable to have some intuitive understanding of the main factors affecting the rate constants of ISC (see Sections 4.8 and 5.4.4).

Caution is necessary when using Equation 2.23 to calculate singlet to triplet ISC rate constants, because the transition from $S_1$ may well occur to an upper triplet state $T_x$ rather than to $T_1$, as long as $E(T_x) \leq E(S_1)$. This will reduce the energy gap and may at the same

---

[j] The symmetry labels $\sigma$ and $\pi$ are used to designate *molecular orbitals* (MOs) that are symmetric and antisymmetric, respectively, with respect to a plane of symmetry of the molecular structure (Section 4.4). Nonbonding $\sigma$-orbitals (lone pairs) are designated with the label n.

time be favoured by a change in orbital type. For example, the rate constants for ISC from the $S_1(n,\pi^*)$ states of benzophenone or 1,4-naphthoquinone are on the order of $2 \times 10^{11}\,s^{-1}$, $\log(k_{ISC}/s^{-1}) \approx 11.3$, presumably because the $S_1(n,\pi^*)$ state couples to a nearly isoenergetic $T_x$ state of $\pi,\pi^*$ character, rather than directly to the $T_1(n,\pi^*)$ state.[42]

### 2.1.7 Quantum Yield: Definition

The quantum yield $\Phi_x(\lambda)$ is equal to the amount $n_x$ of photochemical or photophysical events x that occurred, divided by the amount $n_p$ of photons at the irradiation wavelength $\lambda$ that were absorbed by the reactant, $n_x/n_p$ (Equation 2.24).[k] Both $n_x$ and $n_p$ are measured in moles or einstein (1 einstein = 1 mol of photons) and the dimension of $\Phi_x$ is unity.

$$\Phi_x(\lambda) = n_x/n_p$$

***Equation 2.24*** *Quantum yield of a process x*

An explicit definition of the process x is essential to avoid confusion. For example, when x refers to a photochemical reaction A $\rightarrow$ B, $n_x$ may be defined by the amount of reactant consumed (quantum yield of disappearance, $\Phi_{-A} = -n_A/n_p$),[l] or by the amount of product formed (quantum yield of product formation, $\Phi_B = n_B/n_p$). These will not be the same, $\Phi_B < \Phi_{-A}$, when products other than B are also formed. In general, quantum yields lie in the range $0 \leq \Phi_x \leq 1$ and represent the probability that a molecule undergoes the defined process after absorption of a photon. However, quantum yields may exceed unity, depending on the definition of process x. For a photoreaction A $\rightarrow$ 2B, for example, the quantum yield of product formation, $\Phi_B = n_B/n_p$, spans the range $0 \leq \Phi_x \leq 2$. If the photoreaction generates radicals or other intermediates that start a chain reaction of a monomer C, $R^\bullet + C \rightarrow R - C^\bullet \xrightarrow{C} R - C - C^\bullet \xrightarrow{C} \dots$, then the quantum yield for the disappearance of C is not bounded, $\Phi_{-C} \geq 0$. When C is different from the absorbing species, $\Phi_{-C} = -n_C/n_p$ is not strictly consistent with the definition given above, because the light is not absorbed by C. By extension of *Kasha's rule*, quantum yields are often independent of the excitation wavelength. There are many exceptions, however. For example, when a molecule has two or more chromophores or when the sample contains a mixture of, e.g., non-equilibrating conformers, the quantum yield and the product distribution may be strongly wavelength dependent (Section 5.5).

The quantum yield of product formation $\Phi_B$ is distinct from the chemical yield of product B. The chemical yield may approach 100% if B is the only photoproduct formed, yet $\Phi_B$ may be low when photophysical processes dominate over product formation. Conversely, the quantum yield of reactant disappearance, $\Phi_{-A}$, may be high, even though the chemical yield of product B is low, when the reaction produces mainly products other than B.

The quantitative analysis of the reaction progress is a task that chemists are well trained to do, but the determination of the amount $n_p$ of photons absorbed is not. Both are needed

---

[k] When other absorbing species are present in the sample, then the fraction of the incident light that is absorbed by the reactant A is $A_A(\lambda,t)/A(\lambda,t)$, where $A_A(\lambda,t)$ is the partial absorbance by the reactant A (Equation 2.4) and $A(\lambda,t)$ is the total absorbance of the sample at the wavelength of irradiation $\lambda$ (Equation 2.5). Some authors define $n_p$ as the amount of photons *absorbed by the sample* (which may contain additional species absorbing at the irradiation wavelength) and consider the internal filter effect separately. By that definition, which is not recommended, the quantum yield is no longer a property of the reactant molecule A.

[l] Because compound A disappears, $n_A < 0$, hence $\Phi_{-A} > 0$.

to define the quantum yield of a photoreaction (Equation 2.24). Experimental methods to determine quantum yields and the fundamental importance of these quantities will be discussed in Section 3.9.

### 2.1.8 Kasha and Vavilov Rules

*Kasha's rule* states that *polyatomic molecules generally luminesce with appreciable yield only from the lowest excited state of a given multiplicity*. The concept has been extended to chemical reactions of excited species, that is, polyatomic molecules react with appreciable yield only from the lowest excited state of a given multiplicity. Kasha's rule is a consequence of the energy gap law (Section 2.1.5); the energy differences between upper excited states of a given multiplicity are usually much smaller than those between the lowest singlet or triplet state and the ground state, so that the rates of IC from upper excited states far exceed the rates of fluorescence, let alone phosphorescence. In the lowest excited states $S_1$ and $T_1$, radiative and nonradiative decay become competitive.

Exceptions from Kasha's rule, in particular fluorescence from $S_2$, are to be expected under three circumstances. First, when the energy gap between the first and second excited singlet states is large, so that IC from $S_2$ to $S_1$ is relatively slow, then $S_2$–$S_0$ fluorescence becomes competitive if the oscillator strength of the $S_0 \rightarrow S_2$ transition is large. Similarly, $S_2 - S_1$ fluorescence may be observed. Examples are azulene (**1**)[43] and cycl[3.3.3]azine (**2**)[44,45] (Case Study 4.1); for these compounds, $S_2$ fluorescence is much stronger than $S_1$ fluorescence. Second, when both the $S_2$–$S_1$ energy gap and the oscillator strength of the $S_0 \rightarrow S_1$ transition are very small but that of the $S_0 \rightarrow S_2$ transition is large, then the $S_1$ state is likely to have a long lifetime and fluorescence from the nearby $S_2$ state is observed due to thermal population of the latter. This type of $S_2$ fluorescence, also called two-level emission, will decrease as the thermal population of $S_2$ is reduced at reduced temperature. The classical example of two-level fluorescence is ovalene (Scheme 2.1), which has an $S_2$–$S_1$ energy gap of $1800\,\text{cm}^{-1}$ in the gas phase and only about $500\,\text{cm}^{-1}$ in toluene solution.[46,47] Third, weak fluorescence from upper excited singlet states can be detected in general with sufficiently high sensitivity when stray light sources are carefully eliminated. This has been elegantly achieved by using triplet–triplet annihilation rather than direct excitation to observe delayed fluorescence (Section 2.2.4) from several aromatic hydrocarbons.[48,49] The total emission spectrum of pyrene is shown in Figure 2.9. It is well worth studying this astounding piece of work that displays fluorescence from several upper excited singlet states in addition to phosphorescence from $T_1$ in solution.

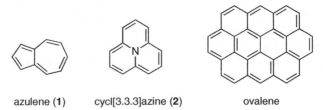

azulene (**1**)        cycl[3.3.3]azine (**2**)        ovalene

**Scheme 2.1**  *Compounds exhibiting $S_2$ fluorescence in violation of Kasha's rule*

*Vavilov's rule* states that *the quantum yield of luminescence is generally independent of the wavelength of excitation*. The quantum yield of luminescence is equal to the

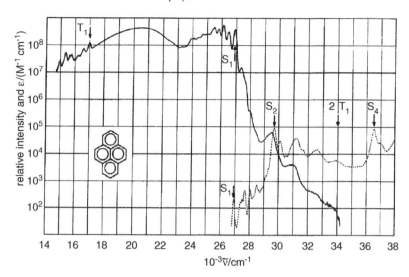

**Figure 2.9** Delayed luminescence (solid line) from a solution of $1 \times 10^{-5}$ M pyrene in methylcyclohexane at 193 K and absorption spectrum (dotted line). Note the logarithmic scale covering eight orders of magnitude in emission intensity. Reproduced from ref. 39 with permission. Copyright 1967, American Institute of Physics

probability that the absorption of a photon by a molecule will be followed by emission of a photon from that molecule as either fluorescence or phosphorescence. Vavilov's rule may be considered as a corollary of Kasha's rule, which implies that molecules that are excited to upper states will quantitatively relax to the lowest excited state by IC and vibrational relaxation. There are exceptions to Vavilov's rule. For example, the fluorescence of benzene vapour suddenly drops to near zero due to some very fast radiationless decay process, the so-called *channel three*,[50] when the photon energy is more than $3000 \, \text{cm}^{-1}$ above the origin of the $S_0 \rightarrow S_1$ transition.

### 2.1.9 Franck–Condon Principle

*The Franck–Condon principle* assumes that *an electronic transition occurs without changes in the positions of the nuclei in the molecular entity and its environment.* Being much lighter than atoms, electrons move much faster. Thus the change in electronic structure (electron redistribution) associated with electronic excitation may be regarded as occurring in the Coulombic field of frozen nuclei. The resulting state is called a Franck–Condon state, an electronically excited state that is born with the structure of the ground state and the transition is called a *vertical transition.*

The quantum mechanical formulation of this principle was given in Section 2.1.5; the intensity of a vibronic transition is proportional to the square of the Franck–Condon integrals between the vibrational wavefunctions of the two states that are involved in the transition. Thus, the band shape of an electronic transition depends on the displacement of the excited electronic state relative to that of the ground state. This is illustrated for one vibrational degree of freedom of a given molecule in the schematic diagram in Figure 2.10.

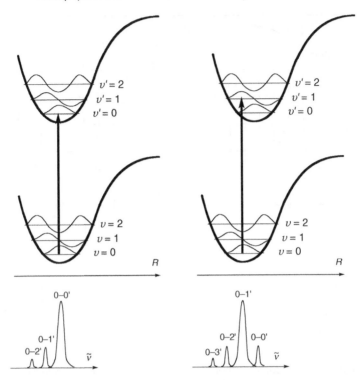

**Figure 2.10**   Illustration of the Franck–Condon principle. The bottom diagrams illustrate the vibrational structure of the absorption bands

When the displacement of the upper electronic PES is small (left), then the overlap between the lowest vibrational levels $\upsilon=0$ in the ground state and $\upsilon'=0$ in the excited state is largest and the 0–0′ transition is by far the most intense. With increasing displacement, the overlap between the ground state vibrational wavefunction $\upsilon=0$ and the higher vibrational levels of the upper state ($\upsilon'=1$ in Figure 2.10, right) becomes largest and the maximum vibronic band intensity moves to higher wavenumbers $\tilde{\nu}$.

The position of the 0–0′ band defines the energy of the excited state relative to that of the ground state, $E_{0-0}=h\nu_{0-0}$. It can usually be located accurately in gas-phase spectra, especially in high-resolution spectra that can be obtained in low-temperature molecular beams. In solution, however, many molecules do not exhibit any vibrational fine structure in their electronic absorption spectra, so that it is difficult to determine $\nu_{0-0}$. Moreover, the intensity dependence illustrated in Figure 2.10 holds only for symmetry-allowed transitions (see Section 4.4). That symmetry-forbidden transitions are observable at all as weak absorptions is due to vibrational borrowing; vibronic transitions to upper (non-totally symmetric) vibrational levels become weakly allowed when the total symmetry of the vibronic transition is considered. Forbidden 0–0 bands are sometimes (barely) detectable in solution spectra due to symmetry perturbations induced by the solvent, but possible contributions from hot bands (Section 2.1.4) must be taken into account.

The spectral distribution of fluorescence usually exhibits an approximate *mirror-image relationship* to the first absorption band when the spectra are plotted on a frequency or wavenumber scale. This is expected from the Franck–Condon principle, when the

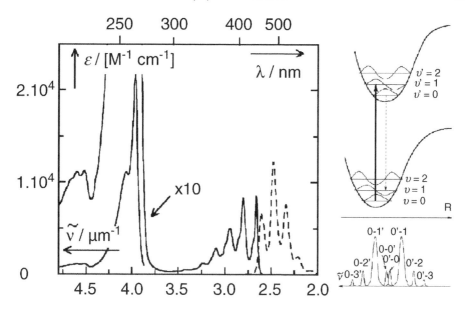

**Figure 2.11** Absorption and fluorescence emission of anthracene

potential energy curves of the ground and excited state are similar in shape, so that the spacing of the dominant vibronic bands is about the same in absorption and emission, as shown in Figure 2.11, right. When diffuse absorption bands do not allow an unambiguous location of the 0–0′ band, then one usually assumes that the crossing point between absorption and emission (normalized to the same height) defines the position of the 0–0′ band and hence the excitation energy (see, e.g., Case Study 5.1).

The spectral positions of the 0–0′ bands of absorption and fluorescence usually do not coincide exactly. The difference is called the *Stokes shift*. The excited state is generally more polarizable and often more polar than the ground state. Solvent relaxation then stabilizes the excited state following Franck–Condon excitation, so that the 0–0′ band of absorption is usually at higher frequency than that of emission. When the excited state is much less polar than the ground state, the 0–0′ band of fluorescence may appear at higher frequency than that of absorption (*anti-Stokes shift*). Unfortunately, the term Stokes shift is also used to refer to the gap between the band maxima of absorption and emission (see the footnote to fluorescence in Section 2.1.1), which may be much larger than the shift between the 0–0 bands.

## Special Topic 2.2: Optical brighteners

Fluorescent whitening agents (FWAs) are important industrial products that are widely used to cover up the yellow tint of laundry, paper and road markings to produce a 'whiter than white' effect.[51] They are also used in polymer sheets operating as concentrators of solar light. FWAs are dyes that absorb in the near-ultraviolet region and fluoresce in the visible region, usually the blue. Fluorescent dyes absorbing and emitting at longer wavelengths are used to enhance the visibility of signposts on ski slopes or as marker inks to emphasize text. For example, a solution of fluorescein appears to be green due to its strong fluorescence; in fact, its colour is yellow when the light source is viewed through the solution. The presence of FWAs on paper and laundry becomes particularly obvious under UV illumination such as under the 'black light' used in bars or for the analysis of thin-layer chromatographic plates. Banknotes usually do not contain optical brighteners except for security marks, so a common method for detecting forged notes is to check for fluorescence.

Desirable properties of FWA are high absorbance in the near-UV region, but not in the visible region, that is, a strong absorption band rising sharply below 400 nm, a high fluorescence quantum yield and therefore low quantum yields of ISC and IC and high photostability in addition to the other properties required for industrial application such as good adsorption and strong adherence to the substrates, low toxicity and rapid degradation in the environment.

The most common classes of chemicals used as FWAs are stilbenes and styrenes, triazines, benzoxazoles, coumarins, naphthalimides and covalently linked or fused combinations thereof (Scheme 2.2). DSBP, a distyrylbiphenyl derivative, is one of the most widely used FWAs in laundry detergents with an estimated worldwide production of 3000 tons per year in 1990. Its photochemical degradation by Fe(III) ions in aqueous media has been studied.[52]

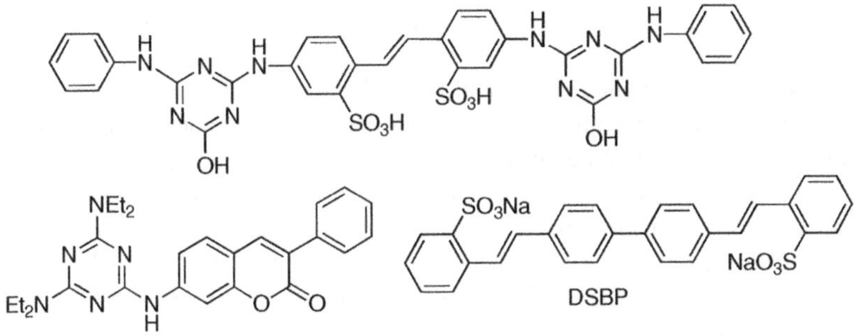

**Scheme 2.2**　*Typical commercial optical brighteners*

A recent development is the use of quantum dots (see Special Topic 6.30) as optical brighteners in fibres.

## 2.2　Energy Transfer, Quenching and Sensitization

### 2.2.1　Diffusion-Controlled Reactions in Solution, Spin Statistics

*Recommended review.*[53]

Electronically excited states are highly reactive and so are many of their primary photoproducts. Therefore, intermolecular processes that occur on every encounter between reactant molecules in solution, also called diffusion-controlled reactions, are fairly common in photochemistry.

Encounter frequencies between solute molecules are lower than collision frequencies in the gas phase, because the translational motion is hindered by the bulk of solvent molecules. Conversely, once reactants have encountered each other in solution, they undergo many collisions, because the surrounding solvent molecules hinder their separation (Franck–Rabinowitch *cage effect*, Special Topic 6.11).

Reactions between molecules A and B via an *encounter complex* A··B may be treated by the kinetic Scheme 2.3, where $k_d$ is equal to the bimolecular rate constant of diffusion, $k_{-d}$ is the first-order rate constant for escape from the encounter complex and $k_P$ is the first-order rate constant of product formation in the encounter complex.

$$A + B \; \underset{k_{-d}}{\overset{k_d}{\rightleftharpoons}} \; A\text{··}B \; \overset{k_P}{\longrightarrow} \; P$$

**Scheme 2.3**

The net rate of change of the encounter complex concentration $c_{A\text{··}B}$ is equal to its rate of formation minus its rate of decay (Equation 2.25).

$$dc_{A\text{··}B}/dt = k_d c_A c_B - (k_{-d} + k_P) c_{A\text{··}B}$$

**Equation 2.25**

The concentration $c_{A\text{··}B}$ will be small compared with $c_A$ and $c_B$ throughout the reaction and after a short induction period its rate of formation will be largely balanced by its rate of decay, that is, we may assume $dc_{A\text{··}B}/dt \approx 0$ (*steady-state approximation*). Rearrangement yields $c_{A\text{··}B} \approx k_d c_A c_B/(k_{-d} + k_P)$ and the rate of product formation, $dc_P/dt = k_P c_{A\text{··}B}$, is given by Equation 2.26.

$$dc_P/dt \approx k_d c_A c_B \frac{k_P}{k_{-d} + k_P}$$

**Equation 2.26**

The partition ratio $k_P/(k_{-d} + k_P)$ defines the *efficiency* of product formation from the encounter complex (see also Section 3.7.4). For the limiting case $k_{-d} \ll k_P$, where most encounters will be reactive, Equation 2.26 reduces to $dc_P/dt \approx k_d c_A c_B$, that is, the overall observed rate constant of reaction approaches the rate constant of diffusion, $k_r \approx k_d$. In 1917, von Smoluchowski derived Equation 2.27 from Fick's first law of diffusion for the ideal case of large spherical solutes.

$$k_d = 4\pi r^* D N_A$$

**Equation 2.27**   *von Smoluchowski equation*

The parameter $r^*$ is the distance at which reaction occurs and $D = D_A + D_B$ is the sum of the diffusion coefficients of A and B. The latter can be calculated from the Stokes–Einstein Equation 2.28:

$$D_A = \frac{k_B T}{6\pi \eta r_A}$$

**Equation 2.28**   *Stokes–Einstein equation*

where $k_B$ is the Boltzmann constant, $\eta$ is the solvent viscosity, which is usually quoted in the non-SI unit poise (1 P $= 0.1$ kg m$^{-1}$ s$^{-1} = 0.1$ Pa s), and $r_A$ is the radius of molecule A, which is assumed to be spherical and much larger than the solvent molecules. The SI units of diffusion coefficients are $[D] = $ m$^2$ s$^{-1}$.

Having adopted various approximations up to this point, we may also assume that $r_A \approx r_B \approx r^*/2$ which, upon replacement of $k_B N_A$ by the gas constant $R$, leads to Equation 2.29.

$$k_d = \frac{8RT}{3\eta}$$

***Equation 2.29*** *Rate constant of diffusion of large spherical solutes*

Note that $r^*$ and the diffusion coefficient $D$ have cancelled from Equation 2.29, because $D$ is inversely proportional to the molecular radii $r^*/2$. Hence the rate constant $k_d$ depends only on temperature and solvent viscosity in this approximation. A selection of viscosities of common solvents and rate constants of diffusion as calculated by Equation 2.29 is given in Table 8.3. The effect of diffusion on bimolecular reaction rates is often studied by changing either the temperature or the solvent composition at a given temperature. For many solvents,[54–56] although not for alcohols,[57] the dependence of viscosity on temperature obeys an Arrhenius equation, that is, plots of log $\eta$ versus $1/T$ are linear over a considerable range of temperatures and so are plots of $\log(k_d \eta / T)$ versus $1/T$.[56]

For solute molecules that are comparable in size to the solvent molecules or even smaller, Equation 2.29 yields values for $k_d$ that are too low, because small molecules can 'slip' between the larger solvent molecules (see Section 2.2.5 on dioxygen). For small molecules, the von Smoluchowski Equation 2.27 gives more accurate predictions than the simplified Equation 2.29, provided that diffusion coefficients and reaction distances $r^*$ can be determined independently.[58]

The lifetime of encounter complexes between neutral reactants is on the order of 0.1 ns in solvents of low viscosity, that is, $k_{-d} \approx k_d$ M. Random diffusive displacements of the order of a molecular diameter occur with a frequency of about $10^{11}$ s$^{-1}$. Subsequently, the fragments from a specific dissociation may re-encounter each other and undergo 'secondary recombination'.[59] If secondary recombination does not take place within about 1 ns, the fragments will almost certainly have diffused so far apart that the chance of a re-encounter becomes negligible. The initial overall electronic multiplicity $2S + 1$ of encounter complexes is thus important in determining the fate of the reactants, because their lifetime is usually insufficient to allow for intersystem crossing during an encounter.

When the multiplicity of at least one of the reactants exceeds unity, we expand Scheme 2.3 to include the multiplicities $m$ and $n$ of the reactants (Scheme 2.4).[53]

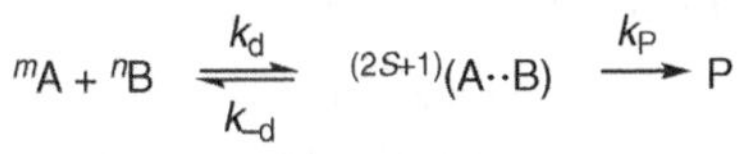

$$^m A + {}^n B \underset{k_{-d}}{\overset{k_d}{\rightleftharpoons}} {}^{(2S+1)}(A{\cdot}{\cdot}B) \overset{k_P}{\longrightarrow} P$$

**Scheme 2.4**

The product $mn = (2S_A + 1)(2S_B + 1)$ is the number of possible total spin angular momentum quantum numbers $S$ of the encounter complex that are allowed by combination of the quantum numbers $S_A$ and $S_B$ of the reaction partners as defined by Equation 2.30.

$$S = S_A + S_B, S_A + S_B - 1, \ldots, |S_A - S_B|$$

***Equation 2.30***

Because the energy differences $\delta E$ between the sublevel states of each multiplet of $^mA$ and $^nB$ are very small, $\delta E \ll kT$ at ambient temperature and above, they are nearly equally populated under equilibrium conditions (Boltzmann's law, Equation 2.9) and the probability of the formation of any given encounter spin state will be equal to all the others; as there are $mn$ choices, it will be equal to the *spin-statistical factor* $\sigma = (mn)^{-1}$.

For example, the multiplicity of radicals with one unpaired electron, $S = \frac{1}{2}$, is $2S + 1 = 2$. Each of four spin states is then expected to form with equal probability upon encounter of two radicals $^2A$ and $^2B$, $\sigma = 1/4$. Three of these are sublevels of the encounter complex with triplet multiplicity, $S = S_A + S_B = 1$, $2S + 1 = 3$, and the fourth is the singlet encounter pair, $S = S_A + S_B - 1 = 0$, $2S + 1 = 1$. Only the latter can undergo radical recombination to form a singlet product $P = A-B$ without undergoing ISC. The above considerations therefore suggest that the rate constant for radical recombination will not exceed one-quarter of the rate constant of diffusion, because only every fourth encounter will lead to recombination.

When two molecules in a triplet state collide, the spin quantum numbers $S_A = S_B = 1$ can combine to give the total spin quantum numbers $S = 2$, 1 or 0 and the corresponding multiplicities of the encounter complex are $2S + 1 = 5$, 3 or 1. If the only possible product P has singlet multiplicity, the statistical factor $\sigma$ for reaction is 1/9 (Scheme 2.5). Thus, observed rate constants for the quenching of triplet states by dioxygen producing singlet oxygen (Section 2.2.5) are generally no more than one-ninth of the diffusional rate constant for oxygen.[60,61] Similarly, the formation of an excited singlet state by triplet–triplet annihilation (Equation 2.51) and the formation of the anthracene dimer[62] (Scheme 6.90) upon encounter of two anthracene triplets are subject to a spin statistical factor of 1/9th.

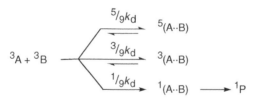

**Scheme 2.5**

## 2.2.2 Energy Transfer

*Recommended textbooks.*[63,64]

This section deals with processes by which the excitation energy of an excited molecule $D^*$, the energy donor, is transferred to a neighbouring molecule A, the energy acceptor (Equation 2.31). The multiplicity of $D^*$ and $A^*$ will be specified as we look at the different mechanisms of energy transfer.

$$D^* + A \rightarrow D + A^*$$

**Equation 2.31** *Energy transfer*

Energy transfer permits electronic excitation of molecules A that do not absorb the incident light. This is exploited, for example, for light harvesting in photosynthetic organisms. The energy absorbed by the antenna, which may consist of tens or even hundreds of chlorophyll molecules, is transported to the reactive centre, which then transforms it into storable chemical energy (Special Topics 6.25 and 6.26).[65] We will consider only the lowest excited singlet and triplet state of the donor D because, under most conditions, *internal conversion* and *vibrational relaxation* (Section 2.1.1) from higher excited states of D will be much faster than any competing intermolecular energy transfer processes. Exceptions may occur in gases at very low pressure, where thermal relaxation is slow or when energy transfer is very rapid in systems containing high local concentrations of A.[66] Energy transfer processes are isoenergetic: the energy lost by the donor reappears in the acceptor. If the electronic excitation energy of A is below that of D, an excited vibrational level of A will be populated initially and the excess energy is then rapidly dissipated to the medium rendering energy transfer irreversible (Figure 2.12, left).

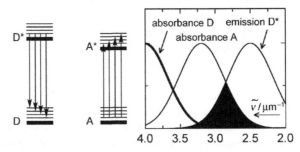

**Figure 2.12**   Resonance condition for energy transfer from D* to A. The thick horizontal bars in the left-hand diagram represent electronic and the thin bars vibrational states. The dark area on the right represents the spectral overlap integral *J*

*Radiative energy transfer* proceeds by way of electromagnetic radiation: a photon is spontaneously emitted by the excited donor D* and subsequently absorbed by the acceptor A. This is by far the dominant process of energy transport in dilute gases as in interstellar space. Radiative energy transfer was considered in the derivation of Planck's law for the frequency distribution of a *black-body radiator* (Special Topic 2.1). It is sometimes referred to as *trivial energy transfer*, because it is conceptually simple and must be taken into account under any circumstances. However, we shall see that it is usually not the dominant energy transfer process, except in very large, dilute systems, where intermolecular interactions are negligible. The absorption spectrum of A must overlap with the emission spectrum of D* for A to absorb a photon emitted by D (Figure 2.12, right).

The probability $p$ that a photon emitted by D* will be absorbed by A is given by Equation 2.32:

$$p = \int_{\tilde{\nu}} \bar{I}_{\tilde{\nu}}^{D^*}[1-10^{-A(\tilde{\nu})}]\, d\tilde{\nu}$$

$$\text{where} \quad \bar{I}_{\tilde{\nu}}^{D^*} = I_{\tilde{\nu}}^{D^*} \Big/ \int_{\tilde{\nu}} I_{\tilde{\nu}}^{D^*} d\tilde{\nu}, \text{ that is } \int_{\tilde{\nu}} \bar{I}_{\tilde{\nu}}^{D^*} d\tilde{\nu} = 1$$

**Equation 2.32**   *Probability of radiative energy transfer from D* to A* where the normalized *spectral radiant intensity*[m,10,22] $\bar{I}_{\tilde{\nu}}^{D*}$ represents the spectral distribution of the fluorescence emission by D* and the term $10^{-A(\tilde{\nu})}$ is equal to the transmittance $T(\tilde{\nu})$ from the point of emission by D* to the bounds of the solution. Equation 2.32 is readily derived by considering the probability for reabsorption of a photon in a given wavenumber interval $d\tilde{\nu}$, $dp = \bar{I}_{\tilde{\nu}}^{D*}[1-T(\tilde{\nu})]d\tilde{\nu}$.

Taylor expansion of the exponential term, $10^{-A} \approx 1 - \ln(10)A$, neglecting higher terms and replacement of $A(\tilde{\nu})$ by $\varepsilon_A(\tilde{\nu})c_A\ell$ gives the approximate expression Equation 2.33, which holds for low absorbances $A$. The probability $p$ for energy transfer is then proportional to the concentration $c_A$, to the average optical pathlength $\ell$ of emitted photons travelling through the solution that depends on the size and the shape of the sample and to the spectral overlap integral $J$ (Figure 2.12). The spectral distributions of the emission by D*, $\bar{I}_{\tilde{\nu}}^{D*}$, and of the absorption by A, $\bar{\varepsilon}_A(\tilde{\nu})$, are normalized by definition, so that *the overlap integrals $J^{\tilde{\nu}} = J^{\lambda}$ do not depend on the oscillator strengths of the transitions involved.*

$$p \cong 2.303 c_A \ell J$$

$$\text{where} \quad J^{\tilde{\nu}} = \int_{\tilde{\nu}} \bar{I}_{\tilde{\nu}}^{D*}\bar{\varepsilon}_A(\tilde{\nu})d\tilde{\nu}; \quad \int_{\tilde{\nu}} \bar{I}_{\tilde{\nu}}^{D*}d\tilde{\nu} \equiv \int_{\tilde{\nu}} \bar{\varepsilon}_A(\tilde{\nu})d\tilde{\nu} \equiv 1$$

$$\text{and} \quad J^{\lambda} = \int_{\lambda} \bar{I}_{\lambda}^{D*}\bar{\varepsilon}_A(\lambda)d\lambda; \quad \int_{\lambda} \bar{I}_{\lambda}^{D*}d\lambda \equiv \int_{\lambda} \bar{\varepsilon}_A(\lambda)d\lambda \equiv 1$$

**Equation 2.33** *Probability of radiative energy transfer from D\* to A*

Radiative energy transfer between identical molecules tends to increase the apparent lifetime of the emitting species D* because photons already emitted are being recaptured. However, it rarely occurs, because the overlap integral $J$ is necessarily small for D=A (Figure 2.12). For D≠A, on the other hand, $J$ may approach unity in favourable cases. Yet *radiative energy transfer from D\* to acceptors A does not affect the lifetime of D\**; it occurs only when D* emits spontaneously and it is irreversible when the excitation energy of A is lower than that of D due to the rapid dissipation of excess vibrational energy on A*.

In practice, it is usually observed that the fluorescence lifetime (Section 3.5) and quantum yield (Section 2.1.7) of D* *decrease* when an acceptor A is added, a phenomenon called *concentration quenching*. Clearly, some interactions between D* and A play a role. Energy transfer processes that reduce the lifetime of D* cannot be attributed to the spontaneous emission of a photon by D* and are thus called *nonradiative energy transfer* processes (stimulated energy transfer would also be an appropriate expression, but is rarely used). Moreover, Pringsheim noted in 1924 that the polarization of fluorescence (Section 3.6) decreases rapidly with increasing concentration of D in solid solutions, much more so than could reasonably be explained on the basis of radiative energy transfer that, moreover, should have depended on the sample size, but did not. In the same year, Perrin attributed fluorescence depolarization to nonradiative energy transfer and offered a classical electrodynamic model to explain energy transfer over large distances. Perrin later provided

---

[m] *Radiant intensity*, $I$, is the radiant power (Equation 2.3), $P$, per solid angle, $\Omega$, emitted at all wavelengths, $I = dP/d\Omega$. If the radiant power is constant over the solid angle considered, $I = P/\Omega$. The unit of $I$ is W sr$^{-1}$. The *spectral radiant intensity* is the derivative of radiant intensity with respect to wavelength or wavenumber, $I_{\lambda} = dI/d\lambda$, $\bar{I}_{\tilde{\nu}} = dI/d\tilde{\nu}$. The SI units are $[I_{\lambda}] = $ W m$^{-1}$ sr$^{-1}$ and $\bar{I}_{\tilde{\nu}} = $ W m sr$^{-1}$.

a quantum mechanical interpretation, but neither model could reproduce the available data accurately.

The analogy to a radio emitter and receiver is sometimes used to visualize energy transfer. This analogy is adequate for radiative but not for nonradiative energy transfer, because in the latter process the emitter is stimulated by the acceptor to forsake its excitation energy.

A quantum mechanical treatment of nonradiative energy transfer is based on *Fermi's golden rule* (Equation 2.21). The interaction term $V_{if} = \langle \Psi_i | \hat{h} | \Psi_f \rangle$, which couples the (initial) zero-order wavefunction $\Psi_i = \Psi_{D^*} \Psi_A$ of the neighbouring molecules prior to energy transfer with the wavefunction $\Psi_f = \Psi_D \Psi_{A^*}$ of the final state, permits energy transfer over extended distances *without* the intervention of a photon. The Coulombic interactions contained in the perturbation operator $\hat{h}$ may be expanded as a multipole series. If the intermolecular distances are much larger than the size of the molecules, the first term, representing the dipole–dipole interaction between the transition moments of the donor $D^*$ and of the acceptor A, predominates for allowed transitions. As the interaction between two point-dipoles falls off with the third power of the distance, the rate constant for *Förster-type resonance energy transfer* (FRET),[n] $k_{FRET}$, being proportional to $V_{if}^2$ (Equation 2.21), falls off with the sixth power of the distance $R$ between $D^*$ and A (Equation 2.34):

$$k_{FRET} = R_0^6 / (R^6 \tau_D^0)$$

*Equation 2.34*  *Förster resonance energy transfer*

where $\tau_D^0$ is the lifetime of $D^*$ in the absence of acceptor molecules A and $R_0$ is called the *critical transfer distance*. When the distance $R$ equals $R_0$, $k_{FRET} = 1/\tau_D^0$, energy transfer and spontaneous decay have equal rate constants and these processes are equally probable. Note that the rate constant $k_{FRET}$ is a first-order rate constant, $[k_{FRET}] = s^{-1}$. It pertains to the rate of energy transfer between $D^*$ and A that are being held at fixed distance and relative orientation.

By applying Fermi's golden rule, Förster derived a very important relation between the critical transfer distance $R_0$ and experimentally accessible spectral quantities (Equation 2.35),[o, 67,68] namely the luminescence quantum yield of the donor in the absence of acceptor A, $\Phi_D^0$, the orientation factor, $\kappa$, the average refractive index of the medium in the region of spectral overlap, $n$, and the spectral overlap integral, $J$. The quantities $J$ and $\kappa$ will be defined below. Equation 2.35 yields remarkably consistent values for the distance between donor and acceptor chromophores $D^*$ and A, when this distance is known. FRET is, therefore, widely applied to determine the distance between markers D and A that are attached to biopolymers, for example, whose tertiary structure is not known and thus

---

[n] The term FRET was first used in papers relevant to life sciences as an acronym for 'fluorescence resonance energy transfer'. This is a misnomer, because fluorescence is not involved in the transfer, which is nonradiative. The term RET (resonance energy transfer) would be more appropriate, but it is also used for 'return electron transfer' (Section 5.4.3). Because FRET is an established acronym, the letter 'F' is now taken to stand for 'Förster, or Förster-type'.

[o] 'For Förster's scholars in Stuttgart this book was so to speak the house bible, a quite appropriate designation, for one because it provided the means of understanding and second because of its precise and concise style, where each word mattered and many sentences required very careful reading and re-reading or – better still – the exegesis by an interpreter, before one could grasp its superb content. For those not conversant in German it must have largely remained a book with seven seals; because it was not translated into English, American colleagues say that it secured a German lead in fluorescence spectroscopy for years.' [Translated from the obituary of Förster given by Albert Weller, 1974].

serves as a *"molecular ruler"* (Special Topic 2.3).

$$R_0^6 = \frac{9\ln(10)}{128\pi^5 N_A} \frac{\kappa^2 \Phi_D^0}{n^4} J$$

**Equation 2.35** *Förster's equation to calculate the critical transfer distance $R_0$ was erroneously printed with $\pi^6$ instead of $\pi^5$ in the denominator in several papers by Förster. He corrected this misprint in a later publication.[69,70,p]*

The spectral overlap integral $J$ can be expressed in terms of either wavenumbers or wavelengths (Equation 2.36). The area covered by the emission spectrum of D* is normalized by definition and the quantities $\bar{I}_{\tilde{\nu}}^{D*}$ and $\bar{I}_{\lambda}^{D*}$ are the normalized spectral radiant intensities of the donor D* expressed in wavenumbers and wavelengths, respectively. Note that the spectral overlap integrals $J$ defined here differ from those relevant for radiative energy transfer (Equation 2.33). Only the spectral distributions of the emission by D*, $\bar{I}_{\tilde{\nu}}^{D*}$ and $\bar{I}_{\lambda}^{D*}$, are normalized, whereas the transition moment for excitation of A enters explicitly by way of the molar absorption coefficient $\varepsilon_A$. The integrals $J^{\tilde{\nu}}$ and $J^{\lambda}$ are equal, because the emission spectrum of D* is normalized to unit area and the absorption coefficients $\varepsilon_A$ are equal on both scales.

$$J^{\tilde{\nu}} = \int_{\tilde{\nu}} \bar{I}_{\tilde{\nu}}^{D*} \varepsilon_A(\tilde{\nu}) \frac{d\tilde{\nu}}{\tilde{\nu}^4}, \quad J^{\lambda} = \int_{\lambda} \bar{I}_{\lambda}^{D*} \varepsilon_A(\lambda) \lambda^4 d\lambda,$$

$$\text{where} \quad \int_{\tilde{\nu}} \bar{I}_{\tilde{\nu}}^{D*} d\tilde{\nu} \equiv 1 \quad \text{and} \quad \int_{\lambda} \bar{I}_{\lambda}^{D*} d\lambda \equiv 1$$

**Equation 2.36** *Spectral overlap integral for nonradiative energy transfer*

Collecting the numerical constants and length conversion factors, we obtain the practical expression Equation 2.37. The proportionality constant will differ for a different choice of units. It is unnecessary and not recommended to convert the emission spectrum of the donor to a wavenumber scale for the calculation of $J$ (see Section 3.4).

$$\frac{R_0}{nm} = 0.02108 \left( \frac{\kappa^2 \Phi_D^0}{n^4} \frac{J^{\lambda}}{mol^{-1} \, dm^3 \, cm^{-1} \, nm^4} \right)^{\frac{1}{6}}$$

**Equation 2.37** *Practical expression for $R_0$*

The variables $\tilde{\nu}$ and $\lambda$ in $J^{\tilde{\nu}}$ and $J^{\lambda}$ are sometimes replaced by constant average values $\bar{\tilde{\nu}}$ and $\bar{\lambda}$, respectively, representing the positions of maximum spectral overlap, so that they can be moved in front of the integrals $J$. The orientation factor $\kappa$ is defined by Equation 2.38:

$$\kappa = \frac{\mu_D \mu_A - 3(\mu_D r)(\mu_A r)}{\mu_D \mu_A}$$

**Equation 2.38** *Orientation factor*

---

[P] In Equation 2.3, the units can be chosen arbitrarily, as long as their product is the same on both sides. Most papers, reviews and textbooks show a factor of 9000 rather than 9 in the numerator. Apparently, this is done to convert $dm^3$ (appearing in the molar absorption coefficient $\varepsilon$) to $cm^3$. Equation 2.35 with a factor of 9000 no longer has equal units on both sides (*caveat emptor!*). Such confusing and unnecessary practice should be avoided. When doing such a conversion, the units must be given explicitly, as in the scaled Equation 2.37.[70]

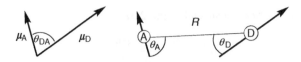

**Figure 2.13**   Angles defining the orientation factor $\kappa$

where $\mu_D$ and $\mu_A$ are the transition moment vectors of D and A and $r$ is the unit vector in the direction of $R$. Possible values for $\kappa^2$ range from 0 to 4 depending on the relative orientation of the transition dipole moments $\mu_D$ and $\mu_A$. The relevant angles are defined in Figure 2.13; $\kappa^2 = 0$ when the electric field generated by $\mu_D$ is perpendicular to $\mu_A$, that is, when $\theta_{DA} = \theta_D = 90°$ or $\theta_{DA} = \theta_A = 90°$ or when the two terms in the numerator of Equation 2.38 cancel; $\kappa^2 = 4$ when all vectors are in line, $\theta_{DA} = \theta_D = \theta_A = 0°$.

Two issues require careful consideration. The first is the effect of diffusion on energy transfer rates in media of low viscosity, which will be considered below. The second is the choice of an appropriate orientation factor $\kappa$. For FRET in solution, it is usually warranted to assume that the molecules D* and A undergo rapid Brownian rotation during the lifetime of D*. Averaging over all possible orientations of the transition dipoles gives $\langle \kappa^2 \rangle = 2/3$. If the dipoles are randomly oriented but do not rotate on the time scale of the donor luminescence in rigid media, other average values must be used, for example $\langle |\kappa| \rangle^2 = 0.476$ for large three-dimensional samples.[71] Concerns[72] that the use of $\langle \kappa^2 \rangle = 2/3$ may be inadequate, when fluorescent labels are attached to specific sites of biopolymers such as enzymes or DNA in order to determine the distance between these sites from observed FRET efficiencies, have been dispelled.[73]

---

**Special Topic 2.3: Energy transfer – a tool to measure distances and to track the motion of biopolymers**

The efficiency (Equation 3.32, Section 3.9.7) of FRET is defined as $\eta_{FRET} = k_{FRET}/(k_{FRET} + 1/\tau_D^0)$. Inserting Equation 2.34 for $k_{FRET}$ one obtains Equation 2.39. This function falls off rapidly as $R$ exceeds the critical Förster radius $R_0$, Figure 2.14.

$$\eta_{FRET} = R_0^6/(R_0^6 + R^6)$$

***Equation 2.39***   *Efficiency $\eta_{FRET}$ as a function of distance*

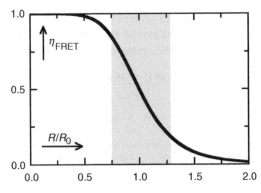

**Figure 2.14**   Efficiency $\eta_{FRET}$ as a function of distance

Having attached a donor chromophore D and an acceptor A to specific sites of a biopolymer such as a DNA hairpin or an enzyme, the efficiency of FRET is measured by comparing the steady-state fluorescence signal intensity of the donor D in the presence and absence of the acceptor A under otherwise identical conditions, $\eta_{FRET} = 1 - I_\lambda^0 / I_\lambda^A$. The critical Förster radius $R_0$ is calculated from the spectral properties of D and A (Equation 2.37). With the experimental for value $\eta_{FRET}$, Equation 2.39 defines the apparent donor–acceptor distance, $R = R_{app}$ (Equation 2.40); hence the expressions '*molecular ruler*' or '*spectroscopic ruler*'.[74,75]

$$R_{app} = R_0 \left( \frac{I_\lambda^A}{I_\lambda^0} \right)^{1/6}$$

**Equation 2.40** *FRET as a molecular ruler*

When the distance $R$ between D and A can be calculated or measured by other methods in relatively rigid systems, it has been amply demonstrated that remarkably consistent values of $R_{app}$ are obtained from FRET measurements. However, $R_{app}$ may differ somewhat from the actual donor–acceptor distance in a flexible macromolecule with a broad distribution of donor–acceptor distances. The distance $R_{app}$ derived from steady-state FRET measurements then considerably overestimates the true mean distance.[76] To treat flexible macromolecules or conformational mixtures, time-resolved measurements of FRET are necessary.[75]

In recent years, single-molecule spectroscopy (SMS) (Section 3.13) has been successfully applied to probe the free-energy surfaces that govern protein folding.[77] Moving away from ensemble measurements is an essential step forward in applications of FRET.[78–82] The conformational changes undergone by a single, denatured protein molecule in search of its native structure are recorded in a 'diary' as FRET is repeatedly switched off and on (Figure 2.15). Ideally, one would like to record the dynamics on the reconfiguration time scale of single polypeptide chains as they wiggle their way through the energy landscape, each molecule taking its own route to the folded state. Similarly, single-molecule FRET can visualize enzymes at work by monitoring conformational changes during substrate conversion.[83] SMS using FRET will continue to attract widespread attention.

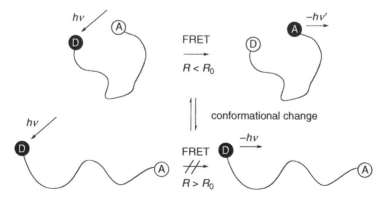

**Figure 2.15** The empty circles represent the ground state and filled circles the excited state of D or A

In 1969, Levinthal posed his famous paradox pointing out that the configurational space of a polypeptide chain is so large that it can never find its native conformation by a random search. Evolution has apparently selected polypeptide chains that find their native fold reliably and quickly. Their free-energy landscapes exhibit minimal ruggedness and the overall shape of a funnel with only shallow local minima.

Rate constants $k_{et}$ for triplet–triplet energy transfer (TET) in flexible bichromophoric systems D–(CH$_2$)$_n$–A were determined from steady-state quenching and quantum yield measurements. The magnitude of $k_{et}$ for $n = 3$ is comparable to that in molecules with a rigid spacer between the chromophores, so that a through-bond mechanism is assumed to remain important. As the connecting polymethylene chain becomes longer, a gradual drop in TET rate constants indicates that through-space interactions, that is, encounter contacts between $^3$D and A, enabled by rapid conformational equilibria, compete and provide the only mechanism responsible for transfer when $n \geq 5$.[84,85]

The elementary steps of peptide folding are single-bond conformational changes, which occur on a picosecond time scale. Intrachain loop formation allows unfolded polypeptide chains to search for favourable interactions during protein folding. The dynamics on the nanosecond time scale represent chain diffusion exploring different local minima on the free energy landscape. Triplet–triplet energy transfer is thus ideally suited to monitor rates of contact formation in polypeptide chains and to explore the early dynamics of peptide folding.[86–88] As an alternative, the long-lived fluorescent state of 2,3-diazabicyclo[2.2.2]oct-2-ene (DBO) was used to monitor contact quenching in oligopeptide chains giving largely consistent results.[89,90] The kinetics of end-to-end collision in single-stranded oligodeoxyribonucleotides of the type 5′-DBO–(X)$_n$–dG (X = dA, dC, dT or dU and $n = 2$ or 4) was investigated by the same method. The fluorophore was covalently attached to the 5′ end and dG was introduced as an efficient intrinsic quencher at the 3′ terminus.[91]

Conditions of FRET from singlet donors $^1$D$^*$ to acceptors A in the singlet ground state are optimal when the excitation energy of the acceptor is somewhat lower than that of the donor, ensuring good spectral overlap. FRET between identical molecules, D = A, is relatively slow, especially at low temperatures, where overlap is restricted to the 0–0 transition. At higher temperatures, the overlap increases somewhat due to thermal population of some excited vibrational levels of $^1$D$^*$. In single crystals, repeated energy transfer can proceed over large distances until the excitation arrives at an impurity trap of lower energy. Fascinating studies of FRET between chromophores in organized systems such as zeolites loaded with chromophores,[92] monolayer assemblies[93] and self-assembling columnar mesophases,[94–96] in dendrimers[97,98] and in other covalently linked chromophores that can be used as molecular switches, multiplexers[99,100] or antenna molecules,[101,102] have been reported. In such closely associated systems of *m* donor and *n* acceptor chromophores, simple models considering only pairwise interactions between nearest neighbours may be inadequate. Rather, all *m* × *n* electronic couplings should be considered simultaneously, which leads to a more delocalized model and even faster rates of energy transfer (*exciton hopping*).[103]

Equation 2.37 predicts very short transfer distances $R_0$ for forbidden transitions of A, such as in singlet–triplet transfer to organic acceptor molecules that have very low singlet–triplet absorption coefficients $\varepsilon$. For $R_0 \leq 0.5$ nm, the dipole approximation is no longer adequate and other interactions come to play (see below). On the other hand, FRET from triplet donors with high quantum yields of phosphorescence and good overlap with the singlet absorption of the acceptor can be fairly efficient,[104,105] notwithstanding the fact that triplet–singlet energy transfer (Equation 2.41) is a spin-forbidden process. The low transition dipole of the triplet emitter $^3D^*$ is compensated by its long lifetime $^3\tau_D^0$.

$$^3D^* + A \rightarrow D + {}^1A^*$$

**Equation 2.41** *Triplet–singlet energy transfer*

Förster radii for a number of common singlet donor and acceptor pairs are given in Table 2.2.

**Table 2.2** *Förster radii determined experimentally for various D–A pairs*

| Donor | Acceptor | Medium | $R_0$/nm | Type | Ref. |
|---|---|---|---|---|---|
| Coronene | Rhodamine-6G | PMMA | 2.84 | $^1D^* \rightarrow A$ | 106 |
| Coronene | Acridine Yellow | PMMA | 3.59 | $^1D^* \rightarrow A$ | 106 |
| 1,12-Benzperylene | Acridine Yellow | PMMA | 3.34 | $^1D^* \rightarrow A$ | 106 |
| Pyrene | Perylene | PMMA | 3.59 | $^1D^* \rightarrow A$ | 106 |
| 9,10-Dichloroanthracene | Perylene | — | 3.78 | $^1D^* \rightarrow A$ | 106 |
| Phenanthrene-$d_{12}$ | Rhodamine B | CA[a] | 4.33 | $^3D^* \rightarrow A$ | 105,106 |
| Perylene | TMPD$^{+b}$ | CH$_3$CN | 4.0 | $^1D^* \rightarrow {}^2A$ | 107 |
| Coumarin 460 | [Ru(bpy)$_3$]$^{2+}$ | CH$_3$CN | 4.4 | $^1D^* \rightarrow A$ | 108 |

[a]Cellulose acetate.
[b]Wurster's blue radical cation.

How does energy transfer affect the decay kinetics and overall intensity of donor fluorescence? In rigid media or highly viscous solvents, where the position of the molecules may be considered as stationary during the lifetime of $^1D^*$ (apart from, possibly, Brownian rotation), the spectral radiant intensity $I_\bar{\nu}^{D^*}$ due to the fluorescence of $^1D^*$ decays non-exponentially with time $t$ after excitation of D by a short light pulse, as predicted by Förster[67,68] (Equation 2.42).[63]

$$I_{\bar{\nu}}^{D^*}(t, c_A) = I_{\bar{\nu}}^{D^*}(t = 0)e^{-\left\{\frac{t}{\tau_D^0} + \frac{2c_A}{c_A^0}\sqrt{\frac{t}{\tau_D^0}}\right\}}$$

**Equation 2.42** *Förster decay kinetics in solid solutions*

The lifetime $\tau_D^0$ is that of the donor fluorescence in the absence of A, $c_A$ is the acceptor concentration and $c_A^0$ is the so-called *critical concentration* of A, defined by Equation 2.43.

$$c_A^0 = \frac{3}{2\pi^{3/2}N_A R_0^3} = 4.473 \times 10^{-25} \left(\frac{R_0}{nm}\right)^{-1} mol \; dm^{-3}$$

**Equation 2.43** *Critical concentration*

Because the decay of $^1D^*$ is accelerated by resonance energy transfer to A, the fluorescence quantum yield of the donor is reduced and that of the acceptor A increases accordingly. Assuming a statistical (Gaussian) distribution of the intermolecular distances, Förster derived Equation 2.44 for the relative fluorescence quantum yield of the acceptor, $\Phi_A(c_A)/\Phi_A{}^{max}$.[68]

$$\frac{\Phi_A(c_A)}{\Phi_A^{max}} = \sqrt{\pi}xe^{x^2}\left\{1 - \frac{2}{\sqrt{\pi}}\int_0^x e^{-y^2}dy\right\}$$

**Equation 2.44**   *Concentration quenching*

The second term in the braces is the Gaussian error function, $x = c_A/c_A^0$, and $\Phi_A{}^{max}$ is the fluorescence quantum yield of A when it is either excited directly or when energy transfer from D to A following excitation of D is complete at high concentration of A. This function and its counterpart, $\Phi_D(c_A)/\Phi_D{}^0 = 1 - \Phi_A(c_A)/\Phi_A{}^{max}$, representing the concentration quenching of $^1D^*$ by A, are plotted in Figure 2.16.

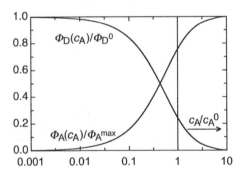

**Figure 2.16**   Relative quantum yields of donor (D) and acceptor (A) fluorescence in rigid solutions. The vertical bar indicates the concentration $c_A = c_A^0$

Thus, in rigid solutions, the critical transfer distance $R_0$ can be determined experimentally either from the observed relative quantum yields of D or A at various concentrations of A (Equation 2.44) or from time-resolved measurement of the fluorescence decay of D (Equation 2.42). The results are in good agreement with those calculated from the Förster Equation 2.37.

In fluid solutions, three regimes can be distinguished. At very high dilution, where donors and acceptors rarely approach to distances below 10 nm, radiative energy transfer dominates and the lifetime of $^1D^*$ is not affected by the presence of A. At moderate acceptor concentrations, fluorescence quenching of $^1D^*$ by A is governed by diffusional approaches of A to $^1D^*$. The mean molecular diffusion distance during the fluorescence lifetime $\tau_D$ of the donor is given by Equation 2.45:

$$\bar{r} = \sqrt{2D\tau_D}$$

**Equation 2.45**   *Mean molecular diffusion distance*

where $D$ is the sum of the diffusion coefficients of donor and acceptor, which can be calculated from the Stokes–Einstein Equation 2.28.

As long as $\bar{r} > 3R_0$, the fluorescence decay is close to exponential, the lifetime of the donor fluorescence decreases linearly with increasing concentration of A and fluorescence quenching obeys Stern–Volmer kinetics (Section 3.9.8, Equation 3.36). However, the bimolecular rate constants $k_{et}$ of energy transfer derived from the observed quenching of donor fluorescence often exceed the rate constants of diffusion $k_d$ calculated by Equation 2.26, because resonance energy transfer does not require close contact between D and A. Finally, when $\bar{r} \leq 3R_0$, at high concentrations and low solvent viscosity, the kinetics of donor fluorescence become complicated, but an analysis is possible,[109,110] if required.

Although triplet–triplet energy transfer (Equation 2.46) is a spin-allowed process, total spin being conserved, it is forbidden in terms of the Coulombic (Förster) formalism, because the oscillator strengths for singlet–triplet absorption by A are extremely small. A quantum mechanical treatment for triplet energy transfer by electron exchange was reported by Dexter; triplet energy transfer is treated as a synchronous, dual exchange of electrons (Section 5.2) between the molecular orbitals of the donor molecule $^3D^*$ and those of the acceptor A (Figure 2.17).

$$^3D^* + A \rightarrow D + {}^3A^*$$

**Equation 2.46** *Triplet energy transfer*

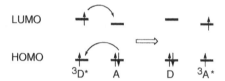

**Figure 2.17** Electron exchange mechanism for triplet energy transfer from $^3D^*$ to A. The horizontal bars represent molecular orbital energies

We need not be concerned with the quantum mechanical treatment of electron exchange. All we need to know is that triplet energy transfer by electron exchange, the *Dexter mechanism*, requires appreciable overlap between the molecular orbitals of D and A, so that the critical transfer distance becomes essentially equal to the sum of the van der Waals radii of D and A.

Triplet energy transfer is readily measured by nanosecond flash spectroscopy (Section 3.7), because molecules in the lowest triplet state have strong and characteristic triplet–triplet absorption spectra commonly extending well into the visible region (Figure 2.18).

Triplet energy transfer is an important method used in preparative photochemistry to generate the triplet state of molecules that either do not undergo spontaneous ISC following direct excitation or that undergo unwanted reactions when excited to the singlet state. The irradiation of suitable donor molecules rather than of the target substrates directly is called *triplet sensitization*. The triplet energy of the donor D, $E_T(D) = E(^3D^*) - E(D)$, should exceed that of the acceptor, $\Delta E_T = E_T(A) - E_T(D) \leq 0$ for efficient energy transfer. Moreover, exclusive excitation of the donor must be possible. This is best achieved if the donor absorbs at longer wavelengths than the acceptor (Figure 2.19).

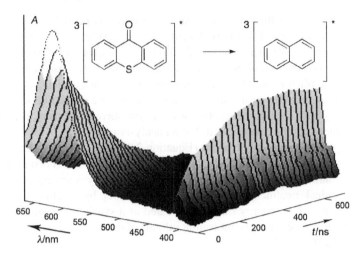

**Figure 2.18**   Energy transfer from triplet thioxanthone ($\lambda_{max}$ 650 nm, $E_T = 265$ kJ mol$^{-1}$) to naphthalene ($1 \times 10^{-3}$ M, triplet–triplet absorption of triplet naphthalene: $\lambda_{max}$ 415 nm, $E_T = 254$ kJ mol$^{-1}$) observed at various delay times with respect to a nanosecond excitation pulse at 351 nm that is absorbed only by thioxanthone

Benzophenone ($E_T = 288$ kJ mol$^{-1}$) and the xanthones ($E_T = 265$ kJ mol$^{-1}$ for thioxanthone, Figure 2.18) are often used as triplet sensitizers, because they absorb at relatively long wavelengths, undergo efficient ISC ($\Phi_{ISC} \approx 1$) and have a small singlet–triplet energy gap. Alternatively, the donor can be used in large excess, for example acetone as a solvent, to ensure that it absorbs most of the light.

If the triplet energy of the acceptor is comparable to or higher than that of the donor, $\Delta E_T = E_T(A) - E_T(D) \geq 0$, then the efficiency for separation of a thermally equilibrated encounter complex with the excitation energy residing on the acceptor A is given by Boltzmann's distribution law, $n_{A^*}/n_{D^*} = \exp(-\Delta E_T/RT)$, so that the rate constant $k_{et} = k_d n_{A^*}/(n_{A^*} + n_{D^*})$ is expected to fall off exponentially as energy transfer becomes strongly endothermic ($\Delta E_T/RT \gg 1$) (Equation 2.47). Such behaviour was observed in pioneering studies for triplet energy transfer from biacetyl ($E_T = 236$ kJ mol$^{-1}$) to a graded

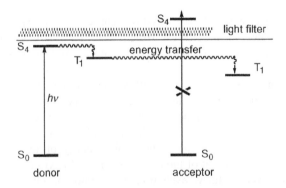

**Figure 2.19**   Triplet sensitization

series of acceptors with higher and lower triplet energies.[111]

$$k_{et} = \frac{k_d}{1 + e^{\Delta E_T/RT}} \approx k_d e^{-\Delta E_T/RT} \text{ for } \Delta E_T \gg 0$$

*Equation 2.47*

Hammond, Saltiel and co-workers subsequently observed that the rate constant of energy transfer becomes slower than that predicted by Equation 2.47 when the structure of the relaxed excited triplet state of the acceptor differs substantially from that in ground state, as is the case for stilbene.[112] These findings initiated extensive investigations on the twisted triplet state geometry of stilbene and related compounds (Section 5.5). The expression *nonvertical energy transfer* (NVET) was coined to describe this situation.

A classical treatment of NVET that was gleaned from electron transfer theory (Section 5.2) was given by Balzani (Equation 2.48).[113]

$$k_{et} = \frac{k_d}{1 + e^{\Delta G/RT} + \frac{k_{-d}}{k_{et}^0} e^{\Delta G^{\ddagger}/RT}}$$

*Equation 2.48*

The free energy difference $\Delta G$ between the donor and the acceptor triplet excitation free energies is usually replaced by the spectroscopic energy difference $\Delta E_T = E_T(A) - E_T(D)$ that is determined from the 0–0 bands of the phosphorescence spectra, ignoring the difference in entropy changes that may be associated with excitation of A and D to the triplet state (see Section 3.11). The free energy of activation $\Delta G^{\ddagger}$ appearing in the denominator represents the excess free energy required for nonvertical energy transfer from D to A. It is treated as a fitting parameter. A value of $\Delta G^{\ddagger} = 1300 \text{ cm}^{-1}$ reduces the rate constant $k_{et}$ by about two orders of magnitude relative to $k_d/2$ for $\Delta G = 0$.

An alternative model, the *'hot-band'* mechanism, found support in studies of the temperature dependence of NVET rates in stilbene sensitization.[114,115] In this model, thermal population of vibrational modes of the ground-state acceptor is assumed to provide access to geometries near to the relaxed excited state, as depicted in Figure 2.20. It is supported by the fact that energy differences between triplet donors are reflected almost entirely in the activation entropy rather than in the activation enthalpy, as would be expected from Balzani's treatment.

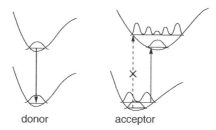

donor          acceptor

**Figure 2.20** The energy of the donor is insufficient for vertical energy transfer to the acceptor (dashed arrow), but sufficient for thermally activated NVET. The squares of vibrational wavefunctions are shown

Other examples of NVET include the triplet energy acceptors 1,3,5,7-cyclooctate-traene,[116] which has a puckered structure in the ground state but relaxes to a planar structure in the triplet state, and a cyclobutadiene derivative,[117] which changes from a rectangular to a square structure.

Lanthanide ions exhibit long-lived, strong luminescence in the visible region. Their f–f absorption bands are, however, very weak and FRET to lanthanide ions is thus inefficient. Efficient FRET and even two-photon absorption[118] can be achieved by complexation with appropriate chromophores as an antenna. Lanthanide ion complexes (Figure 2.21) are used

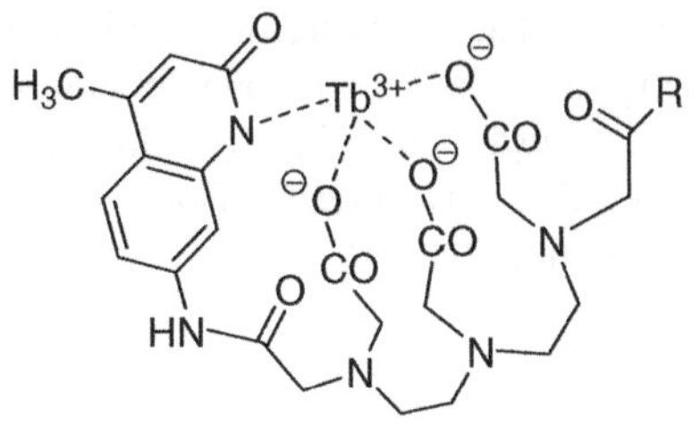

**Figure 2.21** A terbium ion complex used as a probe in biophysical studies

for imaging ion channels in living cells, to study protein–protein interactions or DNA–protein complexes in cells and for high-throughput screening assays to measure peptide dimerization associated with DNA transcription factors and ligand–receptor interactions.[119]

### 2.2.3 Excimers and Exciplexes

The fluorescence spectrum of dilute pyrene ($<10^{-4}$ m) in degassed cyclohexane exhibits vibronic structure and has a maximum at 395 nm. As the concentration of pyrene is increased, the fluorescence quantum yield of pyrene decreases and a broad, structureless emission band with a maximum at about 480 nm gains in intensity (Figure 2.22, left).

This phenomenon was first identified and explained by Förster.[120] The structureless emission is attributed to an excited pyrene dimer ($^1P^*\cdots P$) that is formed by collisional association of singlet excited pyrene $^1P^*$ with a pyrene molecule P in the ground state. It was subsequently found that many aromatic molecules exhibit similar behaviour. The expression *excimer* (*exci*ted di*mer*) was proposed by Stevens to distinguish such species from the excited state of a ground-state complex. Excimer formation is prominent at relatively low concentrations of pyrene (Figure 2.22, left), because of its unusually long fluorescence lifetime, $^1\tau = 650$ ns, which allows for diffusional encounters of $^1P^*$ with P even at low concentration.

The diagram in Figure 2.22 (right) shows that the approach of two pyrenes is repulsive if both molecules are in the ground state, but slightly attractive when one molecule is in the excited singlet state. The vibrational states of pyrene refer to the internal coordinates of the molecule that are orthogonal to the intermolecular distance $r_{P\cdots P}$. They give rise to the structured emission of single pyrene molecules. The excited dimers are in a shallow potential energy minimum that has no counterpart on the ground-state surface.

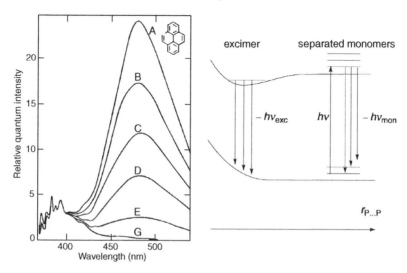

**Figure 2.22** Left: fluorescence spectra of pyrene in cyclohexane. Intensities are normalized to a common value of $\Phi_f$. A, $10^{-2}$ M; B, $7.75 \times 10^{-3}$ M; C, $5.5 \times 10^{-3}$ M; D, $3.25 \times 10^{-3}$ M; E, $10^{-3}$ M; G, $10^{-4}$ M. Reproduced by permission from ref. 109. Copyright 1970, John Wiley & Sons, Ltd. Right: potential energy surfaces for excimer formation; $r_{P...P}$ represents the distance between two pyrene molecules. The vibrational levels shown for the monomers at large separation refer to degrees of freedom other than $r_{P...P}$

Time-resolved measurements of pyrene fluorescence show that only the structured monomer emission is observed immediately after excitation, because the pyrene molecules are not associated in the ground state. The broad emission then grows in as the excimers are formed by diffusional encounter and the equilibrium between monomers and excimers is reached.

Why are many electronically excited molecules prone to associate with one or more ground-state molecules, whereas no such association is observed in the ground state? We first recall that entropy always works against the association of two molecules, because the translational and rotational degrees of freedom of one molecule are lost. From statistical thermodynamics we can easily calculate the translational entropy of pyrene in the gas phase ($p = 1$ bar, $T = 298$ K), $S_{trans} = 174$ J K$^{-1}$ mol$^{-1}$. It is slightly lower in solution due to the restricted mobility, but the entropy change associated with excimer formation, $\Delta_{exc}S$, is still on the order of $-100$ J K$^{-1}$ mol$^{-1}$. Hence the enthalpy of association, $\Delta_{exc}H$, needs to be substantial to ensure exergonic excimer formation (Equation 2.49). The equilibrium constant for the formation of the pyrene excimer amounts to $K_{ass} = 3.3 \times 10^3$ M$^{-1}$ in isooctane at room temperature.[121]

$$\Delta_{exc}G = \Delta_{exc}H - T\Delta_{exc}S < 0 \text{ requires } \Delta_{exc}H < -30 \text{ kJ mol}^{-1}(T = 298 \text{ K})$$

**Equation 2.49** *Thermochemistry of excimer formation*

This condition is relaxed when two pyrene molecules are linked covalently by a relatively short tether such that only some rotational degrees of freedom are lost upon excimer formation (Figure 2.23).[122–124]

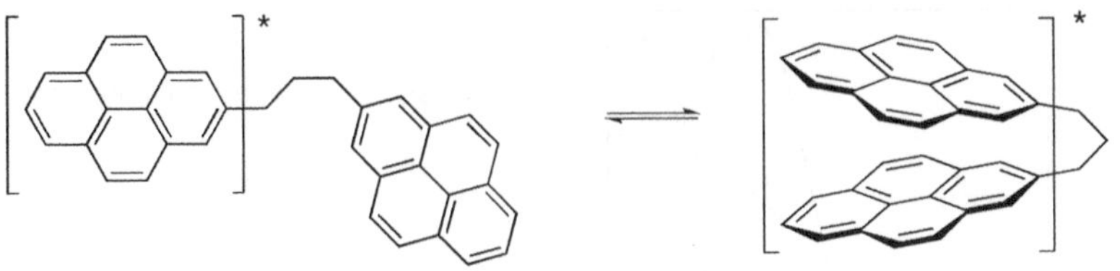

**Figure 2.23**  Excimer formation in a covalently linked bispyrene

Quantum mechanical treatments have indicated three conceptually different types of stabilizing interactions that all contribute to the stabilization of excimers. The most important of these is called *exciton interaction*, which can be viewed as the stabilization energy resulting from the delocalization of the excitation energy between the neighbouring molecules. The same kind of interaction is considered in the Förster treatment of singlet energy transfer that we dealt with in the last section. We use a trial wavefunction $\Psi_{exc}$ for the excimer that consists of an equal mixture of two wavefunctions representing the excitation being localized on one or the other of the two molecules A and B, $\Psi_{exc} = c$ $[\Psi_A \Psi_{B^*} \pm \Psi_{A^*} \Psi_B]$. If the wavefunctions $\Psi_A$, $\Psi_{A^*}$, $\Psi_B$ and $\Psi_{B^*}$ are eigenfunctions of the separate molecules representing the ground and lowest excited singlet state, respectively, then the energy of the excimer can be calculated using the variation theorem (Equation 1.14), which leads to the $2 \times 2$ secular determinant (Equation 2.50).

$$\begin{Vmatrix} H_{11} - \varepsilon & H_{12} \\ H_{21} & H_{22} - \varepsilon \end{Vmatrix} = 0$$

*Equation 2.50*

For $A = B = P$ (pyrene), the energies associated with the two basis functions are equal, $H_{11} = H_{22} = E(P) + E(P^*)$, as are the interaction terms $H_{12} = H_{21}$, which in the simplest approximation are equal to the interaction between the transition moments $M_{0 \rightarrow 1}$ (Equation 2.18) of the two molecules. The secular determinant expands to the quadratic equation $(H_{11} - \varepsilon)^2 - H_{12}^2 = 0$, which yields two exciton states (Figure 2.24), the lower of which is stabilized relative to the energy of the separated molecules by the interaction energy $H_{12}$.

The $S_0$–$S_1$ transition of pyrene is fairly weak, $\log(\varepsilon/\text{M}^{-1}\,\text{cm}^{-1}) \approx 2$ (see Section 4.7), so that only a small exciton stabilization would be predicted by the simple treatment outlined above. In this case, higher excited states and also charge transfer states of the type

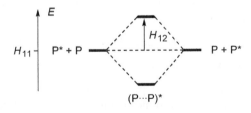

**Figure 2.24**  Exciton interaction

$\Psi_{A+}\Psi_{B-} \pm \Psi_{A-}\Psi_{B+}$ must be included in calculating $H_{12}$. Contributions from charge transfer states will be much more important in encounter complexes of an excited molecule with different ground-state molecules that can act as electron donors or acceptors. Such excited-state complexes are called *exciplexes* and will be considered in Section 5.2. Transition moments for singlet–triplet transitions are essentially zero, so that very little exciton interaction will result in a triplet encounter pair $^3A^* \cdots B$. Indeed, few unambiguous examples of *triplet excimers* are known.[125] The observation of excimer phosphorescence requires either very high concentrations (with the associated danger of impurity emissions) or covalent linking of two chromophores[126] to reduce the entropic handicap of excimer formation.

So far we have assumed that the overlap between the molecular orbitals of the two molecules is negligible in an excimer complex. At short distances, say $r_{P \cdots P} \leq 300$ pm, orbital overlap leads to further stabilization of an excimer. As can be seen from Figure 2.25, first-order perturbation of the degenerate orbitals (Section 4.3) due to

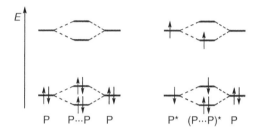

**Figure 2.25** Orbital diagram for the encounter of two pyrene molecules P (left, ground state; right, excimer)

intermolecular overlap does not stabilize the encounter complex in the ground state (in fact it is destabilized, because the antibonding interaction of the doubly occupied orbitals is somewhat stronger than the bonding interaction), but leads to a net stabilization of two electrons in the excimer $(P \cdots P)^*$.

## 2.2.4 Delayed Fluorescence

The fluorescence emission discussed so far is produced by direct excitation of a molecule M to one of its excited singlet states that, after IC to $S_1$ when an upper singlet state was initially populated, emits prompt fluorescence from $S_1$ with a lifetime on the order of nanoseconds. In addition, several processes can be envisaged that permit repopulation of $S_1$ following ISC to $T_1$, which can give rise to emission that has the spectral characteristics of fluorescence, but a lifetime much longer than that of prompt fluorescence, that is, to *delayed fluorescence*. These processes can be sub-classified as '*P-type*' and '*E-type*' *delayed fluorescence*.[32] We do not consider 'artificial' delayed fluorescence due to impurities, or due to ionization followed by recombination with the formation of $^1M^*$.

The classical example of *E-type delayed fluorescence* is that of eosin (4′,5′-dibromo–2′,7′-dinitrofluorescein disodium salt) in degassed solvents. The name E-type refers to *e*osin, which has a high quantum yield of ISC and a small singlet–triplet energy gap, $\Delta E_{ST} = 18\,\text{kJ mol}^{-1}$. Thermally activated repopulation of the $S_1$ state by reverse ISC from $T_1$ is responsible for delayed fluorescence with the lifetime of $T_1$. The intensity of E-type delayed fluorescence decreases strongly at reduced temperature.[32] It is not observed with aromatic hydrocarbons, which have large singlet–triplet energy gaps (Section 4.7).

An unusual example of delayed fluorescence exemplifying El Sayed's rules (Section 2.1.6) was recently reported for the triplet sensitizer xanthone,[127] which undergoes ultrafast ISC within 1 ps. Delayed fluorescence with a lifetime of 700 ps was observed in aqueous solution. Temperature-dependent steady-state and time-resolved fluorescence experiments indicate that the $T_2(n,\pi^*)$ state, which is primarily accessed by ISC from $S_1(\pi,\pi^*)$, is nearly isoenergetic with the $S_1$ state. The delayed fluorescence is attributed to reverse ISC from $T_2(n,\pi^*)$, in competition with internal conversion to $T_1(\pi,\pi^*)$.

*P-type delayed fluorescence* was first observed with *pyrene*[128] (Figure 2.9) and subsequently with many other aromatic compounds.[48,49] It is due to triplet–triplet annihilation (Equation 2.51). According to spin statistics (Scheme 2.5), one in nine encounters of two molecules $^3M^*$ will give an overall singlet pair and the energy gap law will then favour formation of one singlet excited molecule $^1M^*$, provided that the combined energy of the two triplet molecules $^3M^*$ is larger than that of $^1M^*$, $2E_T(M) > E_S(M)$.

$$^3M^* + {}^3M^* \rightarrow {}^1(M\cdots M) \rightarrow {}^1M^* + M$$

**Equation 2.51**    *Triplet–triplet annihilation*

Because two triplets $^3M^*$ are required for triplet–triplet annihilation, the intensity of P-type delayed fluorescence is proportional to the square of the radiant power at moderate intensity of the light source. The decay of P-type delayed fluorescence does not follow a simple rate law. Its decay kinetics are related to those of the triplet state, of mixed first and second order.

## 2.2.5    Dioxygen

The trials and tribulations leading to an eventual understanding of photooxidation processes (Section 6.7) represent a fascinating part of the history of science.[129,130] G. N. Lewis proposed in 1924 that paramagnetic oxygen was a biradical species. Four years later, R. S. Mulliken put forth a molecular orbital description of dioxygen and predicted the existence of two low-lying excited singlet states above its triplet ground state. He assigned the known 'atmospheric absorption bands' of oxygen at '1.62 volts' ($1.31\,\mu m^{-1}$, 762 nm) to the upper of these 'metastable' states.

In 1931, H. Kautsky claimed that the sensitized formation of a short-lived, highly active and diffusive form of molecular oxygen was responsible for the photooxygenation of dyes.[131] He supported that claim with an ingenious experiment: a yellow dye and a colourless oxidation substrate (leuco-malachite) were adsorbed separately on silica gel beads and mixed. Oxidation of the colourless gel particles to develop the colour of Malachite Green took place only when the mixture was irradiated under oxygen and did not occur, when either light or oxygen was absent.

Kautsky's 'three-phase test' indicated that the excited yellow dye generated diffusible excited $O_2$ species that were able to reach the colourless gel particles and to oxidize the leuco dye. In spite of this convincing evidence, his hypothesis was strongly challenged and not generally accepted for more than 30 years. In retrospect, this is a classical example of the Babylonian information barrier between various disciplines of science. Unfortunately, Kautsky had originally attributed the chemically active oxygen species to the then known (upper) excited singlet state of $O_2$. His hypothesis was subsequently buried by the observation that photooxidation could be induced by infrared irradiation beyond $1.31\,\mu m^{-1}$, despite the fact that the existence of a lower singlet state had been predicted both by Mulliken and Hückel and identified by Herzberg in 1934. Kautsky's reassignment of the active species to the lower-lying singlet state in 1939 had little impact. His hypothesis was revived only when Kasha reported on the luminescence of chemically produced singlet oxygen in 1963,[132] and when Foote showed a year later that the chemical reactivity of singlet oxygen generated chemically was identical with that formed by sensitization.[133]

An MO diagram of the dioxygen molecule is shown in Section 4.4, Figure 4.13. The highest occupied orbital is twofold degenerate and these two MOs allocate two electrons in total. Thus, the ground state of molecular oxygen is a triplet state, in accord with *Hund's rule*. The term symbol for triplet oxygen is $^3\Sigma_g^-$. Two low-lying singlet states of oxygen, $^1\Delta_g$ and $^1\Sigma_g^+$, have energies of 95 and 158 kJ mol$^{-1}$, respectively, above the triplet ground state.[134-136] The upper excited $^1\Sigma_g^+$ state is rapidly deactivated in solution to the long-lived $^1\Delta_g$ state, which is commonly referred to as *singlet oxygen*, $^1O_2$. Singlet oxygen is an extremely reactive (Section 6.7) and highly cytotoxic species.

Molecular oxygen (*dioxygen*), $^3O_2$, plays a dominant role in photochemistry. Its concentration is on the order of $2 \times 10^{-3}$ M in air-saturated organic solvents at ambient temperature, $0.27 \times 10^{-3}$ M in water and $5 \times 10^{-3}$ M in perfluoroalkanes, close to that in air, $8 \times 10^{-3}$ M.[137] Oxygen diffusion in low-viscosity organic solvents is very rapid, $k_d \approx 4.5 \times 10^{10}$ M$^{-1}$ s$^{-1}$ in acetonitrile[137] and $2.7 \times 10^{10}$ M$^{-1}$ s$^{-1}$ in cyclohexane.[138] Because triplet oxygen has two low-lying excited singlet states, it is a very efficient quencher of excited states. An excited sensitizer molecule $^1S^*$ can produce two molecules of singlet oxygen by the sequence of reactions shown in Equation 2.52:

$$^1S^* + {}^3O_2 \rightarrow {}^3S^* + {}^1O_2 \quad \text{or} \quad {}^1S^* + {}^3O_2 \rightarrow {}^3S^* + {}^3O_2$$
$$^3S^* + {}^3O_2 \rightarrow {}^1S + {}^1O_2$$

**Equation 2.52**   *Quenching of singlet and triplet states by dioxygen*

Both reactions shown on the first line are spin allowed and may occur on every encounter of $^1S^*$ with $^3O_2$ (singlet quenching). The first reaction is exothermic only if the singlet–triplet energy gap $\Delta E_{ST}$ of the sensitizer is sufficient for energy transfer to $^3O_2$, $E_S - E_T \geq 95$ kJ mol$^{-1}$. In that case, formation of $^1O_2$ will be favoured by the energy gap law. If $\Delta E_{ST}$ is smaller, singlet oxygen will not be formed, but the second reaction called *oxygen-catalysed ISC* is always possible. It is spin allowed, because the nascent encounter complex of $^3S$ and $^3O_2$ can have overall triplet multiplicity. The reaction in the second line (triplet quenching) is spin allowed if the encounter complex of $^3S$ and $^3O_2$ has overall singlet multiplicity, as will be the case in every ninth encounter (see Scheme 2.5, Section 2.2.1). Quantum yields of singlet oxygen formation are usually lower than expected from the above simple considerations due mainly to charge transfer interactions in polar solvents.[139] In practice, *oxygen quenching in aerated organic solvents limits the* *lifetimes of excited organic molecules to about 20 ns for $S_1$ and 200 ns for $T_1$. The quantum yield of any photoreaction that proceeds via an intrinsically longer-lived state will thus be reduced accordingly unless oxygen is removed.*

The quantum yields of formation and the lifetimes of singlet oxygen under various conditions have been investigated in considerable detail. The yield of $^1O_2$ is usually determined by chemical trapping with, for example, 1,3-diphenylisobenzofuran (Section 6.7). The lifetime of $^1O_2$ is measured mostly by its phosphorescence emission at 1270 nm, which has recently been observed in single cells.[140] It can also be detected by its absorption to the upper singlet state, $^1\Delta_g \rightarrow {}^1\Sigma_g^+$, $\tilde{\nu}_{max} = 0.52\,\mu m^{-1}$.[141] Systematic studies have shown that both the quantum yields[142] and the lifetimes[143–145] of $^1O_2$ can be understood in terms of the energy gap law (Section 2.1.5) for the dissipation of excess energy to the solvent. The lifetime of $^1O_2$ in $D_2O$ ($\tau_\Delta \approx 68\,\mu s$) is about 20 times longer than that in $H_2O$ ($\tau_\Delta \approx 3.5\,\mu s$) in the absence of quenchers, because the O–H stretch vibration is a much better accepting mode for the conversion of the electronic excitation energy of $^1O_2$ to vibrational energy of the solvent.

The vibrational deactivation of $^1O_2$ has been elegantly exploited to achieve highly stereoselective photooxygenation of alkenes in which one site is 'protected' by a nearby C–H bond acting as a quencher.[146] Replacement of the protective C–H group by C–D significantly reduces the selectivity of attack. The authors anticipate that this new concept of stereocontrol may prove to be general and may facilitate chiral control in a variety of phototransformations.

---

### Special Topic 2.4: Barometric paint

Pressure-sensitive paint (PSP) provides a visualization of two-dimensional pressure distributions on airfoil and model automobile surfaces for wind tunnel research. The airfoil surface is coated with a polymer film that is permeable to oxygen and contains a luminescent pigment. Platinum porphyrins are mostly used, which emit intense phosphorescence that is strongly quenched by oxygen.[147] The surface is irradiated in the wind channel and monitored by a CCD camera (Section 3.1). Complementary filters are used with the light source and camera to ensure that no light can pass directly from the lamp to the detector. PSP then gives a continuous map of the oxygen pressure on the surface as a function of wind speed using a Stern–Volmer-type relationship for luminescence quenching (Section 3.9.8). A complementary method is based on porphyrin-sensitized singlet oxygen emission at 1270 nm, which can be imaged with an InGaAs near-infrared camera.[148]

The conventional method used to design and optimize the shapes of airfoil and other surfaces is to construct a 'loads model' by drilling several hundred small holes called pressure taps, each connected by its own tube to multiplexed pressure transducers. These pressure taps then provide the map of the lift needed to engineer properly the support structure and other features of an airplane. Such models can cost millions of dollars and are time consuming to build. The goal of PSP is to obtain the same information using highly resolved digital images of luminescence with only few pressure taps to calibrate the PSP emission. Furthermore, PSP can provide data from places where it is impossible to install taps, for example, at very thin parts of a wing model.

## 2.3 A Classification of Photochemical Reaction Pathways

The concept of *potential energy surfaces* (PES) (Section 1.4) allows us to visualize what should properly be described by the abstract quantum mechanical notion of a wave packet moving in the $(3N-6)$-dimensional space of the internal nuclear coordinates. The absorption of a photon by a molecule may initially place it on an elevated point of the excited state surface (*Franck–Condon principle*). Nearby minima on the excited state PES are called *spectroscopic minima*, because their position and shape can be gleaned from the vibronic structure of the absorption spectrum. The energy released when the molecule starts 'rolling down' on the excited-state surface is irreversibly dissipated to the surrounding medium, much like the potential energy of a rolling marble is eventually converted to heat by friction. Like a marble, the molecule is likely to follow the path of steepest descent.[q] Having arrived at a local minimum, it is presented with the choice of various photophysical processes, fluorescence, internal conversion or intersystem crossing. In principle, this choice is always available, but thermal relaxation is fastest on a slope of the PES. Given sufficient time near the minimum, the molecule may be jostled over a barrier by thermal reactivation to find a deeper minimum on the excited PES. Once it has arrived on the ground-state surface, the shape of the latter will govern what becomes of it next. The task of theoretical calculations and models (Section 4.9) will be to

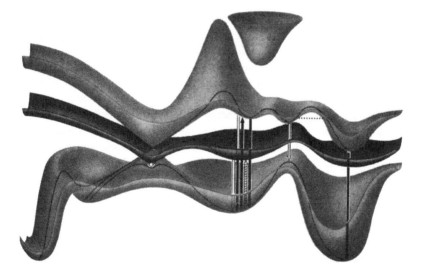

**Figure 2.26** Potential energy surfaces. Reproduced by permission from ref. 16. Copyright 1990, John Wiley & Sons, Ltd

predict or estimate the shapes of the relevant PES (Figure 2.26) as a guideline to understand the photochemical reactivity of organic molecules.

---

[q] This classical analogy should obviously be taken with a pinch of salt. Apart from ignoring wave mechanics, it does not properly account for solvent effects, which may impose a barrier on 'downhill' changes of molecular structure (see, e.g. Section 5.5).

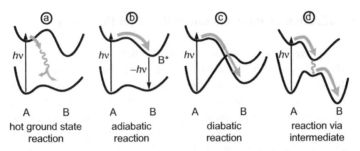

**Figure 2.27**   Four classes of photoreactions

At which point along the reaction coordinate does the molecule jump from the excited state PES to the ground state? Asking this question leads to a classification of photoreactions to four different categories that were originally proposed by Förster. Figure 2.27 shows four two-dimensional cross-sections of an excited- and a ground-state PES along a reaction coordinate leading from reactant A to product B. We do not specify at this point whether the excited-state surface represents a singlet or a triplet state.

*Class* ⓐ Internal conversion from the local ('spectroscopic') minimum in the excited state forms a 'hot' (vibrationally excited) molecule in the electronic ground state. The excess energy initially acquired by IC or ISC is often more than sufficient for some reactions to take place. However, vibrational relaxation will cool the molecule within a few picoseconds in solution and, in practice, hot ground-state reactions leading to a new product B rarely compete with vibrational relaxation back to A. The two-dimensional diagrams shown in Figure 2.27 are somewhat misleading in this respect: Movement along most of the many degrees of freedom other than the reaction coordinate will encounter a high barrier. Reaction towards B therefore requires intramolecular vibrational redistribution to channel a large part of the excess energy into the internal degree of freedom that can heave the molecule towards B. In other words, the density of states is low in the region above a barrier on the ground-state surface, so that the barrier is felt even at excess energies well above its height. Another reason why few hot ground-state reactions have been established to occur may be that it is difficult to obtain unambiguous evidence for such a reaction path. In gaseous molecules at low pressure and in rare gas matrices at low temperature, on the other hand, vibrational cooling is slow and hot ground state reactions are quite likely to occur.

*Class* ⓑ Reactions that proceed entirely on an excited-state surface, initially forming the final product in its excited state B*, are called *adiabatic reactions*.[r] Adiabatic reactions are not unusual in the triplet state, because the competing photophysical decay processes are slow, allowing for the passage across substantial barriers on the triplet surface. Singlet-state adiabatic reactions, on the other hand, are usually associated with only minor structural changes with very low barriers such as valence isomerization, geometric isomerization and hydrogen or proton transfer.

---

[r] The term 'adiabatic' implies that the reaction path does not 'cross a boundary', that is, it does not leave the excited state PES. In thermodynamics, the term 'adiabatic' is used in an entirely different context of system boundaries that are impenetrable to heat and material.

The observation of emission from a product B* (or triplet–triplet absorption by $^3$B* in flash spectroscopy) upon excitation of a reactant A would seem to be clear evidence for an adiabatic reaction, class ⓑ. However, this is true only when excitation of A is possible at a wavelength where B does not absorb. Otherwise, diabatic reactions can easily masquerade as adiabatic ones due to re-excitation of the photoproduct B by the light source used for excitation of A.[149]

*Class* ⓒ *Diabatic reactions*$^s$ proceed directly from the excited state to the ground state of the photoproduct via geometries, at which the two surfaces cross (conical intersections, Special Topic 2.5) or nearly cross (avoided crossings, funnels). In such reactions, no intermediates other than the excited state reached by absorption can usually be detected, because the molecule arrives on or near to a cusp of the ground-state surface and immediately proceeds to a stable product.

Are potential energy surfaces allowed to cross, as shown in diagram ⓒ? This has been under dispute for some time. It was argued that Born–Oppenheimer states belonging to the same irreducible representation, that is, to states with spatial wavefunctions having the same transformation properties under symmetry operations of the molecule (Section 4.4), cannot have exactly the same energy, because an exact Hamiltonian operator would generally give rise to interaction terms that split them apart. In fact, the *noncrossing rule* holds strictly only for diatomic molecules;[150–153] surface crossings may occur in a restricted, $(3N-8)$-dimensional geometric subspace of polyatomic molecules (see Special Topic 2.5). Nevertheless, consideration of a single reaction coordinate with only one degree of freedom prohibits accidental surface crossings at most geometries, so that only avoided crossings will be seen in most cross-sections of the full $(3N-6)$-dimensional geometric space such as diagram ⓒ. Approximate PESs that cross, because the relevant interaction terms are ignored in a particular model used, are called *diabatic surfaces* (Section 4.9). Whether a molecule actually descends through a conical intersection or an avoided crossing in a diabatic reaction is, however, a rather academic issue. PESs become meaningless when two states are very close in energy, because the Born–Oppenheimer approximation fails and energy dissipation is very fast in such situations.[153]

A semiclassical model to calculate the probability $p_{12}$ that a molecule follows a diabatic pathway, by retaining the electronic structure of the initial electronic state as it passes through an avoided crossing, was developed independently in 1932 by Landau and by Zener.[154] In the *Landau–Zener model*, the nuclear movements are described classically, that is, nuclear motion enters parametrically and only a single reaction coordinate, as shown in diagram ⓒ in Figure 2.27, is considered. In the Landau–Zener expression given in Equation 2.53:

$$p_{12} = \exp\left(\frac{-4\pi^2 V_{12}^2}{h\upsilon|s_1 - s_2|}\right) = \exp(-2\pi\omega_{12}\tau)$$

**Equation 2.53** *Landau–Zener model for diabatic reactions*

---

$^s$ Also called *nonadiabatic reactions*, a double negation.

## Special Topic 2.5: Conical intersections

Two conditions must be met at a crossing of two PESs: the energy of the two states must be the same and the interaction between the two electronic states must vanish. The electronic interaction between states of different multiplicity is very small in general, so PESs of different multiplicity can cross freely, but the additional requirement of vanishing interaction between states of equal multiplicity reduces the $(3N-6)$-dimensional space of the nuclear coordinates by a second degree of freedom. Hence PESs are allowed to cross in a $(3N-8)$-dimensional subspace, even if they have the same spatial symmetry. In linear molecules, there are $3N-5$ degrees of freedom and the subspace reduces to $3N-7$. Diatomic molecules have only one degree of freedom, the internuclear distance $r$. Hence both conditions cannot be met simultaneously and states of the same symmetry cannot cross. This is the *noncrossing rule* for diatomics.

If the PESs are plotted against two special internal coordinates $x_1$ and $x_2$, which define the so-called branching plane, they would have the shape of a double cone in the region surrounding the degeneracy. In the remaining $3N-8$ directions the PESs remain degenerate, whereas movement in the branching plane lifts the degeneracy (Figure 2.28).

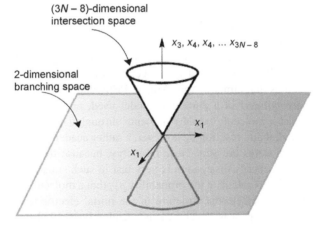

**Figure 2.28**   Conical intersection. Adapted from ref. 22

$V_{12}$ is the electronic coupling term between the zero-order states at the crossing point, $h$ is Planck's constant, $\upsilon$ (not $\nu$!) is the nuclear velocity along the reaction coordinate as the molecule approaches the crossing point and $|s_1 - s_2|$ is the difference between the slopes of the two intersecting PESs at this point. Even in systems for which an accurate calculation of $p_{12}$ is possible nowadays, the Landau–Zener model provides a useful first estimate.

Equation 2.53 makes sense when one realizes that $2\pi V_{12}/h$ is the frequency $\omega_{12}$ with which the system oscillates between the two electronic structures represented by the diabatic states and $V_{12}/|s_1 - s_2|$ represents an 'interaction length' so that the duration $\tau$ of the interaction equals $V_{12}/[\upsilon|s_1 - s_2|]$. Consider the two cases shown in Figure 2.29.

The zero-order diabatic PESs that cross at the origin of the reaction coordinate are shown as dotted lines and the interaction $V_{12}$, chosen at 0.05 eV, leads to an avoided

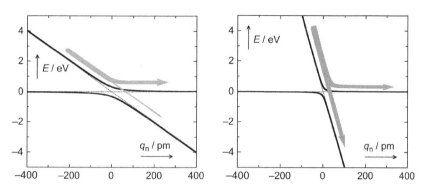

**Figure 2.29** Diabatic and adiabatic reaction paths

crossing of the adiabatic curves (solid lines). Atoms move at speeds on the order of $10^4 - 10^5 \, \mathrm{cm \, s^{-1}}$, depending on their weight. We choose $10^5 \, \mathrm{cm \, s^{-1}}$ for relatively light atoms. The slopes of the diabatic PESs are chosen as 0 and $0.01 \, \mathrm{eV \, pm^{-1}}$ on the left and as 0 and $0.05 \, \mathrm{eV \, pm^{-1}}$ on the right. For a molecule moving from left to right along the reaction coordinate, the Landau–Zener model (Equation 2.53) then predicts $p_{12} = 9\%$, the probability for the diabatic reaction to occur as it passes through the region of the avoided crossing, that is, a 91% probability for staying on the upper PES. When the difference in slope is five times higher (right), $p_{12}$ amounts to 62%.

Diabatic reactions are rarely initiated from triplet states, because the interaction terms $V_{12}$ between states of different multiplicity, which are due to spin–orbit coupling (Section 4.8) are small. Even if surface crossings with the singlet ground-state PES do occur in certain regions of the $(3N - 6)$-dimensional space of the internal nuclear coordinates and the molecule does explore such a region in its lifetime, it is unlikely to find the minute escape hatch to the ground state PES.

*Class* ⓓ The reaction proceeds by formation of reactive intermediates such as a radical pair, a carbene, a biradical or a zwitterion, which have low-lying excited states and are therefore situated near a minimum on the excited state PES. Many photoreactions belong to this class (see Section 5.4).

## 2.4 Problems

1. Using Equation 2.11 and the absorption spectrum of anthracene shown in Figure 2.11, calculate the fluorescence rate constant $k_f$ for anthracene in a solvent of refractive index $n = 1.5$. For a rough estimate of the area of the first absorption band, it is sufficient to sketch a rectangle of about the same area and replace the integral by the product of height and width of the rectangle [$k_f \approx 1 \times 10^8 \, \mathrm{s^{-1}}$].

2. Estimate the rate constants $k_{IC}$ and $k_{ISC}$ from $S_1$ and $k_{ISC}$ from $T_1$ for anthracene. Refer to Equation 2.22, Equation 2.23 and Table 8.6. [$k_{IC} \approx 5 \times 10^6 \, \mathrm{s^{-1}}$, $k_{ISC}(S_1) \approx 6 \times 10^7 \, \mathrm{s^{-1}}$, $k_{ISC}(T_1) \approx 1 \times 10^4 \, \mathrm{s^{-1}}$].

3. Referring to the spectra of anthracene (Figure 2.11) and biacetyl (Figure 6.5), give a quick-and-dirty estimate of the critical Förster distance $R_0$ (Equation 2.37) for FRET

from anthracene to biacetyl. Is Förster theory adequate to predict $R_0$ in this case? [$R_0 \approx$ 50 pm. No, the calculated distance is too short to warrant using the dipole approximation].

4. IR-absorbing dyes with high quantum yields of fluorescence are hard to come by. Explain. [Low rate constants of fluorescence (Equation 2.11), fast competing IC (Equation 2.22)].

5. Give an adequate name to the processes shown in the Jabłoński diagram (Figure 2.30) and discuss the probability of these processes or any special conditions that may be required to permit their observation. [①: Singlet–triplet absorption; very weak, heavy-atom solvent or oxygen pressure, phosphorescence excitation. ②: Triplet–triplet absorption; flash photolysis. ③: Triplet–singlet phosphorescence; unlikely, because IC is fast. ④: Triplet-Singlet ISC: unlikely, because IC is fast. ⑤: Fluorescence; very weak, violation of Kasha's rule].

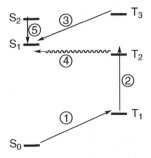

**Figure 2.30**   Photophysical processes

6. In a solution containing $1 \times 10^{-5}\,\mathrm{M}$ of a dye, $\varepsilon(\lambda_{irr}) = 2.5 \times 10^4\,\mathrm{M}^{-1}\,\mathrm{cm}^{-1}$, what percentage of radiation at $\lambda_{irr}$ is absorbed by the dye, if the absorbance of the solvent in the absence of dye in a 1 cm cell is $A(\lambda_{irr}) = 0.4$? [38.5%].

7. Calculate the probability $p_{12}$ for diabatic reaction (Equation 2.53) given that $v = 10^5\,\mathrm{cm\,s^{-1}}$, $V_{12} = 0.02\,\mathrm{eV}$ and $|s_1 - s_2| = 0.02\,\mathrm{eV\,pm^{-1}}$. [$p_{12} = 83\%$]

8. On the basis of Arrhenius' empirical law for first-order rate constants, $k_r = A\exp(-E_a/RT)$, calculate the maximum activation energies $E_a$ that permit reactions initiated on a singlet and on a triplet state PES to compete with the rate constants, $k_d$, of the unavoidable photophysical decay processes occurring from these states. Assume $k_r(S_1) = k_d(S_1) \approx 1 \times 10^8\,\mathrm{s^{-1}}$, $k_r(T_1) = k_d(T_1) \approx 1 \times 10^3\,\mathrm{s^{-1}}$, $A = 1 \times 10^{13}\,\mathrm{s^{-1}}$ and $T = 300\,\mathrm{K}$. [$S_1$: $E_a = 30\,\mathrm{kJ\,mol^{-1}}$. $T_1$: $E_a = 60\,\mathrm{kJ\,mol^{-1}}$].

9. The quantum yield of some photoreaction increases 100-fold when the solution is degassed. Which excited state is likely being quenched by dioxygen and what is its lifetime in degassed solution? [$T_1$, $\sim$20 µs].

# 3

# Techniques and Methods

*Recommended textbooks.*[137,155–159]

## 3.1 Light Sources, Filters and Detectors

The spectrum of the main light source on Earth, the Sun, was shown in the Introduction (Figure 1.1). It contains only a small amount of UV radiation. Photolyses in the laboratory are carried out using artificial sources of ultraviolet or visible light, combined with optical filters, which confine multiwavelength radiation to the desired spectral region. Compilations of the spectral distributions of various light sources, of cut-on, cut-off, narrow-band or solution filters and other optical components are available in handbooks[1,137,156–158] and from many commercial providers (Oriel). Here, we cover only a selection of the material.

The most widely used sources of UV–VIS light for continuous irradiation in laboratory experiments are xenon and mercury arc lamps, whereas deuterium and tungsten lamps are utilized for spectrophotometers. Various mercury lamps (Figure 3.1) are commercially available. Their spectral irradiance is strongly dependent on the mercury vapour pressure. Figure 3.2 shows the line spectrum of a typical *low-pressure mercury arc* (the vapour pressure of mercury is about $\sim 10^{-3}$ mbar). That of a *medium-pressure* ($\sim 1$ bar) lamp is shown in Figure 3.3. The former emits primarily two bands of radiation centred at 253.7 and 184.9 nm, due to $Hg(^3P_1)$ and $Hg(^1P_1)$ de-excitation, respectively. Depending on the transmittance of the quartz envelope, the short-wavelength line may be partially filtered out. These lamps are often utilized in Rayonet reactors (see below). The excited mercury atoms in medium-pressure arc lamps, undergoing more frequent collisions with electrons, are in part excited to higher states. As a result, more emission lines, such as 313.9 or 365.4 nm, useful for photochemical experiments, become available. The lamp envelope's temperature can reach 600 °C and produce a considerable amount of infrared radiation and heat. Therefore, cooling-water circulation or water cuvette filters must be utilized to prevent the sample from becoming heated.

---

*Photochemistry of Organic Compounds: From Concepts to Practice*   Petr Klán and Jakob Wirz
Copyright © 2009 P. Klán and J. Wirz

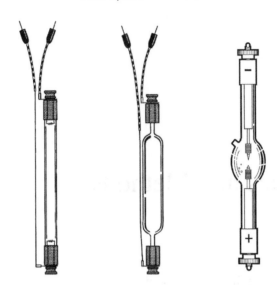

**Figure 3.1** Examples of low- (left), medium- (middle) and high-pressure (right) mercury arc lamps. Reproduced by permission of Ace Glass Inc. and Newport Corp, Oriel Product Line

The most intense sources of UV radiation are the *high-pressure* (~100 bar) *mercury* arcs. The spectral lines are broadened due to the high pressure and temperature and they are superimposed on a continuous background of radiation (Figure 3.4). While common mercury–xenon [Hg(Xe)] lamps still produce significant mercury emission bands, especially in the UV region, the smoother xenon lamp spectrum finds application in environmental photochemistry experiments because of its resemblance to solar radiation (Figure 1.1).

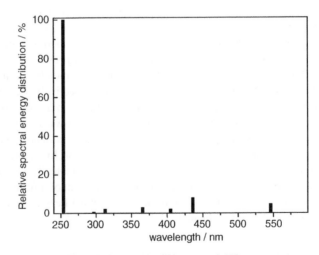

**Figure 3.2** Relative spectral energy distribution of a 15 W low-pressure Helios Italquartz Hg lamp. Adapted from ref. 157

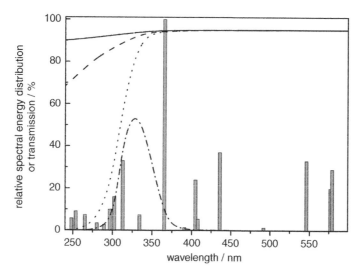

**Figure 3.3** Relative spectral energy distribution of a 125 W medium-pressure Helios Italquartz Hg lamp and transmission curves of glass cut-off filters: quartz (—), Vycor (- - -), Pyrex (· · ·) and a band-pass filter [a solution of $Cr(SO_4)_2 \cdot 12H_2O$ (15% w/v) in 0.5 M $H_2SO_4$ (– · –). Adapted from Ace Glass Inc. Catalogue and ref. 157

The working temperature of tungsten-filament incandescent lamps lies between 2200 and 3000 K. Therefore, they emit light mostly in the visible and infrared parts of the spectrum. Such a source of radiation may be useful in photoreactions of coloured chromophores, for example in the photodissociation of bromine or chlorine molecules to initiate photohalogenation reactions (Section 6.6.1).

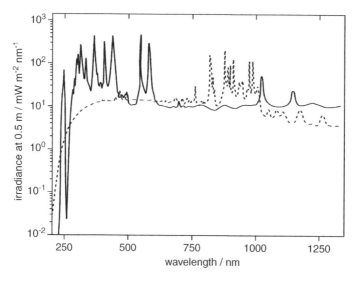

**Figure 3.4** Typical spectral irradiance of 200 W Hg(Xe) (—) and 150 W Xe (- - -) lamps. Adapted by permission from Newport Corp. Oriel product line

*Lasers* (see Special Topic 3.1) are mostly used in photochemistry for time-resolved measurements (Sections 3.5 and 3.7), but can be highly useful for preparative experiments or for the irradiation of samples in remote places such as NMR or EPR cavities, because they produce highly intense, strictly monochromatic and parallel beams. Pulsed lasers can give rise to absorption of two or more photons by a single molecule or to re-excitation of short-lived intermediates, thereby forming products that are different from those obtained by continuous irradiation. Pulsed lasers may also be advantageous when the photoproduct is also sensitive to irradiation and secondary photolysis of the product cannot be avoided by choosing an appropriate wavelength. If the photoproduct is not formed within the duration of the laser pulse, a condition often met when the reaction proceeds through reactive intermediates or a triplet state, secondary photolysis can be completely avoided by irradiating each sample with only a single laser pulse. This is easily realized on a preparative scale with the aid of the simple device shown in Figure 3.5: the cylinder dips into a solution containing the substrate and, when rotated, picks up a film of the solution. The speed of rotation is adjusted to the repetition frequency of the laser. A collector arm skims off the film with a thin Teflon sheet before the cylinder surface dips back into the bath. The whole device may be kept under a Plexiglas box equipped with a quartz window for irradiation under an inert gas atmosphere. A DC motor, such as available in toy model building shops, drives the cylinder and allows for simple adjustment of the rotating speed by changing the applied voltage.

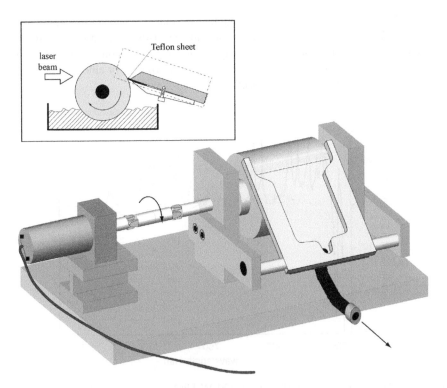

**Figure 3.5**  Apparatus for preparative photolysis with pulsed lasers

## Special Topic 3.1: Lasers

Lasers emit coherent electromagnetic radiation in a monochromatic beam of low divergence. Theodore Maiman built the first working laser in 1960.[160] Currently, the most widespread use of lasers is in optical storage devices such as CD and DVD players, bar code readers, laser pointers and laser printers. Lasers are used for cutting and carving, for target identification and weapons delivery, for surgery and cosmetic applications and, of course, in photochemistry. They have become the foremost tool in photophysical and spectroscopic work, because of their well-defined wavelength and short pulse durations.

Lasers consist of a gain medium surrounded by a resonant optical cavity. The medium is pumped by an external energy source, usually by light from a flash lamp or from another laser or by an electric discharge. Laser emission occurs only when the number of particles in an excited state exceeds that in some lower energy state (*population inversion*, Figure 2.4) so that light amplification takes place in the cavity.

In the *continuous wave* (CW) mode of operation, the population inversion is maintained by a steady pump source and the laser output is essentially constant in time. In the *pulsed mode*, the output consists of short pulses; the length of the cavity defines the time elapsing between sequential pulses. In many applications, very short laser pulses are required. In a *Q-switched laser*, the population inversion is allowed to build up by keeping the cavity output mirrors reflective. Once the pump energy stored in the cavity is at the desired level, the reflectance of the exit is adjusted by an electro- or acousto-optical Q-switch to release the light until the population inversion is depleted. Pulse lengths achieved in this way are on the order of 3–20 ns. *Mode-locked lasers* emit pulses of about 30 ps duration, which, using various tricks of the art, can be shortened to a few femtoseconds. Due to the energy–time uncertainty principle (Equation 2.1), a pulse of such short duration must contain a broad spectrum of wavelengths and the laser medium must have a broad gain profile to amplify them all. A suitable material is titanium-doped, artificially grown sapphire (Ti:sapphire).

*Population inversion* cannot be achieved in a two-level system, a material with two electronic states. At best, a nearly equal population of the two states is reached, resulting in optical transparency, when absorption by the ground state is balanced by stimulated emission from the excited state. An indirect method of populating the emitting excited state must be used. In a *three-level laser* (Figure 3.6, left), irradiation of the laser medium pumps an upper level 2, which is rapidly depleted by a nonradiative

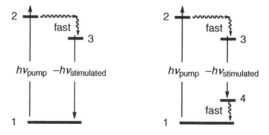

**Figure 3.6** Three-level (left) and four-level (right) laser systems

process populating a long-lived level 3. Thus stimulated emission induced by further pumping is avoided and the population of level 3 will eventually exceed that of the ground state. Benzophenone, for example, would fulfil these requirements, but its radiative rate constant of phosphorescence is very low. The chromium(III) ions in the ruby laser introduced by Maiman do the job. More efficient lasers can be built based on *four-level* systems. Here, the stimulated emission from level 3 reaches a low-lying level 4 that is rapidly depopulated to the ground level 1 by a second nonradiative process. Because only few atoms need be excited into the upper laser level to form a population inversion between levels 3 and 4, a four-level laser is much more efficient than a three-level laser and most practical lasers are of this type.

Classical optics, so-called linear optics, assumes a linear response of transparent matter to electromagnetic radiation, as expressed in the refractive index. This description of the interaction light and matter retains only the first (linear) term of a Taylor expansion. Linear laws describing physical responses hold only up to a point; a well-known example is Hooke's law, describing the elasticity of a spring. The power of a single 1 mJ pulse released within a picosecond is 1 GW, equal to that of a modern nuclear power plant. With the high radiative powers of lasers, nonlinear responses necessitate the inclusion of higher (quadratic, cubic, etc.) terms. *Nonlinear optics*[161] can produce unusual, but very useful, phenomena such as the frequency doubling (second-harmonic generation) in a BBO crystal, parametric down-conversion, optical parametric oscillators and the like, but also unwanted phenomena such as dielectric breakdown. Current activity to achieve light amplification in quantum dots holds promise for exciting new technical developments.[162,163]

As a consequence of *Heisenberg's uncertainty principle*, (Equation 2.1), ultrashort laser pulses are no longer highly monochromatic. Ti:sapphire femtosecond lasers can produce an octave-spanning spectrum directly from the oscillator and have led to the development of *optical frequency combs*.[164] A mode-locked femtosecond laser maintains a short pulse circulating inside the optical cavity. With each round trip, an attenuated copy of the light pulse escapes so that the laser emits a regular train of ultrashort pulses. To measure the unknown frequency of a laser wave, the beam and the pulse train are superimposed with a beam splitter and a detector registers an interference signal. Optical frequency combs from mode-locked femtosecond lasers extend the limits of time and frequency metrology. Precise comparisons of optical resonance frequencies of atomic hydrogen and other atoms with the microwave frequency of a caesium atomic clock are establishing sensitive limits for conceivable slow variations of fundamental constants. Optical high harmonic generation is extending frequency comb techniques into the extreme ultraviolet region. Frequency comb techniques are also providing a key to attosecond science.

Some lasers that are frequently used in the photochemical laboratory and their properties are listed in Table 8.4.

*Optical filters* can modify the spectral output from spectrally broad light sources and essentially monochromatic light can be obtained from light sources with several sharp but widely separated emission lines, such as the low-pressure mercury arcs. Commercially available glass filters and readily prepared filter solutions are inexpensive alternatives to monochromators. The most common cut-off (long-wavelength pass) filters are the glass of which the photochemical apparatus (Section 3.2) consists, such as lamp envelopes, immersion wells, test-tubes, and so on. An additional glass filter can be introduced between the source of the radiation and the sample. Fused quartz and fused silica have a very high UV transmission; they cut off below 180 nm and, depending on thickness and quality, also partly below 250 nm, while they are almost fully transparent above 250 nm (Figure 3.3). *Vycor*, a glass with good temperature and thermal shock resistance, can also be used for irradiation near and above 250 nm, while Pyrex is commonly utilized when wavelengths >290 nm are required. Many different short- or long-wavelength pass and bandpass filters are available commercially and their optical parameters can be found in the manufacturers' catalogues. *Dichroic filters* are thin-film dielectric mirrors, which selectively allow light of a narrow range of wavelengths to pass. Because unwanted wavelengths are reflected rather than absorbed, dichroic filters do not absorb much energy and consequently do not heat up or age much during prolonged operation.

*Solution filters* are readily prepared in any photochemical laboratory. Their compositions and optical properties are available throughout the literature.[137,157,158] Figure 3.3 shows, for example, the transmission curve of a band-pass solution filter made of $Cr(SO_4)_2$ dissolved in dilute aqueous $H_2SO_4$, which transmits radiation at $325 \pm 20$ nm. As a result, the 313.9 nm band, available in the emission spectrum of medium-pressure mercury lamps, can be isolated. However, some of the reported solutions have limited photochemical stability and should be regularly checked for transmittance.

Pure solvents may also act as optical filters; for example, methanol and benzene cut off most of the radiation below 205 and 280 nm, respectively (Table 3.1). Their (internal) filter effect must be considered in planning photochemical experiments with solutions, because it may lower the reaction efficiencies.

*Photomultipliers* are sensitive to light in the ultraviolet, visible and near-infrared regions, allowing for the detection of single photons. They consist of a glass or quartz vacuum tube housing a photocathode, a sequence of electrodes called dynodes and an anode (Figure 3.7). When an incident photon of sufficient energy, $E_p \geq h\nu_{th}$ (*photoelectric effect*; see Figure 1.5), strikes the photocathode material, an electron is ejected and directed towards the electron multiplier, which consists of about eight dynodes, each of which is held at a more positive voltage than the previous one. The electron is accelerated

**Table 3.1** *Cut-off wavelengths of common solvents[a]*

| Solvent | Cut-off wavelength/nm |
|---|---|
| Water | 185 |
| Acetonitrile | 190 |
| n-Hexane | 195 |
| Methanol | 205 |
| Diethyl ether | 215 |
| Ethyl acetate | 255 |
| Carbon tetrachloride | 265 |
| Benzene | 280 |
| Pyridine | 305 |
| Acetone | 330 |

[a]Adapted from ref. 1.

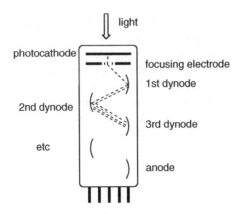

**Figure 3.7**    Schematic diagram of a head-on photomultiplier tube

by an electric field. On striking the first dynode, more electrons are emitted and these in turn are accelerated towards the second dynode. The dynode chain produces an increasing number of electrons at each stage. Finally, the anode is reached, where the accumulation of charge results in a sharp current pulse indicating the arrival of a photon at the photocathode. Photomultipliers produce up to about $10^8$ electrons per incident photon. The temporal half-width of a single pulse of electrons reaching the anode is on the order of 1 ns, which limits the time resolution of photomultipliers.

In *photodiodes*, light absorption promotes an electron from the valence band to the conduction band of a semiconductor, thereby creating a mobile electron and a positively charged site, an 'electron hole'. When a voltage is applied (reverse bias, photoconductive mode), this will give rise to a photocurrent through the diode. A diode array consists of thousands of photosensitive diodes in a one- or two-dimensional matrix. Each diode may span only $100 \, \mu m^2$. The readout to store a digitized image can be done by measuring the charge required to reload a capacitor behind each irradiated diode [*charge coupled device* (CCD)] and is usually clocked to 30 Hz. Diode arrays have largely replaced light-sensitive films in photographic devices.

*Image intensifiers*, also called microchannel plate (MCP) detectors, are a two-dimensional array of tubes (e.g. $1000 \times 1000$ on a chip of $2 \times 2 \, cm$) that also operate by the photoelectric effect. An image is projected on to the MCP. As a photon hits a photocathode on the detector plate, the metal ejects an electron that is accelerated in an electric field and amplified. The anode is a phosphorescent screen that emits up to $10^4$ photons upon impact of the fast electrons generated from a single photoelectron. A green phosphor is generally used because the human eye is most sensitive to that colour. An amplified, but monochromatic, image is thereby recreated on the phosphor screen. MCPs are used in night vision devices and are sensitive up to the near-infrared region. The phosphor is often combined with a CCD camera to record the image. For time-resolved applications, MCPs can be gated with rise-times of about 1 ns by applying a counter voltage near the photocathode so that photoelectrons do not reach the accelerating field and the device remains opaque as long as the counter voltage is applied. This can be used to capture objects moving at high speed or for spectrographic flash photolysis. In spectrographic applications, the output of a polychromator (essentially a monochromator with a wide exit hole instead of the slit selecting a single colour) is focused on to the MCP. The spectrum is separated in one dimension, so that the green phosphor shows red light on one side and blue light on the other. Images so obtained can be read by a CCD camera. The time resolution is limited by the gating of the MCP, not the slow read-out process of the CCD camera. *Microchannel plate photomultipliers* can achieve response times of 30 ps.

A *streak camera* transforms the temporal profile of a light pulse into a spatial profile on the detector. For example, a spectrally resolved light pulse from a fluorescent sample that is excited by a short pulse of light hits a one-dimensional photocathode array through a narrow slit along one direction. The photoelectrons are accelerated in a cathode ray tube and pass through a pair of plates that produce a rapidly rising electric field that deflects the electrons orthogonally to the slit. The phosphor screen at the end of the tube then shows the spectral distribution of the light pulse in one direction and the time-varying deflection of the electrons in the other. A CCD array is used to record the streak pattern on the screen, displaying both the temporal and the spectral profile of the light pulse. In order to record periodic phenomena, the streak camera needs to be triggered, like an oscilloscope. The time resolution of the fastest optoelectronic streak cameras is about 0.2 ps in single-pulse mode. Jitter of the source trigger and of the instrument response will, however, limit the time resolution of accumulated periodic signals to 10 ps or more.

---

### Special Topic 3.2: Organic Light-Emitting Diodes (OLEDs)

*Recommended textbook.*[165]

Light-emitting diodes (LEDs) accompany our daily lives. They are often used as small indicator lights on electronic devices and increasingly in higher power applications such as flashlights and area lighting. The colour of the emitted light depends on the composition of the semiconducting material used and can range from the IR to the near-UV. Red LEDs became commercially available in the late 1960s and were commonly used in seven-segment displays. Later, other colours became widely available and appeared in many appliances. As the LED materials technology became more advanced, the light output was increased and LEDs became bright enough to be used for illumination. Low-cost efficient blue LEDs did not emerge until the early 1990s, completing the desired RGB colour triad (Section 1.3). High-brightness colours gradually emerged in the 1990s, permitting new designs for outdoor signage and huge video displays for billboards and stadiums.

In recent years, organic light-emitting diodes (OLEDs) have started to be commercially applied in bright displays and the future holds tremendous opportunity for their low cost and high performance. They may eventually replace the liquid crystal displays (LCDs) that require backlight, which are used in mobile devices such as cellular phones and laptop computers, and perhaps even the fluorescent and incandescent light sources used for illumination. LEDs and OLEDs emitting in the near-UV region have been developed and they are increasingly employed as light sources in photochemical equipment for preparative irradiation and for quantitative work such as quantum yield measurements.

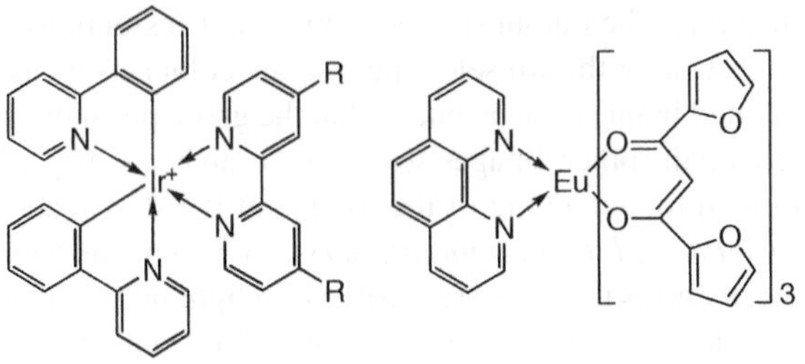

**Figure 3.8**   Organic transition metal complex dopants used in OLEDs

The basic mechanism of electroluminescence involves charge carrier recombina-tion in semiconductor materials forming an electron–hole pair (exciton; see Special Topic 6.29). Due to spin statistics (Section 2.2.1), the recombination of charge carriers in semiconductor materials leads to the formation of 25% singlet and 75% triplet excitons. The triplet excitons in LEDs are generally wasted as heat. In OLEDs, the energy is transferred to organic transition metal complex (Section 6.4.4) dopants containing heavy atoms that exhibit strong spin–orbit coupling (SOC) (Section 4.8). This opens a radiative path for $T_1 \rightarrow S_0$ emission. SOC also enhances $S_1 \rightarrow T_1$ ISC. Thus, all excitons can be converted to triplet states of transition metal complexes (triplet harvesting) with high quantum yields and sub-microsecond lifetimes of emission (Figure 3.8). As a result, OLEDs can reach a four times greater efficiency than LEDs or OLEDs built with purely organic fluorescent emitters.

## 3.2   Preparative Irradiation

Whereas elaborate equipment and tools are needed to study the mechanisms of photochemical reactions, techniques to achieve preparative irradiations are relatively simple and cheap. Nevertheless, when photochemical transformations in solution are used for organic synthesis, several specific requirements have to be fulfilled. High chemical yields and reduction of side-product formation can be achieved, but photochemists have to be careful in choosing proper irradiation wavelengths and reactant concentrations, to consider internal filter effects and whether oxygen needs to be removed during irradiation.[156,158] The chemical reagent (initiator) is light; therefore, the reaction vessel and any optical components (lenses) should be transparent to the wavelength range of interest. The products or intermediates formed in the course of the reaction may absorb at the wavelengths of irradiation. When their concentrations increase during the conversion, reaction efficiency may decrease and eventually stop. Side-reactions may be eliminated in some cases as shown in Figure 3.5 or by decreasing the reactant concentration, which also improves light penetration within the sample, but as a result larger volumes of the reaction mixture will be necessary. Since *dioxygen* is an efficient triplet quencher (Sections 2.2.5 and 6.7.1), it should usually be removed from the solution prior to irradiation. The most

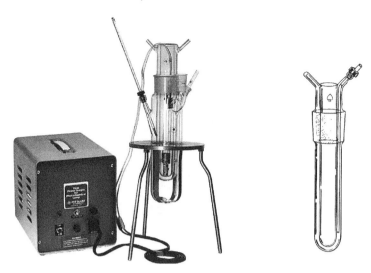

**Figure 3.9** Left: a photochemical reactor with immersed configuration with the lamp and a power source. Right: detail of the immersion well. Reproduced by permission of Ace Glass Inc

common procedures are purging the solution with argon or nitrogen for several minutes before or during irradiation (which, at the same time, serves to mix the solution constantly) or the more rigorous but demanding *freeze–pump–thaw method*. The latter process requires several cycles of freezing the solution in liquid nitrogen, evacuation using a high-vacuum pump and subsequent thawing when the vacuum inlet is closed, in order to remove completely all dissolved gases. The glass vessel with the frozen solution can then be torch sealed or closed with a good PTFE stopcock.

*Photochemical reactors* with immersed configuration (Figure 3.9) are the most common photochemical equipments in preparative irradiation.[158] They usually consist of double-walled immersion wells that provide water cooling of the lamp and filtering of the incident IR radiation. The sample solution, stirred by a magnetic bar typically under an argon atmosphere, is placed outside the lamp finger. The reactors are not ideal for photolysis experiments, in which a narrow excitation band is required. The construction material (Pyrex, quartz, etc.; Section 3.1) limits the wavelength range of irradiation, but a sleeve made of a filtering glass can be inserted between the lamp and the well wall. Furthermore, despite vigorous stirring, polymeric products may precipitate on to the reactor walls and absorb some of the light. This can be avoided by using falling-film or microstructured photochemical reactors,[166] in which the reactant solution falls down along the irradiation well as a liquid film. The photolysed solution is continuously removed via an opening in the bottom of the reactor. Another advantage of falling-film reactors in photochemical reactions is that the thin layers allow extensive penetration and thus a more efficient use of light.

Photochemistry can also be carried out in a photoreactor of *external configuration*.[158] This arrangement is very simple; the solution is held outside the source of irradiation. One option is to use an immersion well as shown in Figure 3.9, but the sample is placed in a separate vessel nearby. The whole setup can be immersed in a thermostated water or methanol bath to control the temperature of the reaction mixture. An external

**Figure 3.10** Rayonet photochemical reactor (*external configuration*). Reproduced by permission of Southern New England Ultra Violet Company

configuration reactor consisting of the reaction sample placed in the middle of an array of UV lamps is also commercially available (Figure 3.10).

A special photochemical reactor, utilizing electrodeless discharge lamps (EDLs), has been designed that generates UV radiation when placed in a microwave (MW) field.[167–169] The EDL consists of a glass tube ('envelope') filled with an inert gas and an excitable substance and sealed under a lower pressure of a noble gas. A low-frequency electromagnetic field (300–3000 MHz) triggers gas discharge, causing emission of UV–VIS radiation. An arrangement in which an EDL is placed inside the reaction solution has several advantages, such as simultaneous UV and MW irradiation of the sample, and is therefore capable of performing photochemistry at high temperatures and low cost.

Heterogeneous transition metal photocatalysis[170] (Section 6.8.1) involves insoluble semiconductor metal oxides or sulfides which, upon irradiation, undergo interfacial electron transfer between the excited semiconductor surface and the reactants. The photolysis reactor usually consists of a thermostated transparent vessel containing a reaction mixture (e.g. $TiO_2$ suspended in an aqueous solution of the reactant) under constant vigorous stirring. The source of visible or UV light is placed above or next to the reactor, like the photoreactor with an external configuration arrangement described above. Such devices are used for large-scale waste-water cleaning.

Organized and constrained media may provide cavities and surfaces, sometimes called microreactors or nanoreactors,[171] that can control the selectivity of photochemical reactions of reactants.[172,173] There are many types of *microreactors*, for example, molecular aggregates of micelles or monolayers, macrocyclic host cavities of crown ethers or cyclodextrins and microporous solid cavities and/or surfaces of zeolites, silica or

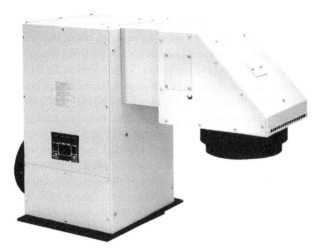

**Figure 3.11** Solar simulator. Reproduced by permission of Newport Corp, Oriel Product Line

alumina.[171,172] The arrangements of the photolysis experiments can be similar to those of photocatalysis. A solid with absorbed reactants can also be externally irradiated in the absence of any solvent.

Many organic compounds undergo photochemical transformations in the crystalline state[172] (see Special Topic 6.5). Intermolecular forces restrict the rotational and translational motions of the molecules in the solid state; therefore, only specific (favourable) conformations are available for the reactions. In a typical photochemical experiment, the crystalline material is ground into a fine powder using an agate mortar and pestle and then deposited in a thin layer between two quartz plates. The plates are sealed inside a nitrogen/argon-purged polyethylene bag and irradiated using an external source of irradiation.

Several commercial providers offer continuous and pulsed *solar simulators* (Figure 3.11) that imitate the emission spectra of the Sun under various conditions, such as those at sea level or in outer space. The wavelength output is adjusted by optical filters. The simulators are typically used in scientific and process control applications such as photobiology, material stability testing, photolithography, solar cell testing and environmental studies. Toughening or hardening of a polymer material by cross-linking by UV irradiation (Section 6.8.1) is called *curing*.[174] The technology, based on an external UV irradiation of a solid surface, is a popular, environmentally friendly processing method for inks and coatings.

## 3.3 Absorption Spectra

UV–VIS absorption spectra are called electronic spectra, because the observed absorption bands indicate the frequencies of the electromagnetic radiation at which the energy of a photon, $h\nu$, matches the energy difference between the electronic ground state and an electronically excited state of the absorbing molecules, $\Delta E = h\nu$. *The absorption and emission spectra* (Section 3.4) *of a substrate should be determined at the outset of any photochemical investigation.* These simple measurements immediately reveal the energy of the lowest excited singlet state and hold a lot of information about its lifetime and reactivity. The vibronic structure of absorption bands tells us about the displacement of the excited state potential energy surface with respect to the ground state (Section 2.1.9). The predictions of quantum mechanical models for electronic excitation (Chapter 4) can be directly compared with the total intensity of an absorption band, quantified by the *oscillator strength* (Section 2.1.4), with solvent shifts and substituent effects on the transition energy (Section 4.4) and with the orientation of the transition moment, obtained by spectroscopy with polarized light (Section 3.6). Based on these criteria, a well-founded assignment of excited states is possible. The quantum mechanical models can then be used to extract information on the electronic structure, dipole moment and reactivity of these excited states. Next, one should analyze (Section 3.7.5) progressive absorption changes resulting from irradiation.

Conventional absorption spectrophotometers are double-beam instruments, in which the intensity of light passing the sample is repeatedly compared with that of a reference beam. This continuously provides accurate correction for temporal fluctuations of the intensity of the monitoring light source, and also for its variation with wavelength, due also to the wavelength dependence of the monochromator transmission and detector sensitivity. Although photomultipliers are very sensitive, approaching the limit of single photon detection, the intensity ratio of probe and reference beam can only be determined to about 0.01% at best.

Most spectrometers plot absorption spectra versus the wavelength, $\lambda$. Chemists have thus become accustomed to seeing absorption spectra on a wavelength scale and to quote absorption maxima as $\lambda_{max}$. Spectroscopists have long preferred scales that are proportional to energy rather than wavelength, such as the frequency, $\nu$, or the wavenumber, $\tilde{\nu}$. This has many advantages, for example the vibronic features are then equally spaced and the first absorption band often exhibits a mirror-image relationship (Section 2.1.9) with the fluorescence spectrum. Modern instruments are equipped with a computer and one can easily convert wavelengths to wavenumbers. For optical spectroscopy, these are conveniently measured in units of $\mu m^{-1}$, where $1\,\mu m^{-1} = 10^6\,m^{-1}$; the wavenumbers then range from $1\,\mu m^{-1}$ for $\lambda = 1000\,nm$ to $5\,\mu m^{-1}$ for $\lambda = 200\,nm$. The non-SI unit $cm^{-1}$ ($1\,\mu m^{-1} = 10^4\,cm^{-1}$) is mostly used in IR, Raman and microwave spectroscopy. In this text, absorption spectra are shown on a wavenumber scale $A(\tilde{\nu})$, $\tilde{\nu} = \nu/c = 1/\lambda$, $[\tilde{\nu}] = \mu m^{-1}$, running from right to left, so that the wavelengths, shown on top of the diagrams on a nonlinear scale, still increase to the right as is more customary to the chemist. To convert to a molecular energy scale we use the relation $E = N_A hc\,\tilde{\nu}$ to obtain the practical Equation 3.1.

$$E = 119.6 \frac{\tilde{\nu}}{\mu m^{-1}}\,kJ\,mol^{-1}$$

**Equation 3.1** *Conversion of wavenumbers to molar energies*

For the determination of molar absorption coefficients, $\varepsilon$, care must be taken to use clean glassware and solvents and the instrumental baseline must be recorded or calibrated prior to measurement. Absorbance measurements are most accurate in the range $0.1 \leq A \leq 1.5$; lower absorbances are prone to baseline errors and those exceeding 1.5 should be avoided, because the amount of light passing the sample cell becomes very small and the readings are easily distorted by stray light, especially at wavelengths below 250 nm. Three independent measurements should be performed, which consist of weighing several milligrams into a small vessel (platinum or disposable aluminium), dissolving the sample in a volumetric flask (20 ml) and preparing several dilutions using volumetric pipettes and further volumetric flasks. When done with care, the three measurements should agree to within $\pm 3\%$. Because the absorption coefficients of organic molecules often vary by many orders of magnitude at different wavelengths, several dilutions are often required to determine absorption bands of very different intensity accurately. The digitized spectra can then be combined using a computer. This is especially important for compounds such as ketones, which exhibit very weak $n,\pi^*$ transitions (Section 4.4 and Figure 6.5) that are easily overlooked at concentrations of $10^{-4}$ M or less.

## 3.4  Steady-State Emission Spectra and their Correction

*Recommended textbooks*[32,175]*and reviews.*[176,177]

Fluorescence spectrophotometers are single-beam instruments. The basic setup of a conventional instrument with a continuous light source (usually a xenon arc) is shown in Figure 3.12. Emission spectroscopy is highly sensitive because it is basically a zero-background technique and because photomultipliers are able to detect single photons with high probability under good operating conditions. A photoelectron released at the photocathode by a single incident photon produces an avalanche of electrons (typically $>10^6$ electrons). The quoted time resolution of photomultipliers refers to the width of a single pulse (usually about 1–2 ns), not the time elapsing between the ejection of a photoelectron at the photocathode and the arrival of the pulse at the anode. Cooling the multiplier housing can reduce thermal release of electrons. Moreover, thermal electrons usually give pulses of lower intensity. Electronic gates that suppress signals below a certain threshold can be used to suppress the background noise further. This technique is called *photon counting*; only the pulses that exceed a certain threshold, say $10^6$ electrons, are counted. Fluorescence spectra are recorded by setting the monochromator M1, which selects the excitation wavelength, to an absorption maximum of the sample and scanning the monochromator M2.

In the geometric arrangement shown in Figure 3.12, the emission is collected at an angle of approximately 90° to the excitation beam (right-angle arrangement). To avoid distortions of the emission spectrum by the reabsorption of emitted light (*inner filter*

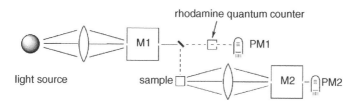

**Figure 3.12** Basic components of a steady-state fluorescence spectrophotometer. The boxes M1 and M2 are monochromators and the detectors PM1 and PM2 are photomultipliers

*effect*), optically dilute solutions should be used. For measurements that must be done on optically dense samples, most instruments provide a front-face arrangement (the sample is irradiated at an angle of 45° and the emission is observed from the front face at an angle of about 30° to avoid reflections of the excitation beam).

Good practice in recording and correcting fluorescence spectra is mandatory and less straightforward than in absorption spectroscopy. The high sensitivity of emission spectroscopy permits the detection and quantification of extremely small amounts of luminescent compounds such as benzopyrene in industrial wastewater or in a river. Conversely, emission spectroscopy of weakly luminescent compounds is extremely sensitive to impurities with strong luminescence. Such impurities can be present in the sample or solvent, the glassware used to prepare the solution or the sample cell itself. A blank spectrum should be recorded with the pure solvent to correct for small impurity emissions and also for Rayleigh scattering and Raman emission by the solvent. Even when these 'trivial' sources have been eliminated or at least quantified for later correction, a photoreaction of the sample that produces a fluorescent compound upon exposure to daylight or to the relatively strong excitation beam of the spectrometer can vitiate the results.

*Fluorescence excitation spectra* are recorded by setting the monochromator M2 to an emission maximum of the sample and scanning the excitation wavelength with the monochromator M1 (see also Special Topic 3.3). Excitation spectra are action spectra[a]; the emission intensity is proportional to the amount of light absorbed by the fluorescent sample at the excitation wavelength (Figure 4.23) provided that Vavilov's rule (Section 2.1.7) holds. The right-angle arrangement must be used. An excitation spectrum will faithfully reproduce the absorption spectrum of the fluorescent compound only when the intensity of the excitation beam is calibrated by the quantum counter (see below) and when the absorbance of the sample is less than 0.05 (see Section 3.9.1, Equation 3.18). At higher absorbances, the amount of light will be attenuated substantially before it reaches the central part of the sample, which is focused on to the slit of the monochromator M2.

The choice of slit widths at the monochromators is important. Small slits give better spectral resolution, while reducing the amount of light transmitted. Hence a compromise must always be made. The low light intensities associated with small slits may be compensated by increasing the sensitivity of the detector (higher voltage) and the integration times (slower scan speed). However, these will also increase the background. Whereas the slit widths at the entrance and exit of a monochromator should be the same for best performance, the slit widths at M1 and M2 can and usually should, be chosen differently. Small slit widths are chosen at M2 to improve the resolution of emission spectra, while small widths at M1 are required to achieve adequate resolution of excitation spectra. Similar arguments apply when double monochromators are used for M1 or M2.

Excitation spectra are especially useful for analysing mixtures. If the emission spectrum is (in part) due to an impurity, the excitation spectrum is likely to differ markedly from the absorption spectrum of the sample. This may be used in low-temperature work (Section 3.10) to determine whether an absorption spectrum produced by irradiation is due to a single photoproduct.

---

[a] An action spectrum is a plot of a biological, chemical or physical response per number of *incident* (prior to absorption) photons versus wavelength or wavenumber.

Few compounds emit substantial *phosphorescence* in solution; biacetyl is a well-known exception to the rule. Phosphorescence lifetimes tend to be much longer (milliseconds) than fluorescence lifetimes (nanoseconds; see Section 2.1.1), so that highly purified solvents must be used and thoroughly degassed to reduce quenching by dioxygen and other adventitious impurities. To prevent diffusional quenching, the compounds are generally dissolved in solvent mixtures[157,159] that freeze to clear glasses when cooled to 77 K, such as diethyl ether–isopentane–ethanol (5 : 5 : 2) or alkane–solvent mixtures. The sample cell (usually a quartz tube) is dipped into liquid nitrogen contained in a transparent quartz Dewar vessel.

Alternatively, glassy solutions that are solid at room temperature [borax or poly(methyl methacrylate, PMMA) glass] can be used for phosphorescence measurements of organic and inorganic solutes. Freshly distilled methyl methacrylate solutions are polymerized in stoppered glass tubes by warming at $\sim$80 °C for a few days. The resulting mould is readily cut to any desired shape and polished to produce transparent blocks. Oxygen is mostly consumed during the polymerization and the diffusion of air into the moulds is fairly slow ($<$1 mm per month); such samples can be used as reference samples to set up a phosphorescence spectrometer or for classroom demonstration of the slow phosphorescence decay of naphthalene, for example, which lasts for many seconds.

Phosphorescence spectra are commonly weaker than fluorescence spectra, even when solid sample solutions are probed. For unperturbed detection of phosphorescence, the fluorescence must, therefore, be eliminated. This is readily achieved by surrounding the sample with a metal cylinder that is rotated at variable speed by a motor ("rotating can"). A hole in the cylinder exposes the sample to the excitation beam for a short period, during which the detector does not 'see' the sample. Rotation of the can subsequently opens the sample to the detector, long after the fluorescence has decayed. The phosphorescence lifetime can then be determined from the intensity dependence of the phosphorescence on the rotational speed. In modern instruments, the elimination of fluorescence is commonly achieved by replacing the continuous light source with a pulsed source (a stroboscope or a pulsed laser) and the photomultiplier is electronically deactivated during and shortly after the excitation pulse.

---

### Special Topic 3.3: Phosphorescence excitation spectra

Singlet–triplet absorption spectra are very difficult to measure, because the absorption coefficients for these forbidden transitions are extremely weak in organic molecules. In principle, one could resort to using very high concentrations and pathlengths, but in practice this requires materials of extremely high purity and the method fails at shorter excitation wavelengths where singlet–singlet absorption sets in. Few reliable results have been obtained in this way with organic compounds (for an example, see Figure 6.15).

The intensity of singlet–triplet absorption can be increased by using heavy-atom media ($CBr_4$ or liquid xenon) and particularly by paramagnetic enhancement under high pressures of dioxygen (oxygen pressure method). Most of these measurements were done by Evans around 1960 using oxygen pressures up to 130 atm. To put reactive organic molecules and solvents under high oxygen pressure is, however, a dangerous venture. Evans reported: 'Only one measurement was made on an acetylene–oxygen mixture at high pressures (acetylene 20 atm, oxygen 100 atm). When the needle valve was opened to release the pressure, a violent explosion occurred which reduced the quartz windows of the cell to fine dust and blew off the stainless steel end-plates at high velocity.'[178]

The more convenient method of *phosphorescence excitation* was introduced by Kearns in 1965. Because emission spectroscopy is much more sensitive than absorption, particularly for phosphorescence that can efficiently be separated from stray light and fluorescence, the determination of singlet–triplet absorption spectra by monitoring the resulting phosphorescence emission has often been successful with comparatively little effort.

Because fluorescence spectrometers are single-beam instruments, a number of corrections are required to obtain reproducible and meaningful emission spectra. Temporal fluctuations of the light source, and also the variations of its intensity at different excitation wavelengths, are corrected by means of a *quantum counter* that consists of a triangular cell containing a solution of a fluorescent dye and a photomultiplier, which constantly monitors the total fluorescence emission from that cell (PM1, Figure 3.12). The fluorescent dye, usually Rhodamine 6G, must be sufficiently concentrated to absorb the incident light completely at all wavelengths of excitation $\lambda$ and it must have a fluorescence quantum yield that is independent of $\lambda$. The signal from PM1 is then proportional to the number of photons per unit time in the excitation beam, part of which is focused on to the quantum counter cell. Fluctuations of the intensity of the light source in time, and also its variations at different excitation wavelengths when scanning monochromator M1 to record excitation spectra, are automatically corrected by dividing the output signal of the fluorescence detector PM2 by that of PM1 (ratiometric measurement). Obviously, the correction of excitation spectra works only up to the red edge of the absorption by the fluorescent dye.

Current spectrometers use a grating, so that the spectral dispersion is linear in wavelength and the spectral bandpass (the width $\delta\lambda$ of light from a 'white' source passing the monochromator) is independent of wavelength. Therefore, spectrofluorimeters generally plot spectra on a wavelength scale. If an emission spectrum is converted to a wavenumber scale, it is essential to convert the instrument readings, the emitted *spectral radiant power* per wavelength interval $P_\lambda^{em}$ according to $P_{\tilde{\nu}} = P_\lambda/\tilde{\nu}^2$ (see footnote to Equation 2.3), which results in substantial changes of the spectral shape. For the determination of fluorescence quantum yields (Section 3.9.5) or of Förster overlap integrals (Section 2.2.2), it is unnecessary, and therefore not recommended, to convert spectra from a wavelength to a wavenumber scale.

*Correction of emission spectra* is required, because the transmission efficiency of the monochromators and particularly the sensitivity of the detectors depend on the wavelength and the polarization of the emitted light. A magic-angle arrangement of polarizers may be used to avoid distortions due to polarized emission from solid samples.[175] Commercial instruments may be delivered with a sensitivity function $S(\lambda)$ that allows correction of the recorded emission spectra by multiplication with the recorded signal, $P_\lambda^{em} = P_\lambda^{obs}S(\lambda)$. However, the function $S(\lambda)$ usually changes substantially as the instrument ages, so it should be calibrated at regular intervals. This can be done either using standard fluorescent dyes[179] such as quinine sulfate, the true emission spectra of which have been determined accurately, or with a calibrated tungsten lamp. For the conversion of an emission spectrum to a wavenumber scale, the sensitivity function $S(\lambda)$ must be replaced by $S(\tilde{\nu})$ (Equation 3.2).

$$S(\tilde{\nu}) = S(\lambda)|d\lambda/d\tilde{\nu}| = S(\lambda)/\tilde{\nu}^2$$

**Equation 3.2**

## 3.5 Time-Resolved Luminescence

*Recommended reviews.*[177,180] See also Section 3.13.

Recent developments of pulsed light sources, optical components, fast and sensitive detectors and electronic equipment for data collection and analysis have permitted the construction of numerous instruments, often commercially available, for the collection of luminescence data with excellent resolution in time, spectral distribution and space. The sensitivity has reached the ultimate level that allows the characterization of such properties for single molecules (see Section 3.13). Only an overview of some of these techniques is given here.

In *pulse fluorimetry*, the sample is excited by a short pulse of light from a laser or from a spark gap (similar to the spark gaps used to ignite combustion in motor engines). Spark gaps are inexpensive and produce highly reproducible excitation pulses of about 2 ns width. Spark gap instruments are operated at repetition rates of $10^4$–$10^5$ Hz and the fluorescence signals are accumulated using single photon counting (Section 3.4) and an electronic time-to-amplitude converter. If the duration of the pulse is comparable to or longer than the lifetime of the fluorescence decay, as is often the case, the fluorescence intensity rises in time with the excitation pulse, passes through a maximum and reflects the true decay function of the sample only when the intensity of the pulse becomes negligible. However, the excellent statistics of the averaged signal permit a mathematical deconvolution of the decay signal from the excitation pulse shape. In this way, a time resolution of about 200 ps can be achieved for fluorescence lifetimes. This can be improved by an order of magnitude by using short excitation pulses from mode-locked lasers and fast microchannel plate photomultipliers as detectors (Section 3.1).

In *phase-modulation fluorimeters*, a continuous excitation beam is modulated at a gigahertz frequency using a Pockels cell and the fluorescence signal is detected with a lock-in amplifier. Time resolutions achievable depend on the instrumentation and are comparable to those in pulse fluorimetry. *Streak cameras* (Section 3.1) offer the muliplexer advantage of recording the time evolution of complete emission spectra. The time resolution reaches a few picoseconds on the fastest instruments. However, signal averaging is usually required to achieve an adequate signal-to-noise ratio and jitter of the trigger signal may then blur the accumulated signal on the time axis.

Ultrashort fluorescence lifetimes are best determined by *photon upconversion*. An excellent description of the technique for non-specialists wishing to set up an experiment is in preparation as part of a current IUPAC project on 'ultrafast intense laser chemistry'.[181] The time resolution is basically limited by the temporal width of the laser pulses that are being used. An intense gate pulse at frequency $\nu_G$ is mixed with the fluorescence at a frequency $\nu_F$ in a nonlinear optical crystal (Special Topic 3.1), to create a short pulse at the sum frequency $\nu_U$. The gate pulse thus represents a time window for the fluorescence during which the latter is upconverted and the intensity of the upconverted light is measured. By controlling the optical delay between the fluorescence and the gating pulse, kinetic traces of the fluorescence can be obtained. Because the upconverted signal appears far to the blue of the fluorescence band, it is easily separated from the signals at frequencies $\nu_G$ and $\nu_F$ by a monochromator or an optical filter.

## 3.6   Absorption and Emission Spectroscopy with Polarized Light

*Recommended textbook.*[182]

The orientation of transition moments in the molecular framework can be predicted by quantum mechanical calculations (Sections 2.1.4 and 4.4). Experimentally, the direction of transition moments can be assessed by studying the absorption of linearly polarized light by samples in which the molecules are preferentially oriented. Complete orientation is available in single crystals. However, optical measurements on single crystals are demanding and rarely practicable, not least because extremely thin crystals are required for absorption studies.

Partial orientation is much easier to achieve and generally sufficient to determine the direction of the transition moments, particularly with respect to any axes of symmetry that the molecules may have. Partial orientation can be achieved by *photoselection* using a linearly polarized excitation source, by application of an electric field or by dissolving the solute in a transparent anisotropic medium (liquid crystals, stretched polymers).[182] Photoselection is the basis of emission anisotropy measurements discussed below. It is also used for molecules that can be either generated or destroyed photochemically in a rigid medium such as poly(methyl methacrylate) or glassy solvents at low temperature. Preferential alignments of dipolar molecules that are achievable by electrostatic fields are unfortunately fairly small.

By far the simplest method for the preparation of partially oriented samples is the doping of stretched polyethylene sheets, which was developed by Thulstrup, Eggers and Michl.[182,183] It can be applied in any laboratory and requires no special equipment other than a UV–VIS or IR spectrometer and linear polarizers. Commercial polyethylene usually contains various additives such as UV stabilizers, which must be removed by extraction, but optically clean sheets are available. The sheets are doped either by swelling them in a solution of the probe and evaporating the solvent or by keeping them under vacuum under vapours of the substrate. The sheets are then stretched to about six times their original length and their absorption spectrum is recorded twice with linearly polarized light beams with the electric vector parallel and perpendicular to the stretching direction. An example taken from the literature [182] is shown in Figure 3.13. A series of linear combinations of the two spectra is generated by computer (centre diagram in Figure 3.13), two of which are chosen such that some spectral features disappear from one or the other of the two combinations. Cooling of the sample in the spectrometer improves the resolution, but is not required. It is well established that the above procedure for spectral reduction reliably recovers the polarized spectra of molecules with a plane of symmetry just as they would be obtained with a fully oriented sample. For details of the procedure and analysis of polarized spectra of partially oriented samples, the reader is referred to the original literature.

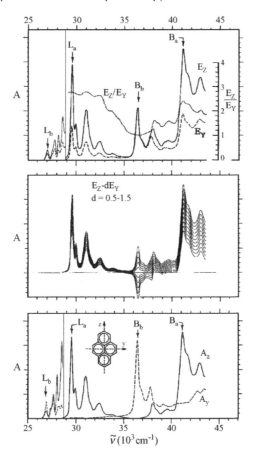

**Figure 3.13** Pyrene in stretched polyethylene at 77 K. Top: baseline-corrected polarized absorption spectra $E_Z$ (parallel) and $E_Y$ (perpendicular to the stretching direction). Centre: linear combinations of the polarized spectra $E_Z$ and $E_Y$. Bottom: reduced UV spectra of pyrene. $A_z = E_Z - 1.0E_Y$ and $A_Y = 1.75(E_Y - 0.36E_Z)$. Adapted by permission from ref. 182. The Platt symbols $L_b$, $L_a$, $B_b$ and $B_a$ labelling the electronic transitions are explained in Section 4.7. The polarization of the weak $L_b$ band shown on the left ($<29\,000\ \mathrm{cm}^{-1}$, spectra recorded at higher concentration) is of mixed polarization

Most light sources, lasers in particular, are linearly polarized to some degree. As a result, the emission from the sample may also be polarized. The *degree of anisotropy R* is defined by Equation 3.3:

$$R = \frac{I_{\mathrm{II}} - I_{\perp}}{I_{\mathrm{II}} + 2I_{\perp}}, \quad -0.2 \leq R \leq 0.4$$

**Equation 3.3** *Degree of emission anisotropy*

where $I_{\mathrm{II}}$ and $I_{\perp}$ are the radiant intensities measured with the linear polarizer for emission parallel and perpendicular, respectively, to the electric vector of linearly polarized incident electromagnetic radiation. The quantity $I_\parallel + 2I_\perp$ is proportional to the total emission intensity.

In an isotropic sample with randomly oriented chromophores, the emission anisotropy arises from preferential excitation (photoselection) of those molecules that happen to be oriented such that their transition moment of absorption is oriented parallel to the polarization of the exciting light. The theoretical values of $R$ depend on the angle $\phi$ between the transition moments of absorption and emission, $\boldsymbol{\mu}_a$ and $\boldsymbol{\mu}_e$: $R = \langle 3\cos^2\phi - 1\rangle/5$, where $\langle\rangle$ denotes an average over the orientations of the photoselected molecules. This leads to the range of values given in Equation 3.3, that is, $-1/5$ for $\phi = 90°$ and $2/5$ for $\phi = 0°$. In practice, $R$ is often reduced by (a combination of) the following depolarizing events: rotational diffusion of the emitter during its lifetime, overlap of differently polarized transitions in the range of excitation or emission wavelengths measured, energy transfer between the emitting chromophores (Section 2.2.2) and adiabatic reactions of the emitter such as proton transfer (Section 5.3). Time-resolved measurements of anisotropy can be used to determine the rate of rotational diffusion of large molecules such as tagged biopolymers, when the rotational time constant is comparable to the lifetime of emission. Rotational diffusion can be reduced or suppressed in highly viscous media.

## 3.7   Flash Photolysis

*Recommended reviews* are mentioned in the second paragraph below.

Since its invention by Norrish and Porter in 1949,[184] flash photolysis is the most important tool to produce transient intermediates in sufficient concentration for time-resolved spectroscopic detection and for the identification of elementary reaction steps (see Section 5.1). The term *photolysis* strictly implies the light-induced breaking of chemical bonds (the Greek expression *lysis* means dissolution or decomposition). However, since its inception, the term flash photolysis has been used to describe the technique of excitation by short pulses of light, irrespective of the processes that follow.

Transient intermediates are most commonly observed by their absorption (transient absorption spectroscopy; see ref. 185 for a compilation of absorption spectra of transient species). Various other methods for creating detectable amounts of reactive intermediates such as stopped flow, pulse radiolysis, temperature or pressure jump have been invented and novel, more informative, techniques for the detection and identification of reactive intermediates have been added, in particular EPR, IR and Raman spectroscopy (Section 3.8), mass spectrometry, electron microscopy and X-ray diffraction. The technique used for detection need not be fast, provided that the time of signal creation can be determined accurately (see Section 3.7.3). For example, the separation of ions in a mass spectrometer (time of flight) or electrons in an electron microscope may require microseconds or longer. Nevertheless, femtosecond time resolution has been achieved,[186,187] because the ions or electrons are formed by a pulse of femtosecond duration (1 fs $= 10^{-15}$ s). Several reports with recommended procedures for nanosecond flash photolysis,[137,188–191] ultrafast electron diffraction and microscopy,[192] crystallography[193] and pump–probe absorption spectroscopy[194,195] are available and a general treatise on 'ultrafast intense laser chemistry' is in preparation by IUPAC.

The duration of the pump pulse is the main factor that limits the time resolution of flash photolysis. The commonly used specification of full width at half-height (FWHH) is incomplete, since the shape of many pulses is far from Gaussian. Tailing and afterglow of, for example, conventional discharge flash lamps and excimer lasers can severely reduce the time resolution achievable. Only a few groups still use the so-called 'conventional' apparatus, that is, flash photolysis systems pumped by a discharge flash lamp of microsecond duration. Yet these instruments still give the best performance on the millisecond time scale and the whole setup costs a fraction of that of a laser. Tremendous developments in the production of short laser pulses, in the time resolution and sensitivity of detectors, the speed and dynamic range of digitizers and the speed and memory capacities of computers (statistical averaging) have improved time resolution down to femtoseconds.

Most flash photolysis systems with excitation pulses of nanosecond and longer duration operate in the kinetic mode, that is, the transient absorption is monitored at a single wavelength as a function of time. Spectrographic detection can be done using gated microchannel plates as light shutters (and amplifiers) and diode-array detectors, which provide digitized transient absorption spectra at a given time delay with respect to the laser pulse. Basically, both techniques probe one-dimensional slices of the same physical information that consists of a two-dimensional array $A$ of absorbances $A(\lambda,t)$ as a function of wavelength $\lambda$ and time $t$ after excitation (Figure 2.18). Systems with shorter excitation pulses and subnanosecond time resolution usually operate in the spectrographic mode (pump–probe spectroscopy). The analysis of kinetic and spectrographic data is described in Sections 3.7.4 and 3.7.5.

Beware of artefacts! This cautionary remark should always be kept in mind when carrying out flash photolysis. Stray light and fluorescence, acoustic shock waves and inhomogeneous transient distributions produced in the sample by focused laser pulses, extraneous electronic pulses from flash lamps or Q-switches, signal echoes and the like often distort the transient waveforms or spectra.

### 3.7.1 Kinetic Flash Photolysis

Flash photolysis setups (Figure 3.14) are single-beam instruments that measure absorbance changes in time.

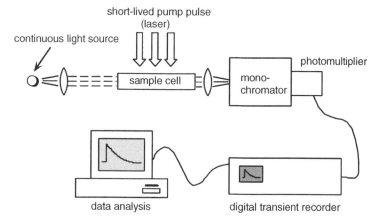

**Figure 3.14** Kinetic setup for flash photolysis

Depending on the pulsed excitation source, but also on the optical arrangement (excitation parallel or vertical to the monitoring light) and focusing optics, the concentrations of generated transients may differ widely. Strictly parallel orientation of the pump and probe beams is avoided to inhibit the strong pump pulse from hitting the detector of the probe beam. Both orientations have some advantages and it is useful to devise the detection system such that it can be switched from one geometry to the other. An advantage of right-angle excitation with diffuse pump beams is the relatively low concentration of transient intermediates produced in the sample. This reduces the half-life of bimolecular reactions between transient intermediates such as triplet–triplet annihilation or radical recombination. Sensitivity can be maintained by collecting transient absorbance over an extended optical pathlength; pathlengths of 4 cm are conveniently used with excimer lasers and 10–20 cm with conventional discharge flashlamps. An advantage of parallel pump and probe beams is that the sample absorbance can be raised to a point where essentially all of the exciting light is absorbed.

With nonexponential decay traces, it may be difficult to distinguish between second-order reactions and two overlapping first-order decays with similar rate constants. In such cases, it is helpful to change the initial concentration of the transient intermediates either by reducing the concentration of the reactant or the intensity of the excitation pulse or by switching from parallel to right-angle excitation. Second-order decays will then exhibit longer half-lives, while the sum of first-order decays remains the same.

Flash photolysis commonly produces transient concentrations that are unevenly distributed. Inhomogeneity arises from hot spots in the laser beam profile, focusing optics, internal filter effects in the sample solutions and imperfect overlap between the volume excited by the pump beam and that analysed by the probe beam. Inhomogeneous sample distributions distort kinetic traces[196] and should be minimized, even at the expense of some loss in the signal intensity. A gradient in the concentration at right-angles to the monitoring beam will distort even the traces of first-order reactions, since the average absorbance of such a solution will not be proportional to the average transient concentration. The absorbance at the excitation wavelength should be kept below about 0.5 at the excitation wavelength with vertical excitation. The effect of inhomogeneous sample distributions for a given setup can be determined by observing a strongly absorbing transient intermediate such as the triplet of 9,10-dibromoanthracene in aerated solution that is known to give a clean first-order decay. Inhomogeneous transient distributions are manifested by a sigmoid distortion of the decay traces. When the transient absorption is monitored parallel to the excitation beam, inhomogeneous transient distributions are less distorting, so that much higher sample concentrations can be used.

A common procedure in mechanistic studies is to follow the growth or decay rate of a transient intermediate as a function of the concentration of some other species such as a sensitizer, quencher, trapping agent, acid or buffer. Bimolecular rate constants are then obtained from the concentration dependence of the observed first-order rate constants, $k_{obs}$ (see Section 3.9.7).

Good kinetic data are usually reproducible within a few percent. Standard errors determined from fitting of a single trace are commonly even smaller, but reproducibility should be judged from the analysis of several decay traces, because systematic errors, such as instability of the monitoring light source over the period of observation or variations of temperature, do not show in a single trace. The temperature dependence of the logarithm of first-order rate constants obeying Arrhenius' law is proportional to the activation energy $E_a$, $\mathrm{d}\log(k/s^{-1})/\mathrm{d}T = E_a/(2.3RT^2)$. Hence fast reactions having low activation energies are generally not very sensitive to temperature changes. Nevertheless, assuming $A = 10^{13}\,s^{-1}$, one predicts a 5% increase for a rate constant of $10^6\,s^{-1}$ when the temperature increases by $1\,°C$ around ambient temperature. Hence sample cells should be held in a thermostat if highly accurate kinetic data are required.

Digitizing oscilloscopes with sampling rates of $\geq 1$ gigasample $s^{-1}$ and high-bandwidth amplifiers ($\geq 500$ MHz) nicely match the time resolution required to see the fastest processes resolvable by excitation with Q-switched lasers, excimer lasers or dye lasers, which have pulse widths on the order of 2–20 ns. The high storage depth of modern digitizers provides access to long delay times even at high time resolution so that both fast and slow processes may be captured in a single trace. Furthermore, the pre-trigger capturing facility of these devices obviates the need for trigger cascades. Jitter is largely eliminated when the trigger is generated directly from the laser pulse with a photodiode.

### 3.7.2 Spectrographic Detection Systems

The photographic plates or film used in the early days to record the absorption spectra of transient intermediates have been replaced by diode arrays. The time window during which the transient spectra are collected can be determined by employing a short probe pulse of light. Alternatively, microchannel plate image intensifiers placed in front of the diode arrays may be used in combination with continuous monitoring light sources to achieve nanosecond time windows; image intensifiers not only enhance the intensity of the monitoring light during the active time window, but also act as shutters with a (de-)activation rise time of about 1 ns and an on:off transmission ratio $>10^6$. Such an effective shutter is required if the monitoring light source is continuous during the long integration times of the detectors ($\sim 30\,ms$). Cut-off filters that do not transmit below the cut-off wavelength should be used for protection of the detector from the excitation pulse and to eliminate second-order reflections from the grating. When excess monitoring light is available at some wavelengths, compensating filter solutions may be adjusted with appropriate dyes to equalize the spectral distribution of the source. A commercially available setup for nanosecond laser flash photolysis (LFP) is shown in Figure 3.15.

Fluorescence will interfere with kinetic absorbance measurements for the duration of the fluorescence or of the pump pulse, whichever is longer. Scattered light and

**Figure 3.15** Spectrographic setup for nanosecond LFP. Reproduced by permission of Applied Photophysics Ltd

fluorescence can easily drown the monitoring light source because the intensity of the pump pulse is commonly orders of magnitude higher than that of the probe pulse or of the continuous monitoring light source. 'Stray light' problems are reduced by using appropriate filters, by pulsing the monitoring lamp and by taking advantage of the different angular dispersion of scattered light and the monitoring beam.

When the observation wavelength lies within the absorbance of the reactant, negative absorption may be due to bleaching. Transient kinetics should always be determined at several wavelengths and transient spectra at various delay times, $t$.

Streak cameras allow one to probe the entire two-dimensional array $A(\lambda,t)$ following excitation with a single pulse. They are, however, rarely used for pump–probe flash photolysis because of their low precision and high price. Repeated excitation is necessary to probe different wavelengths on a kinetic apparatus and different time delays on spectrographic devices. Variations of the pump pulse intensity will affect the spectral information determined point-by-point in the kinetic mode and the kinetic information derived from sequential spectra. Systematic variations of the instrument performance or decomposition of the sample with time can easily produce artificial kinetics in the spectrographic mode and artificial spectral features in the kinetic mode, respectively. It is, of course, best to determine and eliminate the source of such problems. Notwithstanding, probing at different wavelengths or time delays should always be done at random rather than in an ordered sequence; if systematic variations occur during data collection (commonly 10 min to 1 h), they will produce an increase in random noise rather than artefacts such as distorted kinetics.

Fluctuations of the laser pulse can be eliminated by calibration, that is, by independent monitoring of the pulse intensity for each individual shot. Sample degradation can be avoided by using flow systems.

### 3.7.3  Pump–Probe Spectroscopy

Lasers with ultrashort pulse lengths down to a few femtoseconds ($1\,\mathrm{fs} = 10^{-15}\,\mathrm{s}$) are now available commercially. However, photomultipliers and the associated electronic digitizers are not sufficiently fast to follow waveforms much below 1 ns. Therefore, devices with pico- and femtosecond time resolution use optical delay lines to define the time delay between the excitation pulse and the probe pulse (Figure 3.16). By focusing part of the

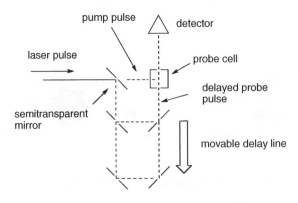

**Figure 3.16**  Conceptual design of a pump–probe apparatus

pump pulse into a solution or a $CaF_2$ plate, a broad supercontinuum (300–700 nm) is generated, which can be used as the probe pulse in *pump–probe spectroscopy*.[194] A pathlength difference of 0.3 mm between the pump and the probe pulse amounts to a time difference of 1 ps. It is essential to ensure that the spatial overlap of the two pulses on the sample does not change with the time delay setting. Proper alignment should be tested routinely by monitoring some absorption which is generated by the pump pulse and which is known to persist throughout the time range covered by the delay line. Any seeming growth or decay of that absorption would then indicate misalignment of the two pulsed beams. This is the reason why delay lines are usually limited to about 0.3 m in length, spanning a time scale of 2 ns in two passes.

An elegant way to study the dynamics from excited vibrational levels of the electronic ground state is to make use of the femtosecond pump–dump–probe scheme in transient absorption experiments. The molecular system is initially excited to the $S_1$ state and, using a second pulse of longer wavelength (the dump pulse), excited vibrational levels of the $S_0$ state are populated via stimulated emission.[197]

### 3.7.4 Analysis of Kinetic Data

An *elementary process* is a reaction involving a small number of molecules (usually one or two, ignoring the surrounding solvent molecules) that is *assumed* to proceed in a single step, that is, over a single barrier. The differential rate law for a monomolecular elementary reaction of a compound A is $-dc_A/dt = -kc_A$ and is said to be *first order* (there is a single concentration associated with the rate constant $k$), that for an elementary bimolecular reaction is either $-dc_A/dt = -2kc_A^2$ for a reaction involving two molecules of A or $-dc_A/dt = -kc_Ac_B$ for a reaction of A with a different molecule B. Both of the latter are *second order*. Zero-order rate laws, $-dc_A/dt = -k$, cannot represent an elementary reaction and are not encountered in fast reaction kinetics. The units of a first-order rate constant are $s^{-1}$ and those of a second-order constant are commonly $M^{-1}s^{-1} = dm^3 mol^{-1}s^{-1}$. Consider the simple reaction scheme of an intermediate or excited state A undergoing two parallel, irreversible, first-order reactions to yield the two products B and C (Scheme 3.1), left.

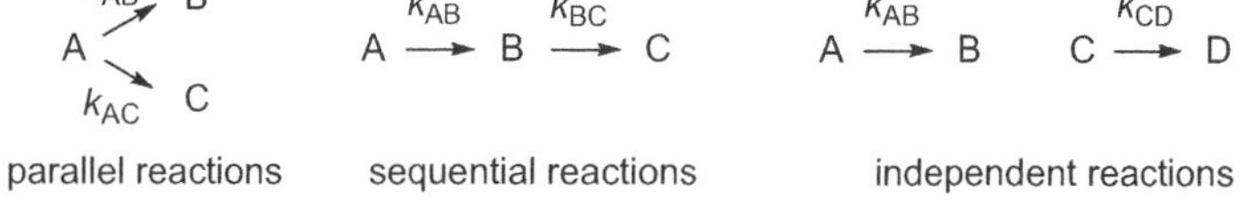

parallel reactions    sequential reactions    independent reactions

**Scheme 3.1** *Parallel, sequential (consecutive) and independent reactions*

The rate constant for the reaction A → B is $k_{AB}$ and that for the reaction A → C is $k_{AC}$. Given that $k_{AB}$ is smaller than $k_{AC}$, say $k_{AB} = 0.2\,s^{-1}$ and $k_{AC} = 0.8\,s^{-1}$, one may be tempted to conclude that the observed rate constant for the formation of C is larger than that of B. This is not the case. The differential rate law for the disappearance of A is given by Equation 3.4:

$$-dc_A(t) = (k_{AB} + k_{AC})c_A(t)dt$$

**Equation 3.4** *Rate law for two parallel reactions of compound A*

This is a first-order rate law with a single observable rate constant $k_{obs} = k_{AB} + k_{AC}$. Integration gives

$$c_A(t) = c_A(t = 0)e^{-k_{obs}t}$$

**Equation 3.5**

The differential rate law for the appearance of compound B is $dc_B(t) = k_{AB}c_A(t)dt$. Insertion of $c_A(t)$ from Equation 3.5 and integration assuming $c_B(t = 0) = 0$ gives

$$c_B(t) = c_A(t = 0)\frac{k_{AB}}{k_{obs}}(1-e^{-k_{obs}t})$$

**Equation 3.6**

Hence the observed rate constant for the disappearance of A, $k_{obs} = k_{AB} + k_{AC}$, is the same as that for the appearance of B (and of C). The difference lies in the pre-exponential partition ratios, $k_{AB}/k_{obs} = k_{AB}/(k_{AB} + k_{AC})$ for B, Equation 3.6, and correspondingly $k_{AC}/k_{obs}$ for C, which define the *efficiency* of the two reactions, as shown in Figure 3.17. Extension of this argument to more than two parallel first-order reactions from a given reactant A is straightforward. The efficiency of a given elementary reaction step is equal to its partition ratio, the rate constant of the process considered, $k_x$, divided by the sum of the rate constants of all processes competing for the depletion the reactant A (Equation 3.7).

$$\eta_x = k_x/\Sigma k_i$$

**Equation 3.7** *Efficiency of a single-step, first-order reaction*

Many reaction schemes involving parallel first- and second-order reactions also have straightforward analytical solutions.[198] Hence numerical integration methods are rarely required for the fast reactions of interest here. When the reactions are monitored by absorption spectroscopy, the total absorbance of the solution at an observation wavelength $\lambda$, $A(\lambda,t)$, is related to the concentrations $c_i$ of the reacting species $B_i$ by the Beer–Lambert law (Equation 2.5). Thus two parallel reactions, A $\rightarrow$ B and C $\rightarrow$ D (Scheme 3.1, right) will, in the limit $k_{AB} = k_{CD}$, give rise to a single exponential obscuring the existence and the relative amplitudes of the two individual steps.

A special problem arises with consecutive reactions, A $\rightarrow$ B $\rightarrow$ C (Scheme 3.1, centre), that have similar rate constants.[199] The integrated rate law for the time-dependent

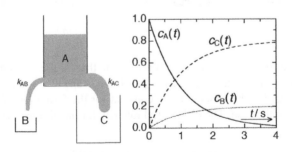

**Figure 3.17** Kinetics of two parallel reactions A $\rightarrow$ B and A $\rightarrow$ C. Left: illustration with a bucket with two holes of different sizes. Right: concentration profiles of compounds A, B, and C

concentration of B is given by Equation 3.8 (concentration profiles for $c_A$ to $c_C$ are depicted in Figure 3.20). If only B absorbs at the wavelength of observation, then a rise in absorbance will be followed by a decay and it is immediately apparent that a consecutive reaction is being observed. Nonlinear least-squares fitting of an observed kinetic trace using Equation 3.8 will nevertheless be difficult or impossible if $k_{AB} \approx k_{BC}$, because the expression is indeterminate for $k_{AB} = k_{BC}$. Moreover, it should be noted that Equation 3.8 is unaffected by an interchange of the microscopic rate constants $k_{AB}$ and $k_{BC}$, so that the observed rate constants cannot be assigned to the individual steps in the reaction sequence, unless other information, such as the absorption coefficients of intermediates A, B and C, is available. The growth part of a growth–decay curve always corresponds to the faster reaction, but it does not follow that it also corresponds to the first reaction in the sequence.

$$c_B(t) = c_A(t = 0) \frac{k_{AB}}{k_{AB} - k_{BC}} [\exp(-k_{BC}t) - \exp(-k_{AB}t)]$$

**Equation 3.8**

When the decay of intermediate B is faster than its formation, $k_{BC} > k_{AB}$, the absolute value of the pre-exponential term will become small and thus the transient concentration $c_B$ will be small at all times. For $k_{BC} \gg k_{AB}$, the appearance of product C will approach the first-order rate law, Equation 3.6 (replace the index B by C), because the transient concentration of intermediate B becomes negligible. This shows that observation of a first-order rate law for the reaction A → C does not guarantee that there is no intermediate involved, that is, that the observed reaction A → C is an elementary reaction.

Rate laws derived for reaction schemes consisting of any number and combination of consecutive, parallel and independent first-order reaction steps can be integrated in closed form to give a sum of exponential terms, so that the total absorbance $A$ of a solution in which these reactions proceed can also be expressed as a sum of exponential terms (Equation 3.9), but see last paragraph of this Section.

$$A(t) = A_0 + A_1 \exp(-\gamma_1 t) + A_2 \exp(-\gamma_2 t) + \cdots$$

**Equation 3.9**

First-order decay rate constants of transients that are also subject to second-order reactions cannot be determined reliably, even when attempts are made to correct for the observable second-order contributions. A typical example is the triplet state half-life of anthracene in degassed solution, which appears to be somewhere in the range $20\,\mu s$–1 ms on most instruments (the values are reproducible on a given instrument but vary widely between different setups and solvent purification methods). The lifetime of triplet anthracene in solution rises to 25 ms when measures are taken to reduce quenching by diffusional encounters with the parent molecule (self-quenching) with other molecules in the triplet state (triplet–triplet annihilation) and with impurity quenchers such as dioxygen.[200] Hence lifetimes of transients that are prone to second-order decay contributions, such as long-lived triplet states or radicals, should always be considered as lower limits.

The response of photomultipliers, photodiodes and diode arrays is generally linear in light flux within a specified range. Within this range, the readings will be proportional to the transmittance $T$ of the sample. Noise associated with the dark current is commonly negligible, because integration times are short. Under these circumstances, the *relative* error of the transmittance and the *absolute* error of the sample absorbance, calculated as $A = \log(1/T)$, will be constant. Minimization of the squared residuals, $\chi^2$, with respect to a model function such as Equation 3.9 should therefore be done with absorbance data where each data point properly receives equal weighting. It is not recommended to transform the data points in order to obtain a linear model function, for example, to $\log A$ for first-order reactions or to $A^{-1}$ for second-order reactions, because the associated errors transform accordingly so that weighting of the data points would be required.

Therefore, the trial model function will in general be a nonlinear function of the independent variable, time. Various mathematical procedures are available for iterative $\chi^2$ minimization of nonlinear functions. The widely used Marquardt procedure is robust and efficient. Not all the parameters in the model function need to be determined by iteration. Any kinetic model function such as Equation 3.9 consists of a mixture of linear parameters, the amplitudes of the absorbance changes, $A_i$, and nonlinear parameters, the rate constants, $k_i$. For a given set of $k_i$, the linear parameters, $A_i$ can be determined without iteration (as in any linear regression) and they can, therefore, be eliminated from the parameter space in the nonlinear least-squares search. This increases reliability in determining the global minimum and reduces the required computing time considerably.

Although analytical model functions (closed integral forms of the differential kinetic equations) are available for most reaction schemes encountered in fast kinetics, it is by no means trivial in general to derive the physically relevant rate constants $k_i$ of the elementary reactions from the observable rate coefficients $\gamma_i$ (Equation 3.9) and to determine the concentration profiles from the model parameters $\gamma_i$ determined by the fit. Moreover, such a mathematical model will be consistent with many mechanisms (combinations of elementary reactions). In such a situation of mathematical ambiguity, additional information is required. It may consist of mechanistic considerations based on chemical intuition or prior independent knowledge of individual molar absorption coefficients and/or rate constants. Distinctions between acceptable and unacceptable mechanisms may also be possible on the grounds that the absorbances of all species must be non-negative throughout the spectrum and that the concentrations of all species must be non-negative at all times. As a rule, *one should never postulate a mechanism that is more complex than is required by the observed rate law*, unless there is other solid evidence requiring a more complex mechanism (Occam's razor).[b]

### 3.7.5 Global Analysis of Transient Optical Spectra

*Recommended textbooks.*[201,202]

Most absorption and emission spectrophotometers are interfaced to a computer, allowing for digital recording of the spectra. If the spectra change in time, for example when a photoreaction is monitored intermittently by absorption spectroscopy, one obtains a series of spectra such as that shown in Figure 3.18, where the absorbance data are stored in a

---

[b] Occam's razor (sometimes spelled Ockham's razor) is a principle attributed to the 14th-century English logician and Franciscan friar William of Ockham. The principle states that the explanation of any phenomenon should make as few assumptions as possible, eliminating those that make no difference in the observable predictions of the explanatory hypothesis or theory. The principle is often expressed in Latin as the *lex parsimoniae* ('law of parsimony' or 'law of succinctness'): *entia non sunt multiplicanda praeter necessitatem*, that is, entities should not be multiplied beyond necessity. This is often paraphrased as 'All other things being equal, the simplest solution is the best.' [Quoted from Wikipedia, 14 November 2007.]

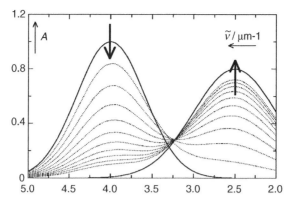

**Figure 3.18** A series of time-dependent spectra

*matrix$^c$A* (Equation 3.10). *Singular value decomposition* (SVD) is the method of choice to analyse such data.[201] Different names such as *principal component analysis* and *factor analysis* are used for essentially the same procedures. These methods are often incorporated in the software provided with the spectrometers. It is important, however, to understand what they do and to what extent they reflect the predilections of the user.

$$\mathbf{A}(t_i, \tilde{\nu}_j) = \begin{pmatrix} A(t_1, \tilde{\nu}_1) & A(t_1, \tilde{\nu}_2) & \cdots & A(t_1, \tilde{\nu}_j) & \cdots & A(t_1, \tilde{\nu}_c) \\ A(t_2, \tilde{\nu}_1) & A(t_2, \tilde{\nu}_2) & \cdots & A(t_2, \tilde{\nu}_j) & \cdots & A(t_2, \tilde{\nu}_c) \\ \cdots & \cdots & \cdots & \cdots & \cdots & \cdots \\ A(t_i, \tilde{\nu}_1) & A(t_i, \tilde{\nu}_2) & \cdots & A(t_i, \tilde{\nu}_j) & \cdots & A(t_i, \tilde{\nu}_c) \\ \cdots & \cdots & \cdots & \cdots & \cdots & \cdots \\ A(t_r, \tilde{\nu}_1) & A(t_r, \tilde{\nu}_2) & \cdots & A(t_r, \tilde{\nu}_j) & \cdots & A(t_r, \tilde{\nu}_c) \end{pmatrix} = (A_{ij})$$

***Equation 3.10*** *Spectral data matrix **A** of dimension* $(r \times c)$

The $r$ rows of the matrix, $i = 1, \ldots, r$, represent the absorption spectra of the sample at a given time $t_i$ and the $c$ columns, $j = 1, \ldots, c$, collect the absorbances at a given wavenumber

---

$^c$ A matrix $A = (A_{ij})$ is a $(r \times c)$ rectangular array of numbers $A_{ij}$ (Equation 3.10); $r$ is the number of rows with index $i = 1 \cdots r$, $c$ the number of columns with index $j = 1 \cdots c$. A *square matrix* has an equal number of rows and columns, $r = c$. The *transpose A′* of a matrix $A$ is obtained by exchanging columns and rows, $A' = (A_{ji})$. In a *diagonal matrix D*, all off-diagonal elements vanish, $A_{ij} = 0$ $(i \neq j)$, hence $D' = D$. A matrix is *symmetrical*, if $A_{ji} = A_{ij}$ for all $i$ and $j$.

Matrix operations are best performed by computer; only some definitions, but not the numerical procedures, are given here. Matrices of equal size $(r \times c)$ can be added (or subtracted) by adding (or subtracting) their corresponding elements, $A \pm B = (A_{ij} \pm B_{ij})$. The product $C = A*B$ of two matrices can be formed only if the number of columns of $A$ equals the number of rows of $B$. The elements $C_{ik}$ of $C$ are then given by

$$C_{ik} = \sum_{l=1}^{m} A_{il} B_{lk}.$$

A square matrix is *singular* if its determinant $\|A\|$ (footnote h at end of Section 1.4) vanishes. The *inverse* $A^{-1}$ of a matrix $A$ is defined by the equation $A^{-1}*A = E$, where $E$ is the *identity matrix*, $E = (\delta_{ij})$ and $\delta_{ij}$ is *Kronecker's delta*, which takes the value of 1 for $i = j$ and 0 otherwise, so that $E*A = A*E = A$. The inverse of $A$ can be determined if and only if, it is non-singular, $\|A\| \neq 0$. Diagonalization of a square matrix $A$ is the equivalent of finding its *eigenvalues* and *eigenvectors* (Section 1.4).

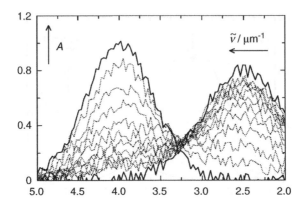

**Figure 3.19**   Time-dependent spectra of Figure 3.18 with random noise

$\tilde{v}_j$ as a function of time. In the spectra shown, the *isosbestic points* of consecutive spectra gradually move from 3.3 to 3.2 µm$^{-1}$, indicating that the overall process observed is not a uniform reaction. The series of spectra was in fact generated artificially assuming a two-step reaction, $B_1 \rightarrow B_2 \rightarrow B_3$, with $k_{1 \rightarrow 2} = 0.5\,s^{-1}$ and $k_{2 \rightarrow 3} = 2\,s^{-1}$. Had they been obtained by spectrographic flash photolysis, the signal-to-noise ratio might be around 10 as in Figure 3.19, where random noise was added to the same spectra. In this case, the systematic deviations from a uniform reaction could easily be missed by visual inspection.

According to the Beer–Lambert law (Equation 2.5), the observed absorbances $A_{ij}$ are linear combinations of the absorbances of the three components $B_k$, $k = 1, 2, 3$. The matrix $A$ of dimension $(r \times c)$ was thus constructed by forming the product $C \cdot \varepsilon$, where the columns of matrix $C$ $(r \times 3)$ contain the concentration profiles of the species $B_1$–$B_3$ given by the assumed rate law and the rows of matrix $\varepsilon$ $(3 \times c)$ contain their spectra (Equation 3.11). To visualize matrix operations, we represent the matrices by rectangles, where the height represents the number of rows and the width the number of columns. For simplicity, the cell length $l = 1$ cm was multiplied into the matrix elements of $\varepsilon$. The concentration profiles and spectra of $B_1$–$B_3$ are plotted in Figure 3.20.

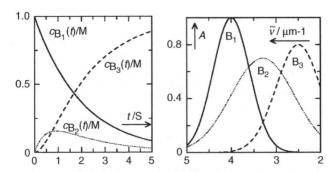

**Figure 3.20**   Concentration profiles of the components $B_1$–$B_3$ (left) and their absorption spectra (right)

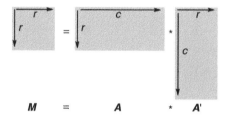

$$A_{ij} = \sum_{k=1}^{3} c_{ik}\varepsilon_{kj}$$

**Equation 3.11**

In real life, we will have spectral data $A$ such as those shown in Figure 3.19, but do not know the underlying matrices $C$ and $\varepsilon$. The goal of global analysis is to revert the process, that is, to recover the information about the number of components, their spectra, $\varepsilon$ and the rate laws with the associated rate constants defining the concentration profiles, $C$, by analysis of the complete set of spectral data, $A$. Global analysis consists of two steps, *factor analysis* and *target factor transformation*. For factor analysis we first form the covariance matrix from the data matrix $A$, $M = A * A'$, where $A'$ is the transpose of the data matrix $A$ (Equation 3.12).[d]

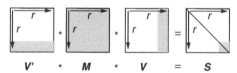

**Equation 3.12** *Covariance matrix*

The covariance matrix $M$ is a symmetrical square matrix, $M_{ij} = M_{ji}$, of size $r \times r$ that can be diagonalized. Solving the secular determinant $\|M - \delta_{ij}\lambda\| = 0$ yields $r$ eigenvalues $\lambda$ with $r$ associated *orthonormal[e]eigenvectors* $v_i$, $i = 1, \ldots, r$ of length $r$.[f] Collecting the eigenvectors in the columns of a matrix $V$ we have found the matrix that diagonalizes $M$ by the operation $V' * M * V$ (Equation 3.13). The eigenvalues appearing on the diagonal matrix $S = (\delta_{ij}\lambda_i)$ are sorted by size.

**Equation 3.13** *Diagonalization of the covariance matrix*

These matrix operations are easily programmed with mathematical software packages. In MATLAB, for example, only two statements, 'M = A*A';' and '[V,S] = eig(M);', are required

---

[d] Another covariance matrix could have been constructed as $A' * A$, which would have had the much bigger size of $c \times c$. As the information content of $A * A'$ and $A' * A$ is the same, we chose the smaller one.

[e] The matrix product $V * V'$ is a diagonal unit matrix $(\delta_{ij})$ of size $r \times r$.

[f] The mathematical procedure is the same as that used in quantum mechanics to optimize LCAO wavefunctions (Sections 1.4 and 4.2).

and with reasonably sized matrices such as $c = 1000$ and $r = 20$ the calculation is completed in seconds on a desktop computer. We are now in a position to determine the number of significant factors, that is, the number of independent components contributing to the spectral matrix $A$. Using the noise-free spectra (Figure 3.18), the diagonal matrix $S$ contains only three non-zero diagonal elements on the bottom right. We have retrieved the information that $A$ was generated using three components, $n = 3$. Having decided on the number $n$ of eigenvalues to be retained, we can reduce the matrices $V$ and $S$ to the shaded areas $V_{red}$ of size $r \times n$ and $S_{red}$ of size $n \times n$ (Equation 3.13) and reconstruct the transpose of the spectral matrix, $A_{reconstr}'$, by forming the product $U_{red}*S_{red}*V_{red}'$ (Equation 3.14). To determine the matrix of the spectral eigenvectors, $U_{red}$, of size $c \times n$, we right-multiply $A_{reconstr}'$ with $V_{red}$ to get $A_{reconstr}'*V_{red} \cong A'*V_{red} = U_{red}*S_{red}$ [the rows of $V_{red}$ are orthonormal, $V_{red}'*V_{red} = (\delta_{ij})$, hence $U_{red} = A'*V_{red}*(S_{red})^{-1}$]. The coefficients in the rows of $V_{red}$ represent the weight of the three spectral eigenvectors $U_{red}(c \times n)$, which is to be multiplied by the corresponding eigenvalue on the diagonal of $S_{red}$ to reconstruct the spectra in the rows of $A$. The spectra reconstructed from the noise-free data are identical with Figure 3.18.

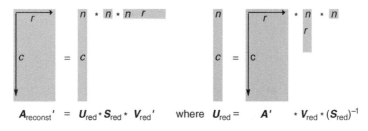

$$A_{reconst}' = U_{red} * S_{red} * V_{red}' \quad \text{where} \quad U_{red} = A' * V_{red} * (S_{red})^{-1}$$

**Equation 3.14**

Using the noisy spectra shown in Figure 3.19, the largest eigenvalues obtained by diagonalization of $M$ are, in decreasing order, 135.1, 43.2, 0.19, 0.12, 0.11, 0.10, etc. The noise has produced 11 nonzero eigenvalues and it is difficult to decide whether the spectra were made up of two or more components. Note that the falloff after the second is 40% and very little thereafter. Using statistical criteria,[201] one would conclude that the third eigenvalue is not significant at the 95% level. However, as chemists, we may have prior knowledge (or belief) that the reaction involves an intermediate and, hence, would want to choose three components. The spectra reconstructed retaining three eigenvalues are shown in Figure 3.21 (top).

At this point, we have achieved two goals. First, the amount of data stored (and subsequently handled) in the matrices $U_{red}$, $S_{red}$ and $V_{red}$ is reduced by at least an order of magnitude. Second, as an additional bonus, much of the random noise has been eliminated (compare Figures 3.19 and 3.21). One should realize, however, that by choosing the number of eigenvalues we introduce some bias in the reconstructed data. This is illustrated in Figure 3.21 (bottom), in which the same data were reconstructed with only two eigenvalues. The data are now forced through an isosbestic point.

Figure 3.22 shows the three abstract eigenvectors (the columns of $U_{red}*S_{red}$) that were used to reproduce the spectra. Clearly they do not correspond to the spectra of the species $B_1$–$B_3$. Rather, the first eigenvector (solid line) represents an 'average' of the whole spectral series and is positive throughout. The second (dotted line) has one node and allows for a 'first correction' in reconstructing the individual spectra and the last (dashed line,

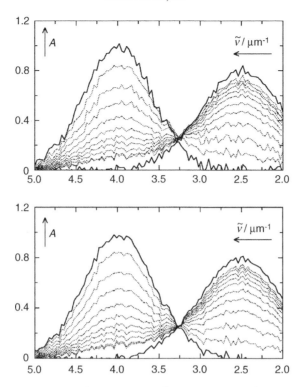

**Figure 3.21** Reconstructed spectra using three (top) or two (bottom) eigenvalues

multiplied by a factor of five) is a 'second correction'. This is reminiscent of the molecular orbitals that are obtained by the same mathematical procedure (Section 4.2); the lowest MO is bonding throughout, the next has one node, and so on.

The row vectors in matrix $V_{red}$ define points in the three-dimensional space of the orthogonal eigenvectors. They lie in (or near, in the presence of noise) a plane, consistent

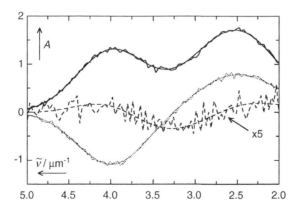

**Figure 3.22** The three eigenvectors of the spectral data. The smooth lines are the eigenvectors obtained from the noise-free spectra (Figure 3.18)

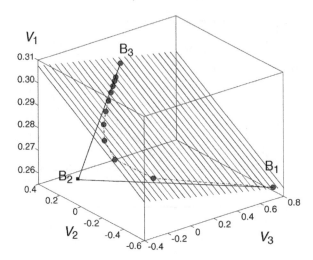

**Figure 3.23**   The reaction $B_1 \rightarrow B_2 \rightarrow B_3$ proceeding in a plane of the eigenvector space

with a reaction involving three components, as shown in Figure 3.23. A one-step reaction $B_1 \rightarrow B_2$ would follow a straight line and a reaction involving more than three species would show systematic deviations from the regression plane.

Having completed factor analysis, we wish to determine the spectra of the chemical species that give rise to the observed spectral data. To achieve transformation of the eigenspectra to the component spectra (*target factors*), additional information is needed. We may know that the first spectrum corresponds to the starting material $B_1$ and the final one to the product $B_3$. The initial spectral changes are then attributed to the reaction $B_1 \rightarrow B_2$ and the final ones to $B_2 \rightarrow B_3$. We can therefore estimate the coordinates of the point in eigenvector space that represents the species $B_2$ as the intersection of two straight lines in the regression plane (Figure 3.23).

Other, more powerful methods are available. For one, we know that the species spectra cannot have negative absorbances. This information may be sufficient to analyse spectra with little overlap, such as highly resolved IR spectra. If we know that the spectra are ordered in a meaningful sequence, for example because a chemical reaction proceeds in time or the pH is changed continuously by titration, then *self-modelling* methods (*evolving factor analysis*) can be used.[203,204] If we know, or have reason to assume, that the species concentrations $C$ follow a certain rate law (as in the present example) or the law of mass action (in a titration), then an initial guess of the $C$ matrix can be made using trial parameters in the model function, for example the rate or equilibrium constants of the assumed model. The model parameters are subsequently optimized by nonlinear least-squares fitting.[201–203,205,206] All of the above methods aim at decomposing the spectral data $A$ to chemically meaningful concentration profiles $C$ and species spectra $\boldsymbol{\varepsilon}$; (Equation 3.11). The best choice depends on the additional knowledge that is available about the system. It is recommended to test any chosen model by unbiased self-modelling methods.

To complete the global analysis of our example, we construct a trial matrix $C$ $(r \times 3)$ using the integrated rate laws for the species $B_1$–$B_3$ (Equation 3.8, Section 3.7.4) with trial parameters $k_{AB}$ and $k_{BC}$. From this guess of $C$, a first guess for the species spectra $\boldsymbol{\varepsilon}$ is

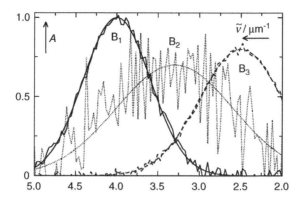

**Figure 3.24** Recovered species spectra. The noisy traces were obtained from the spectral data shown in Figure 3.19

obtained without iteration as $\boldsymbol{\varepsilon} = (\boldsymbol{C'}*\boldsymbol{C})^{-1}*\boldsymbol{C'}*\boldsymbol{A}$ (MATLAB command: 'EPS = C\A;'). The matrix $\boldsymbol{\varepsilon}$ is then optimized by minimizing the sum of squares of the residuals, $\boldsymbol{R} = \boldsymbol{A} - \boldsymbol{C}*\boldsymbol{\varepsilon}$, upon variation of the nonlinear parameters $k_{AB}$ and $k_{BC}$ that determine $\boldsymbol{C}$. The final species spectra contained in the rows of $\boldsymbol{\varepsilon}$ are shown in Figure 3.24. The rate constants obtained from the noisy spectra are 0.49 and $2.2 \, \mathrm{s}^{-1}$.

In conclusion, the global analysis of spectral data is a very useful tool to validate a proposed model, when used with proper understanding and caution. However, statistical analysis in general is not able to eliminate any systematic errors that may be hidden in experimental data. On the contrary, it will emphasize any such deviations from a chosen model and might thereby insinuate false complexity of the system investigated. No mathematical treatment can ever make up for less than optimal methods of data collection.

## 3.8 Time-Resolved IR and Raman Spectroscopy

Before the advent of NMR, infrared (IR) absorption and the largely complementary Raman spectroscopy were the most important tools to identify functional groups by their characteristic vibrational frequencies and they still hold their place in analytical chemistry. Both IR and Raman spectroscopy provide fingerprints that permit an unambiguous identification of simple molecules and even of specific sites in biopolymers. Due to the low absorption coefficients of vibrational bands, time-resolved IR spectroscopy was hampered by a lack of sensitivity and was originally useful only for the detection of very strong vibrational bands such as those of metal carbonyls. This has changed dramatically over the last 20 years with the advent of Fourier transform instruments and amplified Ti:sapphire lasers that produce broadband femtosecond light pulses. Vibrational spectroscopic methods are now extensively used in structural studies of the dynamics of chemical and biological reactions.

In picosecond time-resolved Raman spectroscopy, the sample is pumped and probed by energetically well-defined optical pulses, producing a full vibrational spectrum over a 1000–2000 cm$^{-1}$ window.[207] One would expect vibrational spectroscopy to be restricted to the picosecond time domain and above by the Heisenberg uncertainty principle (Equation 2.1), because a 1 ps transform-limited pulse has an energy width of

$15\,\text{cm}^{-1}$ FWHM and a 10 fs pulse has an intrinsic energy bandwidth of $1500\,\text{cm}^{-1}$. This impasse is elegantly circumvented by ultrafast femtosecond stimulated Raman spectroscopy, which can provide $<100$ fs temporal and $<35\,\text{cm}^{-1}$ spectral resolution by exploiting coherence.[208,209] An additional femtosecond pump pulse (460–670 nm, $<30$ fs) that is significantly shorter than the period of the vibrationally active modes of the molecule is added to initiate a photochemical reaction. It produces a localized wave packet that evolves on the excited state surface. The spectral resolution is mainly limited by the bandwidth of the Raman pulse ($<10\,\text{cm}^{-1}$), while the time resolution is determined by the cross-correlation of the pump and Raman probe pulses (typically $<50$ fs). This technique has the additional advantages of an improved signal-to-noise ratio, short data acquisition times and insensitivity to background fluorescence. It was used to monitor the primary events of vision.[210]

To obtain IR spectra on a time scale of nanoseconds, the sample cell in conventional spectrometers is usually excited by an Nd:YAG laser. Flow cells with a pathlength of at least 0.1 mm must be used for photoreactive samples and the pulse repetition frequency is then limited to $\sim$1 Hz. In *step–scan FTIR* spectroscopy,[211] the time evolution is collected at single points of the interferogram, which is then reconstructed point-by-point and subsequently transformed to time-resolved IR spectra. Alternatively, dispersive instruments equipped with a strong IR source can be used.[212] The time resolution of both methods is about 50 ns. FTIR instruments provide a triggerable fast-scan mode to collect a complete spectrum within a few milliseconds.[213]

## 3.9 Quantum Yields

Quantum yields are fundamental quantities that define the photonic economics of processes induced by light absorption. They are required to determine rate constants of photophysical and photochemical processes (Section 3.9.7). Many different techniques are used to measure quantum yields depending on the process studied. In the following, we describe some procedures commonly used in the chemical laboratory. The measurement of quantum yields is an art that has a number of pitfalls. The experimenter has few options to double-check his or her own results other than reproducibility, which will not reveal any repeated systematic errors. Therefore, it is prudent to reproduce the quantum yield of a related, well-known process in the laboratory before determining an unknown one.

### 3.9.1 Differential Quantum Yield

The definition of a *quantum yield* for a given process $x$, $\Phi_x(\lambda) = n_x/n_p$, was given in Equation 2.24, Section 2.1.7. Here $n_x$ is the amount of photochemical or photophysical events $x$ that occurred during irradiation and $n_p$ is the amount of photons at the irradiation wavelength $\lambda$ that were absorbed by the reactant. Both $n_x$ and $n_p$ are measured in moles or einsteins (1 einstein $=$ 1 mol of photons) and the dimension of $\Phi_x$ is unity.

With photoreactive systems, the amount of light absorbed by the reactant changes as the reaction proceeds and the amount of product formed may change further when irradiation of the photoproducts leads to secondary photoreactions. To deal with this, the quantum yield $\Phi_x(\lambda) = n_x/n_p$ must be defined differentially (Equation 3.15). As infinitely small conversions cannot be measured, one is forced to use finite doses of light to determine quantum yields. Exact integration of Equation 3.15 is possible in certain cases and will be dealt with below (Section 3.9.3). In any case, small conversions (<10%) are preferable to avoid corruption by secondary photoreactions. The best compromise depends on the sensitivity of the method of analysis used (GC, absorbance, luminescence, etc.).

$$\Phi_x(\lambda) = dn_x/dn_p$$

**Equation 3.15** *Differential quantum yield*

For monochromatic light, the *molar photon flux* $q_{m,p}^0 = n_p/t$, the amount (in moles or einsteins) of photons incident on a sample cell per unit time, is proportional to the incident spectral radiant power $P_\lambda^0$ (Equation 3.16). The unit of $q_{m,p}^0$ (Equation 2.3) is mol s$^{-1}$.

$$q_{m,p}^0 = P_\lambda^0/N_A hc\tilde{\nu}$$

**Equation 3.16** *Molar photon flux*

When only the reactant absorbs at the wavelength of irradiation $\lambda$, the amount $dn_p$ of photons absorbed by the sample in a short time interval $dt$ is equal to the absorbed molar photon flux $q_{m,p}$ (Equation 3.17).

$$dn_p = q_{m,p}^0[1-T(\lambda)]dt = q_{m,p}^0[1-10^{-A(\lambda,t)}]dt$$

**Equation 3.17**

If the fraction of light transmitted is substantial, a small correction should be applied to account for the light reflected back into the liquid from the rear face of the cell (about 4%). When other species that absorb at the wavelength of irradiation are present or when the photoreaction produces absorbing photoproducts, the right-hand side of Equation 3.17 must be multiplied by the fraction of light absorbed by the reactant A, $A_A(\lambda,t)/A(\lambda,t)$, Equation 3.18.

$$dn_p = q_{m,p}^0\left[1-10^{-A(\lambda,t)}\right]\frac{A_A(\lambda,t)}{A(\lambda,t)}dt$$

**Equation 3.18**

For small absorbances, $A(\lambda,t) \ll 0.1$, the quotient $[1-10^{-A(\lambda,t)}]/A(\lambda,t)$ approaches $\ln(10) = 2.303$, so that $dn_p/dt \approx 2.303 q_{m,p} A_A(\lambda,t)$. For large absorbances, $A(\lambda,t) > 2$, the term $10^{-A(\lambda,t)}$ may be neglected. If the reaction progress is measured by analytical methods such as GC, HPLC or NMR, then the quantum yield is usually calculated by Equation 3.15, simply replacing the differential terms $dn_x$ and $dn_p$ by small increments $\Delta n_x$ and $\Delta n_p$ measured after short time intervals $\Delta t$ of irradiation. Average values for the absorbances $A(\lambda,t)$ and $A_A(\lambda,t)$ during the irradiation interval should then be used to calculate the amount of light $\Delta n_p$ absorbed by component A (Equation 3.18) and the conversion per time interval $\Delta t$ should be kept well below 10%. Quantum yields determined in this way for consecutive irradiation periods should remain constant; a decreasing trend indicates that secondary photoreactions are occurring.

### 3.9.2   Actinometry

*Recommended review.*[214]

To determine a quantum yield using Equation 3.18, the *radiant power* $P_\lambda^0 = q_{m,p}^0 N_A hc\,\tilde{\nu}$ incident on the sample cell must be known. Absolute measurements of radiant power are difficult to perform with precision and are rarely done in the chemical laboratory. In practice, radiant power is measured by *actinometry*. An actinometer is a chemical system that undergoes a light-induced reaction, for which the quantum yield $\Phi_{act}(\lambda)$ is known accurately. Equations 3.15 and 3.18 are then used in reverse, that is, the radiant power of the light source is determined by measuring the conversion of the actinometer per unit time, $dn_x/dt = dn_{act}/dt$, and inserting the known value of $\Phi_{act}(\lambda)$. Requirements for a good chemical actinometer are that its quantum yield is well established and preferably constant over a wide range of wavelengths, radiant power and total radiant energy, and also high sensitivity and precision coupled with simplicity of use and ready availability of the photosensitive material.

Spectrophotometric monitoring of the reaction progress in both the sample and the actinometer solution (Section 3.9.3) is convenient and gives accurate results. Due to fluctuations of most light sources in time, the irradiation of sample and actinometer solution should be done simultaneously (Figure 3.29). Alternatively, standard analytical techniques following irradiation on a merry-go-round apparatus (Figure 3.30) are frequently used in the organic laboratory. A comprehensive list of established chemical actinometers and recommended standard procedures is available.[214] The ferrioxalate actinometer is probably the most widely used.[214–216] It has been calibrated repeatedly by comparison with a thermopile and also against other actinometers. It is prepared from readily available chemicals (Case Study 3.1) and can be used for a wide range of irradiation wavelengths (205–509 nm, Table 3.2). Its disadvantages are that it must be

*Table 3.2*   *Quantum yields of the ferrioxalate actinometer at different wavelengths.*

| Wavelength/nm | Quantum yield[32] | Quantum yield[216] |
|---|---|---|
| 509 | 0.86 | |
| 480 | 0.94 | |
| 468 | 0.93 | |
| 436 (0.15 M) | 1.01 | |
| 436 (0.006 M) | 1.11 | |
| 405 | 1.14 | |
| 365/366 (0.15 M) | 1.18 | |
| 365/366 (0.006 M) | 1.22 | 1.27 |
| 340 | | 1.23 |
| 334 | 1.23 | |
| 320 | | 1.27 |
| 313 | 1.24 | |
| 297/302 | 1.24 | 1.25 |
| 260–300 | | 1.25 |
| 253.7 | 1.25 | 1.40 |
| 240 | | 1.45 |
| 225 | | 1.46 |
| 220 | | 1.47 |
| 214 | | 1.50 |
| 205 | | 1.49 |

handled in a darkroom under red light and that its quantum yield is wavelength dependent. For details on the convenient, reversible actinometer azobenzene, see Section 3.9.4. Aberchrome 540, another reversible and widely used actinometer, is no longer commercially available and exhibits fatigue and also side-reactions.[217] Its use is not recommended. The self-sensitized photo-oxidation of *meso*-diphenylhelianthrene can be used for irradiation wavelengths in the range 475–610 nm (Scheme 6.259).[218] Several compounds giving standard transient absorbance have been recommended for the actinometry of laser pulses.[214]

---

**Case Study 3.1: Actinometry – Ferrioxalate**

Pure potassium ferrioxalate is prepared by mixing three volumes of 1.5 M potassium oxalate with one volume of 1.5 M ferric chloride, both of analytical quality. The precipitated green potassium ferrioxalate $[K_3Fe(C_2O_4)_3 \cdot 3H_2O]$ should be recrystallized three times from warm water and the crystals dried in a current of warm air (45 °C). The whole procedure must be carried out in a darkroom lit with a red safelight. The crystals can be stored indefinitely in the dark.

The 0.006 M actinometer solution is prepared by dissolving 2.947 g of the ferrioxalate crystals in ~800 ml of water. Sulfuric acid (1.0 N = 0.5 M, 100 ml) is added, the solution diluted to 1 l and mixed. The 0.15 M solution is made in a similar manner using 73.68 g of ferrioxalate. At least 99% of the light is absorbed by the 0.006 M solution up to 390 nm and by the 0.15 M solution up to 445 nm in a cell of pathlength 1 cm. Above these wavelengths, the fraction absorbed at the irradiation wavelength should be measured.

On exposure to light, the following reactions occur:

$$[Fe(C_2O_4)]^+ \xrightarrow{h\nu} Fe^{2+} + C_2O_4^{-\bullet}$$
$$C_2O_4^{-\bullet} + [Fe(C_2O_4)]^+ \rightarrow 2CO_2 + Fe^{2+} + C_2O_4^{2-}$$

After photolysis, the ferrous ion formed is converted to its 1,10-phenanthroline complex and the latter determined by spectrophotometry. The minimal detectable amount of light is about $2 \times 10^{-10}$ einstein $ml^{-1}$ and the maximal light dose that can be measured accurately (±2%) is about $5 \times 10^{-6}$ einstein $ml^{-1}$.

The following procedure is taken from the literature.[32,219] It is recommended to prepare a calibration graph by mixing solutions of (a) $0.4 \times 10^{-3}$ M of $Fe^{2+}$ in 0.1 N sulfuric acid (freshly prepared by dilution from a standardized solution of 0.1 M $FeSO_4$ in 0.1 N sulfuric acid), (b) a buffer solution of 600 ml of 1 M sodium acetate and 360 ml of 1 N sulfuric acid diluted to 1 l and (c) 0.1% (w/v) of 1,10-phenanthroline monohydrate in water. Solution (c) must be kept in the dark and stored for no longer than 3 months. Into a series of 20 ml calibrated flasks, sequentially add the following volumes of solution and mix: $x = 0.0, 0.5, 1.0, \ldots, 4.5, 5.0$ ml of solution (a). Add 5 ml of solution (b) and $(10 - x)$ ml of 0.1 N sulfuric acid. Add 2 ml of solution (c), make up to mark, mix and allow to stand for at least 0.5 h. Measure each absorbance at 510 nm in a 1 cm cell and correct each reading for the value obtained with the solution, to which no ferrous ion was added (should be ≤0.01). The solutions may be kept for several hours in the dark before measurement. The resulting plot of absorbance against the molarity of ferrous ion should be linear with a slope corresponding to $\varepsilon = 11\,050\,\text{M}^{-1}\,\text{cm}^{-1}$.

During irradiation of the actinometer, the solutions should be stirred with a magnetic bar. After irradiation, pipette an aliquot (2 ml) of the solution into a 20 ml calibrated flask. Add a volume of buffer (b) equal to half the volume of photolyte taken (1 ml) and 2 ml of the phenanthroline solution (c). Make up to the mark with water, mix and allow to stand for at least 0.5 h. Measure the absorbance at 510 nm and repeat with the same volume of unexposed actinometer solution. Convert the absorbance difference of ferrous iron using the calibrated slope. Convert the quantity of ferrous ion formed in the total volume of the irradiated solution to a radiation dose (see Equation 2.24) using the recommended quantum yield given in Table 3.2. If necessary, allow for the fraction of light absorbed (Equation 3.17).

### 3.9.3  Spectrophotometric Determination of the Reaction Progress

A convenient, sensitive and highly accurate method of monitoring the progress of photoreactions is to record absorption spectra intermittently between successive periods of irradiation. It works best for uniform reactions; this does not imply that a single photoproduct must be formed, but that the product distribution remains unchanged throughout. Non-uniform reactions arise when transient intermediates have sufficient lifetime ($>1$ min) to interfere with spectrophotometric monitoring or when some photoproducts are unstable either thermally or under the irradiation conditions. When a photoproduct is light sensitive and undergoes secondary photolysis upon prolonged irradiation, then the quantum yield of the initial photoreaction may be estimated at low conversion, provided that the absorption spectrum of the light-sensitive photoproduct is known.

Sensitive tests for the uniformity of a reaction can be done by global analysis of the complete set of spectra recorded during photolysis. These methods, described in Section 3.7.5, provide the best evaluation of the minimum number of spectral components required to reproduce a sequence of spectra within experimental accuracy and the time-dependent species concentrations thus obtained accurately define the reaction progress. Simpler versions use absorbance differences observed at a few selected wavelengths where the changes are largest. Uniform reactions give linear plots of $\Delta A(\lambda_1, t)$ versus $\Delta A(\lambda_2, t)$. For two sequential photoreactions, absorbance difference plots are curved, but plots of absorbance difference quotients, $\Delta A(\lambda_1, t)/\Delta A(\lambda_2, t)$ versus $\Delta A(\lambda_1, t)/\Delta A(\lambda_3, t)$, are linear.[198,220,221] *Isosbestic points* provide the simplest criterion for the uniformity of a reaction: An isosbestic point is a wavelength, wavenumber or frequency at which the total absorbance of a sample does not change, because reactant and product(s) accidentally have the same absorbance (Figure 3.25). A reaction is non-uniform if the intermediate spectra do not cross at the same point, as in Figure 3.18.

We now derive several equations for the spectrophotometric quantum yield determination of unidirectional photoreactions A $\rightarrow$ B. Reversible photoreactions will be treated in Section 3.9.3. The reader should not be deterred by the complex appearance of some of these equations. They are easy to use and give highly reproducible results, because absorbance measurements are precise. The photoreaction is induced by continuous irradiation with a monochromatic light source that exposes the sample

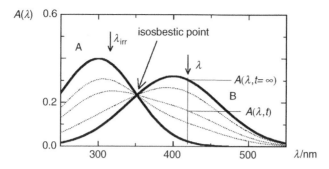

**Figure 3.25** Spectrophotometric observation of a photoreaction A → B. The wavelengths of irradiation and of observation are designated $\lambda_{irr}$ and $\lambda$, respectively

solution of volume $V$ to a constant photon flux $q^0_{m,p}$ (Equation 3.16). The product of $q^0_{m,p}$ and exposure time $t$ is equal to the amount of photons that have impinged on the sample cell at time $t$. Note the distinction between the wavelength of irradiation, $\lambda_{irr}$, and the wavelengths of observation, $\lambda$, which in most cases will not be the same. To simplify the notation, we will abbreviate absorbances $A(\lambda,t)$ at the wavelength of observation by $A_t$ and those at the wavelength of irradiation, $A(\lambda_{irr},t)$, by $A_t'$. Replacement of the index $t$ by 0 or $\infty$ indicates no and exhaustive irradiation, respectively. Similarly, absorption coefficients $\varepsilon$ and $\varepsilon'$ refer to the wavelengths of observation and irradiation, respectively.

In order to determine the quantum yield $\Phi_{-A} = -dn_A/dn_p$ by absorption measurements, the molar absorption coefficients of the starting material A and of the product (mixture) B must be known. The differential amount of light absorbed by the reactant, $dn_p$, during an exposure interval, $dt$, is then given by Equation 3.18. Replacing the partial absorbance by the reactant A at time $t$, $A'_{A,t}$, by $\varepsilon'_A n_A(t)l/V$, where $l$ is the optical pathlength, we obtain

$$dn_p = q^0_{m,p}(1-10^{-A_t'})\frac{\varepsilon'_A n_A(t)l}{VA_t'}dt$$

*Equation 3.19*

The total absorbance at the observation wavelength $\lambda$ is given by $A_t = [\varepsilon_A c_A(t) + \varepsilon_B c_B(t)]l$. With $c = n/V$ and $n_B(t) = n_A(t=0) - n_A(t)$, we can express the amount of component A in the solution as a function of the observed absorbance:

$$n_A(t) = \frac{A_t - A_\infty}{l(\varepsilon_A - \varepsilon_B)}V$$

*Equation 3.20*

The differential of Equation 3.20 is $dn_A = dAV/([\varepsilon_A - \varepsilon_B]l)$ and substitution using Equation 3.20 gives Equation 3.21:

$$dn_A = \frac{n_A(t)}{A_t - A_\infty}dA$$

*Equation 3.21*

We insert Equations 3.19 and 3.21 into $\Phi_{-A} = -dn_A/dn_p$ and replace $\Phi_{-A}\varepsilon_A'$ by $Q(\lambda_{irr})$ to obtain Equation 3.22, the general differential equation used for the spectrophotometric determination of quantum yields; the reason for introducing the so-called *pseudo quantum yield* $Q = \Phi_{-A}\varepsilon_A'$ will become evident in Section 3.9.4.

$$Q(\lambda_{irr})dt = \frac{VA_t'}{lq_{m,p}^0(1-10^{A_t'})(A_t - A_\infty)} dA(\lambda)$$

**Equation 3.22**   *Differential equation for the spectrophotometric determination of quantum yields.*

Approximate integration of Equation 3.22 is possible by linear interpolation of the reciprocal photokinetic factor,[222] $F(\lambda_{irr}, t)$ (Equation 3.23).

$$F(\lambda_{irr}, t) = \frac{(1-10^{-A_t'})}{A_t'}, \quad \bar{F}(\lambda_{irr}) = \frac{F(\lambda_{irr}, t_2) + F(\lambda_{irr}, t_1)}{2}$$

**Equation 3.23**

Integration of Equation 3.22 for a short irradiation interval $\Delta t = t_2 - t_1$, having replaced the variable $F^{-1}(\lambda_{irr}, t)$ with the constant $\bar{F}^{-1}(\lambda_{irr})$, gives the operational Equation 3.24, which is generally applicable.

$$Q(\lambda_{irr}) \cong \frac{V}{lq_{m,p}^0 \bar{F}(\lambda_{irr})(t_2 - t_1)} \ln\left[\frac{A_{t_1} - A_\infty}{A_{t_2} - A_\infty}\right]$$

**Equation 3.24**

When the absorbance at the irradiation wavelength is very large (*total absorbance* $A_t' > 3$), so that $1 - 10^{-A(\lambda_{irr})} \approx 1$, then $\bar{F}(\lambda_{irr})$ can be replaced by $(1/A_{t_1}' + 1/A_{t_2}')/2.$

Exact integration[g] of Equation 3.22 is possible *when the reactant A can be irradiated at a wavelength where product B does not absorb*, $\varepsilon_B' = 0$, yielding Equation 3.25:

$$Q(\lambda_{irr}) = \frac{V\log\left(\frac{10^{A_0'}-1}{10^{A_t'}-1}\right)}{ltq_{m,p}^0}, \quad \text{where} \quad A_{t \text{ or } 0}' = (A_{t \text{ or } 0} - A_\infty)\frac{\varepsilon_A'}{\varepsilon_A - \varepsilon_B}$$

**Equation 3.25**

*When the reaction is monitored at the wavelength of irradiation,* $\lambda = \lambda_{irr}$, *then* $A_{t \text{ or } 0}'$ *in Equation 3.25 is the same as* $A_{t \text{ or } 0}$.

Prior to collecting accurate experimental data for spectrophotometric quantum yield determinations, one should always do a quick-and-dirty test run and decide which case is applicable (Equation 3.24 or 3.25). Note that in these equations the right-hand side must be multiplied by 1000 if the conventional but non-SI units $[V] = \text{cm}^3$, $[\varepsilon] = \text{mol}^{-1} \text{dm}^3 \text{cm}^{-1}$ and $[l] = \text{cm}$ are used. For quantitative work, it will be worthwhile to write a small program that computes the quantum yield from the measured absorbances according to the appropriate equation.

---

[g] Use the integral $\int(1/[a + be^{px}])dx = x/a - \ln(a + be^{px})/(ap)$, with $x = (A_t - A_\infty)$, $a = 1$, $b = -1$ and $p = -\ln(10)\varepsilon_A'/(\varepsilon_A - \varepsilon_B)$.

### 3.9.4 Reversible Photoreactions

In a reversible photoreaction (Scheme 3.2), such as the photoisomerization of azobenzene (Section 6.4.1), the differential rate for the disappearance of the reactant A is given by Equation 3.26, where $dn_{p,A}$ and $dn_{p,B}$ are the amounts of light absorbed by A and B, respectively, as defined by Equation 3.19.

$$A \xrightleftharpoons[h\nu]{h\nu} B$$

*Scheme 3.2*

$$-dn_A = \Phi_{A \to B}(\lambda_{irr})dn_{p,A} - \Phi_{B \to A}(\lambda_{irr})dn_{p,B}$$

*Equation 3.26*

Once a photostationary state (PSS) is reached upon irradiation at the wavelength $\lambda_{irr}$, the amount of reactant A no longer changes, $-dn_A/dt = 0$, and by inserting Equation 3.19 for $dn_{p,A}$ and $dn_{p,B}$ we obtain Equation 3.27:

$$\frac{\Phi_{A \to B}}{\Phi_{B \to A}} = \frac{\varepsilon_B' n_B(\infty)}{\varepsilon_A' n_A(\infty)}$$

*Equation 3.27*

where $n_A(\infty)$ and $n_B(\infty)$ are the amounts of A and B present in the photostationary state and the absorption coefficients $\varepsilon'$ refer to the wavelength $\lambda_{irr}$.

Hence the composition of the photostationary state, $n_A(\infty)/n_B(\infty)$, is defined by the ratio of the quantum yields for the forward and backward reactions and that of the absorption coefficients $\varepsilon'$. For a spectrophotometric determination of that composition, the absorption spectrum of at least one of the compounds A or B must be known. To determine the individual quantum yields $\Phi_{A \to B}$ and $\Phi_{B \to A}$ we define the pseudo quantum yield $Q$[222] (Equation 3.28).

$$Q(\lambda_{irr}) = \varepsilon_A' \Phi_{A \to B} + \varepsilon_B' \Phi_{B \to A}$$

*Equation 3.28* Pseudo quantum yield

Proceeding as in the previous section we obtain the differential Equation 3.22, where $Q(\lambda_{irr})$ is now defined by Equation 3.28. The integral of Equation 3.22 was given by Equation 3.24 and Equation 3.25 in the previous section. The individual quantum yields $\Phi_{A \to B}$ and $\Phi_{B \to A}$ are now obtained from $Q(\lambda_{irr})$ by combining Equation 3.27 and Equation 3.28. These relations are also used for actinometry with reversible systems, for example the carefully calibrated *E–Z* isomerization of azobenzene. Recommended *Q*-values for common irradiation wavelengths are given in Table 3.3.[223] It should be ascertained that the thermal back reaction $Z \to E$ of azobenzene does not interfere with the quantum yield measurements under the reaction conditions used; the sample cell may have to be cooled if it is located near the lamp housing. When *E*-azobenzene is used for actinometry in the visible region (405 or 436 nm), it should be preirradiated at 313 nm in order to yield larger absorbance changes upon irradiation in the visible.

**Table 3.3**  *Recommended Q-values for the azobenzene actinometer[223]*

| $\lambda_{irr}/nm$ | $Q \times 10^{-3}/dm^3\,mol^{-1}\,cm^{-1}$ |
|---|---|
| 254 (Hg) | $2.75 \pm 0.05$ |
| 280 (Hg) | $2.7 \pm 0.2$ |
| 313 (Hg) | $3.70 \pm 0.05$ |
| 334 (Hg) | $2.80 \pm 0.05$ |
| 337.1 ($N_2$ laser) | $2.8 \pm 0.1$ |
| 365/366 (Hg) | $0.13 \pm 0.01$ |
| 405/408 | $0.53 \pm 0.01$ |
| 436 | $0.82 \pm 0.01$ |

In practice, many largely reversible photoreactions do not remain in the photostationary state under prolonged irradiation due to the occurrence of minor side-reactions. The composition of the PSS should then be validated by an approach from both isomers or, preferably, from an authentic mixture near the PSS.

### 3.9.5  Luminescence Quantum Yields

Quantum yields of fluorescence $\Phi_f$ and phosphorescence are conveniently measured by comparison with a suitable standard, a reference compound, for which the quantum yield of luminescence is known.[157,176,179] As mentioned in Section 3.4, the absorbance of both the sample and the reference should be kept below 0.05 to avoid internal filter effects.

By definition of the differential quantum yield (Equation 3.15), the photon flux emitted as fluorescence, $q_{m,p}{}^{em}$, is equal to the fraction of the incident photon flux $q_{m,p}{}^0$ that is absorbed by the fluorescent compound times the quantum yield of fluorescence $\Phi_f$ (Equation 3.29).

$$q_{m,p}^{em} = q_{m,p}^0[1-10^{-A}]\Phi_f$$

**Equation 3.29**

Expanding the exponential function and retaining only the first term of the series we obtain Equation 3.30:

$$q_{m,p}^{em} \approx q_{m,p}^0 \ln(10)A\Phi_f$$

**Equation 3.30**

For small absorbances $A$, the photon flux emitted by the sample will be proportional to $A$, a prerequisite for obtaining faithful fluorescence excitation spectra (Section 3.4). The errors introduced by the approximate Equation 3.30 are 5.5% for $A(\lambda) = 0.05$ and 1% for $A(\lambda) = 0.01$. These errors will cancel when the absorbances of sample and reference are chosen to be equal.

The fluorescence quantum yield of the sample is given by Equation 3.31, where $F(s) = \int I_\lambda(s)d\lambda$ and $F(r) = \int I_\lambda(r)d\lambda$ are the integrated, corrected (Section 3.4) fluorescence bands of sample s and reference r, respectively. Both must be measured with small, preferably identical, absorbances and with the same instrumental settings. It is not recommended to convert emission spectra to a wavenumber scale for quantum yield determinations.

$$\Phi_f(s) = \Phi_f \frac{n^2(\lambda_{em}, s)A(\lambda_{exc}, r)F(s)}{n^2(\lambda_{em}, r)A(\lambda_{exc}, s)F(r)}$$

**Equation 3.31**

The correction for the refractive indices of the sample and reference solutions, $n(s)$ and $n(r)$, allows for the variation in the angles of the luminescence emerging from the solution to air. The correction may be substantial, for example $n_D^2(benzene)/n_D^2(water) \approx 1.27$, and may depend somewhat on optical parameters of the instrument. Moreover, internal reflections within the cell also depend on the refractive index. It is therefore preferable to dissolve both sample and reference in the same solvent to avoid errors from these sources.

Careful work will produce highly reproducible quantum yields with relative standard errors of only a few percent. Such reproducibility is useful for comparing relative quantum yields in a series of related compounds or in quenching studies (Section 3.9.8). It should be realized, however, that absolute quantum yields are usually fraught with systematic errors that do not show up in the statistics. Quantum yields determined in different laboratories or with different actinometers often differ by 10% or more.

### 3.9.6 Polychromatic Actinometry and Heterogeneous Systems

It is not always possible to choose ideal conditions for the determination of quantum yields. For example, the mineralization of organic solutes in waste waters by heterogeneous photocatalysis is done in suspensions of semiconductor particles such as $TiO_2$ with broadband light sources (Special Topic 6.29). A protocol has been proposed for the determination of relative photonic efficiencies allowing for an approximate conversion to quantum yields of such systems.[224,225] A sunlight actinometer has been designed for determining abiotic photodegradation quantum yields in surface waters.[226]

Very small quantum yields on the order of $10^{-6}$ and less are hard to measure accurately, but are nevertheless important, for example with regard to the light fastness of dyes on tissue[227,228] or of fluorescent dyes used for single molecule spectroscopy in heterogeneous systems (Section 3.13). In the latter case, processes induced by the simultaneous or sequential absorption of several photons leading to highly excited states may be largely responsible for photobleaching. One then has to resort to relative measures of stability under the given experimental conditions.

### 3.9.7 Relating Quantum Yields to Rate Constants

In ground-state chemistry, the measurement of reaction kinetics often provides the rate constant of the process studied directly, because a single reaction dominates. Photoreactions, on the other hand, always compete with photophysical processes, as outlined in Section 2.1.1. In Section 3.7.4, we showed that the efficiency of a specified single-step, first-order process $x$ starting from a reactant or excited state A is equal to the partition ratio $\eta_x$, the rate constant $k_x$ of the process considered divided by the sum of the rate constants of all processes competing for the depletion the reactant A, $\eta_x = k_x/\Sigma k_i$ (Equation 3.7). The combined efficiencies of all pathways starting from a given reactant must add up to unity, so that the efficiency of a given reaction step represents the *probability* for that step. *Therefore, the quantum yield of a single-step reaction is equal to its efficiency,* $\Phi_x = \eta_x$.[h] There is no need to invoke the *steady-state approximation* (Equation 2.26) to define the efficiency of a reaction in terms of rate constants. In fact, it was assumed in Section 3.7.4 that the initial concentration of the reactant A, $c_A(t=0)$, is large, as would be the case when a large excited-state population is generated by a strong pulse of light. The efficiencies $\eta_B$ and $\eta_C$ will be the same under continuous irradiation.

Now consider the photochemical reaction sequence shown in Scheme 3.3, which allows for the formation of two different photoproducts B and C, where B is formed from the singlet state $S_1(A)$ and C from the triplet state $T_1(A)$.

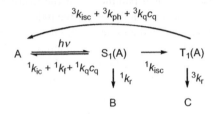

**Scheme 3.3**   *A typical reaction sequence*

The upper left indices of the reaction rate constants shown in Scheme 3.3 indicate whether a reaction step starts from the singlet state, $^1k$, or from the triplet state, $^3k$. For example, the efficiency $^1\eta_B$ of the reaction that forms product B from the excited singlet state $S_1$ is given by Equation 3.32:

$$^1\eta_B = {}^1k_r/({}^1k_r + {}^1k_{ic} + {}^1k_f + {}^1k_{ISC} + {}^1k_q c_q)$$

**Equation 3.32**

where the rate constant $^1k_r$ refers to the reaction from the singlet excited state $S_1$ leading to the photoproduct B, $^1k_{ic}$ to internal conversion to the ground state A, $^1k_f$ to fluorescence, $^1k_{ISC}$ to intersystem crossing to the triplet state and $^1k_q$ is the second-order rate constant for quenching of $S_1$ by a quencher q. The observed decay of $T_1$ will still obey the first-order rate law as long as the concentration $c_q$ of the quencher remains constant, because the product $^1k_q c_q$ then amounts to a first-order rate constant. The formation of B consists of a single reaction step following excitation of the molecule A to its excited singlet state $S_1(A)$, so that the efficiency of this step is equal to the quantum yield of formation of product B, $^1\eta_r = \Phi_B$.

The formation of product C, on the other hand, proceeds in two steps with the efficiencies $^1\eta_{ISC}$ (Equation 3.32) and $^3\eta_C = {}^3k_r/({}^3k_r + {}^3k_{ISC} + {}^3k_{ph} + {}^3k_q c_q)$ where $^3k_r$ refers to the reaction from the lowest triplet state $T_1$ leading to the photoproduct C, $^3k_{ph}$ is the rate constant of phosphorescence, $^3k_{ISC}$ that of intersystem crossing to the ground state and $^3k_q$ is the second-order rate constant for quenching of $T_1$ by a quencher q. The overall quantum yield for the formation of product C is equal to the product of the efficiencies of the two reaction steps, $\Phi_C = {}^1\eta_{ISC}{}^3\eta_C$. Many authors do not distinguish between

---

[h] The expression *quantum efficiency* $\eta$ is sometimes used, which is rather misleading, because the efficiency is not defined in terms of light quanta.

efficiencies and quantum yields and use the symbol $\Phi$ or $\phi$ for both. In that case, the quantum yields representing single steps must be distinguished clearly from overall quantum yields to avoid confusion, for example $\Phi_C = {}^1\Phi_{ISC}{}^3\Phi_C$.

Let us generalize: *the quantum yield of a multi-step process x is equal to the product of the efficiencies of all steps required to complete that process. This allows us to determine rate constants by measuring quantum yields and lifetimes.* Kinetic measurements such as time-resolved fluorescence or kinetic flash photolysis yield observed rate laws for the decay of excited states or of reactive intermediates. When the decay of an intermediate $x$ obeys first-order kinetics, as is frequently the case, then the observed lifetime $\tau = 1/k_{obs}$ is equal to the inverse of the sum of the rate constants of all processes contributing to the decay of the species observed.[i]

For a single-step reaction such as the formation of product B from the excited singlet state $S_1$ (Scheme 3.3), the rate constant ${}^1k_r$ can be determined from the quantum yield $\Phi_B = {}^1\eta_B$ and the fluorescence lifetime $\tau_f$, $\Phi_B = {}^1k_r\tau_f$. In general, for a single-step process $x$, the efficiency and reaction rate of that process are related by Equation 3.33.

$$\boxed{\eta_x = k_x\tau_x, \text{ e.g. } \Phi_f = k_f\tau_f}$$

**Equation 3.33**

Equation 3.33 also holds for the efficiency of the reaction forming product C from the triplet state, ${}^3\eta_C = {}^3k_r{}^3\tau$. However, the rate constant ${}^3k_r$ cannot be determined directly from the quantum yield of that process, $\Phi_C = {}^1\eta_{ISC}{}^3\eta_C$ and the triplet lifetime, ${}^3\tau$, because the efficiency of the first step, ${}^1\eta_{ISC}$, is involved (Equation 3.34).

$$\Phi_C = {}^1\eta_{ISC}{}^3\eta_r = {}^1\eta_{ISC}{}^3k_r{}^3\tau; \quad \text{i.e., } {}^3k_r = \Phi_C/({}^3\tau\,{}^1\eta_{ISC})$$

**Equation 3.34**

As we shall see in the next section, Stern–Volmer analysis of triplet sensitization or quenching experiments can lead to a determination of ISC efficiencies.

### 3.9.8 Stern–Volmer Analysis

In Chapter 2, we discussed several energy transfer processes of excited molecules $M^*$ in solution. Energy transfer from $M^*$ to molecules different from M amounts to *quenching* of $M^*$; the lifetime of molecules $M^*$ is thereby reduced and the luminescence yield is reduced accordingly. Another important quenching mechanism is electron transfer between $M^*$ and electron donors or acceptors (Section 5.2), because electronically excited molecules are, both, strong oxidants and reductants (Equation 4.1). Exergonic electron and energy transfer processes commonly take place whenever $M^*$ collides with an appropriate quencher, that is, with diffusion-controlled quenching rate constants $k_q$ on the order of $10^9$–$10^{10}\,M^{-1}\,s^{-1}$ (Table 8.3).

The purpose of adding quenchers that reduce the quantum yield of a photophysical or photochemical process $x$ is normally to determine the multiplicity and lifetime of the

---

[i] The *fluorescence lifetime* $\tau_f = 1/k_{obs}$ is shorter than the *radiative lifetime*, which is defined as the inverse of the fluorescence rate constant, $\tau_r = 1/k_f$.

excited state from which that process occurs and to obtain information about the rate constant $k_x$. Quenching experiments are easy to perform, do not require any sophisticated equipment and go some if not all the way in that endeavour. A simple case in point is fluorescence. The fluorescence quantum yield $\Phi_f$ is equal to its efficiency (Section 3.9.7) and its dependence on the concentration of some quencher q often obeys Equation 3.35.

$$\Phi_f^q = \frac{{}^1k_f}{\sum {}^1k_i + {}^1k_q c_q}$$

**Equation 3.35**

Here, the term for fluorescence quenching, ${}^1k_q c_q$, was written separately from the sum of all the other processes depleting $S_1$ in the absence of quencher. The ratio of the quantum yield of fluorescence in the absence of quencher, $\Phi_f^0$, and that with various amounts of quencher added, $\Phi_f^q$, then increases linearly with the quencher concentration $c_q$ according to Equation 3.36, which is called the *Stern–Volmer equation*. Note, however, that the kinetics of singlet quenching involving resonance energy transfer may be more complicated; the decay of the donor fluorescence in the presence of a resonant acceptor in viscous or rigid solvents is non-exponential (Section 2.2.2, Equation 2.42). Other deviations from the simple Stern–Volmer expression are discussed below.

$$\frac{\Phi_f^0}{\Phi_f^q} = \frac{{}^1k_f}{{}^1k_f}\frac{\sum {}^1k_i + {}^1k_q c_q}{\sum {}^1k_i} = 1 + \frac{{}^1k_q c_q}{\sum {}^1k_i} = 1 + {}^1k_q\,{}^1\tau_f^0 c_q$$

**Equation 3.36** *Stern–Volmer equation of fluorescence quenching*

The advantage of measuring the ratio $\Phi_f^0/\Phi_f^q$, rather than $\Phi_f$ itself, is that upon insertion of Equation 3.31 most terms cancel, provided the quencher does not absorb at the excitation wavelength, so that $\Phi_f^0/\Phi_f^q$ reduces to $F^0(s)/F^q(s)$, the ratio of the integrated emission spectra in the absence and in the presence of quencher. Moreover, $F^0(s)/F^q(s)$ can be replaced by the ratio of the observed fluorescence signal intensities at any wavelength $\lambda$, $I_\lambda^0/I_\lambda^q$, provided that the shape of the emission spectrum is not affected by quenching. There is no need for calibration of the detector, because the calibration factor would cancel in the ratio. Thus *the ratio of fluorescence intensities $I_\lambda^0/I_\lambda^q$ increases linearly with quencher concentration $c_q$* (Equation 3.37). The essential result of a fluorescence quenching study is the slope, ${}^1K_{SV} = {}^1k_q\tau_f^0$, called the *Stern–Volmer coefficient*. The unit of ${}^1K_{SV}$ is $M^{-1}$.

$$\frac{I_\lambda^0}{I_\lambda^q} = 1 + {}^1K_{SV}c_q$$

**Equation 3.37** *Practical Stern–Volmer equation for fluorescence quenching*

The Stern–Volmer Equation 3.36 can be used for any photophysical or photochemical single-step process $x$. Moreover, because $\Phi_x^0 = k_x\tau^0$ and $\Phi_x^q = k_x\tau^q$, the ratio of the quantum yields, $\Phi_x^0/\Phi_x^q$, is equal to that of the lifetimes of the reactive excited state or intermediate in the absence and presence of quencher (Equation 3.38). This relation provides a stringent test for the assignment of an intermediate $x$ that is observed by time-resolved methods [fluorescence lifetime measurement (Section 3.5) or kinetic flash photolysis (Section 3.7.1)]. *Assignment of an observed intermediate to the one that* *undergoes a single-step process x characterized by the quantum yield $\Phi_x = \eta_x$ requires that the Stern–Volmer coefficients $K_{SV}$ obtained independently by lifetime and quantum yield measurements are the same within experimental error.*

$$\Phi_x^0 / \Phi_x^q = \tau^0 / \tau^q = 1 + K_{SV} c_q$$

**Equation 3.38** General Stern–Volmer relation for a single-step reaction

Time-resolved methods are required to separate the two individual parameters contributing to $K_{SV} = k_q \tau^0$. The inverse of $\tau^q$, that is, the decay rate constant of the observed intermediate, $k_{obs}$, should increase linearly with the quencher concentration $c_q$ (Equation 3.39).

$$k_{obs} = 1/\tau^0 + k_q c_q$$

**Equation 3.39**

Thus $k_q$ and $\tau^0$ are determined as the slope and intercept, respectively, obtained from a series of kinetic experiments with varying $c_q$. In the absence of time-resolved data, the lifetime $\tau^0$ of the reactive state is often estimated from $K_{SV}$ by assuming a rate constant on the order of $10^9$–$10^{10}\,M^{-1}\,s^{-1}$ for near diffusion-controlled quenching, as may be justified by reference to known quenching rate constants of related systems. As the quenching rate constant may in fact be less than diffusion-controlled, $k_q \leq k_d$, the lifetimes so obtained represent an upper limit.

Frequently, the reactant M itself acts as a quencher of its excited molecules M*, M* + M → 2M. It is difficult to determine *self-quenching* accurately from the dependence of the quantum yield on the concentration $c_M$, because the absorbance by M increases as its concentration is raised. However, the effect is readily measured by time-resolved methods. Replacing the index q by M in Equation 3.39 and replacing $\tau^0$ by $1/k_{obs}(c_M \rightarrow 0)$ one obtains Equation 3.40. The observed decay rate constant of the excited molecule M*, $k_{obs}(c_M) = 1/\tau^M$, increases linearly with the concentration of M. The slope $k_M$ is the rate constant of self-quenching and the intercept $k_{obs}(c_M \rightarrow 0)$ is the decay rate constant of M* at infinite dilution.[229]

$$k_{obs}(c_M) = k_{obs}(c_M \rightarrow 0) + k_M c_M$$

**Equation 3.40** Self-quenching

Deviations from the simple Stern–Volmer relations given above can arise for numerous reasons. We have earlier discussed singlet quenching by resonance energy transfer (Section 2.2.2), which may result in non-exponential decay of the excited donor. A trivial, but frequently overlooked, source of error is competing absorption or light scattering by the quencher. Medium effects induced by the addition of large amounts of quencher, such as the change in ionic strength caused by the addition of salts, may lead to deviations from the simple Stern–Volmer equation. Strong deviations from linearity are observed when the quencher forms a complex with the reactants in the ground state.

Multi-step reactions generally do not obey the simple Stern–Volmer Equation 3.38 and plots of $\Phi_x^0 / \Phi_x^q$ versus $c_q$ may show upward or downward curvature. Some examples are discussed below. Many more cases have been worked out in the literature.[230] The derivations given here should enable the reader to determine the appropriate Stern–Volmer equations for any particular reaction scheme. Consider the case depicted in the reaction Scheme 3.3, where a reactant A forms product B via its excited singlet state and product C via the triplet state. The quantum yield of disappearance of A is equal to the sum of the quantum yields for the formation of B (Equation 3.32) and of C (Equation 3.34), $\Phi_{dis} = \Phi_B + \Phi_C = {}^1\eta_B + {}^1\eta_{ISC}{}^3\eta_C$. By inserting the appropriate rate constants defining the efficiencies $\eta$ into the ratio $\Phi_{dis}{}^0/\Phi_{dis}{}^q$ and some algebraic tidying, we obtain Equation 3.41, which is a function of the three parameters, $a = {}^1k_{ISC}{}^3\eta_C{}^0/{}^1k_r = \Phi_C{}^0/\Phi_B{}^0$, ${}^1K_{SV} = {}^1k_q{}^1\tau^0$ and ${}^3K_{SV} = {}^3k_q{}^3\tau^0$.

$$\frac{\Phi_{dis}{}^0}{\Phi_{dis}{}^q} = \frac{(1 + {}^1K_{SV}c_q)(1 + {}^3K_{SV}c_q)}{1 + {}^3K_{SV}c_q/(1 + a)}$$

**Equation 3.41** *Stern–Volmer equation for the disappearance of reactant A by two reaction paths (Scheme 3.3) with quenching of both the lowest singlet and triplet state*

Equation 3.41 also holds for the appearance of product when products B and C are the same or when their yields are combined. For the single-step reaction forming product B from the singlet state, Equation 3.38 holds, $\Phi_B{}^0/\Phi_B{}^q = (1 + {}^1K_{SV}c_q)$. For the two-step reaction forming product C via the triplet state we obtain Equation 3.42.

$$\Phi_C{}^0/\Phi_C{}^q = (1 + {}^1K_{SV}c_q)(1 + {}^3K_{SV}c_q)$$

**Equation 3.42** *Stern–Volmer equation for the appearance of product C by the two-step reaction (Scheme 3.3) with quenching of both the lowest singlet and triplet state*

An example of the three Stern–Volmer functions is shown in Figure 3.26 (left). The parameters were assumed to be ${}^1K_{SV} = 5\,\text{M}^{-1}$, ${}^3K_{SV} = 1000\,\text{M}^{-1}$ and $a = \Phi_C{}^0/\Phi_B{}^0 = 10$. Even when the quenching constants ${}^1k_q$ and ${}^3k_q$ are the same, ${}^3K_{SV}$ is usually much larger than ${}^1K_{SV}$, because the triplet lifetime is much longer. In the Figure 3.26 (right), ${}^1K_{SV}$ was set at zero, that is, we assumed that quenching of the singlet state is negligible, ${}^1K_{SV} \ll 1$. In that case, Equation 3.42 reduces to Equation 3.43. This is a common situation because singlet state lifetimes may be very short and because the singlet energy of the quencher is often higher than that of the quenchee, that is, ${}^1k_q \ll {}^3k_q$. Alternant hydrocarbons have large singlet–triplet energy gaps, in contrast to triplet ketones (see Section 4.7). Thus, piperylene

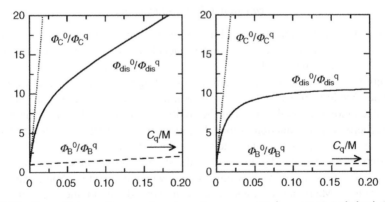

**Figure 3.26** Stern–Volmer diagrams for a two-step process (Scheme 3.3). Left: both $S_1$ and $T_1$ are quenched. Right: only triplet quenching is effective

(penta-1,3-diene) is often referred to as a triplet quencher, because its triplet energy is well below that of most ketones, so that $^3k_q$ is close to diffusion controlled, whereas its singlet energy is usually higher than those of ketones, $^1k_q \ll {}^3k_q$. Piperylene and other polyenes should always be freshly distilled because they form polymers that are poorly soluble, especially in polar solvents, giving turbid solutions. Sorbic acid or inorganic transition metal ions can serve as quenchers in aqueous solutions.

$$\Phi_C^0/\Phi_C^q = 1 + {}^3K_{SV}c_q$$

**Equation 3.43** *Stern–Volmer equation for the appearance of product C by the two-step reaction (Scheme 3.3) with quenching of only the triplet state*

It is worth spending some time to digest the diagrams in Figure 3.26. Begin with the right-hand diagram, which represents the case of triplet quenching only. Consequently, the yield of product B that is formed directly from the singlet state is not affected by the quencher (dashed horizontal line). The straight line representing the formation of the triplet product C (dotted line, Equation 3.43) rises steeply because $^3K_{SV}$ is large. The solid line representing the disappearance of reagent A starts out with the same steep slope, but reaches a plateau at $\Phi_{dis}^0/\Phi_{dis}^q = 11 = 1 + a$. At quencher concentrations $c_q > 0.2$ M, the triplet reaction is completely suppressed, but the formation of the singlet product B remains unaffected. In the left-hand diagram, which displays quenching of both excited states, the dashed line representing the singlet product has a small slope corresponding to $^1K_{SV}$. The dotted curve representing the formation of C looks the same as that on the right-hand side, but is in fact no longer linear, showing upward curvature; at very high quencher concentrations it would rise quadratically, $\Phi_C^0/\Phi_C^q \rightarrow {}^1K_{SV}{}^3K_{SV}c_q^2$ (Equation 3.42). Finally, the bent solid curve for $\Phi_{dis}^0/\Phi_{dis}^q$ no longer saturates completely when the triplet pathway is suppressed, but continues to rise linearly with a slope of $^1K_{SV}(1 + a) = 55$ M$^{-1}$ above $c_q = 0.1$ M, as singlet quenching takes its toll.

Triplet sensitization (Section 2.2.2) can be used to determine $^3K_{SV}$ for the quenching of a triplet reactant. A complete reaction scheme for sensitization of a reactant A yielding product B from the triplet state would lead to a rather complex Stern–Volmer expression, if all steps are treated explicitly. However, if ISC of the sensitizer is very fast and efficient ($\Phi_T = 1$) and if the concentration of A is chosen to be much higher than that of the quencher added, then we may assume that triplet energy transfer from the sensitizer to A is fast and efficient at all quencher concentrations, so that a plot of $\Phi_B^0/\Phi_B^q$ versus $c_q$ will obey the simple Stern–Volmer relation and yield $^3K_{SV}$ as the slope.

A frequently encountered case is depicted in Scheme 3.4, where the singlet state of A, $S_1(A)$, reacts with a substrate B, for example an alkene yielding a cycloaddition product C. In most cases the substrate B will be in large excess, so that $c_B$ can be treated as a constant. Scheme 3.4 includes an intermediate A··B that may represent an exciplex (Section 5.2) or a biradical intermediate (Section 5.4.4), which can proceed either to product C, $^{A··B}k_r$ or

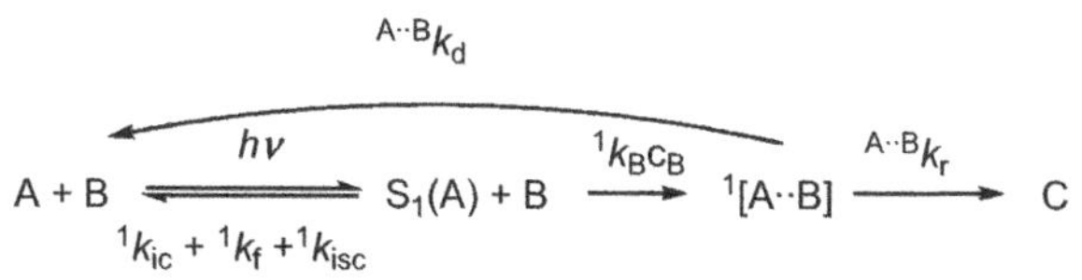

**Scheme 3.4** *Bimolecular reaction*

revert to the starting materials A + B, $^{A\cdot\cdot B}k_d$. All competing decay processes from $S_1(A)$ are assumed to lead back to the starting materials.

The quantum yield of product formation $\Phi_C$ is equal to the product of the efficiencies of the two reaction steps leading to C, $\Phi_C = {}^1\eta_{A\cdot\cdot B}{}^{A\cdot\cdot B}\eta_C$, where ${}^1\eta_{A\cdot\cdot B} = {}^1k_B c_B/(\Sigma^1 k + {}^1k_B c_B)$ and $^{A\cdot\cdot B}\eta_C = {}^{A\cdot\cdot B}k_r/({}^{A\cdot\cdot B}k_r + {}^{A\cdot\cdot B}k_d)$. The inverse of $\Phi_C$ is then given by Equation 3.44.

$$1/\Phi_C = \frac{1}{^{A\cdot\cdot B}\eta_C}\left(1 + \frac{1}{^1k_B\,^1\tau^0 c_B}\right)$$

**Equation 3.44** *Double reciprocal Stern–Volmer plot*

Thus a plot of the inverse of $\Phi_C$ versus the inverse of $c_B$, a so-called double reciprocal plot, should be linear and the intercept/slope ratio gives $^1K_{SV} = {}^1k_B\,^1\tau^0$. See, however, the comments regarding statistical analysis given below.

The trapping of reactive intermediates such as carbenes or radicals by suitable additives may also reduce the quantum yield of product C. The Stern–Volmer treatment applies similarly and may be used to identify transient intermediates observed by flash photolysis; if half of the product is replaced by the trapping product, then the lifetime of the alleged reactive intermediate must be reduced by a factor of two at the same concentration of the trapping reagent.

Finally, a phenomenon called *concentration quenching* or *static quenching* can lead to upward curvature of Stern–Volmer plots even at moderate quencher concentrations ($c_q \geq 0.01$ M). Molecules that are located next to a quencher at the time of excitation will be quenched immediately. Therefore, the fluorescence decay curve will be non-exponential initially, exhibiting a very fast initial component. Moreover, the initial depletion of these molecules will result in an inhomogeneous distribution of the remaining excited molecules around the quenchers. As a result, the diffusion coefficient $k_d$ is no longer a constant, but becomes a function of time, $k_d(t)$, until the statistical distribution of excited molecules is re-established. The impact of these effects has been analysed in detail.[231] Intrinsic rates of electron transfer in donor–acceptor contact pairs can be extracted from the resulting curvature in Stern–Volmer plots.[232]

A general remark regarding the statistical analysis of Stern–Volmer quenching data is in order. The standard errors of the measured quantum yield or lifetime ratios usually increase with increasing quencher concentration. One should either weight each data point by the inverse of its estimated variance or transform the appropriate Stern–Volmer equation to a form in which the distribution of standard errors is *homoskedastic*, that is, independent of the quencher concentration. The variance of the data points could be estimated from repeated measurements or from *a priori* considerations, but it is easier to fit the parameters of the appropriate homoskedastic function by nonlinear least-squares fitting (see Section 3.7.4). It is common, but bad, practice to transform Stern–Volmer relations to a linear but strongly heteroskedastic form and to analyse by linear regression without weighting.

In practice, the *relative* error of $k_{obs} = 1/\tau$ and hence the error of $\log(k_{obs}{}^q/k_{obs}{}^0)$ is often independent of the quencher concentration. Consider, for example, triplet decay rate constants that were determined by flash photolysis in the absence, $k_{obs}{}^0$, and presence, $k_{obs}{}^q$, of various quencher concentrations. We expect Equation 3.38 to hold,

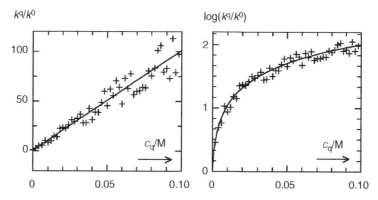

**Figure 3.27** Heteroskedastic (left) and homoskedastic (right) Stern–Volmer plots

$\tau^0/\tau^q = k_{obs}^q/k_{obs}^0 = 1 + K_{SV}c_q$. Figure 3.27 (left) shows artificial data that were generated by assuming $K_{SV} = 1000\,\text{M}^{-1}$ and adding 10% random noise to the values of $k_{obs}$. The slope obtained by unweighted linear regression is $K_{SV} = 965 \pm 43\,\text{M}^{-1}$ and the intercept is ill-defined as $1.7 \pm 2.4$. By taking the logarithm of the rate constants (right-hand diagram), the same data become homoskedastic. Nonlinear least-squares fitting of the function $\log(a + K_{SV}c_q)$ gives a much more accurate intercept, $a = 0.86 \pm 0.12$, and a slope $K_{SV} = 975 \pm 22\,\text{M}^{-1}$. Usually one will have far fewer experimental data than are shown here, so one risks obtaining highly distorted results by linear regression. Having done the proper nonlinear analysis, one can still display the results as shown on the left to emphasize the linear relationship.

### 3.9.9 Quantum Yields of Triplet Formation

The measurement of triplet quantum yields $\Phi_T = {}^1\eta_{ISC}$ is difficult because triplet states are transient intermediates. Phosphorescence quantum yields $\Phi_{ph}$ represent a lower limit for $\Phi_T$, because $\Phi_{ph} = {}^1\eta_{ISC}{}^3\eta_{ph}$, but the measurement of $\Phi_{ph}$ requires low temperatures. Photothermal methods (Section 3.11) can be applied to the determination of triplet quantum yields in solution when the triplet energy $E_T$ is known from phosphorescence measurements.[233]

Molar triplet–triplet absorption coefficients of some important triplet sensitizers with $\Phi_T$ close to unity, such as benzophenone ($\varepsilon_{530} = 7200\,\text{M}^{-1}\,\text{cm}^{-1}$) and xanthone ($\varepsilon_{610} = 5300\,\text{M}^{-1}\,\text{cm}^{-1}$), have been determined reliably,[234] so that monitoring triplet energy transfer by flash photolysis (Figure 2.18) can be used to estimate $\Phi_T$ of other compounds. The triplet absorbance generated by sensitization is then compared with that observed by direct excitation of the probe.[235] The concentrations of sensitizer and probe and of the excitation wavelength must be chosen such that the sensitizer absorbs most of the pump pulse in the energy transfer experiment. Alternatively, energy transfer to β-carotene ($E_T \approx 96\,\text{kJ}\,\text{mol}^{-1}$, $\varepsilon_{515} = 187\,000\,\text{M}^{-1}\,\text{cm}^{-1}$) is very useful, not only to identify transient intermediates as triplets, but also to estimate $\Phi_T$.[236] Although part of the excitation pulse is unavoidably absorbed by β-carotene, spontaneous ISC of β-carotene is inefficient on direct excitation. The near-IR emission of singlet oxygen has been applied to measure $\Phi_T$ of Rose Bengal.[237]

Two self-calibrated methods are available that do not rely on the knowledge of $\Phi_T$ of a reference compound. Horrocks *et al.* described an accurate method based on the measurement of triplet–triplet absorption by flash photolysis, in combination with Stern–Volmer analysis of fluorescence quenching (Section 3.9.8).[238] Bromobenzene was used as a heavy-atom quencher of the fluorescence of 9-phenylanthracene. More recently, time-resolved measurements of delayed fluorescence (Section 2.2.4) were analysed to give accurate triplet quantum yields.[239]

### 3.9.10  Experimental Arrangements for Quantum Yield Measurements

Whenever possible, essentially monochromatic light sources such as low- or medium-pressure mercury arcs equipped with bandpass filters or a monochromator (Figure 3.28), narrow-band photodiodes or lasers should be used for quantum yield determinations, because quantum yields can only be defined for monochromatic irradiation. This can be relaxed if the absorption spectrum of the actinometer is close to that of the sample (Section 3.9.6). One then assumes that the quantum yield is independent of the wavelength of irradiation. The stability of the light source over time is essential. Medium-pressure mercury arcs that have a stable output for many hours after a burn-in period of about 30 min are available. Xenon arcs tend to fluctuate abruptly when the arc between the electrodes jumps from one position to another. Intensity fluctuations of the light source in time can be monitored with photodiodes. This should be routinely done with pulsed lasers.

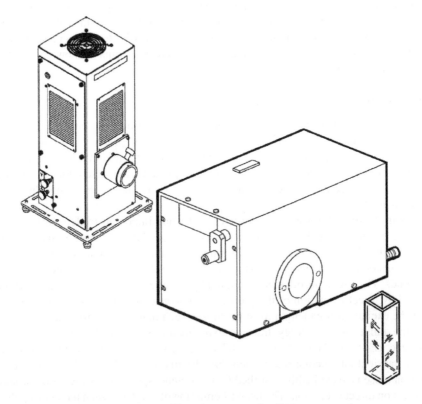

**Figure 3.28**  Components of an optical bench (source of radiation, monochromator, cuvette). Reproduced by permission of Newport Corp, Oriel Product Line

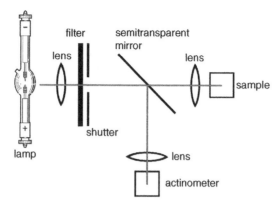

**Figure 3.29** Experimental arrangement for measuring reaction quantum yields

Various arrangements are used for actinometry. In the setup shown in Figure 3.29, the sample and the actinometer are each placed in standard quartz cells ($1 \times 1 \times 4$ cm) that are equipped with a small magnetic stirring bar. Efficient stirring is essential when moderate absorbances, $A < 2$, are used. The sample and actinometer are irradiated simultaneously. First, the relative amount of light reaching the two cells from the beam splitter (a semitransparent mirror) is calibrated by placing an actinometer solution in both cells. Such a setup is insensitive to variations in the intensity of the light source. If the volumes of the sample and actinometer solutions are not the same, correction for volume $V$ ($c = n/V$) is required. For convenience, the actinometer may be replaced by a photodiode to monitor the relative light intensity continuously; the diode output is calibrated by placing the actinometer in the sample compartment. During subsequent irradiation of the sample, the amount of light is calculated from the diode reading.

A convenient, if somewhat less accurate, method for measuring a series of quantum yields of reaction is the *merry-go-round* setup (Figure 3.30). It holds about 20 sample and actinometer solutions with equal absorbance in cylindrical cells (test-tubes), which are

**Figure 3.30** Merry-go-round apparatus. Reproduced by permission of Ace Glass Inc

rotated around a tubular light source such as a low-pressure mercury arc, a medium-pressure mercury arc equipped with appropriate filters or a fluorescent screen. Rotation ensures that all samples receive the same amount of incident light. After irradiation, the sample compositions are analysed using conventional methods such as NMR or GC. Caution is needed when the individual samples have different quantum yields, for example due to the addition of increasing amounts of quencher or reactant, because the amount of light absorbed by the reactant (Equation 3.18) will differ in the individual samples.

## 3.10   Low-Temperature Studies; Matrix Isolation

*Recommended reviews.*[240–244]

Studies involving the irradiation of solid solutions at low temperature have a long history, culminating in the identification of the triplet state of organic molecules by Lewis and Kasha in 1944.[23] Ten years later, the expression *matrix isolation* was coined by George Pimentel,[245] who pioneered the technique of co-condensing a substrate with a large excess of an inert gas on a cold surface, thus preventing any aggregation that may occur when a liquid is frozen. With the advent of commercial closed-cycle cryostats that obviate the need for liquid He and thus offer low-cost operation for matrix isolation in Ar (12 K) and in Ne (two-stage cooling, 4 K), the technique has found widespread use in many laboratories. Nevertheless, the much less demanding method of cooling solutions in (mixtures of) organic solvents that freeze to solid, transparent glasses[157,159] still has its place. Many spectroscopic techniques are used to follow the reactions of matrix-isolated species, mostly IR (high-resolution spectra), optical spectroscopy and ESR, but also mass spectrometry of species that are released from the matrix by laser desorption or ion sputtering.[246] A very powerful variant of matrix isolation is the deposition of mass-selected ions.[247]

Many excellent reviews of matrix isolation work are available,[240,242–244,247] and specific results are mentioned elsewhere in this text. In this section, we restrict ourselves to a few general remarks. Matrix isolation at low temperature mummifies highly reactive intermediates by two distinct effects and thereby permits their spectroscopic characterization without the need for fast detection. First, it isolates the reactive species in cages of an inert medium and thereby prevents diffusion and bimolecular reactions. Nevertheless, such reactions can be initiated at will, either by co-deposition with reactive gases such as dioxygen or carbon monoxide or by softening the matrix by a controlled rise in temperature. Second, monomolecular reactions are suppressed by the low temperature, ensuring that even small activation barriers become insurmountable and possibly by the cage effect inhibiting structural changes.

The reactive target molecules may be formed by rapid deposition of the gaseous eluates from flash vacuum pyrolysis, plasmas, and so on, or by irradiation of stable matrix-isolated molecules with high-energy radiation (UV, X-rays, $\gamma$-rays, etc.). However, rare gas matrices are very poor at conducting heat. Consequently, any excess energy left in the primary products of a photoreaction will dissipate only slowly, allowing for further rearrangement. The same applies to hot molecules formed by internal conversion, so that the products formed by irradiation of matrix-isolated species may well result from hot ground-state reactions, which are rarely seen in solution (Section 2.3). Hence the counterintuitive finding that the product distributions formed by irradiation of molecules isolated in cold matrices may resemble those expected from high-temperature pyrolysis.

## 3.11  Photoacoustic Calorimetry

The measurement of heat deposition following the absorption of light can provide spectroscopic, kinetic and thermochemical information. The technique was invented in 1880 by Alexander Graham Bell as a means for the wireless transmission of speech.[248] A membrane covered by a reflective coating transformed the sound waves of the speaker to an oscillating beam of reflected sunlight. The reflected beam was detected by the distant listener using a kind of stethoscope equipped with a closed transparent detector that was covered with black soot on the inside to reconstruct the sound wave pattern superimposed on the light beam. The oscillating release of heat developed by absorption of the light beam reproduced the sound wave as oscillating volume changes in the enclosed air.

Of course, Bell moved on to develop better methods of sound transmission, but the 'spectrophone' was rediscovered some 80 years later and developed to a highly sensitive technique for trace gas analysis. As photochemists, we are more interested in the time-resolved analysis of heat deposition. A comprehensive review of the applications of photothermal methods in chemistry and biology covering the literature up to 1992 was published,[249] and an excellent overview and discussion of the methodology appeared in 2003.[250] The volume changes initiated in the solvent by the release of heat during photophysical and photochemical processes can be detected either with a microphone directly attached to the sample cell or optically by exploiting the resulting local changes in the refractive index of the solvent. Photorefractive techniques usually employ one of three methods, namely *transient grating* when two light waves with parallel polarization interfere, *thermal lensing* or *beam deflection.*

No attempt is made to analyse directly the complex but reproducible shock waves, which are produced by a short light pulse in the sample cell and recorded by a microphone. Rather, the shock waves produced in the sample are compared with those generated by a chromophore that is known to release the entire absorbed light energy as heat within a short time period (usually within 1 ns). This reference wave, produced for example by a solution of ferrocene, is called the T-wave. Delayed heat release in the sample due to the intervention of reactive intermediates is then reflected by a delay in the temporal evolution of the shock wave and incomplete heat release due to the formation of a (meta-)stable photoproduct or excited state of high energy is expressed in a diminution of the signal amplitude. The pair of signals is then analysed by mathematical deconvolution, which amounts to an iterative fitting of the parameters of a (multi-)exponential decay function to reproduce the observed wave of the sample.[251,252]

For example, a solution of benzophenone in aerated acetonitrile releases part of the absorbed energy very rapidly (<1 ns) due to ISC and thermal relaxation, and the triplet state of benzophenone decays with a lifetime of about 200 ns. The observed signal then consists of a T-wave of reduced amplitude due to the fast process and the delayed heat release due to the decay of the triplet state is modelled by adding small incremental T-waves for the heat released in short time intervals $\Delta t$ as defined by the exponential decay function of the triplet state. The amplitudes and rate constant(s) of the processes occurring in the sample are then determined by iterative least-squares fitting of the nonlinear trial parameter(s) in the exponential function, the rate constant(s), to the observed sample signal (see Section 3.7.4 for the separation of linear and nonlinear parameters in least-squares fitting).

Time-resolved photothermal methods are highly sensitive and cover a huge time range from about 1 ps to 1 s, although the individual methods are restricted to smaller time windows. They can detect reactive intermediates or conformational changes of enzymes that are invisible to other spectroscopic methods if these processes are accompanied by substantial enthalpic or volume changes. The techniques can be implemented at very little cost in any laboratory equipped with pulsed lasers and, when performed with due diligence, will yield highly accurate rate constants. In addition, they are unique in providing thermochemical data (heats of formation) of reactive intermediates.

However, reaction enthalpies obtained by photoacoustic methods must be considered with caution. Although the results are often highly reproducible, with standard errors of only a few kJ mol$^{-1}$, the data may be fraught with hidden systematic errors, which arise mainly from two sources. First, the quantum yield of the process considered must be known with very high accuracy. This is a minor problem in the case of benzophenone considered above, because the triplet yield of benzophenone is known to be very close to unity. For intermediates formed with a smaller quantum yield, however, errors in the quantum yield of only a few percent translate to huge errors associated with the enthalpies of formation and decay of reactive intermediates. Second, the volume changes detected as pressure waves can be of thermal or non-thermal origin. Non-thermal volume changes can arise from Coulombic effects such as a change in the number of charges, dipole moments or hydrogen bonds associated with a reaction and from changes in the molar volume of the reactant associated with conformational changes or fragmentation reactions.

The separation of these cumulative effects is not an easy task, but is necessary for the determination of thermodynamic parameters, such as chemical bond strengths. Measuring very dilute water solutions at 3.9 °C, where the thermal expansion coefficient of water vanishes (or at slightly lower temperatures in more concentrated aqueous solutions, such as buffer solutions) can be used to separate the so-called structural volume changes from the thermal effects due to radiationless deactivation.[253,254] In this way, it is also possible to determine the entropy changes concomitant with the production or decay of relatively short-lived species (e.g. triplet states), a unique possibility offered by these techniques.[254,255]

For reactions in organic solvents, the separation of the thermal and non-thermal effects is more complex. Some methods have been suggested.[256] In view of the much larger value of the thermal expansion coefficient of organic solvents than the value for water, it is expected that the non-thermal volume changes in these solvents are in general much smaller than the thermal effects and can be neglected. However, caution is recommended, considering that some reactions (e.g. electron transfer reactions in polar organic solvents) might be accompanied by substantial solvent reorganization, which would give rise to large structural volume changes.

## 3.12 Two-Photon Absorption Spectroscopy

The interaction of a molecule with a strong laser pulse at a wavelength that is too long to populate an excited state by absorption of a single photon can lead to the simultaneous absorption of two or more photons. The phenomenon is described by nonlinear optics[161] (Special Topic 3.1) and should be distinguished from the sequential absorption of two photons as, for example, in triplet–triplet absorption. Because the cross-sections for two-photon absorption are fairly small even under strong radiation, two-photon absorption is usually measured by fluorescence that is emitted well to the blue of the irradiation wavelength. Two-photon fluorescence excitation spectra differ from conventional absorption spectra because the selection rules for two-photon absorption are different. This permits the identification of excited states that have very low absorption coefficients.

Two-photon absorption is explored for applications in photodynamic therapy, for the release of photoremovable protecting groups (Special Topic 6.18) in living tissue and for stereolithography (3D optical storage). The strong light scattering by biological tissues hampers imaging by confocal microscopy. This is overcome by two-photon fluorescence microscopy because the scattering of near-IR radiation is low so that multiply scattered signal photons can be assigned to their origin allowing for deep-tissue imaging.[257]

## 3.13 Single-Molecule Spectroscopy

*Recommended review articles.*[78,81,258–261]

The detection and manipulation of single molecules became practicable only in the late 1970s[j] thanks to the combination of modern microscopic techniques and lasers and the impact of *single-molecule spectroscopy* (SMS) has grown tremendously over recent decades. The detection of single molecules is not primarily a sensitivity problem; photomultipliers have long been able to detect single photons that reach the photocathode. Rather, SMS is limited by background noise and by the photostability of the fluorescent molecules. Originally, SMS was possible only with a few highly photostable fluorescent molecules isolated in inert matrices at low temperatures. The implementation of clever techniques and statistical analyses to suppress stray light, noise and background signals now permits the detection of single molecules in biological media at room temperature. The signal-to-noise ratio was improved by orders of magnitude thanks to *confocal microscopes* that can probe sub-femtolitre volumes within biological tissue and discard adventitious light emerging from outside that volume.

Apart from the obvious challenge of reaching the ultimate limit of analytical chemistry, why is it of interest to detect single molecules? Conventional measurements done with an ensemble of molecules provide average values for the properties studied, but knowledge about the distribution of a property in time and space rather than its average value can be essential. Single-molecule detection provides information about the temporal fluctuation and local variation of these properties in heterogeneous media. Single molecules can act as

---

[j] '… *we never experiment with just one electron or atom or (small) molecule. In thought experiments we sometimes assume that we do; this invariably entails ridiculous consequences.* . .' E. Schrödinger, *Br. J. Philos. Sci.* **1952**, 3, 233–242.

highly sensitive reporters of their environment and movements. In a pioneering study, up to eight independent parameters (anisotropy, fluorescence lifetime, fluorescence intensity, time, excitation spectrum, fluorescence spectrum, fluorescence quantum yield, distance between fluorophores) of single molecules were simultaneously analysed, allowing for a quantitative analysis of 16 different compounds in a mixture via their characteristic patterns.[262]

*Fluorescence correlation spectroscopy* analyses the temporal fluctuations of the fluorescence intensity by means of an autocorrelation function from which translational and rotational diffusion coefficients, flow rates and rate constants of chemical processes of single molecules can be determined. For example, the dynamics of complex formation between $\beta$-cyclodextrin as a host for guest molecules was investigated with single-molecule sensitivity, which revealed that the formation of an encounter complex is followed by a unimolecular inclusion reaction as the rate-limiting step.[263]

The ultimate goal in biological applications of single-molecule methods is to investigate biological processes in living organisms by tracking the movement of individual molecules in three dimensions and at molecular resolution. Examples of interest abound, such as protein folding and the invasion of a single cell by a virus and its attack on the cell's nucleus.[264] SMS is able to detect rare events such as protein misfolding and reaction intermediates. The spatial resolution of modern optical imaging techniques is approaching 1 nm, far beyond the dispersion limit of a conventional microscope.[265] *In situ* dynamic studies of proteins (enzymes at work)[83] and of intranucleosomal DNA have been reported.[266]

*Green fluorescent protein* (GFP) (Scheme 3.5) and its variants are particularly useful probes for live-cell imaging and to study intracellular dynamics.[267,268] A Chemical Abstracts search for 'GFP' yielded about 35 000 references (June 2007), which appeared mostly during the last 15 years. SMS was used to explore the unfolding pathways of a green fluorescent protein (GFP) mutant called Citrine under nonequilibrium conditions.[269] An intermediate on the unfolding/folding energy landscape was identified and two distinct conformations from which the protein unfolds along parallel pathways were found.

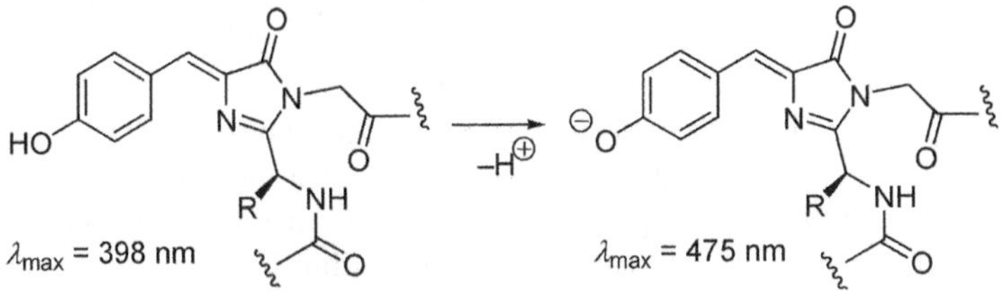

$\lambda_{max} = 398$ nm                    $\lambda_{max} = 475$ nm

**Scheme 3.5**    *The chromophore of green fluorescent protein (GFP)*

## 3.14  Problems

1. Estimate the maximum amount of abiotic photoconversion in mol $m^{-2} h^{-1}$ that can occur in surface waters in bright sunlight (800 W $m^{-2}$). Assume that 5% of the solar spectrum (Figure 1.1) is of sufficiently short wavelength to induce a photoreaction with a quantum yield of 1. [0.5 mol $m^{-2} h^{-1}$]

2. Derive Equation 3.25 using the hint given in the footnote to that equation.

3. A plot of the inverse quantum yield for the photocycloaddition of 1-aminoanthraqui-
   nones (AQ) and (E)-piperylene (Q) was found to be linear and the ratio of the slope,
   $0.20 \, dm^3 \, mol^{-1}$, and intercept, 1.20, of this plot was equal to the Stern–Volmer constant
   for fluorescence quenching of AQ by Q, $K_{SV} = 0.17 \, dm^3 \, mol^{-1}$. It was concluded that
   the reaction proceeds from the excited singlet state of AQ via an exciplex intermediate
   $^1(AQ \cdots Q)^*$. Explain. [Ref. 270]

4. The fluorescence quantum yield of anthracene in degassed acetonitrile is 0.3. It is
   reduced by 30% upon admission of air. Estimate the lifetime of the singlet state of
   anthracene in degassed solution and the rate constant of fluorescence. Refer to Section
   2.2.1. The concentration of dioxygen in air-saturated acetonitrile is $2 \times 10^{-3} \, M$.
   $[\tau_f^0 = 5.6 \, ns, \, k_f = 5.4 \times 10^7 \, s^{-1}]$

5. Relative quantum yields $\Phi^0/\Phi^q$ have been measured as a function of quencher
   concentration, $[q] = 0., 0.02, 0.04, \ldots, 0.2 \, M$. The data, $\Phi^0/\Phi^q = 1.0, 2.1, 3.2, 4.5, 4.8,$
   5.2, 6.0, 8.5, 8.5, 9.0, 9.0, obey a linear relationship (Equation 3.38), within
   experimental accuracy. Determine the intercept and the slope $K_{SV}$ (a) by linear
   regression and (b) by nonlinear least-squares fitting of the function log(Equation 3.38)
   to $\log(\Phi^0/\Phi^q)$; cf. Figure 3.27. Which analysis is more reliable? [(a) intercept
   $1.40 \pm 0.33$, slope $42 \pm 3$; (b) intercept $1.08 \pm 0.10$, slope $46 \pm 2$]

6. Derive Equation 3.41.

# 4

# Quantum Mechanical Models of Electronic Excitation and Photochemical Reactivity

## 4.1  Boiling Down the Schrödinger Equation

The task of finding reasonably accurate solutions to Schrödinger's Equation 1.5 for a medium-sized organic molecule looks like a mission impossible. For measure, consider a modestly sized molecule such as benzene that contains 42 electrons. A table listing only 10 values of its electronic wavefunction on a grid along each of its independent variables would contain $10^{126}$ entries (three coordinates must be specified for each electron). This number is far larger than the number of atoms in the visible universe. Nevertheless, numerical solutions of the Schrödinger Equation 1.5 can nowadays be obtained for reasonably large organic molecules with chemically useful accuracy using approximate *ab initio* or *density functional theory* methods.[a]

Trial wavefunctions are usually constructed by linear combination of Gaussian error functions that are convenient to integrate. The results can be of predictive value and such calculations have become everyday tools for chemists in all branches of chemistry, to guide experiments and not least to rule out untenable hypotheses. This is a remarkable achievement that seemed to be out of reach a few decades ago. Still, simple qualitative models that are amenable to *perturbation theory* are required to understand and predict trends in a series of related compounds. Our goal here is to describe the minimal quantum mechanical models that can still provide a useful qualitative description of electronically excited states, their electronic structure and their reactivity. Such models also provide a language to convey the results of state-of-the-art, but essentially 'black-box' *ab initio* calculations.

---

[a] Closed, exact solutions of the Schrödinger equation not being available for all except the most simple of systems, it is unavoidable to take recourse to approximations. The so-called *ab initio methods* refrain, however, from replacing parts of the calculations by empirical parameters that are optimized by adjustment to experimental data.

---

The electronic structure of molecules is commonly described in terms of *molecular orbitals* (MOs). This is a cavalier attitude because, strictly, orbitals cannot be used to describe many-electron systems. Atomic orbitals (AOs) are exact solutions of the Schrödinger Equation 1.5 only for hydrogen-like atoms. If we attempt to construct an electronic wavefunction for, say, the helium atom by placing both electrons in the lowest AO, $\phi_{1s}$, which was determined for $He^+$ (a hydrogen-like atom), $\Psi_{el}(He) \approx \phi_{1s}(e_1)$ $\phi_{1s}(e_2) = \phi_{1s}^2$, we run into two problems. The first is that the Hamiltonian operator for helium contains the Coulomb term $e^2/r_{12}$ accounting for the repulsion between the two electrons. That term was not included in the determination of $\phi_{1s}$ for $He^+$. If we calculate the lowest expectation value $\langle E_1 \rangle$ for the energy (Equation 1.14) of He using the AO obtained for $He^+$, the resulting energy will be much too high, because the electrons are squeezed into the $\phi_{1s}$ orbitals of $He^+$, which are too small. This can be amended by using some shielding factor that reduces the nuclear charge when determining AOs for He, a semiempirical procedure, or by re-optimizing the AO for He by taking account of electronic repulsion. Both procedures will make the AO for He more diffuse and thereby yield a lower expectation value $\langle E_1 \rangle$.

Nevertheless, the calculated energy will still be substantially higher than that of the He atom in the ground state, unless we resort to an empirical shielding factor that is calibrated to yield the energy of He. This is due to a more fundamental and less tractable problem. By assuming that we can represent the electronic wavefunction for He by a product function of the type $\Psi_{el}(He) \approx \phi_{1s}(e_1)\phi_{1s}(e_2)$, we implicitly assume that the motions of the two electrons are independent of each other: The square of such a wavefunction, which represents the time-averaged spatial distribution of the two electrons (*Born interpretation*, Section 1.4), is the product of their individual distributions, $\Psi_{el}^2 \approx \phi_{1s}(e_1)^2 \phi_{1s}(e_2)^2$. However, the combined probability of two events, such as that of throwing two ones with two dice, is equal to the product of the individual probabilities *only* if the two events are independent. This is clearly not the case for the movements of two electrons in an atom or molecule. Rather, given the position of one electron, it will be unlikely that the other electron is nearby. Electronic motions are correlated, electrons tend to avoid each other; if $e_1$ happens to be on one side of the atom, the chances are that $e_2$ will be on the other. The error associated with calculating time-averaged electrostatic interactions of electrons in stationary orbitals is called *correlation error* and amounts to about $1\,eV$ (or $\sim 100\,kJ\,mol^{-1}$) per electron pair in an organic molecule. Hence the energy calculated in this way will be too high, in accord with the variation theorem (Equation 1.14).

Keeping these cautionary remarks in mind, we nevertheless proceed to build trial wavefunctions for molecules as product functions of molecular orbitals. Moreover, we will construct the *molecular orbitals* as *linear combinations of atomic orbitals* (LCAOs), so that we can take advantage of the efficient Rayleigh–Ritz procedure (Equation 1.6). The idea behind the LCAO method is that electronic motion near a nucleus should be reasonably well described by an AO. Having committed all these sins, we will start by going one step further, namely by ignoring electronic interaction altogether. But do not despair! Although such crude methods can never give accurate energies for molecules and their excited states unless they are parameterized (semiempirical methods), they do retain a fair measure of physical reality, providing insight and allowing us to make useful predictions.

Before we proceed, recall the *Aufbau principle*: Having defined a set of orbitals for a given molecule, we fill the electrons successively into the orbitals of lowest energy, *allowing no more than two electrons of opposite spin per orbital.* Why can we not place all the electrons into the lowest energy molecular orbital $\psi_1$? This is a consequence of the *Pauli principle* (Equation 1.12), which demands a wavefunction to be antisymmetric with respect to electron exchange. Suppose the spin states of two electrons are the same, say $\alpha\alpha$ ($\uparrow\uparrow$). The spin function $\alpha\alpha$ is symmetric (invariant to electron exchange), so the electronic wavefunction must be antisymmetric. A trial electronic wavefunction with both electrons in $\psi_1$, $\Psi \approx \psi_1(e_1)\psi_1(e_2)$ would be symmetric. We could construct an antisymmetric linear combination, $\Psi \approx [\psi_1(e_1)\psi_1(e_2) - \psi_1(e_2)\psi_1(e_1)]$, only to find that this vanishes at all positions in space; no such state exists. Not so if the two orbitals carrying parallel spins are different: $\Psi \approx [\psi_1(e_1)\,\psi_2(e_2) - \psi_1(e_2)\psi_2(e_1)]/\sqrt{2}$ is an acceptable antisymmetric wavefunction; it does change sign upon exchange of the two electrons.

Provided that the electronic part of the wavefunction obeys the Aufbau principle, antisymmetric wavefunctions can always be constructed by using appropriate linear combinations (*Slater determinants* or antisymmetrized wavefunctions) and we can leave this technical problem to the computer programs. An occupation scheme defining the distribution of electrons among a set of given orbitals is called an *electronic configuration.*[b] The ground state of stable organic molecules will in general be represented by a *closed-shell* configuration, in which the lowest available orbitals are doubly occupied with electrons of antiparallel spin ($\uparrow\downarrow$) (Figure 4.1, left). We denote the wavefunction for a configuration by the letter $\chi$, that is, $^1\chi_0 = \psi_1^2\psi_2^2\cdots\psi_i^2$ is the (singlet) ground-state configuration. Electronic excitation can then be described by raising one or more electrons from a bonding orbital $\psi_i$ to an unoccupied orbital $\psi_j$. The resulting (singlet or triplet) excited state configuration becomes $^{1\,or\,3}\chi_{i\to j} = {}^{1\,or\,3}(\psi_1^2\psi_2^2\cdots\psi_i\psi_j)$.

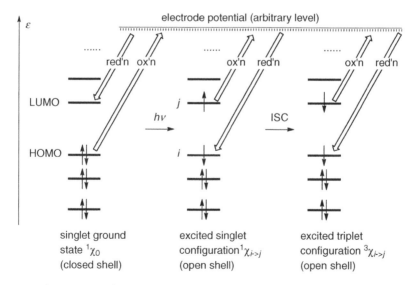

**Figure 4.1** Electronic configurations. The acronym HOMO designates the highest occupied MO and LUMO the lowest unoccupied MO of the ground state

---

[b] The expression configuration used in stereochemistry has a different meaning.

In an excited, open-shell configuration, the Pauli exclusion principle allows us to have an excess of electrons with parallel spin (e.g. ↓↓ in the triplet state of Figure 4.1, right). In two configurations with the same occupancy numbers but different total spin, the configuration with the highest spin will be the lowest in energy. In the present case, the triplet configuration will have a lower energy than the corresponding open-shell singlet. This is an extension of *Hund's first rule*, which was originally formulated for atoms. It is a rule, not a law, but it holds most of the time. Hund's rule may be attributed to the *Fermi hole*, the traffic law that keeps electrons apart in antisymmetric spatial wavefunctions (Figure 1.14). The energy difference between the singlet and triplet state is not manifested in Figure 4.1, because we have decided to ignore electronic interaction. It will show when we reintroduce it as an afterthought (Section 4.7).

Without even doing any MO calculation, we can already conclude from Figure 4.1 that *any molecule M will become both a stronger oxidant and a stronger reductant upon electronic excitation.* Ignoring entropy effects, the difference between the corresponding standard potentials $E°/V$ is equal to the excitation energy $E_{0-0}$ of the excited singlet or triplet state measured in units of electronvolts (Equation 4.1). Thus, if the 0–0 band of $S_0$–$S_1$ absorption lies at $3\,\mu m^{-1}$ (333 nm), both the oxidation and the reduction potential are shifted by 3.72 V! Note the change in sign in the second equation for oxidation.

$$\frac{E^0(M^*/M^{-\bullet})-E^0(M/M^{-\bullet})}{V} = \frac{E_{0-0}(M)}{eV} = 0.01036\frac{E_{0-0}(M)}{kJ\,mol^{-1}} = 1.240\frac{\tilde{\nu}_{0-0}(M)}{\mu m^{-1}}$$

$$\frac{E^0(M^{+\bullet}/M^*)-E^0(M^{+\bullet}/M)}{V} = -\frac{E_{0-0}(M)}{eV} = -0.01036\frac{E_{0-0}(M)}{kJ\,mol^{-1}} = -1.240\frac{\tilde{\nu}_{0-0}(M)}{\mu m^{-1}}$$

**Equation 4.1** *Reduction and oxidation of excited molecules*

## 4.2   Hückel Molecular Orbital Theory

*Hückel molecular orbital* (HMO) theory[271] deals only with the π-electrons of unsaturated systems; the σ-electrons are considered to be part of a frozen core. That is, we use only the $2p_z$-AOs $\phi$ on the unsaturated carbon atoms $C_1$, $C_2$, ..., $C_\mu$, ..., $C_\nu$, ..., $C_\omega$, which are assumed to be orthonormal (orthogonal and normalized) (Equation 4.2).

$$<\phi_\mu|\phi_\nu> = \begin{cases} 1 \text{ for } \mu = \nu \\ 0 \text{ else} \end{cases}$$

**Equation 4.2**

The HMOs are constructed as linear combinations of the $2p_z$-AOs (LCAO):

$$\psi = c_1\phi_1 + c_2\phi_2 + \ldots + c_\mu\phi_\mu + \ldots + c_\nu\phi_\nu + \ldots + c_\omega\phi_\omega = \sum_{\mu=1}^{\omega} C_\mu\phi_\mu$$

In order to find solutions for the eigenvalue problem $\hat{H}^{HMO}\psi = E\psi$, we have to set up the corresponding *secular determinant* (see Equation 1.16). No attempt is made to calculate the matrix elements of the secular determinant; instead, the matrix elements are

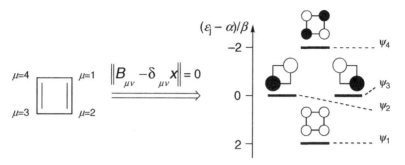

**Figure 4.2** Hückel molecular orbitals for cyclobutadiene (**3**). The $2p$-AOs are viewed from the top. The radius of the rings is proportional to the size of the AO coefficients $|c_{j\mu}|$, so that their area is proportional to $c_{j\mu}^2$. The shading indicates the relative sign, i.e. white is used for $c_{j\mu} > 0$ and black for $c_{j\mu} < 0$ (or vice versa). The coefficients of the NBMOs $\psi_2$ and $\psi_3$ are $|c_{j\mu}| = 1/\sqrt{2}$ or 0 and those for $\psi_1$ and $\psi_4$ are $|c_{j\mu}| = 0.5$

treated as adjustable parameters (Equation 4.3). Thus the HMO model is devoid of any physical input; it makes use only of the topology (connectivity) of the π-system.

$$H_{\mu\mu} = <\phi_\mu|\hat{H}^{HMO}|\phi_\mu> = \alpha \text{ (a constant value for carbon atoms } C_\mu)$$
$$H_{\mu\nu} = <\phi_\mu|\hat{H}^{HMO}|\phi_\nu> = \begin{cases} \beta \text{ for } \mu \text{ bound to } \nu \\ 0 \text{ else} \end{cases}$$

***Equation 4.3*** *List of HMO matrix elements*

The parameters $\alpha$ and $\beta$ are called Coulomb integral and resonance integral, respectively. The resonance integral is a negative (stabilizing) energy. Now, let us set up the HMO secular determinant for cyclobuta-1,3-diene (**3**, Figure 4.2) as an example, using the matrix elements defined by Equation 4.3. The determinant contains the matrix elements $H_{\mu\nu} - \delta_{\mu\nu}\varepsilon$ with index $\mu$ varying in the rows $\mu = 1, \ldots, 4$ and index $\nu$ in the columns $\nu = 1, \ldots, 4$.

$$||H_{\mu\nu} - \delta_{\mu\nu}\varepsilon|| = \begin{array}{c} \\ \\ \begin{matrix} 1 \\ 2 \\ 3 \\ 4 \end{matrix} \end{array} \begin{Vmatrix} \alpha - \varepsilon & \beta & 0 & \beta \\ \beta & \alpha - \varepsilon & \beta & 0 \\ 0 & \beta & \alpha - \varepsilon & \beta \\ \beta & 0 & \beta & \alpha - \varepsilon \end{Vmatrix} = 0$$

***Equation 4.4***

A somewhat simpler form of the determinant is obtained by dividing all elements by $\beta$ and substituting $-x$ for $(\alpha - \varepsilon)/\beta$. Division by a constant is allowed, because the determinant is set equal to zero. We obtain the *Hückel determinant* (Equation 4.5). All diagonal elements $B_{\mu\mu}$ in the Hückel determinant are equal to $-x$ and the off-diagonal elements $B_{\mu\nu}$ are equal to zero, except for atoms $\mu$ and $\nu$ that are connected by a σ-bond ($\mu - \nu$), where $B_{\mu-\nu} = 1$.

$$||B_{\mu\nu}-\delta_{\mu\nu}x|| = \begin{Vmatrix} -x & 1 & 0 & 1 \\ 1 & -x & 1 & 0 \\ 0 & 1 & -x & 1 \\ 1 & 0 & 1 & -x \end{Vmatrix} = 0$$

**Equation 4.5**   *Hückel determinant for cyclobutadiene (3)*

Equation 4.5 is solved in a fraction of a second by a desktop computer using a mathematical program such as MATLAB.[c] A set of four eigenvalues, $\varepsilon_j = \alpha + x_j\beta$, $j = 1, \ldots, 4$, is obtained, together with a set of four HMO coefficients, $c_{j1}, \ldots, c_{j4}$, that is associated with each of the eigenvalues $\varepsilon_j$ and defines the four Hückel molecular orbitals (Equation 4.6). The result is depicted graphically in Figure 4.2. The Hückel calculation for **3** produces two nonbonding orbitals (NBMOs), $\varepsilon_2 = \varepsilon_3 = \alpha$ and $x_2 = x_3 = 0$.

$$\psi_j = \sum_{\mu=1}^{\omega} c_{j\mu}\phi_{\mu}, \ j = 1, 2, \ldots n$$

**Equation 4.6**   *Hückel molecular orbitals ($n = \omega = 4$ for cyclobutadiene)*

In general, any normalized linear combination of *degenerate wavefunctions*, wavefunctions that are associated with the same eigenvalue, is an equally valid wavefunction. Thus, the orthogonal wavefunctions $\psi'_2$ and $\psi'_3$ shown in Figure 4.3, $\psi'_2 = (\psi_2 + \psi_3)/\sqrt{2}$ and $\psi'_3 = (\psi_2 - \psi_3)/\sqrt{2}$ are equally acceptable solutions for the NBMOs, as are the wavefunctions $\psi_2$ and $\psi_3$ shown in Figure 4.2.

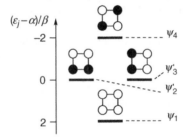

**Figure 4.3**   Alternative set of HMOs for cyclobutadiene (**3**). The size of all coefficients is $|c_{j\mu}| = 0.5$

General solutions exist for the Hückel determinants of special systems with any number $n$ of carbon atoms, namely for the linear polyenes (Equation 4.7) and for the monocyclic systems (Equation 4.8).

$$\varepsilon_j = \alpha + 2\beta\cos\left(\frac{\pi}{n+1}j\right), \ j = 1, 2, \ldots, n$$

**Equation 4.7**   *General HMO energies for linear polyenes with n carbon atoms*

---

[c] A simple HMO program is available on the Internet: http://www.chem.ucalgary.ca/SHMO/

$$\varepsilon_j = \alpha + 2\beta\cos\left(\frac{2\pi}{n}j\right), \; j = 0, 1, 2, \ldots, n{-}1$$

**Equation 4.8** *General HMO solutions for monocyclic polyenes with n carbon atoms*

From Equation 4.8, the energy of the lowest orbital is always $\varepsilon_0 = \alpha + 2\beta$; that of the highest is $\alpha - 2\beta$ if $n$ is even. The other orbitals are twofold degenerate, $\varepsilon_j = \varepsilon_{n-j}$, because the cosine function is symmetric. Equation 4.8 can be recast graphically, by inscribing the $n$-polygon into a circle of radius $2\beta$ (Figure 4.4).

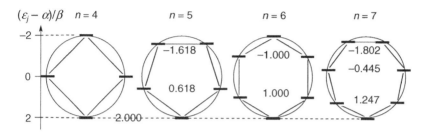

**Figure 4.4** Frost–Musulin diagrams for the HMO energies of monocyclic systems

Systems such as cyclopentadienyl anion, benzene and tropylium cation with $(4N + 2)$ $\pi$-electrons ($N = 0, 1, 2, \ldots$) will thus have a *closed shell*, that is, four electrons in the degenerate pair of the highest occupied MOs in their ground configuration, and that is especially favourable energetically. This is the basis of the well-known *Hückel rule of aromaticity*.

Having determined the HMO orbitals and filled in the appropriate number of $\pi$-electrons, we can characterize the resulting configuration by specifying the *occupation number* $b_j$ of each orbital $\psi_j$; $b_j = 2$ for doubly occupied, $b_j = 1$ for singly occupied and $b_j = 0$ for unoccupied orbitals. The *charge density* $q_\mu$ of the $\pi$-electrons of that configuration on atom $\mu$ is then given by Equation 4.9.

$$q_\mu = \sum_j^n b_j c_{j\mu}^2$$

**Equation 4.9** *HMO charge density*

The *bond order* $p_{\mu\nu}$ between atoms $\mu$ and $\nu$ is given by Equation 4.10.

$$p_{\mu\nu} = \sum_j^n b_j c_{j\mu} c_{j\nu}$$

**Equation 4.10** *HMO bond order*

The bond order between bound pairs of atoms in cyclobutadiene (**3**) is 0.5 and the charge density equals unity on all atoms. On average there is exactly one $\pi$-electron on each carbon atom.

## 4.3 HMO Perturbation Theory

HMO theory gives particularly simple and intuitively appealing results upon application of *Rayleigh–Schrödinger perturbation theory* and we shall take advantage of this to interpret trends and make predictions (see, in particular, Section 4.6).[d] The equations for first- and second-order perturbation given below are derived in the Appendix (Section 4.11).

The Hückel operator $\hat{H}^{HMO}$ of a perturbed system can be expressed as a sum of the operator of the parent molecule, $\hat{H}^0$, and a perturbation operator, $\hat{h}$ (Equation 4.11). As for $\hat{H}^{HMO}$ itself (Equation 4.3), we define the perturbation operator $\hat{h}$ by a list of matrix elements $h_{\mu\nu} = <\phi_\mu|\hat{h}|\phi_\nu>$.

$$\hat{H}^{HMO} = \hat{H}^0 + \hat{h}$$
$$h_{\mu\mu} = \delta\alpha_\mu[\text{for inductive perturbation(s) at atom(s) } \mu]$$
$$h_{\mu\nu} = h_{\nu\mu} = \delta\beta_{\mu\nu}[\text{for resonance perturbation(s) between atoms } \mu \text{ and } \nu]$$
$$h_{\mu\nu} = h_{\mu\mu} = 0 \text{ otherwise}$$

***Equation 4.11*** *A list of suggested HMO perturbation parameters $\delta\alpha_\mu$ and $\delta\beta_{\mu\nu}$ for heteroatoms is given in Table 8.5*

Changes in the resonance integrals between atoms $\mu$ and $\nu$ due to bond length alternation or twisted double bonds can be simulated using Equation 4.12.

$$\beta_{\mu\nu} = \beta e^{-A(r-r_0)}, A \approx 0.02 \text{ pm}^{-1}, r_0 = 140 \text{ pm (for benzene)}$$
$$\beta_{\mu\nu} = \beta\cos(\varphi), \text{ where } \varphi \text{ is the angle of twist around the } \pi\text{-bond}$$

***Equation 4.12*** *Variation of $\beta_{\mu\nu}$ with bond length and twist* [272]

First-order perturbation uses the Hückel orbital coefficients $c_{j\mu}$ of a given parent system to predict the shifts $\delta\varepsilon_j$ on the corresponding orbital energies $\varepsilon_j$. The shift induced by an inductive perturbation $\delta\alpha_\mu$ is determined by the probability $c_{j\mu}^2$ that an electron in orbital $\psi_j$ resides on atom $\mu$ (Equation 4.13).

$$\delta\varepsilon_j^{(1)} = c_{j\mu}^2\delta\alpha_\mu$$

***Equation 4.13*** *First-order perturbation of the MO energy $\varepsilon_j$ by an inductive effect $\delta\alpha_\mu$ introduced on atom $\mu$*

An example of the inductive effect of 2-methyl substitution of buta-1,3-diene is shown in Figure 4.5. For reasons that will become clear shortly, the occupied orbitals are numbered $1, 2, \ldots$ in order of decreasing energy and the unoccupied orbitals are numbered $-1, -2, \ldots$ in order of increasing energy.

A change in the resonance integral between atoms $\mu$ and $\nu$, $\delta\beta_{\mu\nu}$, will change the energy $\varepsilon_j$ by the amount given in Equation 4.14. For operators $\hat{h}$ that contain several elements $\neq 0$, the *first-order perturbations are additive*.

---

[d] In the early days of quantum mechanics, perturbation theory was used mostly because it requires less computational effort than calculating each system separately. In the computer age, this is no longer a valid justification.

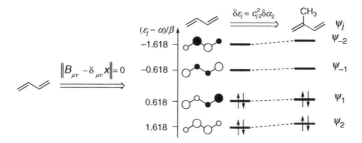

**Figure 4.5** Inductive effect of 2-methyl substitution of buta-1,3-diene using $\delta\alpha_2$ $(CH_3) = -0.4\beta$. The numerical values of the HMO coefficients for butadiene are either $|c_{j\mu}| = 0.372$ or $0.602$

$$\delta\varepsilon_j^{(1)} = 2c_{j\mu}c_{j\nu}\delta\beta_{\mu\nu}$$

**Equation 4.14** First-order perturbation of the MO energy $\varepsilon_j$ by a change in the resonance integral $\beta_{\mu\nu}$ by $\delta\beta_{\mu\nu}$

An example of the conjugative effect of partial 1,4-bonding in cisoid butadiene is shown in Figure 4.6.

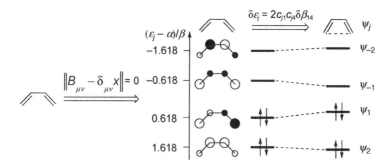

**Figure 4.6** Conjugative effect of 1–4 interaction in cisoid buta-1,3-diene using $\delta\beta_{14} = 0.3\beta$

Perturbation theory gives reliable results only as long as the perturbation $\delta\varepsilon_j^{(1)}$ of a given orbital is small compared with the difference between its energy $\varepsilon_j$ and that of all other orbital energies $\varepsilon_i$, $|\varepsilon_j - \varepsilon_i| \gg \delta\varepsilon_j^{(1)}$. To account for perturbations on degenerate orbitals, $\varepsilon_i = \varepsilon_j$, we need to solve the corresponding secular equation, that is, a $2 \times 2$ determinant for two degenerate MOs. But how small is small when $\varepsilon_i \neq \varepsilon_j$? Moreover, is it justifiable to use the HMO eigenfunctions of the parent system to calculate changes in the energies of the perturbed system? To answer these questions, we must go one step further and allow for variation in the wavefunctions upon perturbation. This requires some more work, but it will expose the limitations of first-order perturbation theory. Being aware of these limitations, one can then in most, but not all, cases go back to use the simpler first-order theory with confidence.

The energy shift of orbital $\psi_j$ calculated to second order is given by Equation 4.15. *Second-order perturbations are not additive for operators $\hat{h}$ that contain several elements $\neq 0$.*

$$\delta\varepsilon_j^{(2)} = \delta\varepsilon_j^{(1)} + \sum_{i \neq j}^{n} \frac{\langle\psi_i|\hat{h}|\psi_j\rangle^2}{\varepsilon_j - \varepsilon_i}$$

**Equation 4.15**   *Second-order perturbation of the MO energy $\varepsilon_j$*

The perturbation matrix elements $<\psi_i|\hat{h}|\psi_j> = h_{ij}$ are determined from the eigenfunctions $\psi_i$ and $\psi_j$ of the unperturbed system, making use of the parameter list given in Equation 4.11 for the elements $h_{\mu\nu}$ in the AO basis. For example, for a single inductive perturbation at atom $\mu$, $\delta\alpha_\mu$, we obtain $h_{ij} = h_{ji} = c_{i\mu}c_{j\mu}\delta\alpha_\mu$ and for a single resonance perturbation between atoms $\mu$ and $\nu$, $\delta\beta_{\mu\nu}$, $h_{ij} = h_{ji} = (c_{i\mu}c_{j\nu} + c_{j\mu}c_{i\nu})\delta\beta_{\mu\nu}$. If required, an improved set of MOs $\psi_j^{(1)}$ for the perturbed system can be determined from Equation 4.16.

$$\psi_j^{(1)} = \psi_j^{(0)} + \sum_{i \neq j}^{n} \frac{\langle\psi_i|\hat{h}|\psi_j\rangle^2}{\varepsilon_j - \varepsilon_i}\psi_i^{(0)}$$

**Equation 4.16**   *First-order perturbation of HMO wavefunctions*

Let us look at a simple case to illustrate the predictions of first- and second-order perturbation and compare them with the full HMO calculation for the perturbed system. We choose ethylene as the parent system and look at inductive perturbation induced by introduction of a substituent at atom 1 or by replacing atom 1 by a heteroatom such N or O. The exact solution is obtained by solving the $2 \times 2$ secular determinant (Equation 4.17).

$$\left|\left|\begin{array}{cc} -x+\delta\alpha_1/\beta & 1 \\ 1 & -x \end{array}\right|\right| = x^2 - x\frac{\delta\alpha_1}{\beta} - 1 = 0$$

$$x_{1,2} = \frac{\delta\alpha_1/\beta \pm \sqrt{(\delta\alpha_1/\beta)^2 + 4}}{2}, x = (\alpha - \varepsilon)/\beta$$

**Equation 4.17**   *Exact HMO calculation for and inductively perturbed ethylene*

To illustrate the accuracy of first- (Equation 4.13) and second-order (Equation 4.15) calculations, their results are compared with the exact solution (Equation 4.17) in Figure 4.7. The perturbation $\delta\alpha_1$ is varied from 0 to $3\beta$. The parameters for N (ethylenimine) and O (formaldehyde) are $\delta\alpha_\mu/\beta = 0.5$ and 1.0, respectively. If the term $(\delta\alpha_\mu/\beta)^2$ in the discriminant of Equation 4.17 is much smaller than 4 and can be neglected altogether, then Equation 4.17 is equal to the result of first-order perturbation theory (Equation 4.13). If $(\delta\alpha_\mu/\beta)^2$ is smaller than 4 but non-negligible, the square root of Equation 4.17 can be expanded, $(1 + x)^{1/2} = 1 + \frac{1}{2}x + \ldots$, and we have the result of second-order perturbation (Equation 4.15). Thus, the results of first-order perturbation are adequate for even a substantial perturbation such as the replacement of $CH_2$ by O in formaldehyde.

However, it must be realized that this result is due to the large difference in energy between the two HMOs of ethylene, $\varepsilon_1 - \varepsilon_{-1} = 2\beta$. The criterion for a 'small' perturbation $h_{ij}$ now becomes clear: In the general case, first-order perturbation is adequate for $(\delta\varepsilon_i)^2/(\varepsilon_i - \varepsilon_j)^2 \ll 1$ and second-order perturbation provided that $(\delta\varepsilon_i)^2/(\varepsilon_i - \varepsilon_j)^2 < 1$.

Clearly, first-order perturbation theory for degenerate orbitals cannot be done as described above, because the choice of any linear combination of degenerate

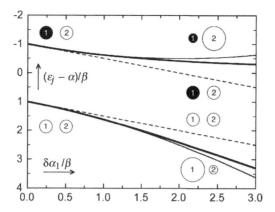

**Figure 4.7** Comparison of first- (dashed line), second-order (thin solid line) and full HMO calculations (thick solid line) for an inductive perturbation of ethylene. The radius of the circles indicates the size of the MO coefficients of $\psi_1$ and $\psi_{-1}$ on atoms 1 and 2, which remain unchanged upon first-order perturbation. The MOs obtained by second-order perturbation or by full HMO are concentrated more on the perturbed atom 1 in the bonding orbital $\psi_1$

wavefunctions is arbitrary (cf. e.g. Figures 4.2 and 4.3). A secular determinant including both degenerate orbitals must be solved. This can be avoided if we select a basis set that is adapted to the reduced symmetry of the perturbed system so that the off-diagonal elements for the interaction between the two degenerate orbitals vanish. That may not be worth bothering with when computers do the job but, more important, the result becomes much more transparent when we start from symmetry-adapted orbitals. We will deal with symmetry considerations in the next section and postpone the discussion of first-order perturbation of degenerate orbitals for now.

An important conclusion from perturbation theory is that *the introduction of a resonance interaction will always destabilize the upper orbital and stabilize the lower one.*[e] *For a given size of the perturbation $\delta\beta_{p\sigma}$, the shifts decrease as the energy gap increases* (Figure 4.8). Note that two basis AOs mix with equal weight when they are degenerate prior to interaction (left) and that the weight of the AO of the basis orbital with lower energy increases in the lower (bonding) MO as the energy gap between the basis orbitals increases (right).

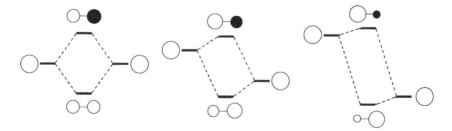

**Figure 4.8** Resonance interaction between two s basis AOs

---

[e] 'Because whoever has, to him will be given and he will have more; but from him who has not, even what he has will be taken away.' (Matthew 13 : 12).

## 4.4  Symmetry Considerations

*Recommended textbooks.*[134,273–275]

Symmetry considerations go a long way in predicting, *inter alia*, properties of wavefunctions, electronic states and transitions between molecular states. The mathematical treatment of symmetry operations is part of a special branch of mathematics known as *group theory*. Only a brief outline of some of its most important applications is given here.

A *symmetry operation* is a movement performed with a molecule – such as the reflection of its nuclear coordinates by a mirror plane $\sigma$ or rotation around an axis of symmetry $C$ – that leaves the molecule unchanged. If a molecule is symmetrical, that is, if it remains unchanged under some symmetry operation $\hat{S}$, its molecular orbitals $\psi_i$ must be either symmetric, $\hat{S}(-\psi_i) = \psi_i$, or antisymmetric, $\hat{S}(-\psi_i) = -\psi_i$, with respect to $\hat{S}$, because the electronic distribution given by $\psi_i^2 dv$ (Born interpretation, Figure 1.13) must remain unchanged, $\hat{S}(\psi_i^2) = \psi_i^2$. For example, seven different symmetry operations can be performed on *ethene* (Figure 4.9). Group theory includes an eighth operation, the *identity operation E*; it does nothing, like the identity operation 'multiplication by 1' in algebra. Any molecule or wavefunction will remain unchanged under the identity operation.

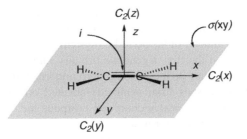

**Figure 4.9**  Symmetry operations

There are three *twofold symmetry axes* denoted $C_2(x)$, $C_2(y)$ and $C_2(z)$, which coincide with the coordinate axes $x$, $y$ and $z$. The origin of the coordinate system shown in Figure 4.9 is an *inversion centre i* of the molecule, that is, the molecule remains unchanged when the nuclear coordinates of the atoms $\mu$, $x_\mu$, $y_\mu$ and $z_\mu$, are replaced by $-x_\mu$, $-y_\mu$ and $-z_\mu$. Finally, there are three planes of symmetry, $\sigma(xy)$, $\sigma(xz)$ and $\sigma(yz)$, which coincide with the three planes spanned by a pair of coordinate axes [$\sigma(xy)$ is marked as a shaded plane]. Any molecule exhibiting the same set of symmetry operations belongs to the same *point group*, which is labelled by the *Schönflies symbol* $D_{2h}$ in the present case. The complete set of symmetry operations of an object defines the point group, to which it belongs. A flow chart to determine the point group of any molecule on the basis of its known (or assumed) 3D structure is shown in Figure 4.10. The combined operation of rotation *and* reflection through a plane perpendicular to an axis is an *improper rotation*; *n*-fold improper rotational axes are designated by the symbol $S_n$. Note that $S_1 \equiv \sigma$ and $S_2 \equiv i$.

The symbol $\sigma$ for symmetry planes is usually characterized by the index $v$ ('vertical', coincident with the rotational axis of highest order $n$) or $d$ ('dihedral' or 'diagonal', coincident with the rotational axis of highest order $n$ and bisecting the angle between

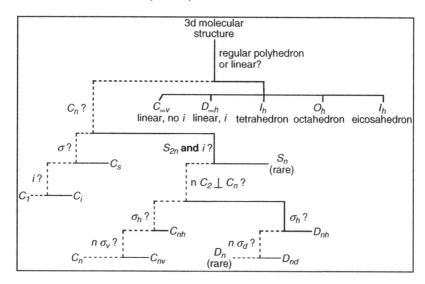

**Figure 4.10** Flow chart for the determination of point groups (Schönflies symbols, $n = 2, 3, \ldots$). To find the point group of a given object, answer questions working downwards; follow the full line to the right if the answer is Yes or the dashed line to the left if it is No

orthogonal axes as in allene) and $h$ ('horizontal', perpendicular to the rotational axis of highest order $n$).

The *character table* of a point group defines the symmetry properties of a (wave) function as either 1 for *symmetric* or $-1$ for *antisymmetric* with respect to each symmetry operation.[f] The first row lists all the symmetry operations of the point group and the first column lists the *Mulliken symbols* of all possible irreducible representations, the symmetry transformation properties that are allowed for wavefunctions. As an example, the character table for the $D_{2h}$ point group is given in Table 4.1. The character tables of all relevant point groups are given in many textbooks.[134,273–275] The last column shows the transformation properties of the axes $x$, $y$ and $z$, which are used to determine electronic dipole and transition moments (Section 4.5).

**Table 4.1**   Character table for the point group $D_{2h}$

| $D_{2h}$ | $E$ | $C_2(z)$ | $C_2(y)$ | $C_2(x)$ | $i$ | $\sigma(xy)$ | $\sigma(xz)$ | $\sigma(yz)$ | |
|---|---|---|---|---|---|---|---|---|---|
| $A_g$ | 1 | 1 | 1 | 1 | 1 | 1 | 1 | 1 | |
| $B_{1g}$ | 1 | 1 | $-1$ | $-1$ | 1 | 1 | $-1$ | $-1$ | |
| $B_{2g}$ | 1 | $-1$ | 1 | $-1$ | 1 | $-1$ | 1 | $-1$ | |
| $B_{3g}$ | 1 | $-1$ | $-1$ | 1 | 1 | $-1$ | $-1$ | 1 | |
| $A_u$ | 1 | 1 | 1 | 1 | $-1$ | $-1$ | $-1$ | $-1$ | |
| $B_{1u}$ | 1 | 1 | $-1$ | $-1$ | $-1$ | $-1$ | 1 | 1 | $z$ |
| $B_{2u}$ | 1 | $-1$ | 1 | $-1$ | $-1$ | 1 | $-1$ | 1 | $y$ |
| $B_{3u}$ | 1 | $-1$ | $-1$ | 1 | $-1$ | 1 | 1 | $-1$ | $x$ |

[f] Character tables of point groups of high symmetry ($n > 2$) have entries other than $\pm 1$, but we need not deal with such groups here.

We now return to *first-order perturbation theory for degenerate orbitals* (Section 4.3). As any linear combination of two degenerate orbitals $\psi_i$ and $\psi_j$ is equally valid, we set up a trial wavefunction function $\psi_k = a_{ki}\psi_i + a_{kj}\psi_j$ and we have to solve the secular Equation 4.18. The eigenvalues of the unperturbed system will be equal for all linear combinations, $\varepsilon_i^{(0)} = \varepsilon_j^{(0)} = \varepsilon_k^{(0)} = \varepsilon^{(0)}$.

$$\left\| \begin{matrix} \varepsilon^{(0)} + \delta\varepsilon_k^{(1)} - \varepsilon & h_{kl} \\ h_{lk} & \varepsilon^{(0)} + \delta\varepsilon_l^{(1)} - \varepsilon \end{matrix} \right\| = 0$$

**Equation 4.18**   *First-order perturbation of two degenerate orbitals*

If we choose the linear combination of the degenerate orbitals such that one is symmetric and the other antisymmetric with respect to a symmetry element that is retained in the perturbed molecule, then the off-diagonal elements will vanish, $h_{kl} = h_{lk} = 0$, because integration of any function over the complete range of its variable $x$ vanishes if the function is *antisymmetric*: $\int f(x)dx = 0$ (Figure 4.11). A function is said to be antisymmetric (*a*) if $f(-x) = -f(x)$. The product of two antisymmetric functions is symmetric (*s*); we write $a \otimes a = s$; similarly, $s \otimes s = s$ but $a \otimes s = s \otimes a = a$.

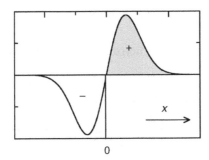

**Figure 4.11**   Integral of an antisymmetric function

Having chosen a symmetry-adapted basis set $\psi_k$ and $\psi_l$, the $2 \times 2$ determinant (Equation 4.18) is diagonal. The diagonal elements are the solutions, $\delta\varepsilon_\pm^{(1)} = \delta\varepsilon_k^{(1)}$ and $\delta\varepsilon_l^{(1)}$, with the associated symmetric and antisymmetric wavefunctions $\psi_k$ and $\psi_l$. The first-order perturbations of the energies $\varepsilon_k^{(0)}$ and $\varepsilon_l^{(0)}$ are thus simply those given by the equations for first-order perturbation of non-degenerate orbitals (Equations 4.13 and 4.14). The same result would have been obtained by solving the $2 \times 2$ non-diagonal determinant (Equation 4.18) obtained with any other linear combination. *The equations for first-order perturbation can be applied to degenerate MOs of the parent system when the linear combination of the degenerate orbitals is chosen such that it is adapted to the symmetry of the perturbed system.*

As an example, consider the effect of bond length alternation in cyclobutadiene (**3**). Let us assume that the molecule relaxes from a quadratic structure with an average bond length of 140 pm to one of the Kekulé structures with the double bonds connecting, say, atoms 1–2 and 3–4 and that the length of the double bonds is reduced to 130 pm and that of the single bonds is increased to 150 pm. From Equation 4.12, we get $\beta_{12} = 1.22\beta$ and $\beta_{23} = 0.82\beta$, that is, $\delta\beta_{12} = \delta\beta_{34} = 0.22\beta$ and $\delta\beta_{23} = \delta\beta_{41} = -0.18\beta$. We choose the set of

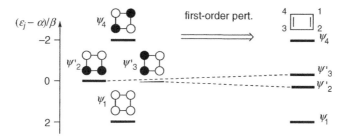

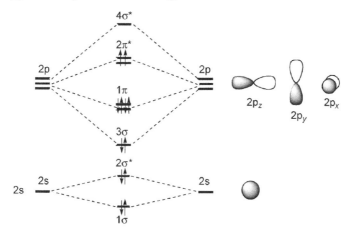

**Figure 4.12** Effect of bond length alternation in cyclobutadiene (3)

nonbonding orbitals $\psi'_2$ and $\psi'_3$ shown in Figure 4.12 (see also Figure 4.3), because these have the proper symmetry to deal with the perturbation. First-order perturbations are additive; adding up the contributions (Equation 4.14) by each bond gives $\delta\varepsilon_+^{(1)} = 0.40\beta$ and $\delta\varepsilon_-^{(1)} = -0.40\beta$.

Let us apply the same procedure to construct a qualitative MO diagram for the dioxygen molecule starting from the valence shell AOs of the separated O atoms, 2s, $2p_x$, $2p_y$ and $2p_z$. The MOs are constructed by choosing symmetry-adapted linear combinations of the basis AOs, for example $\sigma_\pm = (2s_1 \pm 2s_2)/\sqrt{2}$, $\pi_{z\pm} = (2p_{z1} \pm 2p_{z2})/\sqrt{2}$, and so on. The overlap between the $2p_z$ orbitals is largest, leading to the large energy difference between the $3\sigma$ and $4\sigma^*$ MOs. The well-known diagram illustrated in Figure 4.13 results, showing that the HOMO $2\pi^*$ is twofold degenerate and holds only two electrons. Hence the ground state of dioxygen is a triplet state according to Hund's rule.

**Figure 4.13** MO diagram for dioxygen

## 4.5 Simple Quantum Chemical Models of Electronic Excitation

*Recommended textbook.*[15]

Let us assume that organic molecules in their electronic ground state are reasonably well represented by a closed-shell *electronic configuration* built by filling electrons pairwise into the lowest energy HMOs in accord with the *Pauli principle*. Excited configurations can then be constructed by promoting an electron to an unoccupied orbital (Figure 4.1). In terms of Hückel theory, the excitation energy required to promote an electron from an orbital $\psi_i$ to an unoccupied orbital $\psi_j$ is equal to the difference of the corresponding orbital energies (Equation 4.19).

$$\varepsilon_j - \varepsilon_i = (x_j - x_i)\beta = hc\tilde{\nu}$$

**Equation 4.19**   *HMO excitation energy*

Excitation from the HOMO ($\psi_1$), the highest occupied molecular orbital, to the LUMO ($\psi_{-1}$), the lowest unoccupied molecular orbital, of buta-1,3-diene (Figure 4.5) requires $-1.236\beta$ ($\beta$ is a negative energy). We can use this result to obtain a first calibration of $\beta$. The absorption maximum of the first $\pi,\pi^*$ transition of buta-1,3-diene lies at $\tilde{\nu}_{max} = 4.6\,\mu m^{-1}$; hence $-\beta/hc \approx 3.7\,\mu m^{-1}$.

To calculate the transition moment $M_{el,n \to m} = e<\Psi_{el,n}|\Sigma r_i|\Psi_{el,m}>$ (Equation 2.17), we represent the ground-state wavefunction $\Psi_{el,n}$ by the closed-shell configuration $\chi_0 = \psi_1^2\psi_2^2 \ldots \psi_i^2 \ldots \psi_{HOMO}^2$ and the excited-state wavefunction $\Psi_{el,m}$ by $\chi_{i \to j} = \psi_1^2\psi_2^2 \ldots \psi_i \ldots \psi_j$. Integration over the coordinates of those electrons that occupy the same orbital in $\chi_0$ and $\chi_{i \to j}$ gives unity, $<\psi_i|\psi_i> = 1$, so the transition moment simplifies to $M_{el,i \to j} = e<\psi_i|\Sigma r_i|\psi_j>$, which involves only the orbitals $\psi_i$ and $\psi_j$ that participate in the transition. The operator $\Sigma r_i = \Sigma(x_i, y_i, z_i)$ and the electronic transition moment $M_{el,i \to j}$ are vector quantities; the coordinates $x_i$, $y_i$ and $z_i$ define the positions of the electrons $e_i$. To obtain the individual components of the electronic transition moment vector (Equation 4.20), we must integrate the products $\psi_i u \psi_j = u \psi_i \psi_j$, where $u$ stands for $x$, $y$ or $z$. The coordinates $u$ themselves are antisymmetric, so that $M_{el,i \to j}$ will vanish, unless the transition density $\psi_i \psi_j$ is itself antisymmetric with respect to one of the coordinates $u$.

$$M_{el,i \to j} = e(<\psi_i|x|\psi_j>, <\psi_i|y|\psi_j>, <y_i|z|y_j>)$$

**Equation 4.20**   *Electronic transition moment for a one-electron transition $\psi_i \to \psi_j$*

Consider the $\pi,\pi^*$ transition of ethene. The HMOs of ethene are $\pi = (\phi_1 + \phi_2)/\sqrt{2}$ and $\pi^* = (\phi_1 - \phi_2)/\sqrt{2}$. The transition density, shown on the right-hand side of Figure 4.14, is the product $\pi\pi^* = \frac{1}{2}(\phi_1 + \phi_2)(\phi_1 - \phi_2) = \frac{1}{2}(\phi_1^2 - \phi_2^2)$; this is easy to calculate, but the result is hard to swallow. Taken literally, it indicates a 50% chance that an electron is located in the left-hand $p_z$-AO $\phi_1$ and an equal chance that *minus* an electron is in the right-hand $p_z$-AO $\phi_2$. Do not ask about the physical meaning of having 'minus an electron' at

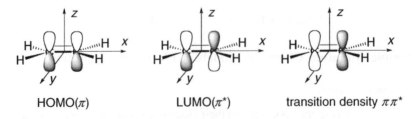

HOMO($\pi$)                    LUMO($\pi^*$)                    transition density $\pi\pi^*$

**Figure 4.14**   HMOs of ethene

some place in space! The transition density results from the quantum mechanical interference of two wavefunctions and has no meaningful classical analogy. The total charge is neutral, as it should be. The transition moment has a finite component in the $x$-direction, but vanishes in the $y$- and $z$-directions. Inserting the transition density into Equation 4.20 gives $M_{el,\pi\pi^*} = e[^1/_2(x_1 - x_2), 0, 0]$, where $x_1$ and $x_2$ are the $x$-coordinates of the carbon atoms 1 and 2. With Equation 2.19, the oscillator strength is predicted to be $f = 0.3$ for the $\pi,\pi^*$ transition of ethene.

Symmetry considerations (Section 4.4) lead to the same conclusion. The transformation properties of the $\pi$- and $\pi^*$-orbitals of ethene are determined by inspection of Figure 4.14. All orbitals are symmetric (1) with respect to the identity operation. Under $C_2(z)$ the $\pi$-orbital is symmetric (1), the $\pi^*$-orbital is antisymmetric $(-1)$, and so on. This leads to the row vectors $(1, 1, -1, -1, -1, -1, 1, 1)$ for the $\pi$-orbital and $(1, -1, 1, -1, 1, -1, 1, -1)$ for the $\pi^*$-orbital. The $\pi$-orbital (Table 4.1) thus belongs to the irreducible representation $B_{1u}$ and the $\pi^*$-orbital to $B_{2g}$. The vector obtained for the transition density $\pi\pi^*$ $(1, -1, -1, 1, -1, 1, 1, -1)$ belongs to the representation $B_{3u}$, which transforms like the $x$-axis (see the last column of Table 4.1), indicating that the transition moment lies parallel to the $x$-axis. The above exercise may appear to be rather futile, given that we had arrived at the same conclusion by simple inspection of the HMO orbitals (Figure 4.14). Note, however, that symmetry arguments do not require any explicit calculation at all. The irreducible representations, to which MOs constructed from a given set of basis orbitals must belong, can be determined from group theory alone.[134,273,274,275] Electronic transitions whose transition moment is zero in all directions because the transition density belongs to an irreducible representation that does not transform like any of the symmetry axes are said to be *symmetry forbidden*.

Is Hückel theory any good at predicting absorption spectra and transition moments? The answer is a mixed bag: yes and no. To begin with, Hückel theory is blind to the energy difference between singlet and triplet configurations because it ignores electron repulsion altogether. Moreover, we shall see that for several classes of molecules the lowest excited singlet state is *not* adequately described by the lowest singly excited configuration. Consider the cyclic systems with $(4N + 2)$ p-electrons $(N = 1, 2, 3, \ldots)$, which are particularly stable ('aromatic') in the ground state (*Hückel rule*, Section 4.2). HMO theory predicts that all four lowest-energy transitions in such systems have the same energy. In fact, these transitions are split into three absorption bands of substantially different energy in benzene (the 'benzene catastrophe') and two in cyclopentadienyl cation. Our goal is to understand where HMO theory works and when we need to upgrade the simple model (Section 4.7).

Let us take the good news first. HMO performs fairly well in predicting the position and intensity of certain types of bands in many classes of compounds. From Equation 4.7, the HOMO–LUMO energy gap of linear polyenes (ethene, buta-1,3-diene, hexa-1,3,5-triene, ...) can be expressed by Equation 4.21.[g]

$$\varepsilon_{LUMO} - \varepsilon_{HOMO} = -4\beta\sin[\pi/(2n+2)] \approx -2\beta\pi/(n+1)$$

***Equation 4.21***

---

[g] Recall that the index $j$ of the LUMO is $n/2 + 1$ and that of the HOMO is $n/2$. Moreover, $\cos\alpha = \sin(\alpha + \pi/2)$, $\sin\alpha = \sin(\pi/2 - \alpha)$ and finally that $\sin\alpha \approx \alpha$ is a good approximation even for the smallest member, ethene, with $n = 2$.

The wavelength of a transition is directly proportional to the inverse of its transition energy; hence HMO theory predicts that the absorption wavelengths of linear polyenes are linearly related to the number of carbon atoms $n$ (Equation 4.22).

$$\lambda_{0-0} \approx -hc\frac{n+1}{2\beta\pi}$$

***Equation 4.22*** *HMO prediction for the wavelength of the HOMO–LUMO transition of linear polyenes*

In Figure 4.15, the observed wavelengths $\lambda_{0-0}$ for eight all-*trans* linear polyenes[276] ($\times$) are plotted versus $n$. Although $\lambda_{0-0}$ does increase with increasing length of the polyenes, the result is disappointing, because the observed relationship is clearly nonlinear; the absorption wavelengths reach a plateau for very long polyenes. Extrapolating linearly from the first three members of the series, we would predict that carrots are blue ($\beta$-carotene is an all-*trans* polyene with 11 conjugated double bonds).

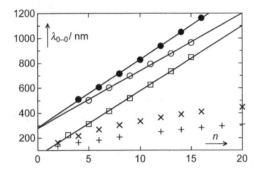

**Figure 4.15** Absorption bands of conjugated linear chains. The compound series are defined in the text together with the corresponding data sources

The deviations from Equation 4.22 can be traced to the alternation of bond lengths of linear polyenes. Even stronger effects of bond length alternation are evident in the series of polyacetylenes[277] ($+$). We could use Equation 4.12 to adjust the resonance integrals $\beta$ to the bond lengths of the essential single and double or triple bonds. Rather than introducing two new parameters using Equation 4.12 – which would fix the problem – let us see whether the HMO prediction of Equation 4.22 works for linear conjugated systems that do *not* exhibit bond length alternation. Indeed it does. The symmetrical cyanines[278] ($\square$), which possess two equivalent resonance structures $Me_2N-(CH=CH)_{(n/2-3)}-CH=N^+Me_2$ and $Me_2N^+=(CH-CH)_{(n/2-3)}=CH-NMe_2$ and thus no alternating bond lengths, accurately obey the predicted linear relationship. The slope corresponds to $-\beta/hc = (3.05 \pm 0.05)\,\mu m^{-1}$. The first absorption band of the linear chains $HC_{2n}H^+$ (radical cations, $\bullet$) and $HC_{2n+1}H^+$ (triplet ground state, $\circ$) and of the linear carbon chains $C_n$ ($n=$ odd, not shown) and $C_n$ ($n=$ even, not shown) also increase linearly with $n$.[241]

Systematic studies of the absorption spectra of benzenoid aromatic hydrocarbons, mostly done by Clar in the 1930s and 1940s,[279] showed that these compounds exhibit four types of UV–VIS absorption bands, which are shifted in a regular way along a homologous series such as the linear acenes (benzene, naphthalene, anthracene, . . .) (Figure 4.16).

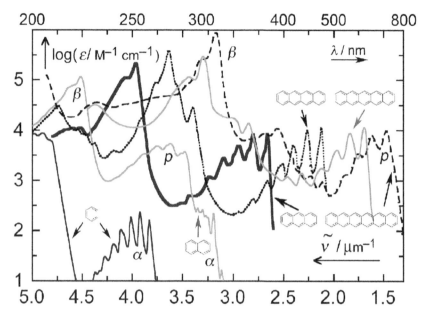

**Figure 4.16** Absorption spectra of the linear acenes[280,281]

Clar labelled the four bands $\alpha$, $p$, $\beta$ and $\beta'$ (in order of increasing energy for naphthalene, the $\beta'$ band of which is slightly above 5 µm$^{-1}$) and empirically assigned them in well over 100 benzenoid hydrocarbons on the basis of their characteristics mentioned below and their regular shifts in homologous series. The $\alpha$ band is weak, $\log(\varepsilon/\,\mathrm{M}^{-1}\mathrm{cm}^{-1}) \approx 2\text{–}3$, with sharp, irregular vibronic structure. Its position is insensitive to temperature, solvent polarity and polarizability. Its intensity, but not its position, increases strongly upon substitution, for example by halogens or alkyl groups. The $p$ band is of moderate intensity, $\log(\varepsilon/\,\mathrm{M}^{-1}\mathrm{cm}^{-1}) \approx 4$. It exhibits a distinct vibrational progression of $\sim$1400 cm$^{-1}$ and is strongly red shifted in polar solvents and with increasing size of the hydrocarbon. The strong $\beta$ band, $\log(\varepsilon/\mathrm{M}^{-1}\mathrm{cm}^{-1}) \approx 5\text{–}6$, is more diffuse and less sensitive to solvent and substitution. Finally, the $\beta'$ band, $\log(\varepsilon/\mathrm{M}^{-1}\mathrm{cm}^{-1}) \approx 4\text{–}5$, is less prominent and not always clearly discernible, because further transitions appear in the far-UV region. The $\alpha$ band is clearly seen as the first absorption band of benzene and naphthalene, then hidden under the stronger $p$ band in anthracene and tetracene, until it resurfaces, showing minute, relatively sharp features in the window between the $p$ and the $\beta$ band in pentacene and hexacene.

The $p$ bands correlate well with HOMO–LUMO energy gaps predicted by Hückel theory (Equation 4.16) (Figure 4.17). Linear regression gives $-\beta/hc = 2.2 \pm 0.1\ \mu\mathrm{m}^{-1}$. The points ($\circ$) for azulene (**1**) and cycl[3.3.3]azine (**2**) fall completely off the regression line (cf. Case Study 4.1). They are shown to emphasize that the correlation holds only for benzenoid hydrocarbons.

Excellent correlation was also found between first ionization potential $I_{1,\mathrm{v}}$ and the HOMO energies obtained by a slightly modified Hückel calculation for a large series of benzenoid hydrocarbons (*Koopmans' theorem*, Section 4.7). The ionization potentials $I_{1,\mathrm{v}}$ are also strongly correlated with the energies of the $^1L_a$ bands, as would be expected on the basis of the pairing theorem (Section 4.6). A few outliers were noted, but a reinvestigation

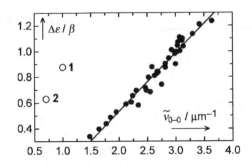

**Figure 4.17**  Correlation of the *p*-band positions of benzenoid aromatic hydrocarbons with HMO HOMO–LUMO gaps. Data taken from ref. 282, except for the outliers azulene (**1**) and cycl[3.3.3]azine (**2**) that do not belong to the same class of compounds

showed that their assumed structure had been incorrect.[283] It is amazing that the structural identification of some of the higher members of these polynuclear hydrocarbons relies entirely (yet convincingly) on their electronic and photoelectron spectra. For example, an earlier X-ray structure of 'circumanthracene' was shown to be incorrect, because the crystals are strongly disordered.[283]

## 4.6  Pairing Theorems and Dewar's PMO Theory

*Recommended reading.*[284,285]

Conjugated hydrocarbons that do not contain an odd-membered ring are called *alternant hydrocarbons* (AHs). The distinction between *alternant* and *non-alternant* hydrocarbons (NAHs) provides a very important classification of conjugated hydrocarbons, especially with regard to excited states. In AHs, the unsaturated C atoms can be assigned to two sets, the starred (*) and the unstarred (o) set, such that no atoms of the same set are bound to each other. This is not possible for NAHs (Figure 4.18).

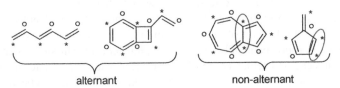

alternant                                          non-alternant

**Figure 4.18**  Alternant and non-alternant hydrocarbons

Coulson and Rushbrooke have derived several theorems and corollaries for AHs,[286] which may be summarized as follows:

1. HMOs with $\varepsilon_j \neq \alpha$ ($x_j \neq 0$) occur in pairs with energies $\varepsilon_{\pm j} = \alpha \pm x_j \beta$. The HMOs of buta-1,3-diene serve as an example. Note that orbital pairing remains intact when a bonding perturbation is introduced between atoms 1 and 4 (Figure 4.6), but not when an inductive perturbation is introduced (Figure 4.5). The HMO-coefficients $c_{j\mu}$ of paired

MOs ($\varepsilon_j \neq \alpha$) are numerically the same, $|c_{j\mu}| = |c_{-j\mu}|$, but they change sign on one set, that is, $c^o_{j\mu} = c^o_{-j\mu}$ and $c^*_{j\mu} = -c^*_{-j\mu}$. Whether the sign is changed in the starred or the unstarred set is irrelevant, because wavefunctions of opposite sign are equivalent. Thus, the energies and HMO coefficients of a bonding orbital $\psi_j$ and its paired antibonding orbital $\psi_{-j}$ are related through Equation 4.23 (cf. e.g. Figure 4.6).

$$\psi_j = \sum_{\mu^o} c^o_{j\mu}\phi^o_\mu + \sum_{\nu^*} c^*_j \phi^*_\nu \quad \varepsilon_j = \alpha + x_j\beta$$

$$\psi_{-j} = \sum_{\mu^o} c^o_{j\mu}\phi^o_\mu - \sum_{\nu^*} c^*_j \phi^*_\nu \quad \varepsilon_{-j} = \alpha - x_j\beta$$

**Equation 4.23**   *Paired orbitals*

2. Nonbonding molecular orbitals (NBMOs) are MOs with an energy $\varepsilon_j = \alpha$, $x_j = 0$. The number of NBMOs is at least equal to the absolute value of the difference between the numbers of atoms in the two sets. Thus odd AHs (AHs with an odd number of atoms) have at least one NBMO.

3. NBMOs must be 'self-paired', i.e., their coefficients must vanish on one set of atoms (the pairing condition $c^o_{j\mu} = -c^o_{-j\mu}$ requires that $c^o_{j\mu} = 0$). Thus an NBMO is confined entirely to one set of atoms, the larger one if they differ in size. The coefficients on atoms that are attached to an atom with $c_{j\mu} = 0$ must add up to zero.

4. The $\pi$-charge densities $q_\mu$ of neutral AHs (Equation 4.9) are unity on all atoms $\mu$.

5. The bond orders $p_{\mu\nu}$ between all atoms of one set of a neutral AH vanish (Equation 4.24).

$$\sum_j b_j c^*_{j\mu} c^*_{j\nu} = \sum_j b_j c^o_{j\mu} c^o_{j\nu} = 0$$

**Equation 4.24**

Rule 4 enables us to determine the coefficients of NBMOs without diagonalizing the Hückel matrix.[287] Start by assigning an arbitrary value of $a$ to one of the atoms with non-vanishing coefficients. Then assign multiples or fractions of $a$ to the other atoms in the same set, using rule 3 that the coefficients on atoms that are attached to an atom with $c_{\text{NBMO},\mu} = 0$ must add up to zero. Finally, the NBMO must be normalized (Equation 4.25), whereby the value of $a$ is defined.

$$\sum_\mu c^2_{\text{NBMO},\mu} = 1$$

**Equation 4.25**

As an example, we show the NBMO of the benzyl radical in Figure 4.19.

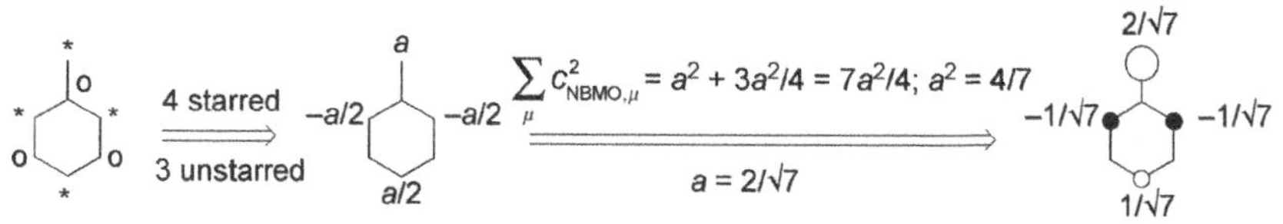

**Figure 4.19**   Back-of-an-envelope determination of the NBMO of benzyl

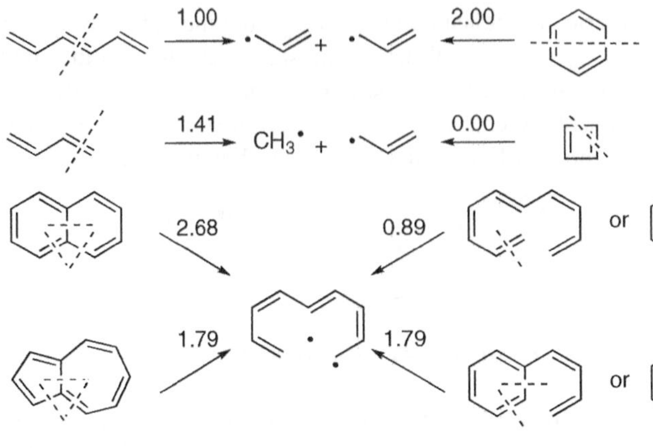

**Scheme 4.1**  *HOMO–LUMO energy gaps in β units determined by PMO theory*

Because the π-charge densities $q_\mu$ of neutral AHs (including odd AH radicals such as benzyl) are unity on all atoms $C_\mu$ (rule 5), the excess charges $q_\mu - 1$ of the benzyl anion and cation obtained by adding or removing an electron to or from the NBMO are equal to $-c_{NBMO,\mu}^2$ and $c_{NBMO,\mu}^2$, respectively.

The Coulson–Rushbrooke theorems have many important consequences that will lead us a long way towards a qualitative understanding of the electronic structure of conjugated molecules, particularly of their excited states. Dewar developed a simple perturbation method (*PMO theory*) to evaluate the HOMO–LUMO gap and the associated excitation energy for π-systems with an even number of conjugated atoms.[284,285,288] Because the NBMO of odd AHs can be determined so easily, the system of interest is dissected into two odd AHs. Some examples are shown in Scheme 4.1.

To calculate the HOMO–LUMO gap of the original system, the coefficients of the two fragment NBMOs 1 and 2 are determined. To recombine the NBMOs through the bonds that were cut, we must first choose the symmetry-adapted linear combinations $\psi_\pm = (\psi_{NBMO,1} \pm \psi_{NBMO,2})/\sqrt{2}$. The resulting energy changes are then calculated by the formula for first-order perturbation (Equation 4.14). First-order perturbations are additive, so if several bonds connecting atoms $\rho$ on fragment 1 with atoms $\sigma$ on fragment 2 are formed upon reunion, all energy changes are summed. The resulting HOMO–LUMO gaps (Equation 4.26) are shown above the corresponding arrows in Scheme 4.1. The HOMO–LUMO energy gaps are also equal to the total π-energy changes $\delta E_\pi$ when two odd AH fragments are united; according PMO theory, $\delta E_\pi$ is approximately equal to the stabilization of the two electrons in the NBMOs of the fragments.[284,285,288]

$$\varepsilon_{LUMO} - \varepsilon_{HOMO} = 2 \sum_{\rho - \sigma} c_{1\rho} c_{2\sigma} \beta_{\rho\sigma}$$

**Equation 4.26**  *HOMO–LUMO energy gaps predicted by PMO theory*

There is an ambiguity in using Equation 4.26 because there are several ways to divide AHs into two odd AHs and the resulting HOMO–LUMO gaps need not be the same. The best first-order estimate will come from the division that minimizes the NBMO splitting. In general, 'aromatic' topologies that strongly stabilize a π-electron system in the ground state feature a large splitting of the NBMOs. Such systems have large excitation energies. *Thus PMO theory* *predicts an approximate mirror-image relationship between the π-electron energies of molecules in the ground state and in the lowest excited state ($S_1$ or $T_1$).*[289]

Perturbation theory is especially well suited to predict the effect of substitution on electronic transitions. The shifts $\delta\varepsilon_j$ of individual orbital energies upon inductive or resonance perturbation are given in Section 4.3. For a transition that is described by single excitation from a bonding orbital $\psi_i$ to an unoccupied orbital $\psi_j$, the transition energy is given by $\varepsilon_j - \varepsilon_i$ (Equation 4.19), so the band shift calculated by first-order perturbation is easily obtained from Equation 4.27. An example is given in Case Study 4.1.

$$\delta\Delta\varepsilon = \delta\varepsilon_j^{(1)} - \delta\varepsilon_i^{(1)}$$

**Equation 4.27** *Inductive effect on transition energies*

If the transition considered is the HOMO–LUMO transition of an alternant hydrocarbon, then first-order theory predicts that inductive perturbation will have no effect at all, because $\delta\varepsilon_j^{(1)} = \delta\varepsilon_i^{(1)}$ as a consequence of the pairing theorem. Small red shifts are in fact observed that can be attributed to *hyperconjugation* with the pseudo-π MO of the saturated alkyl chain.[290] On the other hand, alkyl substitution gives rise to large shifts in the absorption spectra of radical ions of alternant hydrocarbons whose charge distribution is equal to the square of the coefficients of the MO from which an electron was removed (radical cations) or to which an electron was added (radical anions), and these shifts are accurately predicted by HMO theory.[291]

## 4.7 The Need for Improvement; SCF, CI and DFT Calculations

We have started by deleting all interaction terms between the electrons to obtain an 'approximate' Hamiltonian $\hat{H}^0$, because this allowed us to solve the resulting Schrödinger equation easily for fixed nuclear positions (Figure 1.10). The total neglect of electronic repulsion is clearly a poor approximation, but the resulting electronic wavefunctions can be used to calculate electronic repulsion as an afterthought. In the following, we retain the frozen core approximation, that is, we consider only the π-electrons of conjugated molecules. If Hückel MOs are taken as a starting point, reintroduction of the inter-electronic repulsion terms $1/r_{12}$ into the Hamiltonian operator (Section 1.4) results in two types of new contributions to the energy.

Consider the interaction of two electrons, $e_1$ and $e_2$, that are located in the AOs $\phi_\mu$ and $\phi_\nu$. We do not exclude the possibility that the two electrons are in the same AO, $\mu = \nu$, provided that they have opposite spin. The time-averaged distribution of electron 1 is given by $\phi_\mu^2(e_1)dv_1$ and that of electron 2 by $\phi_\nu^2(e_2)dv_2$ (Born interpretation). Therefore, the

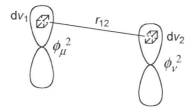

**Figure 4.20** Coulomb repulsion between the electron density in two volume elements

Coulomb repulsion between $e_1$ and $e_2$ is obtained by summing all repulsive contributions for electron $e_1$ in a small volume element $dv_1$ and of $e_2$ in $dv_2$ (Figure 4.20).

Integration over all space ($dv = dv_1 dv_2 = dx_1 dy_1 dz_1 dx_2 dy_2 dz_2$) yields the *Coulomb repulsion integral $J_{ij}$* for two electrons in MOs $\psi_i$ and $\psi_j$ (Equation 4.28).

$$J_{ij} = \left\langle \psi_i^2(e_1) \left| \frac{1}{r_{12}} \right| \psi_j^2(e_2) \right\rangle = \left\langle \left( \sum_\mu c_{i\mu} \phi_\mu(e_1) \right)^2 \left| \frac{1}{r_{12}} \right| \left( \sum_\nu c_{j\nu} \phi_\nu(e_2) \right)^2 \right\rangle$$

**Equation 4.28**   *Coulomb repulsion integral J*

The integral $J$ can be simplified to the double sum of two-centre Coulombic repulsion integrals $\gamma_{\mu\nu}$ (Equation 4.29) by adopting the *zero differential overlap* (ZDO) *approximation*, that is, we assume that atomic orbitals located on different atoms do not overlap, $\phi_\mu \phi_\nu = 0$ for $\mu \neq \nu$.

$$J_{ij} \approx \sum_\mu \sum_\nu c_{i\mu}^2 c_{j\nu}^2 \gamma_{\mu\nu}, \text{ where } \gamma_{\mu\nu} = \left\langle \phi_\mu^2(e_1) \left| \frac{1}{r_{12}} \right| \phi_\nu^2(e_2) \right\rangle$$

**Equation 4.29**

In semiempirical calculations, $\gamma_{\mu\mu} \approx 10.8\,\text{eV}$ is commonly used for two electrons occupying a p-orbital on the same carbon atom $\mu$ and the integrals $\gamma_{\mu\nu}$ for $\mu \neq \nu$ are usually estimated by a point charge approximation, $\gamma_{\mu\nu} \approx e^2/(r_{\mu\nu} + a)$, where $a = e^2/\gamma_{\mu\mu} = 133\,\text{pm}$, that is, $\gamma_{\mu\nu} \approx 1440/(r_{\mu\nu}/\text{pm} + 133)\,\text{eV}$.

We are not completely finished, however, because we have implicitly assumed that electrons are distinguishable (identifiable as $e_1$ and $e_2$), which is not possible in fact and therefore not allowed in quantum mechanics. A correct calculation using *Slater determinants* for electronic configurations (Section 4.1) yields a correction to the Coulomb repulsion $J_{ij}$ for electrons with parallel spin that has no classical analogy, the so-called *exchange integral $K_{ij}$* (Equation 4.30), which represents the Coulomb repulsion between the overlap densities $\psi_i \psi_j dv_1$ and $\psi_i \psi_j dv_2$.[h] Due to the traffic rule arising from the Pauli principle (Fermi hole, Section 1.4), the effective repulsion between electrons of parallel spin, $J_{ij} - K_{ij}$, is less than expected from classical physics. The approximate right-hand expression is again obtained with the ZDO approximation.

$$K_{ij} = \left\langle \psi_i(e_1) \psi_j(e_1) \left| \frac{1}{r_{12}} \right| \psi_i(e_2) \psi_j(e_2) \right\rangle \approx \sum_\mu \sum_\nu c_{i\mu} c_{j\mu} c_{i\nu} c_{j\nu} \gamma_{\mu\nu}$$

**Equation 4.30**   *Exchange integrals*

We have started with HMOs that are not eigenfunctions of a Hamiltonian $\hat{H}$ that includes electronic repulsion terms. However, we can now re-optimize the AO coefficients of the HMOs by solving the secular equations for the operator $\hat{H}$ instead of $\hat{H}^{HMO}$, that is, by setting the secular determinant equal to zero, $\|H_{ij} - \delta_{ij}\varepsilon\| = 0$ (Equation 1.16, Section 1.4). The MOs

---

[h] It is unfortunate that the reverse assignment of symbols has emerged in EPR spectroscopy, i.e. $J$ for the exchange integral and $K$ for the Coulomb integral. We retain the symbols that are generally used in the context of electronic spectra and are recommended by IUPAC. The integrals are given in atomic units (Table 8.3).

so obtained will differ from the HMOs used to calculate the integrals $J$ and $K$ and, conversely, calculation of the latter from the new MOs will give different values for $J$ and $K$. Hence the new MOs are still not proper eigenfunctions of $\hat{H}$ and the procedure must be repeated until the MO coefficients no longer change substantially. Once the orbitals are reproduced essentially unchanged by further reoptimization, we have reached a *self-consistent field* (SCF). The mathematical procedure described above is called the Hartree–Fock (HF) procedure. It must be done iteratively, because one needs some trial MOs beforehand to calculate the integrals $J$ and $K$ that, in turn, are needed to optimize the orbitals.

The orbital energies obtained by SCF theory are given by Equation 4.31, where the sum runs over all orbitals that are doubly occupied in the ground configuration and the so-called *core integrals* $h_{ii}$ are the energies of a single electron in MO $\psi_i$ in the field of the frozen core of nuclei and the σ-electrons (but not the other π-electrons). They correspond to HMO energies.

$$\varepsilon_i = h_{ii} + \sum_{j}^{\mathrm{occ}} (2J_{ij} - K_{ij})$$

*Equation 4.31  SCF orbital energies*

Equation 4.31 can be derived on an intuitive basis. Consider the following simple cases. The energy of an electron in a single doubly occupied orbital $\psi_i$ is $\varepsilon_i = h_{ii} + J_{ii}$, which is the same as $h_{ii} + 2J_{ii} - K_{ii}$ obtained from Equation 4.31, because $J_{ii} = K_{ii}$, as follows from the definitions, Equations 4.29 and 4.30. Noting that that the effective repulsion between electrons of parallel spin amounts to $J_{ij} - K_{ij}$, the energy of an electron in orbital $\psi_i$ in a system containing four electrons in orbitals $\psi_i$ and $\psi_j$ is $\varepsilon_i = h_{ii} + J_{ii} + 2J_{ij} - K_{ij}$.

Orbital energies are related to oxidation and reduction of molecules. Koopmans' theorem states that, on the basis of a single determinant SCF calculation for a closed-shell molecule, the orbital energies defined by Equation 4.31 are the best estimate for the vertical ionization energies, $I_{v,i} = -\varepsilon_i$, of that molecule in the gas phase, which are determined by photoelectron spectroscopy. Similarly, the energies of unoccupied orbitals are a measure of electron affinities, $E_{ea} = \varepsilon_j$, where $E_{ea}$ are the energies required to detach an electron from a singly charged negative ion.[i] SCF (and also HMO orbital energies) also correlate well with standard potentials for oxidation or reduction determined in solution (Figure 4.1).

The total energy of a given electronic ground configuration $\chi_0$ is given by Equation 4.32. Note that it is not simply twice the sum the energies of the occupied orbitals as in HMO theory, because that would count the interaction between each pair of electrons twice.

$$E(\chi_0) = \sum_{i}^{\mathrm{occ}} (h_{ii} + \varepsilon_i) = 2\sum_{i}^{\mathrm{occ}} h_{ii} + \sum_{i}^{\mathrm{occ}}\sum_{j}^{\mathrm{occ}} (2J_{ij} - K_{ij})$$

*Equation 4.32  SCF energy of a ground configuration*

An excited singlet or a triplet configuration can be built by promoting an electron from an occupied orbital $\psi_i$ to a unoccupied orbital $\psi_j$ (Figure 4.1). Their energies above that of the ground configuration are given by Equation 4.33, which follows from Equations 4.31 and 4.32.

---

[i] The energies of unoccupied ('virtual') orbitals obtained by *ab initio* calculations are, however, not suitable to estimate electron affinities and reduction potentials.

$$E({}^1\chi_{i\to j})-E({}^1\chi_0) = \varepsilon_j-\varepsilon_i-J_{ij}+2K_{ij}$$
$$E({}^3\chi_{i\to j})-E({}^1\chi_0) = \varepsilon_j-\varepsilon_i-J_{ij}$$

**Equation 4.33**   *Energies of singly excited configurations ${}^1\chi_{i\to j}$ and ${}^3\chi_{i\to j}$*

Thus, *the energy difference between identical singlet and triplet configurations amounts to twice the exchange integral between the orbitals $\psi_i$ and $\psi_j$ that are involved in the excitation* (Equation 4.34).

$$E({}^1\chi_{i\to j})-E({}^3\chi_{i\to j}) = 2K_{ij}$$

**Equation 4.34**   *Energy difference between singlet and triplet excited configurations*

This is an important result of SCF theory. In fact, we can estimate the size of the integral $K_{ij}$ fairly well using Hückel MOs, so that we usually do not need to do any SCF calculation at all! Equations 4.33 and 4.34 can again be justified intuitively, if one allows for a stabilizing contribution of $-K_{ij}$ for exchangeable electrons of the same spin and a destabilizing one of $+K_{ij}$ for exchangeable electrons of opposite spin, provided that their exchange is allowed by the Pauli principle. For example, the energies of the configurations $\chi_{i\to j} = \psi_i^1\psi_j^1$ are $E({}^1\chi_{i\to j})-E({}^1\chi_0) = h_{ii} + h_{jj} + J_{ij} + K_{ij}$ and $E({}^3\chi_{i\to j})-E({}^1\chi_0) = h_{ii} + h_{jj} + J_{ij}-K_{ij}$, hence $E({}^1\chi_{i\to j})-E({}^3\chi_{i\to j}) = 2K_{ij}$.

Let us note some important qualitative rules regarding the exchange integrals $K_{ij}$ that determine the energy gap $\Delta E_{ST}$ between ${}^1\chi_{i\to j}$ and ${}^3\chi_{i\to j}$. The Coulomb integral $J_{ij}$ falls off roughly as $1/\!<\!r_{ij}\!>$, as the orbitals $\psi_i$ and $\psi_j$ move apart to some 'average' distance $<\!r_{ij}\!>$ (Figure 4.21). The exchange integral representing an overlap repulsion falls of in an exponential manner, $\exp(-\!<\!r_{ij}\!>)$, that is, much more rapidly.

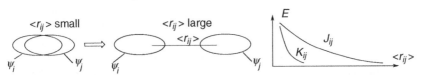

**Figure 4.21**   Distance dependence of Coulomb integrals $J_{ij}$ and exchange integrals $K_{ij}$

*Thus, for a transition with charge-transfer character (small $K_{ij}$), the singlet–triplet splitting will be small and the singlet absorption will be at longer wavelengths than expected from the HMO energy gap $\Delta\varepsilon$.* These conclusions follow immediately from Equations 4.34 and 4.33, respectively. Examples are radical ion pairs or the charge-transfer state in a covalently linked donor–acceptor molecule such as 4-amino-4′-nitrostilbene, and also n,π*-excited states of ketones (acetone or benzophenone), where the n-orbital is localized on the oxygen atom and the π*-orbital is distributed over a conjugated system and has a small coefficient on the oxygen atom. At the other extreme, the HOMO–LUMO π,π*-transitions of alternant hydrocarbons will have a very large singlet–triplet splitting and lie at relatively short wavelengths, because the HOMO and LUMO coefficients at a given atom have the same magnitude (Section 4.6), that is, the HOMO and LUMO are focused on the same atoms and $<\!r_{ij}\!>$ is small. This also explains the 'anomalous' properties (Figure 4.17) of azulene (**1**) and cycl[3.3.3]azine (**2**, Case Study 4.1).

## Case Study 4.1: Spectroscopy – electronic spectra and photophysical properties of azulene (1) and cycl[3.3.3]azine (2)

The absorption and fluorescence spectra of azulene (**1**) and cycl[3.3.3]azine (**2**) are shown in Figures 4.22 and 4.23. The fluorescence emissions are from the second excited singlet state $S_2$, in violation of Kasha's rule (Section 2.1.8), due to the large energy gap between the $S_1$ and $S_2$ states in these compounds.

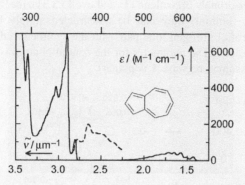

**Figure 4.22**   Absorption (adapted from[280]) and fluorescence (- - -) emission spectra of azulene (**1**)

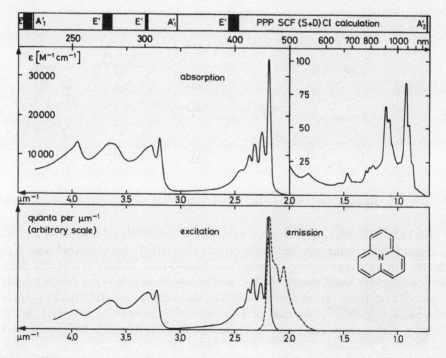

**Figure 4.23**   Absorption, fluorescence emission (- - -) and excitation spectra of cycl[3.3.3] azine (**2**). Reproduced by permission from ref. 36

The first absorption bands of **1** and **2** fall way off the regression line (Figure 4.17), which correlates the $p$ bands of benzenoid hydrocarbons with the corresponding HMO energy gaps $\Delta\varepsilon$: Anthracene is colourless ($\tilde{\nu}_{0-0} = 1.9\,\mu\text{m}^{-1}$), whereas **1** is blue ($\tilde{\nu}_{0-0} = 1.0\,\mu\text{m}^{-1}$),[292] although their HOMO–LUMO gaps are nearly the same ($\Delta\varepsilon = 0.828\beta$ and $0.877\beta$, respectively). Tetracene is yellow, $\Delta\varepsilon = 0.59\beta$, $\tilde{\nu}_{0-0} = 2.1\,\mu\text{m}^{-1}$, whereas the first band of **2** with $\Delta\varepsilon = 0.63\beta$ lies in the near-IR region, $\tilde{\nu}_{0-0} = 0.7\,\mu\text{m}^{-1}$. Why does Hückel theory fail to reproduce the twofold (**1**) and threefold (**2**) change in energy of the lowest excited singlet state?

The HMO frontier orbitals of azulene (**1**) and cycl[3.3.3]azine[44,45] (**2**) are shown in Figure 4.24. For **2** (bottom) they can be constructed simply by considering the interaction of the central nitrogen lone pair with the symmetry-adapted (Section 4.4) NBMOs of [12]annulene. The parameter for the central N atom is $\delta\alpha_{\text{N}} = 1.5\beta$ (Table 8.5; $\delta\beta_{\text{C-N}} = 0$ is assumed because **1** is planar).

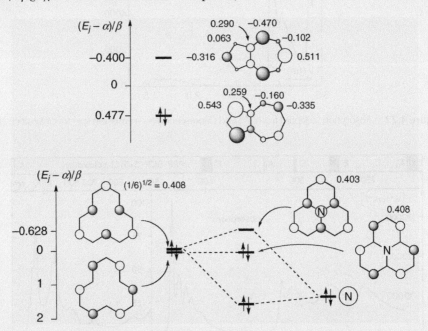

**Figure 4.24**   HMO frontier orbitals of **1** (top) and **2** (bottom)

The HOMO–LUMO gap is obtained as $0.63\beta$ by a full HMO calculation for **2** (first-order perturbation using only the frontier orbitals gives $0.69\beta$). Inspection of Figure 4.24 reveals that the local overlap between the HOMO and LUMO orbitals of these molecules is very small, that is, the MO coefficients are largely (**1**) or completely (**2**) separated to different sets of atoms. Therefore, the exchange integral $K_{\text{HOMO, LUMO}}$ is very small, $0.25\,\text{eV}^{292}$ in the case of **1** and essentially zero in the case of **2**. In the alternant hydrocarbons anthracene and tetracene, the HOMO and LUMO coefficients are large on the same atoms and the exchange integrals $K_{\text{HOMO, LUMO}}$ are large ($\sim 0.7\,\text{eV}$). The 'anomalies' of **1** and **2** (Figure 4.17) may thus be traced to HMO's total neglect of electronic interaction: From Equation 4.33, one would predict that the energies of their lowest excited singlet states would be lower by 0.9 and 1.4 eV or 0.7 and

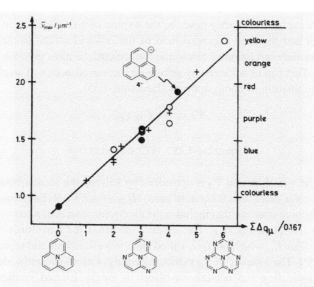

**Figure 4.25** Inductive perturbation of the $S_0 - S_1$ transition of **2**. Reproduced by permission from ref. 34

$1.1\,\mu m^{-1}$, respectively, than predicted by HMO theory. From Equation 4.34, we expect that these compounds have very small singlet–triplet energy gaps, in contrast to the alternant hydrocarbons anthracene and tetracene. Indeed, $\Delta E_{ST}(\mathbf{1}) = 5.4\,kJ\,mol^{-1}$,[293] and $\Delta E_{ST}(\mathbf{2}) \le 2\,kJ\,mol^{-1}$.[34]

Cyclazine is isoconjugate to the phenalenyl anion $\mathbf{4}^-$ (Figure 4.25). Phenalenyl radical $\mathbf{4}^{\cdot}$ is an alternant hydrocarbon with a charge density of 1 on all carbon atoms. The excess charge in $\mathbf{4}^-$ is given by the square of its NBMO, $q_\mu = 1/6 = 0.167$ on atoms $\mu = 1, 3$, and so on. Because HOMO–LUMO excitation is accompanied by substantial charge reorganization, one expects large shifts of the absorption bands upon substitution. For example, many aza-substituted cyclazines with N atoms in positions 1 and 3 are known. Using first-order perturbation theory (Equation 4.13), one predicts that each N atom will stabilize the HOMO by $0.408^2\,\delta\alpha_N = 0.25\beta$. The LUMO will not be affected, because $c_{1,LUMO} = c_{3,LUMO} = 0$. Indeed the position of the first absorption band of azacyclazines correlates well with the number of N atoms. It is shifted all the way from the near-IR region in parent $\mathbf{2}(\tilde{\nu}_{0-0} = 0.7\,\mu m^{-1})$ throughout the visible region up to $(\tilde{\nu}_{0-0} = 2.4\,\mu m^{-1})$ in 1,3,5,7,9,11-hexaazacycl[3.3.3]azine.[34]

SCF calculations provide improved MOs, because the average electronic repulsion is minimized, but the correlation of electronic movements is still neglected. *Ab initio* SCF calculations using the HF procedure do not adopt the ZDO and frozen core approximations mentioned above. Nevertheless, the energies calculated by *ab initio methods* for single-determinant wavefunctions are still much too high, even if an 'infinitely' large basis set of AOs, referred to as the *Hartree–Fock limit*, is used to optimize the molecular orbitals. The excess relative to the true energy is called the *correlation error* and amounts to roughly 1 eV per electron pair. Although we have accounted for electronic repulsion by calculating the best orbitals for a single electron in an averaged field of all the other electrons and nuclei, we have not considered their movements explicitly, having retained the notion of MOs that describe the motion of a single electron. SCF wavefunctions are still essentially single-electron wavefunctions; they describe the motion of a single electron in the field of stationary nuclei and the time-averaged field of the other electrons in the molecule. In order to allow for the correlation of electronic movement, we must give the wavefunction more flexibility. This can be achieved by setting up the wavefunction of an electronic state $J$ as a linear combination of configurations (Equation 4.35).

$$\Psi_J = \sum_K C_{JK}\chi_K$$

**Equation 4.35**  *A CI wavefunction*

Once again, the wavefunction $\Psi_J$ is optimized by solving the secular equations, that is, by setting the secular determinant equal to zero, $\|H_{JK} - \delta_{JK}E\| = 0$ (Equation 1.16, Section 1.4). The Hamiltonian operator that includes the electronic interaction terms $1/r_{ij}$ is used to calculate the matrix elements $H_{JK}$. The index $J$ identifies the wavefunctions associated with each eigenvalue $E_J$, which are now called state wavefunctions and represented by the capital symbol $\Psi_J$. The index $K$ of $\chi_K$ marks the configurations formerly identified by the symbol $i \rightarrow j$ indicating the orbitals involved in singly excited configurations. The procedure is called *configuration interaction* (CI or SCI, if only singly excited configurations are included for optimization). Of course, multiply excited configurations may be included in setting up the trial CI wavefunction (DCI for doubly excited configurations, etc.). A disadvantage of CI wavefunctions is that they can no longer be interpreted in a transparent way unless individual states and transitions are well described by a 'leading' configuration, which has a large coefficient $C_{JK}$ in the CI wavefunction.

The additional freedom given to the CI wavefunction now allows for correlation and, 'in principle' (i.e. if an infinitely large basis set of AOs and an infinitely large set of excited configurations are used in *ab initio* calculations), this method provides exact solutions of the Schrödinger Equation 1.5. The approximations mentioned above are those included in the method developed by Pariser, Parr and Pople (PPP SCF CI method).[294,295,296] Such calculations for large organic molecules are completed in seconds on a desktop computer and are still useful for predicting and interpreting the absorption spectra of planar aromatic molecules and radicals or radical ions.

We have been able to deduce a number of useful results without considering electron correlation or even by disregarding electronic repulsion altogether. Indeed, single-determinant SCF wavefunctions are usually similar to configurations built from simple HMOs. Some of the serious deficiencies of the HMO procedure, such as the degeneracy of singlet and triplet configurations with the same occupation numbers, can be removed by qualitative consideration of the relevant $K$ integral determined from HMOs (Equation 4.34). It turns out, however, that excited state configurations calculated by the HMO or PPP SCF procedure frequently occur in degenerate pairs, not only due to molecular symmetry, but often accidentally as a consequence of the *Coulson–Rushbrooke pairing theorems*. In order to obtain an adequate representation of such excited states, first-order configuration interaction between degenerate configurations must be considered.

Consider the absorption spectrum of naphthalene (Figure 4.16) as an example. To predict the near-UV absorption bands, we need to consider only the four inner HMOs $\psi_2$ to $\psi_{-2}$, which are shown on the left-hand side of Figure 4.26. On the right, the four lowest singly excited configurations are shown and their energies relative to the ground state

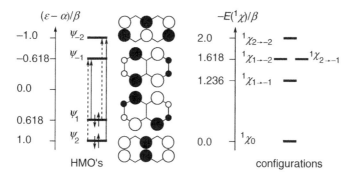

**Figure 4.26** Inner HMOs (left) and energies of lowest configurations (right) of naphthalene. The size of the coefficients is $|c_{j\mu}| = 0.408$ in $\psi_2$ and $\psi_{-2}$ and $|c_{j\mu}| = 0.425$ or $0.263$ in $\psi_1$ and $\psi_{-1}$

$^1\chi_0$, $E(^1\chi_{i \to j}) - E(^1\chi_0) = \varepsilon_j - \varepsilon_i$ are given in units of $\beta$, which is a negative quantity. Note that $E(^1\chi_{1 \to -2}) = E(^1\chi_{-1 \to 2})$ due to the pairing of orbital energies (Equation 4.23). If we calculate the energies of the excited configuration by an SCF method (Equation 4.33), they will be shifted, but the qualitative picture remains the same. In particular, the degeneracy $E(^1\chi_{1 \to -2}) = E(^1\chi_{-1 \to 2})$ will not be removed. It has been shown that the pairing theorems remain valid for PPP SCF calculations.

Because an SCF calculation does not allow for correlation, the wavefunctions of the configurations are not eigenfunctions of the Hamiltonian operator $\hat{H}$. We set up a CI wavefunction (Equation 4.35) considering only the two degenerate wavefunctions, $\Psi_\pm = C_1\chi_{1 \to -2} + C_2\chi_{2 \to -1}$, and we have to solve the $2 \times 2$ secular problem (Equation 4.36).

$$\begin{Vmatrix} E(\chi_{1 \to -2}) - E & H_{12} \\ H_{21} & E(\chi_{2 \to -1}) - E \end{Vmatrix} = 0$$

**Equation 4.36** *First-order CI calculation*

The off-diagonal elements $H_{12} = H_{21}$ calculated by the PPP SCF method are substantial, $H_{12} = 0.8$ eV. The resulting split between the two formerly degenerate configurations is sufficient to move the lower CI state below the energy of the configuration $\chi_{1 \to -1}$. The result of a full PPP SCF SCI calculation (Figure 4.27) is essentially the same.

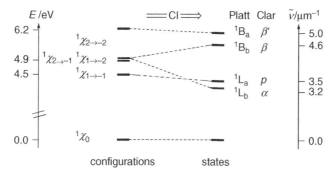

**Figure 4.27** State diagram for naphthalene resulting from a PPP SCF SCI calculation

The state symbols $^1L_b$, $^1L_a$, $^1B_b$ and $^1B_a$ shown in Figure 4.27 were introduced in 1949 by Platt.[297,298] He assumed a free-electron model, similar to the electron-in-a-box, in which the $\pi$-electrons of a cyclic system are confined to a one-dimensional loop of constant potential (a circular wire). The eigenvalues of a single electron in a perimeter of length $l$ are given by Equation 4.37.

$$\varepsilon_q = q^2 h^2 / (2m_e l^2), \quad q = 0, \ \pm 1, \ \pm 2, \ \ldots$$

**Equation 4.37** *Eigenvalues of an electron in Platt's perimeter potential*

The orbital energies $\varepsilon_q$ are all positive because the model does not contain an attractive potential. They are twofold degenerate except for the lowest one with $q = 0$. In the closed-shell configuration $\chi_0$ of a perimeter with $(4N + 2)$ electrons, all orbitals up to $\varepsilon_N$ and $\varepsilon_{-N}$ are doubly occupied (Figure 4.28). The quantum number $q$ represents the orbital angular momentum of an electron in orbital $\varepsilon_q$ and the levels with $|q| > 0$ accommodate electrons moving either clockwise or counter-clockwise around the loop. Platt's original model has been recast in terms of LCAO MO CI theory, refined and extended to cover a wide range of electronic spectral data.[15,299] Thus, the Platt symbols[j] are still in use, although the original model is now only of historical interest.

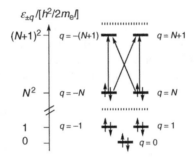

**Figure 4.28**   Closed-shell ground configuration of a perimeter with $4N$ electrons

We return to Figure 4.27. The transition moments to the singly excited configurations are readily calculated using Equation 4.20 and the HMO coefficients given in Figure 4.26. All four transitions are allowed and in-plane polarized ($x$ long axis, $y$ short axis), $M_{el,y}(\chi_1 \rightarrow -1) = 82\,e\,\text{pm}$, $M_{el,x}(\chi_2 \rightarrow -1) = M_{el,x}(\chi_1 \rightarrow -2) = 153\,e\,\text{pm}$ and $M_{el,y}(\chi_2 \rightarrow -2) = 70\,e\,\text{pm}$. First-order mixing of the configurations $\chi_2 \rightarrow -1$ and $\chi_1 \rightarrow -2$ gives the two states $^1\Psi_{\pm} = 2^{-1/2}(\chi_2 \rightarrow -1 \pm \chi_1 \rightarrow -2)$. The transition moments of the two configurations cancel in the lower state and add up in the upper one. Using Equation 2.19 and the calculated transition energies shown in Figure 4.27, which are close to the experimental ones, we predict the following oscillator strengths for the four transitions: $f(^1L_b) = 0$, $f(^1L_a) = 0.25$, $f(^1B_b) = 1.3$, $f(^1B_a) = 0.17$.

---

[j] The orbital angular momentum change is $\Delta q = 1$ (allowed) for excitation to the configurations $^1\chi_N^{N+1}$ and $^1\chi_{-N}^{N-1}$; $\Delta q = 2N + 1$ (forbidden) for the transition to $^1\chi_{-N}^{N+1}$ and $^1\chi_N^{N-1}$. States composed of the first two were given the label B (starting from A for $\Delta q = 0$) and L for the last two. The indices a and b refer to the polarization of the absorption bands. In molecules that contain a plane of symmetry perpendicular to the molecular plane, the subscript a designates states with the transition moment from the ground state directed through one or two perimeter *atoms* and the subscript b those with the transition moment cutting only across *bonds* (cf. e.g. Figure 4.26).

The predicted energies, polarizations and intensities of the first transitions of naphthalene are now in good agreement with experiment.[299] The $^1L_b$ absorption is very weak due to the cancellation of transition moments, $\varepsilon \approx 10^2\,M^{-1}\,cm^{-1}$. Although the excited state belongs to the $B_{3u}$ irreducible representation of the $D_{2h}$ point group (Section 4.4) and the $^1L_b$ transition should, therefore, be x-polarized, as is the $^1B_b$ band, the vibronic bands of the $^1L_b$ transition are actually of mixed polarization due to vibrational intensity borrowing from higher excited states. Such bands, which are common in alternant hydrocarbons as a consequence of the pairing theorem, are said to be *parity forbidden*.

The intensity of parity-forbidden transitions is very sensitive to slight perturbation of the parent compound by substitution. If one calculates the effect of inductive perturbation on the transitions $\chi_{2 \to -1}$ or $\chi_{1 \to -2}$ by first-order perturbation theory (Section 4.4, Equation 4.13), they will be shifted in opposite directions, because the orbitals involved are not paired. Even a minor imbalance in the energy of the two configurations $\chi_{2 \to -1}$ and $\chi_{1 \to -2}$ destroys full cancellation of the two oscillator strengths in the $^1L_b$ transition, leading to an increased intensity. For example, the transition energies in the quinolines are similar to those in naphthalene, as expected, but the intensity of the $^1L_b$ band is increased by over an order of magnitude. The energies are not shifted much, because the charge density at all atoms of an alternant hydrocarbon remains unity in all states, including the mixed $^1L_b$ and $^1B_b$ states. The uneven, but opposite, charge distributions in the configurations $\chi_{2 \to -1}$ and $\chi_{1 \to -2}$ cancel, when they are mixed 1 : 1 by CI. In general, *the energies of $\pi,\pi^*$-transitions of alternant hydrocarbons are fairly insensitive to inductive perturbation*.

Annulenes with $(4N + 2)$ $\pi$-electrons, $N = 1, 2, \ldots$, are the parent representatives of Platt's perimeter model. The absorption spectra of [n]annulenes with $n = 4N + 2$, $N = 1, 2, 3$ and 4, are shown in Figure 4.29.

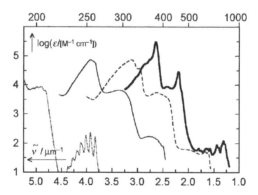

**Figure 4.29**  Absorption spectra of [n]annulenes: benzene = [6]annulene[280] (···), 1,6-methano[10]annulene[300] (—), [14]annulene[301] (---) and [18]annulene[302] (—)

The four lowest singly excited configurations of $(4N + 2)$-membered annulenes are degenerate in the HMO and Platt free electron (Figure 4.28) models. Equation 4.8 gives $\varepsilon_{LUMO} - \varepsilon_{HOMO} = -4\beta\sin(\pi/n)$ for the HMO excitation energies.[k] A PPP SCF CI

---

[k] The indices $j$ and $i$ for the HOMO and LUMO are $n$ and $n + 1$, respectively. Use the trigonometric equation $\cos\alpha - \cos\beta = -2\sin\gamma\sin\delta$, where $\gamma = (\alpha + \beta)/2$ and $\delta = (\alpha - \beta)/2$.

calculation splits them into three states, which belong to the irreducible representations $B_{2u}(^1L_b)$, $B_{1u}(^1L_a)$ and $E_{1u}(^1B_{a,b})$ of the $D_{nh}$ point groups, but the positions of the three bands are still linearly related to $\sin(\pi/n)$.[302] Transitions from the ground state ($A_g$) to the first two are symmetry forbidden; furthermore, that to the $^1L_b$ state is also parity forbidden. The highest energy transition to the degenerate $^1B_{a,b}$ state is allowed. For the smallest member of the series, benzene, only the onset of the transition to the $^1B_{a,b}$ state is seen in the deep-UV region.

More recently, the electronic spectra of the $(4N + 2)$ carbon cycles $C_{10}$, $C_{14}$, $C_{18}$ and $C_{22}$ were reported.[303,304] The spectral pattern of $C_{14}$ differs from that seen in the larger rings, $C_{18}$ and $C_{22}$, indicating some sort of a structural distortion (bond length alternation). The lowest singlet state of fullerene ($C_{60}$) lies at $1.57\,\mu m^{-1}$ (639 nm) and the compound exhibits many transitions throughout the visible and UV region, which have been assigned with the aid of calculations and symmetry arguments.[305,306]

Biphenylene and trisdehydro[12]annulene are representatives of conjugated hydrocarbons with a $4n$-membered ring. The pattern of their absorption spectra is completely different from that of benzenoid aromatics. Their HOMO and LUMO are derived from the two NBMOs of an ideal perimeter and both the lowest excited singlet and triplet state can be described by the configuration $\chi_{HO \rightarrow LU}$. The transition is symmetry forbidden in molecules of $D_{2h}$ symmetry or higher. Nonradiative decay usually dominates their photophysical properties. Quantum yields of fluorescence and intersystem crossing are low.[307] The LCAO version of Platt's perimeter model has been extended to treat conjugated systems with $4N$ $\pi$-electrons derived from [$n$]annulenes.[308,309]

The rich absorption spectra of radical anions and cations extend well into the visible and often into the near-IR region. The radical anion and cation spectra of alternant hydrocarbons are generally very similar, nearly superimposable. This is another remarkable consequence of the pairing theorem (Section 4.6). Sketch, for example, the HMO diagram of naphthalene (Figure 4.26), add or remove an electron, respectively, and draw the arrows leading to the lowest energy excited configurations in both the anion and the cation. It will become clear that for each transition of the cation there exists a transition of equal energy in the anion; for example, $E(\chi_{2 \rightarrow 1})$ of the cation is equal to $E(\chi_{-1 \rightarrow -2})$ in the anion, and so on. Hence the configuration diagrams for the anion and cation are identical. This is also true for the off-diagonal elements of a CI calculation, so that the absorption spectrum predicted by a PPP SCF CI calculation for the anion of an alternant hydrocarbon is identical with that of the corresponding cation. An extensive collection of radical ion spectra is available.[310]

Many linear polyenes exhibit anomalous fluorescence behaviour in the sense that the fluorescence rate constant $k_f$ calculated from the absorption spectrum (Equation 2.11) is much smaller than that determined by lifetime and quantum yield measurements (Equation 3.33). In 1972, Hudson and Kohler reported high-resolution absorption and emission spectra of all-*trans*-1,8-diphenylocta-1,3,5,7-tetraene at low temperature, which proved that the lowest excited singlet state $S_1$ was *not* that reached by the strongly allowed $\pi,\pi^*$ transition ($f \approx 1.5$) that is predicted by MO theory and observed at 410 nm.[311] Rather, very weak ($f \approx 0.06$), structured absorption that had been hidden under the tail of the $\pi,\pi^*$-absorption in solution spectra was detected at slightly longer wavelengths.

The search for a 'forbidden' state was initiated by the finding that the experimentally determined rate constant of fluorescence, $k_f = \Phi_f/\tau_f$, was orders of magnitude less than that calculated by the Strickler–Berg relation (Equation 2.11). This 'forbidden' state was shown to be the $S_1$ state responsible for the long-lived fluorescence emission. The forbidden state was immediately[312] identified on the basis of PPP SCF CI calculations that included doubly excited configurations. The leading configuration of this state is $^1\chi_{HO,HO \to LU,LU}$, that is, the configuration formed by *twofold* $\pi,\pi^*$ excitation from the HOMO to the LUMO. As a configuration with only doubly occupied orbitals, it belongs to the same totally symmetric irreducible representation, $^1A_g$, as the ground state, so that the transition moment leading to the $S_1$ state is zero.

How can a doubly excited configuration, $\chi_{HO,HO \to LU,LU}$, be *lower* in energy than the singly excited one, $\chi_{HO \to LU}$, when, according to simple HMO theory, $E(\chi_{HO,HO \to LU,LU}) = 2E(\chi_{HO \to LU}) = 2(\varepsilon_{LU} - \varepsilon_{HO})$? The doubly excited configuration is stabilized by exchange interaction when the electronic repulsion is included in the Hamiltonian operator: $E(\chi_{HO,HO \to LU,LU}) - 2E(\chi_{HO \to LU}) = -2K_{HO,LU}$. The exchange integral $K_{HO,LU}$ is large in alternant hydrocarbons, whereas the energy gap $\varepsilon_{LU} - \varepsilon_{HO}$ is small in extended polyenes. Thus, PPP SCF CI calculations place the doubly excited state slightly below the singly excited state in linear polyenes.[312] The excited $^1A_g$ state plays an important role in the photochemistry of alkenes, because it is further stabilized when the structure relaxes by twisting a formal double bond, which reduces the energy gap $\varepsilon_{LU} - \varepsilon_{HO}$ (Section 5.5).

The transition moment of the first allowed $\pi,\pi^*$ transition of linear polyenes is largest in the all-*trans* configuration, as predicted by HMO theory: the local charges in the transition density do not depend on the geometry, but the transition dipole is largest in the extended all-*trans* isomers. Isomerization of a central double bond to the Z-configuration is generally accompanied by the appearance of a new $\pi,\pi^*$ absorption band at shorter wavelengths, the so-called *cis*-peak. This phenomenon has been observed in many linearly conjugated systems, including $\beta$-carotene and neolycopene,[313] $\alpha$-carotene[314] and 1,6-diphenylhexa-1,3,5-triene.[315] The appearance of the *cis*-peak is readily understood in terms of an HMO model that includes first-order CI of the degenerate $\chi_{2 \to -1}$ and $\chi_{1 \to -2}$ configurations (cf. the case of naphthalene, Figure 4.27). The lower of the states mixed by CI is forbidden in both isomers, but the higher lying state has a moderate transition moment that is orthogonal to the central double bond in, for example, the 3-Z isomer of hexatriene, but vanishes in the all-*trans* isomer.

We conclude that the HMO model, upgraded when necessary by a qualitative consideration of electronic repulsion (Equation 4.33) and of the correlation of electronic motion (by first-order CI for degenerate excited configurations, Equation 4.36), is able to capture the essence of electronic transitions in conjugated molecules correctly. Clearly, modern, much more sophisticated quantum chemical calculations yield more reliable quantitative predictions, but the HMO model retains its value as a language to project the output of essentially black-box calculations to a lucid formalism that provides insight, systematics and the means to generalize and extrapolate. Both the HMO and the PPP SCF CI methods are, of course, inadequate for the geometry optimization of nonplanar molecules.

Finally, we mention an entirely different quantum chemical approach, known as *density functional theory* (DFT). This method is designed to determine electron densities directly, rather than the electronic wavefunction itself. Knowledge of the electron density is also sufficient to deduce any observable quantity of interest. Exact DFT is an *ab initio* method, although most current implementations introduce some empiricism in the form of functionals such as B3LYP and M06.[316] DFT methods have become very popular, because they achieve the required level of accuracy at the lowest computational cost. Excited-state energies and transition moments can be obtained using time-dependent DFT methods (TDDFT), which determine the poles of the frequency-dependent polarizability of the ground state. For these and other modern methods of computational quantum chemistry dealing with excited states such as CASPT2,[19] and a discussion of their advantages and limitations, the reader is referred to specialized texts.[14,317]

## 4.8   Spin–Orbit Coupling

*Recommended reviews.*[318–320]

Rate constants of *intersystem crossing* (ISC) are calculated by Fermi's golden rule (Equation 2.21, Section 2.1.5). In the coupling term $V_{if} = <\Psi_i|\hat{h}_{magn}|\Psi_f>$, $\Psi_i$ is the wavefunction of the initial state, for example the singlet state $S_1$, $\Psi_f$ is that of the final state, in this case the triplet state $T_1$ to which ISC occurs, and $\hat{h}_{magn}$ is a perturbation operator that accounts for magnetic interactions. Spin–orbit coupling (SOC) is usually the dominant interaction inducing ISC, except when the radical centres are well separated in space as in extended localized biradicals (Section 5.4.4) and in radical pairs. In such systems, the magnetic interactions between electronic and nuclear spins (*hyperfine coupling*) become important. SOC underlies El Sayed's rules (Section 2.1.6 and Figure 4.30), which states that ISC processes involving a change in orbital type (n,$\pi^*$ $\Leftrightarrow$ $\pi,\pi^*$) are generally substantially faster than those that do not. Here, we give only a brief, qualitative description of SOC. Quantitative calculations of SOC are feasible with several quantum chemistry program packages. Practical examples are discussed in Sections 5.4.1–5.4.4.

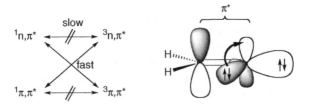

**Figure 4.30**   Left: El Sayed's rule for ISC. Right: the n,$\pi^*$ transition of formaldehyde

The SOC operator $\hat{h}_{SO}$ is a sum of one- and two-electron terms. We consider only the more essential one-electron terms, which represent the interaction of the spin magnetic dipole moment of an electron with the magnetic dipole moment induced by its own orbital motion (Section 1.4). Approximate inclusion of the two-electron part can be done by introducing *effective* SOC constants $\zeta$ for the individual atoms, which correct for the effect of the two-electron part. The atomic SOC constants increase roughly with the fourth power of the atomic number $Z$. Thus, for an unpaired electron occupying an LCAO-MO, the main contributions to SOC originate from the AOs that are located on heavy nuclei and have a large angular asymmetry (p, but not s). The atomic number of protons equals 1 and nearby electrons sojourn primarily in s orbitals, so that H atoms contribute very little. The acceleration of ISC rates in the presence of heavy atoms (either in the molecule itself or in the surrounding medium, e.g. carbon tetrabromide or a xenon matrix) is called the *heavy-atom effect*. Rare cases of inverse heavy-atom effects are, however, also known.[321] Electrons in doubly occupied orbitals, such as inner shells, can be ignored, because the contributions of the two opposed electron magnetic moments cancel. In a biradical that is well described as having two singly occupied MOs, SOC is due to the electrons occupying these two orbitals.

The SOC matrix element $<S_1|\hat{h}_{SO}|T_1>$ represents a vector consisting of three terms, $<S_1|\hat{h}_{SO}|T_{1x}>$, $<S_1|\hat{h}_{SO}|T_{1y}>$ and $<S_1|\hat{h}_{SO}|T_{1z}>$, because the symbol $T_1$ comprises three states that are very close in energy and a reacting molecule can be in any one of them. The terms inducing ISC to the individual triplet sublevels generally differ from each other, so that the triplet sublevels formed by ISC from $S_1$ will initially be unevenly populated. Rates of ISC from $T_1$ to the ground state $S_0$ will also differ for each sublevel. However, equilibration between the three sublevels of $T_1$ is established and maintained within nanoseconds at room temperature and the decay of longer-lived triplet states obeys a single exponential rate law. For equilibrated sublevels, the length of the $S_0$–T spin–orbit coupling vector, $[<S_0|\hat{h}_{SO}|T_{1x}>^2 + <S_0|\hat{h}_{SO}|T_{1y}>^2 + <S_0|\hat{h}_{SO}|T_{1z}>^2]^{1/2}$, determines the overall rate of ISC.

The SOC operator contains the product of a term that acts on the spin part of a wavefunction, converting a triplet function to a singlet function and a term that acts on the space part of the wavefunction, changing one electronic configuration to another. Each component of the SOC vector, $<S_0|\hat{h}_{SO}|T_{1u}>$, $u = x$, $y$ or $z$, is a sum over all atoms and each atom contributes a sum over pairs of all basis set orbitals in the molecule. The terms in the sum are multiplied by numerical coefficients obtained from the CI expansion and from the coefficients of the valence (primarily p-) orbitals in the two singly occupied MOs of $T_1$. The dominant contributors to the sum are those in which both basis set orbitals are located on the same atom ('one-centre terms'). Each term reflects the degree to which a 90° rotation around the axis $u$ through the atom converts one member of the orbital pair into the other. If the space part of the wavefunctions differs only by the occupation number of two MOs $\psi_i$ and $\psi_j$, we need to consider only the three matrix elements $<\psi_i|\hat{h}_{SO}|\psi_j>$ (Figure 4.30).

## 4.9   Theoretical Models of Photoreactivity, Correlation Diagrams

When trying to predict the outcome of a photoreaction, one should certainly not let oneself be guided by the rules of ground-state reactivity. If anything, expect the opposite to happen! This sweeping statement may be justified by Dewar's PMO theory (Section 4.6), which predicts an approximate mirror-image relationship between the π-electron energy of the excited state and that of the ground state. Features, such as 'aromaticity', that are taken as indicators of a stable ground state, tend to be destabilizing factors in the excited state and vice versa. Examples are provided by the heterolytic fission of 9-fluorenol to form the antiaromatic fluorenyl cation (Section 5.4.5, Scheme 5.19) and by the photodecarboxylation of arylacetic acids **5** and **6** (Scheme 4.2).[322,323] The quantum yields and rates for photodecarboxylation are $\Phi(5) = 0.042$, $k(5, \; S_1) = 9 \times 10^6 \, s^{-1}$ and $\Phi(6) = 0.60$, $k(6, \; S_1) = 6 \times 10^9 \, s^{-1}$. The opposite would have been expected for thermal decarboxylation of **5** and **6** to form 'aromatic' and 'antiaromatic' anions, respectively.

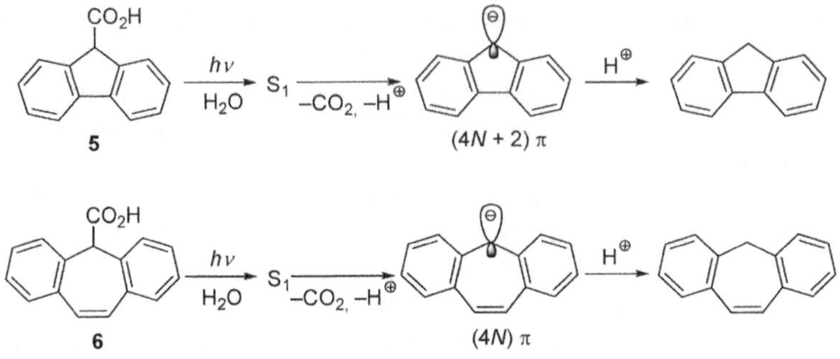

**Scheme 4.2**   *Photodecarboxylation*[322,323]

In Section 2.1, we developed rules of thumb to predict rate constants of photophysical processes for a given molecule (Table 2.1). These unavoidable energy-wasting processes limit the lifetime of a singlet state to nanoseconds or less and that of a triplet state to milliseconds or 200 ns in aerated solutions. For a photoreaction to compete efficiently, Arrhenius' equation indicates that barriers on the excited state PES exceeding $E_a = 30 \, kJ \, mol^{-1}$ for $S_1$ and $E_a = 60 \, kJ \, mol^{-1}$ for $T_1$ will be prohibitive.

Once promoted to an excited singlet state by vertical excitation (Franck–Condon principle), a molecule is likely to follow the path of steepest descent to a local minimum and relax rapidly to the lowest singlet state by internal conversion. Intersystem crossing to the triplet manifold is an option, especially for molecules with low-lying n,π* states or heavy atoms. Now we are looking for guidelines to predict which reaction pathways might compete with the photophysical processes that eventually restore the starting material.

Conical intersections and avoided crossings of excited- and ground-state surfaces correspond to minima on the singlet excited PES and internal conversion to the ground state will be especially fast in such regions. These locations in the space of molecular geometries represent *funnels* that siphon the molecule to the ground state. The shape of the ground-state PES will then govern the subsequent reactions and may lead the molecule to a new structure, a photoproduct. Photoproducts often have a higher heat of formation than the starting material, because funnels on the excited state PES are by definition at an energy maximum of the ground state. Our task is then to predict at which geometries such funnels are located and whether they are accessible from the molecular structure that is initially reached by the absorption of a photon. No funnels are expected on the lowest triplet PES, because there are no lower triplet states and diabatic ISC at crossings with a lower singlet state has a low probability (Equation 2.53). Thus, triplet states are likely to decay more slowly by ISC from a local minimum on the triplet PES.

Modern quantum mechanics can provide accurate excited-state PESs,[14,16,17] but such calculations are demanding and in order to obtain useful answers one must ask the right questions. The chemist's intuition or knowledge from related reactions is instrumental in suggesting conceivable photoproducts of the initial reactant. Qualitative guidelines to predict the initial slope on an excited PES along a chosen reaction coordinate can be obtained by first-order perturbation theory using excited-state charge densities for heterolytic reactions and bond orders for homolytic reactions and rotations around formal double bonds. Thus, the *changes* in charge density and bond order upon excitation are most relevant.[324]

Early examples for the use of excited-state charge distributions to predict regioselectivity are the nucleophilic aromatic substitution reactions of nitromethoxybenzenes (Case Study 6.15)[325,326] and the photosolvolysis reactions of donor-substituted benzyl derivatives (Scheme 6.147).[327] In the ground state, the charge distribution of donor-substituted benzene derivatives corresponds to that of the isoelectronic benzyl anion, which is given by the square of the NBMO [*ortho*- and *para*-activation, Figure 4.31; the saturated side-chain with a leaving group (e.g. $-CH_2OAc$) that may be attached at the *ortho*-, *meta*- or *para*-position is not shown]. HOMO–LUMO excitation accumulates excess charge predominantly in the *meta*- and, to a lesser extent, in the *ortho*-position, which favours heterolytic cleavage of the leaving group.

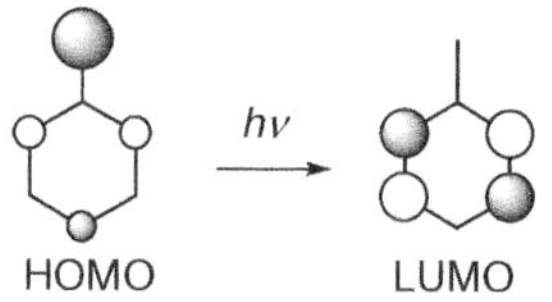

**Figure 4.31** Reorganization of the charge distribution in benzyl anion upon HOMO → LUMO excitation. The charge distribution in the ground state is given by the square of the HOMO and that of the excited state by the square of the LUMO

In a related reaction, the photohydrolysis of trifluoromethylnaphthols, the calculated charge distribution in the excited singlet state was found to correlate remarkably well with the excited-state heterolysis rate constants of eight isomeric compounds.[328,329] The most reactive of these were the 1,8-, 1,5- and 2,3-substituted derivatives. The primary photoproduct formed from 3-trifluoromethyl-2-naphthol (**7**, Scheme 4.3), a naphthoquinone methide, was observed by flash photolysis. Depending on the solvent polarity and on the leaving group, homolytic cleavage may, however, compete and reactions that proceed via the triplet state do not necessarily follow the same pattern.[330]

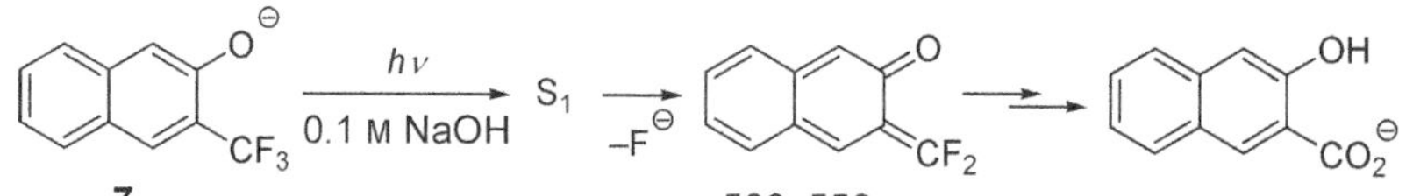

**Scheme 4.3** *Photohydrolysis of 3-trifluoromethyl-2-naphthol (7) in aqueous base*

Various approximations allow one to draw diabatic surfaces that connect the electronic states of the reactant with those of conceivable photoproducts and symmetry considerations (Section 4.4) are extremely helpful in this respect. We first consider state symmetry correlation diagrams.[331] If the molecules of interest are symmetric (perhaps ignoring some substituents), we consider a reaction coordinate leading from reactant to product that preserves one or more of the symmetry elements along the reaction coordinate. We can then connect states of the reactant and the product that belong to the same irreducible representation. We begin by estimating the energies of the lowest electronic states of the reactant and of the possible product(s). This information may be drawn from electronic spectra, calculations or educated guesswork. To be sure, the preferred reaction path may well be one *not* retaining any symmetry. We choose a symmetrical reaction path only because it allows us to construct a correlation diagram. Because potential energy surfaces are continuous, this path will give some indication of the shape of the surfaces nearby. Consider the protonation of a ketone in its lowest $n,\pi^*$ excited triplet state (Figure 4.32). The carbonyl moiety is planar and we assume that the proton attacks the carbonyl oxygen in the plane of symmetry.

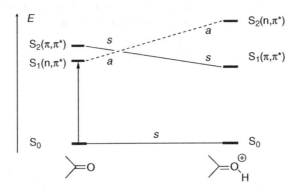

**Figure 4.32**   State-symmetry diagram for the protonation of an $n,\pi^*$ excited triplet ketone

The $n,\pi^*$ state of the ketone is strongly destabilized by protonation of the oxygen atom, so that it will have a higher energy than the $\pi,\pi^*$ state in the conjugate acid. We can now classify the two lowest excited singlet states of reactant and product by simply counting the unpaired electrons. The wavefunctions of the closed-shell ground state and also those of the $\pi,\pi^*$ excited states are symmetric ($s$), because they contain an even number of electrons in the antisymmetric ($a$) $\pi$ and $\pi^*$ orbitals (recall that $a \otimes a = s$, Section 4.4). Those of the $n,\pi^*$ states are antisymmetric ($a$) with respect to the plane of symmetry, because they have a singly occupied $\pi^*$-orbital and the n-orbital is symmetric, $a \otimes s = a$. Connecting the states of the same symmetry, we see that the protonation of an $n,\pi^*$ excited ketone is a state-symmetry forbidden reaction, as long as the plane of symmetry is retained.

Benzophenone is a very weak base in the ground state and its protonation requires very strong acid. The acidity of the conjugate acid, $pK_a(S_0) = -4.7$, is, however reduced by four orders of magnitude in the excited triplet state $pK_a(S_1) = -0.4$ (see also Case Study 5.1). Nevertheless, the protonation of triplet benzophenone is much slower than usual for an oxygen-protonation reaction, $k_{H^+} = 3.8 \times 10^8 \, M^{-1} s^{-1}$, and it is the rate-determining step preceding the faster hydration of the aromatic nucleus, $k_0 = 1.5 \times 10^9 \, s^{-1}$ (Scheme 4.4).[332] The hydrated intermediate regenerates benzophenone after ISC and is responsible for the quenching of aromatic ketones by protons in aqueous acid reducing the quantum yields of their usual reactions. It does, however, open the way to some novel

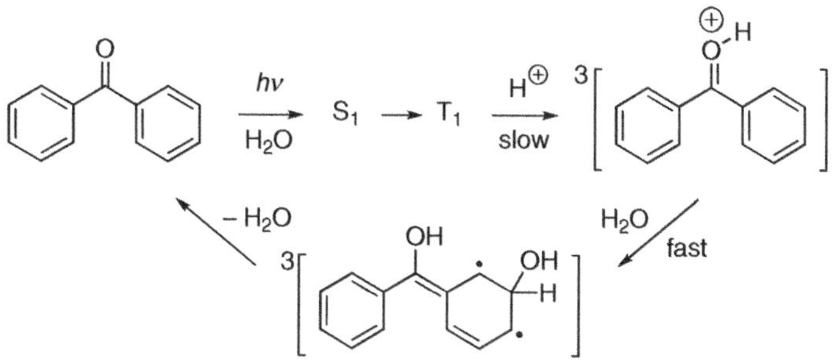

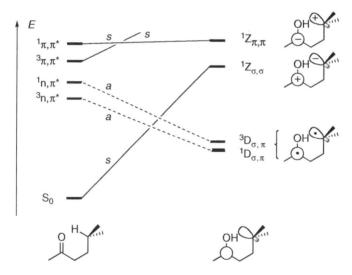

**Scheme 4.4** *Photohydration of benzophenone in aqueous acid*

photoreactions[333–335] of substituted aromatic ketones in aqueous acid (see also Scheme 6.239). The in-plane protonation of phenylnitrene is also state-symmetry forbidden (see Section 5.4.2).

A state-symmetry diagram for the Norrish type II reaction (Section 6.3.4) shows that it is allowed only from the $n,\pi^*$ singlet and triplet states to give the covalent biradicals (Section 5.4.4) $^1D_{\sigma,\pi}$ and $^3D_{\sigma,\pi}$, respectively, but not from $\pi,\pi^*$ states (Figure 4.33).[331]

**Figure 4.33** State-symmetry diagram for intramolecular hydrogen abstraction. The symmetry labels refer to the plane of the carbon chain

The *orbital correlation diagrams* put forth by Woodward and Hoffmann[336] for electrocyclic and sigmatropic reactions (Section 6.1.2) and for cycloaddition reactions (Section 6.1.5) are well known and the details of their construction are not reiterated here. We show only the case of the $[2_s + 2_s]$ cycloaddition of two ethene molecules to cyclobutane as an example (Figure 4.34).

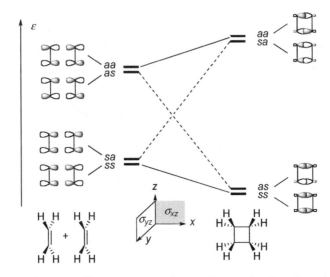

**Figure 4.34** Woodward–Hoffmann MO correlation diagram for [2 + 2] cycloaddition. The labels *s* (symmetric) and *a* (antisymmetric) refer to the symmetry planes $\sigma_{xz}$ and $\sigma_{yz}$ in that order

To predict the shape of potential energy surfaces along the symmetry-retaining reaction coordinate from an MO correlation diagram, one must construct a correlation diagram of the electronic configurations (Figure 4.35).

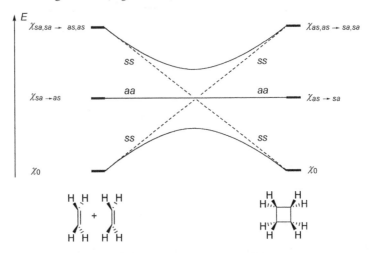

**Figure 4.35** Correlation diagram of the lowest configurations in [2 + 2] cycloaddition

It is seen that the ground configuration of the two ethene molecules correlates with a doubly excited configuration of cyclobutane because the electron pair in the antisymmetric combination of the bonding $\pi$-orbitals of the ethenes (symmetry *sa*) is lifted to the antibonding $\sigma^*$ orbital of the same symmetry in cyclobutane. Starting from the singly excited configuration $\chi_{sa \to as}$ of symmetry *aa*, one ends up in the singly exited configuration

$\chi_{as \rightarrow sa}$ of the same symmetry $aa$ (dashed lines). Configuration interaction between $\chi_0$ of symmetry $ss$ and the doubly excited configuration of the same symmetry splits the two crossing states of symmetry $ss$ apart (full lines), but a 'symmetry-imposed' barrier remains, which classifies the reaction as 'forbidden' in the ground state.

The reaction starting from the lowest singly excited state $S_1$ (represented by the configuration $\chi_{sa \rightarrow as}$ in Figure 4.35) does not encounter a symmetry-imposed barrier, it classifies as 'allowed'. In fact, reactions initiated from $S_1$ usually do not proceed adiabatically to the lowest excited singlet state of the product. Rather, the region near the crossing of the first-order diabatic PESs often represents a funnel, from where the molecule returns to the ground state PES. In short, ground-state reactions that are 'forbidden' by the Woodward–Hoffman rules are generally 'allowed' in photoreactions and vice versa, but exceptions from this general dichotomy do exist.[337]

Simple valence-bond theory provides another approach to construct correlation diagrams and to predict regions of diabatic surface crossings.[20,338]

## 4.10  Problems

1. Predict the effect of methyl substitution of azulene (replace C–H by C–Me individually in positions 1, 2, 4, 5 and 6) on the energy of the lowest excited state $S_1$. Use $\delta\alpha_{C-Me} = -0.333\beta$ to allow for the inductive (destabilizing) perturbation by the methyl group and compare with the experimental shifts of $-790$, 430, 370, $-350$ and 460 cm$^{-1}$, respectively. [0.097, $-0.033$, $-0.065$, 0.034, $-0.087\beta$]

2. Predict the shifts of the first absorption band of anthracene upon methyl substitution by inductive first-order perturbation. [None]

3. Predict the shift of the first absorption band of naphthalene radical cation upon 1-methyl substitution. [$0.425^2$ ($-0.333$)$\beta = -0.06\beta$, hypsochromic shift, ref. 291]

4. Predict the effect of aza substitution on the first absorption bands of naphthalene (quinoline and isoquinoline: energy, band intensity). [No shift, increased intensity of the $^1L_a$ band, cf. Figure 4.27]

5. Predict the shifts of the first ionization potential of naphthalene upon methyl substitution in positions 1 and 2. The coefficients of the HOMO are $c_{HOMO,1} = 0.425$ and $c_{HOMO,2} = -0.263$. [Ref. 291]

6. Explain why the first absorption band of cisoid dienes lies at longer wavelengths than that of transoid analogues. [Figure 4.6]

7. Verify the HOMO–LUMO energy gaps calculated by PMO theory for the systems shown in Scheme 4.1.

8. Why are the absorption spectra of the radical cation and anion of anthracene nearly identical? [End of Section 4.7]

9. Calculate the transition moment and oscillator strength of the $^1L_a$ transition of naphthalene using the HMO coefficients given in Figure 4.26. [$M_{el,y} = 82\,e\,pm$, $f = 0.25$].

10. Derive Equation 4.21 from Equation 4.7.

11. Why do cross-conjugated alkenes absorb at shorter wavelengths than linearly conjugated alkenes? Compare hexa-1,3,5-triene and 1,1-divinylethene. [PMO]

## 4.11   Appendix

### 4.11.1   First-Order Perturbation

Assume that the solution for the Hückel secular determinant of a parent molecule has been determined, that is, that the eigenvalues $\varepsilon_j$ and the associated set of orthonormal HMOs $\psi_j$ are known. We now introduce a 'small' perturbation of the system, say by adding a substituent or by replacing one of the C atoms by a heteroatom, for example N. To this end, we must first express the MO energies $\varepsilon_j$ as a function of the MO coefficients. Inserting Equation 4.6 into $\varepsilon_j = <\psi_j|\hat{H}|\psi_j>$ we obtain Equation 4.38.

$$\varepsilon_j = \left\langle \sum_{\mu=1}^{\omega} c_{j\mu}\phi_\mu |\hat{H}| \sum_{\nu=1}^{\omega} c_{j\nu}\phi_\nu \right\rangle = \sum_{\mu=1}^{\omega}\sum_{\nu=1}^{\omega} c_{j\mu}c_{j\nu}\langle \phi_\mu|\hat{H}|\phi_\nu\rangle$$

**Equation 4.38**

The right-hand expression results from replacing the integral over sums by sums of integrals. We replace the integrals $<\phi_\mu|\hat{H}|\phi_\nu>$ by the parameter list (Equation 4.3), retaining the atomic indices that assign the parameters to specific atoms or bonds, because we will introduce perturbations at some particular atoms, $\alpha_\mu = \alpha + \delta\alpha_\mu$, or bonds, $\beta_{\mu\nu} = \beta + \delta\beta_{\mu\nu}$:

$$\varepsilon_j = \sum_{\mu=1}^{\omega} c_{j\mu}^2\alpha_\mu + \sum_{\mu=1}^{\omega}\sum_{\substack{\mu=1 \\ \nu\neq\mu}}^{\omega} c_{j\mu}c_{j\nu}\beta_{\mu\nu} = \sum_{\mu=1}^{\omega} c_{j\mu}^2\alpha_\mu + 2\sum_{\mu-\nu}^{\omega} c_{j\mu}c_{j\nu}\beta_{\mu\nu}$$

**Equation 4.39**

The last sum in Equation 4.39 runs only over pairs of atoms that are bound ($\mu-\nu$), because $\beta_{\mu\nu} = <\phi_\mu|\hat{H}|\phi_\nu>$ is zero otherwise, and is multiplied by two, because each bond appears twice in the double sum on the left and $\beta_{\mu\nu} = \beta_{\nu\mu}$. Introducing a perturbation at atom $\mu$, $\alpha_\mu = \alpha + \delta\alpha_\mu$, the energy $\varepsilon_j$ will change by $\delta\varepsilon_j = (\partial\varepsilon_j/\partial\alpha_\mu)\delta\alpha_\mu$. The partial derivative of Equation 4.39 is $(\partial\varepsilon_j/\partial\alpha_\mu) = c_{j\mu}^2$, so that we obtain Equation 4.13 for the effect of an inductive perturbation $\delta\alpha_\mu$ on $\varepsilon_j$. Similarly, a change in the resonance integral between atoms $\mu$ and $\nu$, $\delta\beta_{\mu\nu}$, will change the energy $\varepsilon_j$ by $\delta\varepsilon_j = (\partial\varepsilon_j/\partial\beta_{\mu\nu})\delta\beta_{\mu\nu}$. The partial derivative of Equation 4.39 with respect to $\beta_{\mu\nu}$ is $(\partial\varepsilon_j/\partial\beta_{\mu\nu}) = 2c_{j\mu}c_{j\nu}$ (Equation 4.14).

### 4.11.2   Second-Order Perturbation

The operator of the perturbed system is written as $\hat{H} = \hat{H}^0 + \hat{h}$, where $\hat{H}^0$ is the Hückel operator of the parent molecule and $\hat{h}$ the perturbation operator. We could, of course, set up a new Hückel determinant for the perturbed system, but having the eigenvalues $\varepsilon_j^{(0)}$ and the associated eigenfunctions $\psi_j^{(0)}$ with the operator $\hat{H}^0$ of the parent system at hand, we already have a good starting point for the new system. Therefore, the wavefunctions are set up as linear combinations of the functions $\psi_j^{(0)}$ (Equation 4.40).

$$\psi = \sum_{j=1}^{n} a_j \psi_j^{(0)}$$

**Equation 4.40**

Optimized coefficients $a_j$ are determined by the variation method requiring the secular determinant to vanish. The indices of the columns and rows now refer to the basis set defined by Equation 4.40 and not to the AO basis as in Equation 4.5:

$$||H_{ij} - \delta_{ij}\varepsilon|| = \begin{Vmatrix} \varepsilon_1^{(0)} + \delta\varepsilon_1^{(1)} - \varepsilon & h_{12} & \cdots & h_{1n} \\ h_{21} & \varepsilon_2^{(0)} + \delta\varepsilon_2^{(1)} - \varepsilon & \cdots & h_{2n} \\ \cdot & \cdot & \cdots & \cdot \\ \cdot & \cdot & \cdots & \cdot \\ h_{n1} & h_{n2} & \cdots & \varepsilon_n^{(0)} + \delta\varepsilon_n^{(1)} - \varepsilon \end{Vmatrix} = 0$$

**Equation 4.41**

The basis functions $\psi_j^{(0)}$ are eigenfunctions of $\hat{H}^0$, $H_{ij}^0 = 0$, so that the off-diagonal elements of the matrix $H_{ij}$ are equal to $h_{ij}$. In the diagonal elements we have $H_{ii} - \varepsilon = H_{ii}^0 + h_{ii} - \varepsilon = \varepsilon_i^{(0)} + \delta\varepsilon_i^{(1)} - \varepsilon$. Solving Equation 4.41 would provide the exact Hückel eigenvalues of the perturbed system. Instead, we seek an approximation for the first eigenvalue, $\varepsilon_1 \approx \varepsilon_1^{(0)} + \delta\varepsilon_1$. Based on the assumption that the perturbation is small compared with the energy differences between $\varepsilon_1^{(0)}$ and the other eigenvalues of the unperturbed system, $|\varepsilon_i^{(0)} - \varepsilon_1^{(0)}| \gg \delta\varepsilon_1$ for $i \neq 1$, we can simplify Equation 4.41 as follows. (i) For all diagonal elements except the first, $\varepsilon \approx \varepsilon_i^{(0)}$. (ii) Only the off-diagonal elements in the first column and row affect $\varepsilon_1$ substantially; we set all others to 0. (iii) The first-order perturbations $\delta\varepsilon_i^{(1)}$ are neglected in all diagonal elements except for $i = 1$. Equation 4.42 is obtained.

$$\begin{vmatrix} \varepsilon_1^{(0)} + \delta\varepsilon_1^{(1)} - \varepsilon & h_{12} & h_{13} & \cdots & h_{1n} \\ h_{21} & \varepsilon_2^{(0)} - \varepsilon_1^{(0)} & 0 & \cdots & 0 \\ h_{31} & 0 & \varepsilon_3^{(0)} - \varepsilon_1^{(0)} & \cdots & 0 \\ \cdot & \cdot & \cdot & \cdots & \cdot \\ h_{n1} & 0 & 0 & \cdots & \varepsilon_n^{(0)} - \varepsilon_1^{(0)} \end{vmatrix} = 0$$

**Equation 4.42**

A determinant does not change its value when a multiple of any row is added to an other row. Multiply the second row by $h_{12}/(\varepsilon_2^{(0)} - \varepsilon_1^{(0)})$ and subtract the result from the first. The first element becomes $\varepsilon_1^0 + \delta\varepsilon_1^{(1)} - \varepsilon - h_{12}h_{21}/(\varepsilon_2^{(0)} - \varepsilon_1^{(0)})$ and the second element vanishes. Now multiply the third row by $h_{13}/(\varepsilon_3^{(0)} - \varepsilon_1^{(0)})$ and subtract the result from the first. The term $h_{13}h_{31}/(\varepsilon_3^{(0)} - \varepsilon_1^{(0)})$ is subtracted from the first element and the third element vanishes. Repeating up to the last row and evaluation of the diagonal determinant gives Equation 4.43.

$$\left(\varepsilon_1^{(0)} + \delta\varepsilon_1^{(1)} - \varepsilon - \sum_{i=2}^{n} \frac{h_{1i}h_{i1}}{\varepsilon_i^{(0)} - \varepsilon_1^{(0)}}\right)(\varepsilon_2^{(0)} - \varepsilon_1^{(0)})(\varepsilon_3^{(0)} - \varepsilon_1^{(0)})\cdots = 0$$

**Equation 4.43**

*Quantum Mechanical Models*

We have assumed that $\varepsilon_i^{(0)}$ is not degenerate with any other eigenvalue $\varepsilon_i^{(0)}$ so that only the first bracket can vanish and it will do so if

$$\varepsilon = \varepsilon_1^{(0)} + \delta\varepsilon_1^{(1)} - \sum_{i=2}^{n} \frac{h_{1i}h_{i1}}{\varepsilon_i^{(0)} - \varepsilon_1^{(0)}} = \varepsilon_1^{(0)} + \delta\varepsilon_1^{(1)} + \sum_{i=2}^{n} \frac{h_{1i}h_{i1}}{\varepsilon_1^{(0)} - \varepsilon_i^{(0)}}$$

**Equation 4.44**

where the sum represents the second-order perturbation $\delta\varepsilon_1^{(2)}$. The same can be done for any row $i$ of Equation 4.42 to give the general equation for second-order perturbation (Equation 4.15), in which the upper indices (0) marking the eigenvalues $\varepsilon_i$ of the unperturbed system have been dropped.

# 5

# Photochemical Reaction Mechanisms and Reaction Intermediates

## 5.1  What is a Reaction Mechanism?

The term reaction mechanism is part of the everyday language of chemists, yet it conveys different things to different people. The IUPAC Gold book (www.goldbook.iupac.org) defines the mechanism of a reaction as 'A detailed description of the process leading from the reactants to the products of a reaction, including a characterization as complete as possible of the composition, structure, energy and other properties of reaction intermediates, products and transition states. An acceptable mechanism of a specified reaction (and there may be a number of such alternative mechanisms not excluded by the evidence) must be consistent with the reaction stoichiometry, the rate law and with all other available experimental data, such as the stereochemical course of the reaction.' On the basis of Occam's razor (Section 3.7.4), one should always choose the simplest mechanism that is consistent with all available evidence.

A truly mechanistic (in the sense of classical mechanics) description of a molecule's reaction is in fact prohibited by Heisenberg's uncertainty relations (Equation 2.1). Some reaction mechanisms of small molecules in the gas phase have been elucidated in the utmost detail, that is, reaction rate constants have been determined for individual rotational and vibrational quantum states of the reactant. We take a more modest view: a reaction mechanism is the step-by-step sequence of elementary processes and reaction intermediates by which overall chemical change occurs.

The mechanism of a photoreaction should ideally include a detailed characterization of the primary events as outlined by the classification of photochemical reaction pathways in Section 2.3; the lifetimes of the excited states that are involved in the reaction path, the quantum yields and hence the rate constants of all relevant photophysical and photochemical processes, in addition to the information about the structure and fate of any reactive intermediates, their lifetimes and reactivities.

---

Knowledge of reaction mechanisms serves at least two purposes apart from the intellectual satisfaction of 'understanding'. First, it provides a classification that reduces the myriad of chemical reactions of individual molecules to a more manageable set and thereby allows prediction by analogy. Second, it enables chemists to design reaction conditions in order to improve yields, the rate of product formation or selectivity of a reaction, to suppress undesired reactivity or to divert a reaction to form new products.

Before we proceed to discuss the properties of some of the most common reaction intermediates, we take a look at the simplest of elementary processes encountered in photochemistry: electron and proton transfer.

## 5.2   Electron Transfer

The redistribution of electrons associated with electronic excitation of a molecule changes all its physical properties, in particular its acidity (Section 5.3) and its redox properties. Electronically excited molecules are both strong oxidants and reductants (Section 4.1, Figure 4.1). Diffusional encounter of an excited molecule M* with an electron donor D or acceptor A will therefore often result in the formation of a *contact ion pair* or *solvent-separated ion pair*, depending on whether electron transfer occurs at a certain distance with at least one solvent molecule between M* and D or A or only upon direct contact. For example, irradiation of naphthalene in an apolar solution containing *N,N*-dimethylaniline leads to the diffusion-controlled formation of a radical ion contact pair (Scheme 5.1).

The contact ion pair represents an excited supramolecular system called *exciplex*; its multiplicity will initially be that of its precursor contact pair and it has several options for further reaction. It could return to the ground state of the neutral reactants by radiative or nonradiative decay (*return electron transfer*, RET), it could undergo intersystem crossing,[339] it could form a pair of 'free' radical ions by escaping from the solvent cage[340] or it could undergo some reaction such as proton transfer from $D^{\bullet+}$ to $M^{\bullet-}$ or rearrangement of $D^{\bullet+}$ to some product $D'^{\bullet+}$.[341] Its actual fate will depend strongly on solvent polarity and also on the presence of a magnetic field.[339]

The Coulombic attraction between two radical ions of opposite charge amounts to $e^2/(4\pi\varepsilon_0\varepsilon_r a)$ in a solvent of relative static permittivity $\varepsilon_r$ (formerly called dielectric constant), where $\varepsilon_0$ is the electric constant (also called permittivity of vacuum) and $a$ is a parameter for the average distance between the centres of charge. In vacuum, the electrostatic work term to separate two singly charged ions is huge, $e^2/(4\pi\varepsilon_0\varepsilon_r a) = 14.4\,\text{eV}$ or $1.39 \times 10^3\,\text{kJ}\,\text{mol}^{-1}$ for $a = 100\,\text{pm}$ ($\varepsilon_r = 1$), but it is strongly reduced in polar solvents. In water ($\varepsilon_r = 78$), it amounts to only $18\,\text{kJ}\,\text{mol}^{-1}$, so that ion separation occurs on the time scale of nanoseconds.

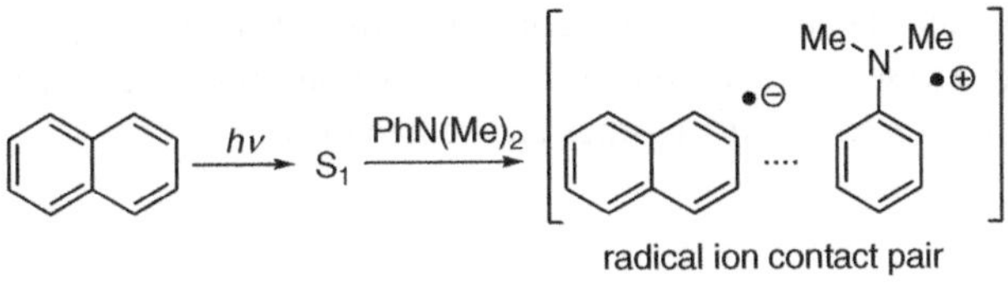

radical ion contact pair

**Scheme 5.1**   *Electron transfer*

The *Gibbs free energy of photoinduced electron transfer*, $\Delta_{et}G°$, in an excited encounter complex $(D{\cdot\cdot}A)^*$ can be estimated from Equation 5.1, where $E°(D^{\bullet+}/D)$ is the standard electrode potential of the donor radical cation, $E°(A/A^{\bullet-})$ that of the acceptor A and $\Delta E_{0-0}$ is the $0-0'$ excitation energy of the excited molecule ($D^*$ or $A^*$) that participates in the reaction.

$$\Delta_{et}G° = N_A\{e[E°(D^{\bullet+}/D) - E°(A/A^{\bullet-})] + w(D^{\bullet+}, A^{\bullet-}) - w(D, A)\} - \Delta E_{0-0}$$

**Equation 5.1**   *Free energy of photoinduced electron transfer*

In the general case, the electrostatic work terms $w$ that account for the Coulombic attraction of reactants (D, A) and products $(D^+, A^-)$ are $w(D, A) = z_D z_A e^2/(4\pi\varepsilon_0\varepsilon_r a)$ and $w(D^+, A^-) = z_{D+} z_{A-} e^2/(4\pi\varepsilon_0\varepsilon_r a)$, where $z_D$ and $z_A$ are the charge numbers of donor D and acceptor A prior to electron transfer and $z_{D+}$ and $z_{A-}$ are those after electron transfer. For neutral species D and A, $z_D = z_A = 0$. Choosing convenient units and collecting the constants we obtain the scaled Equation 5.2.

$$\frac{\Delta_{et}G°}{kJ\,mol^{-1}} = 96.49\frac{E°(D^{\bullet+}/D) - E°(A/A^{\bullet-})}{V} + 138.9\frac{z_{D+}z_{A-} - z_D z_A}{\varepsilon_r a/nm} - 119.6\frac{\Delta E_{0-0}}{\mu m^{-1}}$$

**Equation 5.2**   *Scaled equation for the free energy of electron transfer*

Mixing electron donor and acceptor molecules in apolar solution may be accompanied by the formation of weakly associated charge-transfer or *electron-donor–acceptor(EDA) complexes* in the ground state.[342] In a simple quantum mechanical treatment, the ground-state wavefunction of EDA complexes may be described as a resonance hybrid of a no-bond ground state $(A{\cdot\cdot}D)$ and a charge-transfer state $(A^-{\cdot\cdot}D^+)$ (Equation 5.3), where $c_{0,no\,bond} \approx 1$ and $c_{0,dative} \approx 0$.

$$\Psi_0 = c_{0,no\,bond}\,\Psi(A{\cdot\cdot}D) + c_{0,dative}\,\Psi(A^-{\cdot\cdot}D^+)$$

**Equation 5.3**   *Wavefunction of an EDA ground state complex*

The charge-transfer excited state is then represented by Equation 5.4, with $c_{1,no\,bond} \approx 0$ and $c_{1,dative} \approx 1$.

$$\Psi_1 = c_{1,no\,bond}\,\Psi(A{\cdot\cdot}D) + c_{1,dative}\,\Psi(A^-{\cdot\cdot}D^+)$$

**Equation 5.4**   *Wavefunction of an excited charge-transfer state*

These wavefunctions account for the attractive force between A and D leading to the formation of an EDA complex, for its increased polar character and also for the existence of charge-transfer absorption that is often observed when EDA complexes form. Low-lying excited states of $A^-$ or $D^+$ must be included in the wave function to describe additional charge-transfer bands. These bands appear in addition to the absorption bands of the molecules A and D; they are usually broad and of low oscillator strengths because there is little overlap between the HOMO of the donor and the LUMO of the acceptor (Section 4.4, Equation 4.20). Many examples of the formation of EDA complexes between, for example, tetracyanoethylene and aromatic hydrocarbons are known. The

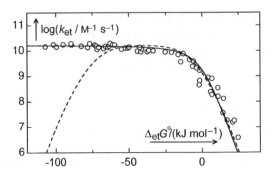

**Figure 5.1**   Rehm–Weller plot; Equation 5.5 was fitted to the data. The dotted line shows a fit of the Marcus Equation 5.7 to the data $\Delta_{et}G° \geq -50\,kJ\,mol^{-1}$ [344]

charge-transfer complexes of tetracyanoethylene with methylated benzenes cover the entire visible region; the maxima are red shifted with increasing methyl substitution, from $\tilde{\nu}_{max} = 2.6\,\mu m^{-1}$ for benzene to $\tilde{\nu}_{max} = 1.9\,\mu m^{-1}$ for hexamethylbenzene.[343] Such solutions can be useful as filters to attenuate the maximum in the spectrum of xenon arcs to produce a more evenly distributed 'white' light source for spectral analysis.

Electron transfer from or to an excited molecule results in fluorescence quenching. In the absence of EDA complex formation, bimolecular rate constants of electron transfer quenching can be determined from the slope of a Stern–Volmer plot of fluorescence quenching combined with measurements of the fluorescence lifetime of M* (Section 3.9.8, Equation 3.36). In a seminal paper, Rehm and Weller[344] reported fluorescence quenching rate constants $k_q$ for four excited aromatic compounds (3,4-benzacridine, anthracene, 1,2-benzanthracene and 1,12-benzperylene) by a series of amino- and methoxy-substituted benzene derivatives as donor molecules in acetonitrile solution. When $\log(k_q/M^{-1}\,s^{-1})$ was plotted against the free energy of electron transfer (Equation 5.2), the data followed a curve approaching the diffusion rate constant $k_d \approx 1.6 \times 10^{10}\,M^{-1}\,s^{-1}$ for $\Delta_{et}G° \leq 0$ and falling off at $\Delta_{et}G° \geq 0$ (Figure 5.1).

The mechanism of quenching had previously been established by observing the formation of free radical ions using flash photolysis.[345] Rehm and Weller proposed the empirical Equation 5.5 to fit the data, where $\Delta_{et}G°$ is the free energy of photoinduced electron transfer in the contact pair (Equation 5.1), $\Delta G^{\ddagger}$ is the free energy of activation that accounts for the structural and solvent reorganization required for the transfer of an electron, $k_d$ and $k_{-d}$ are the rate constants for the formation and separation of the encounter complex, respectively, $K_d = k_d/k_{-d}$ is the equilibrium constant of complex formation and $Z$ is the bimolecular collision frequency in an encounter complex, $Z \approx 10^{11}\,s^{-1}$.[346] A value of $k_d/(ZK_d) = 0.25$ was used.

$$k_q = \frac{k_d}{1 + \dfrac{k_d}{K_d Z}\left(e^{\Delta_{et}G^{\ddagger}/RT} + e^{\Delta_{et}G°/RT}\right)},$$

$$\text{where } \Delta_{et}G^{\ddagger} = \frac{\Delta_{et}G°}{2} + \sqrt{\left(\frac{\Delta_{et}G°}{2}\right)^2 + \left(\Delta_{et}G^{\ddagger}(0)\right)^2}$$

**Equation 5.5**   *Rehm–Weller equation*

The two parameters of Equation 5.5, $k_d$ and $\Delta G^{\ddagger}(0)$, the free energy of activation when $\Delta_{et}G^{\circ} = 0$, may be determined by nonlinear least-squares fitting to the data, giving $\Delta G^{\ddagger}(0) = 90 \pm 10\,\mathrm{kJ\,mol}^{-1}$ and $k_d = (2.2 \pm 0.3) \times 10^{10}\,\mathrm{M}^{-1}\mathrm{s}^{-1}$. Clearly, the Rehm–Weller equation fits the data well (Figure 5.1), but it is in disagreement with a bold prediction made earlier by Marcus,[346] namely that electron transfer rates should decrease not only when the reaction is endergonic, $\Delta_{et}G^{\circ} > 0$, but also as it becomes strongly exergonic, $\Delta_{et}G^{\circ} \ll 0$, the so-called *Marcus inverted region*. The authors noted that Marcus's theory (see Equation 5.7 below) would fit the data for $\Delta_{et}G^{\circ} \geq -50\,\mathrm{kJ\,mol}^{-1}$ equally well, but that it deviates by up to four orders of magnitude at more negative values of $\Delta_{et}G^{\circ}$, as shown by the dashed curve in Figure 5.1.

Even when the series was extended to include values of $\Delta_{et}G^{\circ}$ down to $-250\,\mathrm{kJ\,mol}^{-1}$, the observed quenching rate constants remained at the diffusion limit.[347] For many years it was a matter of controversy why the predicted decrease in the inverted region was not observed in the Rehm–Weller experiment nor in related data obtained later. Rehm and Weller considered possible explanations such as H-atom transfer (i.e. essentially simultaneous electron and proton transfer) or the formation of a low-lying excited state of one of the radical ions, resulting in a free energy change $\Delta_{et}G^{\circ}$ less negative than that calculated by Equation 5.1, but these were ruled out for some of the data points. For a personal, historical perspective on Weller's work, see ref. 348.

In simplified form, classical Marcus theory[346] may be recast as follows. Imagine a reaction coordinate $Q$ that reflects the structural changes of an $(A\cdot\cdot D)^*$ encounter pair and of the surrounding solvent when an electron is transferred within $(A\cdot\cdot D)^*$ to give $(A^-\cdot\cdot D^+)$ [Figure 5.2 (left)]. The potential energy surfaces along the reaction coordinate $Q$ are approximated as parabolas with $(A\cdot\cdot D)^*$ on the left-hand side (thick solid line) and several parabolas on the right that are shifted vertically to represent the situation for various $(A\cdot\cdot D)^*$ pairs that have different free energies of electron transfer, $\Delta_{et}G^{\circ}$.

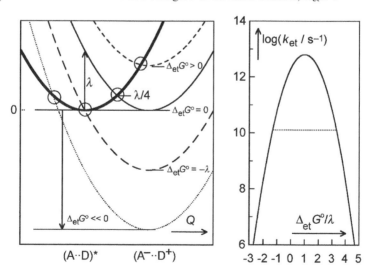

**Figure 5.2** Left: potential energy curves for an electron-transfer reaction in an encounter complex of an acceptor A and a donor D, $(A\cdot\cdot D)^* \rightarrow (A^-\cdot\cdot D^+)$. Right: rate constants predicted by combining Equations 5.6 and 5.7. The diagrams are explained in the text

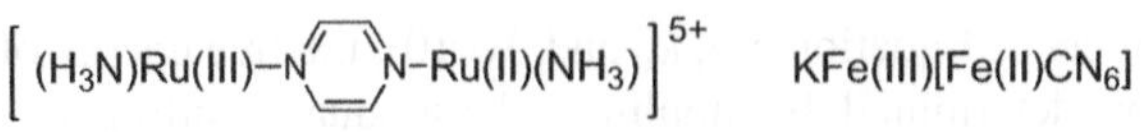

$$\left[ (H_3N)Ru(III)-N\underset{}{\overset{}{\bigcirc}}N-Ru(II)(NH_3) \right]^{5+}$$        $KFe(III)[Fe(II)CN_6]$

8                                        Prussian blue

**Figure 5.3**

Consider first the case of a thermoneutral reaction, $\Delta_{et}G° = 0$ (solid line). It could represent the case of electron transfer between identical molecules, that is, $(M^-\cdot\cdot M) \rightarrow (M\cdot\cdot M^-)$. The reaction coordinate $Q$ accounts for all the structural changes associated with accommodating the excess electron on one or the other molecule. The case $\Delta_{et}G° = 0$ is also realized in the famous Creutz–Taube[349] ion **8** (Figure 5.3) and related so-called mixed-valence compounds.[350] Compound **8** exhibits a broad, moderately intense (oscillator strength $f \approx 0.03$) electronic absorption band in the near-IR region ($\lambda_{max} = 1560$ nm in $H_2O$) that is not observed in model compounds in which both Ru atoms are in an oxidation state of either II or III. The colour of Prussian Blue is also attributed to a charge-transfer absorption, moving an electron from Fe(II) to Fe(III), which are in a different environment, so that $\Delta_{et}G° \neq 0$.

The energy of the IR transition in compound **8**, marked $\lambda$ in Figure 5.2, represents the Franck–Condon energy of electron transfer from Ru(II) to Ru(III) and is called *reorganization energy*. Simple geometric considerations show that the crossing point of the two parabolas lies at $\lambda/4$. Ignoring entropy terms, $\lambda/4$ corresponds to the quantity $\Delta G^{\ddagger}(0)$ used by Rehm and Weller, the activation free energy for thermal self-exchange in **8**. However, the parabolas represent zero-order adiabatic curves, which will be split apart by consideration of the electronic coupling term $V_{12}$ between the zero-order states at the crossing point. The transition probability can then be estimated by the Landau–Zener equation (Equation 2.53). Recent work indicates that the stabilization by $V_{12}$ is in fact comparable to $\lambda/4$ in compound **8**, so that the thermal barrier nearly vanishes.[351] Electron transfer between metal centres sharing a common ligand is called *inner-sphere electron transfer*.

Returning to intermolecular electron transfer (*outer-sphere electron transfer*), we assume that the electronic coupling term $V_{12}$ in the encounter pair $(A\cdot\cdot D)^*$ is sufficient to ensure a high probability of electron transfer at the crossing points, but much smaller than $\Delta G^{\ddagger}(0) = \lambda/4$. The rate of electron transfer is then given by Eyring's transition state theory (Equation 5.6).

$$k_{et} = \frac{k_B T}{h}e^{-\Delta G^{\ddagger}/RT}, \text{ where } \frac{k_B T}{h} = 6.21 \times 10^{12} \text{ s}^{-1} \text{ at } T = 298K$$

**Equation 5.6**

When the curves for the radical ion pair $(A^-\cdot\cdot D^+)$ are moved up or down to represent some change in $\Delta_{et}G°$, the geometric position of the crossing points (marked by circles in Figure 5.2) depends on the reorganization energy $\lambda$ and on the free energy of electron transfer $\Delta_{et}G°$ (Equation 5.7).

$$\Delta G^{\ddagger} = \frac{\lambda}{4}\left(1 + \frac{\Delta_{et}G°}{\lambda}\right)^2$$

***Equation 5.7***   *Classical Marcus relationship for the activation free energy of electron transfer*

By inserting Equation 5.7 into Equation 5.6, one obtains the inverted parabola predicted by classical Marcus theory for the rate constants $k_{et}$ (Figure 5.2, right). The left-hand side of the curve, $\Delta_{et}G° < -\lambda$, where $k_{et}$ decreases with decreasing $\Delta_{et}G°$, has been called the *Marcus inverted region*. Clearly, the rate of diffusion will limit the electron transfer rate constants, as shown by the dotted horizontal line in Figure 5.2. However, for highly exergonic reactions, the decrease should reappear.

The existence (or non-existence) of an inverted region is not a matter of purely academic interest. *Back electron transfer* or *return electron transfer* (RET) presents a challenge to all attempts to store photoinduced charge separation as an electrochemical driving force, for example in photosynthesis (Special Topic 6.25) or in photovoltaic cells (Special Topic 6.31). RET from the initially formed ion pair $(A^- \cdot\cdot D^+)$ to the ground-state contact pair $(A \cdot\cdot D)$ is an energy-wasting process; the electrochemical potential is dissipated as heat. The rate constants of (usually unwanted) RET processes that lie deeply in the inverted region are, therefore, a decisive factor for the efficiency of any solar energy storage device.

Clear-cut experimental evidence for the existence of an inverted region was provided in 1984 by the work Miller *et al.*[352] on long-distance intramolecular electron transfer in the rigidly linked bichromophoric radical anions **9** (Figure 5.4), which were generated by pulsed electron injection. The electrons are initially captured with nearly equal probability by either the 'donor' chromophore biphenyl or the acceptor chromophore mounted on the other side of the saturated 5α-androstane spacer. Electron transfer rates to attain equilibrium were determined by time-resolved observation of the ensuing absorbance changes.

The temperature dependence of the electron transfer rates in compounds **9** was subsequently measured in the range of 170–373 K.[353] The rate constants of the compound with 2-naphthyl as an electron acceptor exhibited an Arrhenius-type dependence, as expected for an essentially thermoneutral reaction with an activation energy of $\lambda/4$, but those with the 2-benzoquinonyl and 5-chloro–2-benzoquinonyl acceptor groups were essentially *independent* of temperature. It became clear that Marcus's prediction of an 'inverted region' had been right, but for the wrong reason.

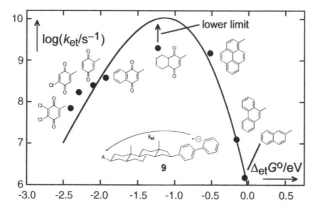

**Figure 5.4** Intramolecular electron transfer rate constants of **9** in 2-methyl-THF. Adapted by permission from ref. 352. Copyright American Chemical Society

The decrease in electron transfer rate constants in the inverted region is *not* due to the reappearance of a barrier, as shown for the lowest (dotted) parabola in Figure 5.2 (left). Rather, exergonic electron transfer should be considered as a radiationless transition. The rate dependence on the free energy difference $\Delta_{et}G°$ in the inverted region is just another manifestation of the energy gap law (Section 2.1.5), which predicts decreasing rates with increasing energy gaps. Theoretical treatments of electron-transfer reactions in the inverted region that in the meantime had been recast[354,355] in terms of Fermi's golden rule (Equation 2.21), are in agreement with these findings and many detailed experimental studies have since established these relationships.[340,356–359]

The connection between electron transfer and triplet–triplet energy transfer is of interest.[360] A quantum mechanical treatment of triplet energy transfer due to Dexter describes triplet energy transfer as a synchronous, dual exchange of electrons (Section 2.2.2, Figure 2.17). This has been 'verified' experimentally:[360] rate constants of the separate electron transfer processes and also of triplet–triplet energy transfer were measured for a series of bichromophoric compounds D–sp–A (sp for spacer) in experiments similar to those described above for compounds **9**. The rate constants for the reactions $D^{\bullet-}$–sp–A → D–sp–$A^{\bullet-}$ (electron transfer, $k_{et}$), $D^{\bullet+}$–sp–A → D–sp–$A^{\bullet+}$ ('hole transfer', $k_{ht}$) and $^3D$–sp–A → D–sp–$^3A$ (triplet energy transfer, $k_{TT}$) with A = 2-naphthyl and various donor chromophores D each varied over several orders of magnitude. Remarkably, the rate constants for triplet energy transfer, which varied over five orders of magnitude ($10^5$–$10^{10}\,s^{-1}$), were in excellent agreement with those calculated as $k_{TT} = Ck_{et}k_{ht}$ with $C = 4 \times 10^{-10}\,s$.

The phenomenon of *intramolecular electron transfer* is observed not only in bi- or polychromophoric systems that are connected through a saturated spacer such as compounds **9**, but also in systems where the donor and acceptor groups are formally conjugated. The first clearly recognized[361] and most investigated[362] example is that of 4-(*N,N*-dimethylamino)benzonitrile (DMABN, Scheme 5.2), which emits two fluorescence bands, whose position and relative intensity depend strongly on solvent polarity (*dual fluorescence*). Absorption initially populates the so-called locally excited (LE) states, because the oscillator strengths of charge-transfer transitions are very low.[a] The subsequent formation of intramolecular *charge transfer* (*CT*) *states* is often associated with small energy barriers indicating some structural changes. Charge separation occurs even in symmetrical compounds such as 9,9′-bianthryl (BA). The dipole moments of the CT states are usually huge, on the order of 10–30 D, consistent with the near-complete charge separation in the CT state. The large dipole moments of CT states are responsible for the large solvatochromic shifts of the fluorescence emission from CT states.

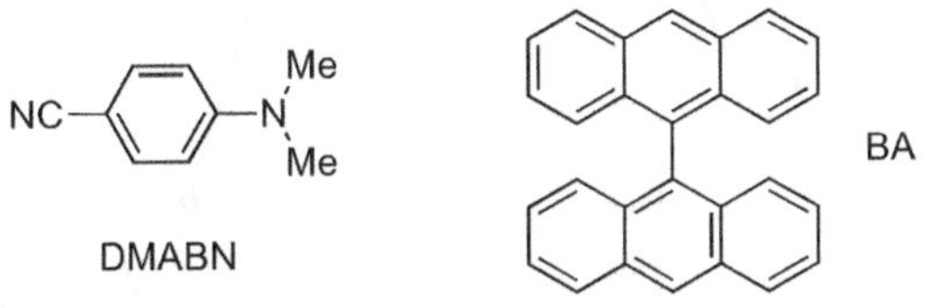

DMABN

BA

**Scheme 5.2**  *Prototype conjugated bichromophoric compounds undergoing intramolecular charge transfer*

---

[a]The transition moment, Equation 4.20, is not related to the change in dipole moment upon exitation. If the orbitals $\Psi_i$ and $\Psi_j$ do not overlap in space, then $M_{el,i\rightarrow j}=0$.

An extensive review of such systems focusing on the structural changes associated with intramolecular electron transfer has appeared.[363] The much-debated structural issue led to a somewhat confusing multitude of acronyms that are used in discussions of these phenomena: Twisted intramolecular charge transfer (TICT), planar intramolecular charge transfer (PICT) and simply intramolecular charge transfer state (ICT). The conversion from LE to ICT states is generally very fast (<1 ps) and, due to the associated sudden increase in polarity, such compounds are useful as probes to study solvent relaxation dynamics.[364] The acronym ICT should be used in the absence of clear-cut information about structural changes associated with CT.

The radical ion pairs generated by photoinduced electron transfer have several options for further reaction (Section 5.4.3).[341,365]

---

## Special Topic 5.1: Electron transfer in biopolymers

Over the last two decades, thousands of studies have been devoted to the biologically important process of electron transfer (ET) in DNA and peptides. Some proteins transport charge over long distances.[366] For example, in the enzyme ribonucleotide, which plays an essential role in DNA replication and repair by providing all of the monomeric precursors (deoxynucleotides) required for these processes, a tyrosine radical is generated 3.5 nm away from the site where ribonucleotide is reduced to deoxyribonucleotide. To measure rate constants of ET, the reaction is typically initiated with a short pulse of light. Site-selective charge injection is required to determine the dependence of ET rates on the structure of the biopolymer. In redox-labelled metalloproteins, a small-molecule oxidant or reductant and a chromophore for photoexcitation are attached covalently to a polypeptide side-chain of a redox protein. In enzymes that include a haem cofactor, Fe may be replaced with Zn, thus permitting the ET reaction to be photoinitiated by production of the excited state of Zn porphyrin of the modified cofactor (see also Section 6.4.4). Another method to study fixed-distance ET in biology relies on the modification of a cofactor in a multicofactor protein or enzyme. The archetype here is the bacterial reaction centre.

DNA damage may be induced by oxidation of a nucleobase due to oxidative stress or UV irradiation. In duplex DNA, the resulting positive charge ('electron hole') can migrate over distances of several nanometres before reaction with water and oxygen takes place yielding mutagenic products such as 8-oxoguanine. This long-distance hole transfer might serve to protect the coding areas of DNA against oxidative stress. The intercalation of a donor and an acceptor along the DNA chain has been used to study ET rates (Case Study 6.31). Alternatively, a DNA base, typically guanine, acts as the electron donor to an excited acceptor such as stilbenedicarbox-amide (two-point attachment, hairpin)[367–370] or anthraquinone[371,372] incorporated at the end of the DNA strand. Tetrahydrofuranyl radicals carrying a phosphate leaving group in the β-position undergo rapid heterolytic cleavage yielding dihydrofur-anosyl radical cations (Scheme 5.3). The system has been utilized for charge injection at defined positions in DNA and peptides to study long-distance electron transfer reactions.[373–376]

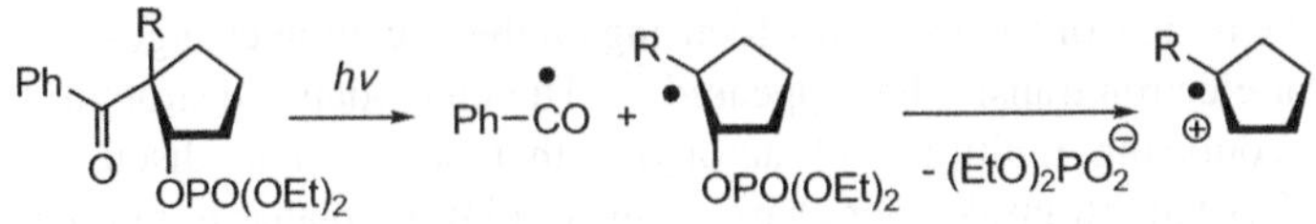

Scheme 5.3

Initial controversies[377] regarding the long-range conductivity properties and wire-type behaviour of DNA have been settled and it is now established that charge transport in biopolymers does occur on fast time scales and over large distances, following different mechanistic regimes (Scheme 5.4).[367,378,379] Over short distances, ET occurs by the so-called *superexchange mechanism*. Here, the rate constant is calculated by Fermi's golden rule (Equation 2.21). The rate constant is expected to fall off exponentially with increasing distance $r$ between the electron donor $D^*$ and the acceptor A, $k_{et} \propto \exp(-\beta r)$. The parameter $\beta$ was calculated to be on the order of $(14 \pm 2)$ nm$^{-1}$. Thus, ET by superexchange becomes negligible for distances on the order of 1 nm. For ET traversing many base pairs, the *hopping mechanism* comes into play. Here, the heterocyclic bases of DNA act as charge carriers for hole migration, particularly guanine, which has the lowest oxidation potential. The distance dependence of the hole-transfer process is significantly influenced by the DNA sequence.[380] Adjacent guanine pairs (GG) and triplets (GGG) act as hole traps. In peptides, the aromatic side-chains of amino acids act as intermediate charge carriers.[373–376] Long-distance electron transport through the biopolymers DNA and proteins occurs in a multistep hopping reaction.

superexchange                      hopping

**Scheme 5.4**    *Superexchange and hopping mechanisms of charge transfer*

The distance dependence of the individual hopping steps is smaller in DNA than in peptides because DNA is more efficient than the peptides as a mediator of the charge transport. On the other hand, peptides are more flexible than DNA double strands and the distances between side-chains of amino acids in peptides can be reduced by conformational changes. The influence of peptide sequence and tertiary structure still remains to be investigated in detail. Investigations of ribonucleotide reductase, DNA photolyase and photosystem II point to the importance of aromatic amino acids such as tyrosine and tryptophan for ET in these proteins.

The electrical current across a few DNA molecules at least 600 nm long has been measured as a function of the potential applied.[381]

## 5.3    Proton Transfer

The acidity of excited molecules often differs substantially from that in the ground state and the ionization of excited acids may occur adiabatically (Section 2.3), yielding the

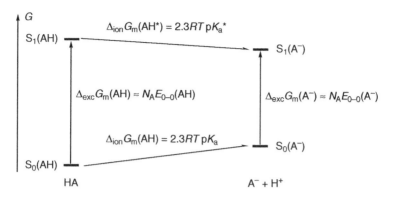

**Figure 5.5** Förster cycle

conjugate base in its excited state. Acidity constants $K_a^*$ of the excited singlet states can be determined from a thermodynamic cycle that makes use of the acidity in the ground state, $pK_a$, and the absorption spectra of the acid and its conjugate base. This cycle is named after Theodor Förster, who first used it to interpret his observations of fluorescence from the conjugate bases of phenols and aromatic amines after excitation of the neutral compounds. The *Förster cycle* for an acid AH and its conjugate base $A^-$ is shown in Figure 5.5.

The molar free energy change for the ionization of the acid AH in aqueous solution is $\Delta_{ion}G_m(AH, aq) = G_m(A^-, aq) + G_m(H^+, aq) - G_m(AH, aq)$. For convenience, we set $pH = 0$, so that $G_m(H^+, aq) \equiv 0$ and drop the label aq (aqueous solution). The molar free energy change for the ionization of the electronically excited acid $AH^*$ is then given by the thermodynamic cycle (Equation 5.8), where $\Delta_{ion}G_m(AH^*) = RT\ln(10)pK_a^*$ and $\Delta_{ion}G_m(AH) = RT\ln(10)pK_a$ are the free energies of ionization of the acid HA in its excited and ground state, respectively, and $\Delta_{exc}G_m(AH)$ and $\Delta_{exc}G_m(A^-)$ are the excitation free energies of AH and $A^-$, respectively.

$$\Delta_{ion}G_m(AH^*) - \Delta_{ion}G_m(AH) = \Delta_{exc}G_m(A^-) - \Delta_{exc}G_m(AH)$$

*Equation 5.8*

In order to place the transition energies $E_{0-0}(AH)$ and $E_{0-0}(A^-)$, which are determined from the absorption spectra of AH and $A^-$, on the free energy scale, we ignore volume and entropy effects, that is, we assume that the changes in volume and entropy associated with ionization in the ground state, $\Delta_{ion}V_m(AH)$ and $\Delta_{ion}S_m(AH)$, are similar to those in the excited state (Equation 5.9).

$$\Delta_{ion}V_m(AH) \approx \Delta_{ion}V_m(AH^*) \text{ and } \Delta_{ion}S_m(AH) \approx \Delta_{ion}S_m(AH^*)$$

*Equation 5.9*

With the general relation $\Delta G = \Delta U + p\Delta V - T\Delta S$, the excitation free energies on the right-hand side of Equation 5.8 can now be replaced by the excitation energies, $\Delta_{ion}G_m(AH^*) - \Delta_{ion}G_m(AH) \approx \Delta_{exc}U_m(A^-) - \Delta_{exc}U_m(AH) = N_A[E_{0-0}(A^-) - E_{0-0}(AH)]$. Insertion of the Einstein Equation 1.4, $E_p = hc\tilde{v}$, yields Equation 5.10.

$$pK_a^* = pK_a + 21.0 \left[ \frac{\tilde{v}_{0-0}(A^-)}{\mu m^{-1}} - \frac{\tilde{v}_{0-0}(AH)}{\mu m^{-1}} \right]$$

where $\dfrac{N_A hc}{RT \ln 10} 10^6 \mu m^{-1} = 21.0$ for $T = 298$ K

**Equation 5.10**   *Förster equation*

The errors associated with the approximations of Equation 5.9 are negligible in practice. Much more important are uncertainties associated with estimating the wavenumbers $\tilde{v}_{0-0}$ from diffuse absorption spectra. In the absence of well-defined 0–0 bands, the positions of the 0–0 transitions of AH and A$^-$ are best estimated as the intersection points of the corresponding absorption and fluorescence spectra when the intensity of the fluorescence spectrum is set equal to the maximum of the absorption spectrum (Figure 5.6).

---

### Case Study 5.1: Mechanistic photochemistry – adiabatic proton transfer reactions of 2-naphthol and 4-hydroxyacetophenone

Förster first observed the adiabatic deprotonation of 2-naphthol (**10**) in the excited singlet state and thereby exemplified the use of Equation 5.10. The absorption and fluorescence emission spectra of **10** in aqueous acid and base are shown in Figure 5.6. The ground state acidity constant of **10** was determined by spectrophotometric titration, $pK_a = 9.5$.

The wavenumbers of the 0–0 transitions for the acid, AH, and the base, A$^-$, estimated from the intersection points of the corresponding absorption and fluorescence spectra, are $\tilde{v}_{0-0}(AH) = 2.95 \, \mu m^{-1}$ and $\tilde{v}_{0-0}(A^-) = 2.65 \, \mu m^{-1}$. By inserting these data into Equation 5.10 we obtain $pK_a^* = 3.2$ for the acidity constant of **10** in the excited singlet state. Thus the acidity of **10** is predicted to increase by over six orders of magnitude upon excitation to the lowest singlet state. The prediction was tested by a fluorescence titration: The relative fluorescence intensity at the emission

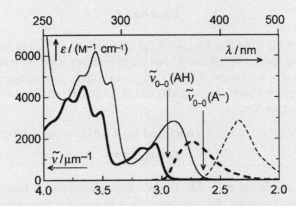

**Figure 5.6**  Absorption (solid) and fluorescence (dashed) spectra of 2-naphthol (**10**) in aqueous acid (0.1 M HClO₄, thick lines) and aqueous base (0.1 M KOH, thin lines)

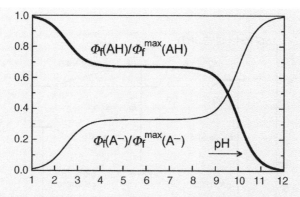

**Figure 5.7** Fluorescence titration of 2-naphthol (**10**)

maximum of the acid AH, $\lambda_{max} = 355$ nm, and of the base A$^-$, $\lambda_{max} = 425$ nm, was measured as a function of pH (Figure 5.7).

The acidity constants calculated by the Förster cycle are thermodynamic values. However, the acid–base equilibrium is rarely established fully during the lifetime of the excited singlet states of acid and base. At pH values well below the ground-state $pK_a$, only AH is excited. The fluorescence of AH drops and that of A$^-$ rises around pH 3, indicating that adiabatic deprotonation of $^1$AH$^*$ forming $^1$A$^{-*}$ occurs (Figure 5.7). However, the fluorescence of AH is reduced by only about 30% at pH $> 3$ and an equivalent amount of fluorescence from $^1$A$^{-*}$ appears. The emission from AH disappears completely only when the pH exceeds the $pK_a$ of AH in the ground state, so that only $^1$A$^-$ is excited. Hence the rate constants of fluorescence and of adiabatic ionization of $^1$AH$^*$ are of comparable magnitude.

The sharply increased acidity of phenols in the excited state can be used to lower the pH of aqueous solutions by a pulsed light source within nanoseconds (*photoacid*). However, the equilibrium is rapidly re-established in the ground state by diffusion-controlled recombination of the released protons with the basic phenolates.

Adiabatic protolytic equilibria in the triplet state are generally fully established due to the intrinsically longer lifetimes of triplets. Soon after Förster's work, Jackson and Porter determined the acidity of 2-naphthol (**10**) in the triplet state by flash photolytic titration.[382] The triplet–triplet absorption of **10** changes from $\lambda_{max} = 432$ to 460 nm as the pH is moved above the triplet $pK_a^*$(**10**) of 8.1. Triplet state acidity can also be predicted using the Förster cycle. The triplet excitation energies $E_T$ of the acid and its conjugate base are determined from the 0–0 bands of their phosphorescence spectra.

4-Hydroxyacetophenone (**11**) is an interesting example of proton transfer in the triplet state,[383] because the basicity of its carbonyl group increases strongly upon excitation. The $pK_a$ of **11** in the ground state is 7.9. Pump–probe absorption spectra of **11** in aqueous solution, pH $< 7$, reveal that ISC to the triplet state $^3$**11** ($\lambda_{max} = 395$ nm) is very fast, $k_{ISC} = 2.7 \times 10^{11}$ s$^{-1}$. Subsequent ionization of $^3$**11** to the anion $^3$**11**$^-$ ($\lambda_{max} = 405$ nm) occurs within 50 ns at pH $> 3.6$. The latter is, however, rapidly re-protonated at the carbonyl oxygen at pH 4. After equilibration (200 ns), one observes the triplet–triplet absorption of the enol **12** ($\lambda_{max} = 360$ nm). The triplet–triplet absorption of $^3$**11**$^-$ ($\lambda_{max} = 405$ nm) is long-lived when the pH is raised above 7, above the $pK_a^*$ of **12**.

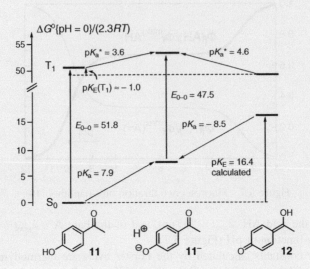

**Figure 5.8**  Protonation equilibria of 4-hydroxyacetophenone (**11**) in the triplet state

Thus, *in the pH range 4–6, the triplet enol tautomer 12 is thermodynamically more stable than the triplet ketone 11.* The spectroscopic and thermodynamic data are summarized in Figure 5.8. When the enol **12** decays to the ground state, it becomes an extremely strong acid and it immediately protonates water to give the anion **11$^-$**. The p$K_a$ of enol **12** in the ground state was estimated at $-8.5$ on the basis of calculations for the equilibrium between **11** and **12** and the measured triplet p$K_a^*$(**12**) $= 4.6$.

---

### Special Topic 5.2: Isotope effects and tunnelling

Isotopes play a multifaceted role in mechanistic chemistry. Isotopic labelling at specific sites of a reagent is often used to support or exclude a hypothetical reaction mechanism based on the location of the isotopic label in the products. *Kinetic isotope effects*, that is, changes in rate constants that are induced by isotopic substitution, may be seen in time-resolved experiments or may express themselves as changes in product quantum yields. They hold important information regarding the nature of the rate-determining step that limits the overall rate in a sequence of reactions.

Deuterium isotope effects are studied most often, because deuterium is relatively cheap, the synthesis of deuterium-labelled compounds is frequently straightforward and the effects are large and thus easy to measure with sufficient accuracy. Kinetic isotope effects can be attributed largely to the difference in the zero-point energies of the C–H and C–D stretching vibrations in the reactant, $E_0 = h\nu/2$. The vibrational frequency $\nu$ is related to the force constant $k$ of the (harmonic) potential and the reduced mass $\mu$ of the oscillator (Equation 5.11).

$$\nu = \frac{1}{\pi}\sqrt{\frac{k}{\mu}}, \quad \mu = \frac{m_X m_Y}{m_X + m_Y}$$

**Equation 5.11** *Vibrational frequency of a harmonic oscillator*

In Section 2.1.5 (Figure 2.7), we encountered kinetic isotope effects on the rates of radiationless decay, which were attributed to the large difference between the frequencies of C–H and C–D stretching vibrations. The hydrogen isotopes X being much lighter than the nuclei Y to which they are attached, one can, to a first approximation, neglect $m_X = 1u$ or $2u$ (X = H or D) relative to $m_Y \geq 12u$ in the denominator of Equation 5.11 to obtain $\mu \approx m_X$ and $\nu_H/\nu_D \approx \sqrt{2}$. For a C–H stretch vibration, $\tilde{\nu} \approx 3000\,cm^{-1}$, the difference in the zero-point energies is substantial, $E_0(C–H) - E_0(C–D) \approx hc\tilde{\nu}_H(1 - 1/\sqrt{2})/2 \approx 440\,hc\,cm^{-1}$ or $\sim 5\,kJ\,mol^{-1}$. On the other hand, the H or D atom will be very loosely bound in the transition state of a hydrogen-transfer reaction such as photoenolization (Section 6.3.6). The activation energy $E_a$ for D transfer will, therefore, exceed that of H transfer by about $5\,kJ\,mol^{-1}$. One can then estimate from the Arrhenius equation, $k = A\exp(-E_a/RT)$, that the ratio of the corresponding rate constants, $k_H/k_D$, may approach 10 at room temperature and higher values below. Isotope effects $k_H/k_D \approx 5{-}10$ are called *primary isotope effects* and they indicate that the proton is 'in flight' and therefore loosely bound in the transition state of the observed reaction step. If the role of the isotopically labelled atom is only that of a spectator located near the reaction centre in the transition state, kinetic isotope effects are much smaller, $k_H/k_D \approx 1.0 \pm 0.1$ (*secondary isotope effects*).

According to the empirical Arrhenius equation, the logarithm of the rate constant $k$ should fall off linearly with a slope of $-E_a/RT$ when plotted against $1/T$. However, when the temperature is strongly reduced, one sometimes finds that $\log(k/s^{-1})$ levels off, as shown in Figure 5.9, where the following Arrhenius parameters were assumed: $A = 1 \times 10^{13}\,s^{-1}$, $E_a(X = H) = 6\,kJ\,mol^{-1}$ and $E_a(X = D) = 10\,kJ\,mol^{-1}$. In addition, a constant term $k_t$ was added to the Arrhenius equation, $k = A\exp(-E_a/RT) + k_t$, namely $k_t(X = H) = 1 \times 10^{12}\,s^{-1}$ and $k_t(X = D) = 1 \times 10^{10}\,s^{-1}$.

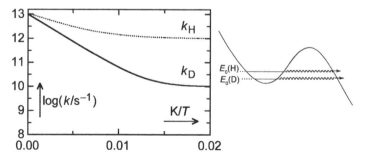

**Figure 5.9** Arrhenius plot for a reaction with tunnelling

The temperature-independent term $k_t$ is attributed to *tunnelling* of the hydrogen or deuterium atom through the barrier, a quantum mechanical phenomenon. The rate of tunnelling depends on the shape (the height and width) of the barrier and on the mass of the particle.[384] Thus electrons tunnel much more easily than nuclei and H tunnels more easily than D and heavier nuclei. Although tunnelling from a given vibrational level requires no activation energy and its rate is therefore independent of temperature, it does proceed more easily from higher levels, where the barrier width and height are reduced (Figure 5.9). The population of upper levers increases at higher temperature. Hence tunnelling contributions to hydrogen atom transfer rate constants may well be substantial even at ambient temperature.[385]

A case in point is the photoenolization reaction of *o*-alkylphenyl ketones (Section 6.3.6) and related compounds. Detailed studies of the temperature dependence of both the primary photoreaction[386–388] and of the back transfer reaction in the ground state[389] have provided convincing evidence for tunnelling contributions in these reactions.

## 5.4   Primary Photochemical Intermediates: Examples and Concepts

*Recommended textbooks.*[390,391]

The wealth of information on reactive intermediates is covered in textbooks of physical organic chemistry and the specialized books quoted above. Many reactions involving such intermediates will be discussed in Chapter 6.

### 5.4.1   Carbenes

Many photoreactions proceed via carbene intermediates (see Schemes 6.169–6.172). Spectroscopic and theoretical attempts to determine the equilibrium bond angle of the parent carbene, methylene ($CH_2$), in its triplet ground state and the energy gap to its lowest singlet state had long given discrepant results that were resolved only in 1984.[392] The triplet ground state of methylene is bent (134°), but the inversion barrier is low and the singlet–triplet energy gap $\Delta E_{ST}$ amounts to $38 \, kJ \, mol^{-1}$.[393] Two valence electrons occupy the two symmetry-adapted NBMOs of $CH_2$, which are a pure p-$\pi$ and an essentially $sp^2$-hybridized $\sigma$ AO (Figure 5.10). The energy of the p-$\pi$ AO is substantially higher than that of the $sp^2$-$\sigma$ AO, which has substantial s-character. The equilibrium bond angle is reduced to 102° in the lowest singlet state. The two unpaired electrons are localized on the same

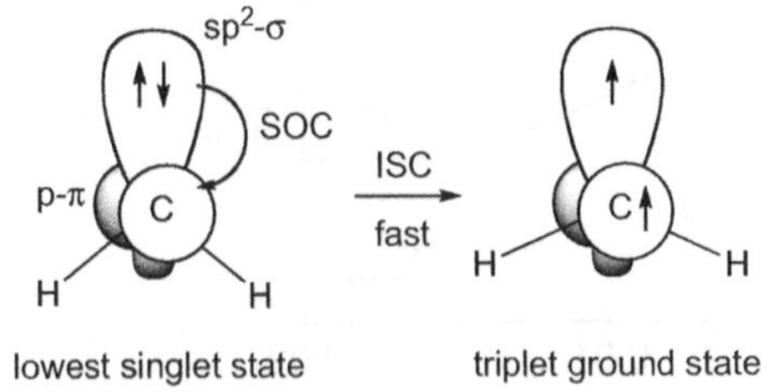

lowest singlet state        triplet ground state

**Figure 5.10**   Ground-state triplet (left) and lowest singlet state (right) configurations of $CH_2$

**Table 5.1** Geometries and singlet–triplet gaps of substituted carbenes X–C–Y[394]

| X | Y | $\Delta E_{ST}/kJ\,mol^{-1}$ | $\phi(X\text{–}C\text{–}Y)$ | |
|---|---|---|---|---|
| | | | Singlet | Triplet |
| F | F | −236 | 104 | 118 |
| Cl | Cl | −84 | 109 | 126 |
| Br | Br | −52 | 112 | 129 |
| H | F | −66 | 102 | 121 |
| H | Cl | −27 | 101 | 124 |
| SiH$_3$ | SiH$_3$ | 82 | 180 | 180 |

atom, so that the exchange integral favouring the open-shell triplet state is large. Nevertheless, the energy gap $\Delta E_{ST}$ of $CH_2$ is rather small, because the closed-shell singlet state is stabilized due to the lower energy of the doubly occupied sp$^2$-σ AO. Thus, Hund's rule favouring the triplet state applies.

Whereas the absolute prediction of singlet–triplet energy gaps in carbenes has long been a challenge for theoretical chemistry, the rationalization of trends in substituted carbenes XCY is straightforward.[394,395] Two kinds of interactions should be considered; inductive effects affect the energy of the σ-AO and conjugative interactions affect the p-π AO. The substituents X and Y may be classified as π-donors ($-NR_2$, $-OR$, $-SR$, halogens) and π-acceptors ($-CO-R$, $NO_2$), which raise and lower, respectively, the energy of the p-π AO. σ-Donors ($-SiR_3$) and σ-acceptors (halogens, especially $-F$) raise and lower, respectively, the energy of the sp$^2$-σ AO. Factors influencing the bond angle such as sterically demanding substituents and geometric constraints in cyclic carbenes should also be considered; bond angles close to 180° favour the triplet state, smaller bond angles near 120° increase the s-character of the σ AO and favour the singlet. Based on these criteria, the trends shown in Table 5.1 are easily rationalized.

*Intersystem crossing in carbenes is extremely fast*, because it is associated with transfer of an electron from the sp$^2$-σ AO to the p-π AO, which is the ideal situation for strong first-order SOC (Section 4.8). It is therefore valid to assume that singlet–triplet equilibration generally precedes intermolecular reactions of carbenes (see Case Study 6.29). Singlet carbenes are frequently more reactive and tend to dominate the chemistry, even when the triplet state is slightly lower in energy. Most bimolecular reaction rates of singlet carbenes are close to diffusion controlled and 'inert' solvents allowing their intermolecular reactions to be studied are not readily available; Freon 113 (1,1,2-trichloro-1,2,2-trifluoroethane) and acetonitrile are often used, but the latter may form ylides with reactive singlet carbenes.[396] Singlet carbenes are readily protonated to give carbocations (Section 5.4.5).

Fast intramolecular reactions of singlet carbenes include 1,2-shifts (predominantly H-shifts) forming alkenes and the Wolff rearrangement forming ketenes (Scheme 6.171). In adamantylidene, the 1,2-H shift is impeded by geometric constraints; it therefore represents a convenient system to study intermolecular reactivity of a simple dialkylcarbene;[397] it has a singlet ground state.[398] Carbenes undergo $C-H$ and $O-H$ insertion (Scheme 6.178) and they add to alkenes, forming cyclopropanes, a synthetically very useful reaction (see Schemes 6.169–6.173). According to the *Skell hypothesis*, also known as the *Skell–Woodworth rule*, singlet carbene additions to alkenes are concerted and stereospecific [addition to a (Z)-alkene gives the *cis*-substituted cyclopropane

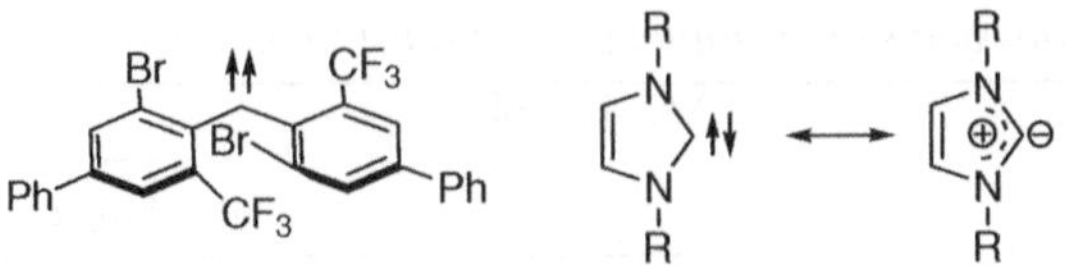

**Scheme 5.5**   *Persistent carbenes*

exclusively], whereas those of a triplet carbene are stepwise, proceeding via a triplet biradical intermediate with sufficient lifetime to permit conformational rotation, which results in scrambling of the stereochemical relationships. Based on this criterion, singlet–triplet energy gaps have been estimated from the product distribution.

Some remarkable persistent carbenes have been synthesized. This was achieved either by substitution with sterically demanding groups that hinder dimerization (triplet carbenes)[399,400] or by substitution with strong π-electron donors (singlet carbenes)[401] (Scheme 5.5). The latter are probably better described as ylides (zwitterions).

A striking, but well-understood, exception from the tenet of fast singlet–triplet equilibration in carbenes is represented by 2-naphthyl(carbomethoxy)carbene **(13)**, Scheme 5.6.[212,402–405] Irradiation of the diazo precursor in an argon matrix at 12 K gave a mixture of two conformational isomers of the triplet carbene $^3$**13**, which were characterized by IR (2120, 1846, 1660 cm$^{-1}$), UV–VIS (sharp peak at 590 nm with a 1380 cm$^{-1}$ progression) and ESR spectroscopy.[402] Upon subsequent irradiation at >515 nm, $^3$**13** was converted to the singlet carbene $^1$**13** that persisted for hours in the dark at 12 K! On standing overnight at 12 K or irradiation at 450 nm, $^1$**13** was reconverted to $^3$**13**. The remarkable persistence of $^1$**13** at 12 K is attributed to an associated change in the conformation of the carbomethoxy group. These findings represent a rare, but convincing, example for conformational control of the spin state. The minima on the singlet and triplet PESs lie at different geometries, so that ISC from $^1$**13** to the triplet ground state is inhibited by an energy barrier. Pump–probe spectroscopy of the diazo precursor in solution has shown that nitrogen loss occurs on a subpicosecond time scale and that the rate constant of ISC $^1$**13** → $^3$**13** is at least $10^{11}$ s$^{-1}$ at ambient temperature.[404] Time-resolved IR studies have further probed the equilibrium between the spin states of

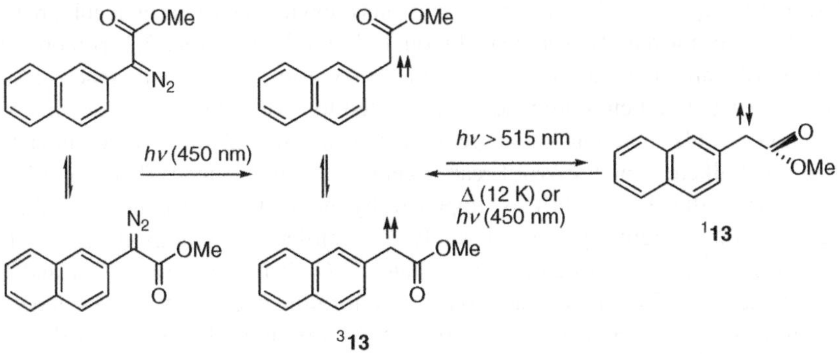

**Scheme 5.6**   *Matrix photolysis of α-diazo(2-naphthyl)acetate in Ar at 12 K[402]*

the carbene.[212,403] The triplet carbene [3]**13** is the ground state in hexane, but the singlet [1]**13** is the ground state in acetonitrile; it is stabilized by about $4\,kJ\,mol^{-1}$ relative to [3]**13** in the more polar solvent. The equilibrated carbenes undergo Wolff rearrangement and C–H bond insertion into the methyl group of the ester.

### 5.4.2 Nitrenes

*Recommended reviews.*[406,407]

Nitrenes (R–N) are hypovalent species that are commonly formed by thermal or photochemical denitrogenation of azides (R–N$_3$), which are readily synthesized and isolated, especially acyl and aryl azides. Simple alkyl azides do not in general produce alkyl nitrenes as detectable or trappable reactive intermediates upon direct irradiation (254 nm) due to rapid 1,2-migration to form the corresponding imines. Trifluoromethylnitrene is, however, formed by irradiation of CF$_3$N$_3$.[408]

A lucid discussion of the electronic structure of nitrenes has been given[407] based on high-quality *ab initio* calculations that are consistent with experimental data. The ground state of the parent diatomic nitrene, imidogen (NH), is a triplet state. The lowest singlet state is twofold degenerate and lies $150\,kJ\,mol^{-1}$ above the triplet ground state. The energy gap $\Delta E_{ST}$ is much larger than in methylene (Figure 5.10) because none of the symmetry-equivalent 2p-AOs of NH has any s-character. Due to the reduced symmetry of phenylnitrene, its lowest singlet state is no longer degenerate and the gap between the triplet ground state and the lowest singlet state is reduced to $75\,kJ\,mol^{-1}$.

The symmetry properties of the spatial wavefunctions describing the lower singlet and the triplet ground state of phenylnitrene are the same (Figure 5.11). The wavefunctions and the relaxed geometries of the two lowest electronic states obtained by *ab initio* calculations are, however, substantially different. In the singlet wavefunction, the highest, singly occupied π-orbital is almost completely localized in the phenyl ring to keep its electron apart from the other unpaired electron in the p-σ AO. This state may be viewed as an iminyl radical attached to a cyclohexadienyl radical. In the triplet state, the same π-orbital is largely localized on the more electronegative N atom because the large exchange integral between the two unpaired electrons keeps the repulsion between electrons of the same spin low. As the lowest singlet and triplet states of phenylnitrene are described by the same orbital occupancy, first-order SOC is zero (Section 4.8). Moreover, the energy gap is

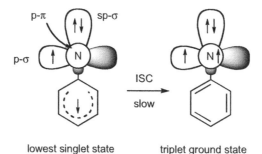

lowest singlet state                    triplet ground state

**Figure 5.11**  Ground-state triplet (left) and lowest singlet state (right) configurations of phenylnitrene

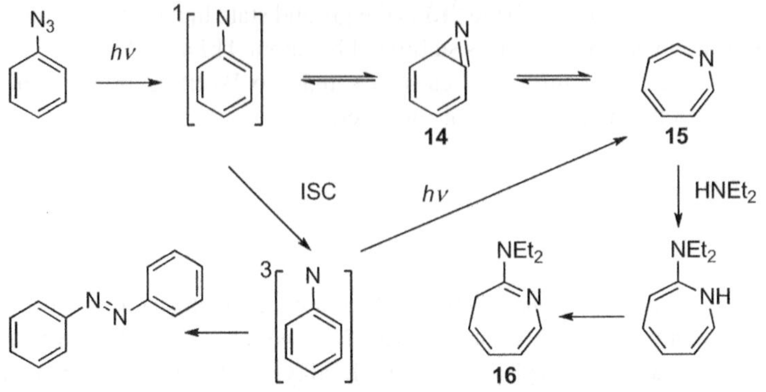

**Scheme 5.7**   *Photochemistry of phenyl azide*

larger, therefore *singlet–triplet ISC in nitrenes is orders of magnitude slower than in carbenes.*

The parent phenylnitrene has been studied in detail.[406] The thermal or photochemical decomposition of phenyl azide and most of its derivatives in solution results in the formation of intractable polymeric tars. Meaningful mechanistic studies became possible only when it was found that amines intercept the intermediate formed by thermal[409] or photochemical[410] decomposition of phenyl azide in solution that was responsible for polymerization. The current knowledge about the mechanism of phenyl azide photochemistry is summarized in Scheme 5.7.

The reactions of the primary photoproduct singlet phenylnitrene are strongly temperature dependent.[411] At room temperature, ISC is too slow to compete with cyclization to the benzazirine **14** followed by ring opening to the 1-aza-1,2,4,6-cycloheptatetraene (**15**). The didehydroazepine **15** is trapped by diethylamine, yielding the isolable azepine **16**. ISC of the singlet nitrene to the triplet nitrene, which is largely independent of temperature, $k_{ISC} \approx 3 \times 10^6 \, s^{-1}$, becomes important only at temperatures below 200 K. All intermediates shown in Scheme 5.7 have been firmly characterized by various spectroscopic techniques, except for the benzazirine **14**.[406] According to calculations, the ring expansion of singlet phenylnitrene does proceed in a two-step mechanism via **14**, but cyclization to **14** is the rate-determining step preceding rapid ring opening to **15**. In glassy solutions at 77 K, ISC is the dominant process. At this temperature, the triplet nitrene is stable in the dark, but it is very light sensitive, yielding **15** upon irradiation. Triplet phenylnitrene dimerizes to azobenzene when the glass is thawed. The effects of various substituents on the phenyl ring and the reactivity of naphthyl and other aryl azides have also been investigated.[406]

Singlet arylnitrenes are strong bases that form nitrenium ions in aqueous solutions.[412] The acidity constant of the conjugate acid of 4-biphenylylnitrene has been estimated as $pK_a = 16$. In-plane protonation of singlet phenyl nitrene is, however, a state-symmetry forbidden reaction (cf. Figure 4.32, Section 4.9), because phenylnitrenium ion has a closed-shell singlet ground state.[413] Indeed, the protonation of singlet nitrenes by water is substantially slower than would be expected for such a strong nitrogen base and it rarely competes efficiently with ring expansion. Even with strong mineral acid, the rate of

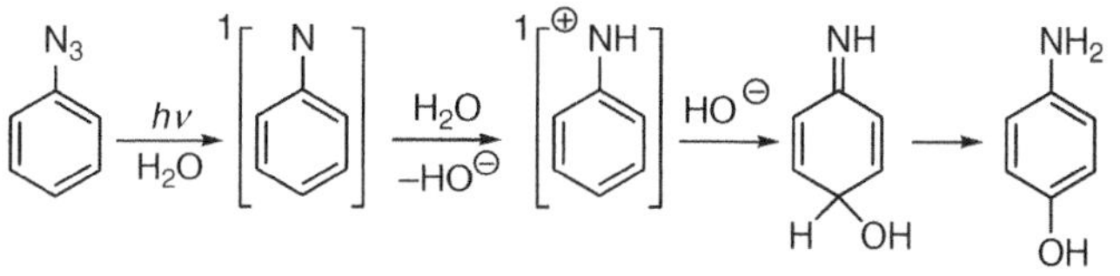

*Scheme 5.8*

protonation of singlet 2,4,6-tribromophenylnitrene is about an order of magnitude below the diffusion-controlled limit despite the huge exothermicity of this reaction.[414] On the other hand, phenylnitrenium ions are extremely electrophilic and, once formed, they rapidly add water to the *para*-position in aqueous solution (Scheme 5.8). The lifetime of parent phenylnitrenium ion has been estimated to be 125–240 ps in aqueous solution using an azide trapping method.[415]

Due to the above-mentioned complications, parent phenylnitrenium has only recently been detected as a transient intermediate using femtosecond pump–probe spectroscopy of phenyl azide in the non-nucleophilic solvent 100% formic acid.[416] Phenylnitrenium ion ($\lambda_{max} = 500$ nm) is formed with a rise time of 12 ps and has a lifetime of 110 ps in this medium. *para*-Substituted nitrenium ions have much longer lifetimes. Thus 4-biphenylylnitrenium and 2-fluorenylnitrenium are readily observed by nanosecond flash photolysis of the corresponding azides in water or aqueous acid with lifetimes of about 300 ns and 30 μs, respectively.[412] Both of these nitrenium ions react with guanine derivatives such as 2'-deoxyguanosine with rate constants close to the diffusion limit.[412,417] These observations may well explain why arylamines that are protected at the *para*-position (e.g. 2-naphthylamine) are carcinogenic. Arylamines are known to undergo a twofold metabolic activation yielding nitrenium ions that will attack DNA given that they are sufficiently persistent in aqueous solution.[418]

Alkylnitrenium ions have also been studied by transient absorption spectroscopy. The antiaromatic carbazolylnitrenium ion is an open-shell singlet that exhibits strong absorption at 620 nm and has a lifetime of 330 ns in acetonitrile solution (Scheme 5.9).[419] Its reaction rate constant with 1,3,5-trimethoxybenzene (TMB) is $9.5 \times 10^9$ M$^{-1}$ s$^{-1}$.

Organic azides are used for *photoaffinity labelling* (see Special Topic 6.16) because the azide group is small and, when attached to enzyme inhibitors, they often retain their activity as inhibitors. Azides are readily synthesized and the light-induced denitrogenation

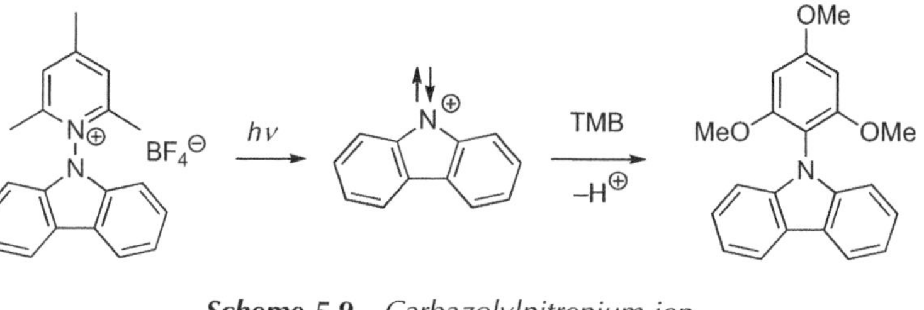

*Scheme 5.9*   *Carbazolylnitrenium ion*

is efficient. Unfortunately, the rapid ring expansion of singlet arylnitrenes is usually too fast to permit the desired cross-linking reactions and triplet nitrenes are not sufficiently reactive to be useful. Various strategies have been followed to increase the lifetime of singlet nitrenes by inhibiting ring expansion. A promising candidate is 2,3,5,6-tetrafluoro-4-azidoacetophenone, which releases a singlet nitrene with a lifetime of 172 ns in benzene solution that reacts with many substrates to form robust cross-linking.[420] Nitrene derivatives that have a singlet ground state, such as acylnitrene,[421] and nitrenes forming didehydroazepines (**15**) that are sufficiently reactive to trap the amine functions of peptides and nucleosides have also been proposed.[422] 2-Fluorenyl azide, which forms nitrenium ions in aqueous solution, could also be considered for photoaffinity labelling.

### 5.4.3 Radicals and Radical Ions

Numerous photoreactions producing radical intermediates and the products resulting therefrom are discussed in Chapter 6. These reactions include homolytic cleavage such as the $\alpha$-cleavage of ketones (Norrish type I reaction, Section 6.3.3), hydrogen abstraction by ketones (Scheme 6.99) and nitrogen elimination following $E$–$Z$ photoisomerization of aliphatic azo compounds (Scheme 6.166). The Norrish type I reaction proceeds preferentially from the n,$\pi^*$ states of ketones. State correlation diagrams,[331] tunnelling calculations, *ab initio* and DFT calculations have been used to interpret the relative reactivity of n,$\pi^*$ and $\pi,\pi^*$ singlet and triplet states, the rates and the preferred sites for $\sigma$-bond homolysis (for further references, see ref. 423). The reactivities of singlet and triplet n,$\pi^*$-excited ketones are comparable, but ISC to the triplet state is very fast in aryl ketones ($10^{11}\,\mathrm{s}^{-1}$ versus $10^8\,\mathrm{s}^{-1}$ in saturated ketones)

The rate constants for the reaction of carbon-centred radicals with various substrates such as alkenes[424,425] and dioxygen[426,427] vary over many orders of magnitude depending on thermodynamic, steric and stereoelectronic effects. Radical recombination reaction rate constants $k_{\mathrm{rec}}$ are often close to the diffusion-controlled limit, $k_{\mathrm{r}} = k_{\mathrm{d}}/4$;[58] the factor of one-quarter is due to spin statistics (Section 2.2.1). The observed second-order rate constant for self-termination reactions is $2k_{\mathrm{r}}$.

The overall multiplicity of geminate radical pairs formed by bond fission is the same as that of the precursor excited state. Remarkably, the multiplicity of the precursor can often be established by NMR spectroscopy of the final products thanks to a phenomenon called *chemically induced dynamic nuclear polarization* (CIDNP, Special Topic 5.3).

---

### Special Topic 5.3: Chemically induced dynamic nuclear polarization

The discovery of CIDNP was marked by the serendipitous observation of emission (negative peaks) in NMR spectra of reacting systems.[428,429] A personal view of the early investigations of the CIDNP phenomenon has appeared.[430]

CIDNP is based on the following principle:[431,432] Initially, the radical pair is born in a spin-correlated state. To form a product in the singlet ground state, the electronic spin state of the radical pair must be a singlet state. Importantly, the electron spins interact with the nuclear spin states. ISC from a triplet to a singlet radical pair is favoured, when the total (nuclear plus electronic) spin is conserved. Therefore, certain nuclear spin states will favour the recombination of triplet pairs. Initially, the nuclear spin states are at Boltzmann equilibrium.[b] However, if reaction pathways starting from an overall triplet pair yield different products than those of singlet pairs, CIDNP is observed. The nuclear spin states are unevenly distributed to the different products and, as a consequence, its NMR signals appear either in enhanced absorption or emission.

The reaction between two spin-correlated geminate radicals leading to the final products is called a *cage reaction*. When the lifetime of a triplet pair is insufficient to permit ISC to the singlet state, the two radicals separate and these free radicals eventually find new reaction partners; this is termed an *escape reaction*. Each of these reaction types leads to differently polarized NMR signals, that is, the resonances appear in enhanced absorption (A) or emission (E). Kaptein and Closs have developed simple rules, which allow one to determine the multiplicity of the precursor radical pair from the sign of the individual CIDNP signals (A or E) and magnetic parameters of the short-lived radical pair (isotropic hyperfine coupling constants and $g$-factors).[433]

---

[b] The energy differences between nuclear spin states are very small compared with $k_B T$ even in strong magnetic fields, so that in thermal equilibrium all nuclear spin states are populated nearly equally in accord with Boltzmann's law (Equation 2.9). This is the reason why NMR spectroscopy, which displays the transitions between different nuclear spin states, is relatively insensitive: One needs milligram quantities for a good spectrum, whereas a few micrograms are usually sufficient to record a UV spectrum. Electromagnetic radiation of the proper frequency produces both absorption and stimulated emission (Figure 2.4) from the sample, and the net absorption is due to the tiny excess of molecules in the lower spin state.

Remarkable effects of nuclear spin on the photochemistry of dibenzyl ketone and its derivatives (see Special Topic 6.11) have been reported and substantial isotopic enrichment of $^{13}C$ in specific products has been achieved under favourable conditions.[434] The influence of magnetic fields and of 'microscopic reactors' (micelles, zeolites, etc.) on these reactions was investigated in detail.[435]

The radical ion pairs generated by photoinduced electron transfer (Section 5.2) have several options for further reaction.[341,365,436] They can undergo return electron transfer (RET) to regenerate the parent systems in the ground state. Due to the near-degeneracy of the overall singlet and triplet state of radical pairs, ISC in the radical pair may be much faster than in the parent aromatic compounds and RET then leads to the formation of one of the reactants in the triplet state within 1 ns ('fast triplet' formation). ISC in radical ion pairs been attributed to hyperfine coupling on the basis of magnetic field effects.[437,438] The radical cations of strained systems can rearrange, because the barriers to rearrangement are frequently much lower in the transient radical ions than in the parent systems (a well-known example is the rearrangement of quadricyclane to norbornadiene, Scheme 6.139). The observation of CIDNP signals (Special Topic 5.3) has proven to be a powerful tool to assess the mechanism of such reactions.

The radical cation of naphthalene has been generated in aqueous acetonitrile both by two-photon ionization and via reaction with $SO_4^{\bullet-}$. Its lifetime in aqueous solution is 25 μs.[439]

The equivalent to the triplet state of molecules with a singlet ground state is the *quartet state* ($S = 3/2$) in radicals. Quartet configurations should, according to Hund's rule, have a lower energy than the corresponding excited doublet configurations with three unpaired electrons. However, such configurations with three unpaired electrons are generally of

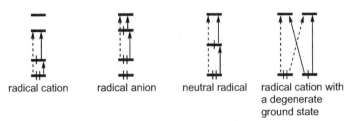

radical cation          radical anion          neutral radical          radical cation with
                                                                        a degenerate
                                                                        ground state

**Figure 5.12**   Quartet states of radicals. Single electron excitations leading to a configuration with a single unpaired electron are marked with a full arrow, those leading to a configuration with three unpaired electrons with a dashed arrow

higher energy than the lowest excited doublet configuration, as indicated by the first three MO schemes on the left of Figure 5.12. Thus ISC to a long-lived quartet state is possible only from upper excited doublet states of radicals, which is unlikely to happen because internal conversion within the doublet manifold will usually be much faster. Therefore, quartet → doublet phosphorescence is difficult to observe and the quartet states of radicals rarely play a role in photochemistry.

The situation is different for radical ions in which the unpaired electron is situated in a degenerate (or nearly degenerate) MO (Figure 5.12, right). Here, the doublet ground state of the radical ion will be degenerate (or nearly degenerate) and the lowest excited state may be one with three unpaired electrons. Indeed, quartet states and quartet → doublet phosphorescence of such systems have been observed, for example, for the decacyclene radical anion,[440,441] or for the radical cations of acetylene and benzene.[442,443]

### 5.4.4   Biradicals

*Recommended reviews and textbooks.*[16,318,444]

This section is intended to provide the organic chemist with relatively simple guidelines, rational structure–reactivity relationships and rules-of-thumb to predict the reactivity of biradicals and their response to changes in manageable parameters such as temperature, solvent polarity and magnetic fields. The same considerations hold, *mutatis mutandis*, for carbenes and nitrenes.

The energy of near-UV photons is on the order of bond energies in organic molecules. Photoreactions are thus frequently initiated by the homolytic cleavage of a covalent bond, which leads to a biradical, if the primary product is still covalently linked (Figure 5.13). *Biradicals*, also called *diradicals*, are usually defined as molecular species that possess two (possibly delocalized) radical centres, that is, one bond less than the number predicted by the standard rules of valence. When the two radical centres are located on the same atom (1,1-biradicals), such hypovalent species are referred to as carbenes (Section 5.4.1), nitrenes (Section 5.4.2), and so on. The two radical centres may be connected through saturated covalent bonds (*localized biradicals*) or they may be delocalized over the same π-system (*conjugated biradicals*).

Chemical transformation is about breaking bonds and forming new bonds. Biradicals are at this interface, so their properties are of fundamental interest. Molecular structures representing biradicals correspond to the funnels connecting excited- and ground-state potential energy surfaces (Figure 2.26). Owing to the near degeneracy of several electronic

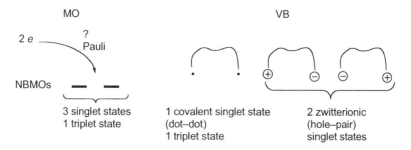

**Figure 5.13**   The four low-lying electronic states of a biradical

states, the electronic structure of biradicals is chameleon-like, which complicates anticipation of their chemical properties, but also opens up the opportunity of influencing their behaviour. Their intrinsic instability has long discouraged attempts to exploit their synthetic potential as bifunctional reagents. Systematic trapping studies, the spectroscopic detection and identification of biradical intermediates in solid matrices and in solution, investigations of their reaction kinetics in solution, combined with the development of theoretical models, have brought some clarification and led to the identification of relatively long-lived biradical intermediates that can be of interest for synthesis.

A molecule with two radical centres will have two nonbonding MOs that are occupied by a total of two electrons; if these electrons have parallel spin, they must occupy one NBMO each, forming a triplet configuration. If they have opposite spin, two additional configurations can be constructed, in accord with the Pauli principle, by placing both electrons into one or the other NBMO. We obtain one triplet state and three singlet states (Figure 5.13), all of which have the same energy in terms of HMO theory. When we consider electronic interaction, the triplet state is likely to be the ground state of the biradical (Hund's first rule) and the two singlet states with localized charges (hole–pair states), represented as *zwitterions* in a valence bond (VB) description, are usually pushed higher in energy than the so-called *covalent singlet state* with one electron on each radical centre.

The definition of a biradical given above is based on G. N. Lewis's rules of valence and would exclude molecules such as cyclobuta-1,3-diene (**3**), which have a structure satisfying these rules, that is, a Kekulé structure. Yet **3** has two NBMOs according to HMO theory (Section 4.2, Figure 4.2) and therefore exhibits the electronic properties of a biradical. One might then define a biradical as an open-shell species with two electrons in two (essentially) nonbonding orbitals. Instead, we prefer the following practical definition of a biradical, which avoids reference to any model (VB or MO): *a biradical is a molecular species having three low-lying electronic singlet states and a triplet state with an energy either below or little above the lowest singlet state.*[445] Because the thermal population of low-lying electronic states depends on temperature, the singlet–triplet energy gap should be judged relative to $k_B T$ ($N_A k_B T = RT = 2.5 \, kJ \, mol^{-1}$ at room temperature). Frequently, the NBMOs are split apart to some extent, for example when the two radical centres interact substantially, when they have intrinsically different electronegativity or when structural relaxation, such as bond localization in cyclobutadiene, stabilizes one NBMO relative to the other. Biradical-like species with a singlet–triplet energy gap exceeding $k_B T$ may be referred to as *biradicaloids*.

A distinction is made between conjugated *Kekulé biradicals* such as **3**, for which at least one Kekulé structure can be drawn, and *non-Kekulé biradicals* such as cyclobutadiene **3**                    trimethylenemethane **17**

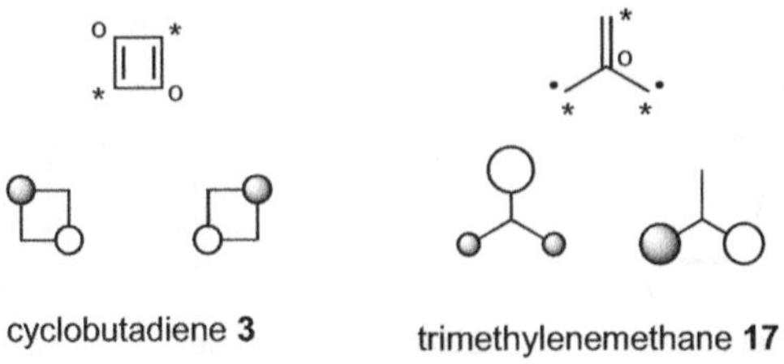

**Figure 5.14**   NBMOs of cyclobutadiene and trimethylenemethane

trimethylenemethane (**17**), which does not have a proper Kekulé structure (Figure 5.14). In non-Kekulé alternant hydrocarbons, the number of starred (*) atoms *s* exceeds that of the unstarred (o) atoms *u*, so that the number of NBMOs must be at least equal to $|s − u|$, that is, $\geq 2$ (Section 4.6). Hence all non-Kekulé hydrocarbons are biradicals. For Kekulé hydrocarbons, $|s − u| \geq 0$; they are normally stable molecules with a closed-shell ground state such as benzene or butadiene. However, some Kekulé hydrocarbons such as cyclobutadiene (**3**) have two NBMOs or a very small energy gap between the HOMO and LUMO; they are called Kekulé biradicals or biradicaloids. Two other Kekulé biradicals (**20** and **21**) are shown in Scheme 5.10.

Another important classification of conjugated diradicals depends on whether their Hückel NBMOs can or cannot be confined to a *disjoint* sets of atoms.[446] As discussed in

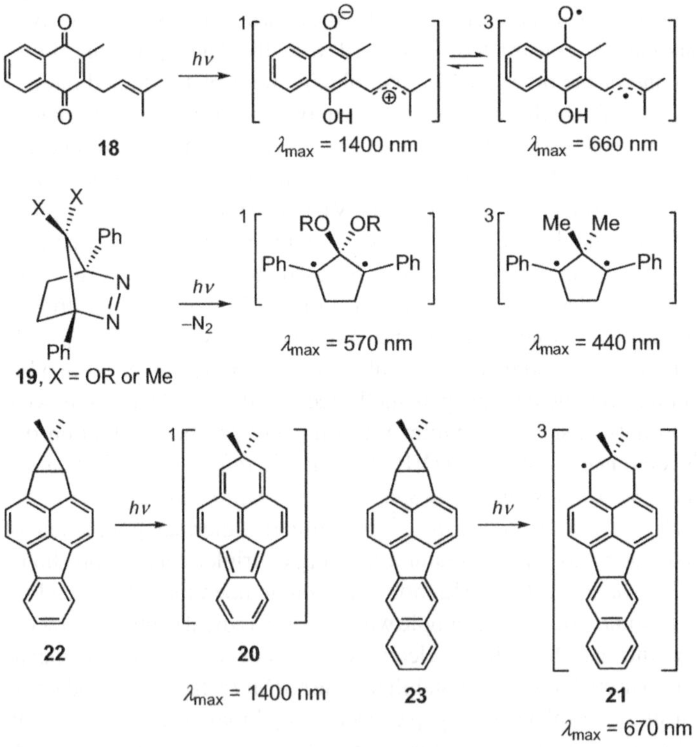

**Scheme 5.10**   *First absorption bands of singlet and triplet biradicals*

Section 4.2, any normalized linear combination of degenerate wavefunctions is an equally valid wavefunction. The linear combinations of the NBMOs with the smallest possible amount of local overlap for **3** and **17** are shown in Figure 5.14. Those of **3** are separated entirely in space to the starred and unstarred sets of atoms. Hence **3** is a *disjoint biradical*. The same cannot be achieved by any linear combination of the NBMOs of **17**; it is classified as a *nondisjoint biradical*.

The distinction between disjoint and nondisjoint conjugated biradicals provides a useful basis for understanding their electronic properties. The class to which a given biradical belongs can be immediately assessed by PMO analysis (Section 4.6). The union of an allyl radical with a methyl radical to form either **3** or **17** will give two NBMOs in either case, because it takes place at a nodal plane of the NBMO of the allyl radical. The NBMO coefficients are then easily obtained from rule 4 (Equation 4.25): the coefficients on atoms attached to an atom with $c_{NBMO,\mu} = 0$ must add up to zero. PMO analysis directly gives the 'localized' NBMOs, which keep the electrons in the open shell apart as much as possible and which are best suited to describe the lowest energy singlet and triplet states by a single configuration.

Starting from this localized basis set, the singlet–triplet gap $\Delta E_{ST} = E_S - E_T$ can be estimated as two times the exchange integral $K_{ij}$ (Equation 4.34) between the two NBMOs $\psi_i$ and $\psi_j$, which will be substantial (on the order of 1 eV) for nondisjoint biradicals such as **17**, but vanishes within the ZDO approximation for disjoint biradicals such as **3**. The triplet state is indeed the ground state of **17** and the $\Delta E_{ST}$ amounts to $67.4 \pm 0.4\,kJ\,mol^{-1}$.[447] *Ab initio* calculations predict that the equilibrium geometry of ${}^3\mathbf{17}$ is planar ($D_{3h}$ symmetry), whereas ${}^1\mathbf{17}$ relaxes to a perpendicular geometry in which one methylene group is orthogonal to the plane of the other atoms ($C_{2v}$).[448] On the other hand, cyclobutadiene (**3**) has a singlet ground state that is stabilized by bond length alternation. Our simple model gives $\Delta E_{ST}(\mathbf{3}) = 0$, but *ab initio* calculations slightly favour the singlet even for planar quadratic **3** due to a small stabilization called spin polarization. Allowing for this minor correction, the classification based on the simple PMO procedure provides robust predictions for the ground state multiplicity of many conjugated biradicals.

The exchange integral falls off rapidly with distance (Figure 4.21) and so does the singlet–triplet gap with increasing separation of the radical centres. In 1,2-biradicals (90° twisted alkenes, Section 5.5) and in localized 1,$n$-biradicals ($n > 2$) the gap is usually within a few $kJ\,mol^{-1}$, that is, comparable to $RT$ at room temperature. The energy gap $\Delta E_{ST}$ plays a decisive role in ISC processes of biradicals.

Because biradicals have three singlet states of low energy, one may predict in general that singlet biradicals will absorb at longer wavelengths than triplet biradicals of the same (or similar) structure. This is illustrated by the MO diagram in Figure 5.15, which shows the two additional low-energy transitions available to singlet biradicals.

Dioxygen is a typical example (Section 2.2.5). In the triplet ground state it absorbs only in the deep UV region, but the lowest *excited* state of singlet oxygen (${}^1\Sigma_g^+$) lies only $62.7\,kJ\,mol^{-1}$ above the lowest singlet state (${}^1\Delta_g$). Other examples of such long-wavelength absorption bands in organic singlet biradicals are shown in Scheme 5.10.

The top row of Scheme 5.10 shows the *m*-quinone methide biradical that is formed by photolysis of menaquinone-1 (**18**), an analogue of vitamin $K_1$.[449] The singlet and triplet state of this biradical are in thermal equilibrium. The zwitterionic singlet is stabilized in polar media and absorbs at 1400 nm. The second row shows 2,2-disubstituted 1,3-

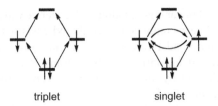

triplet                    singlet

**Figure 5.15**  MO scheme of electronic transitions in singlet and triplet biradicals

diphenylcyclopentane-1,3-diyl biradicals that are generated by photolysis of the corresponding azoalkane precursors **19**. The 2,2-dimethyl derivatives have a triplet ground state and their absorption spectra are similar to those of benzyl radicals, showing a moderately strong band at 320 nm and a very weak, parity-forbidden band at 440 nm.[450] The singlet state is the ground state in the 2,2-difluoro- and 2,2-dialkoxy-substituted derivatives, which show a broad, strong absorption band in the visible region, $\lambda_{max} = 500$–600 nm.[451] The bottom row shows two nonalternant hydrocarbons, 2,2-dimethyl-2*H*-benzo[*cd*] fluoranthene (**20**) and its benzannellated analogue 2,2-dimethyl-2*H*-dibenzo[*cd,k*]fluoranthene (**21**), that were formed by photolysis of their cyclopropane precursors **22** and **23**, respectively.[452] Compound **21** is the only known Kekulé biradical with a triplet ground state. The singlet–triplet gap for **20**, which has a singlet ground state, was estimated to be $-10 \, \text{kJ mol}^{-1}$ and that of **21** (triplet ground state) about $+5 \, \text{kJ mol}^{-1}$. Although $^3$**21** is more extended, its first absorption band lies at much shorter wavelength than that of $^1$**20**. The saturated singlet carbene adamantylidene exhibits a broad, weak absorption band at 620 nm.[398]

   Scheme 5.11 shows a general blueprint that serves as a useful framework for considering reaction mechanisms that involve biradical intermediates B. The scheme also applies to carbenes (Section 5.4.1) and nitrenes (Section 5.4.2). When formed by thermal or photochemical homolytic bond cleavage from the reactant molecule M, biradicals will initially have the multiplicity of their immediate precursors. Singlet biradicals ($^1$B) tend to be short-lived because spin-allowed bond formation leading back to the reactant M or to a

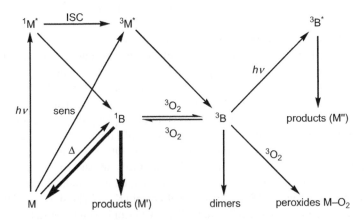

**Scheme 5.11**   *A general reaction scheme for reactions via biradical intermediates*

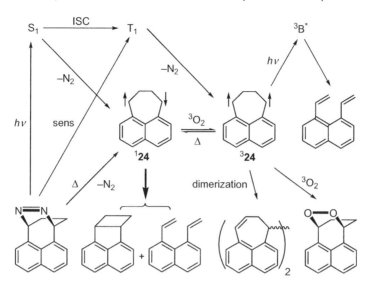

**Scheme 5.12**    *Singlet- and triplet-state reactivity of a conjugated hydrocarbon biradical[453]*

new product is generally fast (picosecond time scale). ISC to the triplet state of the biradical ($^3$B) is unlikely to compete in hydrocarbon biradicals, except in carbenes (Section 5.4.1) or at low temperatures, if bond formation requires some activation energy.

The triplet biradicals may be generated from the triplet state of the reactant M. They are generally longer lived (nanoseconds) because they need to undergo ISC to return to a singlet ground-state product. In aerated solution, they can be trapped by dioxygen to form peroxides (or carbonyl compounds in the case of carbenes). Similarly, dimers of $^3$B may be formed under high-intensity irradiation. These are spin-allowed processes for one in nine encounters (Scheme 2.5). At low temperature, triplet biradicals may be persistent and can be identified by ESR spectroscopy. Light absorption by $^3$B in flash photolysis or at low temperature may lead to new products via an excited-state $^3$B*. Practical examples are provided by the photoreactions of the azo precursor of 1,4-perinaphthadiyl (**24**)[453], Scheme 5.12, and Case Study 5.2.

---

**Case Study 5.2: Mechanistic photochemistry – cyclopentane-1,3-diyl biradicals**

1,3-Diphenylclopentane-1,3-diyl[454,455] (Scheme 5.13) can be generated in its singlet state by either direct irradiation or thermolysis of its azo precursor and also by thermal homolysis of the central bond of 1,4-diphenylbicyclo[2.1.0]pentane (housane). Homolysis of this bond releases a large amount of strain energy. Moreover, the resulting biradical is stabilized by the phenyl groups. Thus the central bond of housane is extraordinarily weak and the degenerate ring-flip reaction, which must proceed via the CP biradical either as the transition state or as an intermediate, occurs about twice per

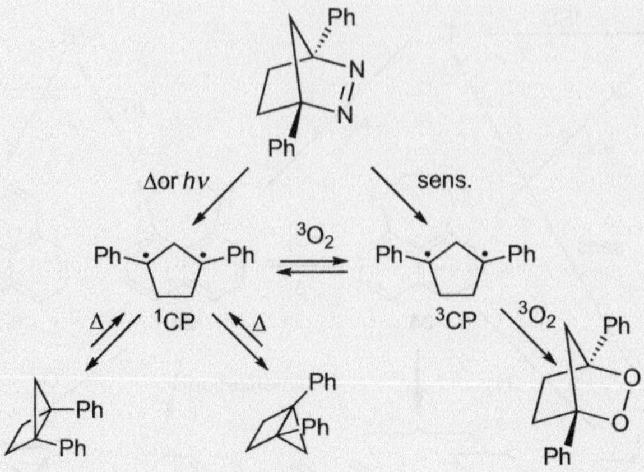

**Scheme 5.13**

second at ambient temperature. The activation parameters for ring-flip were determined by H-NMR as $\Delta H^{\ddagger} = 51.0 \pm 0.4\,\text{kJ mol}^{-1}$ and $\Delta S^{\ddagger} = -69 \pm 6\,\text{J K}^{-1}\,\text{mol}^{-1}$.[455]

Does $^1$CP lie in a potential well and does it undergo spontaneous ISC to $^3$CP? The answer to the second question is no.[454] This was established by flash photolysis and by oxygen trapping. The triplet state $^3$CP was identified by its ESR spectrum, which persists down to 3.8 K, indicating that the triplet state is the ground state of the biradical. Flash photolysis of benzophenone to sensitize the diazo precursor forms $^3$CP, which has a lifetime of 27 μs ($\lambda_{\text{max}} = 320$ nm) in degassed solution. Oxygen quenching of $^3$CP is fast, $k = 7.5 \times 10^9\,\text{M}^{-1}\,\text{s}^{-1}$, and yields the endoperoxide. Upon direct flash photolysis of the azo compound, $^3$CP is not observed and the housane is formed in high yield. The latter is stable in degassed solution, but it reacts with oxygen forming the endoperoxide in aerated acetonitrile. The rate law for oxidation is first order in both oxygen and housane concentration, $k_{\text{ox}} = 2.1\,\text{M}^{-1}\,\text{s}^{-1}$ at 24 °C. Thus, at an oxygen concentration of $2 \times 10^{-3}$ M (air-saturated) or less, the ring-flip reaction, $k_{\text{flip}} = 2\,\text{s}^{-1}$, is much faster than that of oxidation, $k_{\text{ox}}[\text{O}_2] = 4.3 \times 10^{-3}\,\text{s}^{-1}$. These observations exclude oxygenation occurring by spontaneous ISC of $^1$CP followed by trapping of $^3$CP, because any $^3$CP formed would be quantitatively trapped and the rate law would be zero order in oxygen concentration.

Returning to the first question, the answer is yes. $^1$CP is a metastable intermediate, as is already indicated by the fact that oxygen trapping does occur. Any encounter of $^1$CP with $^3$O$_2$ will induce *oxygen-catalysed ISC*, a spin-allowed process (Section 2.2.5). Once formed, $^3$CP is likely to leave the encounter complex with $^3$O$_2$, which has overall triplet multiplicity and cannot form the endoperoxide directly. However, formation of $^3$CP is irreversible, as $^3$CP will rapidly be trapped by another oxygen molecule. Assuming from many precedents that catalysed ISC occurs at the diffusion limit, $k_{\text{cat}} \approx 2.8 \times 10^{10}\,\text{M}^{-1}\,\text{s}^{-1}$, the observed oxidation kinetics indicates that the lifetime of $^1$CP is at least 20 ps and that the barrier protecting it from cyclization amounts to about 12 kJ mol$^{-1}$.[454] These results are illustrated by the PES for the ring-flip reaction shown in Figure 5.16.

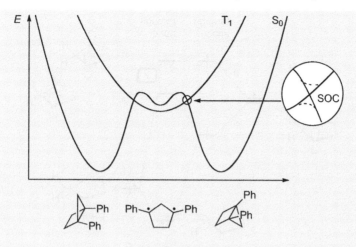

**Figure 5.16**

How does a triplet biradical $^3$CP return to the singlet ground state? As shown in Figure 5.16, it will have to pass a small barrier to the intersection with the singlet PES and the magnitude of spin–orbit coupling (SOC) (Section 4.8) at the crossing point will determine the small probability for diabatic reaction (Landau–Zener model, Equation 2.53), that is, that the molecule will find the escape hatch to the singlet ground state. Application of a general analysis of SOC terms in biradicals[320] to the case of $^3$CP shows that SOC of the triplet state to the lowest singlet state is zero for a planar $^3$CP biradical, because the lowest singlet state wavefunction (covalent, dot–dot) cannot mix with the zwitterionic wavefunctions (hole–pair) that are needed to induce SOC with the covalent triplet state (Section 4.8). This already explains the remarkably long lifetimes of the geometrically constrained $^3$CP biradicals. However, these lifetimes are fairly sensitive to the structural environment, as indicated by the examples shown in Scheme 5.14. Clearly, 2,2-dimethyl substitution strongly reduces the lifetimes of these geometrically constrained triplet 1,3-biradicals.

To understand this trend, we need to look at the requirements permitting SOC through admixture of the lower zwitterionic (hole–pair) state to the covalent singlet ground state wavefunction. Two conditions must be met:[320] the molecule must bend

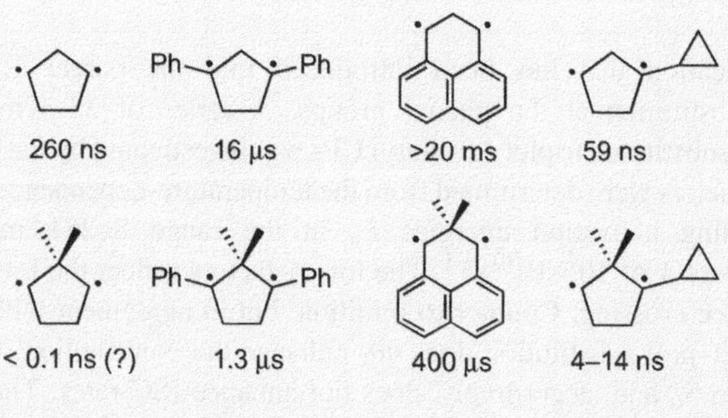

260 ns          16 µs          >20 ms          59 ns

< 0.1 ns (?)     1.3 µs         400 µs          4–14 ns

**Scheme 5.14**   *Lifetimes of some triplet 1,3-biradicals in degassed solution*[450,456–459]

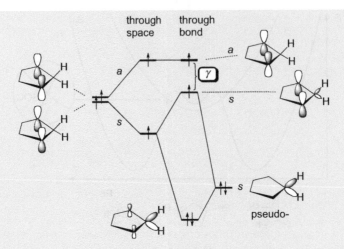

**Figure 5.17**   MO interaction diagram for cyclopentane-1,3-diyl (CP). The symmetry labels *a* and *s* refer to reflection through the plane defined by the $CH_2$ group at position 2

along the reaction coordinate leading to housane (Figure 5.16) and the two NBMOs must be split by a 'covalent' interaction. Consider the simplified MO diagram for planar CP shown in Figure 5.17.

The interaction between the two symmetry-adapted NBMOs *s* (symmetric with respect to the plane of the $CH_2$ group at position 2) and *a* (antisymmetric) shown on the left is considered in two stages. The direct through-space interaction stabilizes the *s* and destabilizes the *a* NBMO. This is counteracted by through-bond interaction involving the pseudo-$\pi$ MO of the 2-$CH_2$ moiety, resulting in a small energy gap $\gamma$ separating the two NBMOs, which house one electron each in the triplet state of the biradical. The matrix element that induces mixing of the lower-lying, closed-shell zwitterionic (hole–pair) wavefunction with the covalent wavefunction is $\gamma$. Hence the rate constant of ISC will be proportional to $\gamma^2$.[450] By replacing the $CH_2$ group at position 2 with a $CMe_2$ group, the basis energy of the pseudo-$\pi$ orbital interacting with the *s*-combination of the $\pi$-NBMO will be raised by about 1.5 eV. As a result, the symmetric NBMO is raised well above the antisymmetric one and the energy gap $\gamma$ is increased. This will increase the contribution of the hole–pair wavefunction to $S_0$ at nonplanar geometries leading to increased rates of ISC in 2,2-dimethyl-substituted CP biradicals.[457]

Another modification that has been introduced into the parent 1,3-diphenyl CP biradicals is substitution at the phenyl groups. A series of 33 symmetrically and asymmetrically substituted triplet 1,3-diaryl CPs was investigated by flash photolysis.[450] Arrhenius parameters were determined from the temperature-dependence of the ISC rate constants, yielding activation energies $E_a$ in the range 8–25 kJ mol$^{-1}$ and pre-exponential factors $A$ of $10^7$–$10^{10}$ s$^{-1}$. The low $A$-factors reflect the low probability of ISC at the surface crossing. Counter to intuition, but in agreement with the model,[320] asymmetric push–pull substitution does *not* enhance the contribution of the hole–pair wavefunctions to $S_0$ and, accordingly, does not enhance ISC rates. The observed rate constants correlate well with calculated values of $\gamma^2$, except for bromo- and iodo-substituted compounds, which exhibit enhanced ISC rates attributable to the heavy-atom effect (Section 4.8).

Several singlet biradicals were generated by femtosecond-pulsed decarbonylation of cyclic ketones and their femtosecond dynamics were investigated by time-resolved mass spectrometry.[460] Very short lifetimes were reported for singlet trimethylene ($\cdot CH_2CH_2CH_2\cdot$, 120 fs), tetramethylene ($\cdot CH_2CH_2CH_2CH_2\cdot$, 700 fs) and parent CP (190 fs). It should be noted, however, that these biradicals were formed with substantial excess energies as isolated species in the gas phase (the CP biradical was formed by absorption of two 307 nm photons). Several singlet 1,4-biradicals were formed by the Norrish type II reaction (Section 6.3.4) of pentan-2-one and their lifetimes in the range 0.5–0.7 ps were determined by the same method.[461] The lifetimes of these biradicals, in particular that of CP, might be substantially higher in solution. A few singlet biradical lifetimes have been estimated by oxygen trapping, for example 1,3-diphenyl-CP (Case Study 5.2) and 1,4-perinaphthadiyl (**24**, Scheme 5.12),[453] or measured by pump–probe spectroscopy [the 1,3-biradical intermediate in the photocyclization of 2,7-dihydro-2,2,7,7-tetramethylpyrene, Figure 6.11], Scheme 5.15.

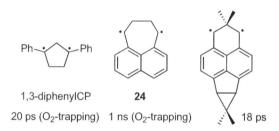

1,3-diphenylCP     **24**

20 ps ($O_2$-trapping)    1 ns ($O_2$-trapping)      18 ps

**Scheme 5.15**  *Singlet biradical lifetimes*

A series of localized triplet 1,*n*-biradicals was generated by laser flash photolysis of the cyclic ketones shown in Scheme 5.16 (Norrish type I cleavage, Section 6.3.3).[40] For small chain lengths, the main products were formed by disproportionation, but for $n \geq 10$, paracyclophanes were obtained in high yield. The lifetimes of the triplet biradical intermediates ($\lambda_{max} = 320$ nm) increased with chain length from 45 ns ($n = 6$) to 80 ns ($n = 8$) and then decreased again to 50 ns ($n = 12$ and 15). ISC to the singlet state is thought to be the rate-determining step in their product forming reactions. Both the singlet–triplet energy gap $\Delta E_{ST}$ and SOC fall off exponentially with the distance between the radical centres. ISC rates will be different for each conformer of these flexible biradicals and subsequent product formation occurs from a small subset of singlet conformers having a short end-to-end distance. Even though the biradicals will spend most of their time in extended conformations, ISC induced by SOC will take place from a few conformers with a relatively small distance separating the radical centres. Such conformations are sparse for $n = 8$ and these biradicals have the longest lifetime.

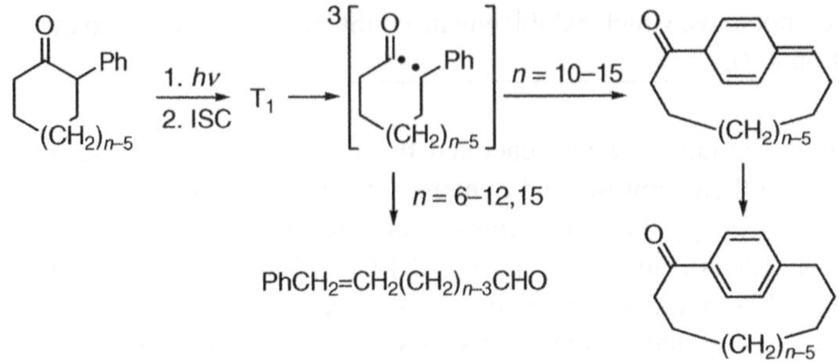

*Scheme 5.16*

Most interesting results were obtained by measuring the lifetimes of these biradicals in magnetic fields $B$ ranging from 50 μT (Earth's magnetic field) to 0.1 T. The ratio of the ISC rate constants in the presence and absence of a magnetic field, $k_{ISC}^B/k_{ISC}^0$, increased from 1 ($B = 50$ μT) to a maximum of 1.13 and then decreased to an asymptotic value of 0.8–0.9 for $B = 0.1$ T. The maximum was reached at 3 mT for $n = 12$, at 12 mT for $n = 11$ and at 60 mT for $n = 10$. The analysis of these and other data[40] proved that ISC is dominated by SOC (76% for $n = 12$ and 88% for $n = 11$), but that hyperfine coupling (HFC) gains importance for the longest chain lengths. It was also shown that for 1,$n$-dibenzyl biradicals HFC becomes dominant for $n > 6$, because SOC is smaller in the absence the acyl group (recall that SOC increases with the fourth power of the atomic number Z, Section 4.8).

---

**Case Study 5.3: Mechanistic photochemistry – the photo-Favorskii reaction of *p*-hydroxyphenacyl compounds**

*p*-Hydroxyphenacyl (*pHP*) is an effective *photoremovable protecting group* for anionic substrates X⁻ (see also Special Topic 6.18). The release of, for example, diethyl phosphate from compound **25** [X = OPO(OEt)₂, Scheme 5.17] is accompanied by a photo-Favorskii rearrangement of *pHP* to *p*-hydroxyphenylacetic acid (**26**), which absorbs only below 300 nm. Following fast and efficient ISC, the reaction is initiated

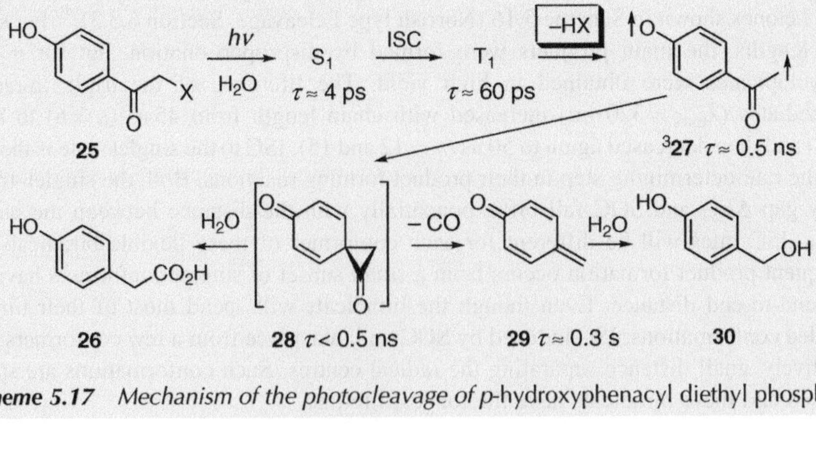

*Scheme 5.17    Mechanism of the photocleavage of p-hydroxyphenacyl diethyl phosphate*

from the lowest the triplet state $T_1$ of **25** ($\lambda_{max}$ 390 nm). It proceeds only in aqueous solvents and the lifetime of $T_1$ decreases in a quadratic fashion with increasing water concentration in wet acetonitrile;[383] with diethyl phosphate as the leaving group X it amounts to several microseconds in degassed, dry $CH_3CN$, but is reduced to 60 ps in wholly aqueous solution.[462] Thus water plays an essential role in the elimination of the leaving group.

Time-resolved resonance Raman spectroscopy of **25** in 50% aqueous $CH_3CN$ proved that the final product **26** appears with a rate constant of $2.1 \times 10^9$ s$^{-1}$ following pulsed excitation of **25**.[207] The appearance of **26** was slightly delayed with respect to the decay of $T_1$(**25**), $k = 3.0 \times 10^9$ s$^{-1}$, that was determined independently by optical pump–probe spectroscopy in the same solvent. The intermediate that is responsible for the delayed appearance of **26**, $\tau \approx 0.5$ ns, is attributed to the triplet biradical $^3$**27**.[462] It shows weak, but characteristic, absorption bands at 445 and 420 nm, similar to those of the phenoxy radical. ISC is presumably rate limiting for the decay of $^3$**27**, which cyclizes to the spiro-dienone **28**. The intermediate **28** is not detectable; its decay must be faster than its rate of formation under the reaction conditions. Decarbonylation of **28** to form *p*-quinone methide (**29**) competes with hydrolysis to **26** at low water concentrations. Hydrolysis of **29** then yields *p*-hydroxybenzyl alcohol (**30**) as the final product.

Detailed knowledge of this mechanism was needed to establish the release rate of the leaving group. With its fast and efficient release of diethyl phosphate, $\tau = 60$ ps, $\Phi \geq 0.4$, the *p*-hydroxyphenacyl protecting holds promise for use in studies of fast reactions such as the primary steps in protein folding (random coil to $\alpha$-helix formation) (Figure 5.18).

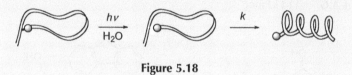

**Figure 5.18**

## 5.4.5   Carbocations and Carbanions

*Recommended reviews.*[391, 463]

Carbocations have held an important place as reactive intermediates in physical organic chemistry since the 1920s following pioneering studies of H. Meerwein, C. Ingold, S. Winstein and others. G. Olah showed that many carbocations are persistent in superacid solvents. Vinyl cations were long considered to be extremely short-lived or even nonexistent as reactive intermediates, but they are now readily accessible for study by light-induced heterolytic fission of halides, acetates and alcohols in polar solvents.[464] The decay of carbocation intermediates is usually first order due to solvent addition, the rate of which increases with increasing nucleophilicity of the medium. 2,2,2-Trifluoroethanol (TFE) and 1,1,1,3,3,3-hexafluoroisopropanol (HFIP) are especially useful for the investigation of carbocations by laser flash photolysis, because these solvents favour the efficient formation of carbocations, yet they are slow to trap them. The lifetime of carbocations is insensitive to the presence of dioxygen in solution.

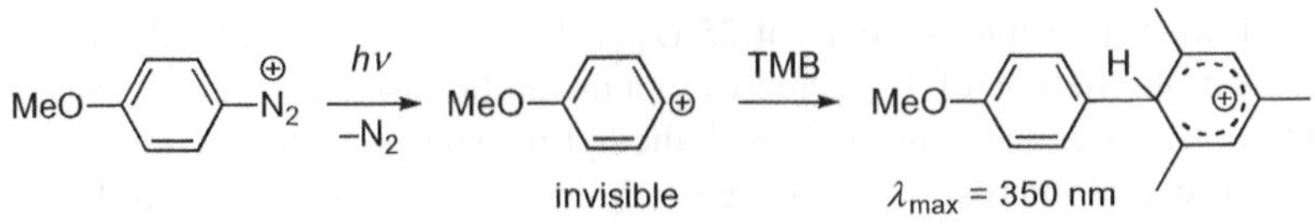

*Scheme 5.18   Making photochemically generated phenyl cations visible by trapping with TMB*

The highly reactive phenyl cation has long escaped detection because it does not absorb in the near-UV region; it has been identified by 'titration' with 1,3,5-trimethylbenzene (TMB) (Scheme 5.18).[465]

Fluorenyl cation (**31**), an 'antiaromatic' carbocation, is formed by irradiation of 9-fluorenol in protic solvents (Scheme 5.19).[466–469] The cation **31** exhibits a narrow absorption band at 515 nm. The lifetime of the precursor singlet state of 9-fluorenol ($\lambda_{max,abs} = 630$ nm, $\lambda_{max,em} = 370$ nm) is strongly reduced in polar solvents, ranging from 1.7 ns in cyclohexane to <10 ps in water. The acceleration of heterolytic cleavage of the C−OH bond in protic solvents must be attributed to fast proton transfer to the *incipient* OH$^-$ leaving group, because fission of the C−OH bond occurs on a time scale that is faster (<10 ps) than that required for solvation of the ions (25 ps).[468] The lifetime of the cation **31** varies from 20 ps in water and TFE,[468] to hundreds of nanoseconds in alkali metal zeolites[469] and to 30 µs in HFIP.[467]

The equilibrium constant for the hydration of **31** has been determined as $pK_R = -15.9$.[470] Hence the equilibrium constant for the heterolytic dissociation of 9-fluorenol in water is $K_{diss} = K_R/K_w = 10^{19.9}$, which amounts to a ground-state free energy difference of $\Delta_r G^\circ = 114$ kJ mol$^{-1}$.

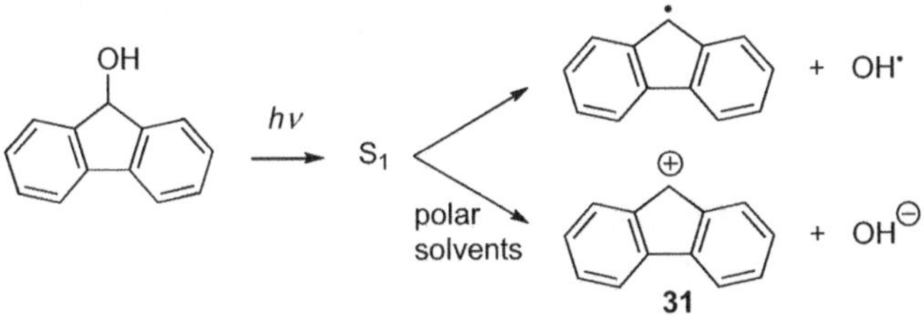

*Scheme 5.19*

### 5.4.6   Enols

*Recommended reviews.*[471–473]

Flash photolysis has contributed much to our understanding of enol chemistry in aqueous solution. The equilibrium constant for the enolization of simple ketones, $K_E = c_{enol}/c_{ketone}$, is usually too small permit a direct determination of enol concentrations $c_{enol}$ at equilibrium. For example, the enolization constant of acetone is $pK_E = 8.3$; the relative amount of 2-hydroxypropene is thus only 5 ppb in aqueous solution. Nevertheless, most reactions of carbonyl compounds proceed via enol intermediates, which, once formed, are highly reactive towards electrophilic reagents. The rate law for the acid-catalysed bromination of acetone is *independent* of the concentration of bromine; this was observed at the beginning of the last century by Lapworth, who concluded that enolization is the rate-determining step of this reaction. Numerous kinetically unstable enols, ynols and ynamines and also keto tautomers of phenols were generated in higher than equilibrium concentrations by flash photolysis of suitable precursors and the kinetics of equilibration was monitored as a function of pH.[471–473] These reactions generally obey a first-order rate law, where the observed rate constant is the sum the rate constants of ketonization and enolization, $k_{obs} = k^K + k^E$. The technique was first applied[474] to study the enol of acetophenone using the Norrish type II reaction of valerophenone (Section 6.3.4, Case Study 6.20). The same intermediate is formed by photoinduced protonation of phenylacetylene (Section 6.1.4, Scheme 6.38). Other methods that were used to generate enols include the photoreduction of ketones by hydrogen abstraction (Section 6.3.1) followed by disproportionation of the resulting ketyl radicals, the photoenolization of *o*-alkylphenyl ketones (Section 6.3.6) and the Wolff-rearrangement of α-diazocarbonyl compounds (Section 6.4.2, Scheme 6.171) followed by hydrolysis of the resulting ketenes via ene-1,1-diols, the enols of carboxylic acids. Some further reactions are shown in Scheme 5.20, together with examples for the generation of keto tautomers of phenols.

Rate constants for the reverse tautomerization reaction can be measured by thermal halogenation of ketones or isotope exchange reactions. Combination of the rate constants of ketonization, $k^E$, and enolization, $k^K$, provides accurate equilibrium constants of enolization, $K_E = k^E/k^K$. The acidity constants of ketones, $K_a^K$, and of the corresponding enols, $K_a^E$, are related to the enolization constant $K_E$ by a thermodynamic cycle, $pK_E = pK_a^K - pK_a^E$ (Scheme 5.21).

Scheme 5.21 also shows the three pathways for keto–enol tautomerization which, depending on pH, are catalysed by acid or base. The 'uncatalysed' pathway is attributed to

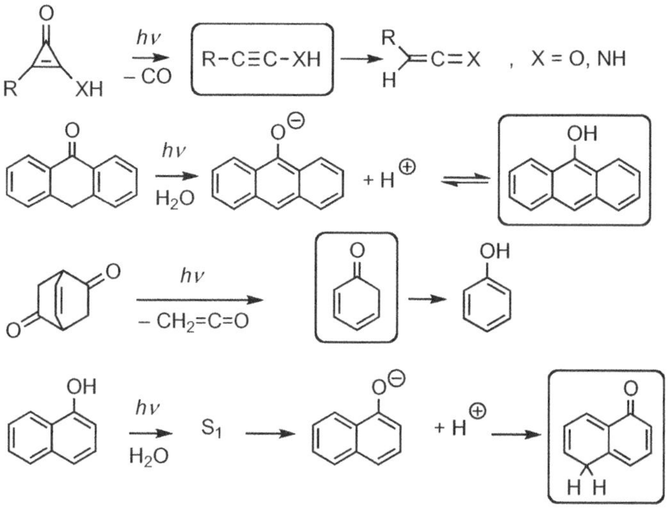

**Scheme 5.20**  *Photochemical generation of enols, ynols, ynamines and keto tautomers of phenols*

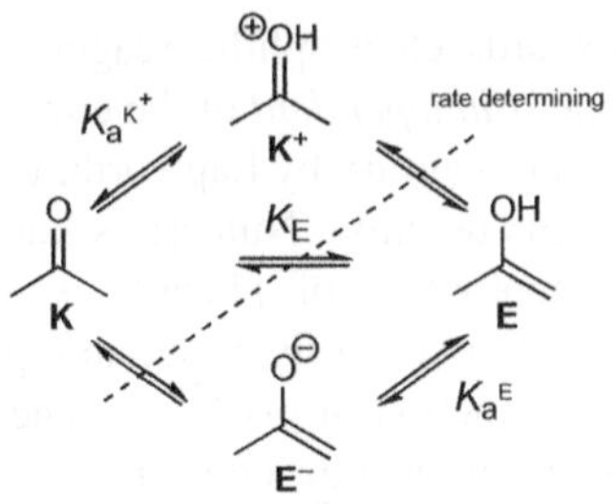

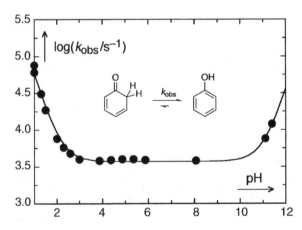

**Scheme 5.21** *Acid- and base-catalysed enolization and ketonization*

water acting as a base, that is, $E + H_2O \rightleftharpoons E^- + H_3O^+ \rightarrow K + H_2O$. The activation energies for unimolecular 1,3-hydrogen shifts are prohibitive, so that thermodynamically unstable enols can survive indefinitely in the absence of water and other general acid or base catalysts. The three independent reaction pathways are manifested in so-called *pH–rate profiles*, such as that shown for the enolization of 2,4-cyclohexadienone in Figure 5.19.[475] Because 2,4-cyclohexadienone is a strong acid that rapidly protonates the solvent water, the 'uncatalysed' reaction dominates over a large range of pH values around 7. That pathway is far less prominent in most keto–enol reactions.

The rate-determining step always corresponds to protonation or deprotonation of a carbon atom, while equilibration of oxygen acids with their conjugate bases is established rapidly. This fact can be used to determine the acidity constants of enols, ynols and ynamines by flash photolysis, $K_a^E$, either kinetically, from downward bends in the pH–rate profiles indicating a pre-equilibrium, or from the changes of the transient absorption in solutions of different pH (spectrographic titration). Such studies have provided some remarkable benchmark numbers, such as the acidity constant of phenylynol ($pK_a^E \leq 2.7$),[476] phenylynamine ($pK_a^E \leq 18.0$)[477] and its pentafluoro derivative ($pK_a^E = 10.3$),[478] and of the carbon acid 2,4-cyclohexadienone, $pK_a^K = -2.9$.[475] The enolization constant of 2,4-cyclohexadienone is $pK_E = -12.7$.

**Figure 5.19** pH–rate profile for the enolization of 2,4-cyclohexadienone to phenol in wholly aqueous solution at 25 °C[475]

## 5.5 Photoisomerization of Double Bonds

The *E–Z* photoisomerization of stilbene has been studied in considerable detail. We use the customary labels 't' (for *trans*) to designate structures near that of (*E*)-stilbene, 'c' (for *cis*) for those near (*Z*)-stilbene and 'p' (for *perpendicular*) for those near the transition-state geometry for *E–Z* isomerization in the ground state. Figure 5.20 shows schematic potential energy surfaces (PESs) for $S_0$, $S_1$ and $T_1$ that are based largely on the experimental data discussed below.

Irradiation of solutions of pure c affords the cyclization product dihydrophenanthrene (DHP) with a quantum yield of about 0.10 (Scheme 6.4) in addition to t ($\Phi_{c\rightarrow t}=0.35$).[481] The Franck–Condon region reached by excitation of c to $^1c^*$ does not correspond to a minimum on the excited singlet PES. Nevertheless, solvent friction presents an effective barrier to its rearrangement, so that the excited state $^1c^*$ persists for 0.3–2 ps depending on solvent viscosity.[479,482] The initial decay of $^1c^*$ is nonexponential, exhibiting pronounced quantum beats due to coherent wavepacket motion with a damping time of about 200 fs.[483] Weak fluorescence is observed from both $^1c^*$ ($\Phi_f=8\times10^{-5}$) and $^1t^*$ ($\Phi_f=7\times10^{-5}$) upon excitation of c in hexane at 30 °C; the fluorescence from $^1t^*$ is due to adiabatic photoisomerization $^1c^*\rightarrow{}^1t^*$.[484] The diffuse emission spectrum of $^1c^*$ develops vibrational structure in rigid alkane glasses at 77 K and its lifetime and quantum yield increase to 4.7 ns and $\Phi_f\approx0.8$.[485,486]

Excitation of t generates $^1t^*$, which yields c and t in approximately equal amounts, $\Phi_{t\rightarrow c}=0.54$.[487] The temperature and solvent viscosity dependence of the fluorescence quantum yields and lifetimes indicates that the reaction from $^1t^*$ proceeds over a medium-enhanced barrier in the excited singlet state; the intrinsic barrier amounts to 12 kJ mol$^{-1}$.[488] The fluorescence quantum yield and lifetime of t depend strongly on solvent viscosity;[489] they increase from $\Phi_f=0.035$, $\tau(^1t^*)=59$ ps (pentane), $\Phi_f=0.040$, and $\tau(^1t^*)=64$ ps (hexane) to $\Phi_f=0.077$, $\tau(^1t^*)=110$ ps (*n*-tetradecane) at 30 °C,[479] and to $\Phi_f=0.95$ and $\tau(^1t^*)=1.6$ ns in a rigid alkane glass at 77 K.[488] The first absorption bands of c and t (Figure 6.1)

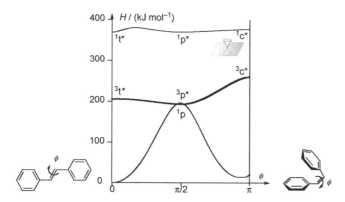

**Figure 5.20** Potential energy surfaces[479] for the *E–Z* isomerization of stilbene. The conical intersection[480] is on the cyclization path to DHP, outside of the torsional reaction coordinate shown

correspond to the HOMO–LUMO transition, but the forbidden, doubly excited state of $A_g$ symmetry (see end of Section 4.7) lies only $1000 \, cm^{-1}$ above $S_1(t)$.[490]

ISC to the triplet state is inefficient in media of low viscosity at ambient temperature; the triplet state must be populated by sensitization. Pulsed laser irradiation of a sensitizer in the presence of high concentrations of either c or t produces the same transient intermediate exhibiting a broad absorption band rising in intensity from 400 to 360 nm.[491] The intermediate decays with a lifetime of 60 ns to give nearly equal amounts of the c and t isomers and is attributed to the twisted triplet state $^3p^*$. The energy of $^3p^*$ was determined by photoacoustic calorimetry to lie $195 \pm 4 \, kJ \, mol^{-1}$ above the ground state of t.[251] The reaction enthalpy for the isomerization of c to t in solution is $\Delta_{c \to t} H = -19.2 \pm 0.4 \, kJ \, mol^{-1}$ and the activation enthalpy for this reaction in the ground state is $182 \pm 8 \, kJ \, mol^{-1}$.[492] Hence the $T_1$–$S_0$ energy gap nearly vanishes at perpendicular geometries, with $^3p^*$ probably lying slightly below $^1p$.

The comparable yields of c and t formed by direct excitation of either isomer have long been taken as evidence that the *E–Z* photoisomerization proceeds via a common intermediate $^1p^*$, which acts as the funnel for IC to the ground state. However, femtosecond laser studies do not appear to be consistent with such an intermediate possessing a lifetime of more than 150 fs because there is no measurable delay between the disappearance of $^1c^*$ and the appearance of ground-state product absorption.[482] Theoretical calculations show that the reaction cannot be described solely by the torsional angle $\phi$.[493–495] A conical intersection (Special Topic 2.5) has been located along a cyclization reaction coordinate leading to DHP.[493,496] DHP opens up very efficiently on irradiation, presumably through the same conical intersection. Contrary to an early suggestion,[497] the minimum at $^1p^*$, probably arising by admixture of the doubly excited $^1A_g$ wavefunction (see end of Section 4.7), is shallow and the energy gap to the ground state is fairly large at this geometry. It is therefore unlikely that $^1p^*$ represents the dominant funnel for IC to the ground state. However, in spite of the extensive work done, this issue is still under discussion.

The ground- and triplet excited-state PES of the homologues 1,4-diphenylbuta-1,3-diene (DPB)[498] and 1,6-diphenylhexa-1,3,5-triene (DPH)[499] have also been elucidated in detail by both experiment and calculation. Stilbene and DPH represent two extreme cases because the extended conjugation in DPH lowers the triplet energies of planar conformers more than those of twisted ones. All three have rather flat triplet PESs with respect to torsional motion about the formal double bonds. The perpendicular geometry $^3p^*$ represents an energy minimum in stilbene. In contrast, the perpendicular geometry corresponds to an energy maximum in $^3$DPH. Decay from the $^3p^*$ state of stilbene and from roughly planar triplets in DPH accounts for the 1000-fold difference in triplet lifetimes (60 ns for stilbene, 60 µs for DPH). The roughly equal amount of (*E*)- and (*Z*)-stilbene formed by sensitization of either isomer is consistent with decay from a perpendicular triplet, whereas the partitioning from $^3$DPH reflects an equilibrium distribution of the planar conformers that strongly favours the all-*trans* triplet. Conformational equilibration is also fast in $^3$DPB, which decays mostly from a perpendicular minimum that lies slightly above the global all-*trans* minimum. Quantum chain processes of triplet–triplet energy transfer may lead to quantum yields of photoisomerization quantum yields far in excess of unity.[500]

The short lifetimes of excited diaryl polyene singlet states precludes the equilibration of conformers following excitation and, as a result, the conformational distribution in the ground state controls the product distribution. Because the absorption spectra of different conformers usually differ, this results in an excitation-wavelength dependence of the product distribution (*NEER principle*;[501] cf. Section 6.1.1 and Case Study 6.3).

## 5.6  Chemiluminescence and Bioluminescence

*Recommended reviews.*[502–509]

*Chemiluminescence* is a fascinating phenomenon that is often displayed in demonstrations ('cold light'), seen in Nature (*bioluminescence*) and is widely used for analytical purposes (drug screening, DNA sequencing, capillary electrophoresis, immunoassays, forensics, etc.),[506] thanks to the utmost sensitivity of light detection in a dark environment. The generation of electronically excited molecules from ground-state reagents is called *chemiexcitation*; it opens up the possibility of performing photochemistry without light.

Chemiexcitation is by definition a *diabatic reaction*. The minimal conditions for chemiexcitation are not easily met. First, the energy released must be fairly large; the energy difference between the transition state of the thermal reaction and the final product must exceed the excitation energy of the product. The energy required to produce an excited state emitting light in the visible region of 400–700 nm amounts to 300–170 kJ mol$^{-1}$. Second, an efficient pathway is needed to store the available excess energy in an electronically excited state of the product, rather than wasting it as heat, which is favoured entropically. This requires that *diabatic potential energy surfaces* cross along the reaction coordinate between reactant and product, such that the wavefunction of the ground-state reactant changes smoothly to that of the excited state of the product (Figure 5.21). In terms of the Landau–Zener model (Section 2.3), formation of the excited state of the product is favoured when the slope of the two zero-order PESs differs strongly at the diabatic crossing point and the electronic interaction term $V_{if}$ between the two zero-order surfaces is small.

The energy requirement is an obstacle to the physical characterization of chemiluminescent compounds because the availability of a reaction of high exothermicity renders most such reagents unstable. They are usually generated *in situ* by mixing stable precursors with appropriate reagents. The observation of blue chemiluminescence from a solution of oxalyl chloride and hydrogen peroxide was first reported in 1963,[510] now well

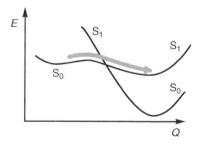

**Figure 5.21**  Shape of a diabatic surface crossing favouring chemiexcitation

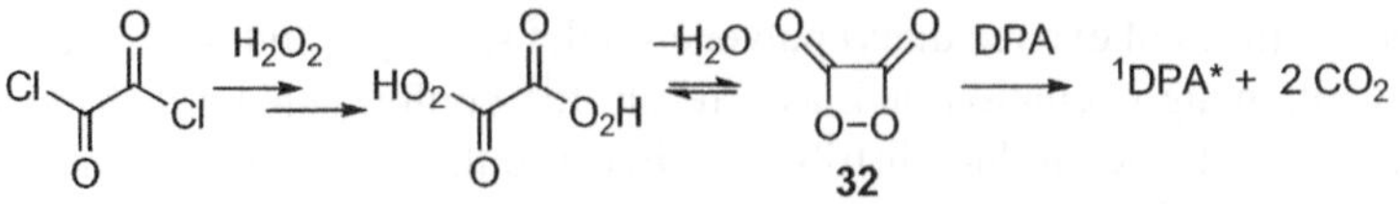

**Scheme 5.22**  *The peroxyoxalate chemiluminescence reaction*

known as the *peroxyoxalate chemiluminescence reaction*. The authors noted: 'The vapors produced by the reaction have the unusual property of inducing the fluorescence of suitable indicators, e.g., a filter paper impregnated with anthracene. These observations are suggestive of a metastable excited electronic state or some other highly energetic species.' The luminescence was strongly enhanced by the addition of fluorescent chromophores such as 9,10-diphenylanthracene (DPA) and the emission corresponded to the fluorescence of these substances. It took over 40 years to provide evidence that the energetic species formed in this reaction is indeed the long-suspected[511] 1,2-dioxetanedione (**32**, a dimer of carbon dioxide!) (Scheme 5.22),[512] and it remains to be shown that the rise and fall of this intermediate are kinetically related to the observed rise and fall of the chemilumines-cence.[509, 513] High-quality *ab initio* calculations have shown that structure **32** represents a minimum on the PES[514] but, surprisingly, a reliable calculation of the barrier to its decomposition is not available at present.

This reaction is the active principle of the 'cold light' sticks originally marketed by American Cyanamid, now OmniGlow. The last step in Scheme 5.22 cannot be due to a simple cleavage of **32** yielding an excited $CO_2$ molecule, the excitation energy of which is far too high. In Section 5.2, we saw that the encounter of an excited molecule with an electron donor or acceptor may lead to the formation of radical ions when the free energy of electron transfer is negative (Equation 5.1). Naturally, the reverse process may occur when the recombination of radical ions is sufficiently exergonic to form an excited state of one of the neutralized reactants. Thus, the reaction of DPA and other activators with **32** is thought to produce a radical ion pair $^1(DPA^{\bullet+} \cdots CO_2^{\bullet-} + CO_2)$, which forms $^1DPA^*$ by return electron transfer.[515] Such processes were named *chemically initiated electron exchange luminescence* (CIEEL);[516] they operate in many other chemiluminescent systems. The radical ions may also be produced by electrochemical oxidation and reduction processes, resulting in *electrochemiluminescence* (ECL).[508] Essentially the same process is exploited in LEDs and OLEDs (Special Topic 3.2) and excimer lasers (Special Topic 3.1). Four important isolable chemiluminescent species are shown in Scheme 5.23.

The cleavage of tetramethyldioxetane (**33**) to two acetone molecules represents a [2 + 2] cycloreversion reaction that classifies as ground-state forbidden by the Woodward–Hoffmann rules (Section 4.9). Hence one may expect a diabatic crossing of zero-order PESs in this reaction (Figure 4.35). Sophisticated calculations indicate that the weak O–O bond breaks initially and that ISC to the triplet state occurs in the resulting biradical intermediate.[517] Exothermic triplet–singlet energy transfer from the chemiex-cited triplet acetone to a heavy-atom substituted chromophore (9,10-dibromoanthracene) then promotes efficient fluorescence emission.[507]

The majority of bioluminescent organisms live in the ocean, but there are many terrestrial forms, notably beetles, that exploit bioluminescence. The bioluminescence of fireflies

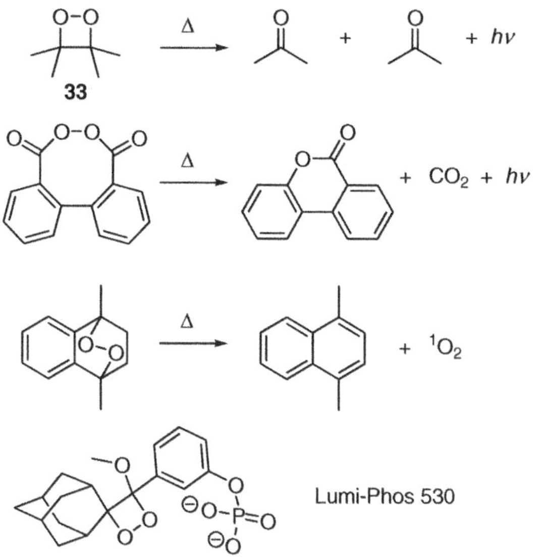

**Scheme 5.23** *Isolable chemiluminescent reagents*

proceeds by oxidation of D-luciferin (**34**) to oxyluciferin (**35**) that is assumed to involve a highly energetic dioxetanone intermediate (Scheme 5.24).[504] The emission wavelength depends on the medium, pH and temperature. The origin of the pH dependence is far from understood.[518] An enol tautomer of **35** may be involved.

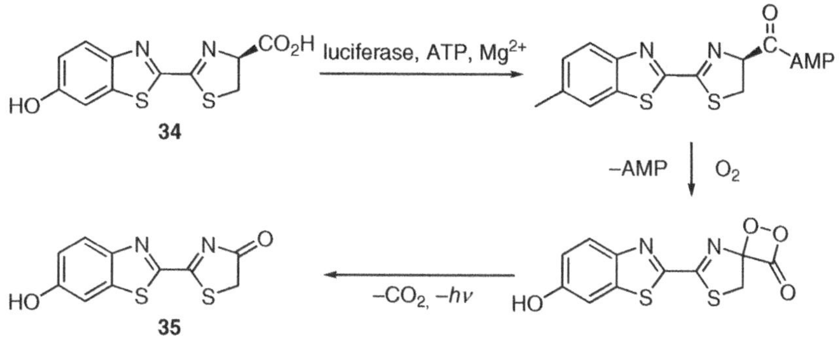

**Scheme 5.24** *Luminescent reaction of fireflies*

## 5.7 Problems

1. The $pK_a$ of phenalene ionizing to its anion **4**$^-$ (Case Study 4.1) is about 20. The first absorption band of phenalene lies at $\tilde{\nu}_{0-0} = 2.9\,\mu m^{-1}$. Predict the acidity constant of phenalene in the lowest excited singlet state. [$pK_a^* = -1$]

2. Identify the symmetry point groups of formaldehyde $[C_{2v}]$, ammonia $[C_{3v}]$, phenol $[C_s]$, gyloxal $[C_{2h}]$ and allene $[D_{2h}]$.

3. Verify Equation 4.31 for a configuration of six electrons in three orbitals by considering all electronic interactions.

4. Predict the ground-state multiplicity of the conjugated biradicals shown in Scheme 5.25. [T, S, S, T, T, T][519]

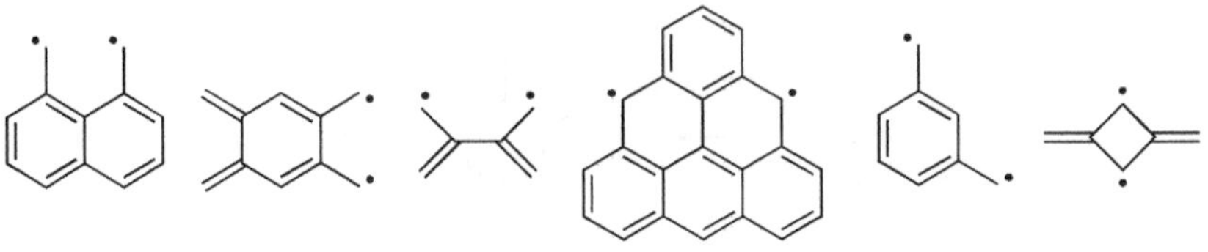

**Scheme 5.25**   *Non-Kekulé biradicals*

5. Calculate the energy gap between the lowest excited singlet and triplet configurations of ethene, $E(^1\chi_{1 \rightarrow -1}) - E(^3\chi_{1 \rightarrow -1}) = 2K_{1-1}$, using $r_{12} = 134\,pm$ and the PPP SCF parameters given below Equation 4.29 [5.407 eV].

6. Calculate the concentration required to quench 99% of singlet excited naphthalene, given that the lifetime of the naphthalene singlet state is 100 ns in the absence of dioxygen and N,N-dimethylaniline and that the rate constant of diffusion $k_{diff}$ is $5 \times 10^9\,M^{-1}\,s^{-1}$. [0.2 M]

7. Explain why the oscillator strengths of charge-transfer transitions in EDA complexes are low and why the absorption maximum of the charge-transfer band of tetracyanoethylene complexes with methylated benzenes is red shifted upon increasing methyl substitution (from $\tilde{\nu}_{max} = 2.6\,\mu m^{-1}$ for benzene to $\tilde{\nu}_{max} = 1.9\,\mu m^{-1}$ for hexamethylbenzene).[343]

8. Rationalize the singlet–triplet energy gaps of the carbenes listed in Table 5.1.

# 6

# Chemistry of Excited Molecules

## 6.1 Alkenes and Alkynes

The absorption spectra of unsaturated hydrocarbons are dominated by $\pi,\pi^*$ transitions, which were discussed in Sections 4.5 and 4.7. Some representative spectra are shown on a logarithmic scale in Figure 6.1. The first strong absorption band, $\log(\varepsilon/[\text{M}^{-1}\,\text{cm}^{-1}]) \approx 4$, of unsaturated hydrocarbons usually corresponds to the HOMO–LUMO transition that is well described by Hückel theory. However, two types of weak transitions may in fact appear at lower energies, which can easily pass unnoticed upon cursory inspection of the corresponding absorption spectra, but may determine the photoreactivity. The first are the parity-forbidden transitions (Section 4.7), $\log(\varepsilon/[\text{M}^{-1}\,\text{cm}^{-1}]) \approx 2.5$, which occur in all *alternant hydrocarbons*. They represent the lowest singlet state $S_1$ in important chromophores such as benzene, naphthalene and their derivatives, as exemplified by the spectra of styrene and phenylacetylene (Figure 6.1, bottom). The second are the symmetry-forbidden $^1A_g$ bands, $\log(\varepsilon/[\text{M}^{-1}\,\text{cm}^{-1}]) \approx 1$ in linear polyenes and their phenyl derivatives (Section 4.7). In terms of MO theory, the lowest $^1A_g$ state is described by a doubly excited configuration with two electrons promoted from the HOMO to the LUMO. Even when the $^1A_g$ state is not the lowest excited singlet state, it may nevertheless be important for photoreactivity, as in polyenes, because the energy of the $^1A_g$ state is lowered along reaction coordinates of double bond twisting (Section 5.5). Donor and acceptor substituents, on the other hand, affect mostly the strong HOMO–LUMO transition, shifting it to lower energies.

Twisting of essential single bonds in the ground state due to steric interactions is associated with a hypsochromic shift of the HOMO–LUMO absorption maximum. This effect is seen in the absorption spectra of (*E*)- and (*Z*)-stilbene (Figure 6.1, bottom). On the other hand, the red edge of the (*Z*)-stilbene absorption extends to lower energies; the non-vertical transition to the relaxed excited state requires less energy, because twisting destabilizes the ground state. The $\pi$-systems of 1,3-dienes (1,1′-bicyclohexene, Figure 6.1,

*Photochemistry of Organic Compounds: From Concepts to Practice*   Petr Klán and Jakob Wirz
Copyright © 2009 P. Klán and J. Wirz

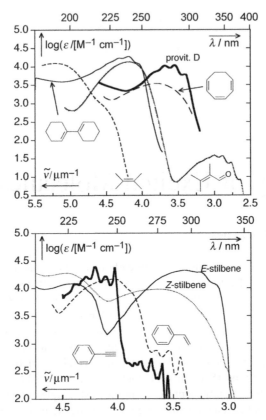

**Figure 6.1** Absorption spectra of prototype alkenes and alkynes. Top: 1,1'-bicyclohexene[280] (—), provitamin D[520] (7-dehydrocholesterol; —), cycloocta-1,3,5-triene[520] (– – –), tetramethylethylene[280] (- - -), and 2,3-dimethylbut-2-enal[280] (····). Bottom: (*E*)-stilbene[521] (—), (*Z*)-stilbene[521] (····), phenylacetylene[522] (—), styrene[280] (- - -)

top) and α,β-unsaturated ketones or aldehydes (2,3-dimethylbut-2-enal) are isolectronic and exhibit similar π,π* bands. However, ketones and aldehydes exhibit an additional, weak n,π* transition at longer wavelengths.

The singlet lifetimes of unsaturated hydrocarbons vary from picoseconds (stilbenes) to nearly 1 μs (pyrene). Radiative lifetimes $\tau_r = 1/k_f$ range from about 1 ns for compounds with a strong $S_0 \rightarrow S_1$ absorption band to many microseconds with compounds having a forbidden $S_0 \rightarrow S_1$ absorption band. The photophysical properties of some prototype unsaturated hydrocarbons are listed in Table 8.6. Radiationless decay associated with *E–Z* (or *trans–cis*) isomerization is usually very fast in polyenes (see below). Intersystem crossing is generally slow – too slow to compete efficiently with these deactivation processes, because *spin–orbit coupling* is weak and the singlet–triplet energy gap is large in alternant hydrocarbons. Therefore, unsaturated hydrocarbons that do not undergo *E–Z* isomerization – due to steric constraints or in extended systems – have high fluorescence quantum yields.

Table 6.1 lists the principal photoreactions of alkenes and alkynes that we describe in this Section. It is structured according to the types of photoreactions that are characteristic for an initial reacting chromophore. Many practical examples and special topics will expand on this general information and a number of references lead the reader to review articles or the original reports.

Most molecules containing a C=C bond are predestined to undergo photoinduced geometric *E–Z* isomerization (Table 6.1, entry 1), unless the double bond is embedded in a small ring or the molecule is constrained by a rigid environment. Since the *E*- and *Z*-isomers have different absorption properties, monochromatic light may be used to enrich the one that absorbs less at the excitation wavelength. The photoisomerization mechanism of singlet and triplet excited alkenes differs (Section 5.5); however, the reactant multiplicity has only a minor impact on the overall reaction. Direct irradiation of monoalkenes is difficult because they absorb at short wavelengths; therefore, photosensitization is a practical technique to carry out the reaction. When the double bond is conjugated to an auxochrome, such as a carbonyl group, the triplet state can be readily formed via intersystem crossing.

*E–Z* isomerization, including chemically nonproductive isomerization about terminal double bonds, will tend to lower the quantum yields of competing photoreactions such as *electrocyclization* or *sigmatropic shift* (entries 2 and 3) of dienes, respectively. The cyclization products absorb at shorter wavelengths than the reactant diene, so that high chemical yields of cyclobutenes may nevertheless be achieved by suppressing secondary photoreactions of the photoproduct such as *cycloreversion* (reverse opening) with appropriate cut-off filters. When two double bonds are connected through a methylene bridge, the *di-π-methane rearrangement* (entry 4) can be the principle reaction for the same reasons.

Excited alkenes and alkynes are highly reactive towards nucleophiles, acids and electron donors. *Nucleophilic addition* and *photoreduction* (entries 5 and 8) predominate with alkenes carrying electron-withdrawing substituents. Some electron-rich alkenes or alkynes readily undergo *photoprotonation* (entries 6 and 7).

Bimolecular *photocycloaddition* (entry 9) of an excited alkene to a ground-state alkene is clearly influenced by alkene concentration. However, close proximity of the reactants can also be promoted by complexation or by constrained environments.

## 6.1.1 Alkenes: *E–Z* Isomerization

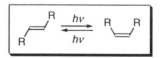

*Recommended review articles.*[500,523–533]
*Selected theoretical and computational photochemistry references.*[16,534–552]

*E–Z* (or *trans–cis*) isomerization (Section 5.5) of alkenes, involving a 180° rotation about a C=C bond, is known to be initiated both thermally and photochemically.[524–526,530,553] After equilibrating, the thermal process produces a mixture of the two isomers in a ratio that reflects their relative thermodynamic stabilities. In contrast, photoisomerization of an alkene (Scheme 6.1) is governed by the ratio of the $E \rightarrow Z$ and

**Table 6.1**    *Photoprocesses involving excited alkenes and alkynes*

| Entry | Starting material[a] | Product(s) | Mechanism | Section |
|-------|----------------------|------------|-----------|---------|
| 1 | | | E–Z isomerization | 6.1.1 |
| 2 | | | Electrocyclization | 6.1.2 |
| 3 | | | Sigmatropic shift | 6.1.2 |
| 4 | | | Di-π-methane rearrangement | 6.1.3 |
| 5 | + Nu⊖ | | Photoinduced nucleophile addition | 6.1.4 |
| 6 | + H⊕ | | Photoinduced proton addition | 6.1.4 |
| 7 | + H₃O⊕ | | Photoinduced proton addition | 6.1.4 |
| 8 | + [e⊖] | | Photoreduction | 6.1.4 |
| 9 | + ‖ | | Photocycloaddition | 6.1.5; 6.1.6 |

[a]Nu⁻ = nucleophile; [H] = hydrogen atom donor; [e⁻] = electron donor.

$Z \rightarrow E$ isomerization quantum yields ($\Phi_{E-Z}$ and $\Phi_{Z-E}$, respectively) and the molar absorption coefficients $\varepsilon$ of the corresponding isomers. The resulting isomer ratio is reached when the rates of formation and disappearance for each isomer become equal (photostationary state; PSS). A large excess of the thermodynamically less stable Z-isomer can be produced when favourable absorption properties (higher $\varepsilon$ values) of the other E-isomer at certain wavelengths allow it (Figure 6.1, bottom). Other competing photoprocesses, such as radiation or radiationless decays (Section 2.1), sigmatropic shifts (Section 6.1.2), carbene formation, [2 + 2] cycloadditions (Section 6.1.5) or electro-cyclization reactions (Section 6.1.2), can also compete with this reaction.

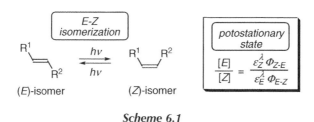

**Scheme 6.1**

E–Z isomerization of alkenes initiated by direct irradiation usually proceeds from the excited singlet state. Because of the existence of two or more close-lying excited singlet states, including the Rydberg [$\pi$,R($3s$)] states,[554] slow mutual interconversion between the excited states is probably involved.[555,556] The presence of a triplet sensitizer is required to carry out the isomerization from a low-lying excited triplet state ($\pi$,$\pi^*$). In such a case, the photostationary alkene concentrations depend on the ratios of the rate constants of energy transfer to both isomers, the magnitudes of which are associated with the relative sensitizer triplet energies (Section 2.2.2).[525,553]

*Direct Irradiation*

Absorption spectra of simple aliphatic alkenes in solutions, generally having their maxima ($\lambda_{max}$) below 200 nm, imply that direct irradiation must be conducted with *vacuum UV* sources (Section 3.1). However, a medium-pressure mercury lamp with a quartz filter can be applied for highly substituted alkenes, possessing $\lambda_{max}$ over 200 nm. Direct irradiation of (Z)-oct-2-ene (**36**, R$=$C$_4$H$_9$) at 185 nm leads to a nonselective excitation and consequently the E–Z isomerization and the 1,3-hydrogen migration products **37** and **38** via carbene intermediates (Scheme 6.2). This rearrangement possibly does not involve the $^1\pi$,$\pi^*$ state but occurs via a *Rydberg* excited state and carbene intermediates.[523,556] E–Z isomerization is, however, exclusively observed upon irradiation above 200 nm.[556]

Rearrangement and decomposition are the major reactions observed when small-ring cycloalkenes are photolysed.[523,524,530] E–Z photoisomerization becomes an important process in medium- and large-ring cycloalkenes, although it is usually accompanied by side-reactions. Although isomerization in cyclopentene is still not geometrically feasible, (Z)-cyclohexene (**39**) gives the highly strained and unstable E-isomer by both direct (Scheme 6.3) and sensitized irradiation.[557] This short-lived intermediate undergoes

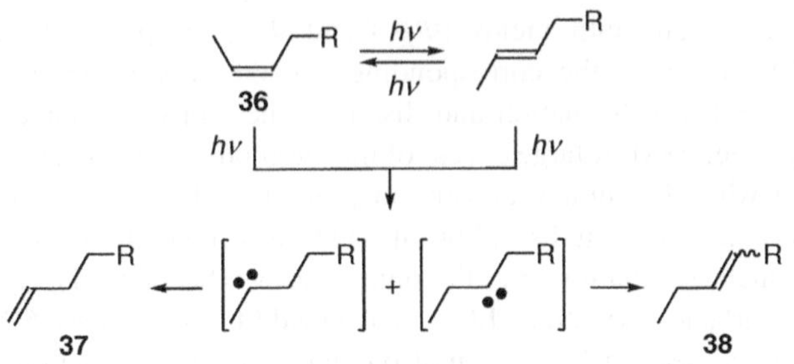

*Scheme 6.2*

non-stereospecific [2 + 2] addition or radical side-reactions in aprotic non-nucleophilic solvents (Section 6.1.5). In contrast, cycloheptene shows little tendency to undergo photodimerization and (Z)-cyclooctene is the smallest cycloalkene producing the corresponding isolable E-isomer upon irradiation at 185 nm.[558]

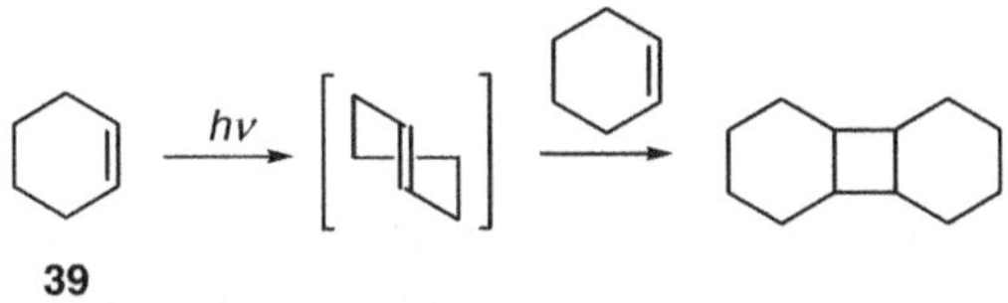

*Scheme 6.3*

*Arylalkenes*, such as stilbene derivatives, are important model compounds for the study of the E–Z photoisomerization.[529,559,560] The compounds absorb significantly over 250 nm, therefore direct irradiation is technically simple. Photolysis of unsubstituted stilbene in aliphatic hydrocarbons at 313 nm affords a photostationary state (PSS) consisting of 93% of (Z)-stilbene and 7% of (E)-stilbene (Scheme 6.4).[112] In addition, a photoinduced 6$\pi$-electrocyclic (Section 6.1.2) formation of dihydrophenanthrene with a quantum yield of $\Phi \approx 0.10$ competes with (Z)-stilbene isomerization ($\Phi_{Z-E} = 0.35$).[487,561]

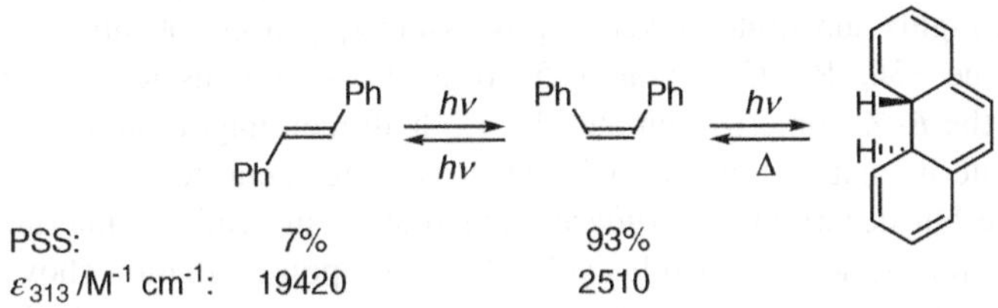

| PSS: | 7% | 93% |
| --- | --- | --- |
| $\varepsilon_{313}$ /M$^{-1}$ cm$^{-1}$: | 19420 | 2510 |

*Scheme 6.4*

The *multiplicity* of the excited state that is responsible for photoisomerization may vary with the nature of the stilbene phenyl ring substituents (Table 6.2). When the coupling

**Table 6.2** *Photolysis of the stilbene derivatives* $Ph-HC=CH-C_6H_4X$[529]

| X | $\Phi_{E-Z}$[a] | $\Phi_{Z-E}$[a] | Mechanism[b] | [E]:[Z] (PSS) |
|---|---|---|---|---|
| H | 0.52 | 0.35 | $S_1$ | 93:7 |
| 4-Cl | 0.60 | 0.42 | $S_1 + T_1$ | 91:9 |
| 4-Br | 0.52 | 0.35 | $S_1 + T_1$ | 88:12 |

[a]$\Phi_{E-Z}$ and $\Phi_{Z-E}$ are the quantum yields for $E \rightarrow Z$ and $Z \rightarrow E$ isomerization, respectively.
[b]Isomerization from the singlet state ($S_1$) or/and the triplet state ($T_1$).

between the singlet and triplet manifolds increases, for example due to the heavy-atom effect (Section 4.8) of a bromine atom, a less selective *triplet* isomerization pathway becomes competitive.[529]

---

### Case Study 6.1: Supramolecular chemistry – photoresponsive stilbene dendrimers

The water-soluble (*E*)-stilbene dendrimer **40** was found to undergo an unusual *one-way E–Z isomerization* to give the thermodynamically less stable (*Z*)-**40** nearly exclusively at the photostationary state upon irradiation in water (Scheme 6.5).[562] Monitoring the absorbance at 330 nm in the course of irradiation showed that the system reaches a photostationary state (Figure 6.2).

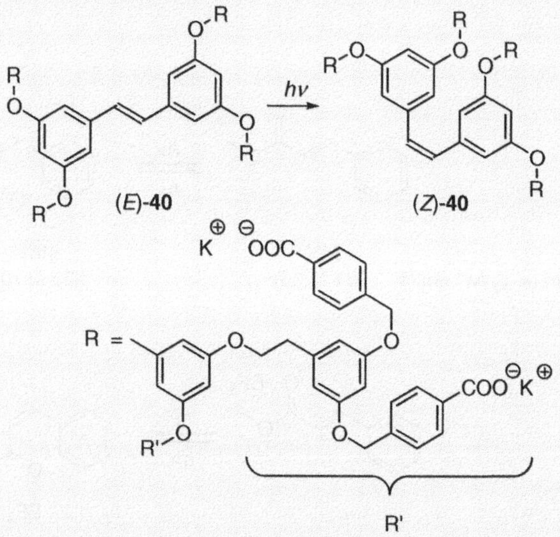

**Scheme 6.5**

*Experimental details.*[562] A KOH aqueous solution ($2 \times 10^{-3}$ M) of **40** was irradiated at 330 nm (Figure 3.9) at 20 °C and product formation was followed by UV spectroscopy.

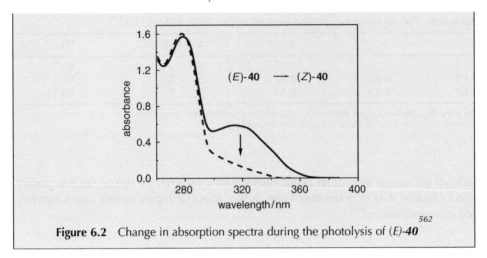

1.6
1.2
0.8
0.4
0.0
absorbance
(E)-40 → (Z)-40
280
320
360
400
wavelength/nm

**Figure 6.2** Change in absorption spectra during the photolysis of (*E*)-**40**[562]

Lewis acid ($BF_3$ or $EtAlCl_2$) complexes of α,β-unsaturated esters can shift the photoequilibrium (PSS) toward the thermodynamically less stable *Z*-isomer even more and may inhibit other competing unimolecular photochemical processes.[563] Such enhanced isomerization results are explained by selective excitation of the ground-state Lewis acid–ester carbonyl complex, which exhibits a red shift in the long-wavelength π,π* absorption band ($\lambda_{max}$) and higher molar absorption coefficients ($\varepsilon_{313}$) (Scheme 6.6).

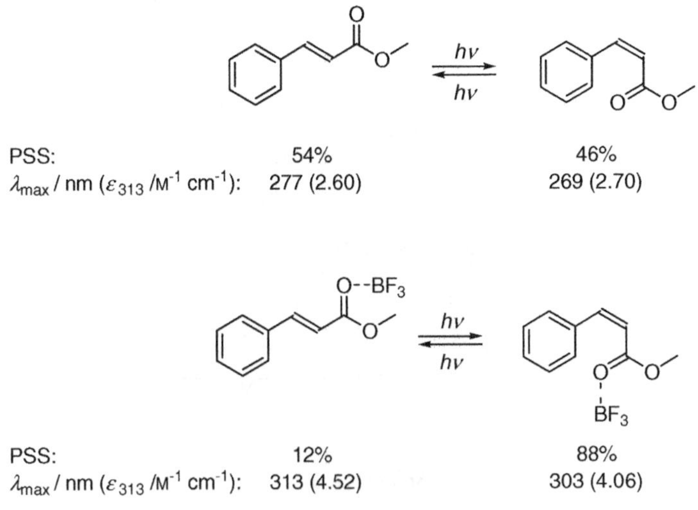

hν
hν
PSS:
54%
46%
$\lambda_{max}$/ nm ($\varepsilon_{313}$ /M$^{-1}$ cm$^{-1}$):
277 (2.60)
269 (2.70)
O--BF$_3$
hν
hν
BF$_3$
PSS:
12%
88%
$\lambda_{max}$/ nm ($\varepsilon_{313}$ /M$^{-1}$ cm$^{-1}$):
313 (4.52)
303 (4.06)

***Scheme 6.6***

Extended conjugation shifts the strong $\pi,\pi^*$ absorption bands of alkenes to higher wavelengths (Figure 6.1). An explanation of their photoisomerization is based on the idea of ground-state conformational control of photoreactivity, termed the *NEER principle* (non-equilibration of excited rotamers), according to which the excited species of the various ground-state rotamers (Section 5.5) will not equilibrate during their short singlet excited-state lifetime.[564] The increased bond order between the unsaturated units of polyenes inhibits *s-cis* ↔ *s-trans* interconversion in the excited state. As a result, the photoproduct concentrations reflect the composition of the ground-state conformational equilibrium.

Three different photoisomerization mechanisms of singlet excited polyenes have been proposed: (1) *one-bond twist*, a typical process observed in *fluid solutions* but also in glassy media,[525,553,565] (2) the *bicycle-pedal mechanism* involving simultaneous rotation about two original double bonds, assumed to occur in a constraining environment,[566–568] and (3) a volume-conserving two-bond '*hula twist*'[531,569,570] (Scheme 6.7). In some cases, the existence of the last mechanism has been ruled out.[567,571]

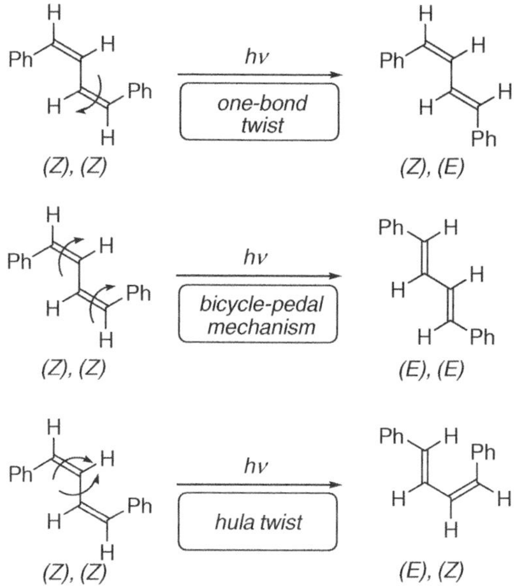

*Scheme 6.7*

Photochemistry of previtamin D (**41**; see also Special Topic 6.4) is an example of *E–Z* isomerization in trienes. Thermal (Δ) interconversion of its conformers **41a,b** is suppressed in a cold, rigid matrix of a solidified medium (92 K) (Scheme 6.8). Upon irradiation, **41a** isomerizes to the thermodynamically more stable tachysterol conformer **42a**, whereas **42b** is formed exclusively from the conformer **41a**.[572,573]

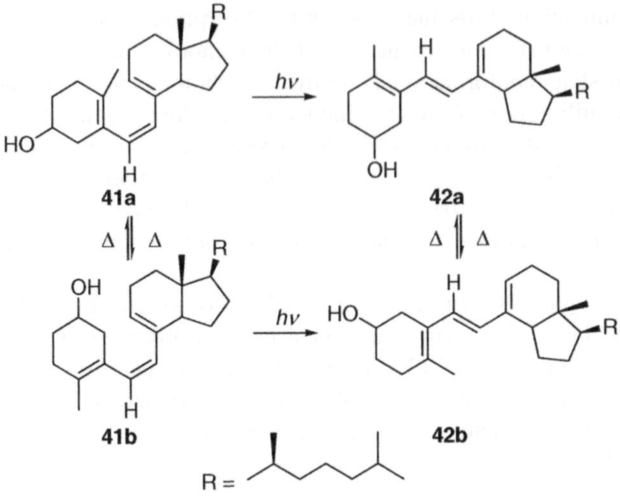

**Scheme 6.8**

---

## Special Topic 6.1: Vision

The retina of the eye consists of a number of *photoreceptor* cells containing the visual pigment *rhodopsin*, a bundle of seven transmembrane helices that enclose the photochemically active chromophore, the imine of 11-*cis*-retinal (**43**),[574] which is formed by oxidation of the dietary supplement vitamin A (all-*trans*-retinol, **44**; Scheme 6.9). Retinal is covalently linked to the opsin receptor as an iminium salt (**45**). This system is capable of *phototransduction*, a process by which light energy is converted into electrochemical potential. The primary photochemical process is *E–Z* photoisomerization ($\Phi \sim 0.65$), which changes the conformation of the opsin molecules and leads to signal transduction cascades.[210,575–577] The visual process has an extraordinarily high sensitivity; a retinal *rod* registers the absorption of a single photon in a dark-adapted eye, while the response of *cones*, used for daytime vision, is typically 10–100 times smaller. Enzymes mediate a reverse *Z–E* isomerization of retinal that subsequently regenerates 11-*cis*-rhodopsin.

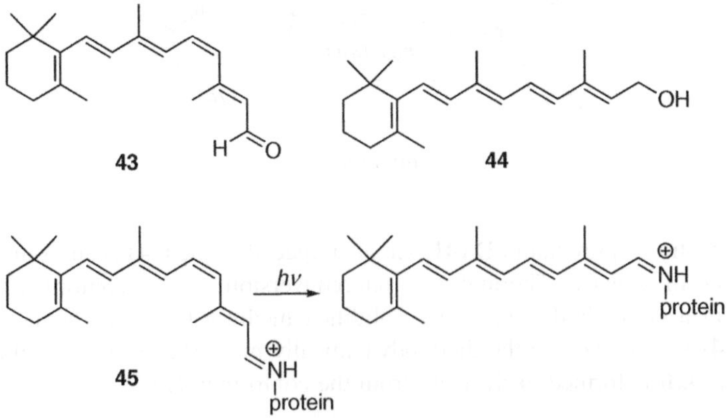

**Scheme 6.9**

## Special Topic 6.2: Phototherapy – treatment of hyperbilirubinaemia

*Phototherapy* or light therapy is exposure to specific wavelengths of visible light to treat health problems, such as acne vulgaris, seasonal and non-seasonal affective disorders or delayed sleep phase syndrome. It is also routinely used in hospitals for treating neonatal jaundice (hyperbilirubinaemia), which is due to an accumulation of the neurotoxic metabolite bilirubin in blood.[578] (Z,Z)-Bilirubin (**46**), a yellow–orange tetrapyrrole derivative, is a product of haem degradation catalysed by haem oxygenase in mammals. This compound is insoluble in water due to intramolecular stabilization by six strong hydrogen bonds (Scheme 6.10); therefore, its direct excretion is impossible. It can be excreted into bile only via a process known as glucuronidation. Formation and excretion are balanced in healthy persons; however, after birth, the high rate of red cell degradation may not be compensated by the initially low liver activity and, as a result, neonatal jaundice occurs. During phototherapy, young patients are exposed to blue or white light to expedite bilirubin elimination. Two major photoprocesses are believed to be involved in the mechanism. The first is the configurational *E–Z* photoisomerization of **46** to form the 4Z,15 *E*-isomer **47**;[579] the second is the less efficient formation of a structural isomer lumirubin (not shown). Because both products are water soluble (the intramolecular hydrogen bonding present in the parent compound is now geometrically suppressed), which permits their direct excretion, both photoreactions help to reduce the bilirubin concentration in blood plasma. It has been found that the initial twisting of the double bond in the excited **46** takes place on a time scale from several hundreds of femtoseconds to a few picoseconds.[580]

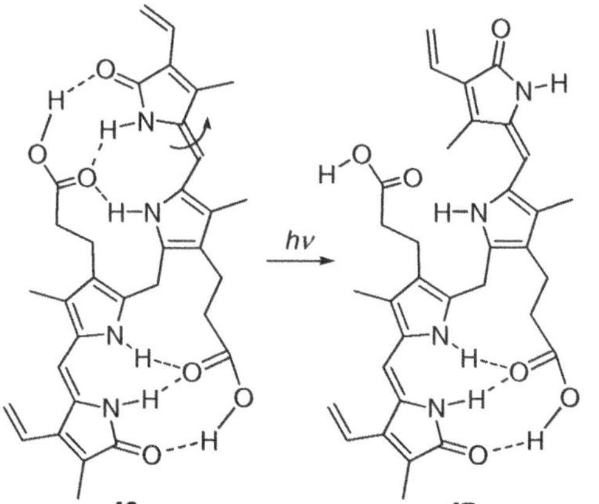

**46**      **47**

*Scheme 6.10*

## Photosensitized Isomerization

Sensitization is often indispensable for inducing reaction from the triplet state and it is practical in the case of aliphatic non-conjugated alkenes, which absorb weakly over 200 nm. The triplet energy ($E_T$) of the sensitizer should be higher than that of the acceptor to afford efficient energy transfer (Section 2.2.2). Nevertheless, triplet sensitizers such as toluene ($E_T = 346 \, kJ \, mol^{-1}$) and $p$-xylene ($E_T = 337 \, kJ \, mol^{-1}$) are known to induce endothermic isomerization of aliphatic alkenes, such as ($E$)-3,4-dimethylpent-2-ene, the $E_T$ of which is higher than $\sim345 \, kJ \, mol^{-1}$ (Table 6.3).[581] When $E_T$ of the sensitizer is too low, the isomerization is inefficient or absent. The dependence of PSS on $E_T$ is interpreted in terms of a selective, energetically more efficient energy transfer to one of the ground-state isomers – the thermodynamically less stable Z-isomer.

**Table 6.3**  *Sensitized E–Z isomerization of (E)-3,4-dimethylpent-2-ene*

| Sensitizer | $E_T/ kJ \, mol^{-1}$ | [E]:[Z][a] |
|---|---|---|
| Benzene | 353 | 61:39 |
| p-Xylene | 337 | 66:34 |
| Pentamethylbenzene | 333 | —[b] |

[a]PSS concentration ratio.[581]
[b]PSS has not been reached after extended irradiation.

Irradiation of ($Z$)-cyclooctene in the presence of a triplet sensitizer, such as xylene ($E_T \approx 337 \, kJ \, mol^{-1}$), affords the $E$-isomer, although in a very low chemical yield.[582] In contrast, the singlet state sensitized isomerization by benzene polycarboxylate derivatives ($E_S \approx 420 \, kJ \, mol^{-1}$) leads to ($E$)-cyclooctene in up to 26% chemical yield (Scheme 6.11).[583]

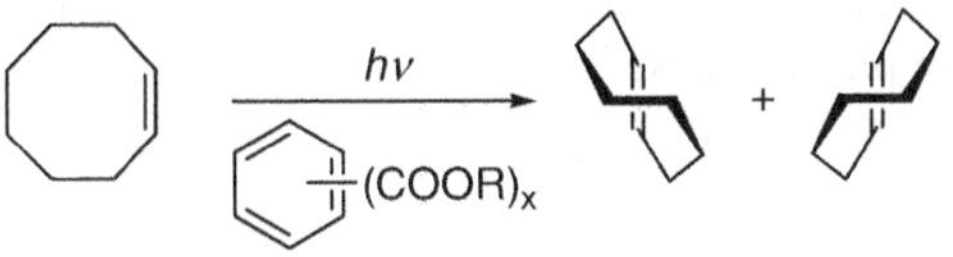

*Scheme 6.11*

## Special Topic 6.3: Asymmetric photochemical synthesis

*Asymmetric* synthesis introduces one or more new features of chirality in a molecule. Several approaches are possible. In general, preferential formation of an enantiomer or diastereoisomer is achieved as a result of the influence of a chiral element present in the substrate, a reagent, catalyst or the environment. Chirality control is also possible in the electronically excited state,[584] as demonstrated in the following examples.

Excitation with circularly polarized light (CPL) is an interesting direct method of asymmetric synthesis, which is based on preferential excitation of one of the enantiomers thanks to a difference in the molar absorption coefficients towards left or right CPL at a given wavelength.[585] Unfortunately, only a small enantiomeric excess (ee) can be generated by this method. Some scientists, nevertheless, believe that asymmetric induction by CPL may have sparked the abiotic synthesis of optically active molecules in nature. For example, chiral induction can enhance the concentration of one enantiomer by photochemical deracemization of sterically crowded ethenes (Scheme 6.12; ee $= \pm0.07\%$),[586] or nonracemic 6$\pi$-electrocyclization (see also Scheme 6.23) of **48** to **49**, followed by oxidation to hexahelicene (Scheme 6.13; ee was below 0.5%).[587]

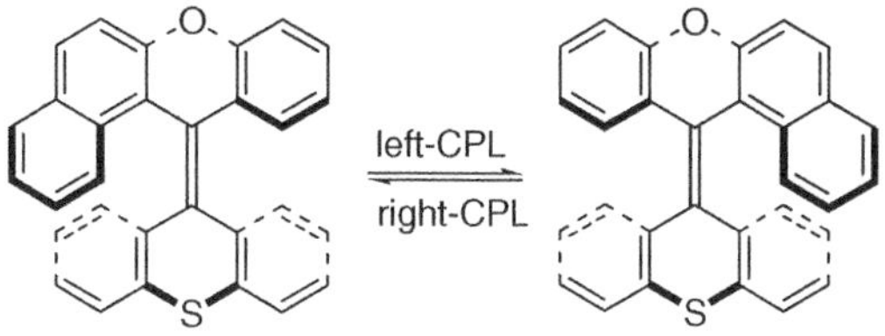

*Scheme 6.12*

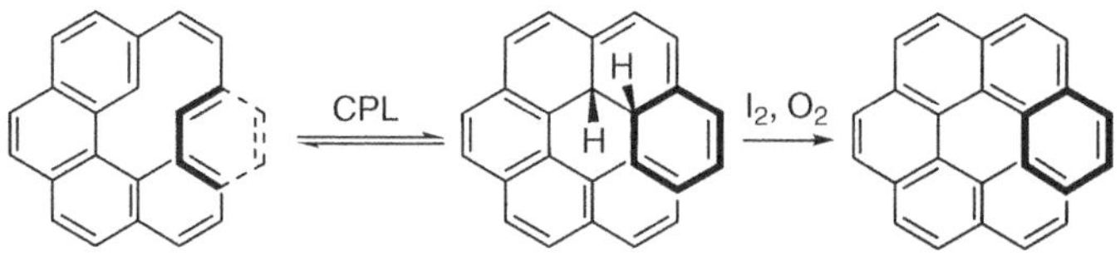

*Scheme 6.13*

*Enantiodifferentiating photosensitization* involves a prochiral or racemic substrate interacting with an optically active sensitizer via quenching (Section 2.2.2) or exciplex formation (Section 2.2.3),[585] rather than in the subsequent thermal process. Asymmetric photosensitization requires only a catalytic amount of the chiral sensitizer. An example is provided in Case Study 6.2 (*E–Z* photoisomerization).

Achiral molecules can crystallize in a chiral space group or form inclusion crystals of achiral guest and chiral host molecules in the absence of any external source of chirality.[588–590] Irradiation of such solid-state samples may then generate optically pure compounds in high chemical yields. This type of asymmetric induction is introduced later in Special Topic 6.5 and Case Study 6.21.

Many other methods of asymmetric induction in photochemistry, such as *template-induced enantioselective photoreactions*,[591] *asymmetric photochemistry in zeolites*,[592] or simple photoreactions of a chiral starting material, parallel their ground-state chemistry counterparts. Diastereoselective photocycloaddition in the Paternò–Büchi reaction of chiral reactants (Case Study 6.17) may serve as an additional example.

## Case Study 6.2: Asymmetric synthesis – enantiodifferentiating photoisomerization

Both singlet and triplet photosensitization of (Z)-cyclooctene (**50**) produces the (E)-cyclooctene enantiomers [(−)-(R)-**51** and (+)-(S)-**51**; Scheme 6.14]. High enantio-differentiation (Special Topic 6.1) has been achieved with singlet excited chiral polyalkyl benzenepolycarboxylates, whereas triplet sensitization (e.g. with chiral alkyl benzyl ethers) gives low optical yields only.[583,593] For example, when a chiral tetra[(−)-bornyl]oxycarbonyl singlet sensitizer **52** was used at −88 °C, (−)-(R)-**51** was obtained in 9% chemical yield and 41% optical purity at 12% conversion. The effect is attributed to specific interactions within an exciplex formed between sensitizer and alkene.

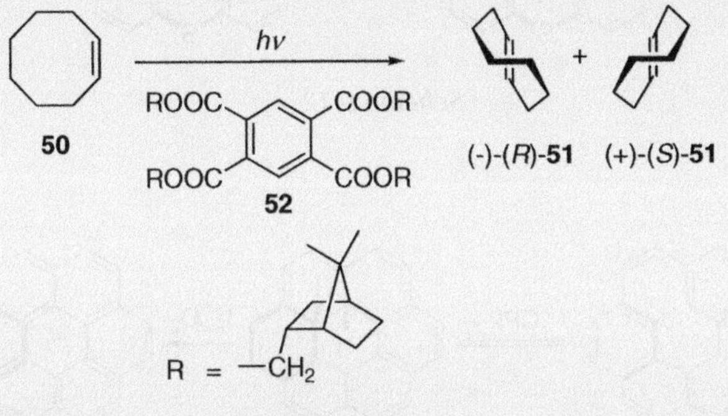

*Scheme 6.14*

*Experimental details.*[583] A pentane solution (200–300 ml) of **50** (200 mM) and the sensitizer **52** (5 mM) was irradiated in a quartz vessel with a low-pressure mercury lamp (30 W) through a Vycor sleeve (>250 nm) (Figure 3.9) under an argon atmosphere at −88 °C for 72 h. After irradiation, the product was extracted from the solution with aqueous silver nitrate at 0 °C and the combined aqueous extracts were washed with pentane and then added dropwise into concentrated ammonia solution to liberate **51**, which was subsequently extracted with pentane.

An alternative mechanism for photosensitized isomerization, termed the *Schenck mechanism*,[594] involves the formation of an adduct between an alkene and a triplet sensitizer [a Paternò–Büchi-like (Section 6.3.2) biradical], in which rotation about the central bond is fast relative to bond breaking. The corresponding alkenes formed after elimination of the sensitizer molecule are in a concentration ratio close to that obtained by the thermal process. This mechanism was thought to be involved when the triplet energy transfer is endergonic (i.e. not efficient), as is the case with the sensitization of an alkene $(E_T > 325 \text{ kJ mol}^{-1})$ by an excited acetophenone $(E_T = 310 \text{ kJ mol}^{-1})$ (Scheme 6.15).[595] In contrast, a *conventional* photosensitized E–Z isomerization mechanism dominates in the presence of acetone $(E_T = 332 \text{ kJ mol}^{-1})$, which transfers energy exergonically.

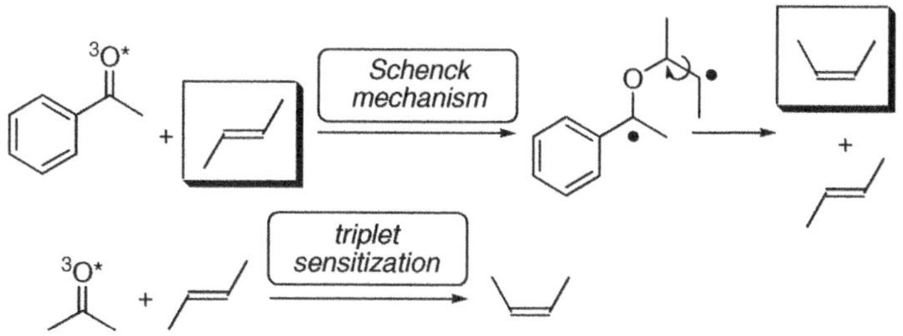

*Scheme 6.15*

## 6.1.2 Alkenes: Electrocyclic and Sigmatropic Photorearrangement

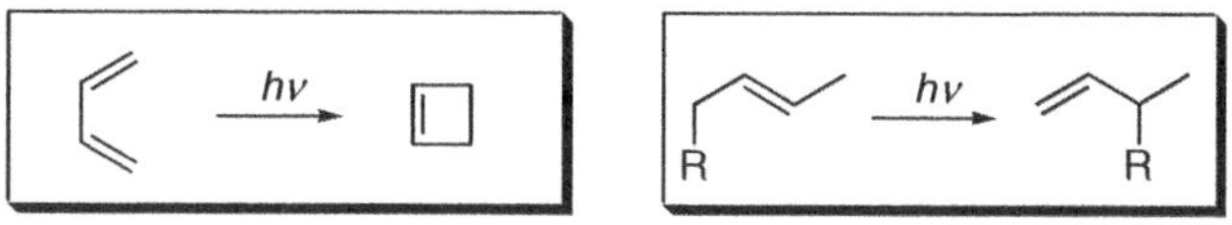

*Recommended review articles.*[487,523,528,530,532,564,596–602]
*Selected theoretical and computational photochemistry references.*[16,480,534,535,538,603–610]

*Electrocyclic Photorearrangements*

*Electrocyclic* ring-opening and ring-closure reactions are unimolecular pericyclic processes which involve a change of the π- and σ-bond positions within a conjugated system in a cyclic transition state. According to the Woodward–Hoffmann orbital symmetry rules,[336] the $4n$ electron systems in the *excited state* react through a disrotatory mode, whereas those with $4n + 2$ electrons react through a conrotatory pathway (Section 4.9; Scheme 6.16). Such a stereospecific process is known to be the opposite of that observed for reactions initiated thermally.

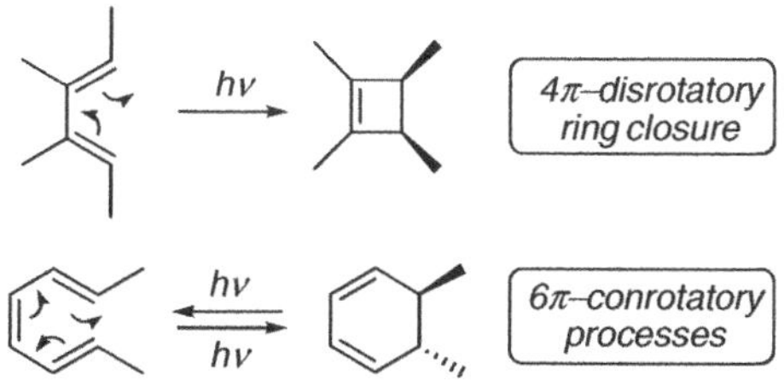

$4\pi$–disrotatory
ring closure

$6\pi$–conrotatory
processes

*Scheme 6.16*

Only the singlet state interconversions of conjugated dienes and cyclobutenes are taken into consideration here, because the triplet state is generally not involved in this process.[596] There are two important factors that affect electrocyclization: (1) reaction efficiency is usually lowered by a concurrent $E$–$Z$ isomerization (Section 6.1.1) along the double bond and (2) $s$-$cis$-diene conformation must be achieved before cyclization proceeds. Direct irradiation of dienes may result in wavelength-dependent photoproduct formation, which is related to different absorption properties of the diene conformers. For example, ($E$)-penta-1,3-diene (**53**), irradiated at 254 nm, gives 3-methylcyclobut-1-ene (**54**) in a low quantum yield ($\Phi = 0.03$) and two other products; however, when **53** is irradiated at 229 nm, where the $s$-$trans$-conformation absorbs predominantly, mostly $E$–$Z$ isomerization is observed (Scheme 6.17).[611]

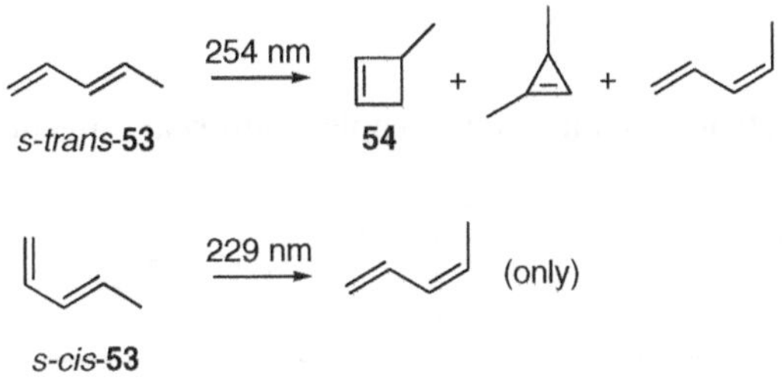

Scheme 6.17

In another example, $cis$-bicyclo[4.3.0]nona-2,4-diene (**55**) undergoes either the intra-molecular $4\pi$-ring closure or $6\pi$-ring opening depending upon the wavelength of excitation (Scheme 6.18).[612] Irradiation at 254 nm results in a 7:3 bicyclononadiene **55**–cyclonona-triene **56** photostationary mixture, reflecting the molar absorption coefficient ratio ($\varepsilon_{254}^{55} = 3800\,\mathrm{M}^{-1}\,\mathrm{cm}^{-1}$; $\varepsilon_{254}^{56} = 1000\,\mathrm{M}^{-1}\,\mathrm{cm}^{-1}$) to some extent, while the production of a small amount of the tricyclononene **57** is observed at 300 nm. Since the triene **56** absorbs considerably at this wavelength ($\varepsilon_{300}^{56} = 2000\,\mathrm{M}^{-1}\,\mathrm{cm}^{-1}$; $\varepsilon_{300}^{55} = 50\,\mathrm{M}^{-1}\,\mathrm{cm}^{-1}$), the photostationary ratio shifts in favour of **55**, thus promoting non-efficient production of **57**.

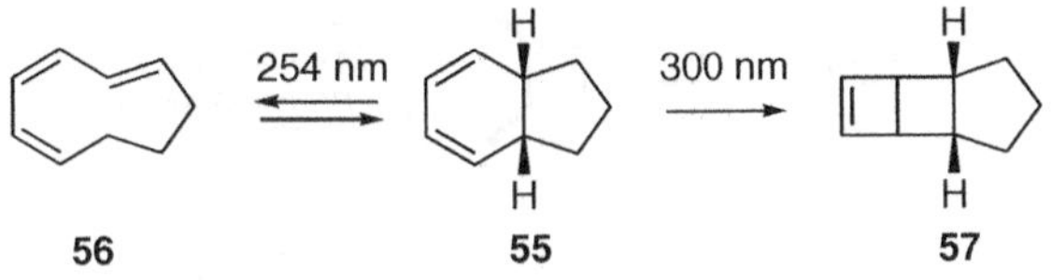

Scheme 6.18

Apart from photocyclization reactions, simple trienes undergo many types of other phototransformations, including $E$–$Z$ photoisomerization (Section 6.1.1), sigmatropic shifts (see later) or photoadditions (Section 6.1.4). For example, conformational control and possibly equilibration between the excited conformers result in a complicated mixture of photoproducts obtained upon irradiation of 2,5-dimethylhex-atriene (**58**) at 313 nm, although 6π photocyclization to **59** is still the major reaction pathway (Scheme 6.19).[613]

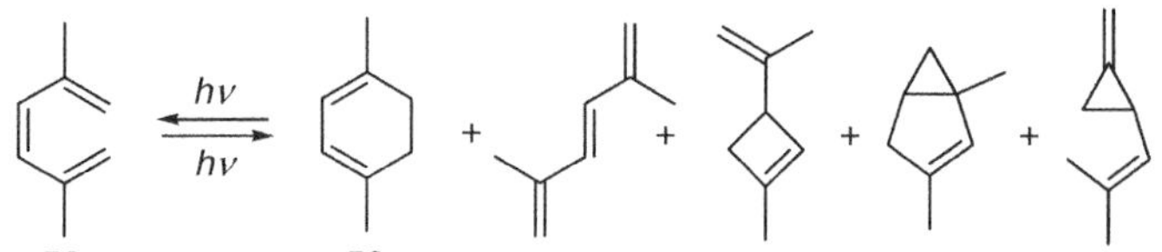

**58** ⇌ (hv) **59**

**Scheme 6.19**

Scheme 6.20 shows an example of photochemical electrocyclization of the enantiomerically pure acrylamide **60** in the presence of NaBH₄, which reduces the immonium functionality in the primary product **61** to give the lactam **62** in 75% chemical yield.[614]

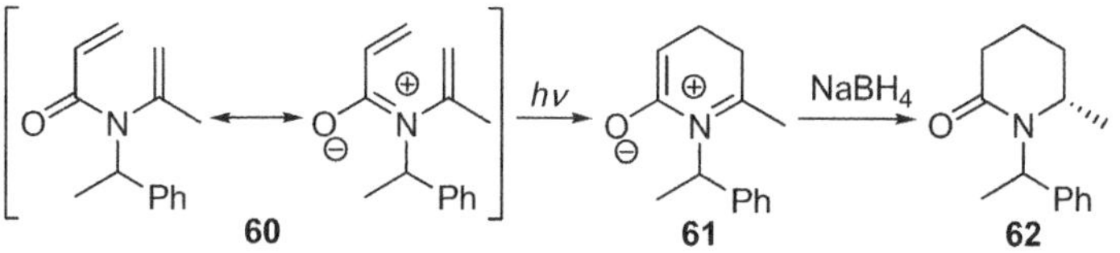

**60**          **61**          **62**

**Scheme 6.20**

## Special Topic 6.4: Photoproduction of vitamin D

*Vitamin D* is associated with biological functions, such as bone formation, immune system responses, cell defences and anti-tumour activity.[615,616] Vitamin D comes in two closely related forms, vitamin $D_2$ (ergocalciferol) and vitamin $D_3$ (cholecalciferol), and their metabolites. Both vitamin $D_2$ and $D_3$ occur naturally in some foods. However, vitamin $D_3$ (**63**) can also be synthesized in skin cells called keratinocytes from 7-dehydrocholesterol (provitamin D; **64**), which undergoes a photochemical six-electron conrotatory electrocyclic ring opening at ~280 nm to previtamin $D_3$ (**41**; see also Scheme 6.8), which spontaneously isomerizes to **63** in a thermal antarafacial hydride [1,7]-sigmatropic shift (Scheme 6.21). Both vitamin $D_2$ and $D_3$ are subsequently converted to active hormone 1,25-D by enzymes in several steps. The recommended daily intake of vitamin D for humans is 5–10 µg per day. For example, 15 ml of fish liver oils and 100 g of cooked salmon contain approximately 35 and 10 µg of vitamin D, respectively, and full-body summer exposure to sunshine at midday for 15–20 min produces 250 µg of vitamin D.

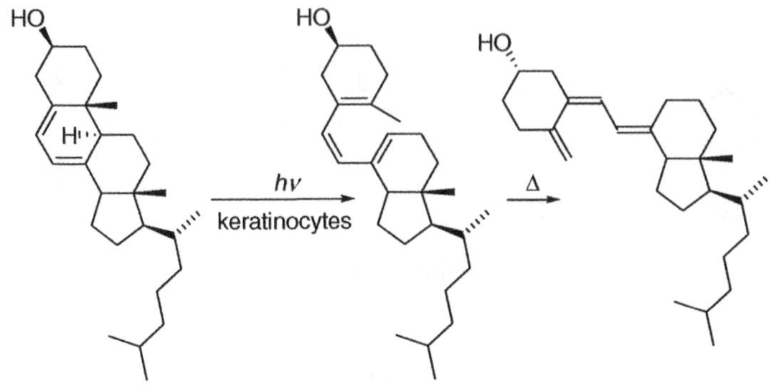

**Scheme 6.21**

## Case Study 6.3: Mechanistic photochemistry – previtamin D photochemistry

The photochemistry of vitamin D (see also Special Topic 6.4 above and Scheme 6.8) has also played a central role in the development of modern organic photochemistry.[564,598,617,618] The concept of *non-equilibration of excited rotamers* (NEER; Section 6.1.1) has been used to explain the excitation-wavelength dependence of *E–Z* isomerization (Section 6.1.1) of previtamin $D_3$ (**41**).[619] Whereas the quantum yield for *E–Z* isomerization decreases with increasing wavelength, the formation efficiencies of the $6\pi$-electron conrotatory ring-closure products, diastereomeric 7-dehydrocholesterol (provitamin D) (**64**) and lumisterol (**65**) (Scheme 6.22), increase dramatically. This was found to occur on the basis of a participation of both the $S_1$ and $S_2$ excited states in the photoreaction.[620] For example, the quantum yields of **64** and **65** formation were: $\Phi_{64} = 0.01$ and $\Phi_{65} = 0.05$ at 285 nm, but $\Phi_{64} = 0.08$ and $\Phi_{65} = 0.21$ at 325 nm. Vitamin $D_3$ is naturally synthesized in *human skin* by a spontaneous thermal antarafacial hydrogen [1,7]-sigmatropic shift in previtamin $D_3$, which is produced from **64** upon solar UVB irradiation in the first step.[621]

*Experimental details.*[620] Sample solutions of previtamin $D_3$ ($2 \times 10^{-3}$ M) in a 1 cm pathlength UV cell were purged with nitrogen and irradiated using a pulsed UV dye laser at $\sim$300 nm (Special Topic 3.1), containing Rhodamine 6G as a sensitizer in methanol or a methanol–water mixture (1:1, v/v). After irradiation, an aliquot of the sample solution was analysed by HPLC.

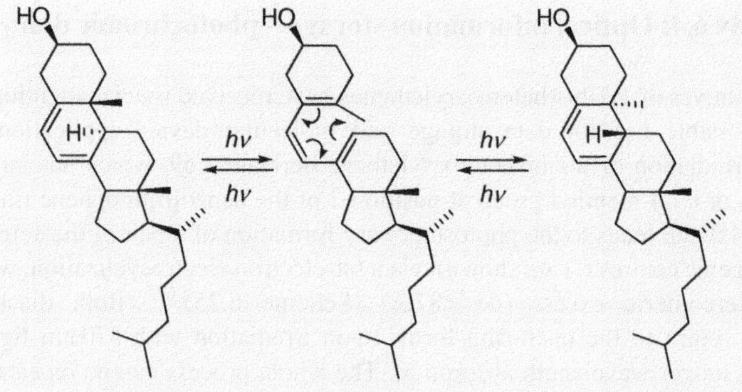

**Scheme 6.22**

(Z)-Diarylethenes, such as stilbene, also undergo a 6π-electrocyclization reaction upon irradiation from the singlet π,π* state, but not upon triplet sensitization.[487,529,599,622] The photochemical orbital symmetry-allowed conrotatory closure leads to the *trans*-dihydro derivative **66** (Scheme 6.23), which can subsequently be oxidized by oxygen or iodine to the corresponding arene (phenanthrene) in nearly quantitative yield.[561]

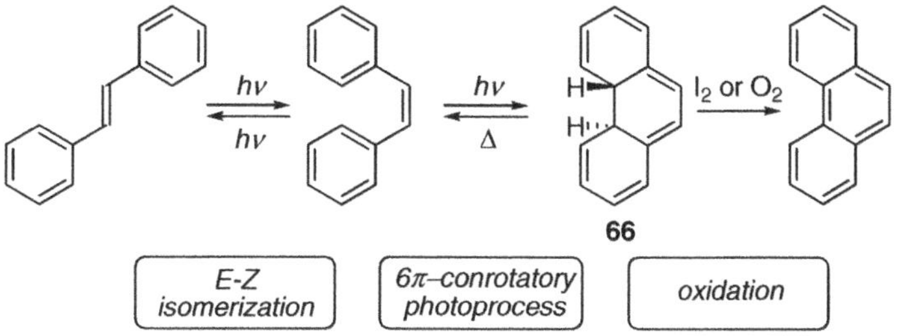

**Scheme 6.23**

This reaction has been exploited as a convenient way of producing *helicene* derivatives (see also Special Topic 6.3).[487,599,622] [7]Helicene (**67**) is, for example, produced from 1,4-bis[(E)-2-(naphthalen-2-yl)ethenyl]benzene (**68**) in 14% overall chemical yield (Scheme 6.24).[623]

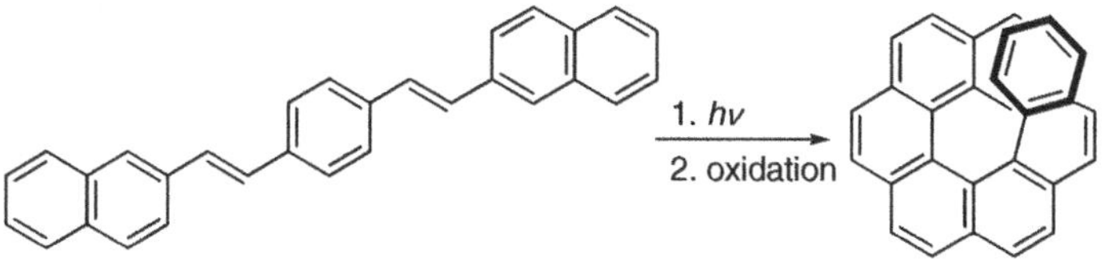

**Scheme 6.24**

---

### Case Study 6.4: Optical information storage – photochromic diarylethenes

Many derivatives of 1,2-bis(hetero)arylethenes have received special attention because of their possible use for data storage and molecular device applications.[624] For example, irradiation of the bisheteroarylethene derivative **69**, which has an optically active (−)- or (+)-menthyl group at position 2 of the benzo[*b*]thiophene ring, in cold toluene at 450 nm leads to the photostationary formation of a pair of diastereomers **70** (the stereogenic centres (*) are shown), via a 6π-electron electrocyclization, with a very high diastereomeric excess (de <87%) (Scheme 6.25).[625] Both diastereomers selectively return to the open-ring forms upon irradiation with 570 nm light due to their much longer wavelength absorption. The whole process can be repeated several times; thus **69** displays *photochromic* (Special Topic 6.15) properties.

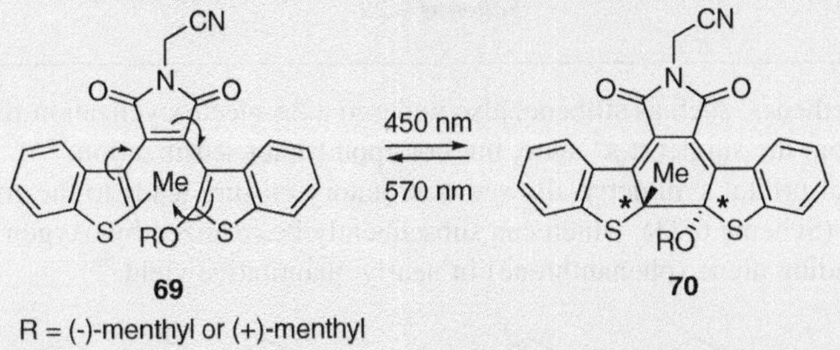

450 nm
⇌
570 nm

**69**

R = (-)-menthyl or (+)-menthyl

**70**

*Scheme 6.25*

*Experimental details.*[625] A toluene solution of **69** (not degassed) was irradiated with a mercury lamp (500 W) or a xenon lamp (Figure 3.9). The desired wavelengths were obtained by passing the light through cut-off and interference filters. The optical purity of the products was determined using HPLC with a chiral column.

---

### Sigmatropic Photorearrangements

Pericyclic intramolecular reactions, involving both the formation of a new σ-bond between atoms previously not directly linked and the breaking of an existing σ-bond via a cyclic transition state, are called *sigmatropic rearrangements*. Concerted[607] photorearrangements of the 4*n* electron systems, according to the Woodward–Hoffmann orbital symmetry rules,[336] are preferred because they involve *supra-supra* (suprafacial) bonding; only longer 4*n* + 2 electron systems may allow the *supra-antara* mechanism. This is demonstrated by the photochemistry of 2-(2-methyl-3-phenylcyclohexylidene)malononitrile (**71**), which gives **72** as the exclusive product by a 4π-electron [1,3]-allylic shift in 25% isolated chemical yield (Scheme 6.26).[626]

An interesting example of a 6π-electron [1,5]-hydrogen shift in the diisopropylidenecyclobutane **73** to give the cyclobutene **74** is believed to occur in an *antarafacial* fashion as predicted by the orbital symmetry rules (Scheme 6.27).[627]

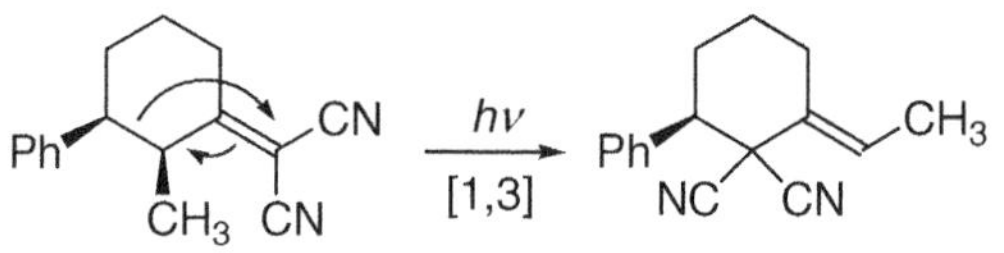


**71** **72**

*Scheme 6.26*

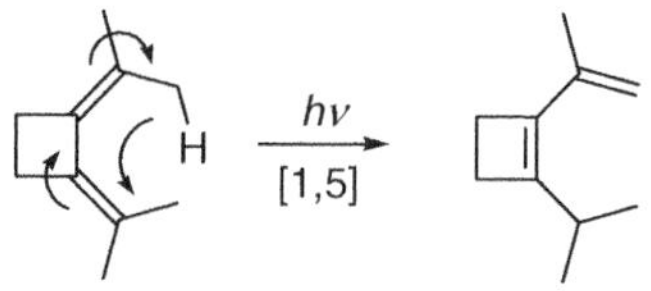


**73** **74**

*Scheme 6.27*

**Case Study 6.5: Organic synthesis – a stepurane derivative**

The tricyclic compound **75**, a derivative of the fungal metabolite stepurane, was obtained by irradiation of the bicyclo[2.2.2]octenone **76**, obtained in good yield from **77** in a five-step synthesis (Scheme 6.28).[628] The reaction proceeds by a 4π-electron suprafacial sigmatropic 1,3-acyl shift.

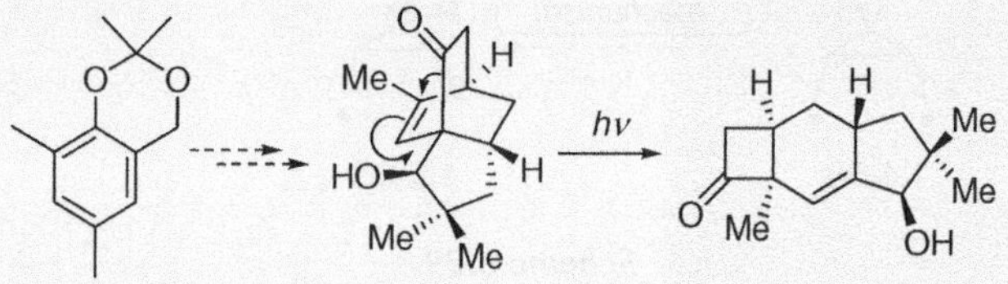


**77** **76** **75**

*Scheme 6.28*

*Experimental details.*[628] A solution of **76** in benzene was irradiated with a medium-pressure mercury lamp (125 W) (Figure 3.9) for 30 min to 65% conversion. The product was isolated in 55% yield.

### 6.1.3 Alkenes: di-π-Methane Rearrangement

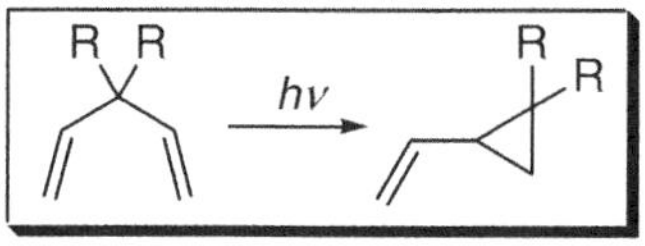


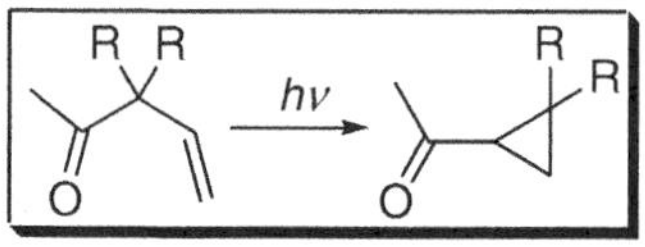

*Recommended review articles.*[532,629–643]
*Selected theoretical and computational photochemistry references.*[534,535,644–647]

The photochemical production of vinylcyclopropane derivatives from compounds having two π-moieties bonded to an sp³-hybridized carbon[648] is termed the *di-π-methane rearrangement*, also known as the *Zimmerman reaction.*[649] A very broad spectrum of di-π-systems can lead to photoproducts that are usually not obtainable by alternative routes.[632,633] The reaction may be classified *formally* as a [1,2]-shift but, according to the proposed stepwise *biradical* mechanism,[650,651] 1,3- and 1,4-biradical (BR) intermediates and also the second π-bond may be involved[652] (Scheme 6.29). A *concerted* (pericyclic) pathway for the di-π-methane reaction from the excited singlet state is, however, not excluded.[629,630,653] Typically, the singlet state reaction occurs upon direct irradiation, whereas the triplet pathway is accessible only using triplet sensitizers due to the poor intersystem crossing efficiencies of alkenes. The di-π-methane rearrangements often show a high degree of diastereoselectivity and/or regioselectivity.

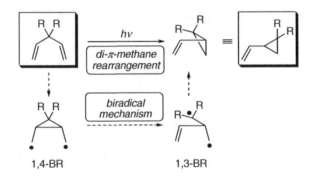

1,4-BR                                    1,3-BR

*Scheme 6.29*

Flexible acyclic di-π-methane systems tend to undergo rearrangement from the lowest singlet state, whereas the triplet state undergoes radiationless decay, because the twisted triplet is efficiently converted to the ground state, often accompanied by *E–Z* isomerization (Scheme 6.30).[654]

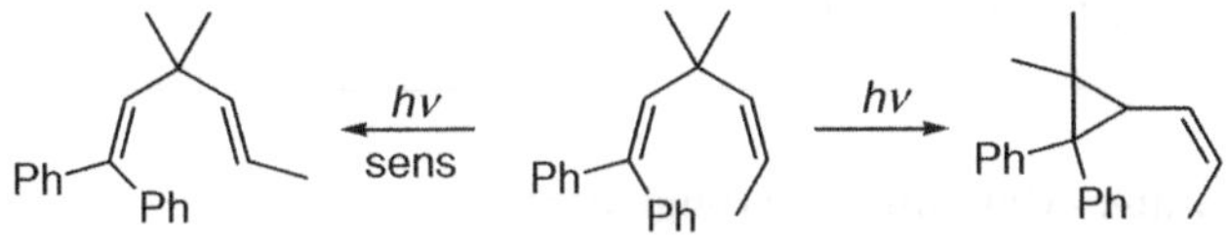

*Scheme 6.30*

In contrast, *cyclic* (less flexible) di-π-methane systems also rearrange from the triplet state. For example, triplet sensitization of bicyclo[2.2.2]octa-2,5,7-triene (barrelene, **78**) provides semibullvalene (**79**) in ~40% chemical yield, whereas direct irradiation produces a small amount of cyclooctatetraene (**80**) via the intramolecular [2+2] cycloaddition (Section 6.1.5) (Scheme 6.31).[648]

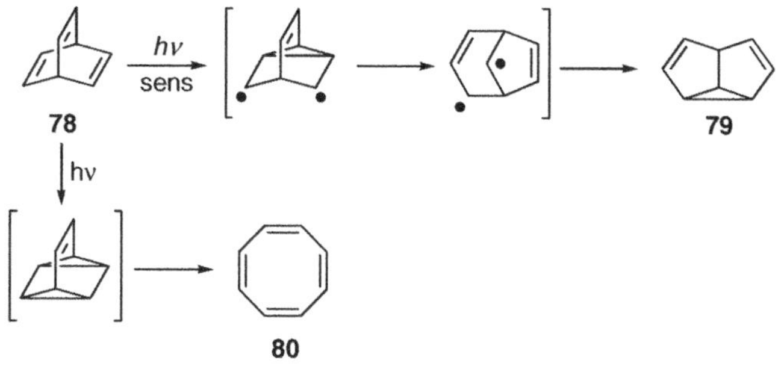

**Scheme 6.31**

The biradical mechanism is commonly proposed to explain the reaction regioselectivity.[629,630,632,633] Two alternative ring-opening processes for the 1,4-biradical **81**, obtained from the photolysis of **82**, may lead to two 1,3-biradicals, **83** and **84** (Scheme 6.32). The latter, which is thermodynamically more stable because the unpaired electron is delocalized on the benzhydryl moiety, should be a precursor to the diphenylvinylpropene **85**. Indeed, the irradiation of **82** afforded **85** as a sole product.[655]

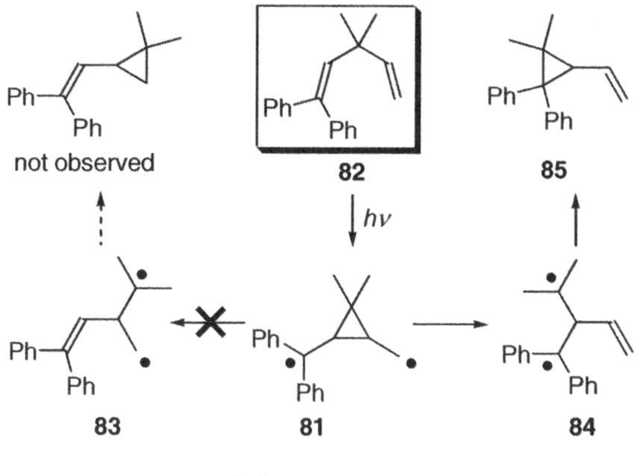

**Scheme 6.32**

**Case Study 6.6: Photobiology – natural photoproduction of erythrolide A**

Erythrolide A (**86**) has been thought to be produced naturally from the diterpenoid **87** by a light-induced di-π-methane rearrangement (Scheme 6.33), because both compounds were isolated from the Caribbean octocoral *Erythropodium caribaeorum*.[656] In order to confirm this hypothesis, the potential precursor **87** was irradiated under various conditions and **86** was obtained in relatively high chemical yields.

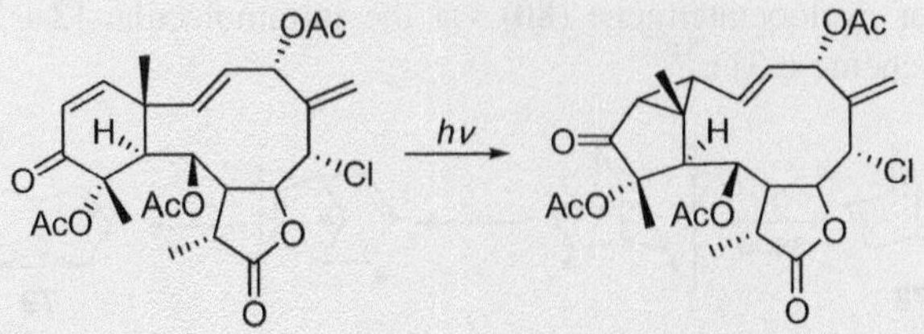

**Scheme 6.33**

*Experimental details.*[656] A benzene solution of **87** in a quartz vessel was irradiated with a medium-pressure mercury lamp (benzene filters to cut off the wavelengths below 280 nm) (Figure 3.9) to give an 87% yield of **86** in 3 h. In comparison, *sunlight* irradiation of **87** in 5% methanolic seawater in a glass vessel produced **86** in 37% yield in 8 days.

The following reactions formally belong to Sections 6.3 and 6.4, because absorption by the C=O and C=N chromophores is largely responsible for the photochemistry; however, discussion of their mechanistic pathways is better suited here. An analogous photochemical rearrangement of $\beta,\gamma$-enones, involving 1,2-acyl migration to give cyclopropane derivatives,[657] is termed the *oxa-di-$\pi$-methane* rearrangement.[631,632,635,643,658,659] The reaction generally occurs from the lowest excited triplet state (T$_1$, $\pi,\pi^*$), possibly via two biradical intermediates (Scheme 6.34). A competing sigmatropic (Section 6.1.2) 1,3-acyl shift occurs from the excited singlet (S$_1$, n,$\pi^*$) state; therefore, direct excitation to the singlet state should be avoided.

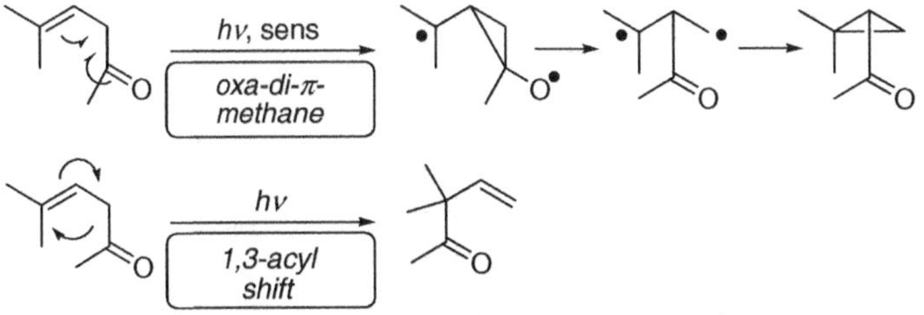

**Scheme 6.34**

## Case Study 6.7: Organic synthesis – substituted cyclopropanes

Oxa-di-$\pi$-methane rearrangement leads to cyclopropane derivatives, compounds that are otherwise difficult to synthesize. The diphenylenal **88**, for example, is converted to the cyclopropyl aldehyde **89** by triplet sensitization (Scheme 6.35).[660] The photoproduct can be further transformed to other compounds, for example a diphenylvinylcyclopropane derivative **90**.

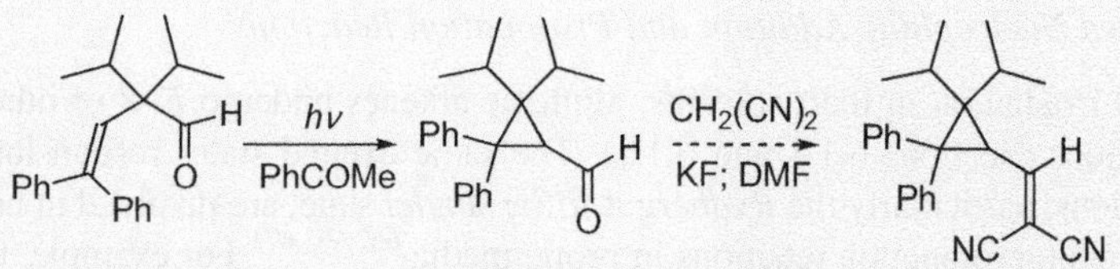

*Scheme 6.35*

*Experimental details.*[660] A benzene solution of **88** (0.653 mmol) and acetophenone (sensitizer; 65 mmol) was irradiated with a medium-pressure mercury lamp (450 W) through a Pyrex filter ($\lambda_{irr} > 280$ nm) (Figure 3.9) for 30 min. The reaction mixture was concentrated under reduced pressure and acetophenone was then removed by bulb-to-bulb distillation at 40 °C under reduced pressure to give **89** as a yellow oil in nearly quantitative chemical yield.

The di-π-methane rearrangement has also been observed in other 1,4-unsaturated systems, such as nitrogen-containing compounds.[630,642] The corresponding *aza-di-π-methane* triplet-sensitized (sens) rearrangement of the β,γ-unsaturated imine **91** produces, inefficiently ($\Phi \approx 0.01$), a single product, cyclopropylmethanimine (**92**), which can subsequently be easily converted to an aldehyde by acid hydrolysis (Scheme 6.36).[661]

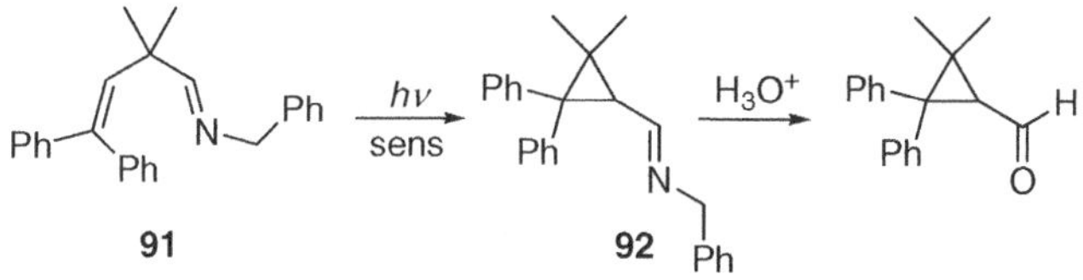

*Scheme 6.36*

## 6.1.4 Alkenes and Alkynes: Photoinduced Nucleophile, Proton and Electron Addition

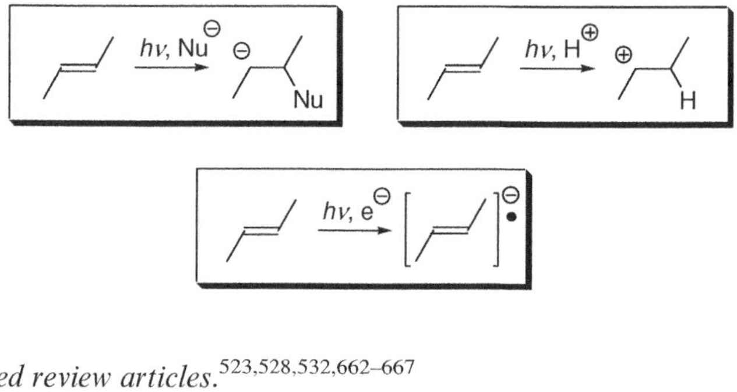

*Recommended review articles.*[523,528,532,662–667]
*Selected theoretical and computational photochemistry references.*[16,668–670]

## Photoinduced Nucleophilic Addition and Protonation Reactions

Upon direct irradiation in inert solvents, aliphatic alkenes undergo *E–Z* or other isomerization reactions (Sections 6.1.1 and 6.1.2). The same excited states responsible for such transformations, particularly the *Rydberg* $\pi,R(3s)$ *singlet* state, are involved in nucleophilic addition or photoprotonation reactions in protic media.[662,663,671] For example, the $\pi,R(3s)$ state of tetramethylethene (**93**), having cation-radical character of the central bond, is readily attacked by a nucleophile (such as methanol) to form a solvated electron and a radical intermediate **94** that disproportionates to a mixture of the ethers **95** (30%) and **96** (37%) (Scheme 6.37).[671] The solvated electrons produced are trapped by the solvent molecules.

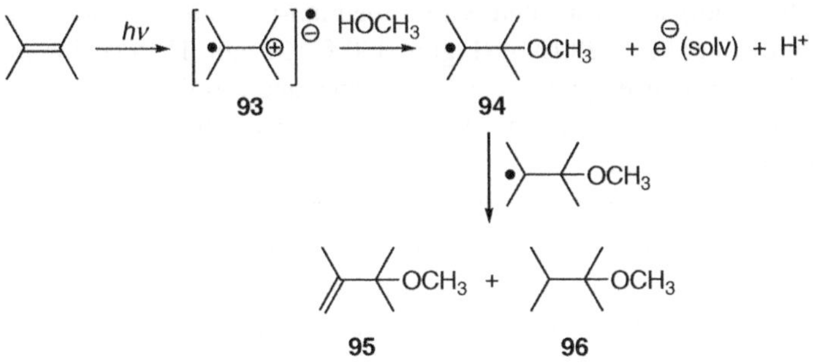

Scheme 6.37

Similarly, direct irradiation of cycloalkenes results in nucleophilic trapping of the $\pi,R(3s)$ excited state.[662] For example, photolysis of dimethylcyclobutene (**97**) in methanol at 228 nm affords three methoxy-substituted products via disproportionation reactions (Scheme 6.38).[672]

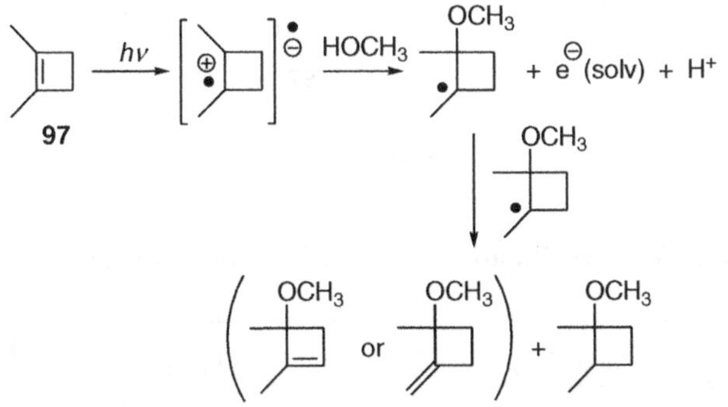

Scheme 6.38

In contrast, arylalkenes or -alkynes readily undergo acid-catalysed *Markovnikov addition* on direct irradiation in water to form the corresponding alcohols and ketones, respectively (Scheme 6.39).[673,674] The initial and at the same time rate-limiting step is the protonation of a more electron-rich $\pi,\pi^*$ excited singlet state (in contrast to aliphatic alkenes), occurring over 10 orders of magnitude more rapidly than protonation of the corresponding ground-state molecule. This step is followed by hydration to form the corresponding alcohols in the case of alkenes or enols[675] in the case of alkynes.

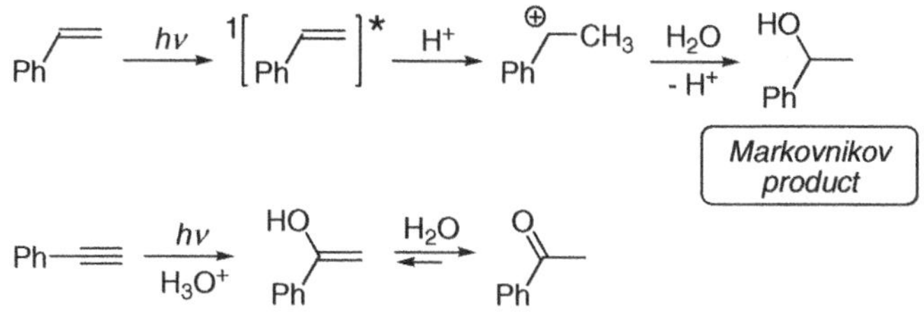

*Scheme 6.39*

Interestingly, photohydration of the nitro-substituted styrenes **98** involves initial formation of benzyl carbanion intermediates via nucleophilic attack of water at the β-carbon of the triplet state (nitrobenzenes generally have high singlet to triplet intersystem crossing quantum yields) and gives an anti-Markovnikov adduct (Scheme 6.40).[676] Here the nitro group enhances the electron-deficient character of the $T_1$ state, thus facilitating the nucleophilic attack on the carbon atom bearing a partial positive charge.

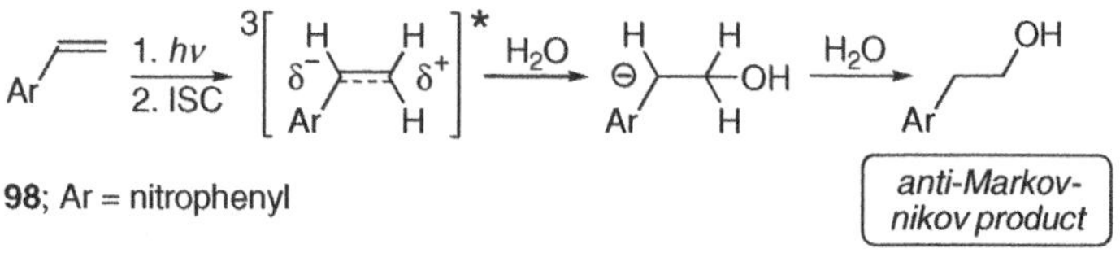

**98**; Ar = nitrophenyl

*anti-Markov-nikov product*

*Scheme 6.40*

Upon sensitized irradiation, the triplet excited acyclic alkenes and large-ring cycloalkenes undergo *E–Z* isomerization in both aprotic and protic media. Medium-ring cycloalkanes – cyclohexenes, cycloheptenes or cyclooctenes – can, however, be protonated via the corresponding thermodynamically unstable (strained) *E*-isomers formed initially by a *Z* → *E* photoisomerization step (Section 6.1.1).[662–664] For example, acid-catalysed irradiation of (*Z*)-1-methylcyclohexene (**99**) in the presence of *p*-xylene as a triplet sensitizer in methanol affords the Markovnikov adduct 1-methoxy-1-methylcyclohexane (**100**) and the elimination product methylenecyclohexane (**101**), both in approximately 40% chemical yield (Scheme 6.41).[671] The *E*-isomer intermediate exhibits extensive incorporation of deuterium in the presence of CH$_3$OD.

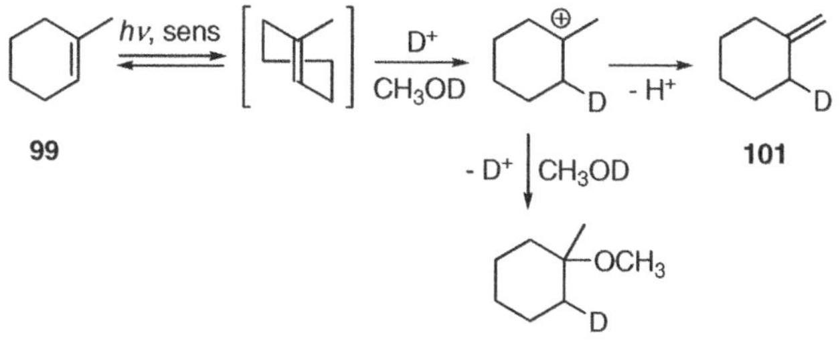

**99**

**101**

**100**

*Scheme 6.41*

## Case Study 6.8: Asymmetric synthesis – diastereoselective photosensitized polar addition

Singlet photosensitized polar addition of methanol to $(R)$-$(+)$-limonene (**102**) in nonpolar solvents afforded a mixture of the diastereomeric ethers **103** and **104** and the rearrangement product **105** (Scheme 6.42).[677] The diastereomeric excess (de) of the photoadduct was optimized by varying the solvent polarity, reaction temperature and nature of the sensitizer. The first step of the reaction is the $Z$–$E$ photoisomerization (Section 6.1.1) of **102** to a highly strained $E$-isomer, followed by protonation and methanol addition. The initial formation of a carbocation via the protonation step has been excluded under those reaction conditions. The Markovnikov-oriented methanol attack on the less-hindered $(R_p)$-$(E)$-**102** compared with that of $(S_p)$-$(E)$-**102** explains why **103** can be obtained in up to 96% de upon sensitization with methyl benzoate in a methanol solution. The hypothesis that $Z$–$E$ isomerization of the cyclohexene moiety affords a strained (reactive) alkene, whereas isomerization of the exocyclic double bond does not, was supported by the observation of an exclusive nucleophilic addition to the cyclohexene double bond.

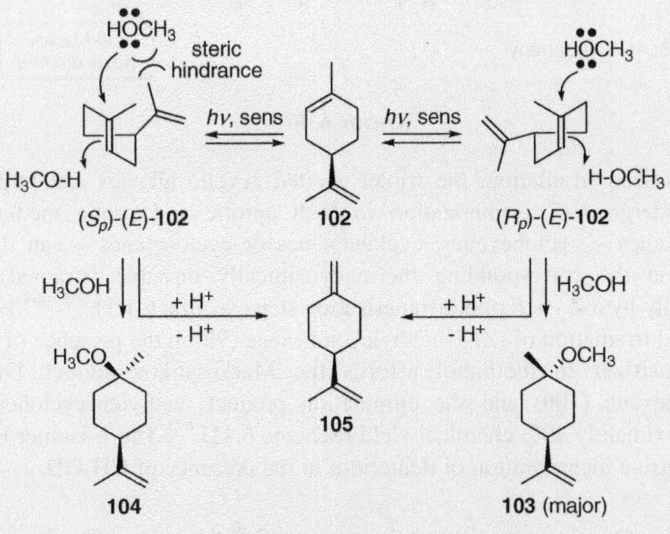

Scheme 6.42

*Experimental details.*[677] A diethyl ether solution containing limonene (**102**; 5 mM) and a sensitizer (2 mM) was irradiated under an argon atmosphere in a quartz tube immersed in a cooling (methanol/ethanol) bath at $-75\,°C$ using a high-pressure mercury lamp (300 W) through a Vycor sleeve (>250 nm) (Figure 3.9). The reaction mixture was evaporated *in vacuo* to give a residue, from which the products were isolated by preparative gas chromatography in about 10% chemical yield.

## Electron Transfer to Excited Alkenes

Quenching of singlet state alkenes (as electron acceptors) by amines (as electron donors) may proceed via the formation of exciplex or radical ion pair [see the photoinduced electron transfer (PET) process in Section 5.2] intermediates, which may regenerate the ground-state reactants via back electron transfer or undergo various chemical reactions.[665–667] For example, singlet excited stilbene in the presence of trimethylamine forms an exciplex–radical ion pair intermediate in polar aprotic solvents, followed by proton transfer (oxidation of an alkylamine, i.e. formation of the radical cation on the nitrogen atom, significantly increases the acidity of the $\alpha$-CH bonds) and radical coupling reactions (Scheme 6.43).[678]

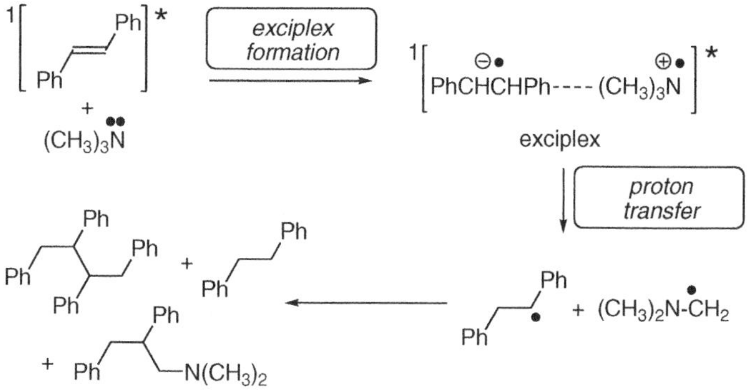

Scheme 6.43

## Case Study 6.9: Mechanistic photochemistry – photocyclization of *N,N*-dimethylaminoalkylstyrenes

The photochemical intramolecular cyclization of *N,N*-dimethylaminoalkylstyrenes in nonpolar solvents was found to be affected by the alkyl chain length.[679] Whereas **106** (having a dimethylene interchromophore spacer) undergoes efficient intramolecular addition to form a single five-membered ring adduct **107** via a 1,5-biradical, **108** (having a tetramethylene interchromophore spacer) produces the six-membered ring diastereomers **109** and **110** in an 8:1 molar ratio via a 1,6-biradical intermediate (Scheme 6.44). A mechanistic study indicated that highly regioselective intramolecular H-atom transfer occurs via least-motion pathways from the lowest-energy folded conformations of the singlet exciplex intermediates.

*Experimental details.*[679] Hexane solutions of a styrylamine (**106** or **108**; 0.01 M) in Pyrex (transparent over 280 nm) test-tubes were irradiated to >95% conversion (GC) under nitrogen using a Rayonet reactor fitted with 16 lamps (21 W; $\lambda_{irr} = 300$ nm; Figure 3.10). Products were isolated in >80% chemical yield by either preparative thick-layer chromatography or column chromatography followed by bulb-to-bulb distillation.

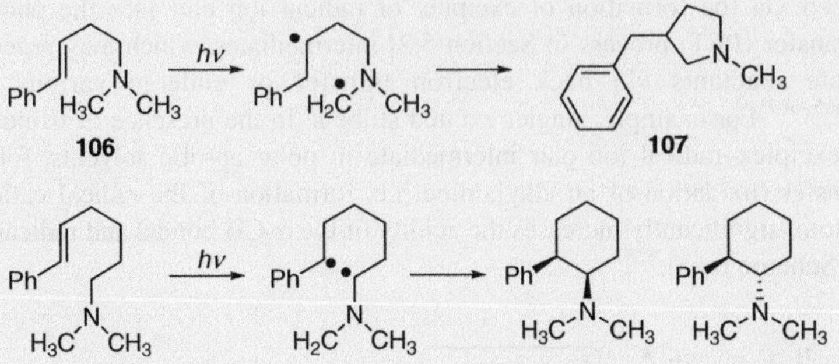

**Scheme 6.44**

### 6.1.5 Alkenes and Alkynes: Photocycloaddition Reaction

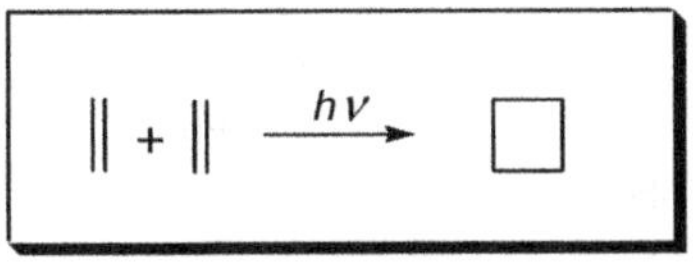

*Recommended review articles.*[532,585,588,602,680–694]
*Selected theoretical and computational photochemistry references.*[16,480,534,535,695–698]

Photoinduced [2 + 2] *cycloaddition* (Section 4.9) of alkenes (alkynes) to form cyclobutane (cyclobutene) derivatives is one of the best studied reactions in photochemistry.[680,682] According to the Woodward–Hoffmann orbital symmetry rules,[336] the cycloaddition of one singlet excited (S$_1$) and one ground-state alkene is allowed by a *suprafacial–suprafacial* concerted stereospecific pathway (Scheme 6.45).[695,699,700] Rare concerted [4 + 2] and [4 + 4] photocycloadditions of conjugated singlet excited dienes must occur in a suprafacial–antarafacial and suprafacial–suprafacial manner, respectively.[690] Since the suprafacial–antarafacial reactant approach is geometrically difficult to achieve, [4 + 2] reactions usually proceed stepwise (involving biradical intermediates). [2 + 2] or [4 + 4] photocycloadditions can occur in either a concerted or stepwise fashion.

Intersystem crossing from the excited singlet state of simple alkenes is inefficient. Triplet state cycloadditions, therefore, are usually achievable via triplet sensitization rather than direct irradiation.[701,702] Such a process often involves exciplexes, formed between electron-poor and electron-rich alkenes,[703] or 1,4-biradical intermediates (Scheme 6.45).[704] Thus the tendency to achieve loose geometries in such species then favours a nonconcerted (stepwise) pathway, in which rotation about the central C–C bond occurs, eventually leading to loss of reaction stereospecificity. In general, the cycloaddition reaction of excited alkenes can be accompanied by other processes, such as radiationless decay or *E–Z* isomerization (Section 6.1.1). In contrast, alkenes conjugated to a carbonyl group (enones) undergo the [2 + 2] cycloaddition with ground-state alkenes to form cyclobutanes via the triplet state available by direct (unsensitized) irradiation (see the following Section 6.1.6).

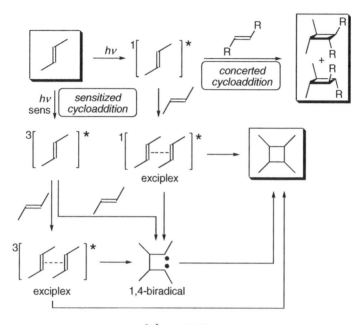

*Scheme 6.45*

Transition metals, especially copper(I), can catalyse intermolecular or intramolecular photocycloaddition reactions of alkenes.[684,688] It has been suggested that the reaction involves coordination of both the reacting alkene molecules in their ground states with a single copper(I) cation. The subsequent photoexcitation of this complex leads to cyclobutane with regeneration of the Cu(I) catalyst, probably via either metal to alkene or alkene to metal charge transfer (Scheme 6.46).[705] A concerted reaction mechanism has also been proposed.[706] Alkene–Cu(I) complexes have been detected and isolated: whereas many aliphatic alkenes and also Cu(I) salts are almost transparent above 240 nm, intense UV absorption bands in this region are observed for their mixtures.[707]

There are only a limited number of *intermolecular* [2 + 2] photocycloaddition reactions known to occur in solution because alkene excited singlet state lifetimes are very short (on the order of 10 ns).[708] Direct irradiation of neat but-2-ene, for example, yields tetramethylcyclobutane with stereochemistry suggesting the concerted mechanism (Scheme 6.45).[709] Inefficient dimerization ($\Phi = 0.04$), however, competes with efficient *E–Z* isomerization ($\Phi = 0.5$).

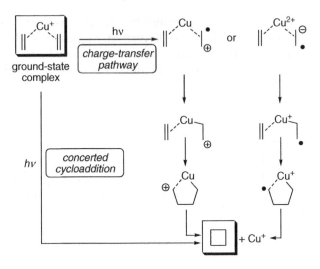

*Scheme 6.46*

In contrast, copper(I)-catalysed *intramolecular* cycloaddition of the triene **111** affords 1-vinyl-3-oxabicyclo[3.2.0]heptane (**112**) in >80% chemical yield.[710] Only double bonds coordinated to the catalyst participate in the photocycloaddition; the bridged arrangement of parallel constrained C=C bonds shown is preferred over all other possible modes of coordination (Scheme 6.47).

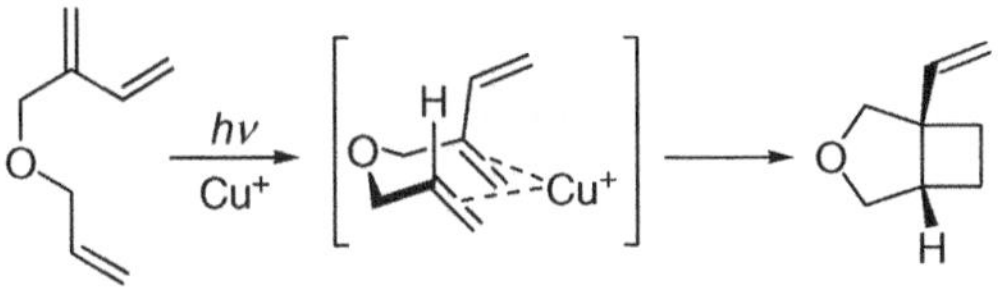

**111**                                    **112**

*Scheme 6.47*

### Case Study 6.10: Organic synthesis – copper(I)-catalysed photocycloaddition

The synthesis of cyclobut-A (**113**), a nucleoside analogue of the potent antiviral antibiotic oxetanocin, was accomplished using intramolecular [2 + 2] cycloaddition catalysed by Cu(I) ions.[711] An inseparable mixture of four (*E*, *Z*, *syn*, *anti*) isomers of **114** was obtained in 70% chemical yield by irradiation of a triene **115** (Scheme 6.48). The nucleoside connection was then established via nucleophilic substitution of an acetate group by adenine and other transformations in several steps.

*Experimental details.*[711] A diethyl ether solution of the triene **115** (10.8 mM) and copper(I) triflate (0.94 mM) was irradiated with a medium-pressure mercury lamp

(450 W) through a double-walled water-cooled quartz ($\lambda_{irr} > 250$ nm) immersion well (Figure 3.9) for 5 h. The reaction mixture was then washed with ice-cold ammonia solution, dried and concentrated under reduced pressure. The residual oil was purified by column chromatography to give a mixture of the isomers **114**.

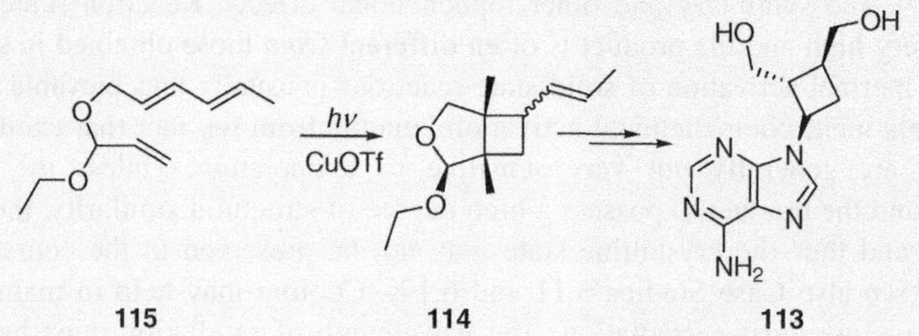

**115**          **114**          **113**

*Scheme 6.48*

Arylalkenes undergo cycloaddition upon photolysis to form various stereoisomers in concentration ratios that are influenced by solvent polarity and constraining environment.[172,685,686] For example, the photodimerization reaction of the stilbene derivative **116** is not efficient in organic solvents, whereas it successfully competes with geometric isomerization in aqueous solutions (Scheme 6.49).[712] In this case, hydrophobic association alters the course of the reaction by aggregating the reacting molecules (increasing their local concentrations) and, as a result, enhances the efficiency of bimolecular reactions. This reaction is presumed to occur only from the excited singlet state of (*E*)-stilbene.

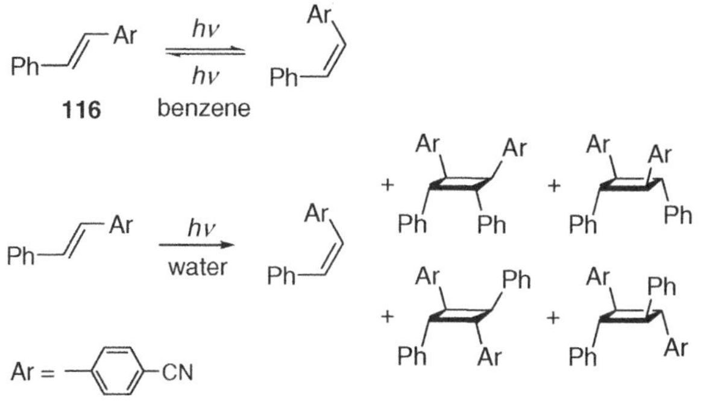

*Scheme 6.49*

## Special Topic 6.5: Photochemistry of organic crystals

Reaction media play an important role in *controlling the nature of reaction pathways* and therefore the reaction products and yields, and also regio- and

stereochemistry (see also Special Topic 6.11). In organic crystals, large conformational, translational or rotational changes are prevented by crystal lattice
constraints.[172,713] The intrinsic reactivity of a molecule is often of secondary or no
importance; the eventual chemical transformations are based on the proximity of
reactive groups to one another (a least-motion reaction pathway; see also Case
Study 6.9), site symmetry and other topochemical effects. Reaction selectivity is
usually very high and the product is often different from those obtained in solution.
Whereas thermal activation of solid-state reactions is usually unachievable because
the crystals melt, photochemical activation benefits from the fact that excited state
reactions are generally not very sensitive to temperature. Unless the starting
material and the product(s) possess a high degree of structural similarity, the crystal
structure and thus the crystalline state may not be preserved in the course of the
reaction (see also Case Studies 6.11 and 6.19). Cooling may help to maintain the
crystal structure during irradiation. The wavelength of irradiation must be chosen
carefully. If the products absorb incident radiation, an optical filter layer consisting
of product molecules is formed on the crystal surface, preventing further irradiation
of the layers beneath.

Following pioneering work on [2 + 2] cycloadditions of cinnamic acid in crystals,[714]
photocycloaddition in the solid state has become another systematically investigated
reaction parameter.[172,588,686] Crystallographic studies have shown that the distance
between the potentially reacting double bonds must be less than ~0.4 nm, although the
course of such photoreactions can also be influenced by other factors. Furthermore,
*optically active* compounds can be produced in the absence of any external chiral
source (such as chiral host molecules in inclusion crystals with achiral guests; see Case
Study 6.21), when achiral reactants crystallize in a chiral space group.[588]

For example, crystallization of the achiral ester **117** (Scheme 6.50) from melt or from
solution affords two chiral enantiomorphous crystal forms.[715] Irradiation of a solid
sample containing both forms then yields two dimers with the either (*S,S,S,S*)- or (*R,R,
R,R*)-**118** absolute configurations, which is an interesting example of *asymmetric
photochemical synthesis.*[588]

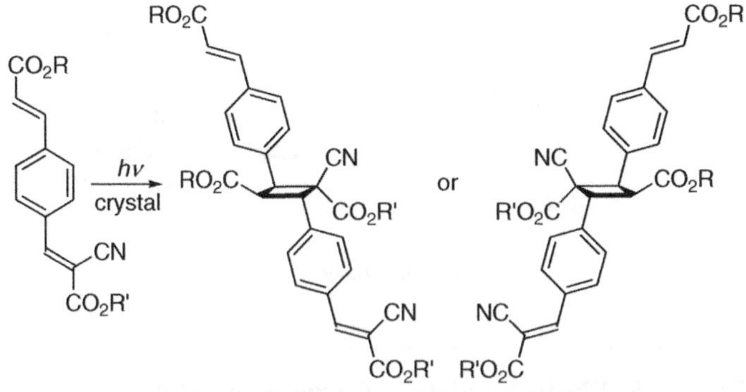

**117**              (*S,S,S,S*)-**118**              (*R,R,R,R*)-**118**

*Scheme 6.50*

**Case Study 6.11: Hologram production – single-crystal-to-single-crystal photoreaction**

A photochemical forward and thermally reversible single-crystal-to-single-crystal [2 + 2] photodimerization reaction (such a process is termed *photochromic*; Special Topic 6.15) was observed on the styrylpyrylium triflate **119** (Scheme 6.51).[716] The choice of the irradiation wavelength was the key parameter to a successful experiment. Intensive photolysis of a crystal using $\lambda_{max}$ close to the absorption maximum (~420 nm) led to its breakdown; microcrystalline dimeric aggregates quickly appeared on the surface and the reaction environment turned out to be heterogeneous. However, using wavelengths within the absorption tail ($\lambda_{irr} > 570$ nm) at 20 °C allowed the single crystalline character of the sample to be retained during the photodimerization. At temperatures above 100 °C, the dimerized compound quantitatively reverted back to the original monomer, again retaining its crystal structure (see Special Topic 6.5). It was demonstrated that thick volume holograms can be reversibly written into such single crystals. Single-crystal-to-single-crystal reactions are feasible only when the transformations do not distort the original reactant crystal lattice too much.

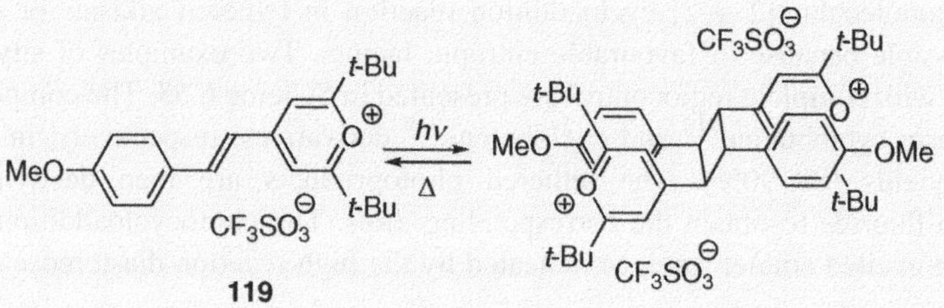

*Scheme 6.51*

*Experimental details.*[716] Grating of a hologram on a single crystal made of **119** was achieved using an He–Ne laser (632.8 nm; 50 mW) (Special Topic 3.1) for both writing and reading. Crystallographic data were obtained for monomer, dimer and mixed crystals that were isolated after different irradiation times.

Aromatic alkynes, such as l-naphthylbuta-1,3-diyne (**120**) in the presence of 2,3-dimethylbutene, undergo a [2 + 2] photocycloaddition reaction to form a cyclobutene derivative **121** in 77% chemical yield (Scheme 6.52). This particular addition occurs exclusively on the C1−C2 triple bond. Quenching, fluorescence and exciplex emission studies suggested that this reaction proceeds from both singlet and triplet excited states.[717]

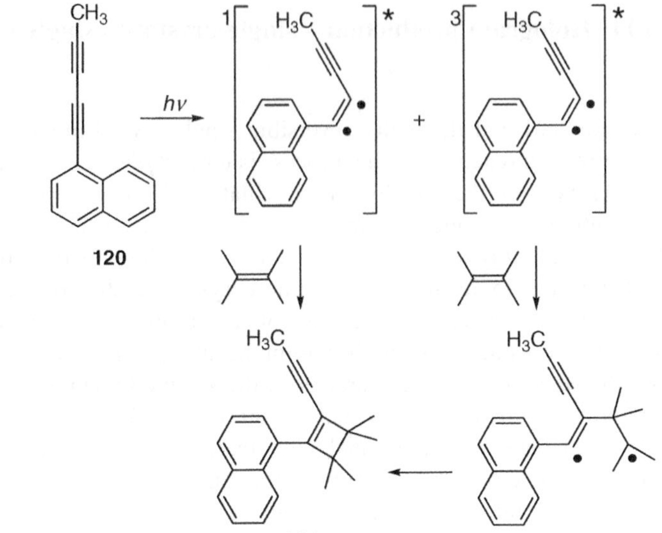

**Scheme 6.52**

The intramolecular [2 + 2] cycloaddition reaction in tethered alkenes or alkynes is more accessible because of favourable entropic factors. Two examples of silyl-tethered derivatives with complete regiocontrol are presented in Scheme 6.53. The compounds **122** and **123** form cyclobutane[718] and cyclobutene[719] derivatives, respectively, in very high chemical yields (80–90%). The tethered photoproducts are then desilylated with ammonium fluoride to obtain the corresponding diols. The photocycloaddition proceeds through the excited singlet state, as indicated by the high reaction diastereoselectivity.

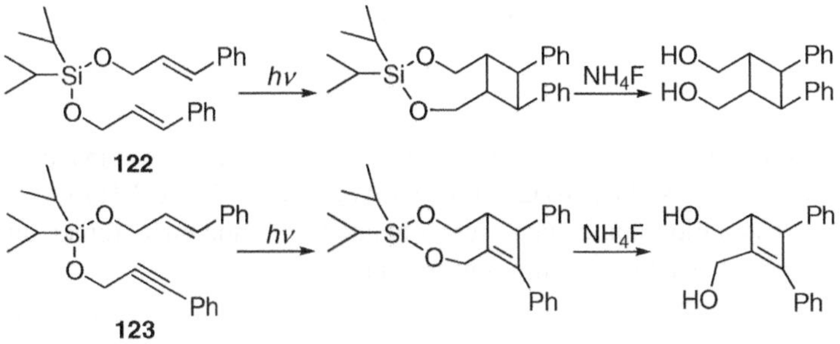

**Scheme 6.53**

The [2 + 2] cycloaddition reaction is an excellent method for the synthesis of cyclophanes.[691] For example, irradiation of the commercially available *m*-divinylbenzene (**124**) in dry benzene yields the cyclobutane **125**, which then slowly photocyclizes to isomeric cyclophanes in a low chemical yield (<10% each) (Scheme 6.54).[720]

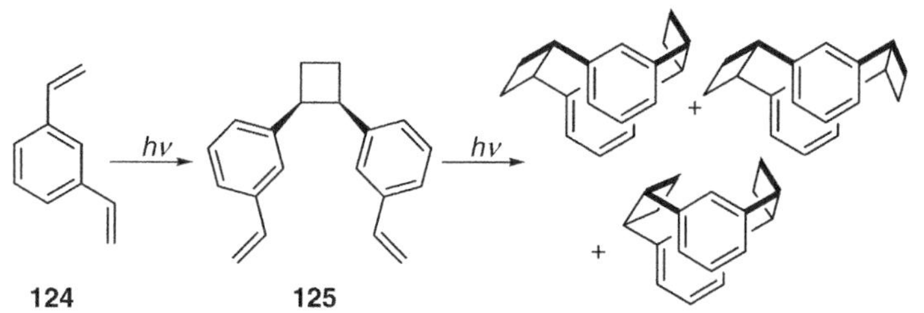

**Scheme 6.54**

Triplet sensitization of alkenes, such as norbornene ($E_T = 314\,kJ\,mol^{-1}$), having triplet energies lower than or close to that of a sensitizer, such as acetophenone ($E_T = 10\,kJ\,mol^{-1}$), leads to the [2 + 2] cycloadduct **126** (Scheme 6.55).[721] In contrast, oxetane **127** formation (Section 6.3.2) is observed when benzophenone ($E_T = 288\,kJ\,mol^{-1}$), being unable to transfer energy efficiently, is used.[722]

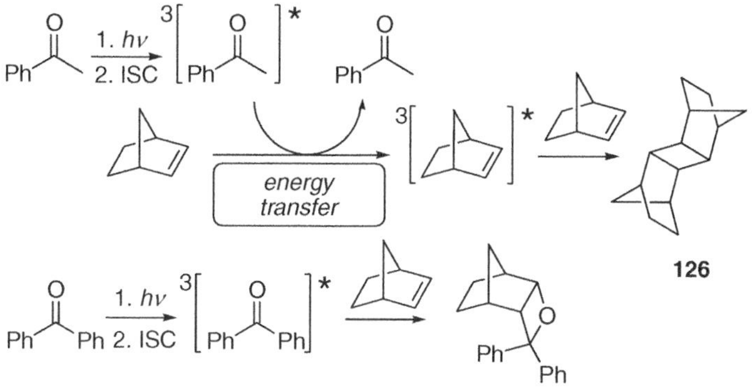

**Scheme 6.55**

Isoprene (**128**) undergoes [2 + 2], [4 + 2] and [4 + 4] photocycloaddition reactions in the presence of a sensitizer (sens, Scheme 6.56). The product distribution is strongly dependent on the $E_T$ of the sensitizer, which is explained by the fact that the *s-cis-* and *s-trans*-conformers of **128** differ in their triplet energies and their subsequent radical (stepwise) reaction preserves their original geometries.[723] A higher population of the excited *s-cis*-**128** is then reflected in higher production of [2 + 2] and [4 + 4] cycloadducts.

Either direct or triplet-sensitized irradiation of cyclohexene produces a stereoisomeric mixture of [2 + 2] dimers in nearly quantitative chemical yields (Scheme 6.57).[557] Dimerization is interpreted in terms of initial photoinduced *E–Z* isomerization of cyclohexene (Section 6.1.1), followed by a non-stereospecific ground-state addition of (*E*)-cyclohexene to the *Z*-isomer. In contrast, copper(I)-catalysed photodimerization of cyclohexene produces only one major stereoisomer **129** (Scheme 6.57).[706] According to the orbital symmetry rules, formation of this product necessitates a formal twisting of the C=C bonds of *both* of the cycloalkene monomers. It has been suggested that this photodimerization involves an initial copper(I)-catalysed $E$–$Z$ isomerization producing the coordination-stabilized ($E$)-cyclohexene **130** and supra-antara stereoelectronic control in the following addition step.

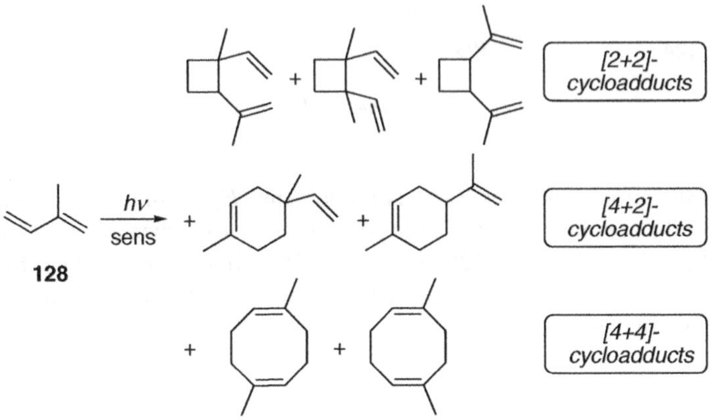

*Scheme 6.56*

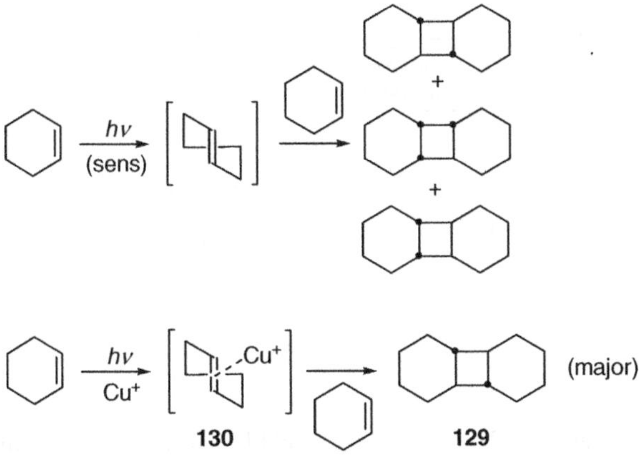

*Scheme 6.57*

---

**Special Topic 6.6: Photochemical synthesis of cage compounds**

Highly strained *cage compounds* have long attracted considerable attention from both experimentalists and theoreticians.[724] The most important structural component of cage compounds is very often a highly strained cyclobutane ring, which is easily accessible by a photochemical route.[725] Cubane for its aesthetic beauty counts as one of the milestones in synthetic (photo)chemistry and structural organic chemistry.[726] A small amount of octamethylcubane (**131**) was, for instance, obtained by direct irradiation of *syn*-octamethyltricyclo[4.2.0.0$^{2,5}$]octa-3,7-diene (**132**) via cycloaddition in *n*-pentane along with the major product octamethylcuneane (**133**), formed possibly through a stepwise radical head-to-tail bond closure (Scheme 6.58).[727]

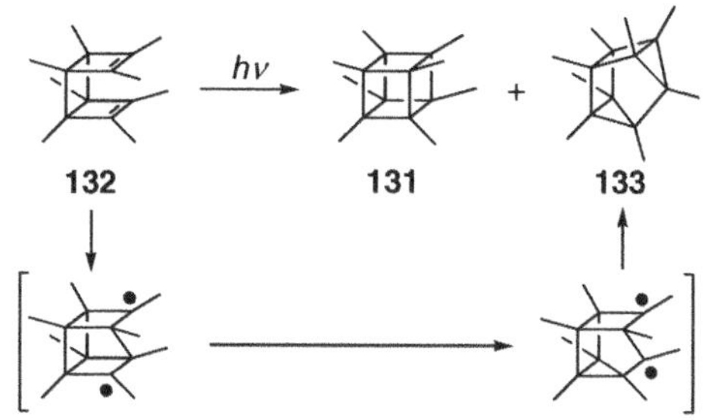

132     *hv*     131     +     133

**Scheme 6.58**

## Special Topic 6.7: Photochemistry of DNA

The UVB region of the solar spectrum (290–320 nm) is effectively filtered out by the ozone layer in the stratosphere; nevertheless, residual radiation that penetrates to Earth's surface still remains harmful to all organisms, increasingly so as the ozone layer is partially depleted. Nucleic and amino acids present in the skin are considered important UVB chromophores due to their absorbance below 320 nm. DNA bases have short $S_1$ excited-state lifetimes (1 ps or less), which largely protects them against greater photochemical damage, which would occur particularly when they undergo intersystem crossing to the reactive triplet states.[728,729] Instead, they rapidly relax to the ground state by nonradiative transition.[730,731] Prebiotic chemistry most probably took place in an oxygen-free atmosphere and, hence, ozone-free stratosphere. Therefore, under conditions of intense short-wavelength irradiation, the photochemical stability of key building blocks probably played an important role in early chemical evolution.[732]

Photodimerization of adjacent pyrimidine bases (cytosine and pyrimidine) is undoubtedly the most common photoreaction resulting from UV irradiation of DNA.[728,729,733,734] This reaction has been correlated with cell survival, mutagenesis and the development of skin cancer. The major reaction is a [2 + 2] photocycloaddition to form a cyclobutane pyrimidine dimer (e.g. the cytosine dimer **134**) via the excited triplet state[735] (Scheme 6.59) or the Paternò–Büchi reaction (Section 6.3.2) to give the dimer **135** (Scheme 6.60). The formation of dimers causes a structural distortion of a double-helical DNA and a considerable destabilization of the duplex around the lesion, which has to be repaired to avoid cell death or mutagenesis.[728] A number of different enzymatic pathways of repair have evolved. In a process called photoreactivation, the enzyme photolyase binds, for example, to the dimer and catalyses the cleavage of the corresponding bonds upon exposure to near-UV and visible light (300–600 nm) to restore the original DNA molecule.

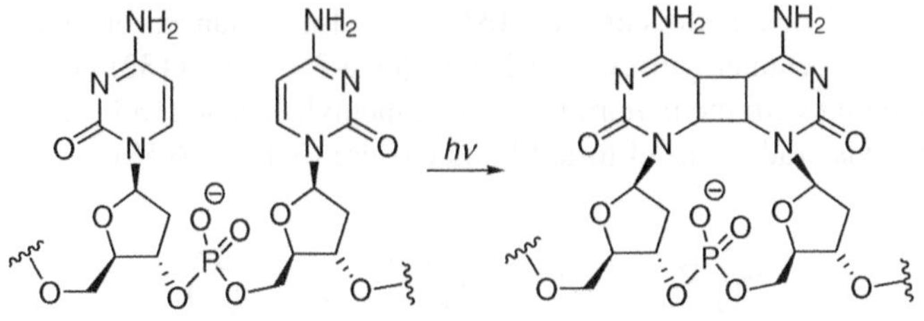

**Scheme 6.59**

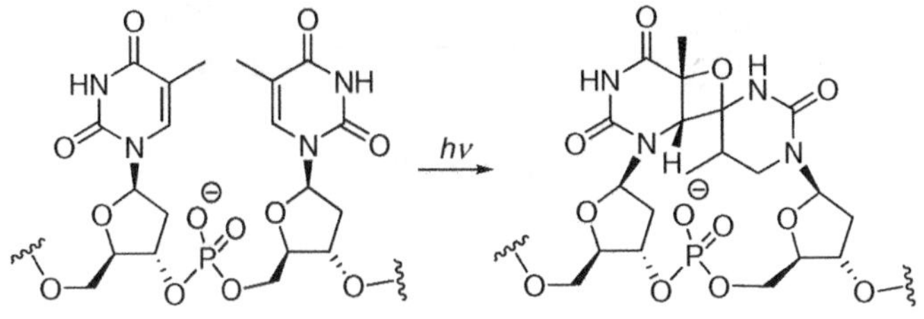

**Scheme 6.60**

In addition, UVA radiation can also cause DNA damage in an indirect process that apparently involves singlet oxygen production[733] (Section 6.7.1), due to some yet unidentified endogenous photosensitizers present in the cells.

---

### Special Topic 6.8: Photochemotherapy – treatment of psoriasis

Psoriasis (hyperproliferation of skin cells), one of the least understood skin illnesses of humans, can be treated by a variety of methods of questionable reliability. Aside from conventional topical treatment or medications taken internally, phototherapy (non-burning exposure to sunlight; see also Special Topic 6.2) and photochemotherapy, the oral or topical administration of psoralen (**136**; Scheme 6.61) or related compounds and subsequent exposure of the skin to UVA radiation (315–380 nm),[736,737] can be successful in healing patients. Several different mechanisms by which photochemotherapy normalizes psoriatic skin have been proposed but their relative importance is not known. The intrinsic photoreactivity of psoralen derivatives is determined by the electronic structure of their lowest singlet and triplet $\pi,\pi^*$ excited states. They may be responsible for the production of reactive singlet oxygen (Section 6.7.1), intercalate with the pyrimidine bases of DNA or react with proteins, lipid membranes, enzymes and other biologically important molecules. Such processes then slow down the abnormally rapid production of psoriatic skin cells. [2 + 2] Photocycloadditions are the most common psoralen photoaddition reactions, such as the formation of two different dimers with thymine of DNA. Unfortunately, long-term photochemotherapy treatment was found to be associated with the development of some types of skin cancers (Special Topic 6.22).

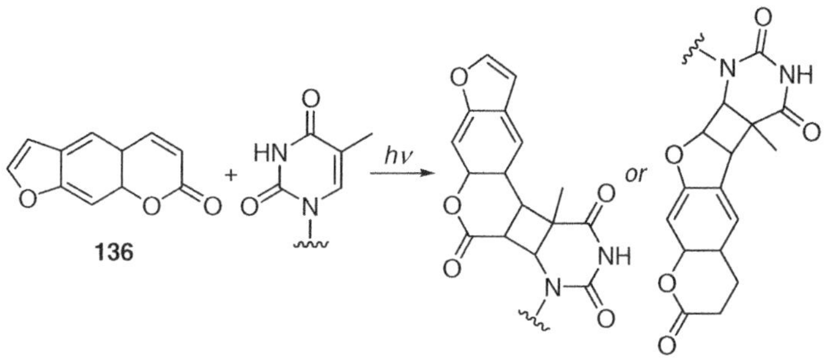

*Scheme 6.61*

The photo-Bergman reaction (cycloaromatization),[738] a photochemically initiated intramolecular reaction of enediynes, consisting of two alkyne moieties connected via an unsaturated bond, is an interesting reaction producing 1,4-dehydrobenzene systems with biradical character.[739] For example, irradiation of 1,2-di(pent-1-ynyl)benzene (**137**) at >313 nm in the presence of propan-2-ol as a hydrogen donor ([H]) gives 2,3-dipropylnaphthalene (**138**) in 25% chemical yield (at 50% conversion; Scheme 6.62).[740] The proposed radical mechanism of this transformation was based on triplet sensitization studies and laser flash photolysis experiments. The cycloaddition – formation of the bond between the radicals in 1,4-dehydrobenzene – cannot be completed because of extreme steric demands. A transition metal-catalysed cycloaromatization is also shown in Scheme 6.289.

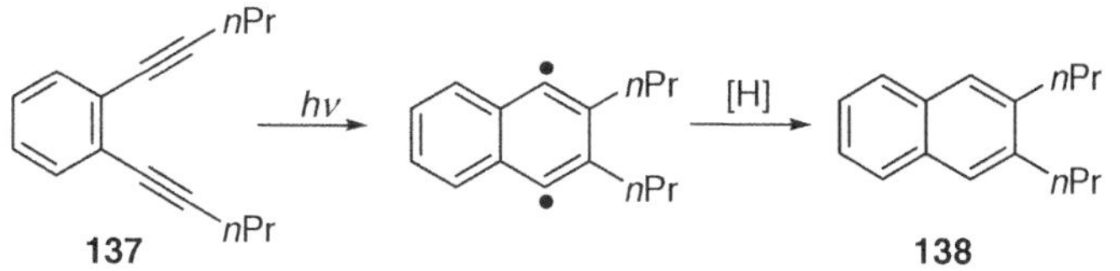

*Scheme 6.62*

### 6.1.6   α,β-Unsaturated Ketones (enones): Photocycloaddition and Photorearrangement

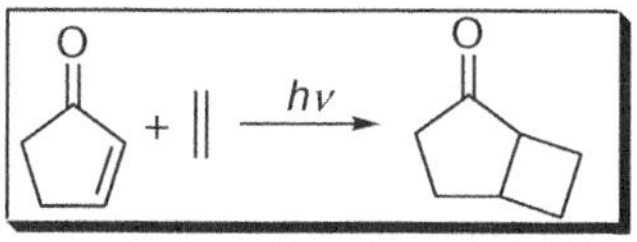

*Recommended review articles.*[692,693,741–751]
*Selected theoretical and computational photochemistry references.*[16,534,535,752–757]

Analogously to photocycloaddition of simple alkenes (Section 6.1.5), excited α,β-unsaturated carbonyl compounds (enones) and alkenes undergo [2 + 2] photocyclization to form cyclobutanes. The reaction pathway generally involves the excited triplet state of an enone (e.g. cyclopentanone in Scheme 6.63) forming an exciplex with a ground-state alkene, which decays to one or more isomeric ground-state triplet 1,4-biradical species or regenerates the starting material by Grob[758] fragmentation. The triplet biradicals must intersystem cross to the singlet biradicals to cyclize to products or eventually revert to ground-state reactants.[741–744]

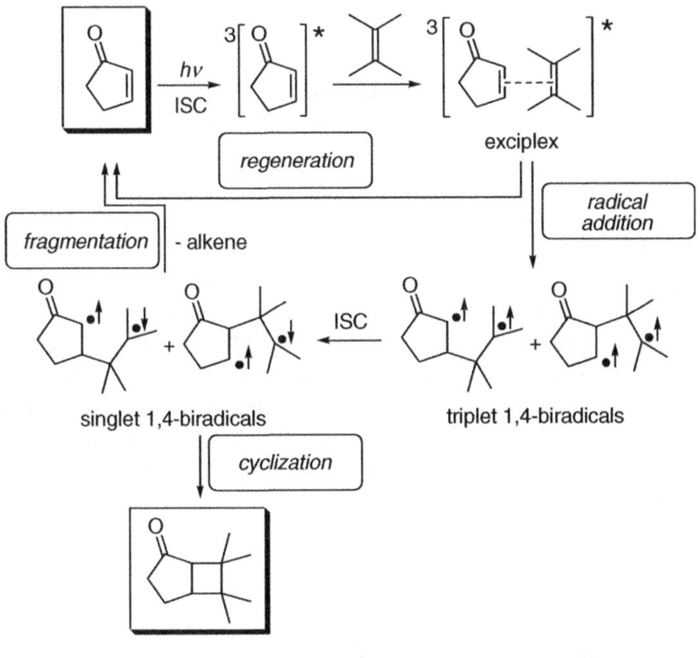

**Scheme 6.63**

The regioselectivity observed in cyclic enone–alkene intermolecular photocyclization reactions does not follow a simple pattern.[741] It was originally suggested that it is associated with the primary binding step, involving interaction of a polarized enone (e.g. cyclopentanone) triplet n,π* excited state (having a higher electron density at $C_\beta$ rather than at $C_\alpha$) with the ground-state alkene (e.g. ethyl vinyl ether) to form a preoriented exciplex (Scheme 6.64).[759,760]

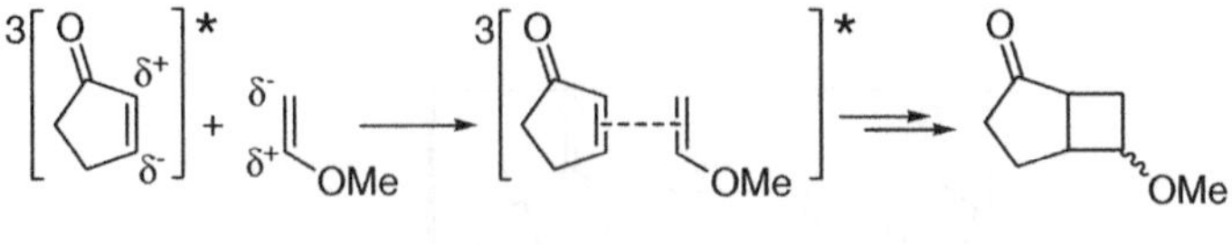

**Scheme 6.64**

However, the lowest energy triplet state of cyclic enones is a $^3\pi,\pi^*$ state, not a $^3n,\pi^*$ state,[761] and it is now believed that the regioselectivity in adduct formation rather reflects differences in the efficiencies of cyclization to products and the efficiencies of regeneration of the starting material via fragmentation of the singlet 1,4-biradicals (Scheme 6.63).[741,743,762] For example, [2 + 2] photocycloaddition of cyclopentanone to ethyl vinyl ether in benzene gives two products, head-to-tail (HT; **139**) and head-to-head (HH; **140**) adducts, in a ~3:1 concentration ratio (Scheme 6.65),[762] which is well in accord with the above-mentioned preoriented exciplex model.[759] As expected, no products derived from the 1,4-biradicals containing a poorly stabilized primary radical centre (**141** and **142**) are obtained. However, radical trapping experiments revealed that the HT and HH 1,4-biradicals are produced in an equimolar ratio. It was concluded that the cyclization of the HT biradical must be more efficient than fragmentation ($\Phi_c > \Phi_f$), in contrast to that of the HH biradical, which possibly relates to different populations of extended versus closed conformations of the biradicals. This mechanistic explanation of regioselectivity could be applied to many other systems.[741,743]

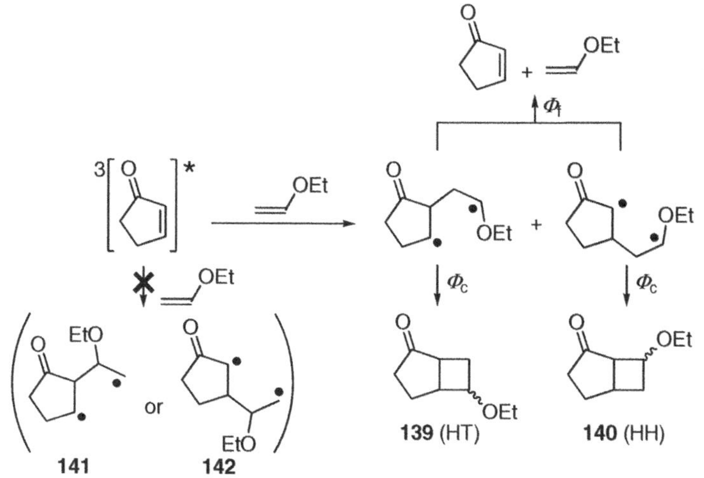

*Scheme 6.65*

---

**Case Study 6.12: Organic synthesis – construction of the AB-ring core of paclitaxel**

A new method for the construction of the AB-ring core of paclitaxel (Taxol), an anticancer drug, was developed utilizing the cyclopentanone (**143**)–allene (**144**) photocycloaddition reaction to give the bicyclic product **145**, which was subsequently transformed to a bicyclic diketone **146** in several steps (Scheme 6.66) in 42% overall chemical yield.[763]

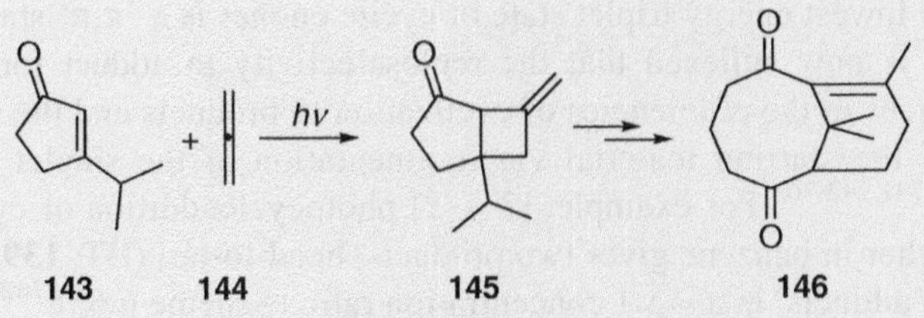

**Scheme 6.66**

*Experimental details.*[763] A solution of **143** and **144** in dichloromethane was irradiated by an unspecified source of UV radiation at −78 °C (Figure 3.9) to give a head-to-head adduct **145** in 84% chemical yield, along with a head-to-tail minor adduct. The product was used in the following step without further purification.

The course of intramolecular enone–alkene photocycloaddition is dependent on the number of atoms between the two reactive C=C bonds. For example, *E*- and *Z*-isomers of 1-acylhepta-1,6-diene (**147**) form a 1:1 mixture of stereoisomeric cycloadducts **148** and **149** upon irradiation, while no *E–Z* isomerization occurs (Scheme 6.67a).[764] The initial bonding takes place between the C2 ($C_\beta$) and C6 atoms, in agreement with the empirical *rule of five*,[765] the regioselective, kinetically preferred formation of five-membered ring biradical intermediates over larger rings due to the entropies of cyclization. As a result, the biradical **150** is not observed. For comparison, the acylhexadiene **151** photolysis also proceeds via a 1,4-biradical (**152**) formed by an initial 1,5-cyclization (Scheme 6.67b).[766]

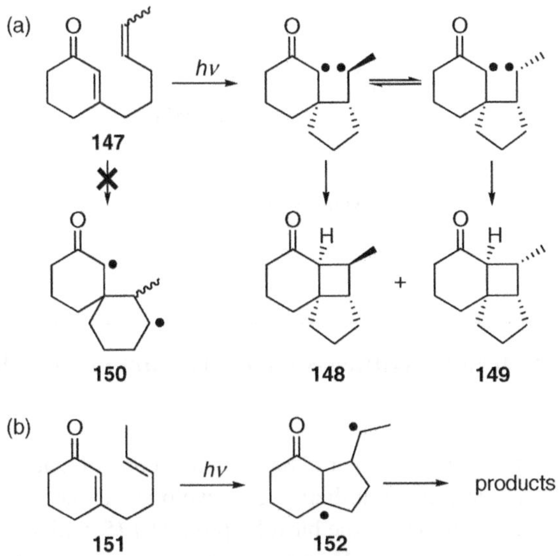

**Scheme 6.67**

The photocycloadditions of alkenes with the *enol* form of 1,3-diketones or 1,2-diketones, referred to as the *de Mayo reaction*,[760,767] also gives cyclobutane derivatives.[742,749] It is assumed that the mechanism of this reaction involves a triplet excited dione, possibly an exciplex, and 1,4-biradical intermediates (see also Scheme 6.63).[749] Intramolecular regioselective photocycloaddition of the enol acetate **153**, as an example of the enolate derived from a 1,3-diketone, leads to the tricyclic adduct **154** in quantitative yield, which undergoes annulation to form a large (eight-membered) ring (see also Special Topic 6.14) of bicycloundecanedione (**155**) in the presence of a base (Scheme 6.68).[768] In another example, an enolized 1,3-diketone group undergoes intramolecular [2 + 2] photocycloaddition to an enamine moiety of the isoquinolone **156** to form **157** via an unstable adduct **158** in 35% overall chemical yield (Scheme 6.69).[769]

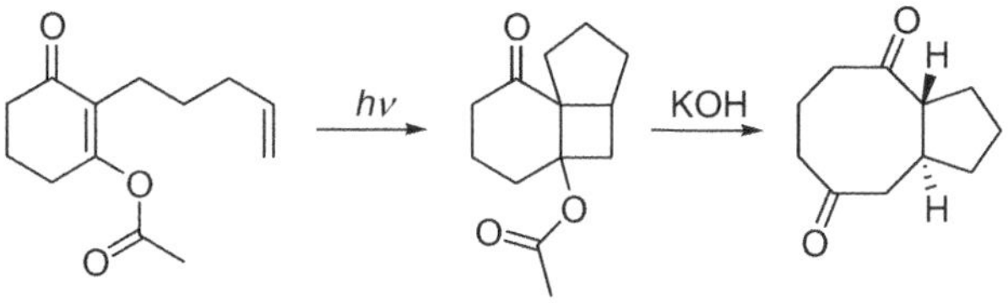

**153**     **154**     **155**

*Scheme 6.68*

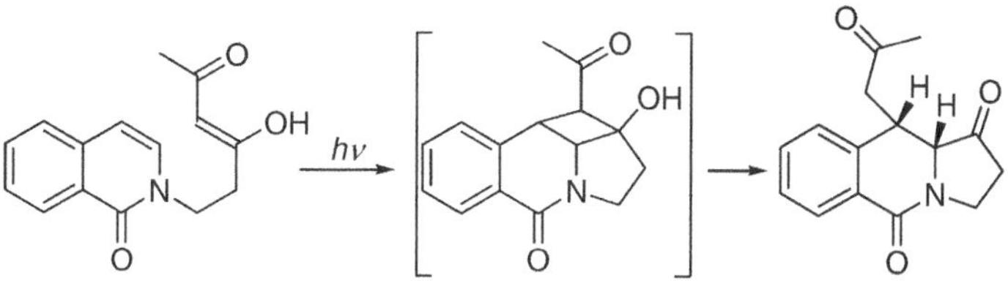

**156**     **158**     **157**

*Scheme 6.69*

Cyclic enones, such as substituted cyclohex-2-enones or cyclohexa-2,5-diones, also undergo sigmatropic photorearrangement to form bicyclo[3.1.0]hexanones (lumiketones) or bicyclo[3.1.0]hex-3-en-2-ones, respectively, for which both concerted and stepwise (biradical) reaction mechanisms have been proposed.[640,641,770] For example, a [1,2]-shift concurrently with the ring contraction (termed the *type A* reaction) is observed upon irradiation of the methylphenyl derivative **159** in polar solvents, whereas phenyl migration (termed the *type B* reaction) predominates in nonpolar solvents (Scheme 6.70).[771,772] The reactions are believed to proceed via both the $\pi,\pi^*$ and $n,\pi^*$ triplet ketone states. In the presence of alkenes, cyclic enones may readily undergo a competitive photocycloaddition reaction (Section 6.1.5).

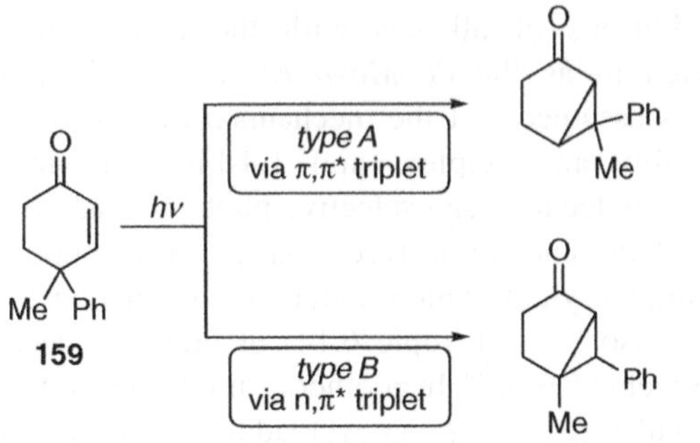

*Scheme 6.70*

Photorearrangements in the crystalline state[773] usually afford products with very high selectivity (Special Topic 6.5). Whereas irradiation of 4,4,5-triarylcyclohex-2-enone (**160**) in benzene solution affords a 1:1 ratio of phenyl to *p*-cyanophenyl migration to form **161** and **162**, the former product is produced exclusively in the solid state (Scheme 6.71).[755]

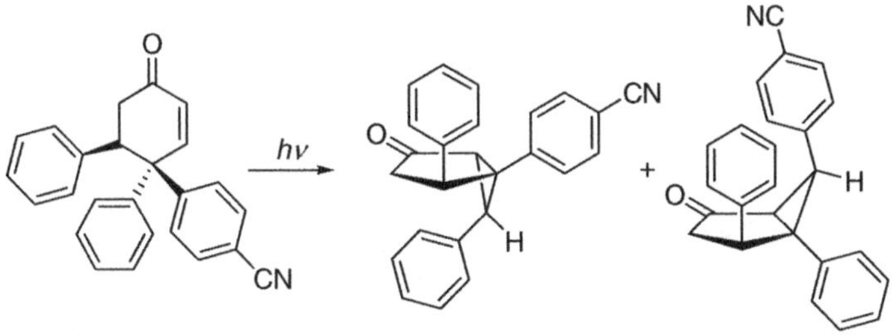

| 160 | 161 | 162 |
|---|---|---|
| benzene solution | 50% | 50% |
| solid state | 99% | not observed |

*Scheme 6.71*

Photochemistry of cross-conjugated cyclohexa-2,5-diones has historically served to explain the mechanism of some primary photochemical processes[774] and later it has been successfully utilized in organic synthesis.[775–777] In general, bicyclo[3.1.0]hex-3-en-2-ones **163** are formed from cyclohexa-2,5-diones **164** via the triplet n,π* excited states of biradical character and the ground-state zwitterion intermediate.[775] This photoprocess can be designated a sigmatropic [1,4]-shift (Scheme 6.72), the equivalent of the type A rearrangement of cyclohex-2-enones (Scheme 6.70).

Santonin (**165**), a well-studied photochemically active compound, undergoes a number of photorearrangements. The [1,4]-shift product lumisantonin (**166**) is obtained in 64% chemical yield in dioxane, whereas a subsequent rearrangement product, photosantonic acid (**167**), is formed in protic media (Scheme 6.73).[778,779]

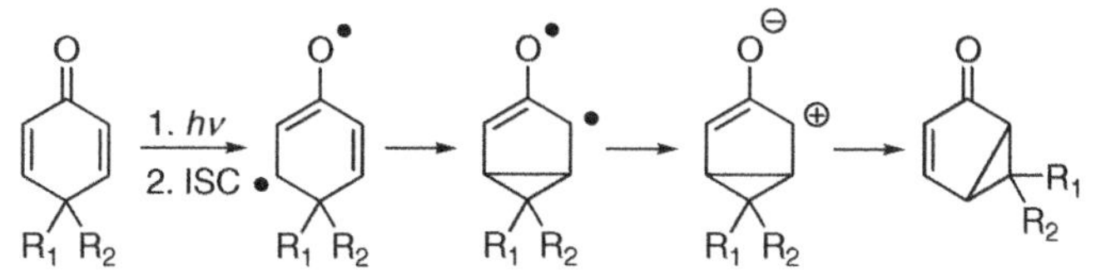


164 163

*Scheme 6.72*

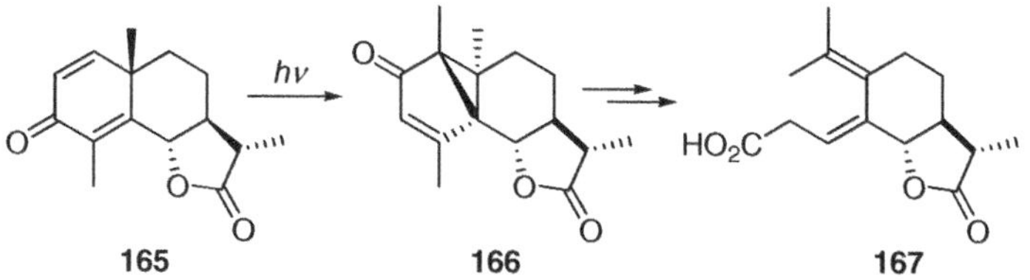


165 166 167

*Scheme 6.73*

### 6.1.7 Problems

1. Explain the following concepts and keywords: photostationary state; *Rydberg* state; NEER principle; bicycle-pedal mechanism; phototransduction; phototherapy; photochemical deracemization; enantiodifferentiating photosensitization; photochromism; di-π-methane rearrangement; asymmetric photochemical synthesis; single-crystal-to-single-crystal photochemistry; photochemotherapy; rule of five.
2. Suggest the mechanisms for the following reactions:

(a)

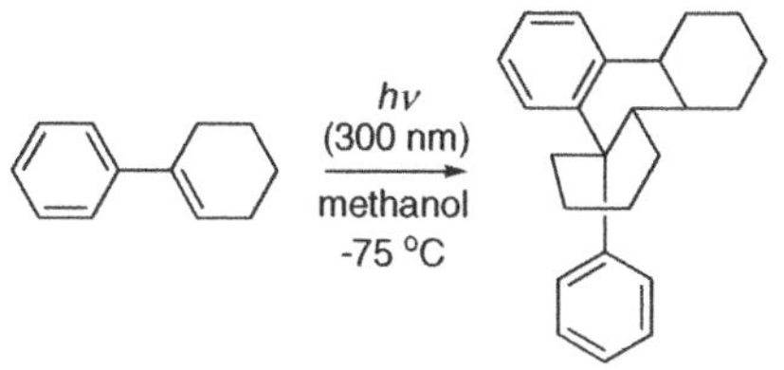


[ref. 780]

(b)

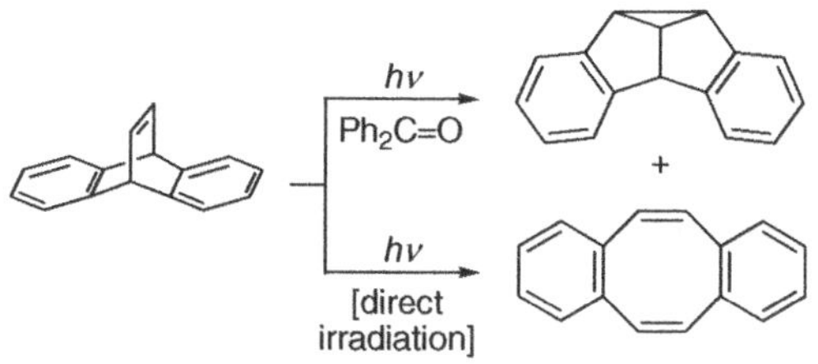


[ref. 638]

(c)

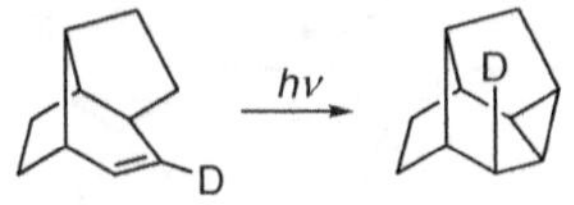

(hint: reaction from a Rydberg excited state)

[ref. 781]

3. Predict the major photoproduct(s):

(a)

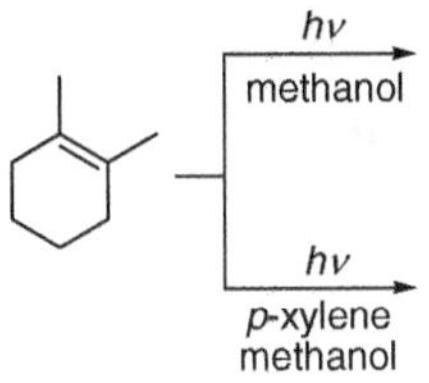

[ref. 671]

(b)

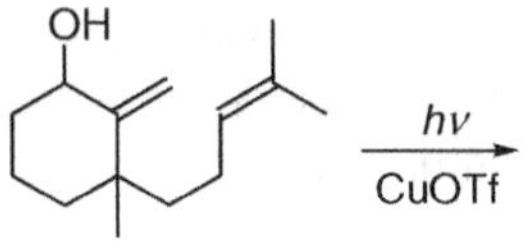

[ref. 782]

(c)

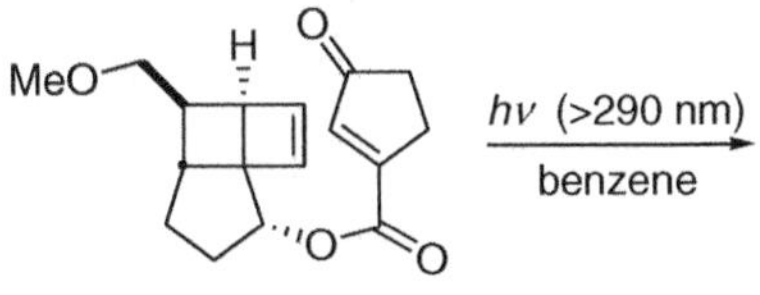

[ref. 783]

## 6.2   Aromatic Compounds

The electronic structure of the lowest excited states of planar aromatic compounds is well described by MO theory (Sections 4.5 and 4.7). Benzenoid aromatic hydrocarbons exhibit three or four $\pi,\pi^*$ absorption bands in the near-UV region, which are labelled $^1L_b$, $^1L_a$, $^1B_b$

and $^1B_a$ (in order of increasing energy for benzene and naphthalene). Because the lowest excited singlet state mostly determines the photophysical and photochemical properties (Kasha's rule, Section 2.1.8), it is essential to note that the HOMO–LUMO transition ($^1L_a$) does *not* correspond to the lowest excited singlet state ($^1L_b$) in benzene and naphthalene and many of their derivatives. The absorption spectra of the linear acenes (benzene, naphthalene, anthracene, etc.) are shown in Figure 4.16. The $^1L_a$ bands of larger systems (tetracene, pentacene, etc.) lie in the visible region, giving rise to the characteristic colour of these compounds. The lowest triplet state is of $^1L_a$ character in all benzenoid hydrocarbons.

The fluorescence rate constants of aromatic compounds predicted by Equation 2.11 differ by about two orders of magnitude depending on the nature of the lowest singlet state; the smaller representatives with $S_1 = {}^1L_b$ have $k_f \approx 2 \times 10^6\,s^{-1}$ and the larger ones with $S_1 = {}^1L_a$ have $k_f \approx 3 \times 10^8\,s^{-1}$. Rate constants of IC and ISC are small in the parent hydrocarbons. Using the rules-of-thumb given by Equations 2.22 and 2.23, one predicts $\log(k_{IC}/s^{-1}) \approx \log(k_{ISC}/s^{-1}) \approx 6$. Therefore, benzenoid aromatic compounds exhibit substantial fluorescence, with quantum yields approaching unity in compounds with $S_1 = {}^1L_a$, and appreciable triplet yields. The quantum yields of fluorescence and ISC add up to less than unity in the larger members, because the rate constant of nonradiative decay, $k_{IC}$, increases as the energy gap $E(S_1) - E(S_0)$ decreases. Due to the long lifetimes of $S_1$ in the smaller benzenoid compounds, $^1\tau \approx 10^2\,ns$, ISC may be accelerated by diffusional encounters with oxygen.

Rates of ISC are increased upon substitution with heavy atoms (Br, I) or with functional groups that have low-lying n,π* states (carbonyls, nitro groups, diazines). The absorption spectra of some nitrogen-containing benzene derivatives are shown later in Figure 6.8.

The first absorption band of nonalternant hydrocarbons and the band shifts induced by substitution are generally well described by HMO theory (Section 4.7). Absorption to $S_1$ corresponds to the HOMO–LUMO transition. Nonradiative decay often dominates the photophysical properties of nonalternant hydrocarbons and also alternant hydrocarbons with a 4*n*-membered ring (biphenylene), so that they generally have short singlet lifetimes and low triplet yields and are less prone to undergo photoreactions upon direct irradiation.

Table 6.4 shows the principal photoreactions of aromatic compounds that we discuss in this chapter. Upon irradiation, aromatic compounds, such as benzenes, naphthalenes and some of their heterocyclic analogues, undergo remarkable rearrangements that lead to some non-aromatic highly strained products, such as benzvalene and Dewar benzene (entry 1), which can be isolated under specific conditions. Quantum and chemical reaction yields are usually low; however, photochemistry may still represent the most convenient way for their preparation. While bulky ring substituents usually enhance the stability of those products, aromatic hydrocarbons substituted with less sterically demanding substituents exhibit ring isomerization (phototransposition) (entry 2).

Excited aromatic compounds are also capable of undergoing bimolecular reactions, which are not observed in the ground state. Stepwise regioselective photocycloadditions to alkenes, for example [2 + 2] photocycloaddition (entry 3), involving short-lived intermediates, provide access to various bicyclic and tricyclic unsaturated hydrocarbons. When an aromatic moiety bears a leaving group, its substitution by nucleophiles (entry 4) is readily available upon excitation. Variations in the nature of the electronically excited state, directing and activating the effects of the ring substituents, and also experimental

**Table 6.4**  *Photoprocesses involving excited aromatic compounds.*

| Entry | Starting material[a] | Product(s) | Mechanism | Section |
|-------|---------------------|------------|-----------|---------|
| 1 | | | Photorearrangement | 6.2.1 |
| 2 | | | Phototransposition | 6.2.1 |
| 3 | | | Photocycloaddition | 6.2.2 |
| 4 | | | Photosubstitution | 6.2.3 |

[a]$Y^-$ = nucleophile.

conditions, are responsible for the great diversity of reaction mechanisms and formation of specific products.

### 6.2.1   Aromatic Hydrocarbons and Heterocycles: Photorearrangement and Phototransposition

*Recommended review articles.*[784–788]
*Selected theoretical and computational photochemistry references.*[16,534,535,789–794]

Benzene, the archetypical aromatic compound, possesses three absorption bands, at $\lambda = 254$ ($S_1$), 203 ($S_2$) and 180 nm ($S_3$).[785,788] Photolysis of benzene in the gas phase at 254 nm into the excited singlet state $S_1$ ($^1B_{2u}$ state; $E_S = 459$ kJ mol$^{-1}$)[157] produces two non-aromatic highly strained products, benzvalene (**168**) (the limiting concentration attainable by irradiation of liquid benzene at higher temperature is only 0.05%) and fulvene (**169**), via a biradical intermediate **170** (prefulvene) with low quantum yields ($\Phi = 0.01$–$0.03$) (Scheme 6.74).[795] The highly energetic non-aromatic Dewar benzene (**171**), along with **168** and **169**, is obtained when benzene is irradiated with wavelengths at $\lambda < 200$ nm. It has been shown that the $S_2$ ($^1B_{1u}$) state and the cyclohexa-2,5-dien-1,4-diyl

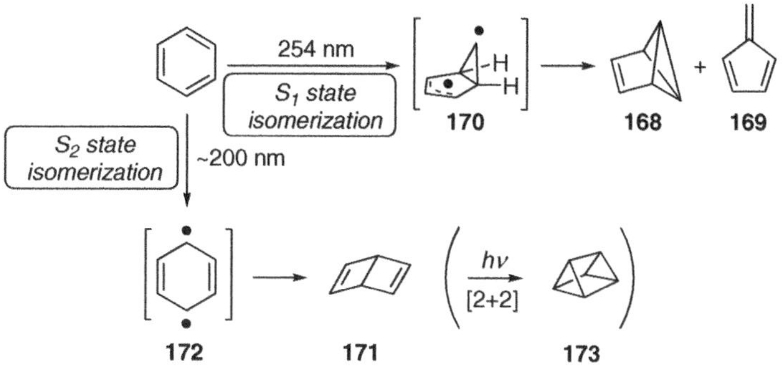

**Scheme 6.74**

biradical intermediate **172** are precursors of this product.[796] Dewar benzene may further undergo [2 + 2] intramolecular photocycloaddition (Section 6.1.5) to form prismane (**173**). Although intersystem crossing in benzene is relatively efficient ($\Phi = 0.23$ in hexane solution), the excited *triplet* state $T_1$ ($E_T = 353$ kJ mol$^{-1}$)[157] is usually not involved in the photoisomerization.

Photochemical transposition (ring isomerization) of the carbon atoms in benzene is known to involve a benzvalene intermediate.[784,785] Gas-phase photolysis of *o*-xylene (**174**) at 254 nm, for example, leads to the benzvalene **175** and subsequently to *m*-xylene (**176**) at low conversion (<10%; Scheme 6.75).[797] A *para*-isomer is formed upon prolonged irradiation.

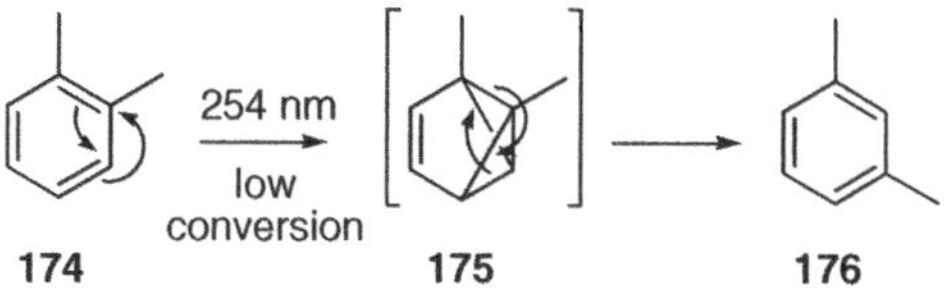

**Scheme 6.75**

Isolation of reactive non-aromatic intermediates can be less problematic when the starting arenes are substituted with bulky substituents. Their stabilization is achieved especially by impeding the re-aromatization process through steric interactions. For example, 1,3,5-tri-*tert*-butylbenzene (**177**) produces 1,2,4-tri-*tert*-butylbenzvalene (**178**) as the sole product with $\Phi = 0.12$ upon photolysis at 254 nm (Scheme 6.76).[798] Re-aromatization and other processes become important after exhaustive irradiation, producing benzvalenes (**179** and **180**), fulvene (**181**), Dewar benzene (**182**) and prismane (**183**). The prismane derivative is obtained as the major product in 65% chemical yield at the photostationary state (PPS) (Section 6.1.1) because it does not absorb significantly at 254 nm. The $S_2$ excited state is not available as a precursor to Dewar benzene (and its photocycloaddition product prismane) at this irradiation wavelength. It has therefore been suggested that the $T_1$ excited state may be involved.[785]

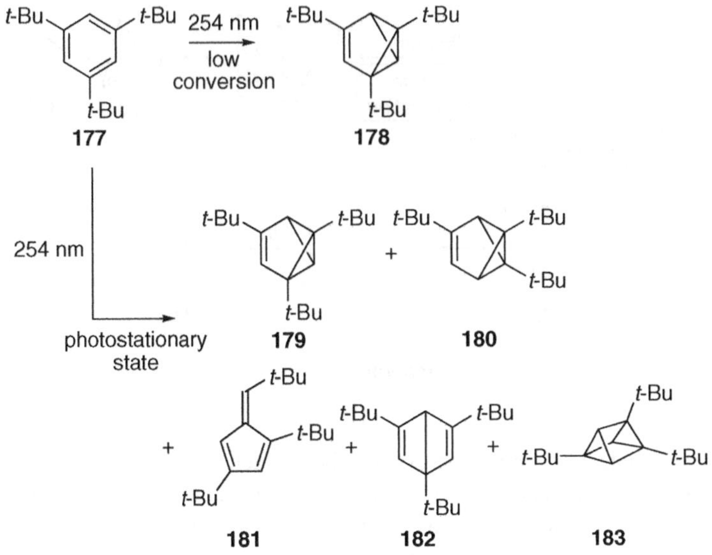

*Scheme 6.76*

These photorearrangements are known to occur not only in substituted benzenes but also in substituted naphthalenes,[785] pyridines,[799] pyrylium cations,[787] and pyridinium salts.[786] For example, irradiation of 2,4,6-trimethylpyrylium perchlorate (**184**) results in the formation of the oxabenzvalene cation **185**, which rearranges to the 2,3,5-trimethyl phototransposition isomer **186** (Scheme 6.77).[800]

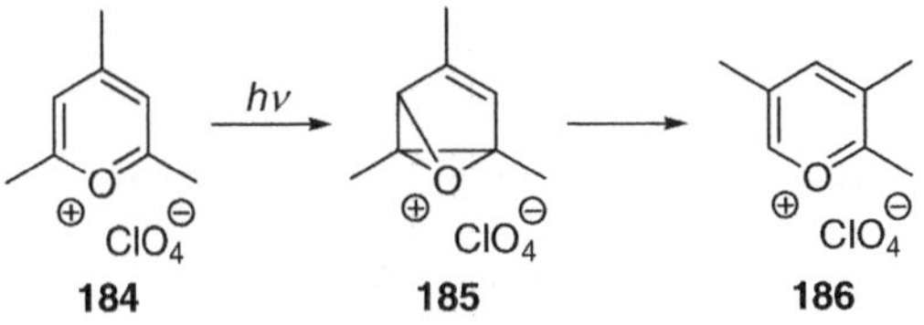

*Scheme 6.77*

## Case Study 6.13: Organic synthesis – photo-ring contraction

4-Hydroxypyrylium salts, produced *in situ* by protonation of 4-pyranones in aqueous $H_2SO_4$, undergo phototransposition and photo-ring contraction in high chemical yields.[787] For example, photolysis of the 2,3-dimethyl-4-hydroxypyrylium cation **187** in 50% sulfuric acid was suggested to form the oxabenzvalene cation **188**, which is trapped by water as a nucleophile, and the adduct **189** is subsequently hydrolysed to give the cyclopenten-2-one **190** (Scheme 6.78).[801] At high $H_2SO_4$ concentrations,

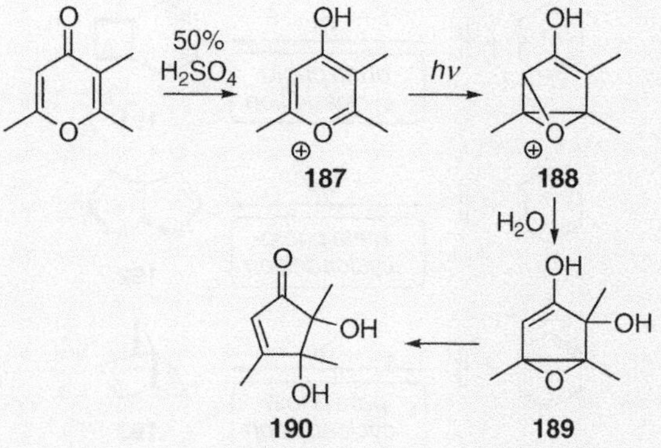

**Scheme 6.78**

where the activity of nucleophilic water is low, the oxabenzvalene cation yields phototransposition products only.

*Experimental details.*[801] 4-Hydroxypyrylium cation (**187**) was prepared by dissolving the corresponding 4-pyrone (1.74 mmol) in 50% aqueous $H_2SO_4$ (8 ml). The acid solution in a quartz tube, placed in a quartz Dewar flask, was irradiated with six low-pressure Hg lamps (8 W; $\lambda = 254$ nm) arranged in a circular array around the Dewar flask (Figure 3.10). The sample temperature was maintained at 0 °C during the irradiation by passing a stream of nitrogen through a heat exchanger coil immersed in a dry-ice–acetone bath and then through the quartz Dewar flask holding the sample. After irradiation, the acid solution was immediately neutralized and washed using diethyl ether. The aqueous layer was concentrated to dryness and a white solid was recrystallized from diethyl ether to give the product **190** in 63% chemical yield.

## 6.2.2 Aromatic Hydrocarbons and Heterocycles: Photocycloaddition

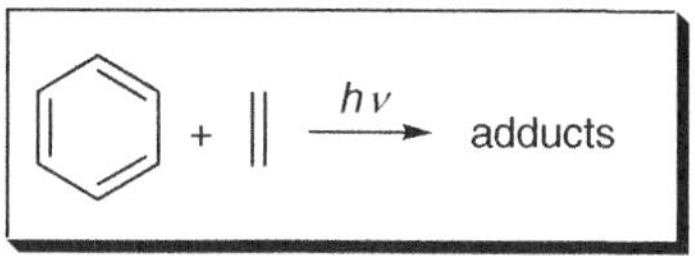

*Recommended review articles.*[602,802–813]
*Selected theoretical and computational photochemistry references.*[814,815]

Benzene in the $S_1$ ($\pi,\pi^*$) excited state is no longer aromatic and is capable of undergoing various chemical reactions not observed in the ground state. Three basic types of photocycloaddition reaction of an alkene to the excited benzene can be described:[802,804,809,816] (a) bicyclo[4.2.0]octa-2,4-dienes (e.g. **191**) are formed by *ortho*-photocycloaddition (1,2- or [2 + 2] photocycloaddition), (b) tricyclo[3.3.0.0$^{2.8}$]oct-3-enes (e.g. **192**) are obtained by *meta*-photocycloaddition (1,3- or [3 + 2]

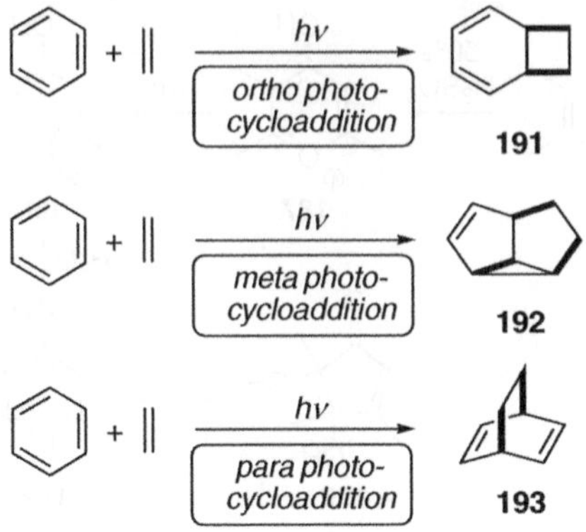

**Scheme 6.79**

photocycloaddition) and (c) bicyclo[2.2.2]octa-2,5-dienes (e.g. **193**) are produced via *para*-photocycloaddition (1,4- or [4 + 2] photocycloaddition) (Scheme 6.79; newly formed σ-bonds are shown in bold).

The photocycloaddition mechanism, and consequently the reaction selectivity, may vary considerably depending on the structure of the initial material and reaction conditions. In general, an excited arene and a ground-state alkene may react with initial polarization to form an exciplex.[802] In [2 + 2] photocycloaddition reactions, biradical intermediates are often involved (Scheme 6.80a), although excitation of a ground-state charge-transfer (CT) complex (Section 2.2.3) has also been discussed in some cases, such as the [2 + 2] photocycloaddition of benzene with maleic anhydride (Scheme 6.80b).[817] Here a zwitterion intermediate **194** collapses to the adduct **195** only in the absence of an acid.

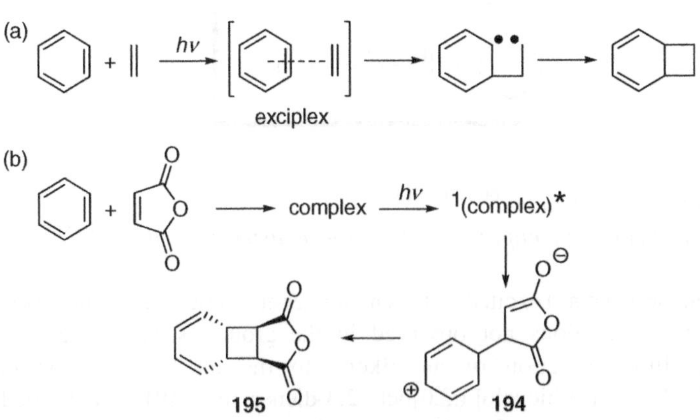

**Scheme 6.80**

## Ortho-Photocycloaddition

This reaction pathway is usually favoured when an aromatic moiety and an alkene bear electron-withdrawing and electron-donating substituents, respectively (or vice versa).[804,809] This addition involves a charge transfer and the course of the reaction is sensitive to the solvent polarity. Such a mechanism may resemble that of [2 + 2] photocycloaddition of alkenes to α,β-unsaturated carbonyl compounds (Section 6.3.2). Scheme 6.81 shows examples of two intermolecular processes and one intramolecular [2 + 2] photocycloaddition reaction: (a) crotononitrile (**196**) is added to anisole (**197**) to yield several stereoisomers of **198** in 38% chemical yield and with high regioselectivity, which is linked to bond polarization in the exciplex;[818]  (b) hexafluorobenzene (**199**) reacts with 1-ethynylbenzene (**200**) to form the bicyclo[4.2.0]octa-2,4,7-triene **201** in 86% yield;[819] and  (c) irradiation of **202** in methanol leads to the single photoproduct **203**. [820]

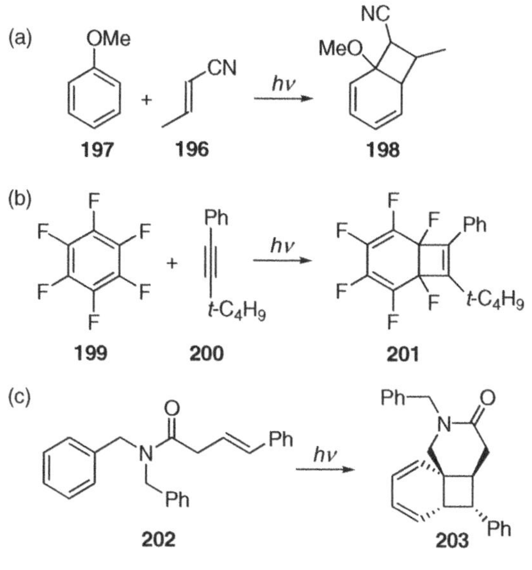

Scheme 6.81

Polycyclic aromatic hydrocarbons are also known to undergo *ortho*-photocycloadditions.[805] For example, the reaction of the chrysene derivative **204** with the electron-deficient methyl cinnamate **205** affords the adduct **206** as the major product (Scheme 6.82).[821] The high stereoselectivity observed has been explained by the formation of an electronically favourable sandwich-type singlet exciplex **207**.

*ortho*-Photocycloaddition has also been observed in aromatic heterocyclic compounds, such as substituted pyridines. For example, irradiation of 2-methoxy-4,6-dimethylnicotinonitrile (**208**) in the presence of methacrylonitrile (**209**) in benzene leads to the addition product **210** that thermally converts to the 3,4-dihydroazocine **211**, which then undergoes electrocyclization in a second photochemical step to give **212** in 45% chemical yield (Scheme 6.83).[822]

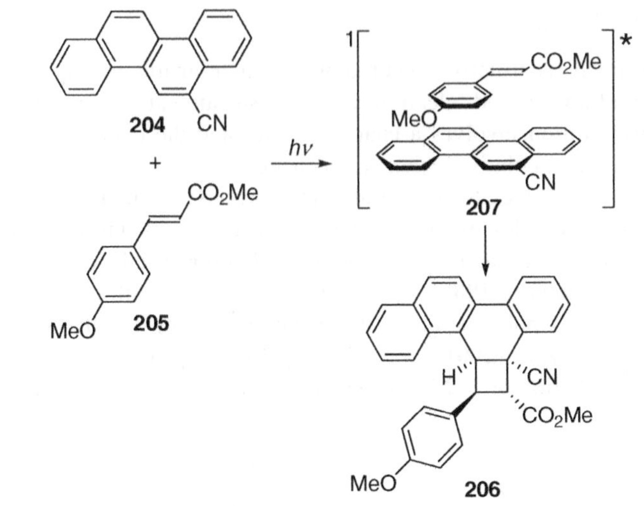

**Scheme 6.82**

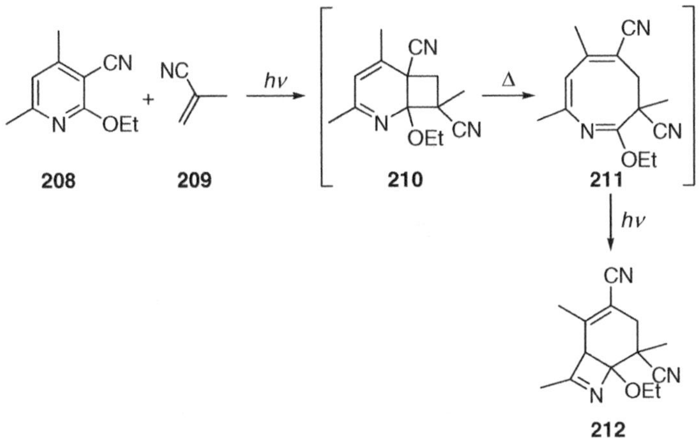

**Scheme 6.83**

---

## Special Topic 6.9: Photochemistry of fullerenes

The (photo)chemical and physical properties of fullerenes, spherical carbon molecules, are an important topic in current research, especially in nanotechnology.[823,824] The molecule with 60 carbons, $C_{60}$, having properties of an electron-deficient alkene, undergoes for example [2 + 2] photocycloaddition similar to aromatic moieties. The absorption spectrum of $C_{60}$ exhibits three principal maxima ($\lambda_{max} = 211$, 256, 328 nm) and a weak band at $\lambda = 540$ nm.[813] Very weak fluorescence from the lowest excited singlet state ($S_1$; $E_S = 193$ kJ mol$^{-1}$) competes with fast and efficient (close to 100%) intersystem crossing to the excited triplet state ($E_T = 157$ kJ mol$^{-1}$). For example, the cycloaddition of both (Z)- and (E)-1-(4-methoxyphenyl)prop-1-ene (**213**) to triplet excited $C_{60}$ (**214**) leads to an adduct **215** through a biradical intermediate **216**

(Scheme 6.84).[825] Fast rotation of the aryl moiety around the former double bond in the biradical and photocycloreversion controls the product stereochemistry.

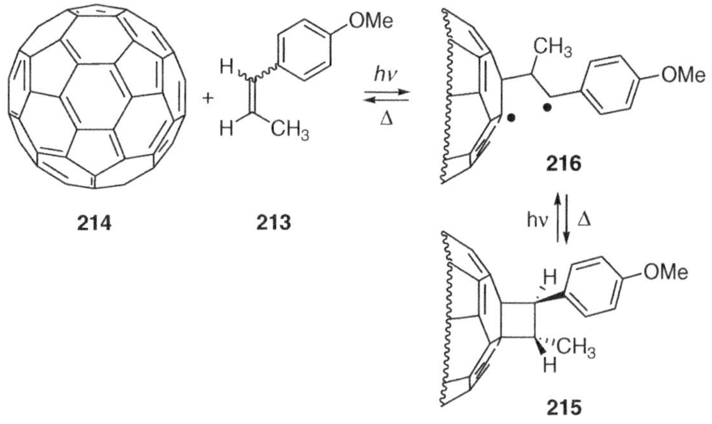

**Scheme 6.84**

Another example of photoaddition to fullerene is shown in Scheme 6.85.[826] In the first step, $C_{60}$ is excited and accepts an electron from an alkaloid scandine (**217**), followed by proton transfer. The resulting pair of radicals couple to give **218**. The second carbon–carbon bond is subsequently formed via an analogous photoinduced electron and proton transfer. The product **219** was obtained in 37% chemical yield.

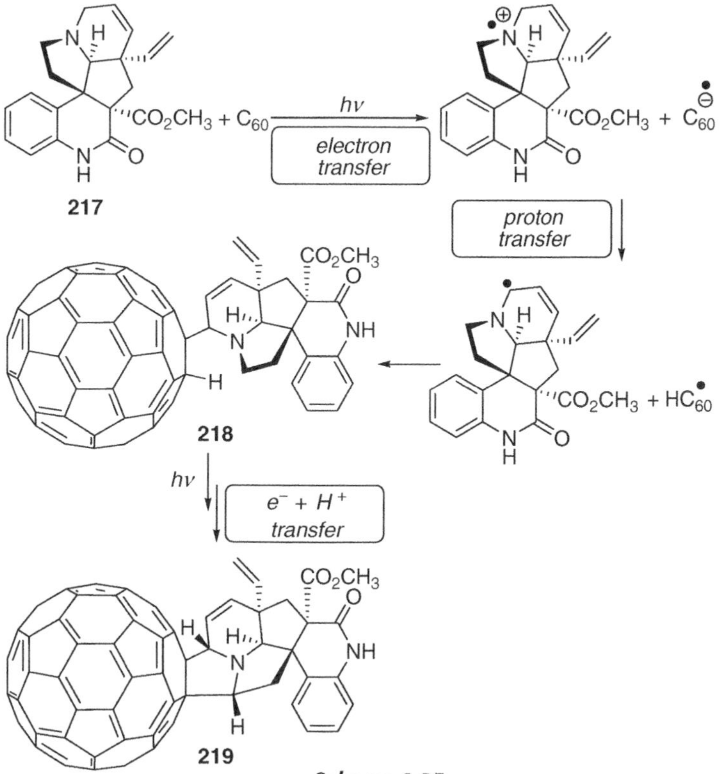

**Scheme 6.85**

## Special Topic 6.10: Cyclopropyl group as a mechanistic probe

The cyclopropyl moiety is often utilized as a mechanistic probe to trap a radical formed on the vicinal carbon (Scheme 6.86). In such a case, the ring is subsequently opened to give a relaxed but-3-en-1-yl (allylcarbinyl) intermediate, which undergoes various reactions.

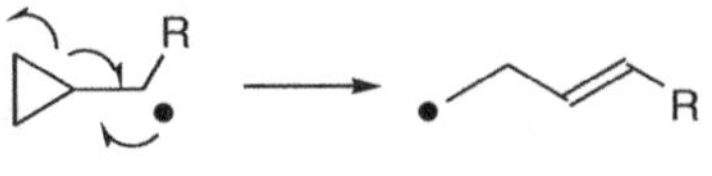

*Scheme 6.86*

This approach has been used, for example, to find whether the intramolecular photocycloaddition reaction of the triplet excited cyclopropyl-substituted 4-(butenyloxy)acetophenone **220** proceeds via the 1,4-biradical **221** (Scheme 6.87).[827] This presumption was confirmed by identifying the three rearrangement cyclization products **222–224**. Because the rate constant of the cyclopropylcarbinyl radical opening to the allylcarbinyl radical is known to be $7 \times 10^7 \, s^{-1}$,[828] it was suggested that the rate constant for the formation of the (not observed) *ortho*-photocycloaddition adduct (**225**) must be less than $3 \times 10^6 \, s^{-1}$. This technique – comparing the rate constants of two parallel processes, of which one is known – is often referred to as a *kinetic* (or *radical*) *clock*.[829]

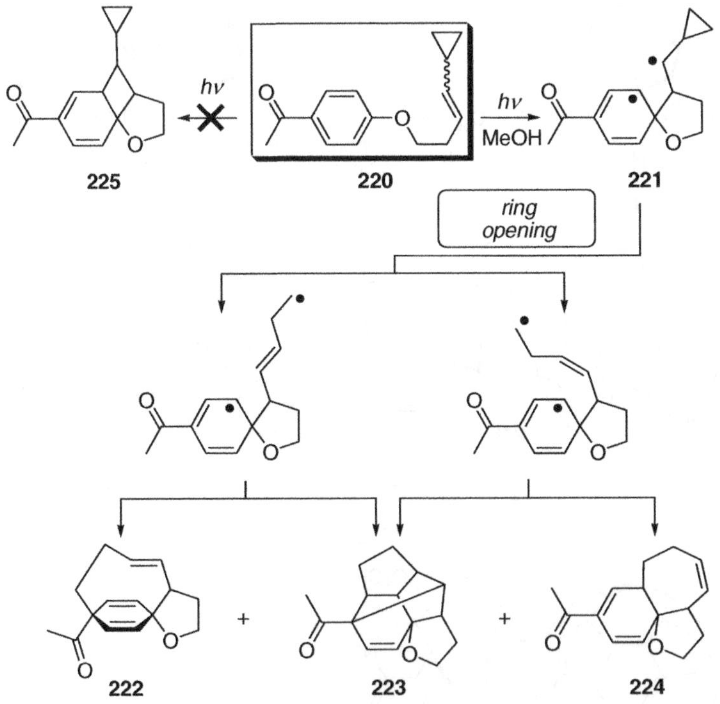

*Scheme 6.87*

*meta-Photocycloaddition*

The [3 + 2] photocycloaddition (Scheme 6.79) usually involves the ground-state alkene and the $S_1$ excited state of an electron-donor substituted benzene derivative, often via an exciplex intermediate.[807,809,811,816] The discrimination between the *ortho-* and *meta-* cycloaddition pathways is dependent on the electron donor–acceptor properties of the reaction partners and the position and character of the reactants' substituents.[807] The reaction typically produces many regio- and stereoisomers; however, a suitable structure modification can reduce their number. Intermolecular and intramolecular versions of the reaction are presented in Scheme 6.88: (a) photolysis of the mixture of anisole and 1,3-dioxole (**226**) leads to the formation of two stereoisomers, *exo-* and *endo-* **227**, in mediocre (~50%) chemical yields;[830] (b) four different isomers are obtained in the intramolecular photocycloaddition of an anisole derivative **228**. [831]

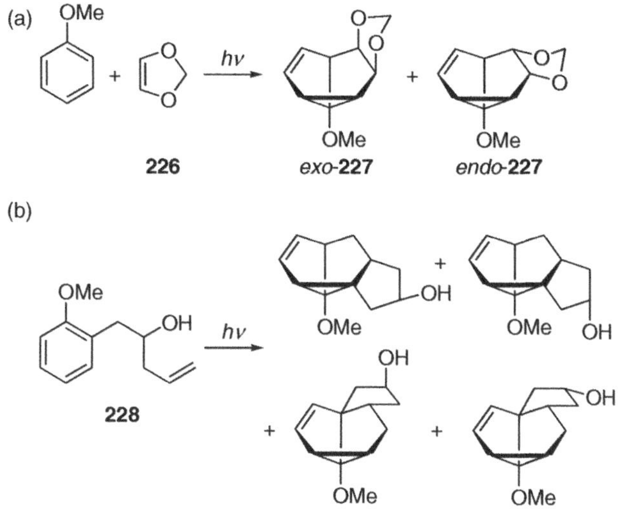

*Scheme 6.88*

*para-Photocycloaddition*

Photo-Diels–Alder ([4 + 2]) photocycloadditions (Scheme 6.79) are rare.[809,816] The asymmetric intramolecular cyclization of anthracen-9-ylmethyl (−)-menthyl fumarate (**229**) producing **230** in 56% diastereomeric excess[832] is one example (Scheme 6.89).

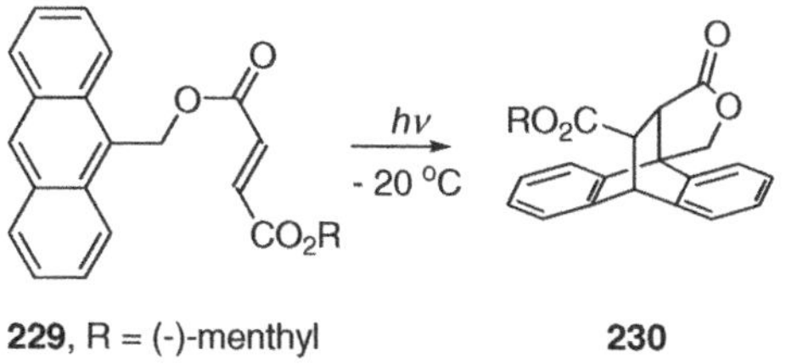

**229**, R = (-)-menthyl          **230**

*Scheme 6.89*

*Photocycloaddition Reactions of Two Aromatic Moieties*

Although intermolecular photocycloaddition of two benzene rings in the condensed phase has not been observed, this reaction is common for polycyclic aromatic hydrocarbons.[805,812] For example, anthracene-9-carbonitrile (**231**) in acetonitrile undergoes efficient [4 + 4] photocycloaddition with anthracene to give an adduct **232** in 94% chemical yield (Scheme 6.90).[833] This process is thermally reversible.

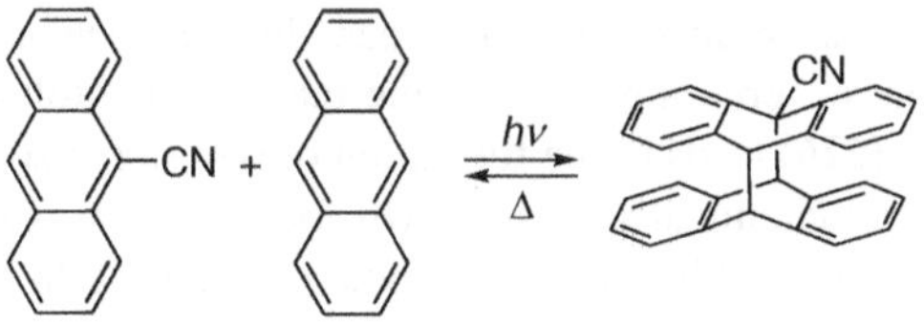

**Scheme 6.90**

However, when benzene or other aromatic rings are constrained in close proximity, photocycloaddition may result in the formation of unusual *cage compounds* (see also Special Topic 6.6).[725,805] For example, [3₄](1,2,3,5)cyclophane (**233**) undergoes [6 + 6] photocycloaddition to give the hexaprismane derivative **234** in 7% chemical yield[834] (Scheme 6.91).

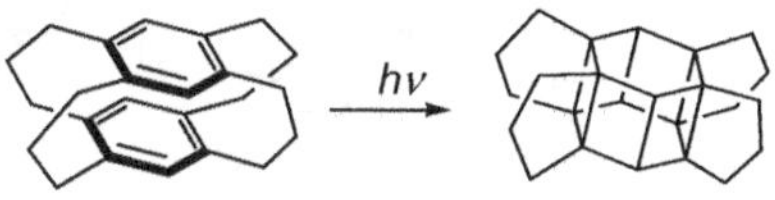

**Scheme 6.91**

---

## Case Study 6.14: Synthesis of cage compounds – octahedrane

Whereas [2.2]paracyclophane (**235**) is almost insensitive to irradiation, two benzene units in the diazacyclophane derivative **236** were found to photocyclize to the octahedrane **237** in one step (Scheme 6.92).[835] The nitrogen atom is apparently the key to the starting compound reactivity here. The bridges hold the aromatic rings closer together and through-bond coupling between the benzene π-orbitals and the C–N σ-orbitals are thought to facilitate an enhanced interaction between the rings in the excited state.

*Experimental details.*[835] A solution of diazacyclophane (**236**; 50 μmol) in benzene (10 ml) in a quartz tube was purged with nitrogen and irradiated in a Rayonet photoreactor (Figure 3.10) equipped with 16 fluorescent lamps ($\lambda_{irr} = 300$ nm, 25 W each) at room temperature for 36 h. The solution was concentrated under reduced pressure and the residue was separated repeatedly by preparative TLC to afford octahedrane in 33% chemical yield.

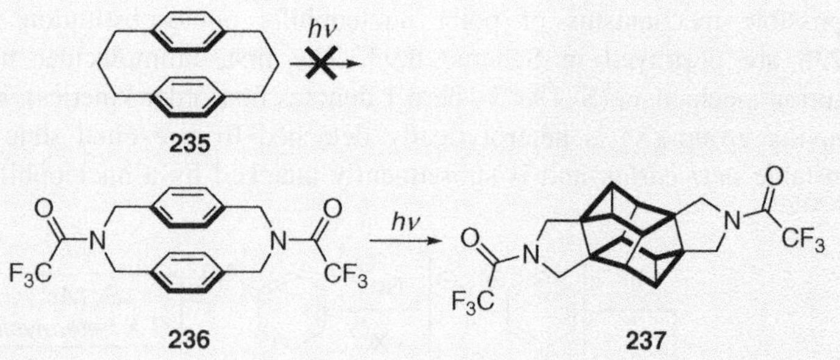

**Scheme 6.92**

### 6.2.3 Substituted Benzenes: Photosubstitution

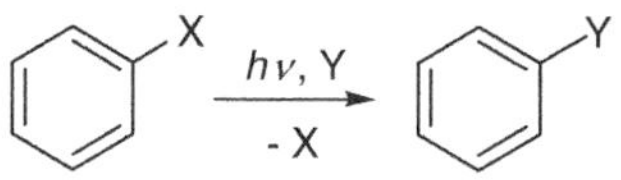

*Recommended review articles.*[836–842]
*Selected theoretical and computational photochemistry references.*[843–846]

Whereas electrophilic aromatic substitution typically occurs in ground-state aromatic compounds, nucleophilic substitution is the most common substitution reaction in excited aromatic compounds (generally depicted as $S_N Ar^*$, where $S =$ substitution, $N =$ nucleophilic and $Ar^* =$ excited aromatic).[836,838,839] Such a behaviour is related to the nature of the electronically excited state, in which one electron is promoted from HOMO to LUMO to form a new electrophilic site (half-filled HOMO) that can be attacked by a nucleophile or can accept an electron from a good electron donor. In contrast, half-filled LUMO has an electron donating character (Figure 6.3). Typical aromatic substrates in $S_N Ar^*$ reactions, such as aryldiazonium salts, aryl halides, aryl alkyl or aryl sulfonates, aryl nitriles or alkoxy/aryloxy arenes, bear good leaving groups. Photoinduced electron transfer (Section 5.2) is often involved in this process.

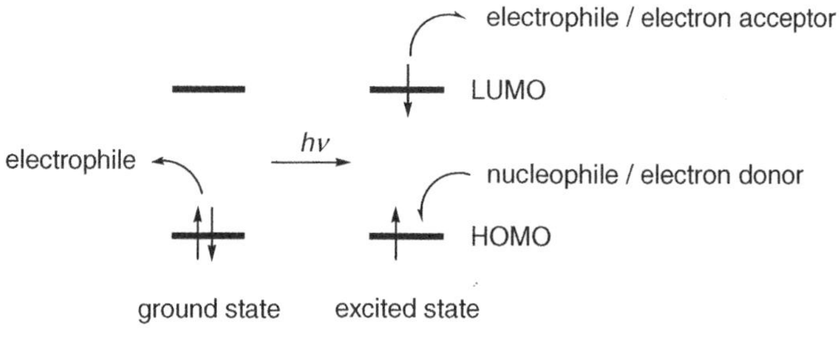

**Figure 6.3** Electrophilic and nucleophilic character of the excited state

Several possible mechanisms of polar nucleophilic photosubstitution in an aryl derivative **238** are portrayed in Scheme 6.93. The first, unimolecular nucleophilic photosubstitution mechanism ($S_N1Ar^*$, where 1 denotes first-order kinetics), in which an excellent leaving group (X) is heterolytically detached from excited state to form a relatively unstable *aryl cation* and is subsequently attacked by a nucleophile, is rarely observed.[836,838]

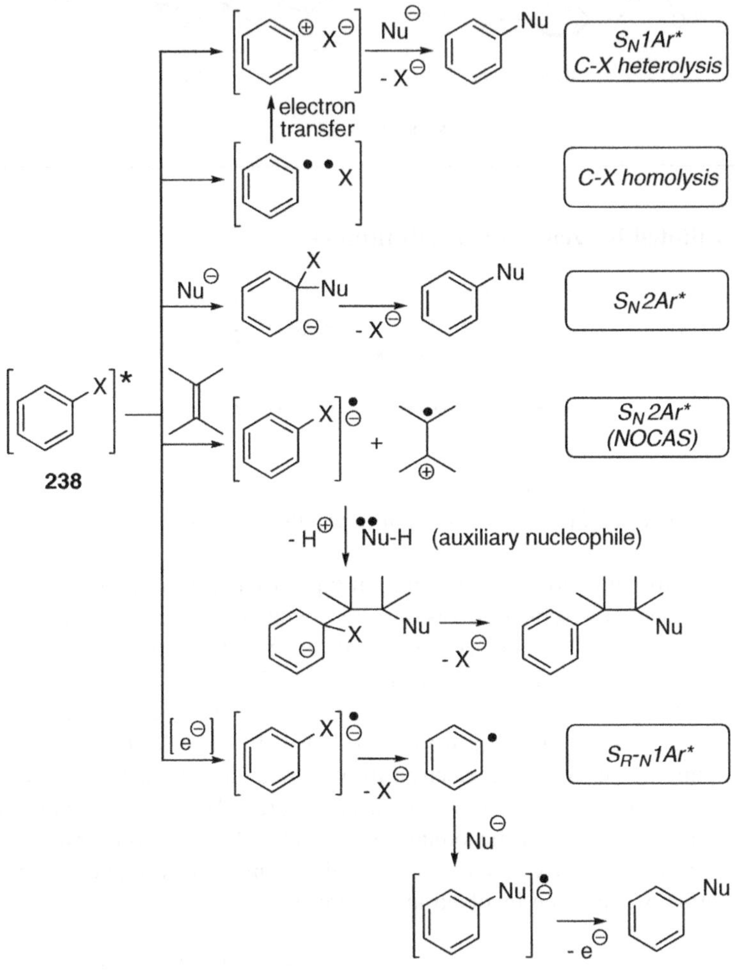

*Scheme 6.93*

Such a mechanism has been hypothesized for the reaction of aryl halides substituted by strong electron-donating groups.[838,845] 4-Chloroaniline (**239**), for example, reacts in the triplet state to give a phenyl cation **240**, apparently of triplet character ($\pi^5\sigma^1$), which has a selective reactivity toward π-, but not n-, nucleophiles (Scheme 6.94), in contrast to the unselective reactions of common singlet aryl cations.[847] Interestingly, the cation is added to an alkene to yield **241** even in nucleophilic methanol.

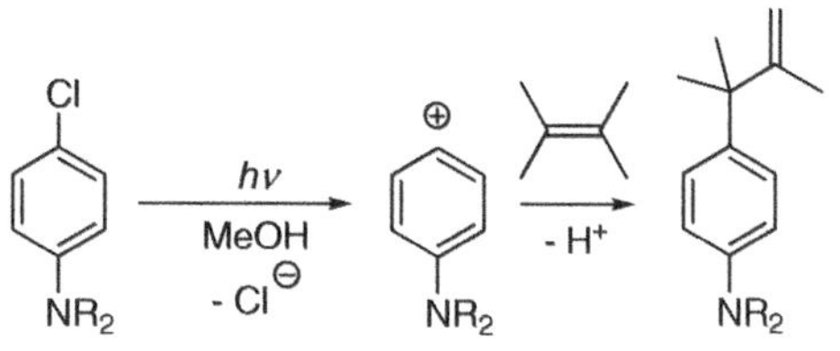

*Scheme 6.94*

In-cage ion pairs can also be formed by initial carbon–halogen photoinduced homolysis (see also Section 6.6.2), followed by an electron transfer step (C–X homolysis, Scheme 6.93). Chlorobenzene, for example, is converted photochemically to phenol in aqueous solutions (Scheme 6.95).[848,849] The hydrogen atom cannot be abstracted from water because of the high hydrogen–oxygen bond dissociation energy ($D_{O-H} = 498\,kJ\,mol^{-1}$); nevertheless, the high solvent dielectric constant promotes an in-cage electron transfer pathway.

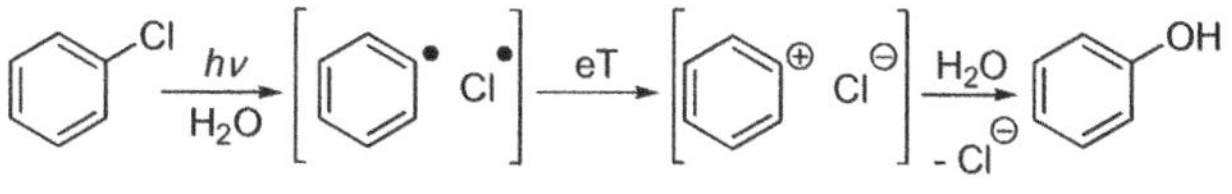

*Scheme 6.95*

Photosubstitution may proceed by direct attack of a nucleophile on the singlet or triplet excited state of an aromatic molecule to form a σ-complex ($S_N2Ar^*$; Scheme 6.93),[836,838] analogous to the Meisenheimer complex intermediate recognized in thermal $S_N2Ar$ reactions. A σ-complex is also known to be formed in the photochemical nucleophile olefin combination aromatic substitution process (*photo-NOCAS*).[837,850] This reaction involves a regioselective interconnection of three reactants: an aromatic electron acceptor (usually an aryl nitrile), an electron donor [olefin (alkene); acting as a π-nucleophile] and an 'auxiliary' nucleophilic species such as methanol. For example, excitation of 1,4-dicyanobenzene (**242**) to its lowest excited *singlet* state, which can also be co-sensitized, promotes electron transfer from an alkene **243** to give a contact radical ion pair (Scheme 6.96).[851] The alkene radical cation is then attacked by methanol to form the corresponding β-methoxyalkyl radical **244** after deprotonation. In the last step of the mechanism, this radical adds itself to the 1,4-dicyanobenzene radical anion at the *ipso* position, forming a σ-complex **245** and finally **246** in 17% chemical yield.

Irradiation of aromatic compounds in the presence of a good electron donor (nucleophile) may promote electron transfer from this species to the excited aromatic substrate in order to form an anion-radical intermediate, which releases the leaving group ($S_{R-N}1Ar^*$, where $R$ = radical, $_-$ = anion; Scheme 6.93).[836,838,840,852] In contrast, electron transfer from the excited aromatic substrate to good electron acceptors, followed by the reaction of the cation-radical intermediate thereby formed with a nucleophile, is possible when an electron-donating group is present on the aromatic substrate ($S_R+_N1Ar^*$, where $_+$ = cation; not shown).[836]

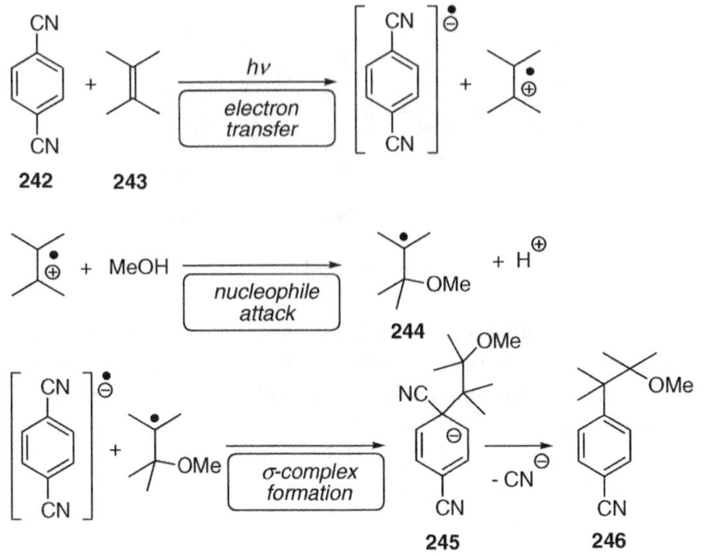

**Scheme 6.96**

Scheme 6.97 presents an example of a photoinduced substitution reaction of iodoben-zene (**247**) with a good electron donor, such as an acetone enolate (**248**), affording the product **249** via the $S_{R-N}1Ar^*$ mechanism in 88% chemical yield.[853] The radical anion species (PhI)$^{\cdot-}$ (**250**) formed in the first step is short lived and it readily releases the

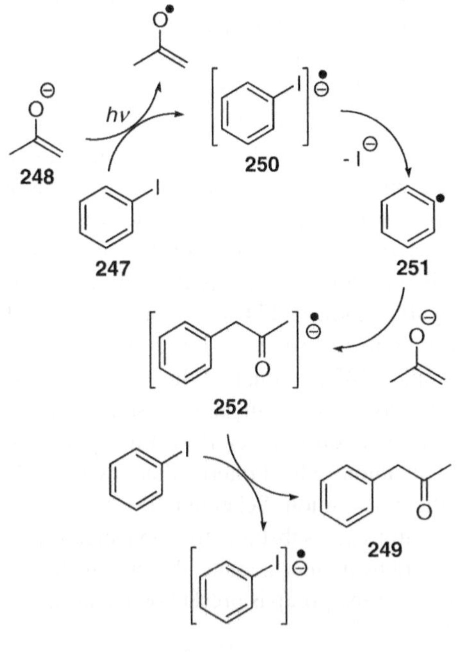

**Scheme 6.97**

halogen ion.[849] The resulting phenyl radical **251** couples with **248** (as a nucleophile) to form the radical anion **252**, which transfers an electron to iodobenzene in the propagation step of the chain mechanism. Aryl iodides are more susceptible to such reactions than aryl bromides or chlorides.

---

**Case Study 6.15: Mechanistic photochemistry – regioselectivity of photosubstitution**

The possibility of aryl–nitrogen bond formation by photosubstitution reaction has been investigated for use in photoaffinity labelling[854] experiments (see also Special Topic 6.16 in Section 6.4.2).[838] 4-Nitroanisole (**253**) presents an interesting model of an aromatic substrate, which could, in the excited triplet state, undergo nucleophilic substitution in the presence of an amine (Scheme 6.98).[855] The reaction was highly regioselective; *n*-hexylamine (**254**) gave rise to methoxy substitution with a maximum quantum yield of $\Phi = 0.018$, while ethyl glycinate substituted the nitro group (**255**) with $\Phi = 0.008$. Experiments with radical scavengers provided evidence that the former reaction occurs via electron transfer from the amine to the excited triplet state of 4-nitroanisole to form a radical ion pair, which undergoes a bimolecular $S_N2Ar^*$ reaction. Compared with alkylamine, the ionization potential of ethyl glycinate is too high to allow for efficient electron transfer to the triplet nitroanisole; furthermore, $S_N2Ar^*$ reactions are known to prefer the transition state that leads to the least stable σ-complex,[325] which is $NO_2$ group substitution in the present case.

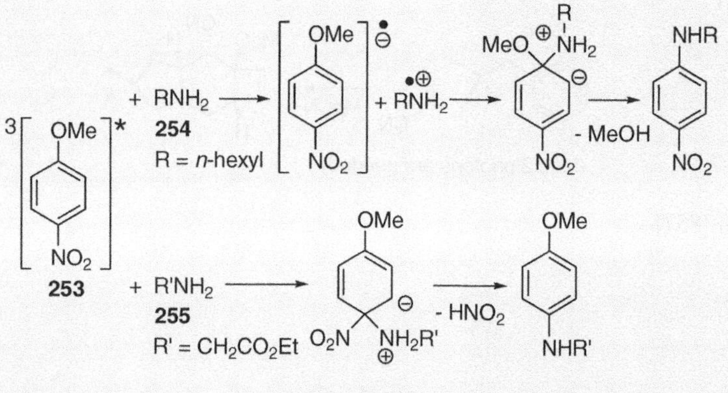

*Scheme 6.98*

---

*Experimental details*.[855] The quantum yields of the reactions shown in Scheme 6.98 were obtained by simultaneous irradiation of the corresponding 4-nitroanisole–amine solutions in methanol–water (20:80, v/v) and actinometer (Section 3.9.2) solutions (aqueous potassium ferrioxalate) in UV cells, placed in a merry-go-round apparatus (Figure 3.30). The samples were irradiated by passing the light from a medium-pressure mercury lamp (250 W) through a monochromator ($\lambda_{irr} = 366$ nm). The conversions were kept below 5% in all cases to avoid photoproduct interference. The 4-nitroanisole concentrations were determined by GC; the concentration of ferrous ions formed in the actinometer samples was determined using absorption spectroscopy after conversion to the coloured tris-phenanthroline complex.[157,214]

### 6.2.4  Problems

1. Explain the following concepts and keywords: phototransposition; photo-ring contraction; *meta*-photocycloaddition; photosubstitution; photo-NOCAS.
2. Suggest the mechanisms for the following reactions:

(a)

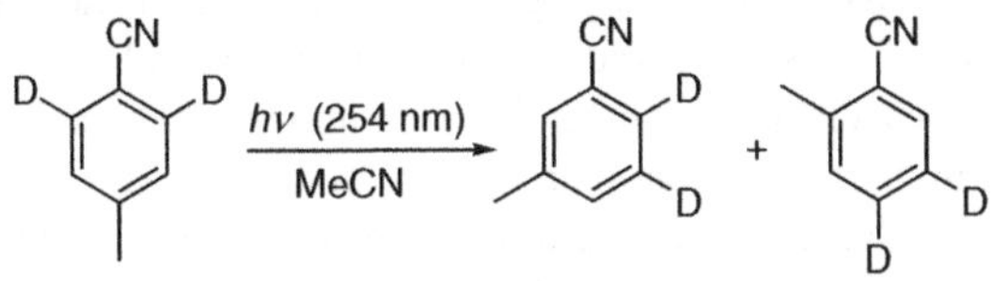

[ref. 856]

(b)

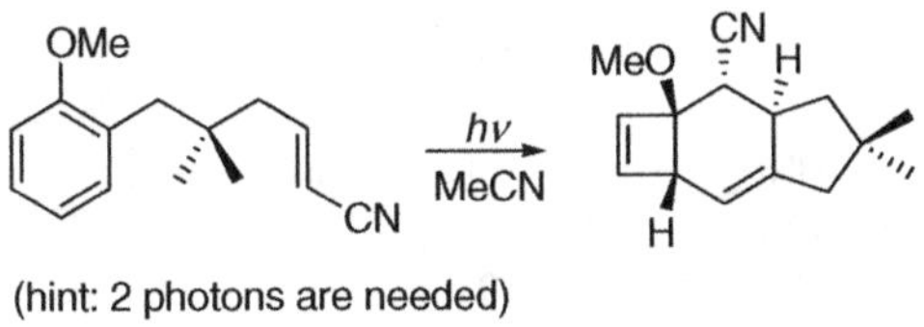

(hint: 2 photons are needed)

[ref. 857]

(c)

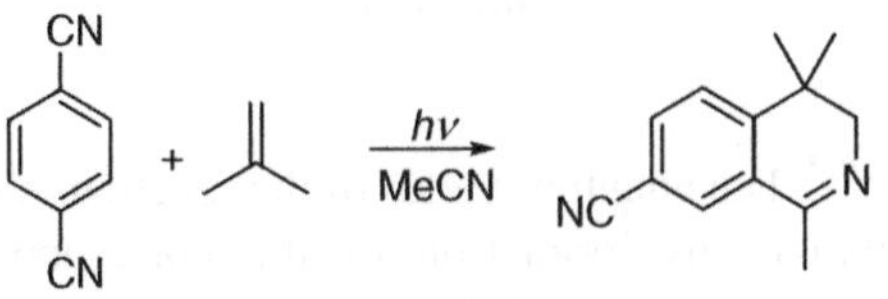

[ref. 858]

3. Predict the major photoproduct(s):

(a)

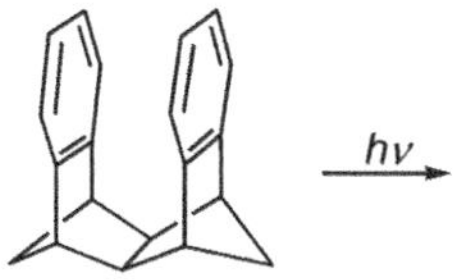

[ref. 859]

(b)

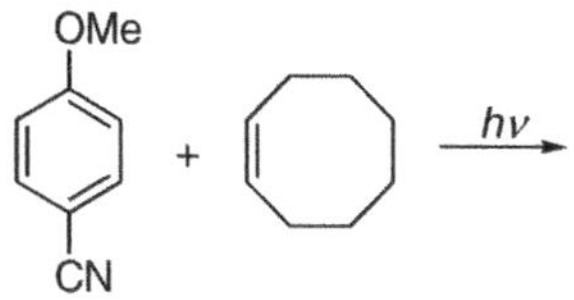

[ref. 860]

(c)

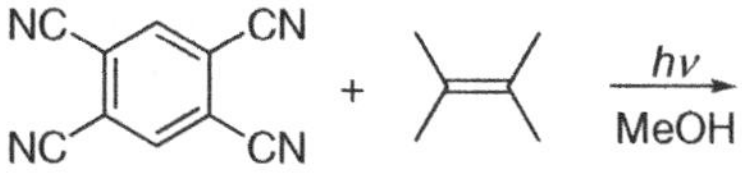

[ref. 861]

## 6.3 Oxygen Compounds

Oxygen is more electronegative than carbon but the nonbonding lone pairs (doubly occupied $n_\pi$-orbitals) of oxygen substituents ($-OH$, $-OR$, $-O^-$) act as a mesomeric electron donor. The absorption spectra of alternant hydrocarbons are not much affected by the inductive effect, but the conjugative interaction shifts their $\pi,\pi^*$ transitions to longer wavelengths. The $^1L_b$ bands are shifted more strongly than the $^1L_b$ bands and the shifts increase along the series $-OH$, $-OR$, $-O^-$. Electronic excitation of phenols involves a substantial amount of electron transfer from oxygen to the aromatic ring, particularly to the *meta*-positions.

Carbonyl compounds also have two nonbonding lone pairs on the oxygen atom. In organic chemistry texts, these are sometimes shown as two equivalent sp$^2$-hybridized lobes (rabbit's ears). While hybridization has no effect on the total energy, the two degenerate $n_{sp^2}$ orbitals are inappropriate as a basis set to discuss one-electron properties such as ionization potentials or n,$\pi^*$ transitions. Rather, the symmetry-adapted lone pairs

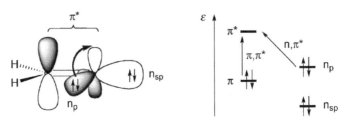

**Figure 6.4** The n,π* transition of formaldehyde

(see Section 4.4) $n_{sp}$ and $n_p$ should be used (Figure 6.4). Although situated on the electronegative oxygen atom and therefore at lower energy than a nonbonding orbital on carbon, the $n_p$ lone pair is higher in energy than the bonding π-MO. The first ionization potential of simple ketones is attributed to electron ejection from the $n_p$-orbital that is situated well above the $ns_p$-orbital having 50% s-character.

The lowest excited state of simple ketones and aldehydes corresponds to excitation of an electron from the $n_p$ lone pair to the π*-MO. The transition is forbidden in compounds of $C_{2v}$ symmetry. The 'local symmetry' is the same in compounds such as acetaldehyde, so that n,π* transitions of ketones and aldehydes are generally weak, $\varepsilon \approx 20\text{–}50\,M^{-1}\,cm^{-1}$, and they are easily overlooked in absorption spectra or hidden by the red edge of stronger π,π* absorption. The nature of the lowest excited state is, however, decisive for the photophysical properties and the photochemical reactivity of carbonyl compounds; the reactivity of n,π* excited ketones is comparable to that of alkoxy radicals (see below).

Solvent shifts are useful as a criterion to identify n,π* transitions in absorption spectra because hydrogen bonding of protic solvents with the carbonyl oxygen stabilizes the $n_p$ lone pair and gives rise to a hypsochromic shift of the n,π* absorption bands; see the positions of the n,π* absorption band of acetone in heptane and water (Figure 6.5, top). This contrasts with π,π* transitions that tend to be shifted bathochromically in polar solvents. Also, the photophysical and photochemical properties often serve to identify the nature of the lowest excited state. Lone-pair interaction in biacetyl splits the two $n_p$-orbitals giving rise to two n,π* transitions at $\tilde{v} = 2.23$ and $3.54\,\mu m^{-1}$. In the spectrum of 1,4-naphthoquinone in methanol, the n,π* band is barely detectable as a shoulder on the red edge of the π,π* absorption.

The position of the π,π* transitions in conjugated ketones may be estimated from the corresponding bands in hydrocarbons of the same topology; the inductive effect of oxygen on π,π* transitions in alternant hydrocarbons is small (Equation 4.26). Thus the π,π* transitions of acetophenone are at about the same position as those of styrene (Figure 6.1). However, the intensity of the first, parity-forbidden absorption band of styrene is increased by the inductive effect in acetophenone (Section 4.7) and the n,π* band is, of course, missing in the isoelectronic hydrocarbon.

The symmetry of n,π* excited configurations is different from that of the π,π* configurations (the wavefunctions are antisymmetric and symmetric, respectively, with respect to the plane of the $R_2CO$ moiety), so the two sets do not interact in CI calculations (Section 4.7) for planar molecules. The shifts of n,π* transitions with increasing conjugation tend to be smaller than those of the π,π* transitions. This is readily understood in terms of HMO or PMO theory: increased conjugation reduces the HOMO–LUMO gap by lowering the π*-MO and raising the π-MO, but only the lowering of the π*-MO affects

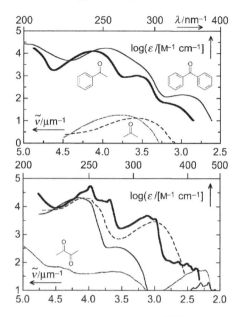

**Figure 6.5** Absorption spectra of prototype ketones.[280] Top: benzophenone (EtOH, —), acetophenone (EtOH, —), acetone (n-heptane, – – –), acetone (water, ····). Bottom: 1,4-benzoquinone (hexane, —), 1,4-naphthoquinone (MeOH, - - -), anthraquinone (cyclohexane, —), biacetyl (hexane, ····)

the n,π* transition. Hence the lowest singlet state is generally of π,π* character in large systems.

Intersystem crossing (ISC) to the triplet manifold is especially fast and efficient when a π,π* triplet state is energetically close to or slightly below the lowest singlet state of n,π* character (El Sayed's rules, Figure 2.8). This is the case for acetophenone, benzophenone, 1,4-benzoquinone and 9,10-anthraquinone. In these compounds, ISC takes place within a few picoseconds so that their quantum yields of ISC are near unity. ISC in saturated ketones is much slower ($\sim 10^{-8}\,s^{-1}$), because the lowest $^3\pi,\pi^*$ state lies well above the $^1 n,\pi^*$ state. The energy gap separating singlet and triplet π,π* configurations is much larger than that for n,π* states, especially in molecules of alternant topology (Section 4.7, Equation 4.33). Therefore, the situation $S_1(n,\pi^*)$ but $T_1(\pi,\pi^*)$, as in donor-substituted acetophenone and benzophenone derivatives, is not uncommon.

The shifts of n,π* transitions in related chromophores can be estimated using perturbation theory (Equations 4.13 and 4.14). The $n_p$-orbital is sensitive mostly to inductive perturbations. The π*-orbital is shifted both by inductive and mesomeric interactions; its AO coefficient at the carbonyl C atom is large (Figure 6.4). These qualitative considerations are supported by the data given in Table 6.5.

Table 6.6 lists the most important phototransformations discussed in this section. The carbonyl compounds (entries 1–6) are typical representatives of the photolabile compounds. Their reactions played an essential role in revealing the mechanisms of some primary photochemical steps. Thanks to their excellent absorption properties, thermal stability, usually uncomplicated synthesis and reaction diversity, they represent popular starting material in applied synthetic or material photochemistry and in photobiochemistry. It is

**Table 6.5** *Substituent effects on n,π\* transitions*

| Chromophore | R | Shift of $n_p$ | Shift of $\pi^*$ | $\lambda_{max}$/nm | $\tilde{v}$/μm$^{-1}$ |
|---|---|---|---|---|---|
| RMeC=O | Alkyl or H | Reference | Reference | 280 | 3.7 |
| | OR or NR$_2$ | ↓↓ | ↑↑ | <210 | >4.9 |
| | F | ↓↓ | ↑ | 220 | 4.4 |
| | SR or Cl | ↓ | ↑ | 240 | 4.2 |
| | Si(Me)$_3$ | ↑ | | 310 | 3.2 |
| RMeC=S | Alkyl | ↑↑ | | 500 | 2.0 |
| RMeC=NMe | Alkyl | ↓$^a$ | | 240 | 4.2 |
| RN=O | Phenyl | (↓) | ↓↓ | 650 | 1.5 |

$^a$Bent geometry at the nitrogen atom of imines; the lone pair has s-character.

obvious from Table 6.6 that the course of a photoreaction must be highly sensitive to the structure of carbonyl moiety in addition to the reaction conditions. The type and energy of the carbonyl compound excitation, and also the nature of the reagent, control intermolecular hydrogen (entry 1) or electron (entry 2) abstraction that lead to a variety of products, such as the corresponding alcohols or diols (photoreduction). While hydrogen is readily abstracted by singlet and triplet n,π\* excited states, both n,π\* and π,π\* excited molecules can be reduced in the presence of good electron donors. The excited n,π\* or π,π\* ketones or aldehydes may also undergo a [2 + 2] photocycloaddition (Paternò–Büchi) reaction to give oxetanes in the presence of an alkene (entry 3). When one or both α-bonds (OC–R) in a carbonyl moiety are sufficiently weak, they are homolytically cleaved to produce radical intermediates (entry 4) and eventually carbon monoxide. Alkyl ketones possessing a γ-hydrogen atom exhibit characteristic photoinduced intramolecular 1,5-hydrogen atom abstraction to form 1,4-biradical intermediates (entry 5). In the absence of reactive γ-hydrogens, intramolecular hydrogen abstraction from a non-γ-position may also occur (Section 6.3.5). A specific photochemically induced intramolecular process takes place on 2-alkylphenyl ketones to produce photoenols (entry 6).

The photochemical behaviour of another class of oxygen-containing compounds, carboxylic acids and their derivatives, has received only limited attention because of their usually inadequate absorption above 250 nm. They typically undergo *homolytic cleavage* to give photodecarboxylation products (entry 7).

Photochemistry of transition metal carbonyl complexes as the borderline between organic and inorganic chemistry is mentioned in Section 6.3.9. Since the dissociation energy of a common metal–carbonyl oxide bond is usually low, photodecarbonylation, that is, release of the CO molecule, is the most common photoprocess observed (entry 8).

### 6.3.1 Carbonyl Compounds: Photoreduction

*Recommended review articles.*[751,862–865]
*Selected theoretical and computational photochemistry references.*[16,534,535,866–870]

**Table 6.6** *Examples of photoprocesses involving excited oxygen-atom containing compounds.*

| Entry | Starting material[a] | Product(s) | Mechanism | Section |
|---|---|---|---|---|
| 1 | + [H] | | Intermolecular H-transfer (photoreduction) | 6.3.1; 6.3.7 |
| 2 | + [e⁻] | | Intermolecular e-transfer (photoreduction) | 6.3.1; 6.3.7 |
| 3 | + | | [2 + 2] cycloaddition (oxetane formation; Paternò–Büchi reaction) | 6.3.2; 6.3.7 |
| 4 | | | α-Cleavage (Norrish type I reaction) | 6.3.3 |
| 5 | | | Intramolecular H-transfer (Norrish type II reaction) | 6.3.4; 6.3.5 |
| 6 | | | Photoenolization | 6.3.6 |
| 7 | [R-COOH]* | R• + •COOH | Photofragmentation | 6.3.8 |
| 8 | | + CO | Photodecarbonylation | 6.3.9 |

[a][H] = hydrogen atom donor; [e⁻] = electron donor.

Hydrogen abstraction (Section 4.9) by excited carbonyl compounds is one of the most fundamental reactions in organic photochemistry.[862–864] In a classical[871] photoreduction reaction, an excited carbonyl compound, such as a ketone, undergoes hydrogen abstraction from a hydrogen donor [H] to form the ketyl radicals, which subsequently abstract another

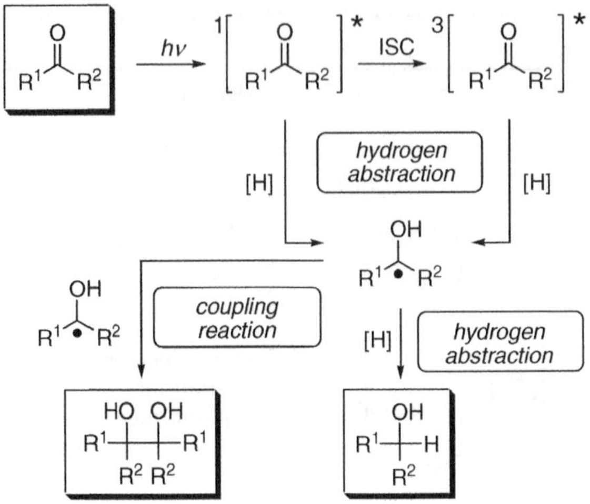

**Scheme 6.99**

hydrogen atom from the environment or recombine to form either alcohols or diols (pinacols, **256**) (Scheme 6.99).

Photoreduction of aliphatic ketones may involve both singlet and triplet excited states, but the quantum yield for product formation via singlet is usually low because other competing processes, such as radical pair recombination, are involved. The rapid intersystem crossing (ISC) in aryl ketones (Section 2.1.6) allows triplet reactivity. Ketones with $n,\pi^*$ lowest triplets, having an unpaired electron localized in an n-orbital on oxygen, are far more reactive than those with $\pi,\pi^*$ lowest triplets.[863]

The rate constants of the hydrogen abstraction by triplet $n,\pi^*$ excited carbonyl compounds are strongly dependent on the reaction thermodynamics. (1) The higher the excitation energy of the hydrogen acceptor, the faster is the reaction. For example, triplet excited $\alpha$-diketones,[872] such as biacetyl ($E_T = 236$ kJ mol$^{-1}$), will generally undergo much slower (less exothermic) hydrogen abstraction than phenyl ketones ($E_T \approx 300$ kJ mol$^{-1}$). (2) Correspondingly, low dissociation energies of the H–S bond ($D_{S-H}$) in the hydrogen-atom donors favour the abstraction, which parallels the energetics of hydrogen abstraction by simple alkoxy radicals.[873] For example, benzophenone triplet ($E_T = 288$ kJ mol$^{-1}$) will abstract hydrogen from the secondary C–H bond of propan-2-ol ($D_{C-H} = 381$ kJ mol$^{-1}$) more rapidly than from the corresponding O–H bond ($D_{O-H} = 442$ kJ mol$^{-1}$) or the C–H bond of neopentane ($D_{C-H} = 420$ kJ mol$^{-1}$), but more slowly than from tributylstannane ($D_{Sn-H} = 326$ kJ mol$^{-1}$) (Table 6.7).

Considerable photoreductive reactivity is also observed for substrates which are poor hydrogen donors but good electron donors (i.e. possessing a low $E_i$) (Table 6.7). Such compounds are able to reduce efficiently both $\pi,\pi^*$ and $n,\pi^*$ triplets via partial or complete electron transfer (Section 5.2) followed by proton transfer. Amines are very common electron donors; depending on their structure, either an N–H or C–H bond is cleaved via a

**Table 6.7** *Rate constants (k) for bimolecular reaction of triplet benzophenone*
*($E_T = 288$ kJ mol$^{-1}$) with various substrates (**H** to be abstracted is shown in bold)*

| Substrate | $k/10^6$ M$^{-1}$ s$^{-1}$ | $D_{S-H}{}^a$/kJ mol$^{-1}$ | $E_i{}^a$/eV |
|---|---|---|---|
| Neopentane | 0.04 | 420 | – |
| **H**–CH$_2$OH | 0.2 | 402 | 10.9 |
| C$_6$H$_5$CH$_2$–**H** | 0.5 | 370 | 8.8 |
| **H**–C(OH)(CH$_3$)$_2$ | 1.9 | 381 | 10.1 |
| $n$-Bu$_3$Sn–**H** | 1300 | 326 | – |
| Ph–O**H** | 1300 | 368 | 8.5 |
| Et$_2$N–CH$_2$CH$_3$ | 3000 | 377 | 7.3 |

$^a D_{S-H}$ are bond dissociation energies; $E_i$ are ionization potentials.[157,863,874]

triplet exciplex or an radical ion pair to form the same primary products [Scheme 6.100; pathway (a) or (b), respectively].[669,875,876] Reduction of aromatic ketones by amines in aqueous media usually yields secondary alcohols rather than pinacol derivatives.

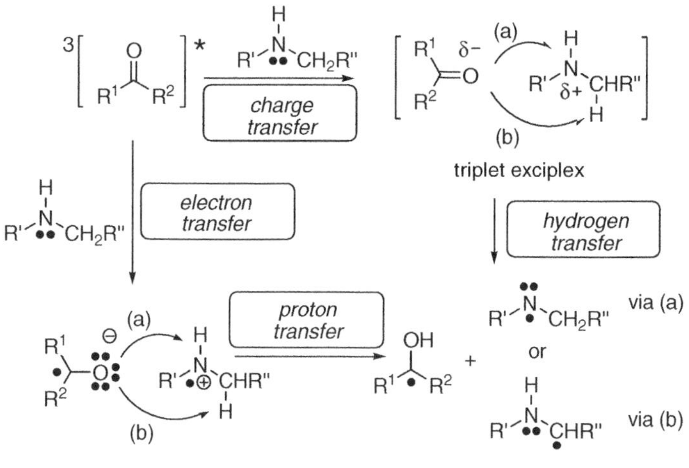

**Scheme 6.100**

---

## Case Study 6.16: Chemistry in ionic liquids – photoreduction

The amine-mediated photoreduction of benzophenones in ionic liquids at 20 °C provides the corresponding benzhydrol derivatives.[877] Unlike the analogous reactions in common organic solvents, where pinacol derivatives are predominately formed, 2-methoxycarbonylbenzophenone (**257**), irradiated in the presence of 2-butylamine in an ionic liquid, 1-butyl-3-methylimidazolium tetrafluoroborate [EMI(OTf)], gave a nearly quantitative isolated chemical yield of methyl 2-[hydroxy(phenyl)methyl] benzoate (**258**) (Scheme 6.101). The authors suggest that benzhydrols are produced via a stabilized ion pair, formed from amine and ketone by single electron transfer. Ionic liquids are considered to be environmentally friendly alternatives to conventional solvents.[878]

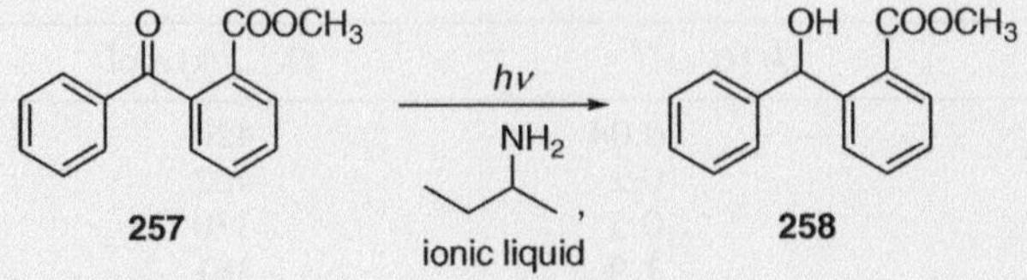

**Scheme 6.101**

*Experimental details.*[877] An EMI(OTf) solution of **257** (25 mM) in a closed Pyrex vessel was purged with argon for 1 h and 2-butylamine solution (1 M) was added via a syringe. The reaction mixture was shaken vigorously, irradiated using a medium-pressure Hg lamp (450 W) at $\lambda_{irr} > 290$ nm (Pyrex optical filter; Figure 3.9) for several hours and the resulting mixture was extracted with diethyl ether. The organic layer was washed with dilute HCl to remove the excess amine. The extracts were dried over MgSO$_4$, concentrated *in vacuo* and the product was purified by column chromatography.

## 6.3.2   Carbonyl Compounds: Oxetane Formation (Paternò–Büchi Reaction)

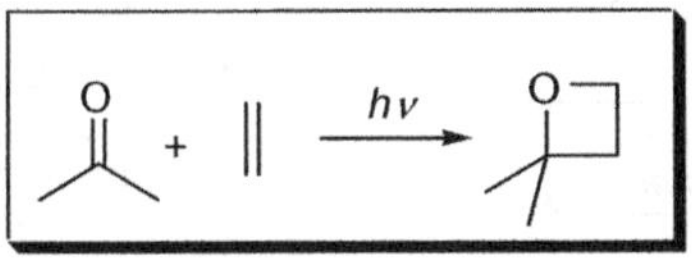

*Recommended review articles.*[584,602,689,751,879–883]
*Selected theoretical and computational photochemistry references.*[16,534,884–886]

The photochemical [2 + 2] cycloaddition of an alkene to an excited carbonyl compound, termed the Paternò–Büchi reaction,[887,888] leads to an oxetane moiety. Although *both singlet* and triplet excited species are known to produce cycloadducts, the scope and regio- and stereoselectivity of this process depend on the reactant multiplicity.[879,880,882] The reaction may involve a concerted reaction or a stepwise mechanism via 1,4-biradical intermediates (BR) and the products are formed from either the C···O or C···C initial attack producing two different triplet biradicals (Scheme 6.102).[879,889] Exciplex, radical ion pair and zwitterion intermediate formation prior to oxetane cyclization have also been noted in some cases.[890,891] This transformation may compete with inter- (Section 6.3.1)/intramolecular hydrogen abstraction (Section 6.3.4) or α-cleavage (Section 6.3.3) of the carbonyl moiety or with the reactions involving the alkene molecule. The course of the reaction then depends on many factors. The carbonyl compound should be the only absorbing chromophore during the irradiation and photostable in the absence of alkene under the given reaction conditions.

The attack of n,π* excited carbonyl compounds on electron-rich or electron-poor alkenes can be qualitatively different.[892,893] Electron-rich alkenes preferentially interact with the electrophilic half-filled n-orbital on the oxygen atom, which is perpendicular to

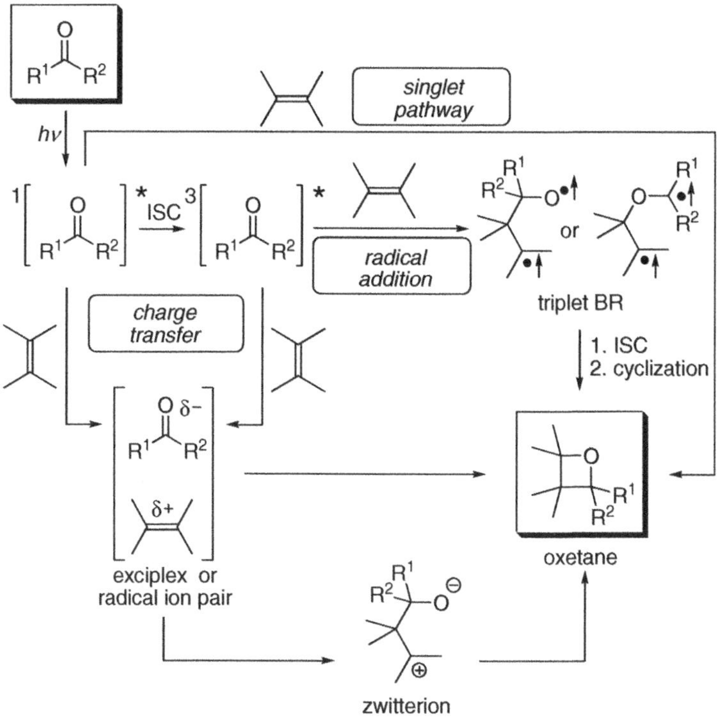

*Scheme 6.102*

the π-plane (perpendicular approach), whereas electron-deficient alkenes attack either the oxygen atom or the nucleophilic carbon atom (parallel approach) (Scheme 6.103). The observed photoproduct stereochemistry cannot, however, distinguish between these mechanisms; but a state correlation diagram analysis (Section 4.9) reveals that the C-atom attack is favoured via a parallel approach.[892] Oxetane formation is also observed in the reaction of π,π* excited carbonyl compounds, such as 4-phenylbenzophenone, with alkenes,[891] although the yields are low (see below).

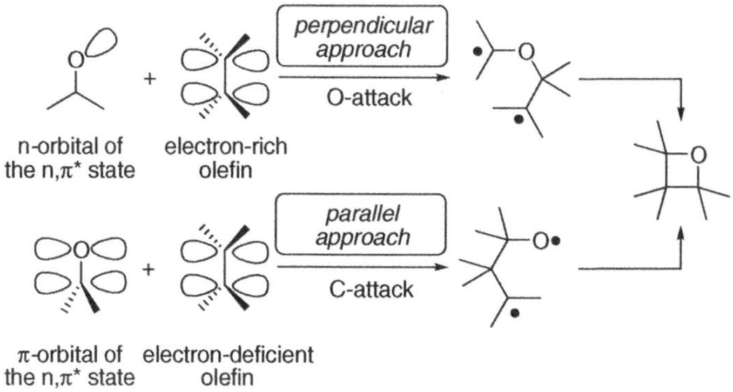

*Scheme 6.103*

**Table 6.8** Fluorescence quenching of norbornan-2-ones by substituted ethylenes

| Ketone | $k_q/(10^9\,M^{-1}\,s^{-1})^a$ | |
|---|---|---|
| | CN / NC (ethylene) | EtO OEt |
| (structure: norbornan-2-one) | 5.1 | 1.2 |
| (structure: methyl norbornan-2-one) | 2.3 | 0.9 |
| (structure: dimethyl norbornan-2-one) | 1.0 | 1.5 |
| (structure: dimethyl norbornan-2-one isomer) | 0.5 | <0.03 |

$^a$Fluorescence quenching rate constants.[892]

The rate constants and selectivities of the Paternò–Büchi reaction are strongly influenced by steric effects.[892] Fluorescence quenching (reflecting the first step to oxetane formation) of singlet excited norbornan-2-one derivatives is reduced by increasing the substitution in the vicinity of the carbonyl group (Table 6.8). The rate constants of quenching ($k_q$) by electron-deficient (E)-1,2-dicyanoethylene decrease when the methyl substituents hinder access to the π-plane of the carbonyl group (C-atom attack), whereas those of electron-rich (Z)-1,2-diethoxyethylene are sterically sensitive to the n-plane approach of the carbonyl oxygen (O-atom attack). Interestingly, a similar 'parallel approach' dependence is observed for the ground-state reduction of norbornan-2-ones by NaBH$_4$, which is known to transfer the hydride to the carbonyl carbon atom in the rate-determining reaction step.[894]

In another example, the reaction selectivity is pronounced by enlarging the *ortho* substituents of the phenyl ring in triplet excited benzaldehyde, which reacts with dihydrofuran (Scheme 6.104).[895] Surprisingly, the cycloaddition produces an excess of the thermodynamically less stable *endo*-diastereomers, which was explained as formation of specific triplet biradical geometries that undergo rapid intersystem crossing.

The lowest excited singlet states of aliphatic aldehydes and ketones have lifetimes on the order of nanoseconds, but they can be trapped by alkenes in a diffusion-controlled bimolecular oxetane formation. According to a theoretical study, a C-atom attack mechanism is either a concerted process producing oxetane directly or it involves a C–C bonded transient singlet biradical intermediate that rapidly cyclizes.[896] The O-atom attack, in contrast, represents a nonconcerted path, allowing conformational motion of the short-lived intermediate thereby formed.

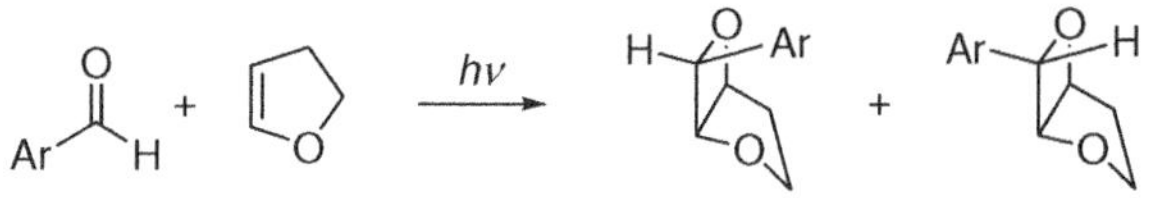

|              | endo | exo |
|--------------|------|-----|
| Ar = Ph | 88% | 12% |
| Ar = 2-methylphenyl | 92% | 8% |
| Ar = 2,4-di-*t*-butylphenyl | 98% | 2% |

*Scheme 6.104*

The triplet biradical intermediates (Scheme 6.103) formed from the triplet excited species have to undergo a spin change to reach the singlet energy surface necessary for the radical recombination. It enhances the intermediate lifetime and bond rotation may be responsible for a considerable lack of reaction selectivity.[895] This is illustrated on the following example of (*E*)-but-2-ene cycloaddition with excited aldehydes (Scheme 6.105). The reaction of singlet excited n,π* acetaldehyde is highly stereoselective (*E*-isomers are preferred), producing a long-lived triplet biradical intermediate, in contrast to that of triplet n,π* benzaldehyde.[897] Interestingly, the singlet state of 2-naphthaldehyde is responsible for a highly stereoselective photocycloaddition, because despite the fact that intersystem crossing to the corresponding π,π* triplet is fast and efficient, the triplet decays nonproductively to the ground state.[898]

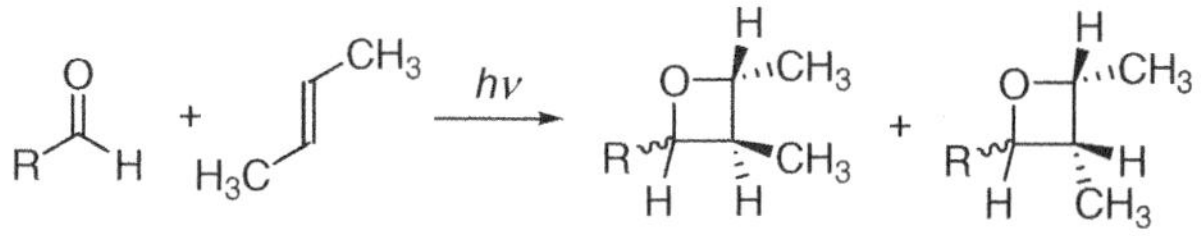

| | | |
|---|---|---|
| R = CH$_3$ ($\Phi$ = 0.13) | 95% | 5% |
| R = Ph ($\Phi$ = 0.53) | 67% | 33% |
| R = 2-naphthyl ($\Phi$ = 0.02) | 94% | 6% |

*Scheme 6.105*

Solvent polarity can trigger a photoinduced electron-transfer (PET) (Section 5.2) step in the Paternò–Büchi reaction of benzaldehyde to dihydrofuran, thus affecting the reaction regio- and diastereoselectivity (Scheme 6.106).[899] Whereas the reaction in benzene proceeds via a triplet biradical intermediate, a radical ion pair is formed in acetonitrile. The electron transfer alters the charge distribution in the reactants, which promotes the formation of two different regioisomers with inverted diastereoselectivity.

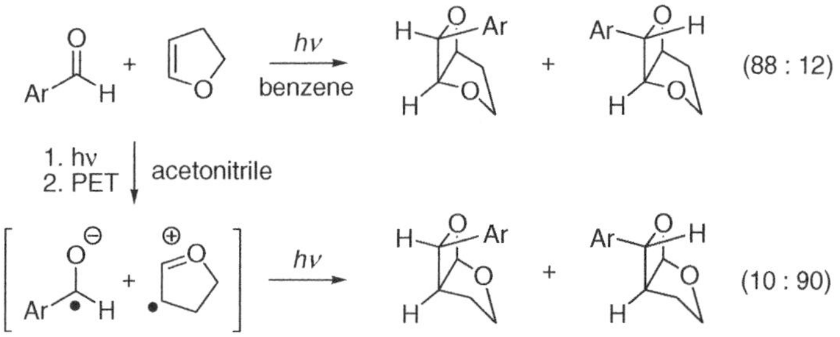

*Scheme 6.106*

## Case Study 6.17: Asymmetric synthesis – photocycloaddition

Hydrogen bonding between the ground-state reactants dihydropyridone (**259**) and an aldehyde **260** was found to be responsible for a high facial diastereoselectivity (>90% de) (see also Special Topic 6.3) in the Paternò–Büchi reaction (Scheme 6.107).[900,901] X-ray analysis, furthermore, revealed that the corresponding lactam moieties in the racemic product **261** are intramolecularly hydrogen-bonded (dashed).

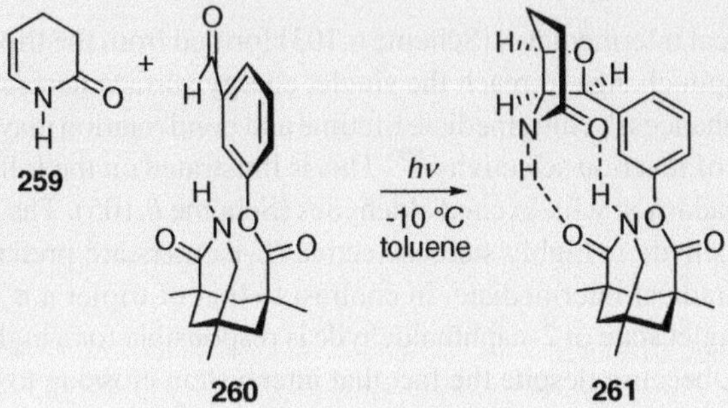

*Scheme 6.107*

*Experimental details.*[901] A solution of **260** (0.24 mmol) and **259** (0.49 mmol) in toluene (20 ml) was irradiated at −10 °C with a high-pressure mercury lamp (150 W) at $\lambda_{irr} = 300$ nm (Duran filter) in an immersion well apparatus (Figure 3.9). The solvent was then removed *in vacuo* and the residue containing the crude product was purified by column chromatography to give **261** in 56% yield. The diastereomeric ratio of oxetane isomers was determined using [1]H NMR.

## Case Study 6.18: Synthesis of cage compounds – merrilactone A analogue

A six-step synthetic approach to the tetracyclic skeleton of neurotrophic sesquiterpene merrilactone A, using intramolecular photocycloaddition to form the oxetane ring, was reported.[902] Irradiation of a degassed acetonitrile solution of **262** gave the product **263** (Scheme 6.108) in a very high yield (93%). This reaction creates three stereocentres in two new rings, forming an oxa[3.3.3]propellane framework.

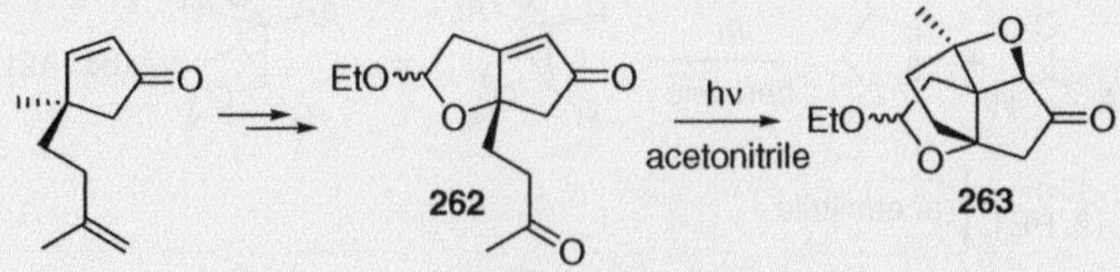

*Scheme 6.108*

*Experimental details.*[902] An acetonitrile solution of **262** (0.54 mм) was purged with nitrogen for 1 h. The solution was then immersed in an ice-bath and irradiated with a medium-pressure mercury lamp (400 W) using a Pyrex filter (Figure 3.9). The reaction was monitored by TLC and was stopped after 3 h. The solvent was evaporated and the product was purified by column chromatography.

### 6.3.3 Carbonyl Compounds: Norrish Type I Reaction

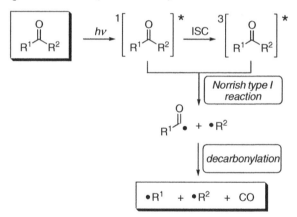

*Recommended review articles.*[751,903–910]
*Selected theoretical and computational photochemistry references.*[16,187,534,535,911–916]

Homolytic cleavage of the α-bond (α-cleavage; Norrish type I reaction[917]), often followed by decarbonylation of the acyl radical intermediate thereby formed[909] (Scheme 6.109), is one of the most common reactions of excited ketones.[903,905] This reaction can be accompanied by competing processes, such as the Norrish type II reaction (Section 6.3.4) or photoreduction (Section 6.3.1).

*Scheme 6.109*

There is a good correlation between the bond dissociation energies ($D_{C-CO}$) and the corresponding rates of α-cleavage and subsequent decarbonylation, so that the quantum yield of the reaction is directly related to the stability of the radicals formed. For example, the excitation energies of acetone ($E_S \approx 373\,kJ\,mol^{-1}$, $E_T \approx 332\,kJ\,mol^{-1}$) and acetophenone ($E_S \approx 330\,kJ\,mol^{-1}$, $E_T \approx 310\,kJ\,mol^{-1}$) are usually sufficient for exothermic release of stabilized benzyl or *tert*-butyl radicals (Figure 6.6, Table 6.9), whereas formation of methyl or phenyl radicals is inefficient.

In general, the n,π* excited ketones undergo much faster cleavage than those with a lowest π,π* excited state, because the σ-orbital of a bond being cleaved overlaps with the half-vacant n-orbital on the oxygen atom.[905,911,919] The cleavage rate constant of the n,π* triplet excited benzyl phenyl ketone (PhCH₂COPh) is, for example, more than three orders

**Table 6.9**  Quantum yields ($\Phi_I$) and rate constants of the Norrish type I cleavage ($k_I$)[a].

| Ketone | $\Phi_I$ | $k_I/10^8\,s^{-1}$ |
|---|---|---|
| $PhCH_2-COPh$ | ~0.4 | 0.02 |
| $t\text{-Bu}-COCH_3$ | ~0.3 | >10 |
| $t\text{-Bu}-CO(4\text{-BP})$[b] | <0.001 | <0.001 |
| Cyclobutanone | ~0.3 | >10 |
| Cyclopentanone | ~0.2 | 2 |

[a]Adapted from refs 905 and 918.
[b]4-BP = biphenyl-4-yl.

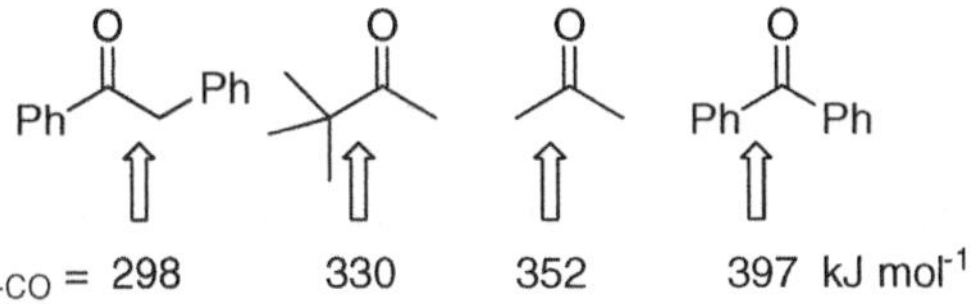

$D_{C\text{-}CO}$ = 298          330          352          397 kJ mol$^{-1}$

**Figure 6.6**  Bond dissociation energies in ketones

of magnitude higher than that of the $\pi,\pi^*$ triplet of biphenylyl *tert*-butyl ketone [*t*-BuCO (4-BP)][919] (Table 6.9). Energy barriers for α-cleavage in aliphatic $n,\pi^*$ ketones in solutions range from nearly 0 to 65 kJ mol$^{-1}$, depending on the ketone structure and the spin multiplicity of the excited state.[920] The cleavage rate constants of singlet $n,\pi^*$ excited ketones are usually high ($k_I = 10^8$–$10^9\,s^{-1}$) and they compare with those of intersystem crossing. In the case of cyclic ketones, the release of the ring-strain energy during α-cleavage increases the exothermicity of the process.[912] Cyclobutanone ring opening is, therefore, faster than that of cyclopentanone (Table 6.9).

Triplet $n,\pi^*$-state cleavage reactions (typically for aromatic ketones, $\Phi_{ISC}$ of which is usually 1.0; Section 2.1.6) are more efficient due to the longer triplet lifetimes and relatively large cleavage rate constants. Furthermore, recombination of the primary triplet radical pair formed is spin forbidden, which allows the radicals to escape the solvent cage. The photochemical *racemization* of the chiral phenylpropiophenone **264**, for example, was found to depend on a partitioning between in-cage radical recombination and diffusion rate constants (Scheme 6.110).[921]

The decarbonylation rates of acyl radicals, the primary α-cleavage intermediates, are also related closely to the stability of the corresponding alkyl radicals formed, and hence to the magnitude of the bond dissociation energies ($D_{C\text{-}CO}$).[903] Fast CO release is observed for the benzyl and *tert*-butyl radicals, whereas the formation of alkyl and particularly unstable aryl radicals is exceedingly slow (Table 6.10). An unsymmetrical ketone $PhCOCH_2Ph$, for example, efficiently produces benzoyl and benzyl radicals upon photolysis, which undergo subsequent reactions. Decarbonylation to release carbon monoxide does not occur in this case because the Ph−CO bond cleavage is energetically unfavourable.

The Norrish type I reaction of acyclic and cyclic ketones in solution typically results in recombination, decarbonylation and disproportionation (hydrogen abstraction) products.[903] For example, irradiation of di-*tert*-butyl ketone (**265**) in hexane solution provides nearly exclusively decarbonylation products from both the singlet and triplet states (>90% chemical yield), whereas the carbonyl group-containing products are produced only in traces (Scheme 6.111).[922,923]

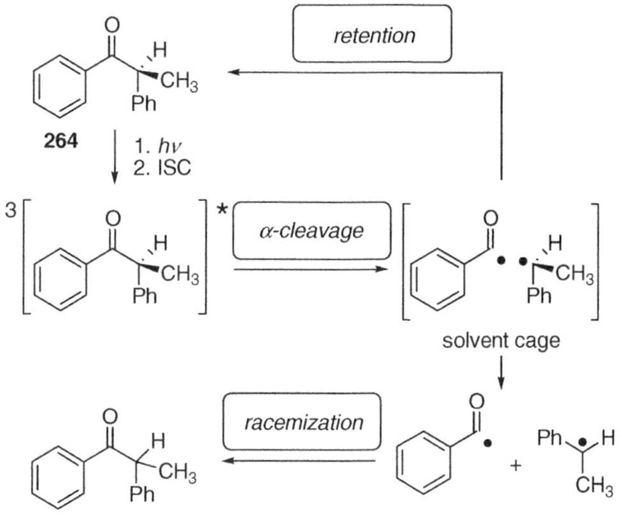

**Scheme 6.110**

**Table 6.10** *Activation energies ($E_a$) and rate constants of decarbonylation ($k_{-CO}$)[a]*

| Ketone | $E_a$/(kJ mol$^{-1}$) | $k_{-CO}$/s$^{-1}$ |
|---|---|---|
| PhCH$_2$–C$^\bullet$(=O) | 29 | $8.1 \times 10^{6}$[b] |
| t-Bu–C$^\bullet$(=O) | 46 | $8.3 \times 10^{5}$[b] |
| Me–C$^\bullet$(=O) | 71 | $4.0$[c] |
| Ph–C$^\bullet$(=O) | ~109 | $1.5 \times 10^{-7}$[c] |

[a]Adapted from ref. 903.
[b]In a nonpolar solvent.
[c]In the gas phase.

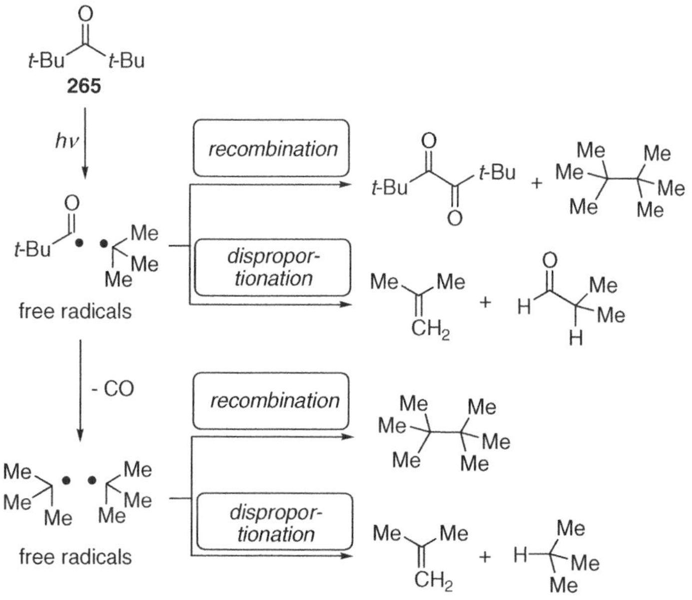

**Scheme 6.111**

## Special Topic 6.11: Cage effects

Organized and constraining media can control the course of photoreactions by affecting the shape and reactivity of guest molecules (cage effects).[172,173,713,924] Zeolites or other nanoporous solid materials, for instance, restrain conformational and translational changes of the reactants and intermediates in their *'hard'* reaction cavities so that the reactions often proceed with high regio- and stereoselectivity (see also Special Topic 6.5). On the other hand, media such as micelles, microemulsions and liquid crystals, the cavity shape of which changes over time (*'soft'* reaction cavities), increase the number of degrees of freedom available to reactant(s) and reaction intermediates. Polar interactions of guest molecules with the walls of the cavities may also alter the course of the reaction. The following example represents a photoreaction, which has been used to study cage effects in many types of constraining media (for other examples, see Case Studies 6.27 and 6.37).

The triplet state of dibenzyl ketone ($PhCH_2COCH_2Ph$) undergoes rapid primary α-cleavage ($k \approx 10^9 \, s^{-1}$) followed by slower decarbonylation ($k \approx 10^7 \, s^{-1}$).[925,926] The triplet benzyl radical pairs thus produced must intersystem cross to singlet radical pairs prior to recombination. The photodecarbonylation of the nonsymmetrical 1-phenyl-3-*p*-tolylpropanone (**266**) in homogeneous solutions provides three different 1,2-diphenylethane products, **267**, **268** and **269**, in a (statistical) 1:2:1 ratio (Scheme 6.112),[925] whereas specific product distribution is usually observed in a heterogeneous environment[172] due to restricted translational diffusion of the radical intermediates. The photochemistry of **266** was studied, for example, in aqueous solutions containing hexadecyltrimethylammonium chloride (CTAC) as a detergent.[927] The detergent molecules are known to self-organize in micelles with hydrocarbon-like interior, capable of embracing hydrophobic organic molecules, while their surface is hydrophilic and highly polar. The relative photoproduct concentrations were found to be dramatically dependent on the detergent concentration and the cage effect was attributed to the mutual recombination of a small number of primary radicals generated inside the micellar aggregates, efficient enough to prevent the radicals from escaping the cage and generating the cross-coupling products **267** and **269**. The average number of ketone **266** molecules per micelle was calculated by the authors for $[CTAC] = 10^{-3}$ and $10^{-2} \, M$ to be 44 and 4, respectively.

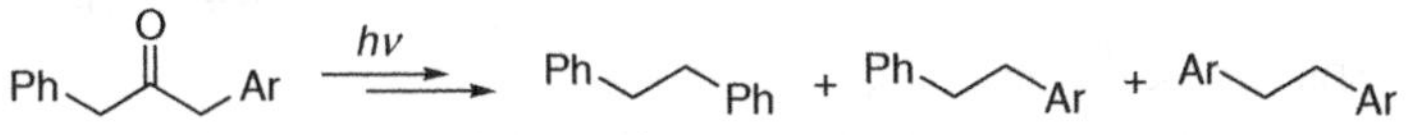

Ph = phenyl; Ar = 4-methylphenyl

| | 267 | | 268 | | 269 |
|---|---|---|---|---|---|
| no CTAC: | 25% | : | 50% | : | 25% |
| [CTAC] = $10^{-3}$ M: | 21% | : | 58% | : | 21% |
| [CTAC] = 5x$10^{-3}$ M: | 6% | : | 88% | : | 6% |
| [CTAC] = $10^{-2}$ M: | 1% | : | 98% | : | 1% |

*Scheme 6.112*

---

## Case Study 6.19: Photochemistry in crystals – solid-to-solid photoreaction

Photochemical reactions in the solid state (see Special Topic 6.5) usually occur with high reaction selectivity and specificity due to conformational bias and the least-motion pathway principle.[172,928] Photochemical decarbonylation of *cis*-2,6-dihydroxy-2,6-diphenylcyclohexanone (**270**) in homogeneous solution results in *cis*- and *trans*-**271** cyclopentanediols in nearly equimolar ratio, whereas its photolysis in single crystals gives predominately the *cis*-isomer in a high chemical yield (∼83%) (Scheme 6.113).[929] Free bond rotation in a biradical intermediate is largely restricted in the solid phase. An X-ray analysis revealed that the starting material and the *cis*-product have a high degree of structural similarity; therefore, the corresponding least-motion pathway reaction preserves shape and volume of the solid (crystalline) environment. By comparison, a small enhancement of *cis*-**271** production was also observed in highly viscous sucrose glasses.

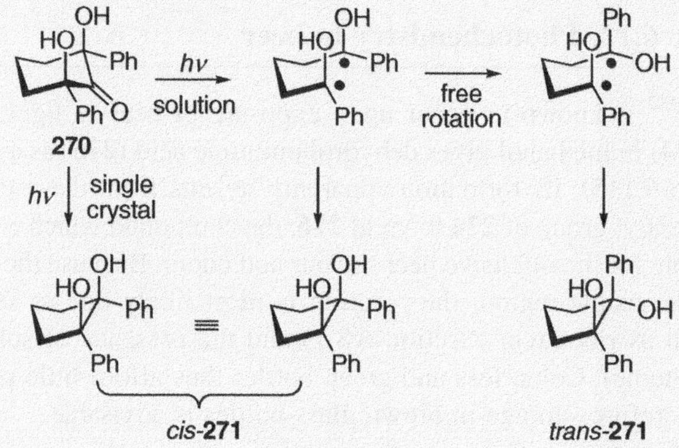

*Scheme 6.113*

*Experimental details.*[929] Single crystals or fine powder of **270**, placed between microscope slides, were irradiated with $\lambda_{irr} = 350$ nm (Figure 3.9) at 20 °C. Dissolution of the irradiated samples led to vigorous gas evolution, indicating that CO remained trapped in the crystal lattice. The ratio of photoproducts was determined by GC and NMR spectroscopy.

---

α-Cleavage in small-ring compounds releases the ring strain in the singlet state and can be sufficiently rapid to compete efficiently with intersystem crossing. Exhaustive photolysis of 2,2,4,4-tetramethylcyclobutanone (**272**) in methanol, for example, affords 2-methoxy-3,3,5,5-tetramethyltetrahydrofuran (**273**) as a major product, in addition to methyl isobutyrate and 1,1,2,2-tetramethylcyclopropane (Scheme 6.114).[930] It has been suggested that the photoinduced ring expansion reaction involves an oxacarbene intermediate via the excited singlet state and a biradical intermediate.[931]

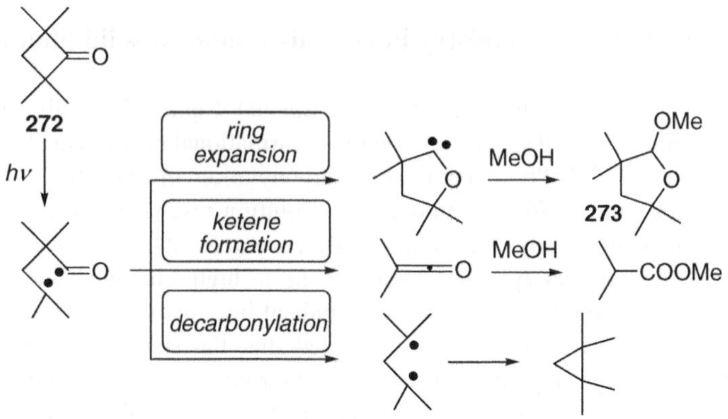

*Scheme 6.114*

---

**Special Topic 6.12: Photochemistry in beer**

A 'skunky' taste[932] is known to occur upon exposure of beer to light. Photolysis of isohumulone (**274**) in methanol gives dehydrohumulinic acid (**275**) as a major reaction product (Scheme 6.115). Its formation apparently results from the α-cleavage of the 4-methylpent-3-enoyl group of **274** to yield **276**, the compound which is believed to be largely responsible for the offensive beer flavour and odour. Because the iso-α-acids do not absorb in the visible region, the reaction is most likely due to sensitization by compounds, such as riboflavin (Section 6.8.1) and the presence of sulfur-containing amino acids (cysteine). Colourless and green bottles thus afford little protection from UV radiation; therefore, storage in brown glass bottles is advisable.

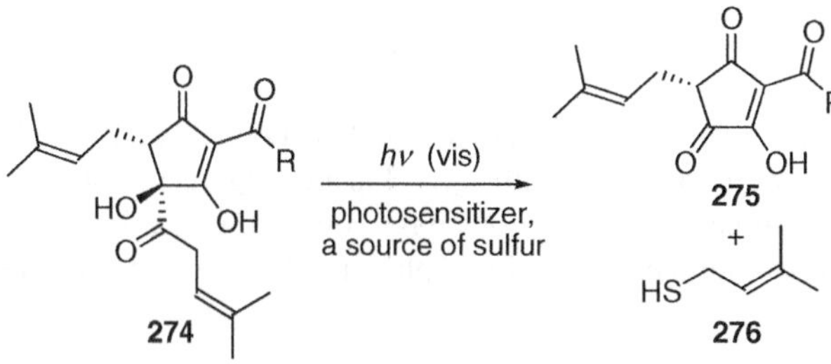

*Scheme 6.115*

---

### 6.3.4 Carbonyl Compounds: Norrish Type II Elimination

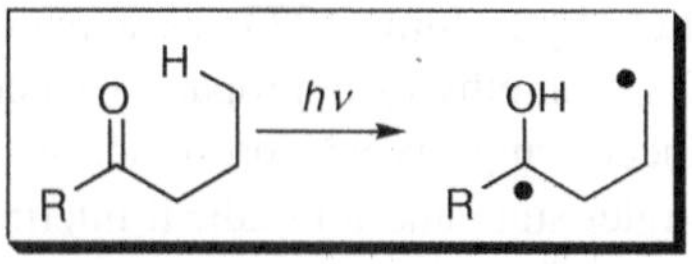

*Recommended review articles.*[751,863,933–939]
*Selected theoretical and computational photochemistry references.*[16,534,535,914,915,940–946]

Many alkyl ketones with a γ-hydrogen atom exhibit photoinduced intramolecular 1,5-hydrogen atom abstraction (Section 4.9) to form 1,4-biradicals (BR), which may undergo subsequent transformations, such as (1) reverse hydrogen abstraction to regenerate the starting material, (2) elimination to form an enol and alkene or (3) coupling of the 1,4-biradical triplet intermediate to produce cyclic alcohols (Yang photocycliza-tion;[947]Section 6.3.5) (Scheme 6.116). [863,933–935,937] The elimination process is termed the Norrish type II reaction.[917] The α-cleavage to acyl and alkyl radicals, that is, the previously discussed Norrish type I reaction (Section 6.3.3), is usually suppressed.

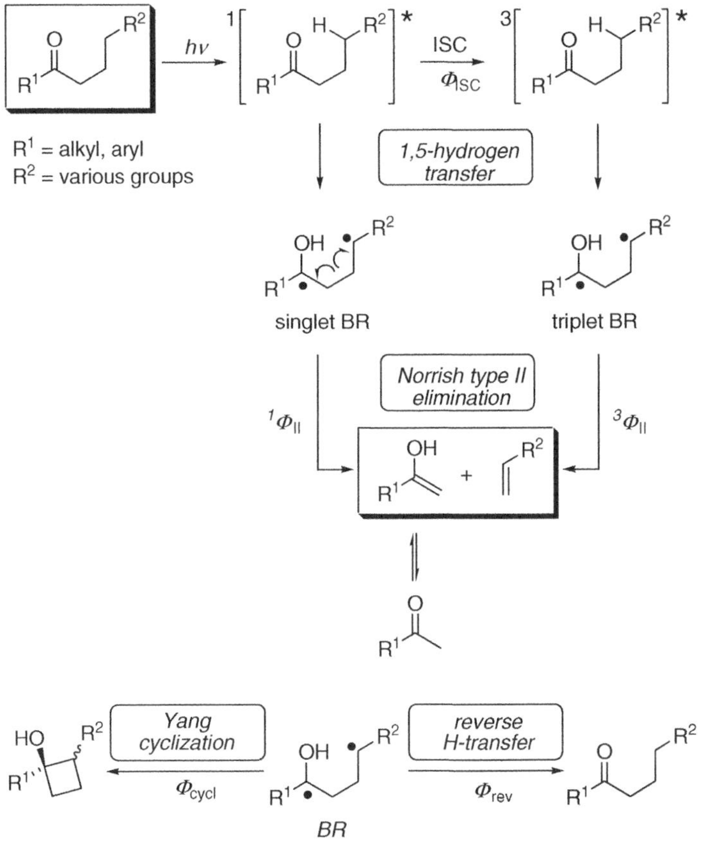

**Scheme 6.116**

Both singlet and triplet excited states and two electronic configurations n,π* and π,π* have been shown to display the type II reactivity.[863,933] Singlet state reactions are fast but their low quantum yield indicates that hydrogen abstraction[866] is accompanied by radiationless decay to regenerate the starting material. Triplet state reactions are usually more efficient. Ketones with the lowest n,π* state are more reactive than those with the lowest π,π* state. The n,π* state, with a localized unpaired electron in a nonbonding orbital of the more electronegative oxygen atom, is responsible for a radical-like reactivity that parallels that of alkoxy radicals.

Aliphatic ketones have $n,\pi^*$ lowest singlets that intersystem cross to their $n,\pi^*$ lowest triplets at rates below $10^8 \, s^{-1}$, which is slow enough to allow singlet reactions to occur.[933,948] Structural changes in an alkyl chain can affect the photokinetic behaviour; the faster the hydrogen abstraction, the less efficient intersystem crossing (Table 6.11).

**Table 6.11**    *Norrish type II elimination of various alkyl ketones ($CH_3COCH_2CH_2R^1$)[a]*

| Ketone | $^1\Phi_{II}$ | $^1k_H/10^8 \, s^{-1}$ | $\Phi_{ISC}$ | $^3\Phi_{II}$ | $^3k_H/10^8 \, s^{-1}$ |
|---|---|---|---|---|---|
| $R^1 = CH_3$ | 0.025 | 1.0 | 0.81 | >0.36 | 0.13 |
| $R^1 = CH_2(CH_3)_2$ | 0.07 | 20 | 0.18 | 0.17 | 3.8 |

[a]In benzene.[933,948] $^1\Phi_{II}$ and $^3\Phi_{II}$ are the quantum yields of the type II elimination from the singlet and triplet state, respectively (Scheme 6.116); $\Phi_{ISC}$ is the intersystem crossing quantum yield; $^1k_H$ and $^3k_H$ are the rate constants of 1,5-hydrogen abstraction.

Intersystem crossing from the singlet to triplet state in alkyl aryl ketones is usually rapid ($<10^{11} \, s^{-1}$) and $\Phi_{ISC}$ is generally close to unity.[863,933] Electron-donating substituents on an aromatic ring of triplet aryl ketones are known to decrease the hydrogen abstraction reactivity by inverting their of triplet $n,\pi^*$ lowest energy levels into those of $\pi,\pi^*$. Table 6.12 demonstrates the percentage of $n,\pi^*$ triplets in equilibrium with $\pi,\pi^*$ triplets and their effect on the observed type II rate constants. In systems where the $\pi,\pi^*$ state is only several kJ $mol^{-1}$ below that of $n,\pi^*$, most of the measured reactivity still arises from low populations of the $n,\pi^*$ triplet. The $\pi,\pi^*$ triplets with a low spin density on oxygen are known to undergo the hydrogen abstraction, but 2–3 orders of the magnitude slower than that observed for $n,\pi^*$ triplets.

**Table 6.12**    *Norrish type II elimination of various aryl ketones ($R^1COCH_2CH_2R^2$)[a]*

| Ketone | $\Phi_{ISC}$ | $^3\Phi_{II}$ | $k_{obs}/10^8 \, s^{-1}$ | $n,\pi^*$ (%) |
|---|---|---|---|---|
| $R^1 = Ph$ <br> $R^2 = CH_3$ | 1 | 0.35 | 0.07 | 99 |
| $R^1 = Ph$ <br> $R^2 = CH_2CH_3$ | 1 | 0.33 | 1.4 | 99 |
| $R^1 = p$-alkylphenyl <br> $R^2 = CH_2CH_3$ | 1 | 0.39 | 0.18 | 18 |
| $R^1 = p$-methoxyphenyl <br> $R^2 = CH_2CH_3$ | 1 | 0.14 | 0.006 | 1 |

[a]In benzene. Symbols as those in Table 6.11 (see also Scheme 6.116); $k_{obs}$ is the observed rate constant of the type II elimination.[933,949]

Several factors affect the partitioning among elimination, cyclization and reversion reactions of the biradical intermediates.[933] Elimination (Grob fragmentation[758]) can occur from either *gauche* or *anti* conformations and this appears to be governed by the stereoelectronic requirement for overlap of the breaking bond with both half-occupied p-orbitals in the 1,4-biradical. Cyclization (and reversion) can take place from the *gauche* conformation only (Scheme 6.117). Singlet biradicals are generally too short-lived to allow for conformational changes; therefore, reversion and elimination are almost exclusively reactions of the singlet excited ketones in solutions. The longer lifetimes of

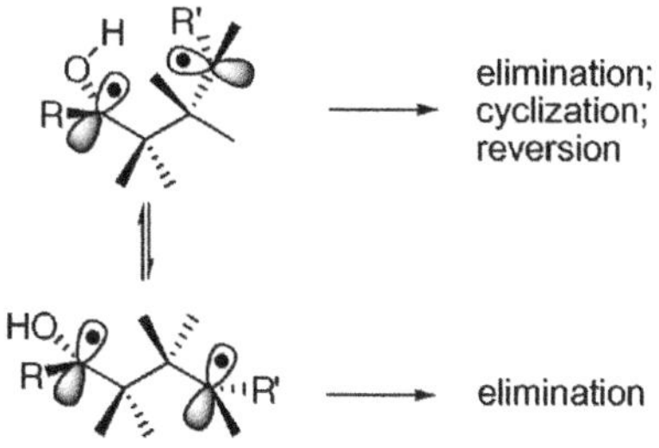

*Scheme 6.117*

triplet biradicals are related to rates of intersystem crossing to the corresponding productive singlets.[950]

The cleavage/cyclization ratio in straight-chain ketones is usually relatively high (>5:1).[933,951] Some bulky α-substituents may increase the probability of cyclization primarily by destabilizing the geometry required for cleavage or by relieving the ring strain in the transition state, for example by the substitution of carbon for oxygen in the alkyl chain (Table 6.13).

**Table 6.13** *Biradical partitioning reactions in aryl ketones* $(PhCOR^1)^a$

| Ketone | $^3\Phi_{II}$ | $^3\Phi_{cycl}$ | $^3\Phi_{rev}$ |
|---|---|---|---|
| $R^1 = CH_2CH_2CH_2CH_3$ | 0.33 | 0.10 | 0.57 |
| $R^1 = CMe_2CH_2CH_2CH_3$ | 0.04 | 0.10 | 0.86 |
| $R^1 = CH_2OCH_2CH_3$ | 0.57 | 0.42 | 0.01 |

$^a$In benzene.[933,951] Symbols as in Scheme 6.116.

---

## Case Study 6.20: Actinometry – valerophenone

Valerophenone (**277**) serves as a reliable actinometer[214] (Section 3.9.2) in quantum yield measurements when monochromatic radiation is used. Acetophenone (**278**), the type II product (Scheme 6.118), is formed in benzene with $\Phi = 0.33$ (conditions: $c_{277} = 0.1$ M; $\lambda_{irr} = 313$ nm; 20 °C).[214,952] In contrast, the quantum yield of valerophenone consumption (type II elimination + Yang photocyclization) in water is close to unity ($\Phi = 0.99$; conditions: $c_{277} = 7 \times 10^{-4}$ M; $\lambda_{irr} = 313$ nm; 20 °C).[953]

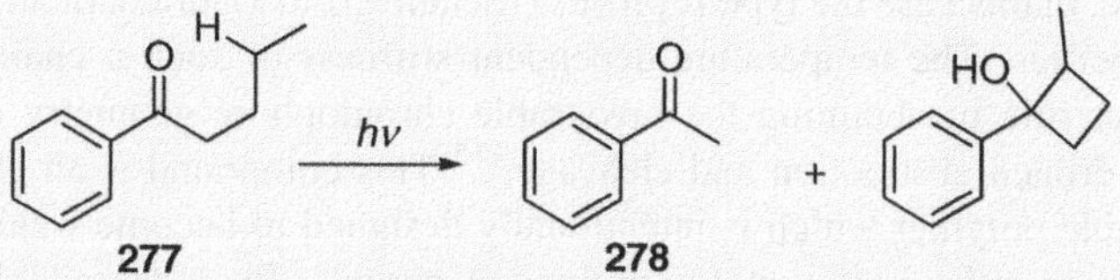

*Scheme 6.118*

*Experimental details.* Actinometry measurements can be performed in a merry-go-round apparatus (Figure 3.30) or on an optical bench (Figure 3.28).

**Table 6.14** *Temperature and environment dependence of the type II reaction in valerophenone*

| Medium | Temperature/°C | $E/C^a$ |
|---|---|---|
| t-BuOH−EtOH (9:1) | 20 | 7.9 |
| t-BuOH−EtOH (9:1) | −30 | 5.9 |
| Silica gel surface | −125 | Only C |
| Zeolite ZSM-5 | 20 | Only E |

$^a$Elimination/cyclization ratio.[937]

The type II reaction of n,π* excited ketones is largely influenced by the reaction environment and experimental conditions.[933,936,937] The initial step requires close proximity of a γ-hydrogen to the oxygen atom, i.e, overlap of the hydrogen s-orbital with the oxygen n-orbital. Flexible ketone chains achieve conformational equilibrium before γ-hydrogen atom is transfered. Conformational (environmental) restrictions can suppress the subsequent elimination reaction and may tolerate other competing reactions, including the Yang photocyclization (Section 6.3.5). Specific inclusion of alkyl aryl ketones in zeolites or cyclodextrins, on the other hand, may prevent the cyclization reaction (Table 6.14).

---

### Special Topic 6.13: Polymer photodegradation

Photodegradation of polymers (*photoageing*), involving chain scission and/or cross-linking, occurs by exposure to solar or artificial radiation and causes structural modifications, usually accompanied by a dramatic deterioration of the physical and mechanical properties of the polymer.[954] Typically, radical intermediates are formed upon excitation, which initiate subsequent (dark) degradation. The presence of other species, such as oxygen, water, organic solvents or additives, and also mechanical stress and heat, may enhance the efficiency of these processes.

   Many polymers, such as polyethylene and polystyrene, do not absorb above 300 nm. Their eventual photosensitivity is then attributed to unwanted incorporation of light-absorbing and photoreactive species during manufacture and processing. When the chromophore is part of the polymer chain, as in poly(phenyl vinyl ketone), then direct photocleavage, in this case the type II process (Scheme 6.119), dramatically affects the polymer properties. The temperature-dependent stiffness of such a chain must then play the major role in obtaining the favourable chromophore geometry required for efficient γ-hydrogen abstraction and cleavage.[955] This compound is an example of a photodegradable polymer which is intentionally designed to become weak and brittle when exposed to solar radiation for prolonged periods. Photodegradable polymers degrade in a two-stage process; the photochemical reactions break specific bonds (some dark propagation steps may follow) to subsequently make the polymer brittle enough to degrade from physical stress. There are several potential applications, such as lowering the chemical persistence of polymers in the environment and photoimaging technology (see also Special Topic 6.27).

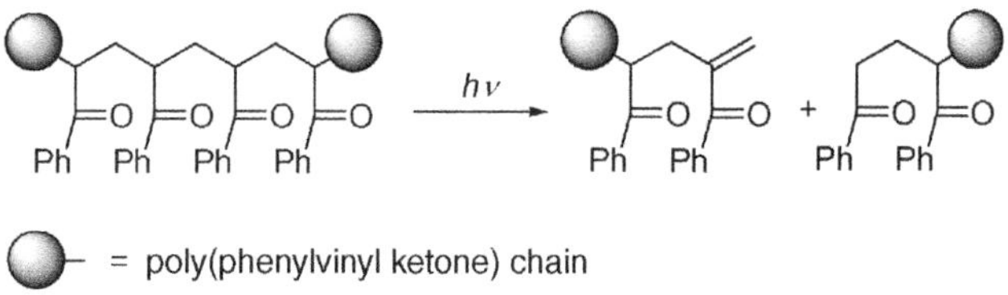
hν
Ph Ph Ph Ph
Ph Ph
Ph Ph
= poly(phenylvinyl ketone) chain

*Scheme 6.119*

Another type of photoageing, perhaps the most common, is *photooxidative degradation*, which involves diffusion of molecular oxygen through the polymer and subsequent formation of reactive singlet oxygen by photosensitization or reaction of photogenerated radicals with ground state oxygen[954,956] (see also Section 6.7). Although stabilizers [UV filters (Section 3.1) or radical traps] are often included in the polymer matrix to provide stability against photooxidation, their effectiveness depends on many factors, including their solubility and concentration in the polymer matrix and physical loss.

## Case Study 6.21: Asymmetric synthesis in crystals – application of chiral auxiliaries

Irradiation of the octahydroinden-2-yl derivatives of acetophenone (**279**) in both solution and the solid state afforded *cis*-hexahydro-1*H*-indene (**280**) and a *para*-substituted acetophenone **281** via a Norrish type II cleavage process (Scheme 6.120).[957] The Yang photocyclization reaction was not observed.

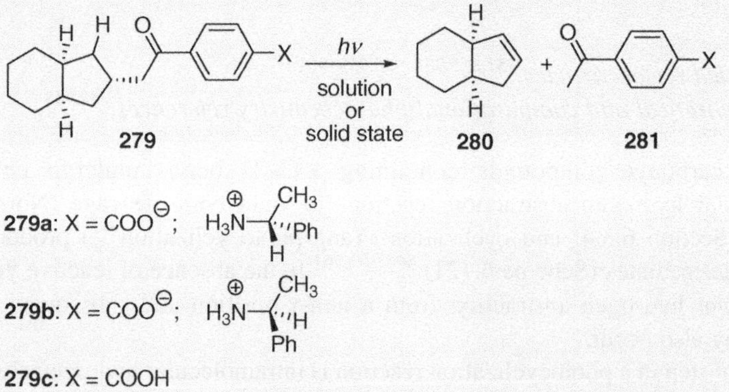
hν
solution
or
solid state
279
280
281
279a: X = COO⊖; H3N⊕–C(CH3)(Ph)H
279b: X = COO⊖; H3N⊕–C(CH3)(H)Ph
279c: X = COOH

*Scheme 6.120*

Molecular motions in confining chiral crystal lattice media are known to be severely restricted. Therefore, high photoreaction selectivities in the solid state can be achieved (see also Special Topic 6.5).[172,173] In the present case, it was shown that salts of achiral carboxylic acids and optically pure amines crystallize in homochiral space groups. The ammonium ions thus act as chiral auxiliaries and the diastereomeric transition states are differentiated by a chiral reaction medium, thereby leading to asymmetric induction.[590,958] For the system shown in Scheme 6.120, irradiation of solid **279a**, employing (R)-(+)-1-phenylethylamine as a chiral auxiliary, afforded an alkene **280** in 32% enantiomeric excess (ee), whereas irradiation of **279b** [employing (S)-(−)-1-phenylethylamine] produced the opposite enantiomer in 31% ee at >99% conversion. Photolysis of both **279a** and **279b** in methanol solutions led, of course, to racemic photoproducts only.

*Experimental details: liquid-state photolysis.*[957] Acetonitrile solutions of **279c** (0.07 M) in Pyrex tubes were purged with nitrogen for 15 min and then irradiated with a medium-pressure mercury lamp (450 W) in a water-cooled Pyrex ($\lambda_{irr} > 280$ nm) immersion well (Figure 3.9) at 20 °C for several hours. The chemical yields of **280** and **281** were 53% and 44% (GC), respectively.

*Experimental details: solid-state photolysis.*[957] A crushed crystalline ketone (**279a** or **279b**) (~5 mg), suspended in hexane (3 ml), was placed between Pyrex microscope slides, sealed in a polyethylene bag under nitrogen and irradiated with a medium-pressure mercury lamp (450 W) at a distance of 10 cm from a water-cooled Pyrex immersion well (Figure 3.9) at either 20 or −20 °C (cryostat ethanol bath). The product, a chiral organic salt, was derivatized to the corresponding methyl ester by treatment with excess diazomethane and purified by column chromatography.

### 6.3.5  Carbonyl Compounds: Photocyclization Following $n$,1-Hydrogen Abstraction

*Recommended review articles.*[751,863,933–935,959–961]
*Selected theoretical and computational photochemistry references.*[942,945]

Excited carbonyl compounds containing γ-C−H bonds undergo characteristic intramolecular hydrogen abstraction reactions to yield both cleavage (Norrish type II reaction;[917]Section 6.3.4) and cyclization (Yang photocyclization[947]) products via 1,4-biradical intermediates (Scheme 6.121).[863,959–961] In the absence of reactive γ-hydrogens, intramolecular hydrogen abstraction from a non-γ-position and subsequent cyclization reaction may also occur.

The initial step in a photocyclization reaction is intramolecular hydrogen abstraction by an excited carbonyl, basic details of which we described in Section 6.3.4. The γ-hydrogen abstraction, proceeding via a six-membered ring transition state, is energetically favoured.[933,938,960,961] In contrast, the β-hydrogen transfer is disfavoured for enthalpic reasons because of the strain that develops in the corresponding five-membered transition state permitting the abstraction, whereas a more negative entropy associated with the larger cyclic transition states (δ- or more distant hydrogen abstraction), reflects the relative

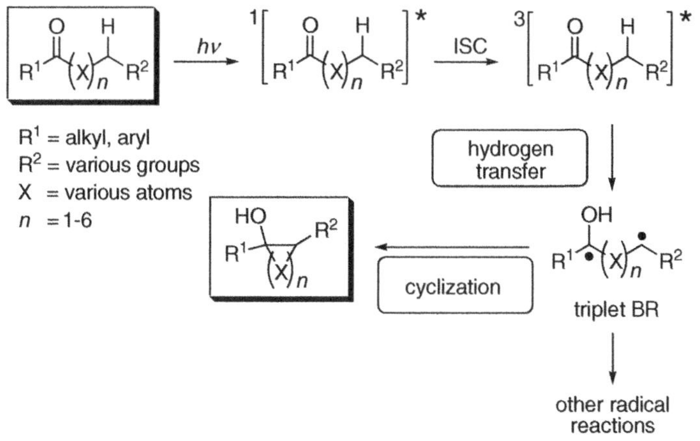

**Scheme 6.121**

improbability of achieving the required geometry. The inherent abstraction and following cyclization efficiencies then depend on many factors, such as the reactive C–H bond dissociation energy, molecule flexibility and solvent polarity. Some photocyclization reactions of ketones can also be accompanied by a photoinduced electron transfer (PET) step (Section 5.2).[960–962]

### β-Hydrogen Transfer: Formation of Three-Membered Rings

Despite the fact that this reaction proceeds via the five-membered transition state, the required alignment for hydrogen abstraction is difficult to achieve. Although some triplet β-dialkylamino ketones produce the 1,3-biradical intermediates due to the rapid initial internal charge transfer step,[959] there are only few examples of direct β-hydrogen abstraction. Irradiation of the 3-hydroxy-2,2-dimethyl-1-(2-methylaryl)alkan-1-ones **282**, for example, produces the dihydroxycyclopropanol derivatives **283** in ~40% chemical yield.[963] Larger R groups apparently cause the ketone to adopt a more favourable conformation for the hydrogen abstraction (Scheme 6.122).

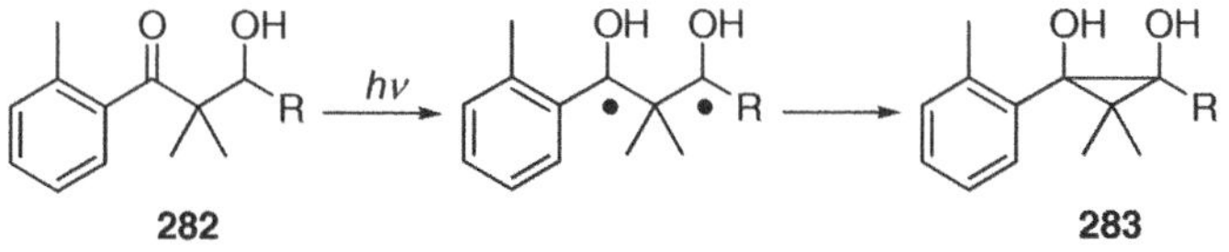

**Scheme 6.122**

### γ-Hydrogen Transfer: Formation of Four-membered Rings (Yang Cyclization)

The cyclization of 1,4-biradicals formed by γ-hydrogen abstraction is discussed in Section 6.3.4, which focuses on the Norrish type II elimination, the most common competing productive reaction. The cyclization efficiency of the triplet-initiated reactions varies greatly with the structure of the biradical, which must attain the *gauche* conformation to cyclize (Scheme 6.117; Section 6.3.4).[959] In acyclic alkyl aryl ketones, substitution on the alkyl chain may have considerable consequences for product concentration ratios (Table 6.12, Section 6.3.4). α-Methylbutyrophenone (**284**), for example, upon irradiation forms racemic 2-methyl-1-phenylcyclobutanol (**285**) exclusively with the methyl and phenyl groups *anti*, which is most probably reflected in a preferred conformation of the triplet biradical[964] (Scheme 6.123).

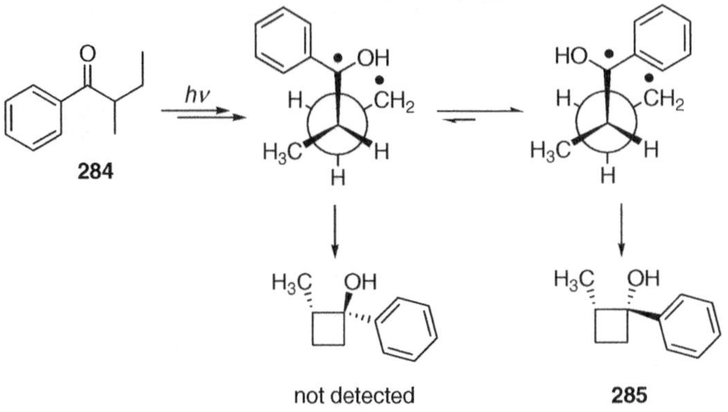

not detected                            285

*Scheme 6.123*

In another example, irradiation of 2-benzoylbicyclo[2.2.2]octane (**286**) in cyclohexane results primarily in the formation of the Norrish type II (Section 6.3.4) photoelimination product **287** (Scheme 6.124).[965] In contrast, 2-benzoyl-2-methylbicyclo[2.2.2] octane (**288**) undergoes complete and stereoselective conversion to the tricyclo [3.3.1.0$^{2,7}$]nonane **289**.[965] The eclipsing interactions between the methyl and phenyl groups in the *exo*-2-methyl-substituted derivative slow the hydrogen abstraction step by two orders of magnitude and presumably the same interactions enhance the probability of cyclization.

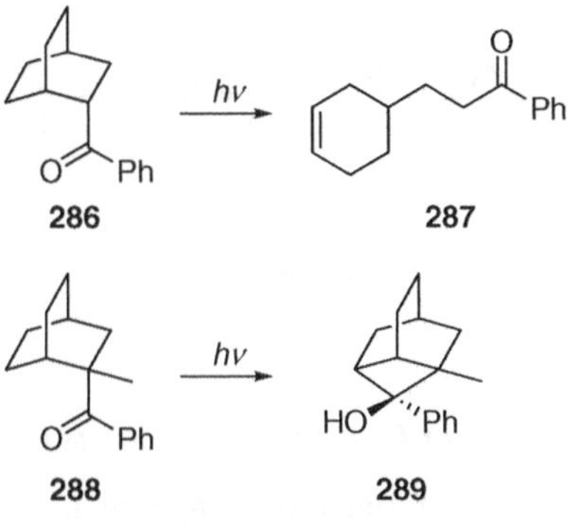

286                            287

288                            289

*Scheme 6.124*

### δ-*Hydrogen Transfer: Formation of Five-membered Rings*

Efficient δ-hydrogen abstraction in alkyl ketones is feasible only if γ-hydrogens are absent, improperly oriented toward the carbonyl group or when the corresponding C−H bond dissociation energy is too high. This can be demonstrated, for example, with β-alkoxy ketones **290** (Scheme 6.125), which produce oxacyclopentanols upon irradiation in nonpolar solvents.[966] Reaction efficiency is lowered by the biradical reversion to ketone involving 1,4-hydrogen transfer.

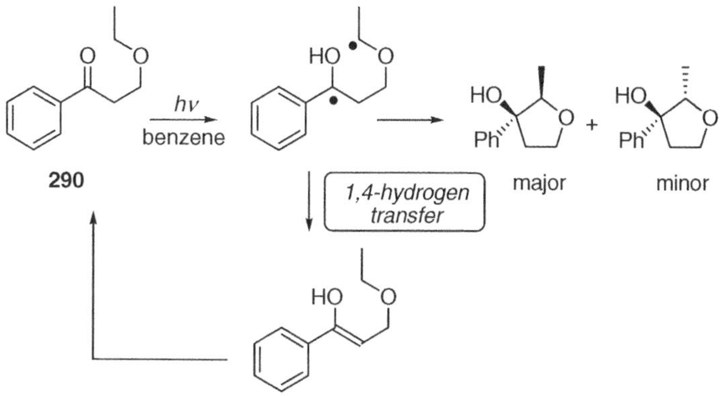

*Scheme 6.125*

Like β-alkoxy ketones, β-amido ketones also undergo photoinduced δ-hydrogen abstraction to give proline derivatives (in <50% chemical yield) (Scheme 6.126).[967] The reaction stereoselectivity depends on the biradical cyclization rate, which competes with conformational changes. Whereas singlet biradicals couple without reaching conformational equilibrium, the triplet biradicals allow bond rotation before the ring forms.

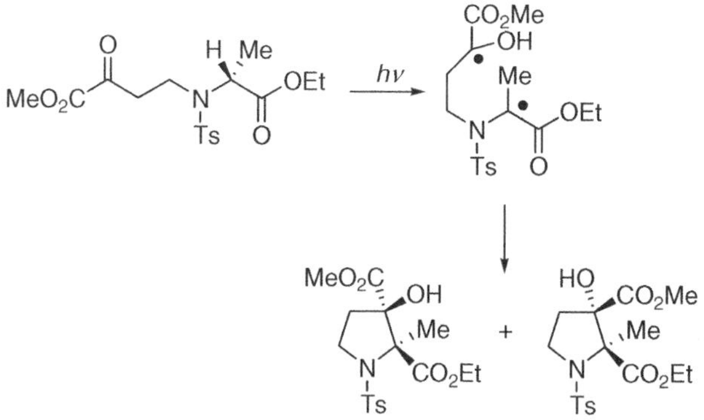

*Scheme 6.126*

*Long-distance Hydrogen Transfer: Formation of Six- and Larger Membered Rings*

There are few experimental examples of longer distance intramolecular hydrogen abstraction. ε-Hydrogen transfer forming 1,6-biradicals generally requires that γ- and δ-hydrogen atoms be unavailable or unreactive [as, for example, in the β-(*o*-tolyl) propiophenone **291** (Scheme 6.127)].[968]

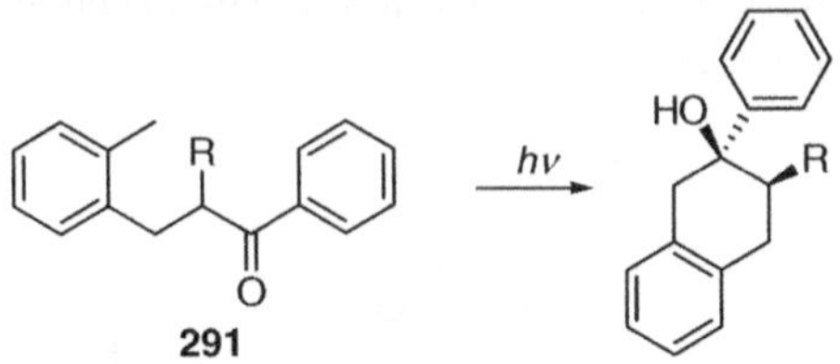

**Scheme 6.127**

The photochemically induced reaction of **292** is an example of seven-membered ring formation, which gives the product **293** in 25% chemical yield (Scheme 6.128).[969] The mechanism involves a spin centre shift[960] approach, which is based on the formation of a 'common' biradical and a subsequent efficient rearrangement bypassing the otherwise favourable cyclization. This reaction, like cyclopropane ring opening (see also Special Topic 6.10), enables one of the radical centres to shift, hence creating a new, more remote biradical that eventually cyclizes.

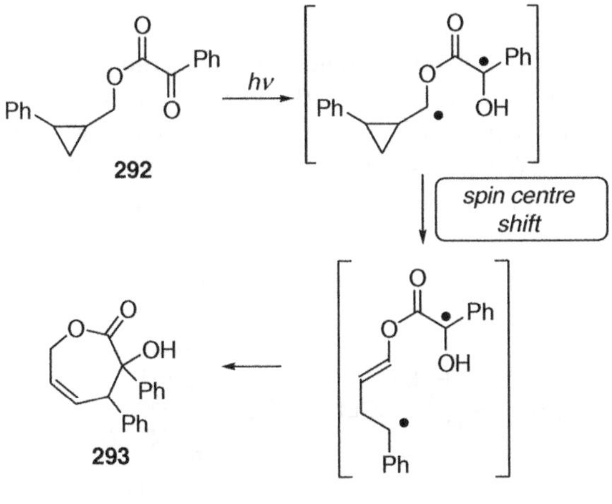

**Scheme 6.128**

## Special Topic 6.14: Photochemical synthesis of large rings

Photocyclization following 1,*n*-hydrogen abstraction seems to be an excellent tool for the synthesis of large macrocycles – one of the great challenges in organic synthesis. Irradiation of benzyloxypentyl phenylglyoxylate (**294**), for example, gives 3-hydroxy-3,4-diphenyl-1,5-dioxacyclodecan-2-one (**295**) in 20% chemical yield (Scheme 6.129), apparently via 1,11-hydrogen abstraction by the excited carbonyl group.[970] The observed selective reactivity is connected to a stabilization of the ensuing radical centre by the alkoxy and phenyl substituents.

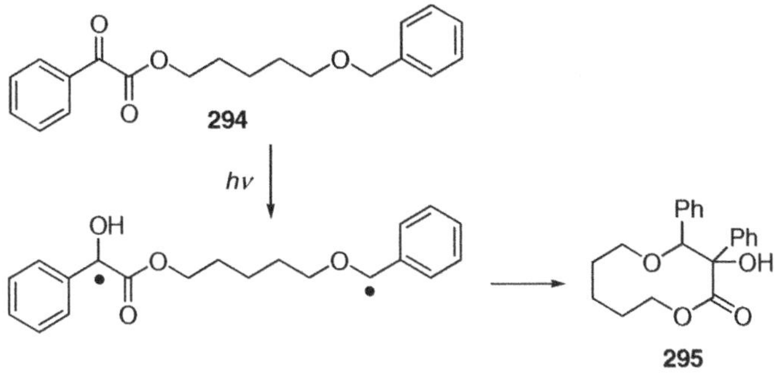

**294**

*hv*

**295**

*Scheme 6.129*

The photocycloadditions of alkenes with the enol form of 1,3-diketones (the *de Mayo reaction*; e.g. Scheme 6.68) or electron transfer-mediated reaction (see below) can also be utilized for synthesis of large rings.

## Cyclization Following Electron Transfer

Carbonyl compounds having a good electron donor group attached on a flexible chain may exhibit intramolecular electron transfer to form radical ions that can undergo a proton migration to produce biradicals similar to those obtained via intramolecular hydrogen abstraction.[960–962] For example, the high oxidizing power of the excited singlet and triplet states of phthalimide moieties can be utilized. The acetone-sensitized triplet state of the phthalimide **296** was proposed to undergo a photoinduced electron transfer, followed by deprotonation and biradical cyclization, to give **297** in 84% chemical yield (Scheme 6.130).[971]

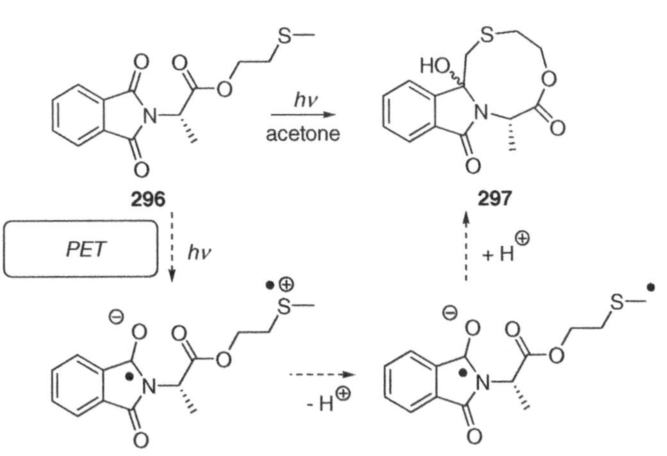

**296**

*hv*
acetone

**297**

PET

*hv*

$+ H^{\oplus}$

$- H^{\oplus}$

*Scheme 6.130*

A similar strategy has also been used for the preparation of cyclic peptides from their noncyclic analogues (**298**) containing *N*-terminal phthalimide as a light-absorbing electron acceptor moiety and a *C*-terminal α-amidocarboxylate centre (Scheme 6.131).[972] The mechanism can be described as intramolecular photoinduced electron transfer from a neighbouring amide donor to the excited phthalimide chromophore, followed by amide radical cation centre migration to the α-amidocarboxylate and decarboxylation to form a 1,ω-biradical intermediate, which subsequently cyclizes to give the product **299** ($n = 1$–3; chemical yields: 38–74%).

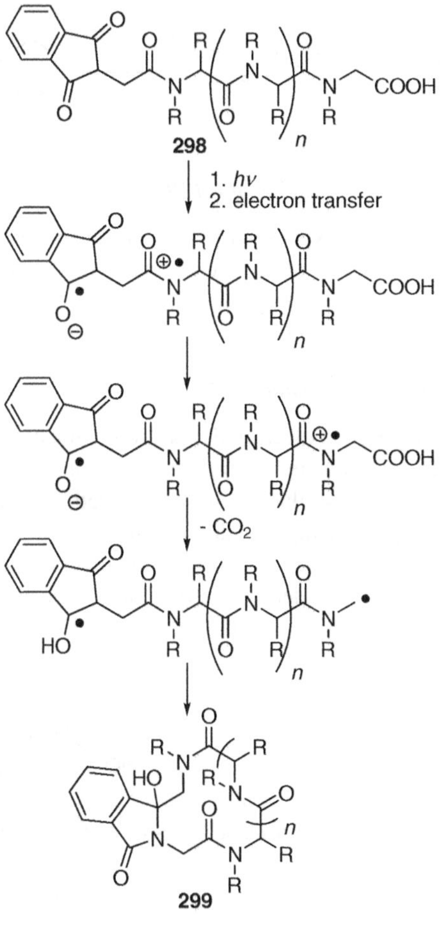

*Scheme 6.131*

## Case Study 6.22: Medicinal chemistry – isooxyskytanthine

Facile synthesis of isooxyskytanthine (**300**), a monoterpene alkaloid, was performed by photoreductive intramolecular cyclization of the 5-oxocyclopentanecarboxamide derivative **301**, followed by reduction of the oxo intermediate **302** (Scheme 6.132).[973] The cyclization step obviously proceeds via the ketyl radical anion formed by electron transfer from triethylamine to the excited ketone.[876]

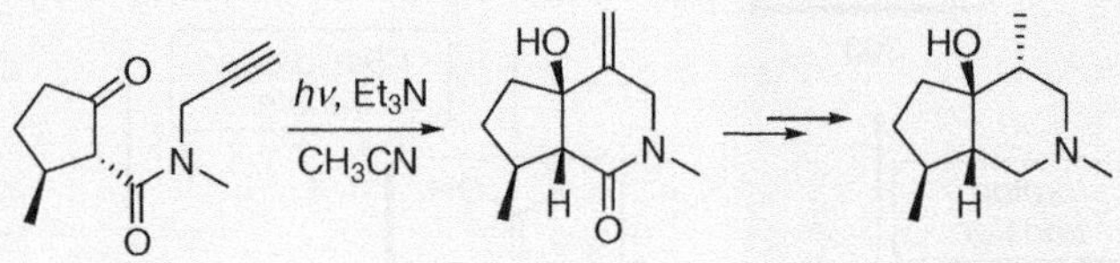

HO
*hv*, Et$_3$N
CH$_3$CN
HO

**301**          **302**          **300**

*Scheme 6.132*

*Experimental details.*[973] An acetonitrile solution of **301** in a 10 mm quartz tube, purged with argon for 30 min, was irradiated in a Rayonet system (Figure 3.10) using 12 low-pressure mercury lamps emitting at $\lambda_{irr} = 254$ nm. The product was obtained in 46% yield.

### 6.3.6 Carbonyl Compounds: Photoenolization

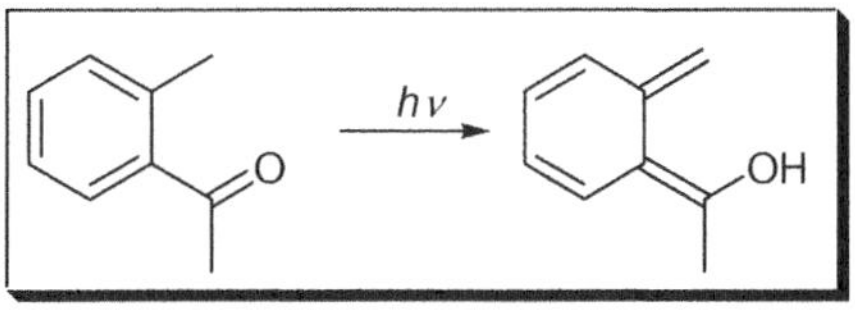

*hv*
OH

*Recommended review articles.*[939,959,974]
*Selected theoretical and computational photochemistry references.*[975–977]

2-Alkylphenyl ketones are known to produce readily the corresponding enols (photoenols) upon photochemical excitation.[974,978] For example, 2-methylacetophenone (**303**) undergoes intramolecular 1,5-hydrogen abstraction via the triplet state to form a triplet 1,4-biradical (triplet enol), yielding two isomeric photoenols, *E*- and *Z*-, whereas fast direct enolization from the lowest excited singlet state produces the *Z*-isomer only (Scheme 6.133).[979,980] The *Z*-isomer, having a lifetime similar to that of the biradical, is converted efficiently back to the starting molecule, but the *E*-isomer may, in the absence of trapping agents such as dienophiles, persist for up to seconds because its reketonization requires proton transfer through the solvent. This reaction can be accompanied by other photochemical reactions typical for the excited ketones, such as hydrogen abstraction (Sections 6.3.1 and 6.3.4).

The cyclization reaction of a biradical intermediate to form cyclobutanol (**304**; see also Yang cyclization; Section 6.3.5) and the Diels–Alder cycloaddition of the photoenols with dienophiles are common subsequent processes (Scheme 6.134).[981,982] Both processes are often stereospecific, typically involving a single, long-lived (*E*)-photoenol.

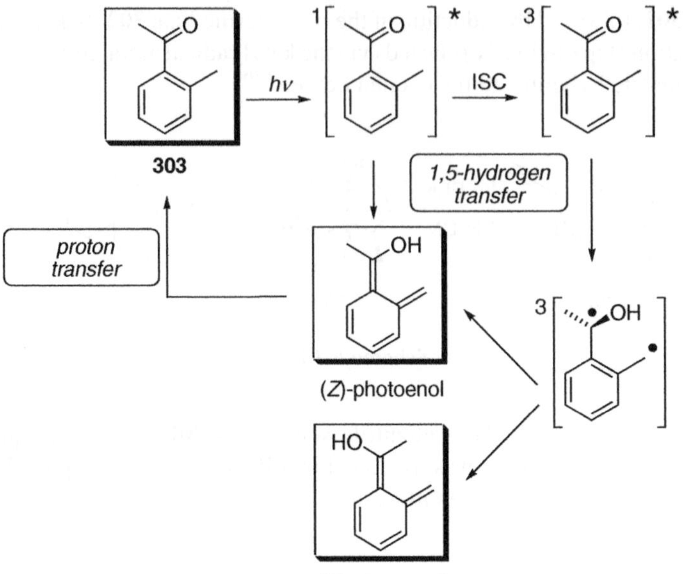

*Scheme 6.133*

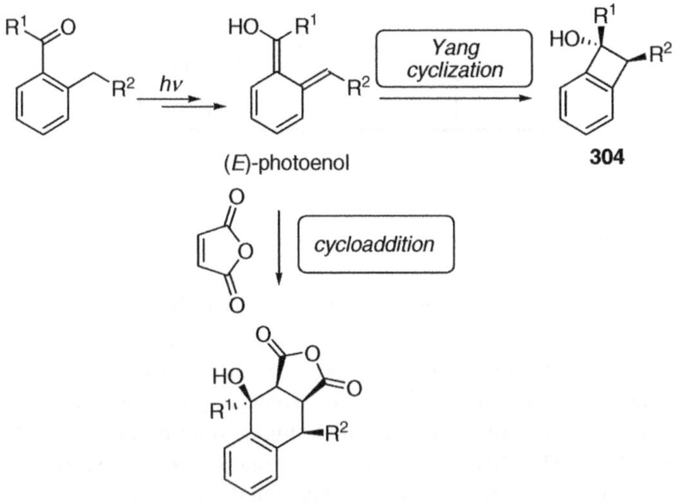

*Scheme 6.134*

## Case Study 6.23: Organic synthesis – Diels–Alder trapping of photoenols

The total synthesis of cytotoxic agents hamigerans was achieved via the photoenolization of substituted benzaldehydes and the subsequent Diels–Alder (dark) reaction.[983]
Two species, *E*- and *Z*-isomers of **305**, while being interconverted by *E–Z*

photoisomerization (Section 6.1.1), also undergo intramolecular cyclization to yield the corresponding product **306** (Scheme 6.135). Irradiation of (*E*)-**305** led to *syn-* and *anti*-**306** in a concentration ratio of 25:1, whereas irradiation of (*Z*)-**305** provided a reverse ratio (1:3). These results understandably depended on the photostationary state (Section 3.9.4) concentrations of the isomers of **305**. It was concluded that the minimization of steric repulsion between the methyl and hydroxymethyl groups *anti* to one another in the transition state (represented by **307**) must be responsible for the formation of a single racemic product.

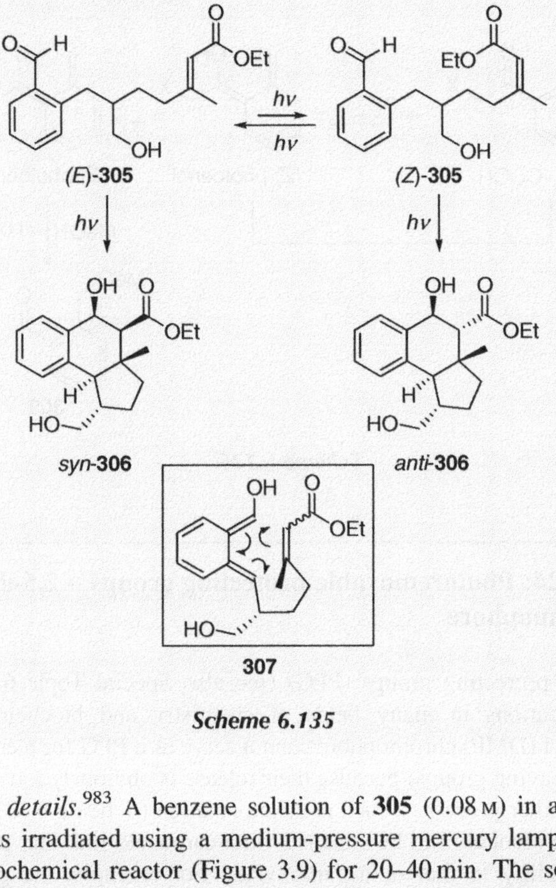

**Scheme 6.135**

*Experimental details.*[983] A benzene solution of **305** (0.08 M) in an argon-purged Pyrex vessel was irradiated using a medium-pressure mercury lamp (450 W) in an immersion photochemical reactor (Figure 3.9) for 20–40 min. The solvent was then evaporated and the reaction mixture was separated by column chromatography to give the product in ~90% chemical yield.

When *leaving groups* are present in an appropriate position, the primary photoenols can undergo elimination reactions. For example, leaving groups such as chloride[984,985] or carboxylate[986–988] on the α-carbon of 2-methylphenacyl compounds are efficiently released to form the indanone derivatives **308** in non-nucleophilic solvents. Furthermore, the acetophenone derivatives substituted on the *o*-methyl group **309** can be produced in the presence of a nucleophile (such as methanol) via (*E*)-photoenol (Scheme 6.136).

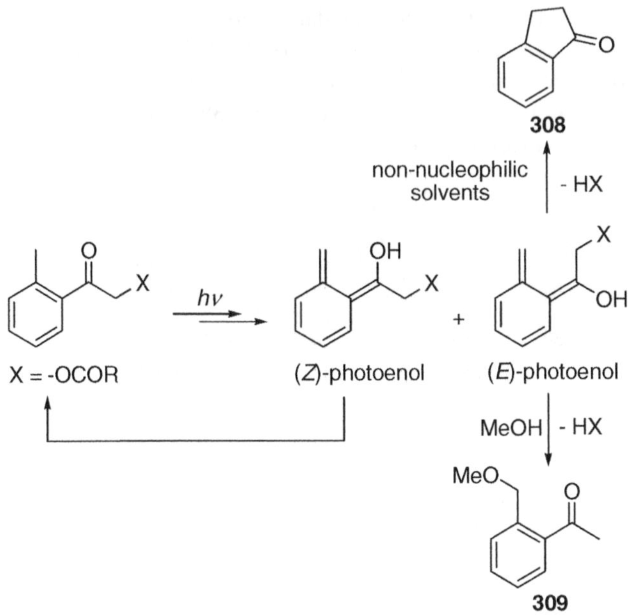

**Scheme 6.136**

---

**Case Study 6.24: Photoremovable protecting groups – 2,5-dimethyl-phenacyl chromophore**

Photoremovable protecting groups (PPG) (see also Special Topic 6.18) have found numerous applications in many fields of chemistry and biochemistry. The 2,5-dimethylphenacyl (DMP) chromophore cannot serve as a PPG for alcohols (which are relatively poor leaving groups) because their release is obviously too slow to compete efficiently with other processes, such as reketonization (Scheme 6.133). To overcome this problem, the alcohols can be attached via a carbonate link that possesses similar leaving group ability to that of a carboxylate. For example, the galactopyranosyl carbonate **310** was found to release the corresponding hydroxy-containing molecules in high chemical yields (<70%) and moderate quantum yields (0.1–0.5) (Scheme 6.137).[989] The primary photoinitiated transformation liberates the carbonic acid **311** from a predominant (*E*)-photoenol, which subsequently slowly decarboxylates in the dark to release the corresponding alcohol **312**.

*Experimental details.*[989] A cyclohexane solution of **310** (0.005 M) in a Pyrex vessel, purged with argon, was irradiated using a medium-pressure mercury lamp (125 W) in an immersion photochemical reactor (Figure 3.9). The irradiation was stopped when the conversion reached at least 95%. The solvent was evaporated and the crude alcohol was purified by column chromatography.

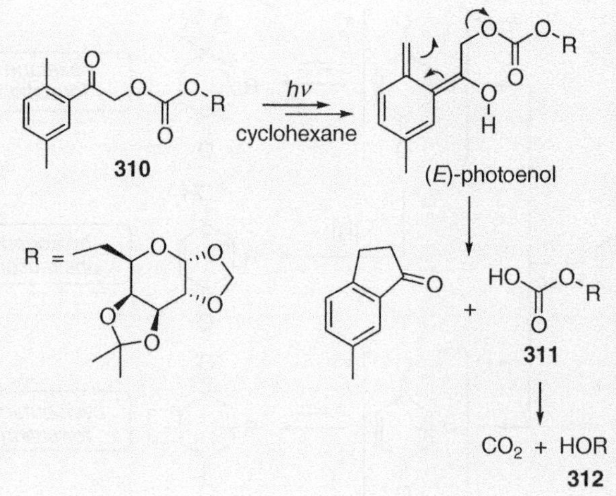

**Scheme 6.137**

### 6.3.7 Quinones: Addition and Hydrogen/Electron Transfer Reaction

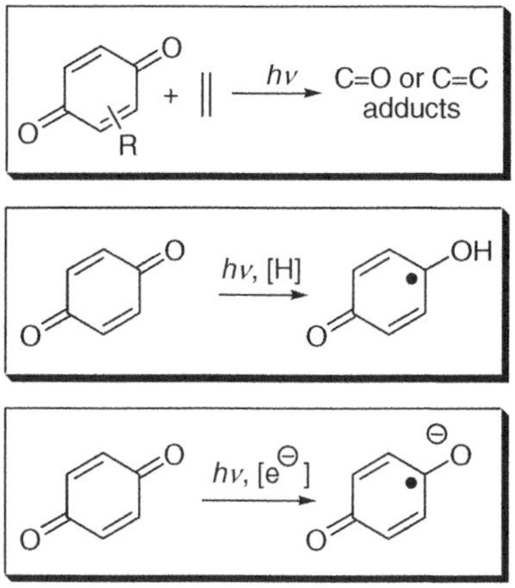

*Recommended review articles*[990–995]
*Selected theoretical and computational photochemistry references.*[254,996–998]

Both 1,2- and 1,4-benzoquinones and naphthoquinones are coloured compounds; 1,4-benzoquinone, for instance, displays two significant absorption band maxima above 250 nm ($\lambda_{max} = 290$ and 362 nm) and a weak absorption tail in the visible region. Intersystem crossing in quinones is generally fast and highly efficient and their photochemistry arises from the lowest excited triplet state in most cases.[990,993,999] There is a significant difference in the reactivity of the $^3$n,$\pi^*$ and $^3\pi,\pi^*$ states. Hydrogen atom abstraction or cycloaddition of an alkene to the carbonyl group (oxetane formation; Section 6.3.2) occurs in quinones with low-lying $^3$n,$\pi^*$ triplet states (Scheme 6.138). In contrast, cycloaddition reactions (Section 6.1.5) to the C=C bond are observed for $^3\pi,\pi^*$ excited compounds. The reactions can proceed by either direct irradiation or triplet sensitization.

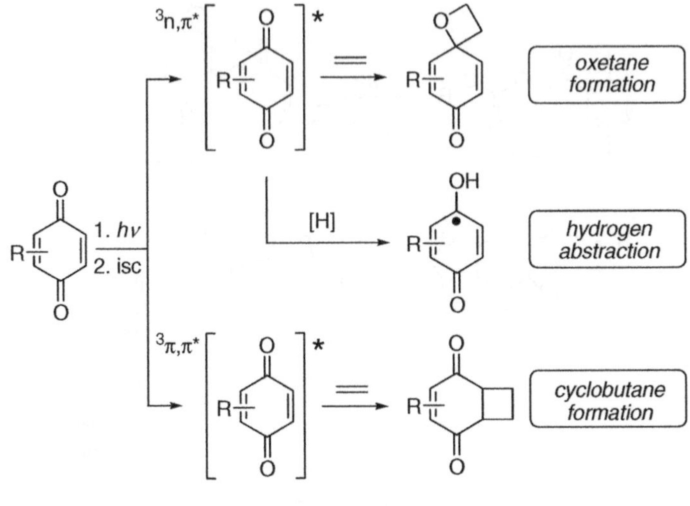

*Scheme 6.138*

For example, $^3$n,$\pi^*$ excited 1,4-benzoquinone affords spiro-oxetanes in the presence of alkenes exclusively, whereas 1,4-naphthoquinone, in which the energies of the $^3$n,$\pi^*$ and $^3\pi,\pi^*$ states are close, produces both spiro-oxetanes and cyclobutanes.[990] Introduction of electron-donor substituents on quinones further destabilizes the $^3$n,$\pi^*$ state.

### [2 + 2] Photocycloaddition Reactions

Cycloaddition reactions of triplet excited 1,4-quinones to ground-state alkenes occur either via a triplet exciplex intermediate, which collapses to a triplet biradical,[1000] or via separated radical ion intermediacy.[990] The existence of biradical intermediates has been proven by measurements of chemically induced dynamic nuclear polarization (CIDNP) (Special Topic 5.3), for example in the reaction of 1,4-benzoquinone (**313**) with norbornadiene (**314**) yielding two products, the spiro-oxetane **315** and the spiro-oxolane **316** (Scheme 6.139).[1001] Interestingly, quadricyclane (**317**) provides the same reaction as norbornadiene.

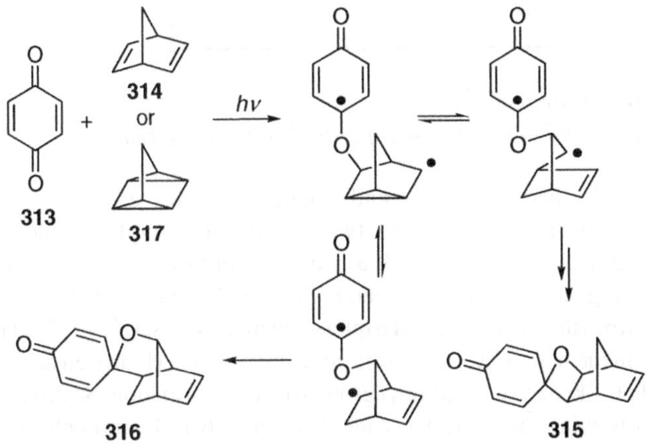

*Scheme 6.139*

The hydroxy group at position 2 in 1,4-naphthoquinone (**318**) is involved in [3 + 2] addition, producing the 2,3-dihydronaphtho[2,3-*b*]furan-4,9-dione **319** in ~50% chemical yield (Scheme 6.140).[1002] The reaction proceeds by a two-step process from a triplet excited quinone via exciplex (Section 2.2.3), intramolecular electron transfer (Section 5.2) and possibly ionic intermediate **320** formation.

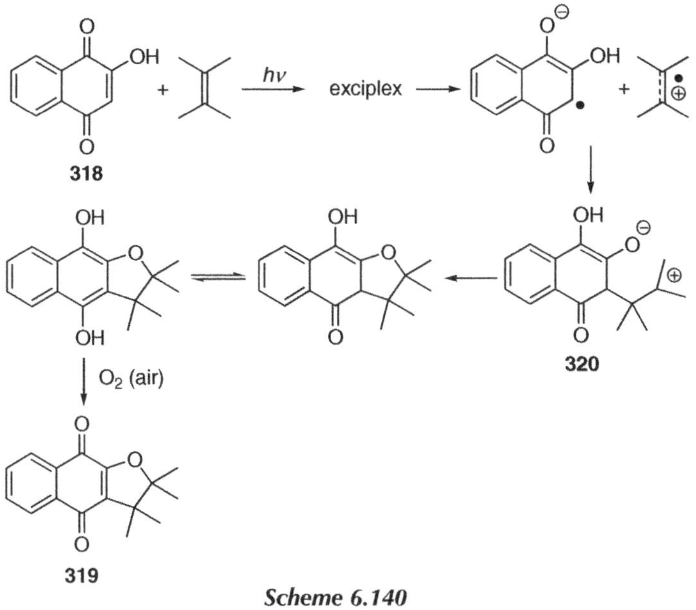

**Scheme 6.140**

*Hydrogen Abstraction*

Hydrogen abstraction is a typical reaction of triplet excited quinones.[994] The first step involves hydrogen atom transfer to form a pair of radicals, similar to that occurring in simple triplet ketones (Section 6.3.1). Alternatively, hydrogen can also be abstracted in a two-step process, which is initiated by electron transfer from the hydrogen atom source to a quinone to form a radical ion pair, followed by proton transfer. The radical mechanism of this reaction is demonstrated in the following example: 9,10-phenanthrenequinone (**321**) in the triplet state abstracts the hydrogen atom from an aldehyde to form a triplet radical pair that recombines to give an acylated product **322** (Scheme 6.141).[1003]

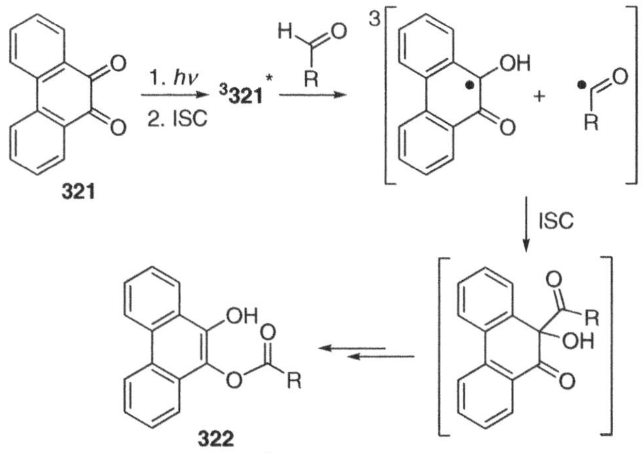

**Scheme 6.141**

## Case Study 6.25: Green photochemistry – photochemical Friedel–Crafts acylation

The synthetic significance of quinone photochemistry is demonstrated with the *'photochemical Friedel–Crafts acylation'*[992,1004,1005] of 1,4-naphthoquinone (**323**) by butyraldehyde, which gives the 1,4-dihydroxynaphthalene derivative **324** via a hydrogen abstraction step and radical recombination (Scheme 6.142).[1006] This reaction was carried out in a solar reactor, which consists of a reaction solution that circulates through a system of parabolic mirrors which collects light ($\lambda_{irr} > 350$ nm) and a heat exchanger to adjust the solution temperature (Figure 6.7). This photochemical reactor (also called a solar plant) has been developed in the context of *green photochemistry*,[1007,1008] which utilizes solar light as a *renewable and cost-free source of energy*.[1009,1010]

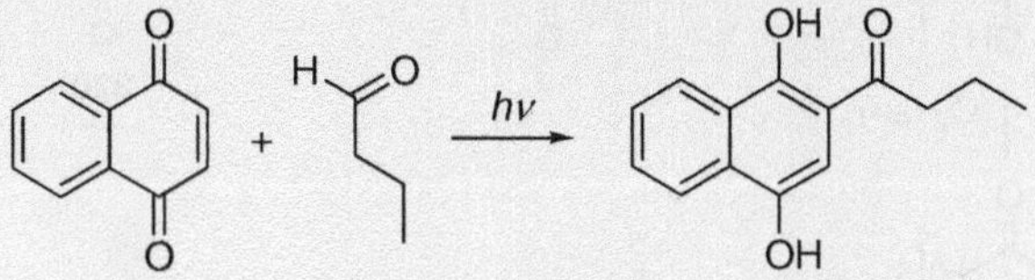

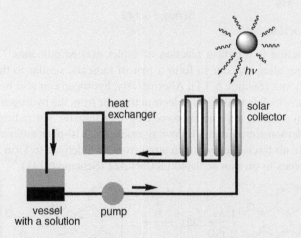

**323**         **324**

*Scheme 6.142*

heat exchanger

solar collector

vessel with a solution     pump

**Figure 6.7**   Solar plant

*Experimental details.*[1006] A solution of **323** (500 g) and an excess of butyraldehyde in a *tert*-butanol–acetone mixture, circulating in a solar reactor (Figure 6.7), was exposed to sunlight for 3 days to give **324** in 90% chemical yield and high purity (GC). The experiment took place in Spain during August and the total illumination time was 24 h.

### 6.3.8   Carboxylic Acids and Their Esters: Photofragmentation and Rearrangement

$$R\text{-}COOH \xrightarrow{h\nu} R^{\bullet} + {}^{\bullet}COOH$$

*Recommended review articles.*[322,323,1011–1018]
*Selected theoretical and computational photochemistry references.*[1019-1024]

*Carboxylic Acids*

The photochemistry of the carboxyl group has attracted only limited attention because it absorbs only at wavelengths well below 250 nm. The principle primary photoprocess is known to be either homolytic cleavage in the case of undissociated acids [for example, the dissociation energy ($D_{C-C}$) in $CH_3-COOH$ is 385 kJ mol$^{-1}$] or heterolytic cleavage in the case of carboxylate anions, followed by a photodecarboxylation step (Scheme 6.143).[322,1025–1027] The latter process may involve the formation of a radical cation precursor. The photoreactions of aliphatic derivatives usually proceed via an excited singlet state (the intersystem crossing efficiency in substituted acetic acids is low) and they are inefficient ($\Phi < 0.05$).

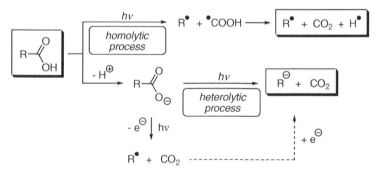

*Scheme 6.143*

For example, the photochemistry of *non-steroidal anti-inflammatory drugs* (NSAIDs) derived from 2-arylpropionic acid, has been studied because these compounds are known to exhibit phototoxicity (see Special Topic 6.22) and skin photosensitivity in some patients.[322,1011,1025] The photodecarboxylation of ketoprofen (**325**), a benzophenone derivative, in neutral aqueous solution proceeds predominately via the excited triplet state (the aromatic ketone is the chromophore in this case) and a carbanion intermediate, formed by intramolecular electron transfer (Scheme 6.144). The subsequent protonation yields the major product **326**.[1028,1029] The degradation quantum yields are higher from the carboxylate form than from the non-dissociated acid. In addition, photoinduced hydrogen abstraction (Section 6.3.1) by the carbonyl group is also observed.

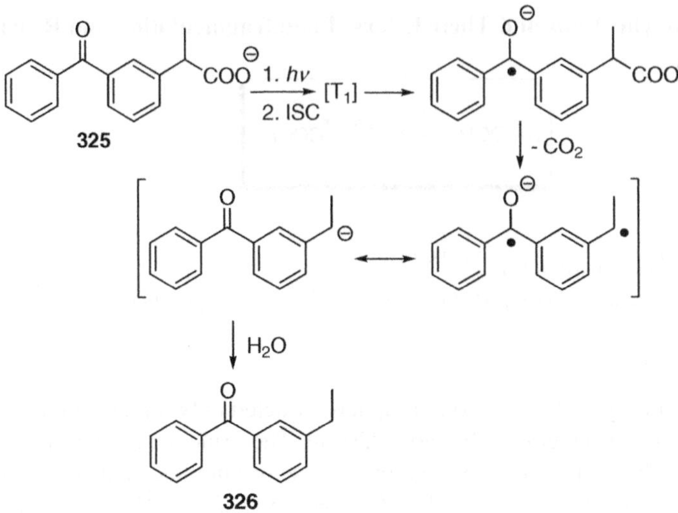

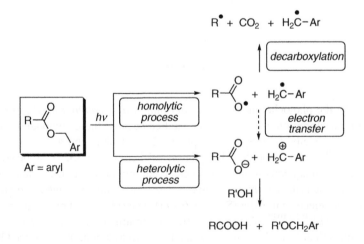

**326**

*Scheme 6.144*

### Esters of Carboxylic Acids and Lactones

Aliphatic esters of carboxylic acids absorb below 200 nm;[1013,1017] therefore, their photochemistry is usually associated with other chromophores. Photolysis of arylmethyl esters affords products derived from either homolytic or heterolytic cleavage via the excited singlet state (Scheme 6.145). The resulting radical or ionic intermediates readily undergo subsequent radical or polar reactions, respectively. The bond dissociation energies of the acyl [O–CC(O)] bond are relatively low (e.g. $D_{O-C} = 355$ or 313 kJ mol$^{-1}$ for ethyl acetate or phenyl benzoate, respectively[874]); therefore the energy of photons with wavelengths below 330 nm is sufficient for σ-bond homolytic fission.

*Scheme 6.145*

Photolysis of the 1-naphthylmethyl ester of phenylacetic acid (**327**) in methanol, for example, affords the 1-naphthylmethyl cation–carboxylate anion pair in addition to 1-naphthylmethyl radical–acyloxy radical pair intermediates, which, after decarboxylation, form an adduct with methanol (**328**, formed along with **329**) or an in-cage radical coupling product **330** (Scheme 6.146).[1030] The competition between the radical and ionic pathways was found to be very dependent upon the substituents on the naphthalene ring.

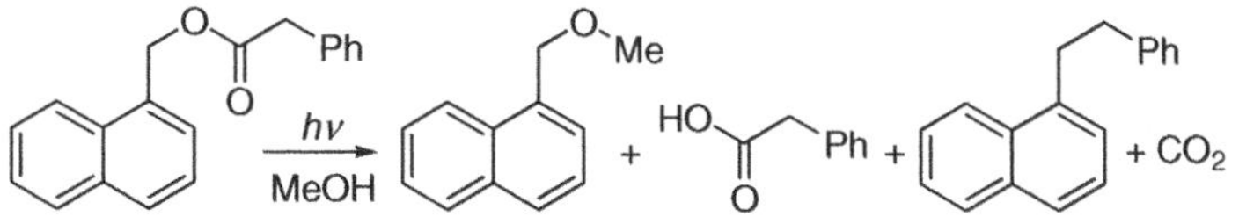

*Scheme 6.146*

It was discussed in Section 6.2.3 that substituted aromatic compounds in the lowest excited singlet state display behaviour contrasting with those in the ground state, which is related to differing charge distributions in the two electronic states. For example, benzene derivatives substituted with an electron-withdrawing group in the *meta* position are more reactive towards photochemical solvolysis than their *para*-substituted analogues (*meta effect*).[1031–1033] Scheme 6.147 shows the difference in photoreactivity of 3-methoxybenzyl acetate (*meta*) (**331**) and 4-methoxybenzyl acetate (*para*) (**332**) in dioxane.[327] Irradiation of the *meta*-derivative leads predominantly (35%) to the product of heterolysis (**333**), whereas the *para*-derivative affords products of homolysis (**334**) only.

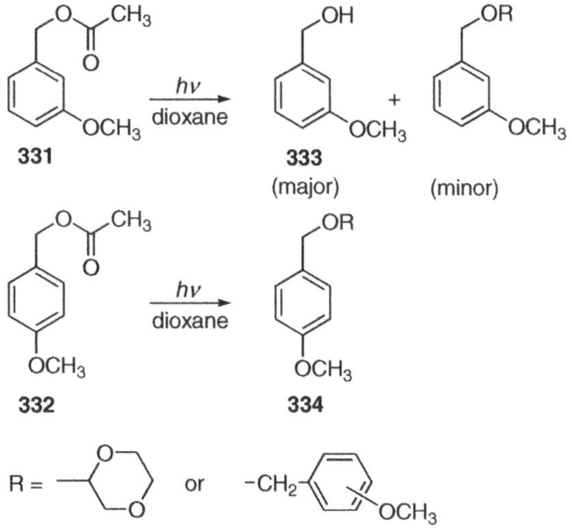

*Scheme 6.147*

## Case Study 6.26: Biology – photoactivatable compounds

Esters of carboxylic acids have been utilized as photoremovable protecting groups (PPGs) (see also Special Topic 6.18) in organic synthesis and biology.[1013,1015,1034,1035] For example, the 4-hydroxyphenacyl protecting group has proven to be an efficient tool for analysis of fast biological processes. Its esters are usually water soluble and stable in aqueous solutions in the dark and the photoproducts are nontoxic. The photolysis of the 4-hydroxy-3-methoxyphenacyl ester of γ-amino butyric acid (GABA) (335) (Scheme 6.148) allows for the spatially and temporally controlled rapid release of GABA (336) in CA1 neurons of acute hippocampal brain slices.[1036] It has been demonstrated that the deprotection mechanism proceeds via the excited triplet state (see Case Study 5.3).

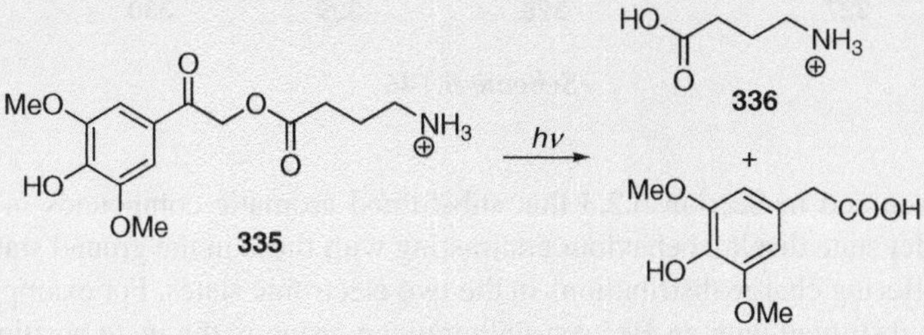

*Scheme 6.148*

*Experimental details.*[1036] CA1 hippocampal neuron from a 7-day-old rat, bathed in a solution of **335** (200 mM), was irradiated by UV flashes from a mercury arc lamp (100 W) coupled to fused-silica optical fibres using transmitting condensers (Figure 3.28).[1037] The light-emitting end of the fibre was positioned above the brain slice, producing an approximately $15 \times 25\,\mu m$ spot. The duration of the light pulse (2 ms) was controlled with an electromechanical shutter. The pulse elicited fast membrane currents due to the specific stimulation of GABA A-receptors (ligand-gated chloride channels) because of the *in situ* GABA photorelease.

The photodecarboxylation of lactones usually involves the homolytic fission of the C−O bond attached to the carbonyl group. Strained ring systems, such as cyclopropane, can be obtained by intramolecular coupling of the resulting biradical intermediate.[1025] Scheme 6.149 shows the photodecarboxylation of the α-santonin derivative **337**, possessing a γ-lactone ring, to give **338** in 56% chemical yield.[1038]

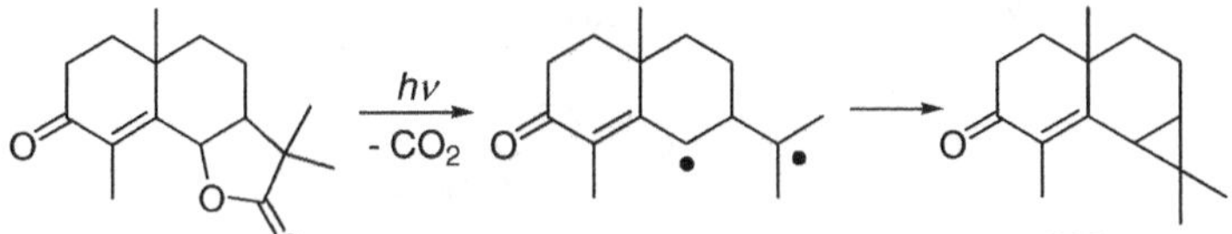

*Scheme 6.149*

Photolysis of aromatic lactones and anhydrides, such as naphthalene-1,2-dicarboxylic anhydride (**339**), gives arynes (in this case **340**)[1039] as reactive intermediates.[1040] Matrix isolation spectroscopy (Section 3.10) then allows their study along with other exotic molecules, which are formed as primary or secondary (photo)products (Scheme 6.150).

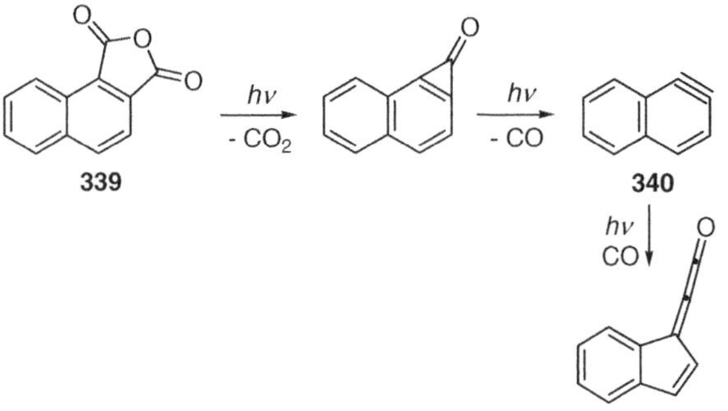

**Scheme 6.150**

The *photo-Fries rearrangement*, analogous to the Lewis acid-catalysed (ground state) Fries reaction, is a typical reaction of the aryl esters of carboxylic acids, but also of aryl carbonates, carbamates, sulfonates and other related compounds.[1012] Scheme 6.151 shows the photorearrangement of phenyl acetate, which cleaves homolytically at the carbonyl–oxygen single bond from the excited singlet state to give a radical pair. Subsequent hydrogen abstraction by phenyloxy radical from a hydrogen atom donor ([H]) affords phenol, while in-cage recombination of the radical pair provides 2- or 4-hydroxyacetophenone. It has also been reported that some aryl esters react from their upper triplet states.[1012,1041]

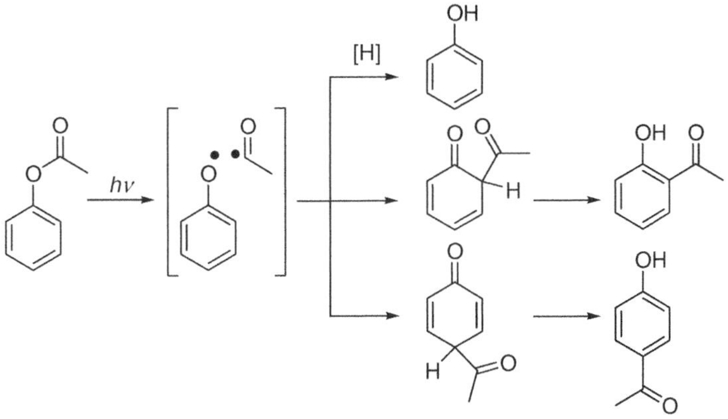

**Scheme 6.151**

## Case Study 6.27: Supercritical CO₂ chemistry – photo-Fries reaction

Supercritical solvents, compounds at a temperature and pressure above their thermodynamic critical points, are interesting reaction media because their properties, such as density, viscosity, diffusivity and dielectric properties, can be controlled by varying the pressure or temperature. A sudden increase in density around the solvent molecule occurring in the near-critical region may cause cage effects (see Special Topic 6.11)[172,173] in chemical reactions. For example, irradiation of 1-naphthyl acetate (**341**) affords a radical pair via an excited singlet state; the radicals can couple to form the acetylnaphthol derivatives **342** and **343** or diffuse apart to form 1-naphthol (**344**) and acetic acid (Scheme 6.152). Supercritical carbon dioxide was found to restrict molecular mobility most efficiently in the region just above the critical pressure due to the formation of specific solvent–solute clusters.[1042] The photo-Fries/1-naphthol formation reaction ratio reached 8.5 in the presence of methanol as a co-solvent.

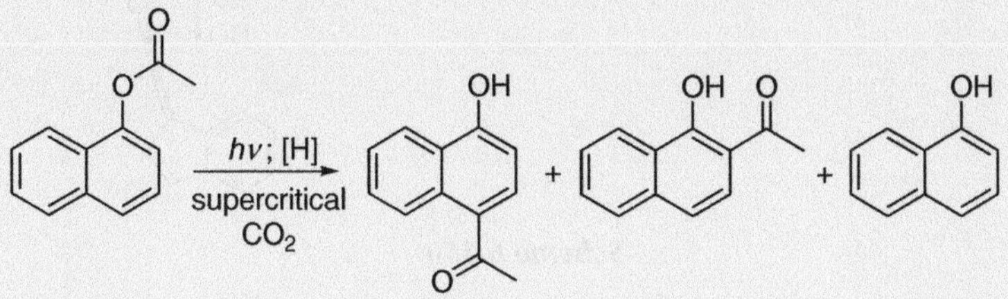

**Scheme 6.152**

*Experimental details.*[1042] 1-Naphthyl acetate (**341**, 0.003 M) in supercritical $CO_2$ (35 °C; 76 bar) was irradiated in a thermostated high-pressure quartz cell by water-filtered light from a high-pressure mercury lamp (100 W) (Figure 3.28). Photoproduct concentrations were determined using GC.

*Photodeconjugation* reactions are known to occur in α,β-unsaturated aliphatic and medium ring-size carboxylic acids or ketones.[1016] In general, the ester **345** isomerizes along the C=C bond upon irradiation (Section 6.1.1); however, after prolonged irradiation, the Z-isomer is able to undergo a photochemical antarafacial [1,5]-sigmatropic hydrogen migration (Section 6.1.2) to give the photodienol **346**, which rearranges to **347** (Scheme 6.153). Such products typically absorb at shorter wavelengths because the double bond is no longer in conjugation with the carboxyl group.

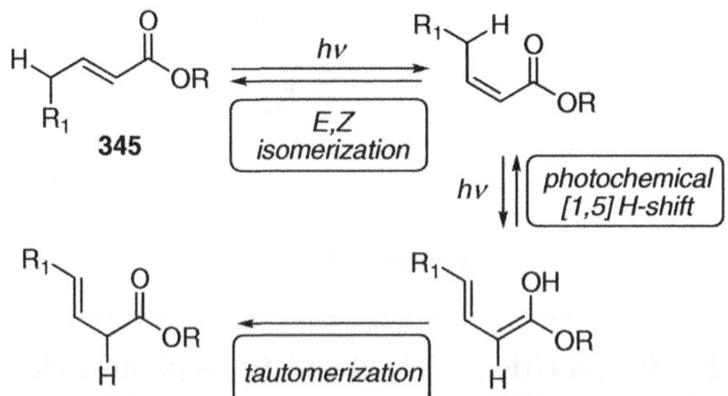

**Scheme 6.153**

### 6.3.9 Transition Metal Carbonyl Complexes: Photodecarbonylation

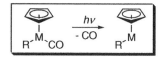

*Recommended review articles.*[1043–1046]
*Selected theoretical and computational photochemistry references.*[1046,1047]

This section briefly discusses photoreactions of transition metal carbonyl complexes, a topic on the borderline between organic and inorganic photochemistry. Such complexes are composed of a transition metal, such as Fe, Rh or W, coordinated with carbon monoxide and often organic ligands, such as cyclopentadienyl or ethylenediamine. The presence of a transition metal in the complexes introduces new types of excited states, which imply unique photophysics and photochemistry.[1046] Their electronic absorption spectra are composed of bands assigned to (often low-lying) metal-to-ligand charge transfer (MLCT), ligand-to-metal charge transfer (LMCT), ligand-to-ligand charge transfer (LLCT), metal-to-metal charge transfer (MMCT) or other transitions.[1047] Electrons in metal-centred molecular orbitals often involve d → d transitions. Therefore, although many organometallic compounds are characterized by extensive covalent bonding, we frequently categorize metals and ligands as charge transfer donors or acceptors.

Both UV and visible light have sufficient energy to initiate many processes in coordination complexes. Decarbonylation is one of the most typical photoreactions, because the dissociation energy of a common metal–carbonyl oxide bond is as low as $\sim$200 kJ mol$^{-1}$.[1048] Scheme 6.154 presents two examples: the fission of (a) metal$-CO$[1049] and (b) metal$-CO-$alkyl[1050] bonds in some carbonyl complexes. In the latter case, irradiation of an enantiomerically pure iron complex **348** leads to decarbonylation, which is followed by alkyl migration.

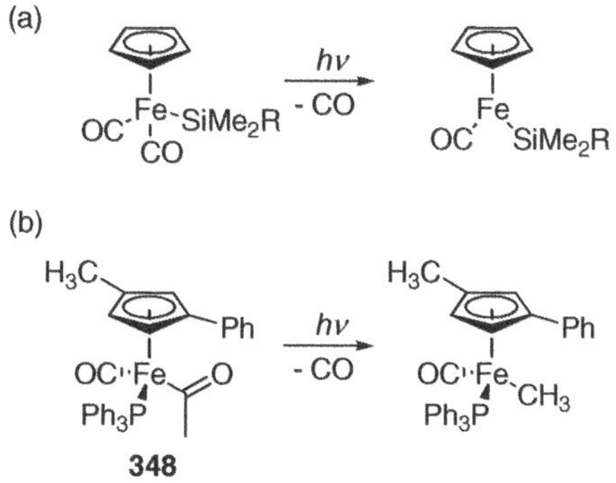

(a)

(b)

**348**

*Scheme 6.154*

The decarbonylation reaction can also be accompanied by oxidative addition, an insertion of the metal into a covalent bond. For example, the tris(dimethylpyrazolyl)borato complex **349** photochemically activates aromatic and saturated hydrocarbons (RH) at room temperature (Scheme 6.155).[1051] The reaction begins by the initial dissociation of a ligand (CO) from the metal complex to produce a coordinatively unsaturated intermediate **350** in less than 100 ps. This reactive species forms an intermediate solvate with RH, which subsequently undergoes C–H bond oxidative addition, yielding **351**. Selective *functionalization of hydrocarbons*, such as carbonylation or oxidative addition, is one of the most important goals of catalysis today.[1043]

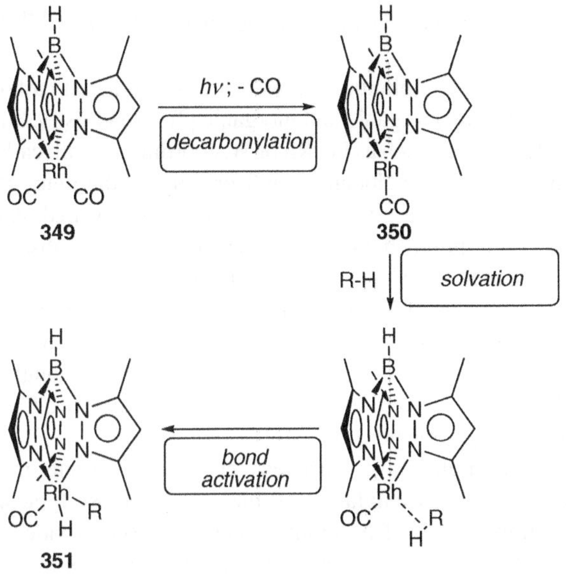

*Scheme 6.155*

### 6.3.10 Problems

1. Explain the following concepts and keywords: photoreduction; perpendicular approach in the Paternò–Büchi reaction; α-cleavage; Norrish type II reaction; photodecarbonylation; cage effect; photoageing of polymers; Yang cyclization; photoenolization; photoremovable protecting groups; photodecarboxylation; phototoxicity.

2. Suggest the mechanisms for the following reactions:

(a)

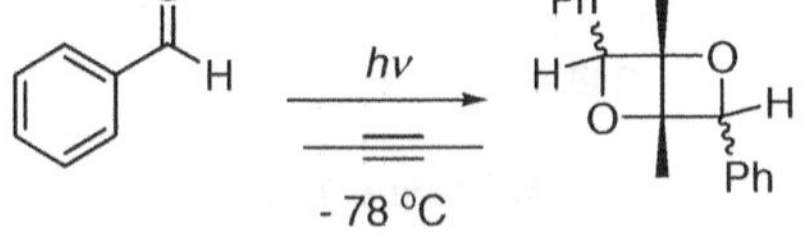

[ref. 1052]

(b)

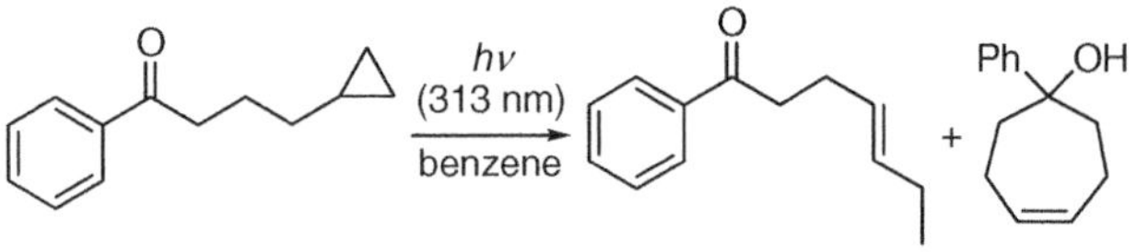

[ref. 1053]

(c)

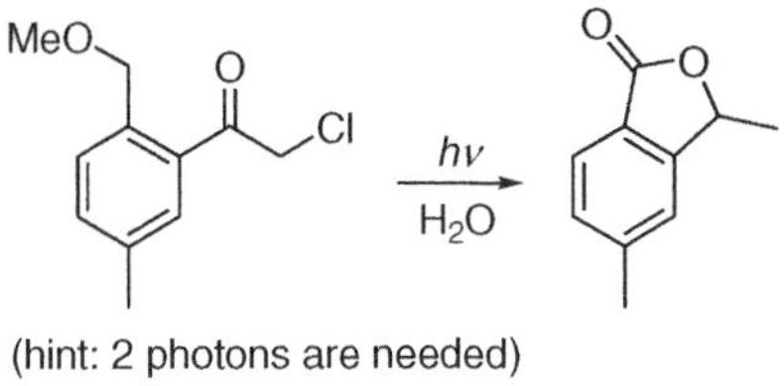

(hint: 2 photons are needed)

[ref. 987]

3. Predict the major photoproduct(s):

(a)

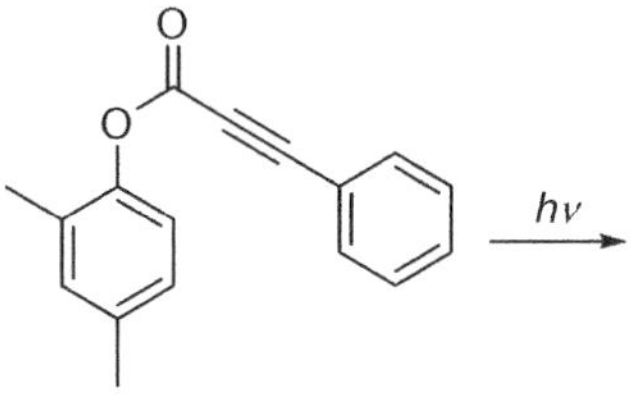

[ref. 1054]

(b)

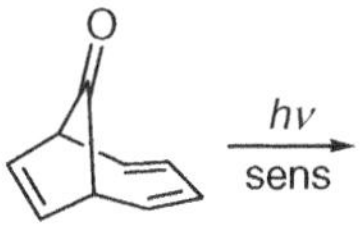

[ref. 1055]

(c)

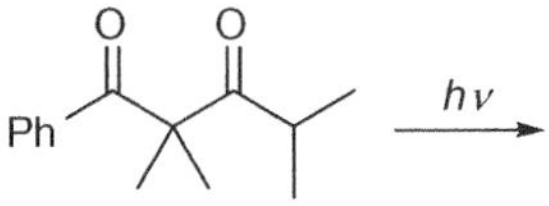

[ref. 1056]

## 6.4   Nitrogen Compounds

The absorption spectra of some nitrogen-containing benzene derivatives are shown in Figure 6.8. The lowest excited singlet state of aromatic nitro compounds is of n,π* character. Aza substitution in the ring, as in pyridine or quinolines, does not shift the absorption bands much with respect to those in the parent hydrocarbons benzene and naphthalene, respectively, but enhances the intensity of the $^1L_b$ bands that now reach $\varepsilon_{max} \approx 10^3\,M^{-1}\,cm^{-1}$ (see Section 4.7). Moreover, the presence of low-lying n,π* bands, where n corresponds to the lone pair orbital of the aza nitrogen, strongly influences the photophysical properties. The weak n,π* bands are usually detectable only as shoulders on the red edge of the first π,π* band. In 1,2-diaza compounds (pyridazine), lone-pair interaction raises the energy of the antisymmetric lone pair orbital. The first n,π* band is then stronger (symmetry allowed) and well separated from the π,π* absorption.

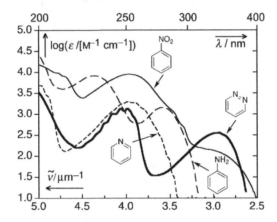

**Figure 6.8**   Absorption spectra of some benzene derivatives: nitrobenzene (——), aniline (– – –), pyridine (·–·–·), pyridazine ( —— ).[280]

The absorption spectra of pyrrole and indole (not shown), in which the nitrogen atom contributes two $p_z$-electrons to the π-system, are also related to those of the isoelectronic hydrocarbons benzene and naphthalene, respectively. π-Donation is somewhat weaker for exocyclic amines such as aniline (Figure 6.8), which is not completely planar in the ground state. Aniline is a weak base; the $pK_a$ of its conjugate acid is 4.6. Protonation in acidic solutions converts the amino group to an inductive acceptor; the absorption spectra of protonated anilines are then similar to those of the corresponding benzene derivatives.

Azoalkanes exhibit an n,π* absorption band in the near-UV region (Figure 6.9). The n,π* transition of the E-isomers is symmetry forbidden, $\log(\varepsilon/[M^{-1}\,cm^{-1}]) \approx 1$; that of the Z-isomers is allowed, $\log(\varepsilon/[M^{-1}\,cm^{-1}]) \approx 2$.[1057] The n,π* transitions are more intense in the two azobenzene isomers and shifted to longer wavelengths, $\lambda_{max} = 450$ nm. Electron-donating substituents such as a p-dimethylamino group strongly shift the π,π* transitions to longer wavelengths. Azo compounds undergo very rapid E–Z isomerization from the lowest n,π*-excited singlet state; hence fluorescence and ISC quantum yields are usually very low. When E–Z isomerization of azo compounds is inhibited by steric constraints, fluorescence, nitrogen elimination or ISC become competitive (see below).

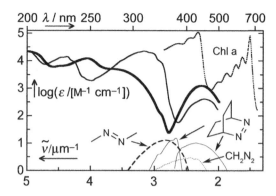

**Figure 6.9** Absorption spectra of azobenzene (hexane, *E*-isomer —, *Z*-isomer —),[280] azomethane (hexane, *E*-isomer, – – –),[280] chlorophyll a (Chl a; in diethyl ether, ....),[280] diazomethane (gas phase, ····);[1058] absorption and fluorescence spectra of 2,3-diazabicyclo [2.2.2]oct-2-ene [in perfluoromethylcyclohexane, — (2 thin lines); shifted downward by 1.5 units for better visibility][1059]

Many important pigments and dyes occurring in nature or used in technical applications are based on the cyclic tetrapyrrole chromophore porphyrin (Figure 6.10). There are 24 atoms participating in the conjugated system, but two double bonds (carbons 2, 3, 12 and 13) are easily hydrogenated, leaving a conjugated 18-membered ring that is bridged by two pyrrole-type nitrogen atoms (bacteriochlorin). Whereas chlorophylls (Figure 6.9) are 2,3-dihydroporphyrin (chlorin) derivatives having a magnesium ion replacing the two central hydrogen atoms, bacteriochlorophylls are 2,3,12,13-tetrahydroporphyrin (bacteriochlorin) derivatives (18-atom conjugated ring).

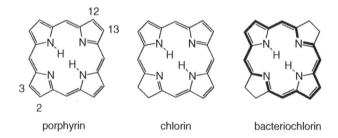

porphyrin          chlorin          bacteriochlorin

**Figure 6.10** Porphyrin and its analogues

All porphyrins exhibit two types of absorption bands, the so-called Q-band(s) of moderate intensity in the visible region and a very strong band in the near-UV, the B-band or *Soret band*. Such a pattern is predicted by simple MO models for cyclic systems with 18 π-electrons (Section 4.7, Figures 4.28 and 4.29). In the highly symmetric zinc tetrakis (perfluorophenyl)porphyrin (TFPP), the symmetry-forbidden Q-band lies at 578 nm, $\log(\varepsilon/[\text{M}^{-1}\,\text{cm}^{-1}]) = 3.7$, with a vibrational progression at 543 nm; the symmetry-allowed B-band at 412 nm is much more intense, $\log(\varepsilon/[\text{M}^{-1}\,\text{cm}^{-1}]) = 5.7$.[1060] Both the Q- and B-bands are due to a degenerate excited state in molecules that belong to the $D_{4h}$ point group. When the symmetry is lowered as in the metal-free H$_2$(TFPP), the Q-band splits into two, $Q_x = 635$ and 582 nm and $Q_y = 535$ and 505 nm. Figure 6.9 shows the B- and Q-bands for chlorophyll a, a magnesium chlorin derivative.

The great structural diversity of nitrogen-containing compounds, ranging from imines and azo compounds to nitro compounds and amines, is connected to its diverse but characteristic photochemical reactivity (Table 6.15). The presence of a lone electron pair on nitrogen in chromophores containing an N=X bond (X=N, C) indicates that both the n,π* and π,π* excited states can be involved in the reactions. While *E–Z* isomerization is a typical reaction for both imines and azo compounds (entry 1), the latter chromophores may additionally

**Table 6.15**    *Examples of primary photoprocesses of excited nitrogen-containing compouunds*

| Entry | Starting material[a] | Product(s) | Mechanism | Section |
|---|---|---|---|---|
| 1 | $\left[ \begin{array}{c} R' \\ N=X \\ R \end{array} \right]^*$   X = C, N | $\begin{array}{c} N=X \\ R \quad R' \end{array}$ | *E–Z* isomerization | 6.4.1 |
| 2 | $\left[ \begin{array}{c} R' \\ N=N \\ R \end{array} \right]^*$ | $R^\bullet + N_2 + R'^\bullet$ | Photofragmentation/ photoelimination | 6.4.2 |
| 3 | $\left[ R_2C=N_2 \right]^*$ | $R_2C{:} + N_2$ | Photofragmentation/ photoelimination | 6.4.2 |
| 4 | $\left[ R-N_2^{\oplus} \right]^*$ | $R^{\oplus} + N_2$ | Photofragmentation/ photoelimination | 6.4.2 |
| 5 | $\left[ R-N_3 \right]^*$ | $R-\ddot{N}{:} + N_2$ | Photofragmentation/ photoelimination | 6.4.2 |
| 6 | $\left[ R-ONO \right]^*$ | $R-O^\bullet + NO$ | Photofragmentation/ photoelimination | 6.4.2 |
| 7 | $\left[ \begin{array}{c} O^\ominus \\ \oplus| \\ R^{\nearrow N} \diagdown \\ R' \end{array} \right]^*$ | $\begin{array}{c} O \\ R^{\nearrow N} \diagdown \\ R' \end{array}$ | Photorearrangement | 6.4.2 |
| 8 | $\left[ R-NO_2 \right]^*$ | $R^\bullet + NO_2$ | Photofragmentation | 6.4.3 |
| 9 | $\left[ R-NO_2 \right]^* + [H]$ | $R-\overset{\bullet}{N}O_2H$ | Photoreduction | 6.4.3 |
| 10 | $R_3N{:} + [A] + h\nu$ | $R_3N^{\oplus\bullet} + A^{\ominus\bullet}$ | Electron transfer | 6.4.4 |
| 11 | $Ar(CN)_x + [D{:}] + h\nu$ | $Ar(CN)_x^{\ominus\bullet} + D^{\oplus\bullet}$ | Electron transfer | 6.4.4 |

[a][H] = hydrogen atom donor; [A] = electron acceptor; [D] = electron donor.

fragment via a homolytic cleavage to release a nitrogen molecule (entry 2) if the corresponding bond dissociation energies are sufficiently low. Fragmentation is the most common photoreaction of other chromophores, such as 3*H*-diazirines, diazo compounds (entry 3), diazonium salts (entry 4), azides (entry 5) and some heteroaromatic compounds (Section 6.4.2). Some of these processes are particularly efficient because thermodynamically stable and photochemically inert molecules, such as $N_2$ or NO, are liberated. The primary photochemical process of organic nitrites is also homolytic fission, in this case of the N–O bond (entry 6). The subsequent isomerization to the corresponding oximes is called the *Barton reaction*. The $\pi,\pi^*$ excited nitrones or heterocyclic *N*-oxides typically undergo *E–Z* isomerization and competing rearrangement to form oxazirine derivatives (entry 7).

Photofragmentation is the principal primary process of excited aliphatic nitro compounds subsequently producing $NO_2$ and the corresponding radicals (entry 8), which may recombine to form nitrites. The excited nitroalkanes or nitroarenes can also be photoreduced in the presence of a hydrogen-atom donor (entry 9).

Amines are generally good electron donors. They readily undergo photoinduced electron transfer (PET) processes, in which amine donates an electron to the reaction partner either in its ground or excited electronic state (entry 10). In contrast, electron-deficient, nitrogen-containing molecules, such as aromatic nitriles, may serve as electron acceptors (entry 11). Many organic metal complexes can also be involved in photochemically initiated redox reactions (Section 6.4.4).

### 6.4.1 Azo Compounds, Imines and Oximes: *E–Z* Photoisomerization

*Recommended review articles.*[1061–1070]
*Selected theoretical and computational photochemistry references.*[16,534,1071–1079]

Like alkenes (Section 6.1.1), chromophores containing the N=N (azo compounds) or C=N (imines, oximes, etc.) bonds can undergo *E–Z* (or *trans–cis*) photoisomerization (Scheme 6.156) and the resulting isomer concentration ratio in the *photostationary state* (PSS) reflects the absorption properties of the isomers and isomerization quantum yields (see Scheme 6.1 in Section 6.1.1). Since conventional (dark) synthesis generally provides access to more stable *E*-isomers, photochemistry is an exceptional tool for preparing sterically hindered *Z*-isomers.[1061,1062] The photoisomerization reaction can be induced by a direct irradiation or by sensitization and it often competes with other phototransformations, such as photofragmentation or photorearrangement (Section 6.4.2).

Unlike alkenes, the unsaturated azo group in N=N or C=N derivatives possess an in-plane lone electron pair in an n-orbital. As a result, both the $n,\pi^*$ and $\pi,\pi^*$ excited states can be involved in two limiting mechanisms of photoisomerization.[1062,1080,1081] The first mechanism is a 180° *rotation* (twist) about the former double bond in conjunction with a reduced bond order, similar to that observed in alkenes (Scheme 6.157). In the second case, an *in-plane inversion* proceeds due to rehybridization of one nitrogen without

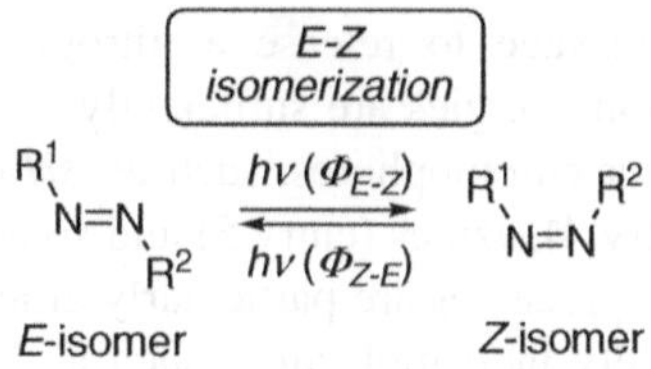

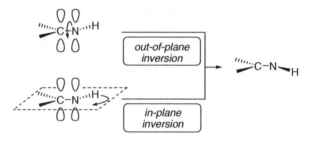

$$R^1 \atop N=N \atop R^2 \quad \begin{matrix} hv\,(\Phi_{E\text{-}Z}) \\ \rightleftarrows \\ hv\,(\Phi_{Z\text{-}E}) \end{matrix} \quad R^1 \quad R^2 \atop N=N$$

*E*-isomer                    *Z*-isomer

**Scheme 6.156**

any significant change in the bond order. The isomerization pathway is dependent on the excitation type. Quantum chemical calculations and also time-resolved experiments suggest that a rotational *E–Z* interconversion in imines and azo compounds occurs from their $^1n,\pi^*$ and $^3n,\pi^*$ states. This remains a subject of intense interest.[1062,1072,1082]

**Scheme 6.157**

*Azo Compounds*

Even in the absence of π-delocalization, azoalkanes absorb in the near-UV region, although the corresponding molar absorption coefficients (n,π*) are small. In the absence of competing photoreactions, some rare sterically hindered Z-isomers can be prepared. For example, (*Z*)-1,1′-azodinorbornane (**352**) was prepared in toluene at 0 °C from its *E*-precursor (Scheme 158).[1083]

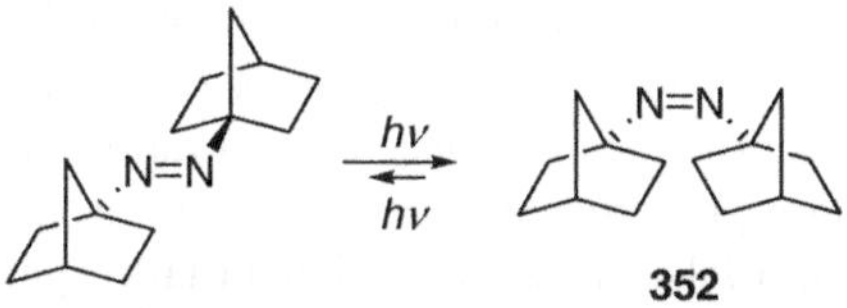

**352**

**Scheme 6.158**

Azoalkanes may readily undergo photoelimination of N$_2$ (Section 6.4.2) when stable radical intermediates are produced. For example, whereas irradiation in the n,π* absorption region (>400 nm) of the cyclic 1,4-dihydronaphthodiazepine **353** induces *E–Z* isomerization, nitrogen eliminates to yield acenaphthene exclusively upon π,π* excitation of the naphthalene moiety at 313 nm (Scheme 6.159).[1084]

The photophysical and photochemical properties of cyclic azo compounds, where *Z* → *E* isomerization is suppressed by geometric constraints, are very sensitive to seemingly minor structural changes (Table 6.16). In contrast to 2,3-diazabicyclo[2.2.1]

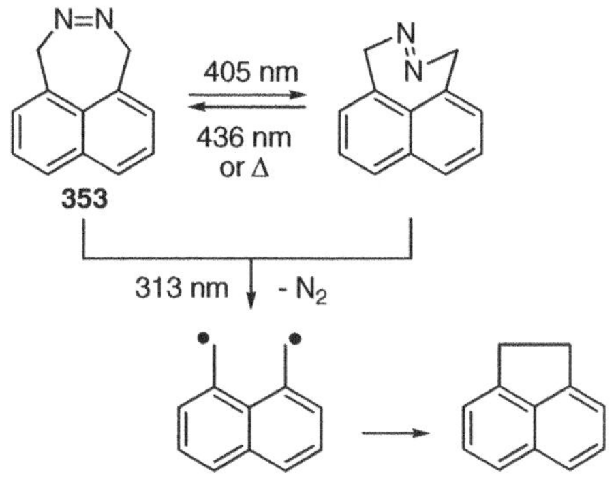

**Scheme 6.159**

hept-2-ene (DBH), 2,3-diazabicyclo[2.2.2]oct-2-ene (DBO) possesses a remarkably long singlet lifetime, a high fluorescence quantum yield and a low quantum yield of denitrogenation, which has been exploited, for example, to measure the kinetics of conformational changes in biopolymers (Special Topic 2.3). The tricyclic diazabicyclo-heptene (fused DBH) is exceptional in exhibiting efficient intersystem crossing, a relatively long triplet lifetime and strong phosphorescence.[1085]

**Table 6.16** *Photophysical properties of bicyclic azoalkanes*[1085]

| Property | 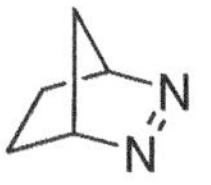 (DBH) | 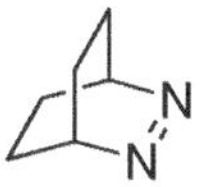 (DBO) | 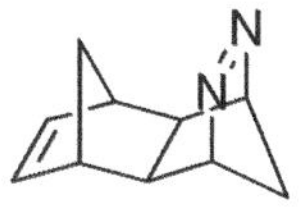 (fused DBH) |
|---|---|---|---|
| $^1\tau$/ns | 0.15 | 690[a] | 2.4 |
| $^3\tau$/ns | 7[b] | 25[c] | 630 |
| $\Phi_{ISC}$ | ~0 | ~0 | ~0.5 |
| $\Phi_f$ | <0.001 | 0.37[d] | 0.02 |
| $\Phi_{-N2}$ | 1.0 | 0.02[d] | 0.59 |
| $\Phi_{ph}$ (77 K) | 0.00 | 0.00 | 0.052 |
| $E_S$/(kJ mol$^{-1}$) | 354[b] | 318[b] | 327 |
| $E_T$/(kJ mol$^{-1}$) | 260[d] | 222[d] | 261 |

[a]Ref. 1059.
[b]Ref. 1086.
[c]Ref. 1087.
[d]Ref. 1088.

Differences in absorption of (*E*)- and (*Z*)-azobenzene (Figure 6.9) enable us to select the irradiation wavelength to obtain the Z-isomer preferentially;[559,1089,1090] however, the photostationary concentration ratio of the isomers is affected by competing thermal $Z \rightarrow E$ isomerization ($E_a \approx 100$ kJ mol$^{-1}$, compared with $E_a \approx 176$ kJ mol$^{-1}$ for stilbene). Many (*Z*)-azobenzene derivatives can be isolated successfully at lower temperatures by chromatographic separation.[1061,1062,1091] Since elimination of nitrogen to give unstable aryl radicals from azobenzene is negligible, *E–Z* isomerization is usually the principal photochemical process observed, which is valuable for many applications.

## Special Topic 6.15: Photochromism

*Photochromism* is a *reversible* chemical transformation, induced in one or both directions by the absorption of light, between two forms having different absorption spectra and other physical properties.[624] Originally, the expression 'photochromic' was used only for compounds exhibiting a light-induced reversible change of colour. In the example shown in Figure 6.11, only the thermodynamically stable isomer 2,7-dihydro-2,2,7,7-tetramethylpyrene absorbs at wavelengths exceeding 330 nm, so that it can be completely converted to the colourless 2,2,7,7-tetramethyldicyclopropa[*a,g*]pyracene by irradiation at $\lambda_{irr} > 330$ nm. Irradiation at 313 nm leads to a photostationary state (PSS) (dashed line), because both isomers absorb at this wavelength. However, complete conversion to the more stable dihydropyrene proceeds in the dark at temperatures above 90 °C.[1092] The reaction medium and the presence of oxygen affect the kinetics and the reversibility of many photochromic systems. Any irreversible (usually photochemical) formation of minor side-products will limit the number of cycles that can be performed and is referred to as *fatigue*.

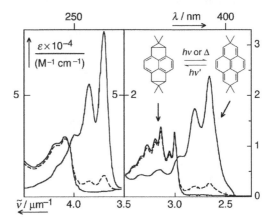

**Figure 6.11**    Photochromism of 2,7-dihydro-2,2,7,7-tetramethylpyrene

There are several families of compounds that display photochromic behaviour involving various mechanisms. The following list of reactions reviews the most common systems (Scheme 6.160). Spiropyrans, spirooxazines, chromenes and fulgides undergo concerted or non-concerted electrocyclization reactions (Section 6.1.2; see also Case Study 6.4), azobenzenes involve *E–Z* isomerization (this section),[624,1093-1101] quinones exhibit a group transfer,[149,1024] and polycyclic aromatic hydrocarbons undergo cycloaddition reactions (Section 6.2.2). Many biological systems are also photochromic; for example, rhodopsin exhibits reversible *E–Z* photoisomerization (Special Topic 6.1).

Photochromism has many potential and existing applications that take advantage of a change in colour or other physicochemical properties during the process, for example variable-transmission optical materials such as photochromic eyeglass or ophthalmic lenses that darken in sunlight (using spiropyran and spirooxazine systems in addition to

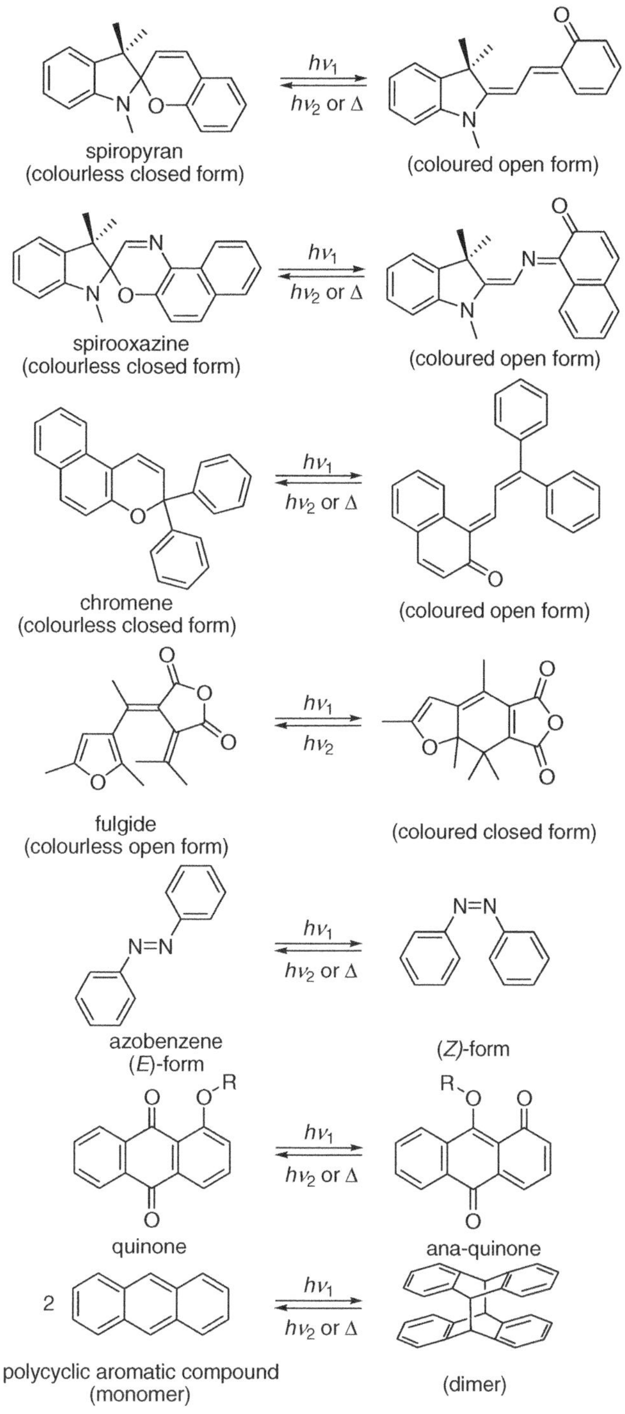

hν1
hν2 or Δ
spiropyran
(colourless closed form)
(coloured open form)
spirooxazine
(colourless closed form)
(coloured open form)
chromene
(colourless closed form)
(coloured open form)
hν2
fulgide
(colourless open form)
(coloured closed form)
azobenzene
(E)-form
(Z)-form
quinone
ana-quinone
2
polycyclic aromatic compound
(monomer)
(dimer)


**Scheme 6.160**

silver halides),[1102] novelty items, such as toys, cosmetics and clothing (involving a reversible change of pigment colour), supramolecular chemistry and nanotechnology (Special Topic 6.19),[1103,1104] biochemistry and biology (Special Topic 6.18),[1034] or liquid crystal alignment.[1105] Although the recording process on CDs or DVDs is based on dye sensitization of polycarbonate layer melting (burning) by laser radiation to change the refractive index of its surface,[1106] utilization of many photochromic systems is being considered for optical information storage.[1100,1103]

For example, reversible photoswitching (Special Topic 6.18) of coordination sites in a self-assembled monolayer containing azopyridine moieties attached to an 8 nm thick film of gold supported on quartz via an $-O(CH_2)_8S-$ linker (Scheme 6.161) has been reported.[1107] Zinc tetraphenylporphyrin (ZnTPP) is coordinated to the nitrogen atom of the pyridine ring of the *E*-form of azopyridine, aiming outwards from the support surface. Waveguided illumination (shown by the arrows) by 365 nm light results in the formation of *Z*-isomers and subsequent, sterically driven release of ZnTPP. Irradiation at 439 nm reverts the azo compound configuration and allows to ZnTPP to re-coordinate. Such a photochromic system presents an opportunity for developing miniaturized (Special Topic 6.19) optical switches controlled by waveguided light.

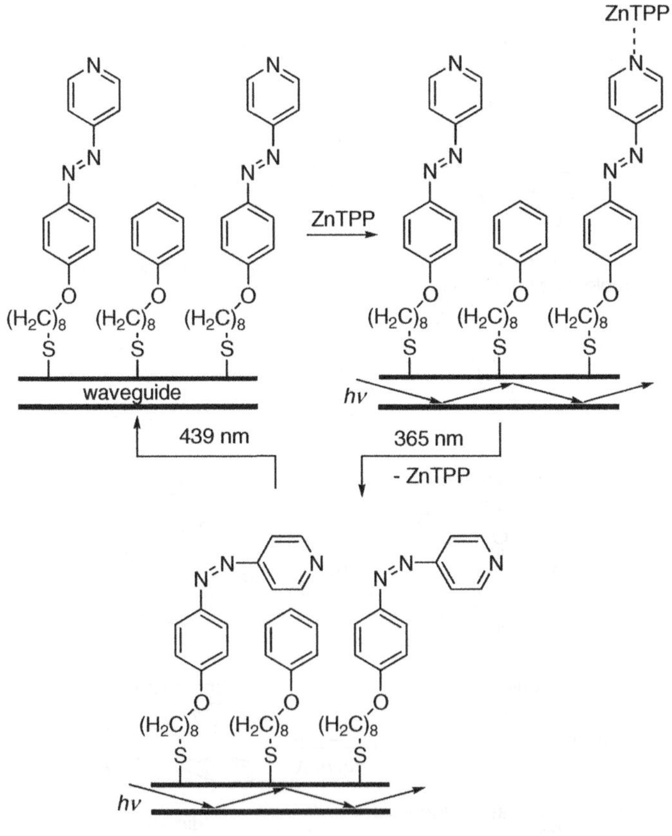

**Scheme 6.161**

---

## Case Study 6.28: Supramolecular chemistry – photoresponsive crown ethers

Photochemical butterfly-like $E \rightarrow Z$ photoisomerization of a bis(crown ether) azobenzene derivative **354** was found to be thermally reversible and the stereoisomers exhibit unique contrasting behaviour in the presence of metal ions.[1108] The concentration of the Z-isomer in the photostationary state was noticeably enhanced by the addition of $K^+$, $Rb^+$ or $Cs^+$, because the corresponding Z-complex achieved a stable sandwich geometry (Scheme 6.162). As a result, the cations could be selectively extracted by the Z-derivative from an aqueous phase to an organic solvent (*o*-dichlorobenzene), whereas no complexation (i.e. no transfer) took place in the case of the *E*-isomer.

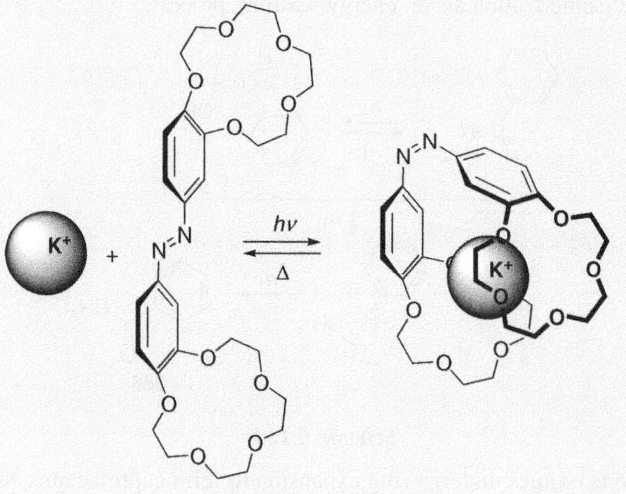

**354**

*Scheme 6.162*

*Experimental details.*[1108] An *o*-dichlorobenzene solution of **354** ($2 \times 10^{-4}$ M; 100 ml) and an aqueous solution of MOH (M = K, Rb or Cs) (25 ml) placed in a U-tube immersed in a thermostated water-bath were irradiated with a high-pressure mercury lamp (500 W) (Figure 3.9). Liquid–liquid phase transfer of cations between the layers was followed by absorption spectroscopy.

---

*Imines and Oximes*

Absorption bands (both $n,\pi^*$ and $\pi,\pi^*$) of *N*-alkylimines are generally below 260 nm, whereas those of aryl derivatives are bathochromically shifted. Efficient production of Z-isomers by photoisomerization reaction [e.g. in *N*-benzalaniline (**355**); Scheme 6.163] is usually feasible only at low temperatures because the thermal reversion has a very low activation barrier ($E_a \approx 65$ kJ mol$^{-1}$).[1061]

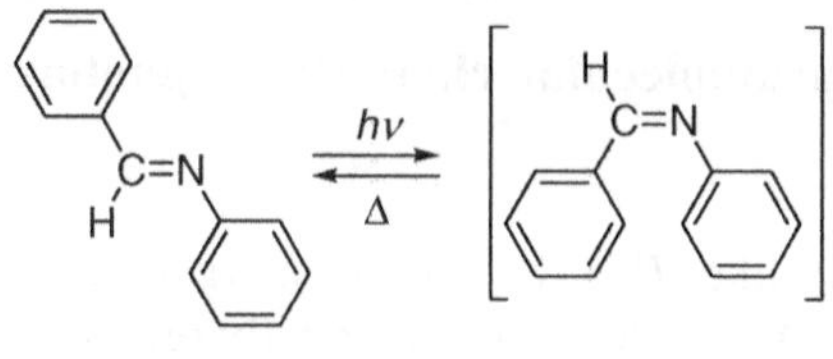

(*E*)-**355**                          (*Z*)-**355**

*Scheme 6.163*

The presence of the hydroxy (alkoxy) group in conjugation with the C=N bond in oximes (alkyloximes) considerably reduces rates of the thermal *E*–*Z* interconversion.[1061] Oximes typically undergo the Beckmann-type photorearrangement to give the corresponding amide **356** in several steps (Scheme 6.164). This reaction is limited, however, by a competing *E*–*Z* isomerization as an 'energy wasting' process.

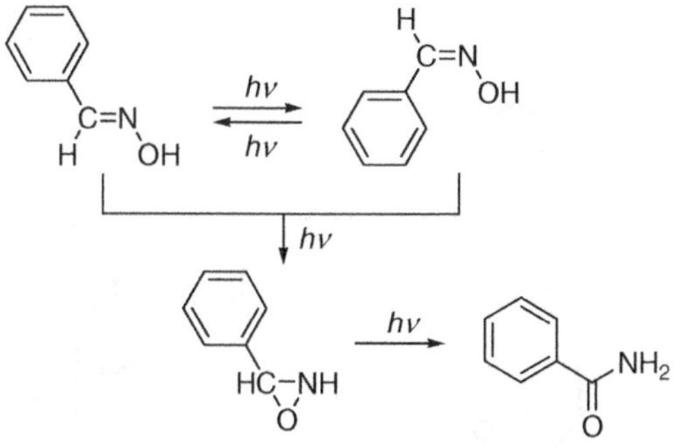

**356**

*Scheme 6.164*

Cyclohexanone oximes undergo ring expansion to form caprolactams, such as **357** and **358** (Scheme 6.165).[1109] The mechanism involves *E*–*Z* photoisomerization, followed by transformation of the singlet excited oximes to an oxaziridine intermediate and subsequent concerted isomerization of singlet excited oxaziridine to the corresponding lactams.[1061]

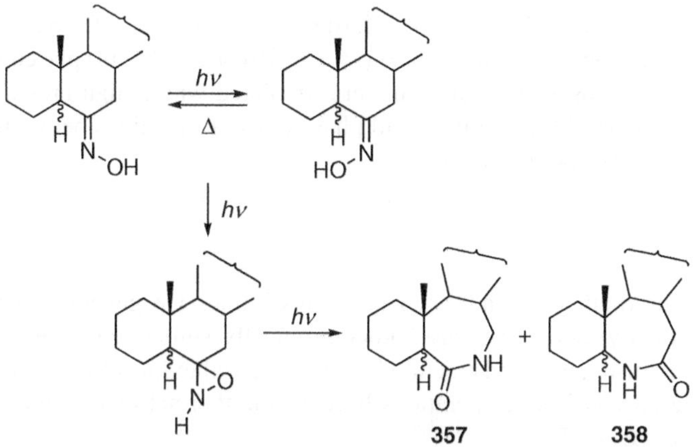

**357**          **358**

*Scheme 6.165*

## 6.4.2 Azo Compounds, Azirines, Diazirines, Diazo Compounds, Diazonium Salts, Azides, *N*-Oxides, Nitrite Esters and Heteroaromatic Compounds: Photofragmentation and Photorearrangement

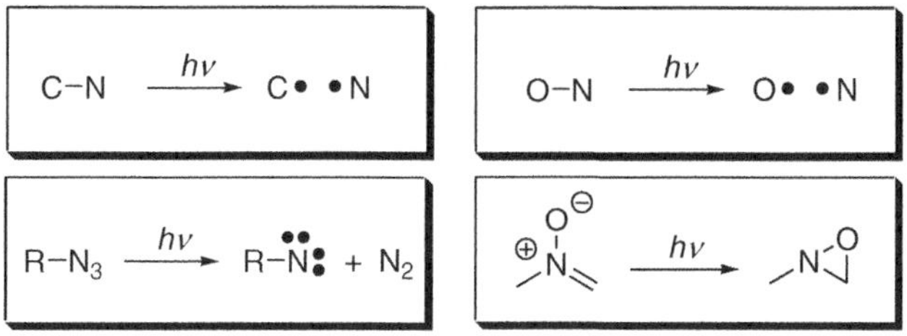

$$C{-}N \xrightarrow{h\nu} C{\bullet} \; {\bullet}N$$

$$O{-}N \xrightarrow{h\nu} O{\bullet} \; {\bullet}N$$

$$R{-}N_3 \xrightarrow{h\nu} R{-}\ddot{N}\colon + \; N_2$$

$$\overset{\oplus}{N}\underset{}{\overset{O^{\ominus}}{\|}}\!\!\diagdown \xrightarrow{h\nu} \diagup N\!\!\diagdown^{O}$$

*Recommended review article.*[1062,1110–1117]
*Selected theoretical and computational photochemistry references.*[16,535,1118–1130]

*Fragmentation* and *rearrangement* are typical primary photoinduced processes in many nitrogen-containing organic compounds. The moieties described in this section typically possess excellent leaving groups (e.g. the nitrogen molecule) and their excitation results in the fission of the weakest bond in the primary step. Many of those photoprocesses have their thermal (dark) counterparts.

*Azo Compounds*

Photochemically or thermally induced elimination of nitrogen from aliphatic azo compounds (see also Section 6.4.1) is accompanied by the formation of alkyl radicals $[D_{C-N}(\text{azoethane}) = 209\,\text{kJ}\,\text{mol}^{-1}]$.[874] For example, photodecomposition of 2,2′-azobisisobutyronitrile (AIBN), a well-known thermally activatable initiator of radical reactions, produces cyanodimethylmethyl radicals (Scheme 6.166).[1131,1132] Similar reactions have been successfully utilized in some radical polymerization reactions, where azo compounds serve as photoinitiators[1114,1115,1133,1134] (see Section 6.8.1).

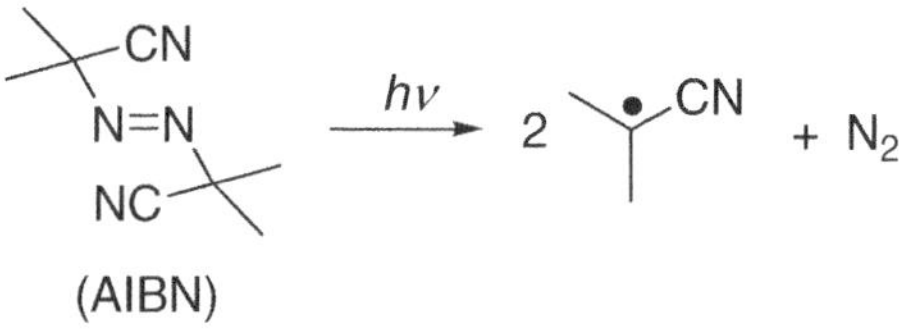

$$\underset{\text{(AIBN)}}{\overset{\displaystyle \overset{\text{CN}}{\diagup}}{\underset{\text{NC}\diagdown}{\text{N=N}}}} \xrightarrow{h\nu} 2 \;\; \underset{}{\overset{\bullet}{\diagup}}\!\!\text{CN} \; + \; N_2$$

*Scheme 6.166*

*2H-Azirines*

Substituted 2*H*-azirines **359** that usually absorb only below 300 nm are known to undergo photochemical irreversible ring opening via C–C bond cleavage to give nitrile ylides (1,3-dipoles) (**360**) as reactive intermediates with strong absorption in the near-UV or visible region, which can be directly observed in frozen matrices (Section 3.10) by steady-state or time-resolved spectroscopy (Scheme 6.167 ).[1116,1135–1139] The [3 + 2] cycloadditions of ylides with dipolarophiles (such as alkenes or carbonyl compounds) provide a convenient method for synthesizing five-membered heterocyclic systems. For example, photolysis of a solution of phenylazirine (**361**) with an excess of methyl acrylate produces the pyrrolinecarboxylate **362** in 80% chemical yield,[1140] and **363** irradiated in the presence of benzaldehyde gives the oxazoline **364** in 20% chemical yield[1141] (Scheme 6.168). The ylide intermediates can also be trapped by various nucleophiles.[1116]

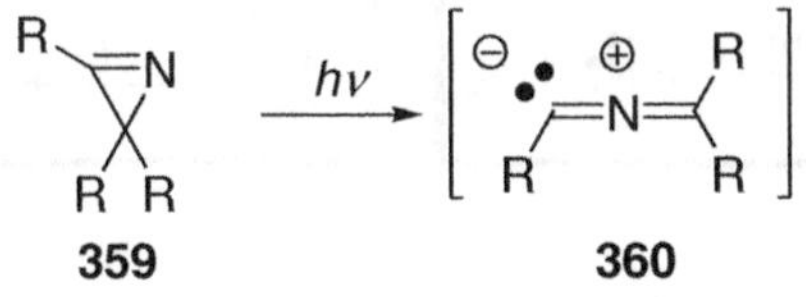

**359**                                              **360**

*Scheme 6.167*

**361**                                              **362**

**363**                                              **364**

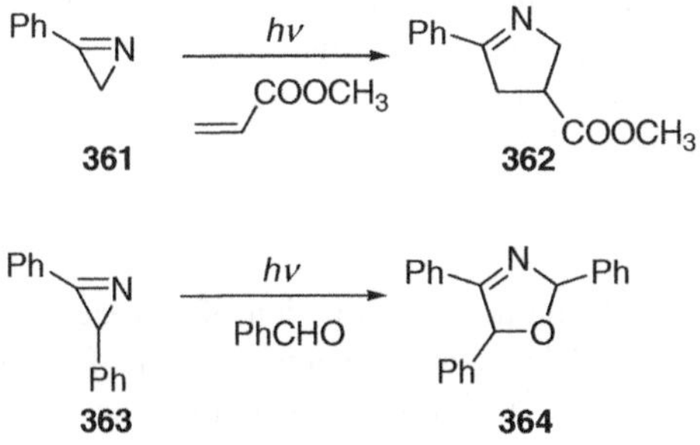

*Scheme 6.168*

### 3H-Diazirines

3*H*-Diazirines have been recognized as important photochemical and thermal precursors to carbenes (Section 5.4.1).[1115,1142–1145] The N=N bond in diazirines, constrained to a three-membered ring, generally displays strong absorption between 310 and 350 nm. Various short-lived intermediates, including singlet excited diazirine, singlet carbene (**365**) and biradical (**366**), may be involved in photolysis of **367** (Scheme 6.169). These reactive species then undergo various rearrangement or bimolecular reactions. Diazirines can also photoisomerize to diazo compounds.[1145] Mechanistic studies of diazirine photochemistry often utilize state-of-the-art methods, such as low-temperature matrix photochemistry (Section 3.10), to trap and detect reactive intermediates.

### Diazo Compounds

Diazoalkanes display a weak absorption band between 300 and 500 nm (Figure 6.9).[1115] An excited singlet state, formed upon irradiation in this region, eliminates nitrogen

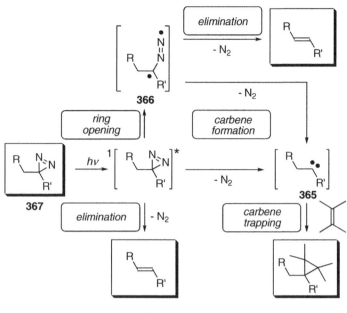

**Scheme 6.169**

$[D_{C=N_2}(\text{diazomethane}) \approx 200 \, \text{kJ mol}^{-1}]^{1146}$ to form a singlet carbene intermediate (Scheme 6.170). Subsequent reactions of the carbene moiety are generally the same as those discussed in the previous paragraph.

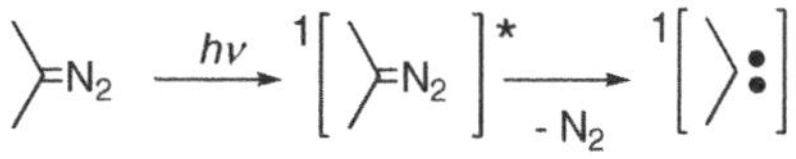

$$\underset{}{\diagup}\!\!=\!\text{N}_2 \xrightarrow{h\nu} {}^1\!\!\left[\underset{}{\diagup}\!\!=\!\text{N}_2\right]^* \xrightarrow[-\,\text{N}_2]{} {}^1\!\!\left[\triangleright\!\!:\right]$$

**Scheme 6.170**

α-Diazocarbonyl compounds represent an important class of photolabile compounds because of their applications in lithographic (see also Special Topic 6.27) production of integrated circuits used by the computer industry, photoaffinity labelling (Special Topic 6.16) and DNA cleavage experiments,[1147] or organic synthesis.[1110,1148] A typical mechanism of the photodegradation of the α-diazocarbonyl compound **368** is the photo-Wolff rearrangement,[1149-1151] which has been suggested to proceed via either simultaneous elimination of nitrogen and rearrangement to a ketene **369** or via a carbene **370** intermediacy (Scheme 6.171).[1152,1153]

For example, steady-state photolysis of 2-diazoindan-1,3-dione (**371**) in alcohol gives a diester **372** in two photochemical steps (Scheme 6.172).[1154,1155]

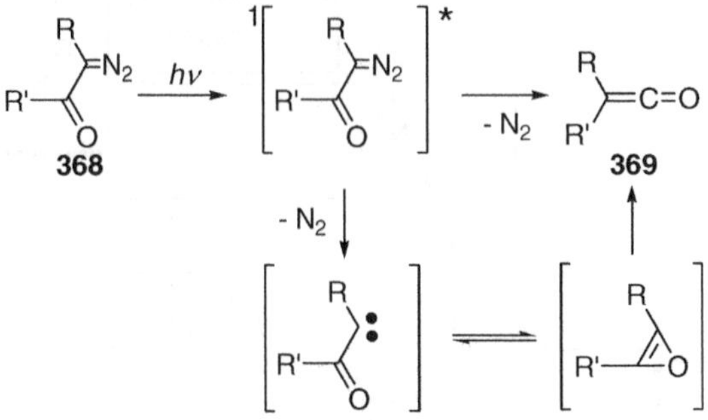

**Scheme 6.171**

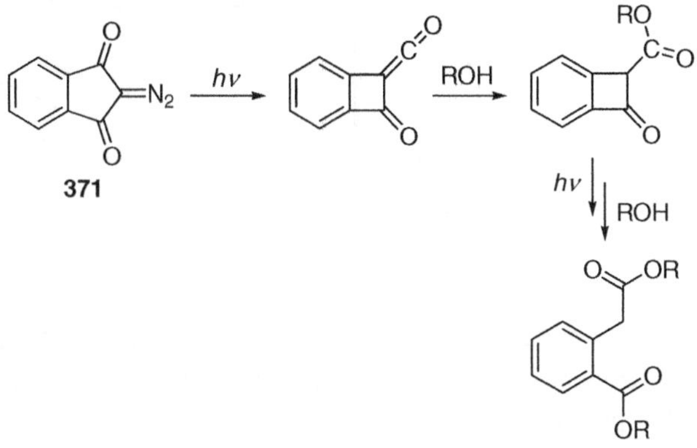

**Scheme 6.172**

## Case Study 6.29: Mechanistic photochemistry – singlet–triplet interconversion of carbenes

Diphenylcarbene (diphenylmethylene) can be generated from diphenyldiazomethane (**373**) by direct irradiation or by triplet sensitization.[1156] The intermediate multiplicity then controls the subsequent reactions: the singlet carbene inserts into the O—H bond of methanol, whereas the triplet carbene adds to an alkene (Scheme 6.173). It has been found that singlet and triplet diphenylcarbenes are in rapid equilibrium relative to the rates of reactions.[1157,1158] Competitive quenching experiments (to obtain $k^1$ and $k_{TS}$) and laser flash spectroscopy (Section 3.7; to obtain $k^2$ and $k_{ST}$) allowed the determination of the free energy difference between the singlet and triplet states of carbene ($\sim$20 kJ mol$^{-1}$).

*Experimental details.*[1157] A solution of **373** in acetonitrile ($2.5 \times 10^{-3}$ M) containing methanol (0.05 M) and isoprene (0.10–10 M) was purged with nitrogen and irradiated with a high-pressure xenon lamp (150 W) through optical filters ($\lambda_{irr} \approx 366$ nm; Figure 3.9). The products shown in Scheme 6.173 were analysed by GC.

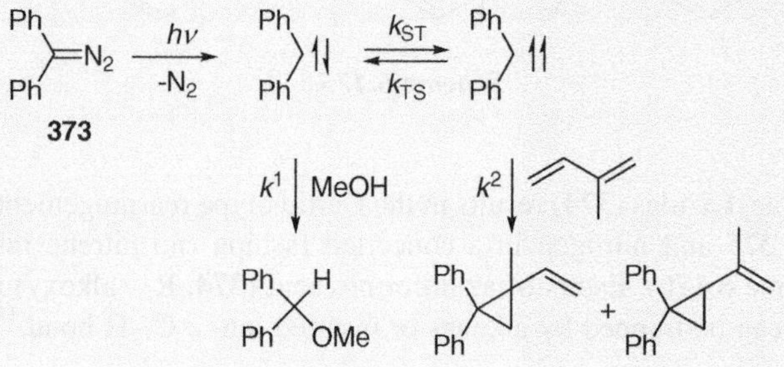

**Scheme 6.173**

## Diazonium Salts

Unsubstituted benzenediazonium ion in solution shows strong absorption maxima at $\lambda_{max} \approx 300$ and 261 nm.[1111] As with thermal decomposition, its principal photoreaction is release of the nitrogen molecule [e.g. $D_{C-N}$(benzenediazonium) $\approx 154$ kJ mol$^{-1}$][1159] to give an aryl cation,[1160,1161] which can be readily attacked by a nucleophile such as water (Scheme 6.174). In contrast, an aryl radical is formed in the presence of an electron donor (such as methanol) by electron transfer, followed by radical reactions, such as hydrogen abstraction from an H-atom donor ([H]).

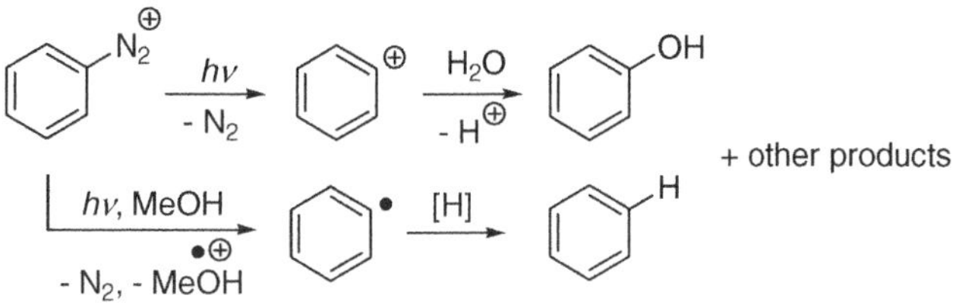

**Scheme 6.174**

## Azides

Unsubstituted organic azides absorb appreciably in the near-UV region (<380 nm). Their direct irradiation leads to the extrusion of molecular nitrogen to form singlet nitrene (Section 5.4.2), which can intersystem cross to triplet nitrene (Scheme 6.175).[1112,1114,1162] Triplet nitrene can also be obtained by photosensitization. In general, nitrenes are reactive intermediates that can undergo various reactions.

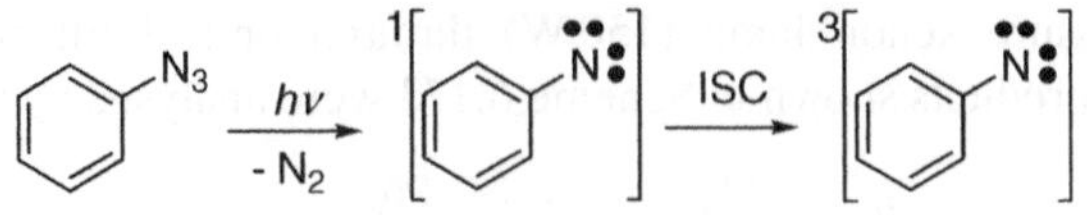

*Scheme 6.175*

Irradiation of acyl azides (**374**) results in the Curtius-type rearrangement[1163] to produce the isocyanates **375** and nitrogen in a concerted fashion (no nitrene intermediates are involved) (Scheme 6.176). Esters of azidoformic acid (**374**, R = alkoxy) produce singlet nitrenes, which can be trapped by alkenes or inserted into a C−H bond.[1112,1164]

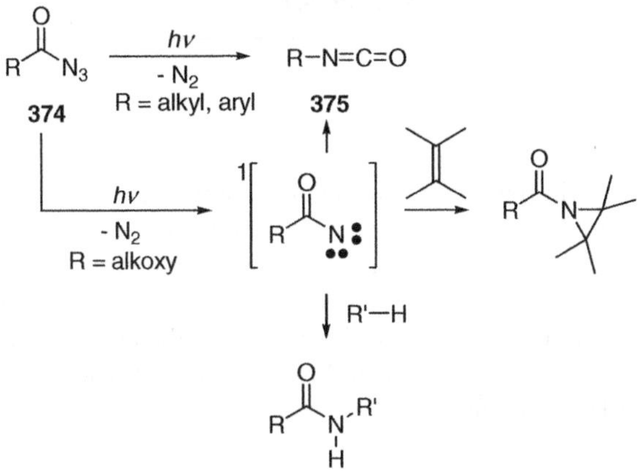

*Scheme 6.176*

Arylnitrenes formed from aryl azides (e.g. **376**) undergo readily photoinduced ring expansion to give a dehydroazepine **377** that can be trapped by nucleophiles (NuH) (Scheme 6.177).[406,411,1113,1164,1165]

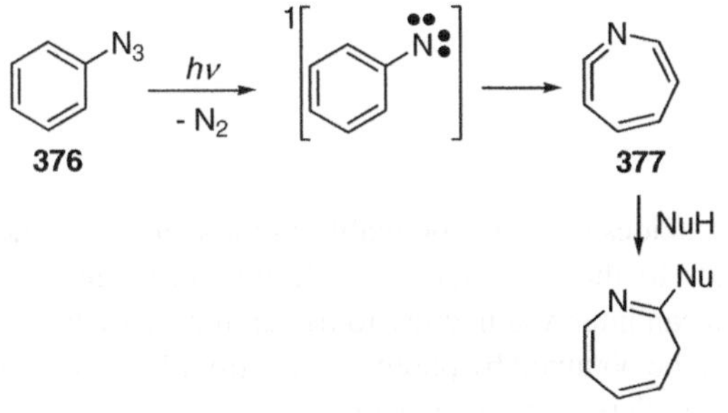

*Scheme 6.177*

## Special Topic 6.16: Photoaffinity labelling

Photoaffinity labelling is a technique for studying the structures of proteins, DNA and other biomolecules, and also biomolecule–ligand and biomolecule–biomolecule transient interactions, in order to understand specific biochemical mechanisms.[854,1147,1166–1171] In this technique, a ligand (label, probe), often fluorescent or radioactive, bears a photoactivatable group (see also Special Topic 6.18), which upon irradiation generates a highly reactive intermediate that appends to the specific site on a biomolecule in its vicinity through a covalent bond (Figure 6.12). Such a modified biomolecule, either directly or after some modifications (such as chemical fragmentation), is examined spectroscopically, radiochemically or by conventional chemical analyses. When the photoactivatable group is a part of another biomolecule, cross-linking between biomolecules may take place.

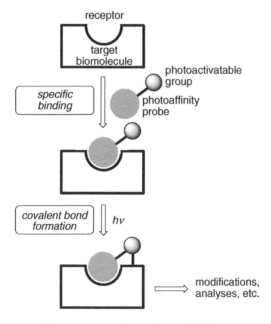

**Figure 6.12** Photoaffinity labelling

Photoactivatable groups should be sufficiently stable under ambient light and the photoreactions should be faster than dissociation of the ligand–receptor complex and be site specific.[1147] The reactions of typical photoaffinity groups have already been described: azides (Scheme 6.175) or 3*H*-diazirines (Scheme 6.169) and diazo compounds (Scheme 6.170) form nitrenes or carbenes, respectively, upon irradiation, whereas excited benzophenone derivatives abstract hydrogen to form ketyl radicals that recombine to form a new bond (Scheme 6.99). Carbenes are generally recognized as being more reactive than nitrenes. A mechanistic study of 3-*p*-tolyl-3-trifluoromethyl-diazirine (**378**) photoreactivity has shown, for example, that *p*-tolyl(trifluoromethyl) carbene (**379**) formed upon irradiation may add to an unsaturated or aromatic system to

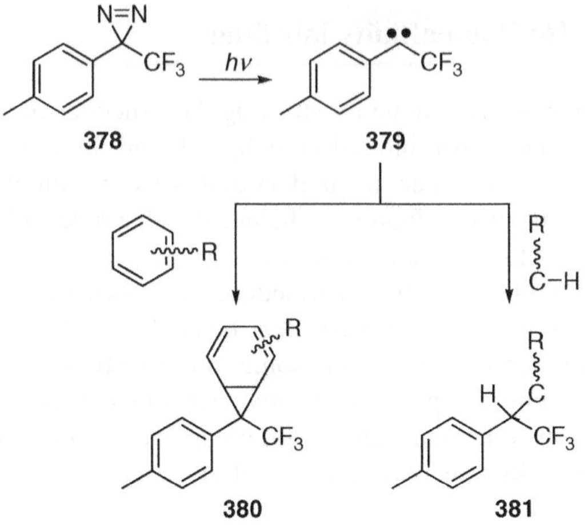

**Scheme 6.178**

form a cyclopropane intermediate **380** (and subsequent rearrangement products) or insert to an inactivated C—H bond to form **381** (Scheme 6.178).[1172]

Here we show two examples of photoaffinity labels. A diazo group containing immunosuppressant cyclosporin A (**382**), which binds to protein cyclophilin and undergoes specific cross-linking upon irradiation due to formation of the corresponding aryltrifluoromethycarbene, was used to study signalling pathways involved in immunomodulation.[1173] The enzymatically non-cleavable azidoanilido guanosine triphosphate analogue **383** was utilized as an efficient label for the G protein (a protein involved in second messenger cascades).[1174] The $^{32}$P tag was used for identification and active site mapping.

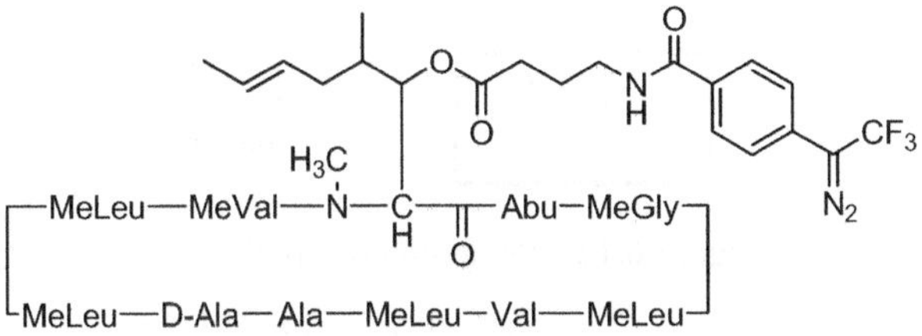

**382**

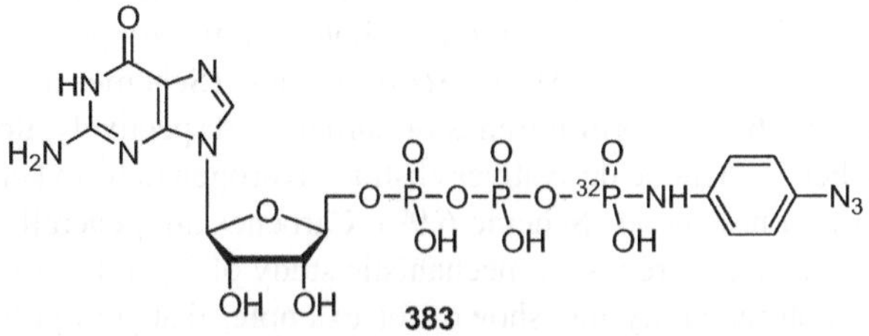

**383**

## N-Oxides

Nitrones or heterocyclic *N*-oxides are examples of photolabile compounds with polarized N—O bonds, where nitrogen is sp$^2$ hybridized. They possess a $\pi,\pi^*$ lowest-energy absorption band with a strong charge-transfer character.[1061,1067] Apart from *E–Z* isomerization, they undergo a characteristic rearrangement to form oxazirine inter-mediates (Scheme 6.179), which are photolabile and may react further. For example, azanaphthalene *N*-oxide (**384**) affords ring enlargement to benzoxazepine (**385**) in aprotic solvents or rearranges in the presence of water to an indoline derivative **386**, possibly via an oxazirine intermediate (Scheme 6.180).[1175]

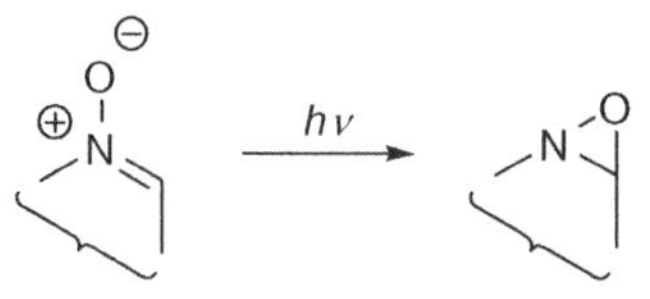

*Scheme 6.179*

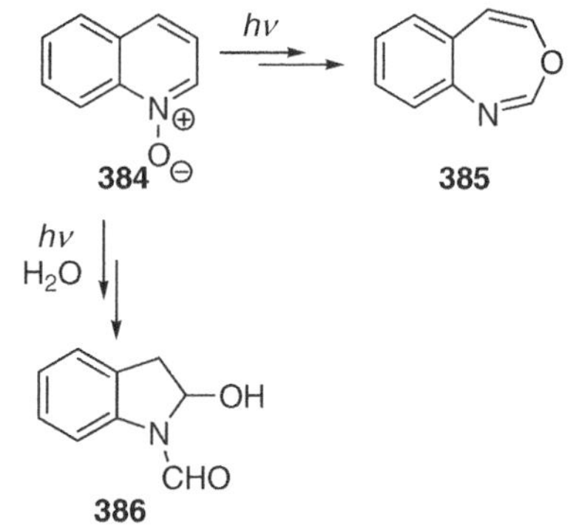

*Scheme 6.180*

## Nitrite Esters

The primary photochemical process of organic nitrites is homolytic fission of the N—O bond[1176] [$D_{O-N}$(ethyl nitrite) $\approx$ 176 kJ mol$^{-1}$].[874] When alkyl δ-hydrogens are available, intramolecular δ-hydrogen abstraction by the resulting alkoxyl radical to generate a carbon radical, which further combines with the nitric oxide hereby formed and isomerizes to the corresponding oxime, is called the *Barton reaction*[1177] (e.g. nitrite ester **387** photolysis[1178] in Scheme 6.181). This reaction provides a unique tool for preparing suitable δ-substituted derivatives in steroids (an oxime group is easily transformed into a carbonyl moiety, for example), because the corresponding methyl groups and alkoxyl radicals with a 1,3-diaxial configuration (**388**) prefer hydrogen abstraction through a six-membered cyclic transition state.[1176]

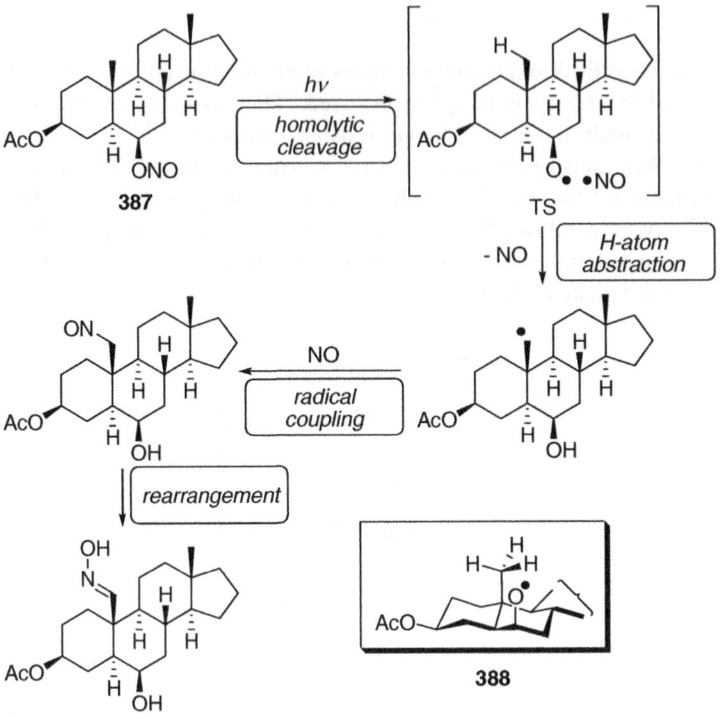

*Scheme 6.181*

*Heteroaromatic Compounds*

Nitrogen-containing heteroaromatic compounds, such as triazoles, tetrazoles,[1111] pyrazoles and 1,2,4-oxadiazoles[1117] undergo various photoinduced isomerization and ring-opening reactions. Photolysis of 1*H*-benzotriazole (**389**), for example, leads to fast and efficient, yet reversible, N–N bond fission to give the diazo compound **390**.[1179] Nitrogen elimination products are obtained only upon prolonged irradiation and typically with low chemical yields (Scheme 6.182).

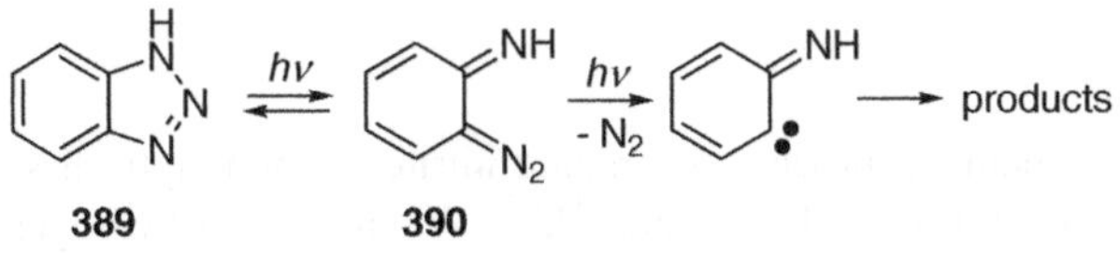

*Scheme 6.182*

Photochemical transposition reactions of some heteroaromatic compounds have already been discussed in Section 6.2.1. Scheme 6.183 shows the photoisomerization of 1-methylpyrazole (**391**), which may involve competition between electrocyclic ring closure and cleavage of the N1–N2 bond to give the same product (1-methylimidazole, **392**).[1117,1180]

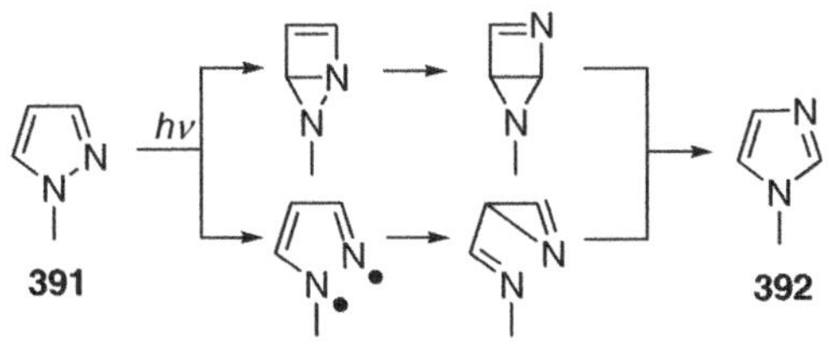

*Scheme 6.183*

## Special Topic 6.17: Photochemistry on early Earth and in interstellar space

The primitive atmosphere on early Earth was composed of nitrogen, methane, ammonia, carbon dioxide and other simple inorganic and organic molecules.[1181,1182] As a heterogeneous system of dust, aerosol particles and water droplets, it was exposed to high-energy radiation (<250 nm) from the young Sun. Complex abiogenic processes presumably produced biologically important compounds essential for emerging life. Some of them, for example guanine and cytosine base pairs of DNA, possess extraordinary photostability (Special Topic 6.7), which could have been an important selective factor in determining the eventual chemical composition of biomolecules.

Laboratory studies have provided evidence that photochemical and photocatalytic (Section 6.8) steps might play an important role in the formation of amino acids or various heterocyclic compounds from very simple molecules. For example, UVC irradiation of acetonitrile–ammonia–water mixture produces hexamethylenetetramine, a potential precursor of amino acids, via two-step photoinitiated fragmentation of acetamide (formed by acetonitrile hydrolysis) to give carbon oxide, which undergoes further photochemical and dark reactions (Scheme 6.184).[1183]

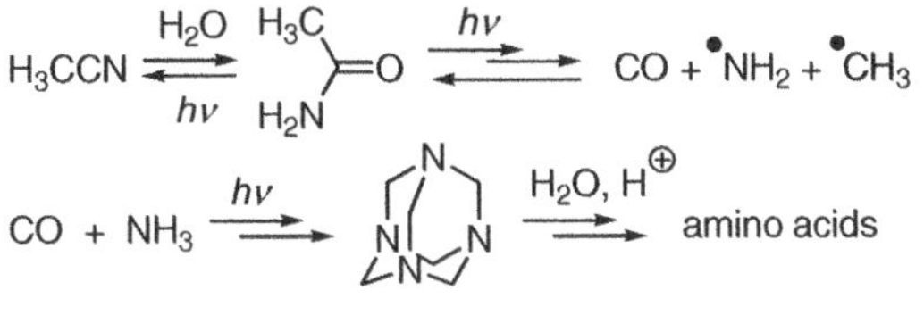

*Scheme 6.184*

Infrared observations, combined with laboratory simulations, have also advanced the understanding of chemical processes occurring in comets in interstellar space.[1184] Comets are ices made of simple molecules, such as $H_2O$, $CH_3OH$, $NH_3$, CO and $CO_2$, although more complex species, including nitriles, ketones, esters or aromatic hydrocarbons, can also be present. Chemical changes can be promoted due to penetrating cosmic radiation or absorbed solar radiation. In the laboratory, UV photolysis (usually by a hydrogen-flow discharge lamp producing Lyman-$\alpha$ emission, $\lambda_{irr} < 200$ nm, in a high vacuum) of cometary ice analogues at temperatures below 50 K gives moderately complex organic molecules, such as ethanol, formamide, acetamide, nitriles,[1185] and even amino acids (Scheme 6.185).[1186] The subsequent delivery of extraterrestrial matter to Earth is suggested to have been an alternative source of prebiotic organic molecules.

$$H_3COH + NH_3 + H_2O \quad \xrightarrow[12\,K]{h\nu} \quad \text{glycine, alanine, serine,}$$
$$+ CO + CO_2 \quad \quad \text{sarcosine, valine, proline, etc.}$$

*Scheme 6.185*

### 6.4.3  Nitro Compounds: Photofragmentation and Photoreduction

$$R-NO_2 \xrightarrow{h\nu} R^{\bullet} + NO_2 \qquad R-NO_2 \xrightarrow{h\nu,\,[H]} R-\overset{\bullet}{N}O_2H$$

*Recommended review article.*[1187]
*Selected theoretical and computational photochemistry references.*[1188–1191]

Simple nitroalkanes absorb below 350 nm and are excited to the lowest singlet n,π* state, which efficiently intersystem crosses to $T_1$. Homolytic photocleavage is the principal primary process in both the gas phase and solution to produce alkyl radicals and $NO_2$ (Scheme 6.186).[1187] Apart from subsequent recombination of the radical intermediates to form nitrites (Section 6.4.2), competing hydrogen abstraction (photoreduction) involving an excited nitro compound and a hydrogen-atom donor may take place.

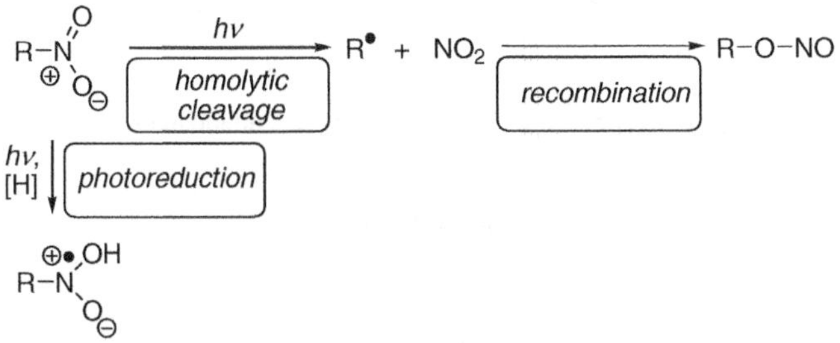

*Scheme 6.186*

Nitroarenes display strong absorption in the near-UV region and are efficiently photoreduced.[1187] An intermolecular version of this reaction is depicted in Scheme 6.187. The photoreduction is initiated by hydrogen atom abstraction. Substituted nitrobenzenes **393**, where X is an electron-donating group (X=p-Me, p-OMe) or nitrobenzene itself (X=H) are photoreduced in the presence of propan-2-ol to the corresponding nitrosobenzene hydrate derivatives **394** as the major product. In contrast, nitrobenzenes with an electron-withdrawing group ($X = p\text{-}NO_2$, $p\text{-}CN$ or $p\text{-}COOH$) give products of two-, four- or six-electron reduction, that is, nitroso compounds, hydroxylamines or anilines, respectively, which are apparently formed sequentially and may be present in the reaction mixture at various stages of conversion.[1187,1192] The photoreduction of unsubstituted nitrobenzene proceeds with a very low quantum yield ($\Phi \approx 0.03$) because of the fast radiationless decay of the triplet state.[1192,1193] The course of the reaction in aqueous solutions is strongly dependent on pH.[1194] Photoreduction of nitrobenzene in concentrated hydrochloric acid proceeds, for example, with $\Phi \approx 0.11$ to form a complex mixture of oligo- and polyanilines.[1195] The $\pi,\pi^*$ triplet excited nitroarenes (e.g. 9-nitroanthracene) are virtually unreactive towards hydrogen abstraction; their principal photoreaction is isomerization to nitrites via homolytic cleavage of the C−N bond (Scheme 6.186).[1187]

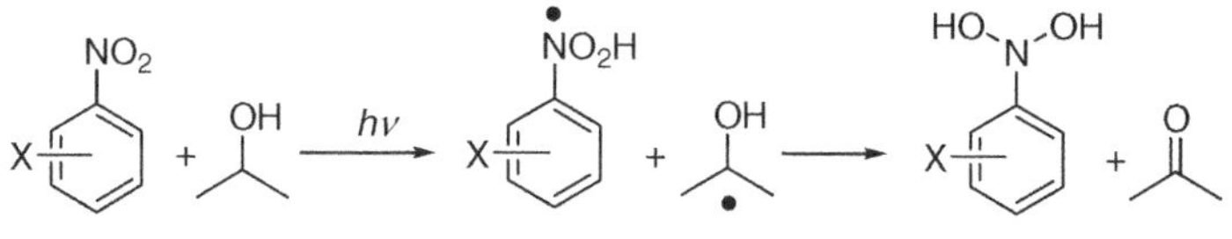

Scheme 6.187

Intramolecular photoreduction of the nitro group in *o*-nitrobenzyl derivatives is one of the most intensively studied reactions in photochemistry.[1035] The excitation of simple 2-nitrotoluene (**395**) rapidly (<1 ns) generates the corresponding *aci*-nitro tautomers (*E*)- and (*Z*)-**396** via hydrogen atom transfer (Scheme 6.188), analogously to the photoenolization reaction (Section 6.3.6). It has not yet been determined whether the reaction proceeds from the excited singlet or triplet state or both or whether H-atom transfer and conformational interconversion occur in an electronic excited state.[1190] The reaction is not very efficient ($\Phi \approx 0.01$ for 2-nitrotoluene) and is largely reversible.[1196] In aqueous solution, the *aci*-tautomers are rapidly equilibrated. The pH–rate profile (Section 5.4.6) for their decay exhibits downward curvature at pH 3–4, which is attributed to pre-equilibrium ionization of the nitronic acid to its anion. Two regions of upward curvature, at pH $\sim 6$ and <0, each indicate a change in the reaction mechanism.

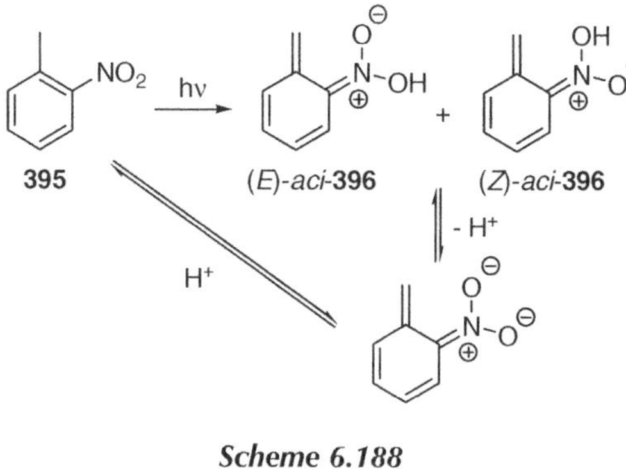

Scheme 6.188

## Special Topic 6.18: Photoactivatable compounds

In general, *photoactivatable compounds* (also called *caged compounds*) are those which, upon photoactivation, either (1) irreversibly release a species (A; Scheme 6.189) possessing desirable physical, chemical, or biological qualities; in such a case, they are called *photochemical triggers*, and the groups that are responsible for the photoprocess are referred to as *photoremovable, photoreleasable* or *photolabile*; or (2) reversibly induce physical or chemical changes in another, covalently or non-covalently bound moiety (B and C in Scheme 6.190a), modify the affinity for another molecule (D in Scheme 6.190b) or exchange protons or electrons; in this case, they are called *photochemical switches* and the process is usually *photochromic* (Special Topic 6.15).

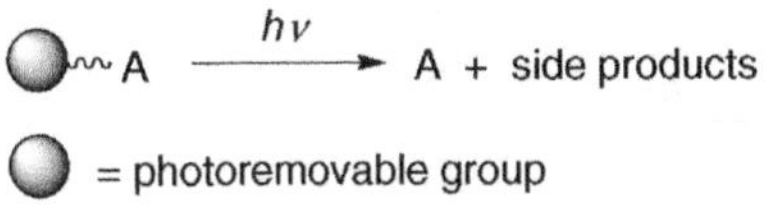

*Scheme 6.189*

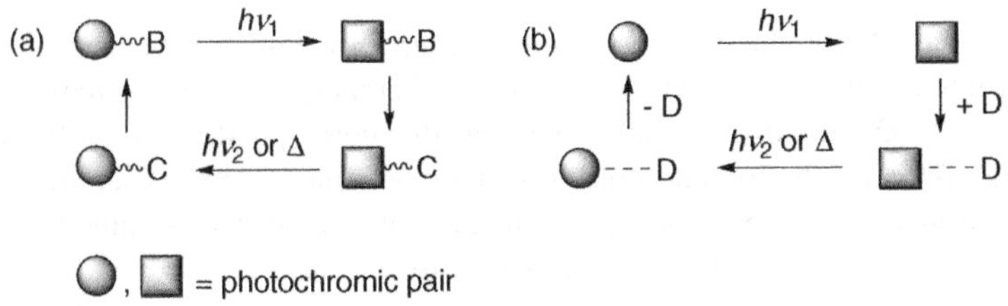

*Scheme 6.190*

Today, photoactivatable compounds are of great interest in connection with biochemical and biological applications (e.g. photoregulation of proteins and enzyme activity, neurotransmitters, ATP and $Ca^{2+}$ delivery or photoactivatable fluorophores),[1015,1034,1035,1197,1198] organic synthesis (e.g. photoremovable protecting groups; solid-phase synthesis; microarray fabrication),[1015,1035,1199,1200] nanotechnology (prospective molecular machines and computers; see Special Topic 6.19),[1103,1104] or even cosmetics (photoactivatable fragrances).[1201] A great advantage of photochemical activation over other stimuli is the ability to control precisely the processes in time and space.

Photochemical triggers irreversibly release free target molecules and also photoproduct(s) formed by the transformation of photoremovable groups (Scheme 6.189). Their design must fulfil several requirements; for example, the side-products should be chemically and photochemically stable and nontoxic (in biological applications) and the photoremovable moiety should absorb at wavelengths where other chromophores present in the system do not.[1035] A further step in this direction is provided by two-photon excitation (Section 3.12) using visible light, which offers excellent three-dimensional control over the localization of target molecule release.[1034,1202]

Irradiation of the most popular photoremovable *o*-nitrobenzyl moiety[1035] liberates a leaving group from the benzylic position in high chemical yields. Scheme 6.191 shows, for example, the 1-(2-nitrophenyl)ethyl ester of ATP (**397**; photoactivatable or caged ATP), which photoreleases the corresponding ATP anion in aqueous solution.[1203] Laser flash photolysis revealed that the release of ATP and the formation of 2-nitrosoaceto-phenone (**398**) occur simultaneously with the decay of the *aci*-anion **399**.[1204] The rate-determining step at pH 7 is cyclization of the *aci* form to **400**.[1205] The aci-dianion **401** (see also Scheme 6.188) does not cyclize, so that its decay is acid catalysed. All the subsequent steps are fast at pH 7. Therefore, the release of the free nucleotide (**402**) and of the side product **398** is indeed synchronous. At lower pH, however, the decay of a hemiacetal **403** becomes the rate-determining step for the release of ATP. In contrast, the release of methoxide (X = Me), a much more nucleophilic leaving group, from 2-nitrobenzyl methyl ether is orders of magnitude slower than the decay of the corresponding *aci*-tautomer. Using 2-nitrobenzyl methyl ether or 2-nitrobenzyl alcohol as precursors, the intermediates shown in Scheme 6.191 have all been identified.[1205–1207] ATP is a multifunctional nucleotide formed as an energy source during the processes of photosynthesis (Special Topic 6.25) and cellular respiration. Its photorelease has already been exploited in several studies of cellular processes.

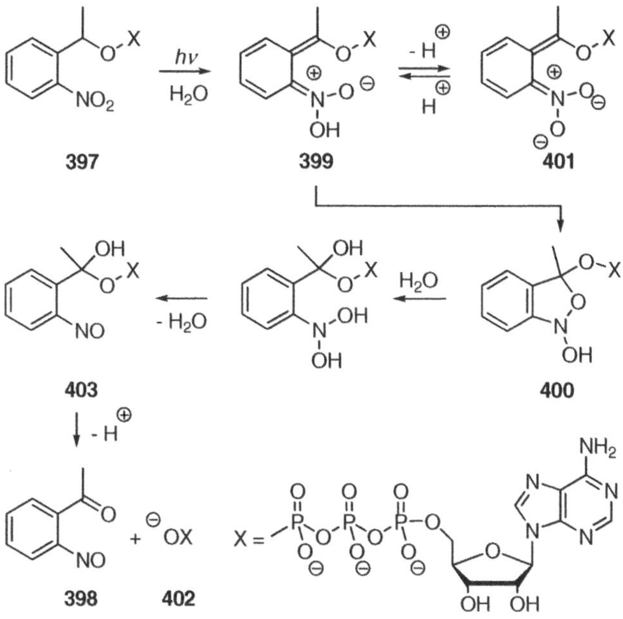

*Scheme 6.191*

Scheme 6.192 shows the *o*-nitrobenzyl moiety used as a photolabile linker in the solid-phase oligosaccharide synthesis on a polystyrene (PS) support.[1208] After the synthesis of the protected oligosaccharide **404**, PS attached to the photoremovable group is removed by photolysis and the final product is obtained by hydrogenolysis. Such a strategy could be promising for combinatorial synthesis of oligosaccharide libraries.

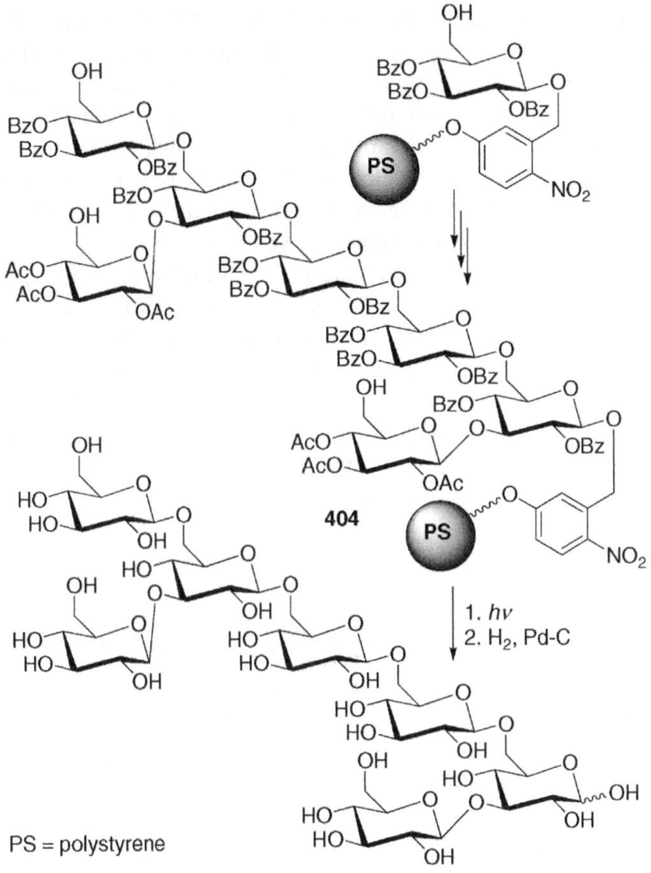

*Scheme 6.192*

Other applications of photoremovable protecting groups are presented in Case Studies 5.3, 6.24, 6.26 and 6.30.

Photochemical switches are also discussed in Special Topic 6.15 (Scheme 6.161) and Special Topic 6.19 (Scheme 6.207). Here we illustrate the photoswitching process, which can control the geometry of biomolecules.[1101] When an azobenzene-derived cross-linker in the DNA-recognition helix of the transcriptional activator MyoD is irradiated at 360 nm, the linker predominantly attains a Z-configuration that significantly stabilizes the helix (Scheme 6.193).[1209] Reverse isomerization can proceed either thermally or photochemically at a different wavelength; therefore, the process is photochromic (Special Topic 6.15).

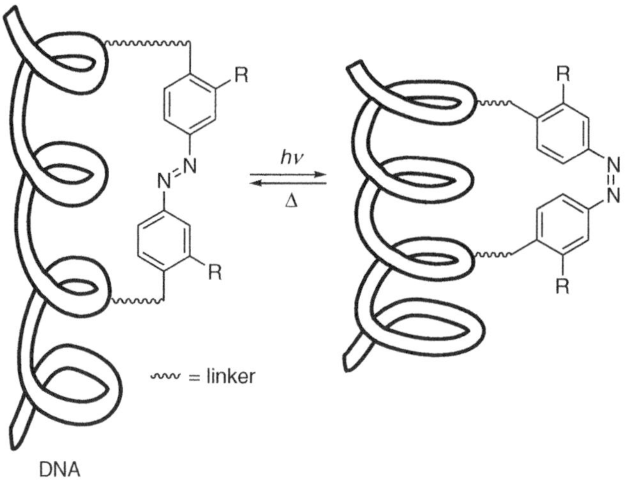

*Scheme 6.193*

Transfer of calcium cations ($Ca^{2+}$) across membranes and against a thermodynamic gradient is important to biological processes, such as muscle contraction, release of neurotransmitters or biological signal transduction and immune response. The active transport can be artificially driven (switched) by photoinduced electron transfer processes (Section 6.4.4) between a photoactivatable molecule and a hydroquinone $Ca^{2+}$ chelator (405) (Scheme 6.194).[1210] In this example, oxidation of hydroquinone generates a quinone to release $Ca^{2+}$ to the aqueous phase inside the bilayer of a liposome, followed by reduction of the quinone back to hydroquinone to complete the redox loop, which results in cyclic transport of $Ca^{2+}$. The electron donor/acceptor moiety is a carotenoid–porphyrin–naphthoquinone molecular triad (see Special Topic 6.26).

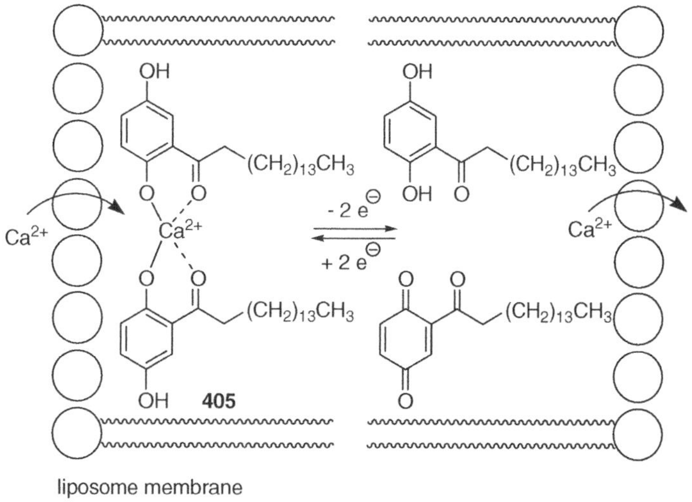

*Scheme 6.194*

## Case Study 6.30: Photoactivatable compounds – chromatic orthogonality

The regioselective control of two independent photochemical processes simply by choosing the wavelength of monochromatic light has been demonstrated on the heptanedioic acid diester **406** (Scheme 6.195).[1211] One carboxylate group is protected by the photoremovable (Special Topic 6.18) 4,5-dimethoxy-2-nitrobenzyl (2-nitroveratryl) group, which is selectively released by 420 nm irradiation to yield **407**. The other carboxylate moiety is protected by the 3′,5′-dimethoxybenzoin group, selectively liberated by 254 nm illumination to give **408**. Both products were obtained as the corresponding carboxylic acids and were methylated (esterified) by (trimethylsilyl)diazomethane in the subsequent step. The 2-nitroveratryl group absorbs at 420 nm exclusively (Figure 6.13); therefore, no photoreaction of the other chromophore can take place by irradiation at this wavelength. This is assuming that energy transfer is not a complicating factor. Since both chromophores absorb at 254 nm (Figure 6.13), differences in the absorbance times excited-state quantum yield determine the selectivity of 3′,5′-dimethoxybenzoin release. The orthogonal approach in the field of protecting groups is a strategy allowing the selective deprotection of multiple protecting groups by changing the reaction conditions. Because the irradiation wavelength is the only variable employed for the deprotection, this method has been termed *chromatic orthogonality*.

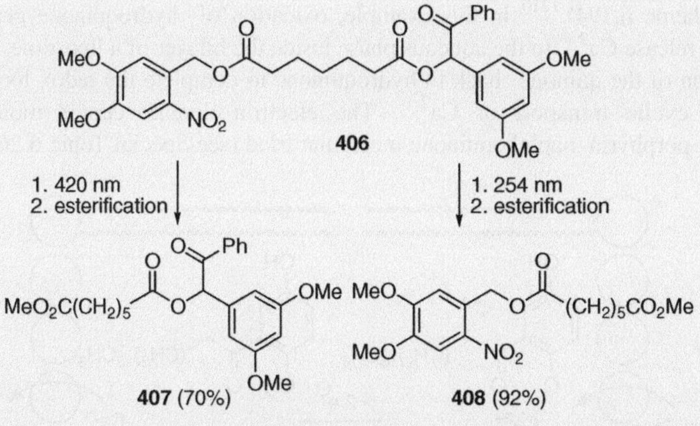

*Scheme 6.195*

*Experimental details.*[1211] A solution of the diester (**406**) (20 μmol) in acetonitrile (10 ml) degassed by purging with nitrogen was irradiated at a selected wavelength (254 or 420 nm) (Figure 3.28) for 24 h. The solvent was evaporated and a crude product (acid) was esterified by (trimethylsilyl)diazomethane (200 μmol in 2 ml of a benzene–ethanol mixture). The chemical yields were determined by NMR spectroscopy.

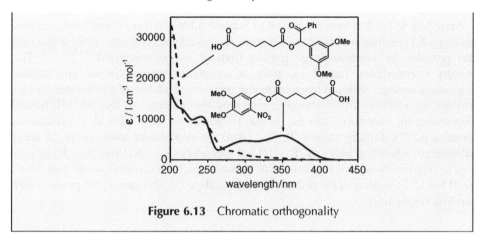

**Figure 6.13**   Chromatic orthogonality

### 6.4.4   Amines, Aromatic Nitriles, Metalloorganic Complexes: Photoinduced Electron/Charge Transfer

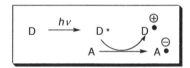

*Recommended review articles.*[602,665,667,669,670,1212–1217]
*Selected theoretical and computational photochemistry references.*[324,1218–1223]

*Amines*

Compared with alcohols, which possess a high standard potential of oxidation, moderate nucleophilicity and weak basicity, amines very often serve as good electron donors and relatively strong bases and nucleophiles in chemical reactions. Photoinduced electron transfer (PET) processes, in which an amine donates an electron to the reaction partner in either its ground or excited electronic state, result in the formation of an amine–substrate exciplex (Scheme 6.196).[670,1224] The driving force for electron transfer is related to the standard potential of oxidation of the donor, the standard potential of reduction of the acceptor and the excited state energy of the absorbing partner (see Chapter 4).

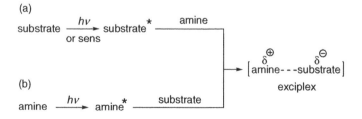

**Scheme 6.196**

According to the first scenario shown in Scheme 6.196a, direct or sensitized (see also Section 6.8.1) excitation of electron acceptors, such as alkenes or arenes, is the initial step that precedes the electron/charge transfer from an amino compound.[665,667,1212] The primary intermediates, radical ion pairs or exciplexes, can regenerate ground-state reactants or undergo chemical reactions to give products. Oxidation of alkylamines, that is, loss of an electron from nitrogen, increases the acidity of the $\alpha$-C$-$H bonds. Deprotonation, therefore, results in the formation of new radical intermediates (Scheme 6.197). Tertiary amines (R, R$'$ = alkyl) can deprotonate from one of the alkyl substituents, whereas primary (R, R$'$ = H) and secondary (R = alkyl/aryl, R$'$ = H) amines tend to deprotonate selectively from the N$-$H bond, which is even more acidic than the $\alpha$-C$-$H bond.[678] Coupling of the radical pairs then leads to the *photoamination* products **409** and **410**, respectively.

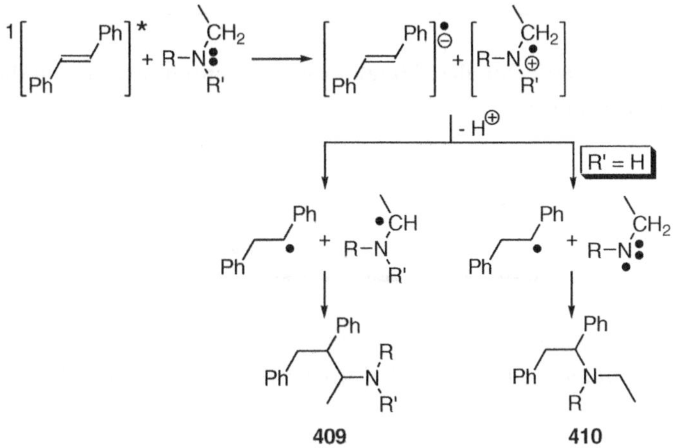

**Scheme 6.197**

The intramolecular version of this reaction can serve, for example, to synthesize medium-sized rings of the azalactam **411** in 43% chemical yield (Scheme 6.198).[1225]

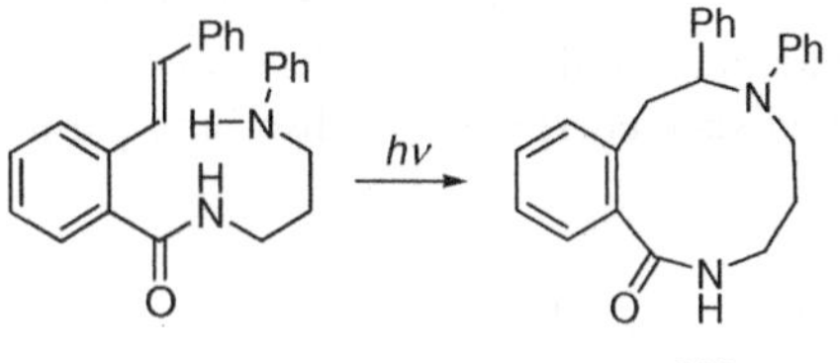

**Scheme 6.198**

Photoamination of arenes[667,1212] – the nucleophilic photosubstitution $S_N2Ar^*$ – was briefly discussed in Section 6.2.3. An intramolecular version of such a reaction, termed *photo-Smiles rearrangement*, is shown in Scheme 6.199.[1226]

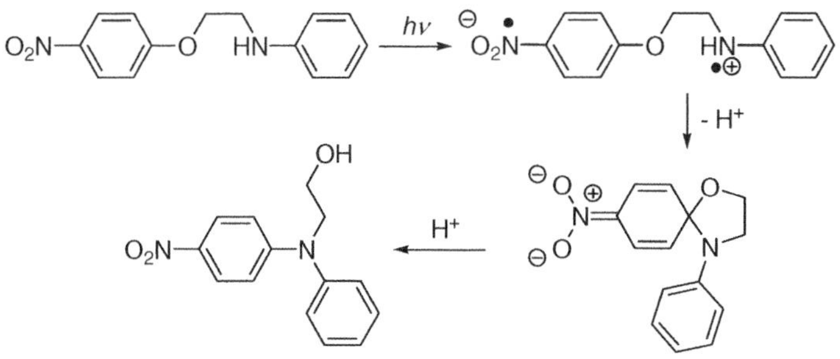

**Scheme 6.199**

Photorelease (Special Topic 6.18) of carboxylates from phenacyl esters represents the reaction, which may involve an electron transfer step induced by amine excitation[1227] (Scheme 6.196b) or hydrogen abstraction from a suitable H-atom donor[1228,1229] (see also Section 6.3.1). For example, upon irradiation, the singlet excited *N,N*-dimethylaniline (DMA) ($E_S = 398 \, \text{kJ mol}^{-1}$)[157] transfers an electron to **412** to form a radical ion pair (Scheme 6.200).[1230] In the subsequent steps, the carboxylate ion **413**, a phenacyl radical (**414**) and a DMA radical cation (**415**) are released and **414** abstracts hydrogen from **415** to form an iminium ion **416** that is hydrolysed by traces of water to give *N*-methylaniline. The electron transfer step was proposed to be exergonic by 60–85 kJ mol$^{-1}$.

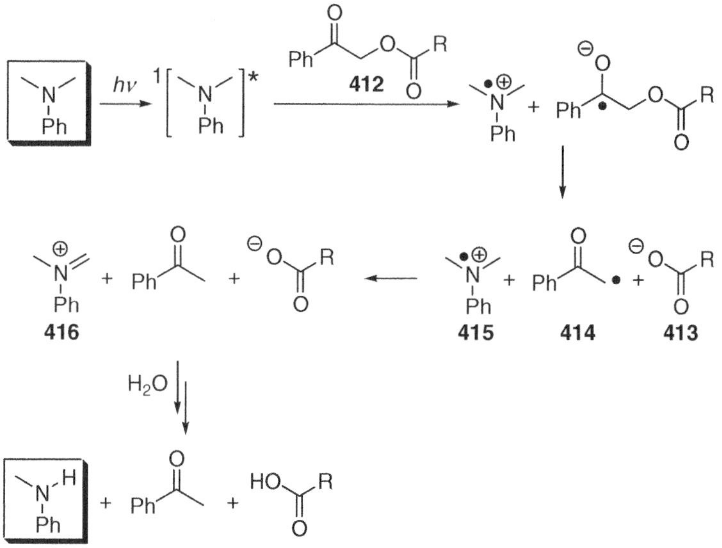

**Scheme 6.200**

*Aromatic Nitriles*

Aromatic nitriles are typical electron acceptors; their electron acceptor abilities increase with increase in the number of electron-withdrawing cyano groups and with increasing size of the aromatic moiety.[670] An analogous model to that described for electron donors (amines) in Scheme 6.196 can also be applied for strong electron acceptors, such as aromatic nitriles. In the first scenario, a ground-state electron acceptor is involved in the electron transfer reaction. For example, an auxiliary electron-donating sensitizer (phenanthrene, **417**; $E_S = 346\,kJ\,mol^{-1}$)[157] is excited to its singlet state, which then transfers an electron to an (auxiliary) acceptor (1,4-dicyanobenzene, **418**) to produce a radical cation **419** (Scheme 6.201).[1231] This species then accepts an electron from the substrate (1,1-diphenylethane, **420**) to give the key intermediate – the radical cation complex **421** – which reacts with a nucleophile (methanol) to afford an anti-Markovnikov product **422** in 70% chemical yield. This very useful approach is termed *redox photosensitization* (see also Section 6.8),[1212] because it permits electron transfer processes in alkene substrates possessing relatively high standard potentials of oxidation.

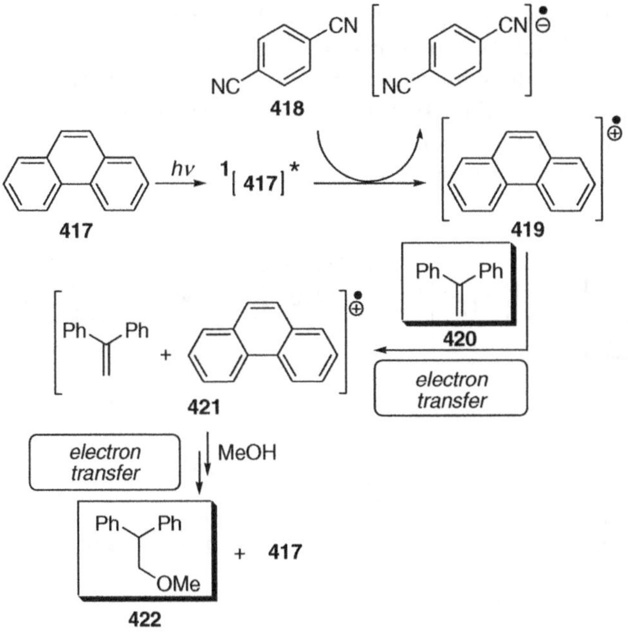

**Scheme 6.201**

According to the second scenario, an auxiliary electron acceptor, for example 9,10-dicyanoanthracene (DCA) in the excited state, is used as a co-sensitizer.[670] DCA absorbs well in the near-UV region and, in the excited singlet state ($E_S = 284\,kJ\,mol^{-1}$), it can accept an electron from *N*-methyl-*N*-(trimethylsilylmethyl)aniline (**423**) to give an amine radical cation **424** (Scheme 6.202).[1232,1233] This intermediate subsequently attacks the C=C bond of 4,4-dimethylcyclohex-2-enone (**425**) and undergoes desilylation upon nucleophilic attack by the solvent. The coupling radical species **426** finally cyclizes to **427** (the overall chemical yield is 10%).

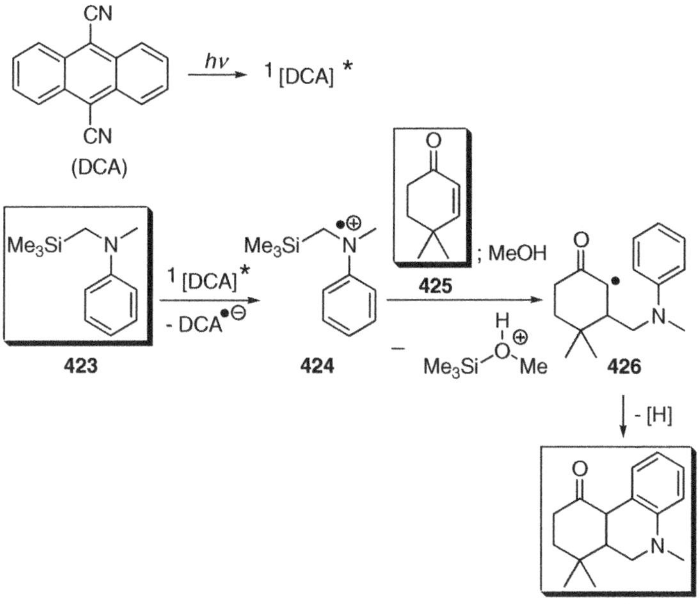

*Scheme 6.202*

Another interesting example is a cascade reaction of the compound **428** in the presence of 2,3,5,6-tetramethylterephthalonitrile as an electron acceptor and a co-sensitizer (biphenyl). The product **429** is obtained via a radical cation **430** in 23% chemical yield (Scheme 6.203).[1234]

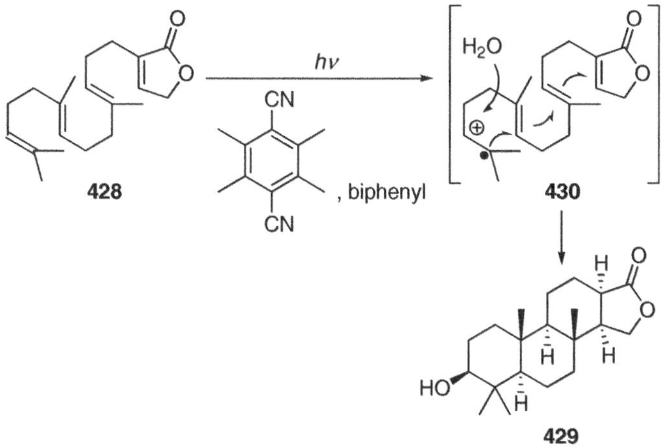

*Scheme 6.203*

*Tris(2,2'-bipyridine)ruthenium(II)*

Some photosensitizers, such as metal organic complexes (see also Section 6.8), can behave as either electron donors or acceptors because of their favourable redox potentials for oxidation or reduction of their excited states.[670] A well-known example is tris(2,2'-bipyridine)ruthenium(II) ion ($[Ru(bpy)_3]^{2+}$; Scheme 6.204).[1046,1235] When a solution of this compound is irradiated with visible light, a triplet charge-transfer excited state is formed, in which a metal-centred electron is promoted to the $\pi^*$-orbital of the bipyridyl ligand [$d,\pi^*$ state; where the metal is the electron-deficient centre (acceptor) and the ligand is the electron rich centre (donor)].

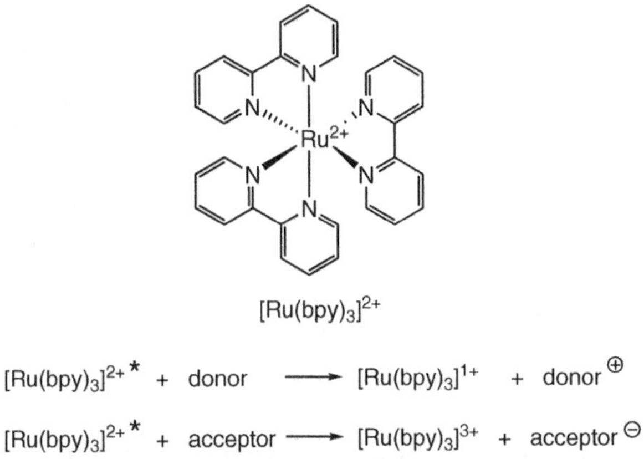

$[Ru(bpy)_3]^{2+}$

$[Ru(bpy)_3]^{2+*}$  +  donor  $\longrightarrow$  $[Ru(bpy)_3]^{1+}$  +  donor$^{\oplus}$

$[Ru(bpy)_3]^{2+*}$  +  acceptor  $\longrightarrow$  $[Ru(bpy)_3]^{3+}$  +  acceptor$^{\ominus}$

*Scheme 6.204*

An interesting application is photosensitized polymerization of pyrrole (**431**) using $[Ru(bpy)_3]^{3+}$ to give a conducting polymer, polypyrrole (**432**), in aqueous solution or in a polymer matrix (Scheme 6.205).[1236] The acting ground-state electron acceptor, $[Ru(bpy)_3]^{3+}$, is obtained in the initial electron transfer step between an excited $[Ru(bpy)_3]^{2+}$ and $[Co(NH_3)_5Cl]^{2+}$ ion.

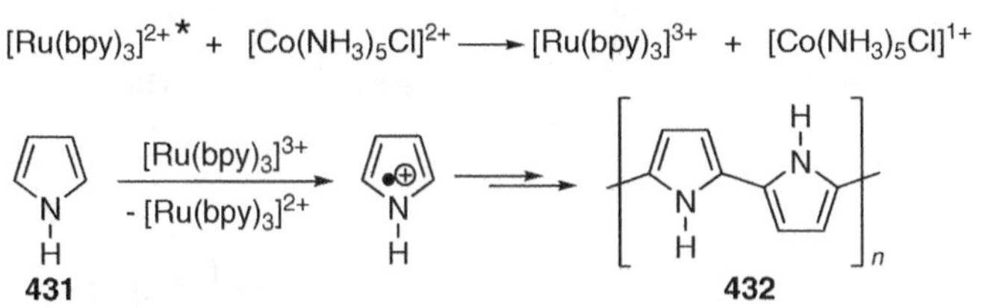

$[Ru(bpy)_3]^{2+*}$  +  $[Co(NH_3)_5Cl]^{2+}$  $\longrightarrow$  $[Ru(bpy)_3]^{3+}$  +  $[Co(NH_3)_5Cl]^{1+}$

**431**                                                                 **432**

*Scheme 6.205*

As part of the search for alternatives to fossil fuel, recent studies have focused on the production of hydrogen, a clean and renewable energy carrier, by direct water splitting

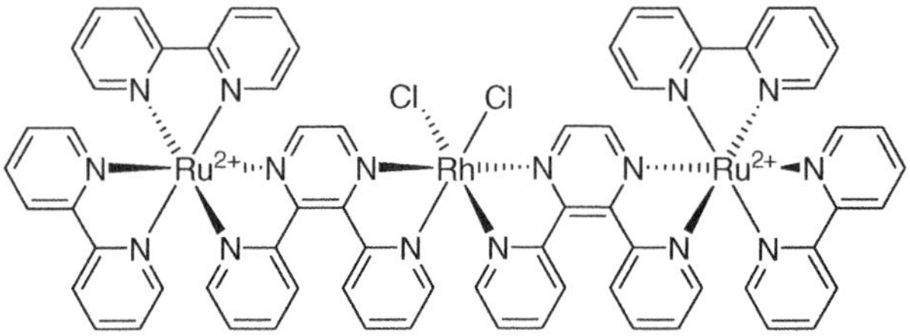

$$[\{(bpy)_2Ru(dpp)\}_2RhCl_2]^{5+}$$

**Figure 6.14** Hydrogen production catalyst

using photocatalysts and solar radiation. Although heterogeneous photocatalysts are relatively effective[1237] (Special Topic 6.26), homogeneous photocatalysts for hydrogen production are also promising. For example, the $[\{(bpy)_2Ru(dpp)\}_2RhCl_2]^{5+}$ (bpy = 2,2'-bipyridine, dpp = 2,3-bis-2-pyridylpyrazine) complex (Figure 6.14) was shown to undergo photoexcitation (Ru → dpp) and electron collection at the rhodium centre, having the ability to be reduced by two electrons by converting $Rh^{3+}$ to $Rh^{+}$.[1238] The complex, in an acetonitrile–water solution and in the presence of $N,N$-dimethylaniline (DMA) (an auxiliary electron donor), was found to produce hydrogen photocatalytically when excited with visible light (470 nm) using an LED array. The quantum yield of the process was 0.01, assuming that two photons were used to produce hydrogen (in general: $2H_2O + 4h\nu \rightarrow 2H_2 + O_2$).

---

### Special Topic 6.19: Molecular machines

The great demand for miniaturization of components in electrotechnical, medicinal or material applications has led to the development of a highly multidisciplinary scientific and technological field called *nanotechnology* to produce devices with critical dimensions within the range 1–100 nm. The ultimate solution to miniaturization is logically a functional molecular machine, an assembly of components capable of performing mechanical motions (rotation or linear translation) upon external stimulation, such as photoactivation.[1103,1104,1239–1244] This motion should be controllable, efficient and occur periodically within an appropriate time-scale; therefore, it involves photochromic behaviour discussed in the Special Topic 6.15. Such devices can also be called photochemical switches (Special Topics 6.18 and 6.15). Here we show two examples of molecular machines: a molecular rotary motor and a molecular shuttle.

Photochemical *E–Z* isomerization (Section 6.1.1) is responsible for continuous unidirectional rotary motion along a carbon–carbon double bond of **433**, a chiral helical alkene mounted on the surface of gold nanoparticles through two octanethiol linkers (Scheme 6.206).[1245] Thanks to the considerable conformational flexibility of the molecule, the position of the methyl group determines its most stable shape (having the methyl substituent in a pseudo-axial orientation), which subsequently undergoes photoisomerization. As a result, it determines whether clockwise or anticlockwise rotary motion occurs.

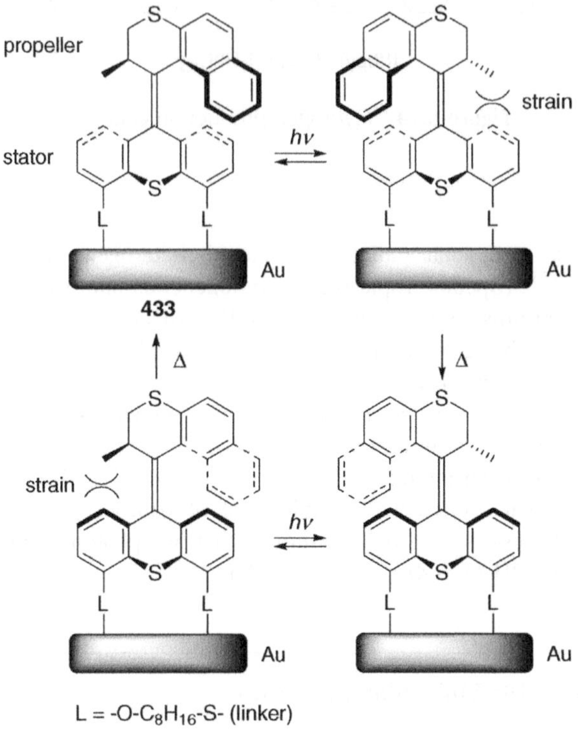

$L = -O-C_8H_{16}-S-$ (linker)

*Scheme 6.206*

A light-driven 'shuttle' is shown in the simplified Scheme 6.207.[1246] The nanomachine **434** ($\sim$5 nm in length) is composed of bis-*p*-phenylene-34-crown-10, an electron-rich macrocycle (M), which can move (shuttle) between two electron-acceptor moieties ('stations'), 4,4'-bipyridinium ($A_1$) and 3,3'-dimethyl-4,4'-bipyridinium ($A_2$), linked to the stoppers, the *p*-terphenyl moiety (S) and the tetraarylmethane group (T). The photoinduced shuttling movement starts with excitation of a $[Ru(bpy)_3]^{2+}$ photosensitizer ($P^{2+}$), covalently attached to the molecule via S, following electron transfer to $A_1^{2+}$ to form $A_1^+$. As a result, the ring M moves to the adjacent $A_2^{2+}$ by Brownian motion, while a reverse electron transfer from $A_1^+$ restores the charge on $A_1^{2+}$, allowing M to move back. The ring-displacement process requires a time frame of $\sim$100 µs and operates with $\Phi \approx 0.12$.

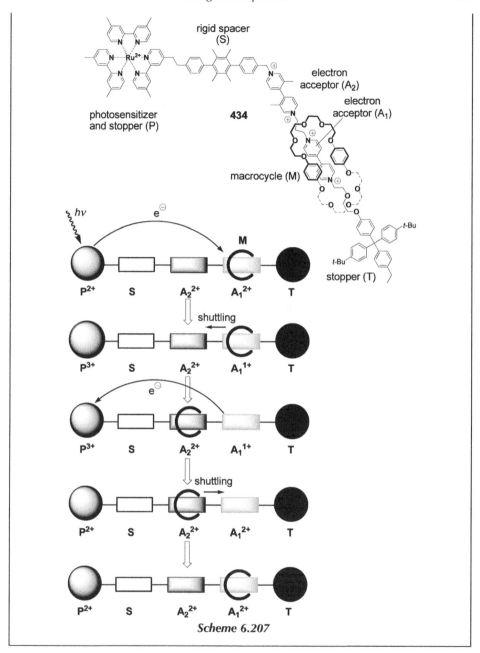

*Scheme 6.207*

## Case Study 6.31: Biochemistry – photocatalytic oxidation of DNA guanine

The photoexcited ruthenium(II) (phenanthroline)dipyridophenazine complex {[Δ-Ru (phen)$_2$(dppz)]$^{2+}$ (**435**)}, an analogue of [Ru(bpy)$_3$]$^{2+}$ (Scheme 6.204), can be oxidized by electron transfer to a weakly bound electron acceptor, such as methyl viologen [paraquat; *N,N'*-dimethyl-4,4'-bipyridinium dichloride (**436**)] or $[Co(NH_3)_5Cl]^{2+}$, resulting in the formation of a $Ru^{3+}$ intercalator (a molecule that binds to DNA and then inserts itself between base pairs) (Scheme 6.208).[1247] This potent ground-state oxidant can be reduced back to the $Ru^{2+}$ complex by reaction with a reduced quencher or by electron transfer from a guanine base of DNA. The resulting guanine radical cation forms a neutral radical **437** by deprotonation, which can irreversibly react, for example, with molecular oxygen. This approach has been developed to explore electron transfer chemistry on double helical DNA (see also Special Topic 6.7).[1248]

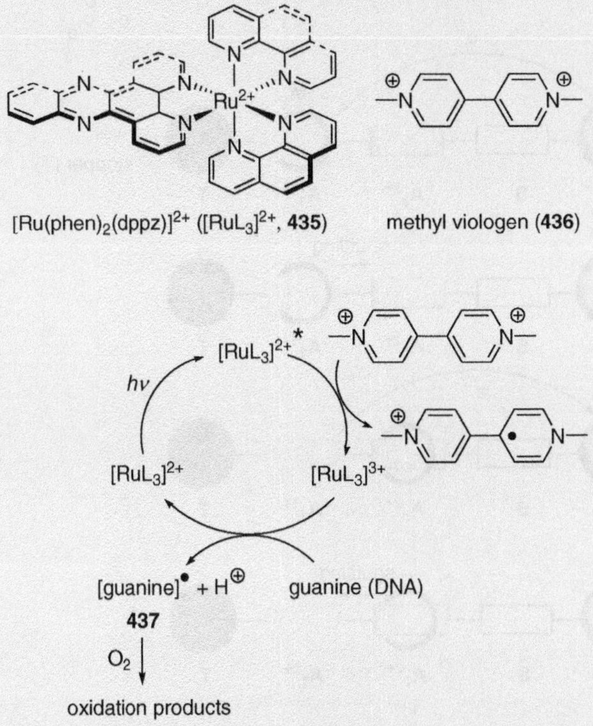

[Ru(phen)$_2$(dppz)]$^{2+}$ ([RuL$_3$]$^{2+}$, **435**)          methyl viologen (**436**)

*Scheme 6.208*

*Experimental details.*[1247] DNA duplexes were formed from oligonucleotides and prepared on a DNA synthesizer by slow cooling of equal concentrations of complementary strands. Solutions of a duplex (8 μM), [Ru(phen)$_2$(dppz)]$^{2+}$ (8 μM) and 10–20 equivalents of a quencher (e.g. methyl viologen), were irradiated at 436 nm with an Hg–Xe lamp (1000 W) equipped with a monochromator (Figure 3.28). After irradiation, the samples were treated with piperidine, dried and electrophoresed through polyacrylamide gel. The extent of DNA damage was evaluated by phosphorimaging.

*Metalloporphyrins*

In addition to $[Ru(bpy)_3]^{2+}$, many other metal complexes can be utilized in photoinduced redox reactions (see also Section 6.8). An important class of such compounds are metalloporphyrins, macrocycles derived from four pyrrole-like subunits with a coordinated metal (haemoglobin or cyanocobalamin (vitamin $B_{12}$) are the best known derivatives). A free base porphyrin, and also some non transition metal ($Al^{3+}$, $Zn^{2+}$, etc.) complexes, are excellent and widely used photosensitizers[1249] to generate singlet oxygen (Section 6.7.1). Many metalloporphyrins can also be involved in photoinduced electron transfer reactions; their redox properties depend on the nature of coordinated metal ion. Complexes of metals possessing a higher positive charge display higher standard potentials of oxidation.[1248] Conjugation of the metal d-electrons with the π-electrons of the porphyrin macrocycle causes strong charge transfer and electrostatic interactions upon excitation (the absorption spectrum exhibits significant bands in the visible part of the spectrum). The central heavy metal ion also enhances intersystem crossing due to spin–orbit coupling (Section 4.8).

The interesting dual behaviour of porphyrin derivatives is demonstrated by intramolecular electron transfer between an electron-rich tris(4-methylphenyl)porphyrin zinc complex (ZnTPP) and an electron-deficient tris(4-methylphenyl)octafluoroporphyrin (TPOFP) (Teflon porphyrin) free base connected via an azobenzene moiety (**438**; Scheme 6.209).[1250] Upon photolysis at >440 nm, $E \rightarrow Z$ photoisomerization (Section 6.4.1) occurs. (*Z*)-**438** can revert to (*E*)-**438** only thermally, because the excited state of the *Z*-isomer is efficiently quenched by intramolecular electron transfer from ZnTPP to TPOFP, whereas no such interaction is possible in the *E*-isomer because the chromophores are too remote. This approach has been suggested for photocontrolled molecular electronics.

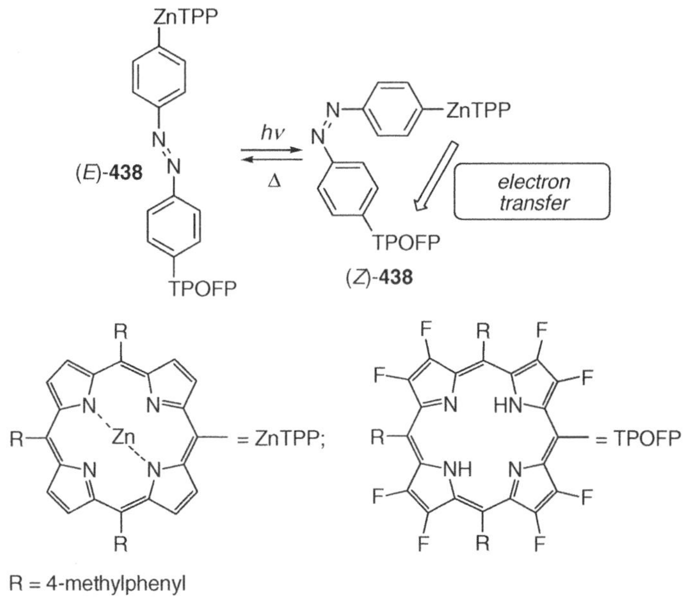

*Scheme 6.209*

### 6.4.5 Problems

1. Explain the following concepts and keywords: photochromism; photoswitching; photoinitiator; photoaffinity labelling; photo-Wolff rearrangement; photoactivatable compound; Barton reaction; photochemical trigger; photolabile linker; photoamination; redox photosensitization.

2. Suggest the mechanisms for the following reactions:

(a)

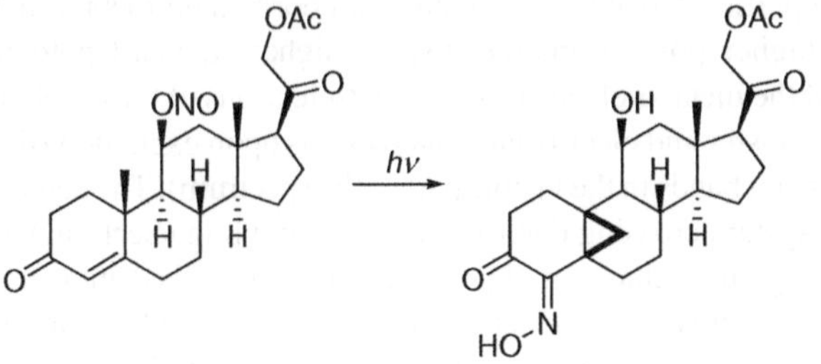

[ref. 1251]

(b)

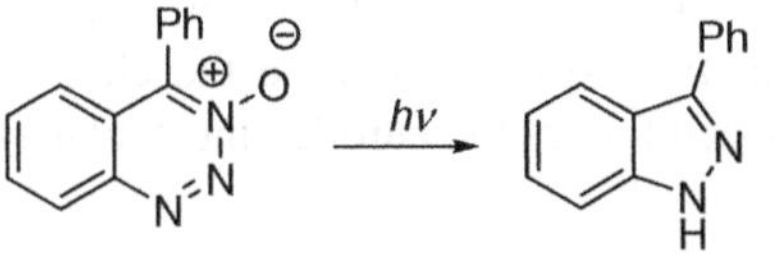

[ref. 1252]

(c)

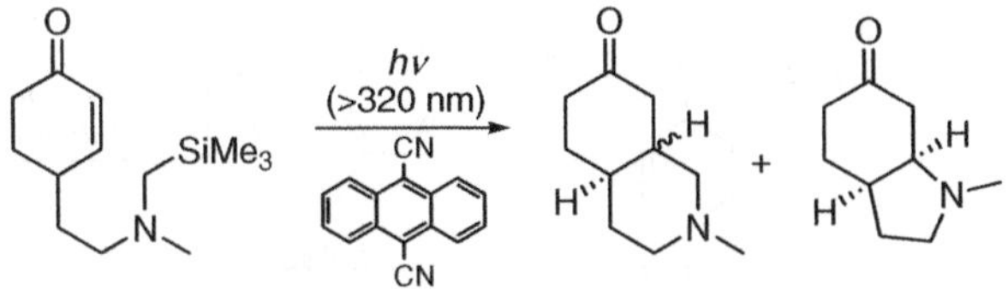

[ref. 1253]

3. Predict the major photoproduct(s):

(a)

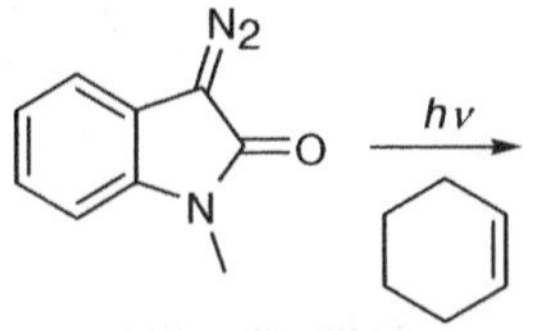

[ref. 1254]

(b)

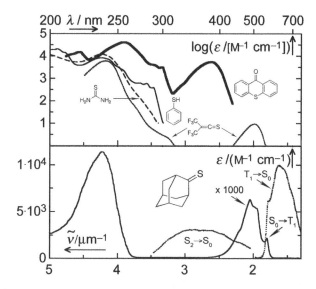

[ref. 1255]

(c)

[ref. 1256]

## 6.5 Sulfur Compounds

Sulfur and oxygen (see Section 6.3) compounds are isoelectronic and the discussion of their excited states is analogous. The differences can be understood by recalling two simple facts: sulfur is less electronegative than oxygen and the C=S π-bond is weaker than the C=O π-bond. The absorption spectra of thiophenols and thioethers are similar to those of the corresponding phenols (Figure 6.15, top).

**Figure 6.15** Absorption spectra of sulfur-containing compounds. Top: thiophenol (in heptane, ——), thioxanthone (in hexane, ——), thiourea (in water, - - -), bis(trifluoromethyl)thioketene (in isooctane, ····).[280] Bottom: absorption (——) and emission (····) spectra of adamantanethione in perfluoroalkane solvents.[1257] The peak at the red edge of the $S_0 \rightarrow S_1$ transition coincides with the 0–0 band of phosphorescence and is attributed to $S_0 \rightarrow T_1$ absorption

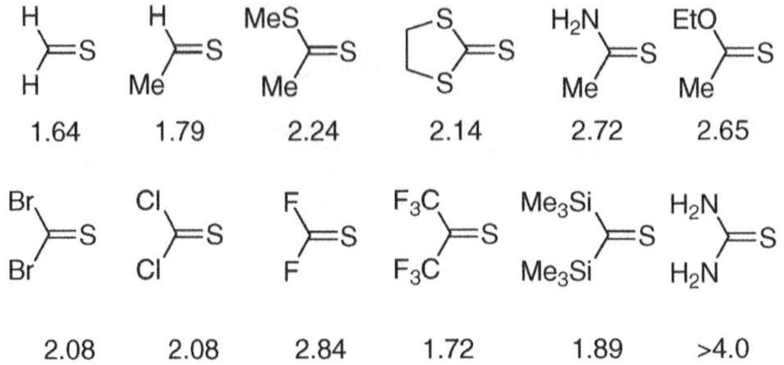

1.64    1.79    2.24    2.14    2.72    2.65

2.08    2.08    2.84    1.72    1.89    >4.0

**Figure 6.16** Positions of the $n,\pi^*$ transitions ($\tilde{v}/\mu m^{-1}$) in thiocarbonyl compounds

The $n,\pi^*$ transitions of thiocarbonyl compounds lie at much longer wavelengths (~500 nm) than those of the corresponding carbonyl compounds, because its $n_p$ lone pair lies at much higher energy. The energy gap between the $S_1(n,\pi^*)$ and $S_2(\pi,\pi^*)$ states of thiocarbonyl compounds is therefore fairly large and the Kasha and Vavilov rules (Section 2.1.8) may be violated (Figure 6.15, bottom). Even inefficient reactions proceeding from the $S_2(\pi,\pi^*)$ state can amount to the dominant photoprocess because the $S_1(n,\pi^*)$ states are short-lived (energy gap law) and not very reactive due to their low excitation energy. Spin–orbit coupling between the ground state and the lowest triplet state of $n,\pi^*$ character is sufficiently large in thiocarbonyl compounds to render singlet–triplet absorption detectable ($\varepsilon \approx 1\,M^{-1}\,cm^{-1}$; Figure 6.15, bottom).

The positions of the $n,\pi^*$ transitions in a series of thiocarbonyl compounds are shown in Figure 6.16.[1258] The shifts between related compounds follow the trends expected from perturbation theory: inductive acceptors lower the energy of the $n_p$ orbital on sulfur and mesomeric interactions raise the energy of the $\pi^*$ orbital.

**Table 6.17** *Photoprocesses involving excited sulfur-containing compounds*

| Entry | Starting material[a] | Product(s) | Mechanism | Section |
|---|---|---|---|---|
| 1 | 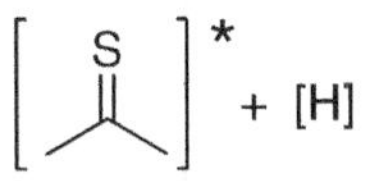 | 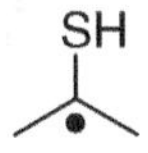 | Photoreduction/Photoaddition | 6.5.1 |
| 2 | 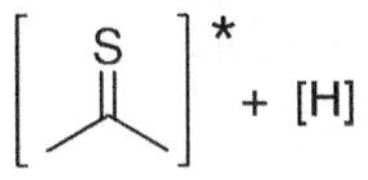 | 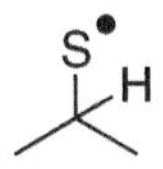 | Photoreduction/Photoaddition | 6.5.1 |
| 3 | 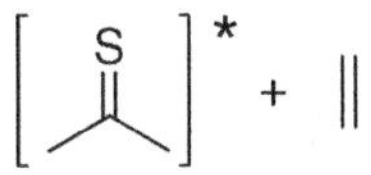 |  | Photocycloaddition | 6.5.1 |
| 4 | 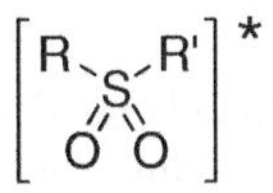 | 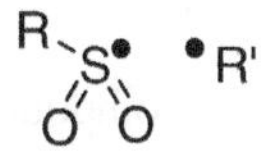 | Photofragmentation | 6.5.2 |

[a][H] = hydrogen atom donor.

Similarly to carbonyl compounds (Section 6.3.1), thiocarbonyl compounds abstract hydrogen upon irradiation; however, both n,π* and π,π* excited states are reactive and the hydrogen atom can be added to either the sulfur (Table 6.17, entry 1) or carbon (entry 2) atoms of the C=S bond. Aliphatic and aromatic thiocarbonyl compounds can also undergo photocycloaddition to unsaturated compounds from both singlet or triplet excited states to form thietanes (analogously to the Paternò–Büchi reaction; see Section 6.3.2) (entry 3) or 1,4-dithianes. On the other hand, fragmentation of the S–C bond is a typical primary process observed in excited sulfones and sulfonates (entry 4), followed by efficient SO₂ extrusion from the radical intermediate.

### 6.5.1 Thiocarbonyl Compounds: Hydrogen Abstraction and Cycloaddition

*Recommended review articles.*[1259–1264]
*Selected theoretical and computational photochemistry references.*[1258,1265–1268]

The differences in the photoreactivity of carbonyl and thiocarbonyl compounds are generally linked to the excitation energies (compare: benzophenone $E_S = 316\,kJ\,mol^{-1}$, $E_T = 288\,kJ\,mol^{-1}$; thiobenzophenone $E_S = 191\,kJ\,mol^{-1}$, $E_T = 165\,kJ\,mol^{-1}$),[157] a high polarizability of the C=S bond and the softer character and lower electronegativity of sulfur compared with oxygen (Section 6.3).[1260,1262,1268] Thiones in solution are known to undergo hydrogen abstraction, cycloaddition, oxidation and fragmentation reactions from both their singlet and excited triplet states.[1261]

*Hydrogen Abstraction by the Excited Thiocarbonyl Group*

In contrast to excited carbonyl compounds, which abstract hydrogen efficiently from the n,π* state (Section 6.3.1), thiocarbonyl compounds react from both n,π* and π,π* states and the hydrogen atom can be added to either sulfur or carbon atom of the C=S bond.[1259,1269,1270] The sulfur atom is larger than oxygen, therefore the reaction can occur over greater C=S····H distances than those involving the carbonyl group.[1271] Singlet π,π* excited (S₂) alkyl aryl thioketone **439**, for example, undergoes unexpected photoinduced intramolecular 1,6-hydrogen atom abstraction to form a short-lived 1,5-biradical intermediate, which cyclizes to the cyclopentanethiol **440** exclusively, although with a low quantum yield ($\Phi = 0.06$) (Scheme 6.210).[1272] The hydrogen abstraction is competitive with back hydrogen transfer and excited state decay. A Norrish type II 1,4-biradical intermediate and subsequent fragmentation yielding elimination products typical of the triplet n,π* excited alkyl aryl ketones are not observed here.

In contrast, hydrogen atom transfer to carbon of the C=S bond occurs in excited (both n,π* and π,π*) 2,4,6-tri-*tert*-butylthiobenzaldehyde (**441**), resulting in the six-membered thiolane derivative **442** formation via a radical mechanism (Scheme 6.211) in nearly quantitative chemical yield.[1273]

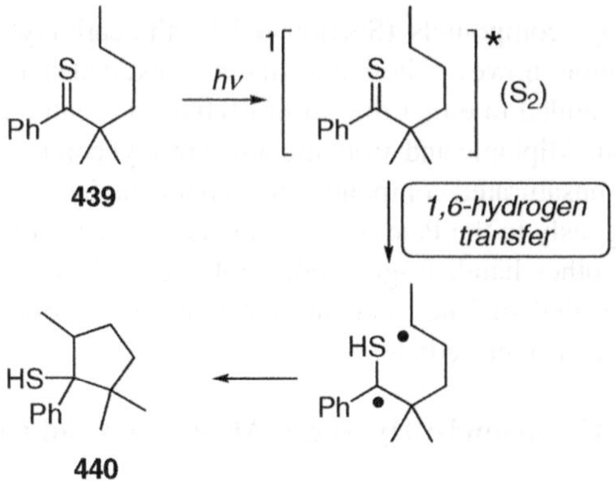

**439**

**440**

*Scheme 6.210*

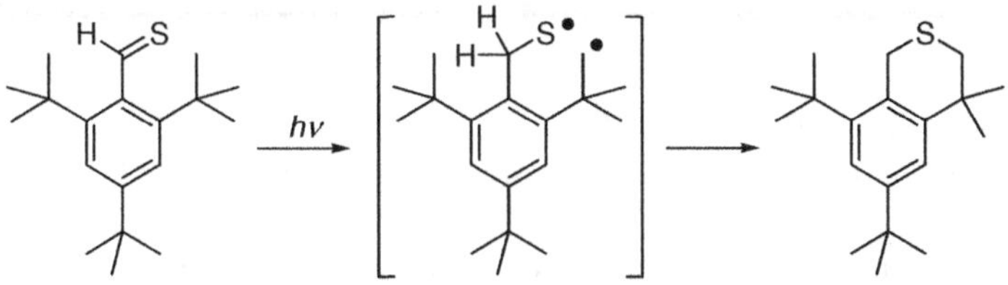

**441**          **442**

*Scheme 6.211*

---

**Case Study 6.32: Mechanistic photochemistry – reactions from different excited states**

Excitation of the alkyl aryl thione **443** with an activated β-hydrogen into either the $S_1$ (n,π*) or $S_2$ (π,π*) state using different irradiation wavelengths was found to produce the cyclopropanethiol **444** via the same 1,3-biradical (Scheme 6.212).[1274] In contrast to the reaction from the $^1$π,π* state, the product formation obtained by excitation into $S_1$ was quenched by a triplet quencher, indicating that the lowest (n,π*) triplet must be the reactive state. A very large isotope effect ($k_H/k_D > 17$) found for this reaction indicates a substantial tunnelling effect (Special Topic 5.2) in the hydrogen abstraction process.

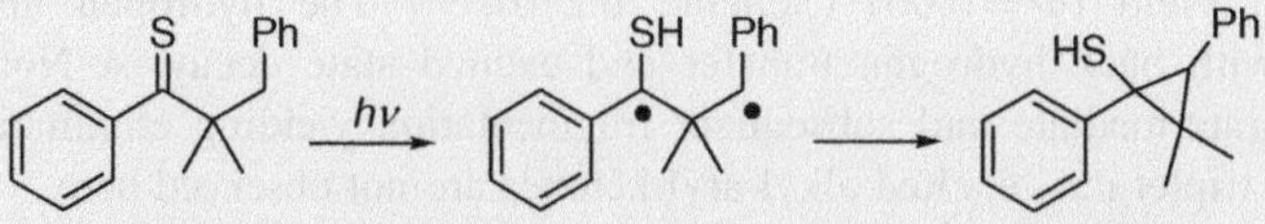

**443**          **444**

*Scheme 6.212*

Experimental details.[1274] A benzene solution of **443** ($5 \times 10^{-3}$ M) was irradiated using a medium-pressure mercury lamp in a Pyrex immersion well through an optical filter ($\lambda_{irr} > 420$ nm; $S_1$ excitation; Figure 3.9) to complete conversion. Preparative TLC and crystallization from *n*-pentane afforded the product in 88% chemical yield.

The same benzene solution was irradiated through a Pyrex filter ($\lambda_{irr} > 280$ nm; $S_2$ excitation) to complete conversion and the product was isolated in 82% chemical yield.

### *Photocycloaddition to Alkenes*

Both aliphatic and aromatic thiocarbonyl compounds can undergo photocycloaddition to unsaturated compounds from either singlet or triplet excited states.[1260,1261,1263] The intersystem crossing quantum yields in aromatic thioketones are usually high and are dependent on the excitation wavelength: $\Phi$ is less than unity ($\sim 0.5$) in the case of excitation to the second excited singlet state, $S_2$ ($\lambda_{exc} \approx 300$–400 nm), but it approaches $\Phi = 1$ when the irradiation is done at $\lambda_{exc} > 500$ nm ($S_1$).[1275] Excited thiobenzophenone (**445**), for example, reacts with an electron-rich enol ether **446** to produce the thietane **447**, analogous to the Paternò–Büchi reaction involving carbonyl compounds (Section 6.3.2) or the 1,4-dithiane **448**, formed by the reaction of two molecules of thiobenzophenone.[1276] Whereas both products are obtained from the n,π* excited triplet state ($T_1$) of thiobenzophenone and electron-rich ethoxyethene via a radical mechanism, the thietane **449** is formed exclusively by a stereospecific cycloaddition from the π,π* excited singlet state ($S_2$) to an electron-deficient *trans*-fumaronitrile (**450**) (Scheme 6.213).

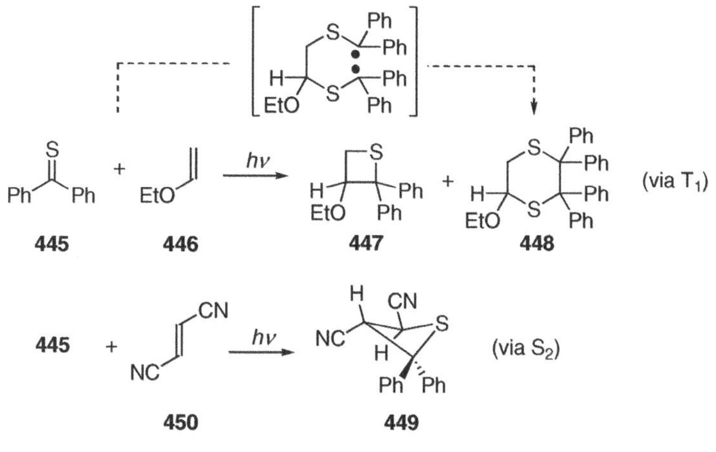

**Scheme 6.213**

### 6.5.2   Sulfones, Sulfonates and Sulfoxides: Photofragmentation

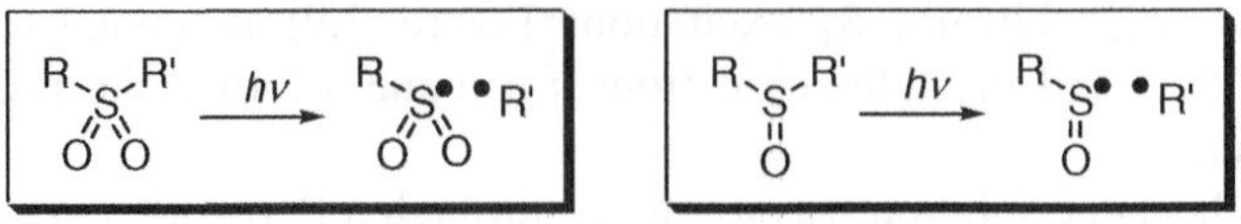

*Recommended review articles.*[1277–1279]
*Selected theoretical and computational photochemistry references.*[1280,1281]

The photochemistry of sulfones (e.g., **451**[1282]) and sulfonates (e.g. **452**[1283]) is dominated by homolytic fissions of the S–C or S–O bonds (Scheme 6.214).[1277,1278] Such bonds are relatively weak; for example, the dissociation energy ($D_{S-C}$) in $CH_3SO_2-CH_3$ is $280\,kJ\,mol^{-1}$ and $D_{S-O}$ in $HO-SO_2CH_3$ is $306\,kJ\,mol^{-1}$.

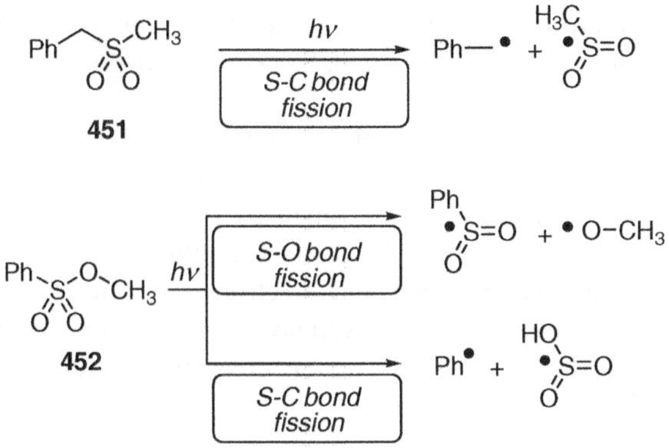

*Scheme 6.214*

Aliphatic sulfones absorb only at $\lambda < 200\,nm$, whereas aromatic sulfones can be irradiated at 250–300 nm. Photolysis of sulfones is a convenient source of alkyl or aryl radicals, which may subsequently undergo various secondary reactions. The $SO_2$ photoextrusion[1284] can originate from both the singlet and triplet excited states depending on the sulfone substituents and the photoreaction can also be triplet sensitized. For example, photofragmentation of **453** to give the cyclophane **454** (54% chemical yield) proceeds primarily from the excited singlet state, whereas that of **455** gives a racemic mixture of the products [(±)-**456**; 64% chemical yield] from the excited triplet state (Scheme 6.215).[1285] This manner of producing cyclophanes is practical because the parent sulfones are easily prepared.

The photochemistry of sulfonates involves the excited singlet state and homolytic cleavage of either the S–C or the S–O bond (Scheme 6.214).[1283] Alkylsulfonates absorb light only at shorter wavelengths ($\lambda_{max} < 200\,nm$), but quartz-filtered irradiation ($\lambda_{irr} > 250\,nm$) is sufficient for excitation of arylsulfonates. Direct irradiation of

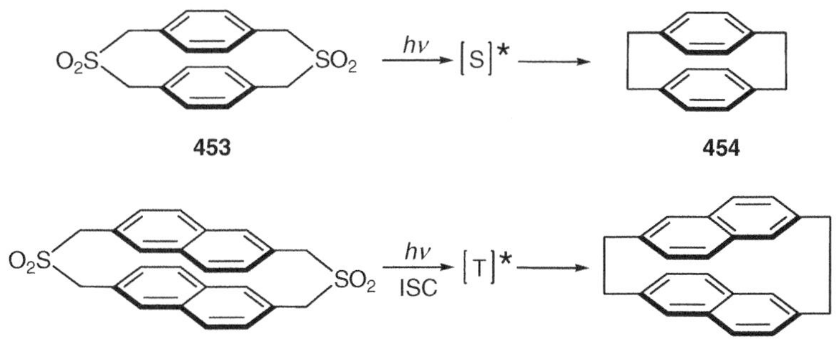

**453** → **454**

**455** → **(±)-456**

*Scheme 6.215*

*p*-toluenesulfonates, for example, leads to S–C bond homolytic cleavage in the first step, followed by $SO_2$ extrusion (Scheme 6.216).[1286] For example, the photoinduced removal of sulfonate moieties has been used in photodeprotection of the hydroxy groups in carbohydrates (see also Special Topic 6.18).[1287]

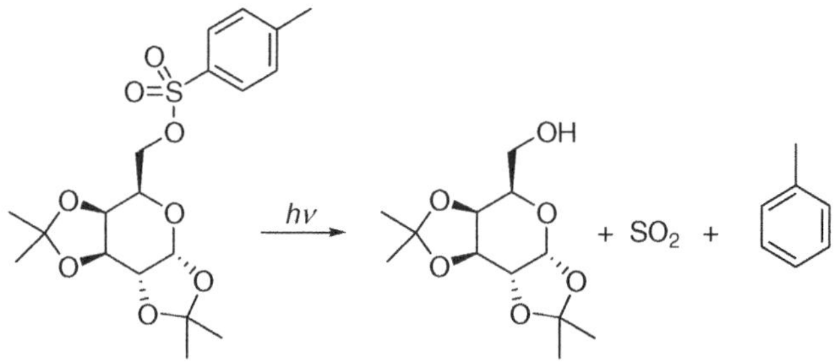

*Scheme 6.216*

## Case Study 6.33: Photoremovable protecting groups – chemistry of carbohydrates

The *p*-tolylsulfonyl group can be utilized as a photoremovable group (Special Topic 6.18) for the protection of carbohydrates in the synthesis of saccharides in the presence of bases, such as the hydroxide anion or amines. No correlation has been observed between the nucleophilicity of the bases and the efficiency of the reaction; however, a qualitative correlation was found with the electron-donating ability of amines.[1288] It has therefore been suggested that the cleavage is induced by an electron-transfer process via the excited singlet state of the sulfonate, the anion-radical $ArSO_2OR^{\cdot-}$, and subsequent release of a leaving group ($RO^-$) (Scheme 6.217).[1288,1289]

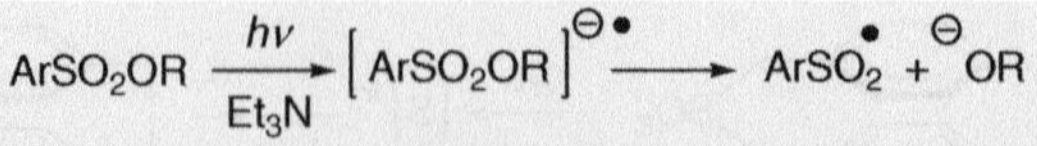

$$ArSO_2OR \xrightarrow[Et_3N]{h\nu} \left[ ArSO_2OR \right]^{\ominus \bullet} \longrightarrow ArSO_2^{\bullet} + {}^{\ominus}OR$$

**Scheme 6.217**

For example, photolysis of methyl 3-*O*-benzyl-2,6-dideoxy-4-*O*-(*p*-tolylsulfonyl)-α-D-*arabino*-hexopyranoside (**457**) in the presence of a high concentration of triethylamine gives the glycoside **458** (Scheme 6.218).[1289]

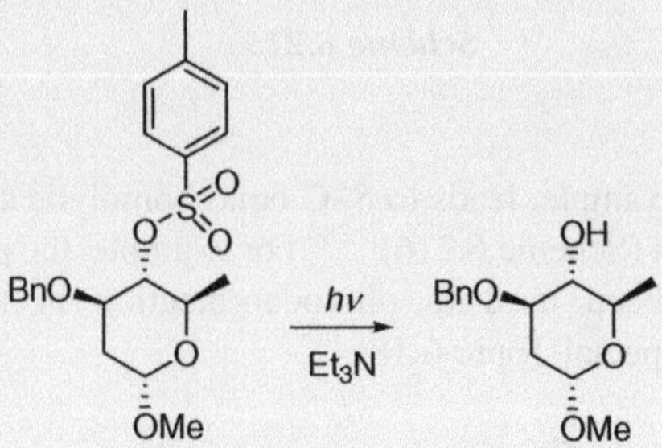

**457**                                    **458**

**Scheme 6.218**

*Experimental details.*[1289] A solution of **457** (0.74 mmol) and triethylamine (2 mmol) in methanol was purged with nitrogen for 1 h and irradiated in a Rayonet photochemical chamber equipped with 16 low-pressure mercury discharge lamps (254 nm) (Figure 3.10) for 12 h. After irradiation, methanol was distilled off under reduced pressure and the product was purified by column chromatography in 90% chemical yield.

Homolytic fission of the C–S bond is a primary process following the excitation of sulfoxides. For example, the sulfoxide **459** produces a 1,5-biradical, which can either rearrange to the sulfenate ester **460** or revert to the starting material (Scheme 6.219).[1290]

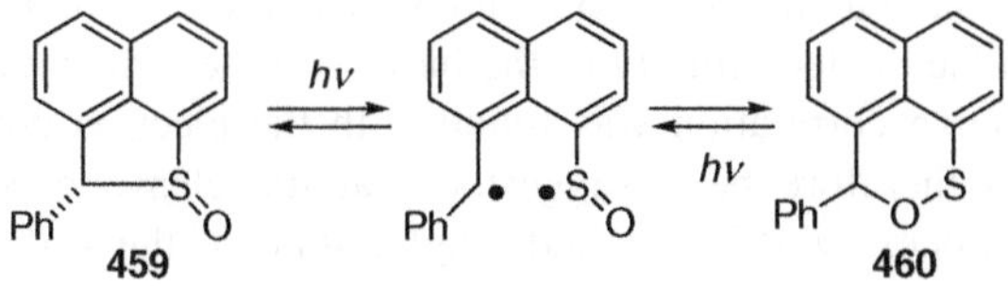

**459**                                    **460**

**Scheme 6.219**

### 6.5.3   Problems

1. Explain the following concepts and keywords: photofragmentation; photoextrusion.
2. Suggest the mechanisms for the following reactions:

(a)

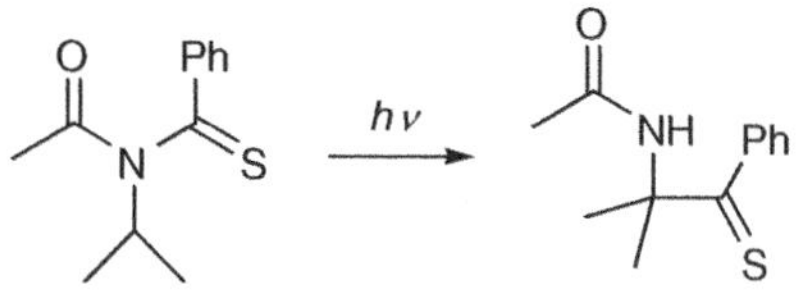

[ref. 1291]

(b)

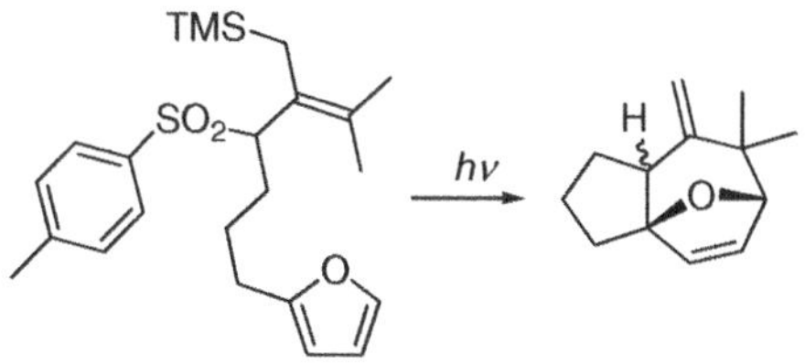

(TMS = trimethylsilyl)

[ref. 1292]

3. Predict the major photoproduct(s):

(a)

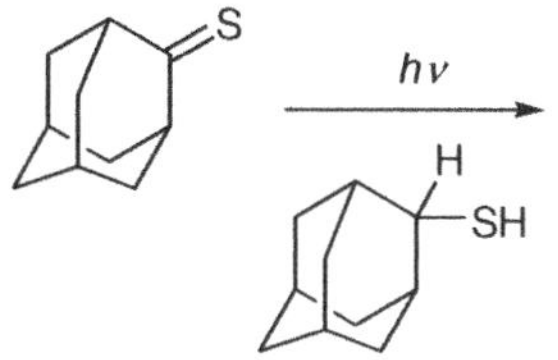

[ref. 1293]

(b)

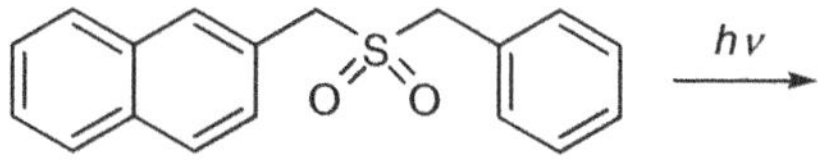

[ref. 1294]

## 6.6 Halogen Compounds

Halogen substitution has little effect on the absorption spectra of conjugated compounds. The heavier halogens bromine and iodine may, however, substantially increase the rate of ISC, especially when the parent compound has a long excited-state lifetime. The $\sigma,\sigma^*$ absorption of haloalkanes reaches into the near-UV region, increasingly with the heavier halogens and geminal di- or trisubstitution. Halogenated aliphatic solvents have relatively low cut-off wavelengths (see also Table 3.1), for example, 2,2,2-trifluorethanol (<190 nm), dichloromethane ($\sim$230 nm), chloroform ($\sim$250 nm), carbon tetrachloride ($\sim$265 nm) and bromoform ($\sim$300 nm).[158]

In this section, we discuss separately the chemical reactions that follow photo-fragmentation of a halogen molecule and the reactions of excited halogen-containing organic molecules. In the first case, a light-absorbing halogen molecule, such as chlorine (Table 6.18, entry 1), is homolytically cleaved to give two halogen atoms, which take part in the subsequent radical chain reactions (halogenations) with, for example, saturated or unsaturated hydrocarbons. Since the dissociation energies of halogen molecules are relatively low, visible light is sufficient to promote the initial step.

**Table 6.18** *Photoprocesses involving halogen compounds*

| Entry | Starting material[a] | Product(s) | Mechanism | Section |
|---|---|---|---|---|
| 1 | $X_2^*$ | $2X^\bullet$ | Photofragmentation | 6.6.1 |
| 2 | $[R\text{-}X]^*$ | $R^\bullet + X^\bullet$ | Photofragmentation | 6.6.2 |
| 3 | $[Ar\text{-}X]^* + Nu^\ominus$ | $Ar\text{-}Nu + X^\ominus$ | Nucleophilic photosubstitution | 6.6.2 |

[a]X = halogen atom; Nu⁻ = nucleophile; Ar = aryl.

Homolytic fission of the carbon–halogen bond is typical of excited organic alkyl and aryl halides (entry 2). Various subsequent reactions may then follow, including radical reactions such as hydrogen abstraction or rearrangements or electron transfer within the caged radical pair. The resulting ion pair intermediates can be attacked by nucleophiles, although direct nucleophilic attack on an excited aryl halide (see Scheme 6.93) is also possible (entry 3).

### 6.6.1 Halogen Compounds: Photohalogenation

$$X_2 \xrightarrow{\;h\nu\;} 2X^\bullet$$

*Recommended review articles.*[155,1295,1296]
*Selected theoretical and computational photochemistry references.*[1297–1301]

Photohalogenation – a photoinitiated reaction between a halogen donor, such as a halogen molecule and a substrate – proceeds typically via a radical chain mechanism; therefore, the reaction quantum yield, closely related to the average chain length, is greater than unity.[155]

The absorption spectra of chlorine and bromine molecules have $\lambda_{max}$ at 330 nm ($\varepsilon = 65 \, l \, mol^{-1} \, cm^{-1}$) and 420 nm ($\varepsilon = 165 \, l \, mol^{-1} \, cm^{-1}$), respectively.[155] Since their dissociation energies are 243 and 192 kJ mol$^{-1}$, which correspond to photons of 492 and 623 nm wavelength, respectively, they readily undergo homolytic fission upon excitation, even when using visible (incandescent) low-power radiant sources.

In the photochlorination of alkanes, chlorine atom formation is followed by various propagation and termination radical reactions (Scheme 6.220). The hydrogen abstraction step is usually exothermic and irreversible. Hydrogen atoms are abstracted at rates that are in the order primary < secondary < tertiary C–H bonds; the activation energies for abstraction may vary from 4 kJ mol$^{-1}$ in the first to 0.4 kJ mol$^{-1}$ in the last case.[155] Such small differences indicate that selectivity disappears at higher temperatures. Hydrocarbon rearrangements are not common, but every possible monochlorinated product is usually formed.[1295] The quantum yields may exceed $10^6$ in the absence of radical trapping agents.

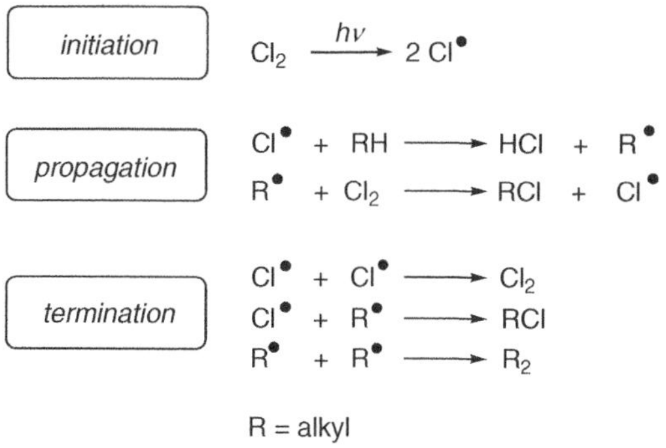

initiation $\qquad$ $Cl_2 \xrightarrow{h\nu} 2 \, Cl^\bullet$

propagation $\qquad$ $Cl^\bullet + RH \longrightarrow HCl + R^\bullet$
$R^\bullet + Cl_2 \longrightarrow RCl + Cl^\bullet$

termination $\qquad$ $Cl^\bullet + Cl^\bullet \longrightarrow Cl_2$
$Cl^\bullet + R^\bullet \longrightarrow RCl$
$R^\bullet + R^\bullet \longrightarrow R_2$

$R = alkyl$

**Scheme 6.220**

Aromatic solvents and carbon disulfide often increase the chlorination selectivity due to the formation of π- or σ-complexes between the solvent molecules and the chlorine atom. Such an interaction decreases the reactivity and hence increases the selectivity toward the hydrogen atoms. For example, photochlorination of adamantane (**461**) gives a higher production of 1-chloroadamantane in CS$_2$ than in benzene (Scheme 6.221).[1302]

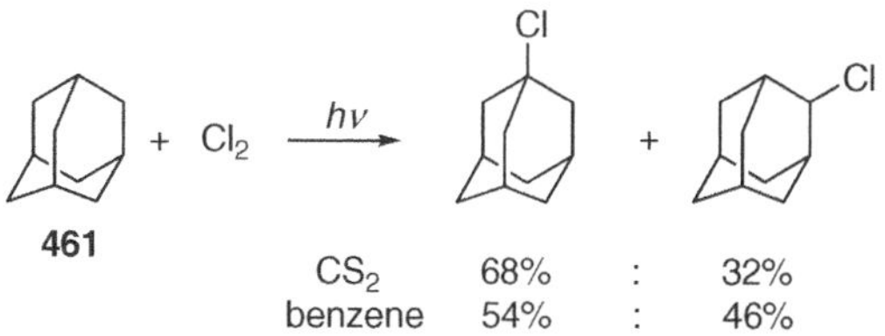

**461** $\qquad$ + $Cl_2 \xrightarrow{h\nu}$

| | | |
|---|---|---|
| CS$_2$ | 68% : | 32% |
| benzene | 54% : | 46% |

**Scheme 6.221**

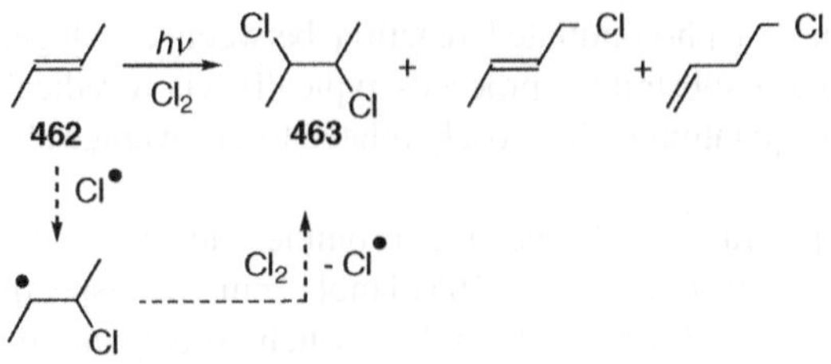

**Scheme 6.222**

Photochlorination of unsaturated hydrocarbons is initiated by exothermic addition of the chlorine atom to a multiple bond to form a chloroalkyl radical and is often accompanied by substitution and rearrangement reactions. This is demonstrated by chlorination of but-2-ene (**462**), where the addition product, 2,3-dichlorobutane (**463**), is obtained as a major product (Scheme 6.222).[1303]

In contrast to photochlorination, the corresponding hydrogen abstraction or addition steps in photobromination are usually reversible, endothermic and more selective.[155] The hydrogen abstraction rates in the allylic or benzylic positions can be relatively high, however. Table 6.19 shows the relative reactivities of the benzylic C–H bonds in a series of alkylbenzenes in the photobromination reaction carried out in CCl₄.[1304] It is apparent that not only electronic but also steric effects control the reaction kinetics.

**Table 6.19**   *Photobromination of alkylbenzenes*

| Alkylbenzene | Relative reactivity [per hydrogen (**H**) atom][1304] |
|---|---|
| 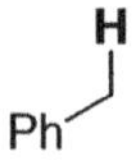 | 1.0 |
| 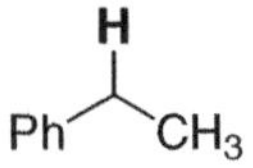 | 17.2 |
| 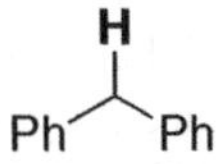 | 9.6 |
| 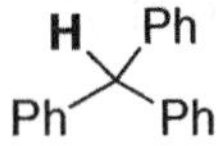 | 17.8 |

---

## Case Study 6.34: Organic synthesis – photobromination of progesterone

The regio- and stereoselective photobromination of progesterone (**464**), an α, β-unsaturated steroid ketone, gave the 6β-brominated product **465** (Scheme 6.223).[1305] The reaction was carried out in the presence of cyclohexene oxide to trap HBr formed during the reaction.

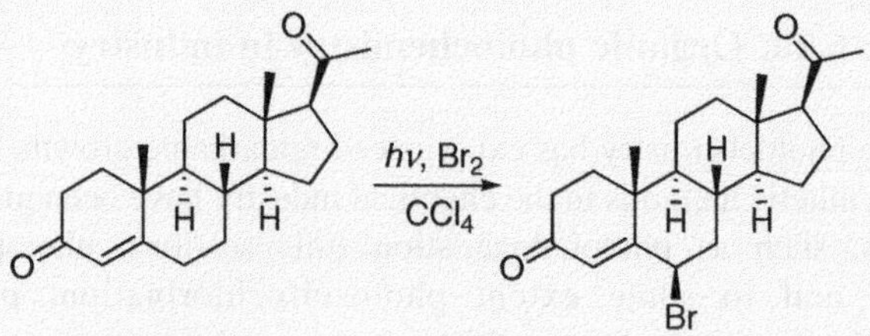

**Scheme 6.223**

*Experimental details.*[1305] A solution of progesterone **464** (6 mmol), bromine (6 mmol) and cyclohexane oxide (9 mmol) in CCl₄ (120 ml) was exposed to light from a tungsten lamp (incandescent; visible light; 100 W) (Figure 3.9). The product precipitated after 15 h and was filtered off. Column chromatography of the remaining filtrate gave an additional amount of the starting material. The overall chemical yield was 87%.

*N*-Bromosuccinimide (NBS) (**466**) is a common bromination agent, used mainly for substitution reactions in the allylic or benzylic positions. The mechanism involves excitation of NBS generating either a bromine atom (Scheme 6.224a) or a succinimidyl radical[1306] (Scheme 6.224b) as one of the chain propagation steps, depending on the solvent and conditions used and, to some extent, on the reactivity of the substrate.[1307]

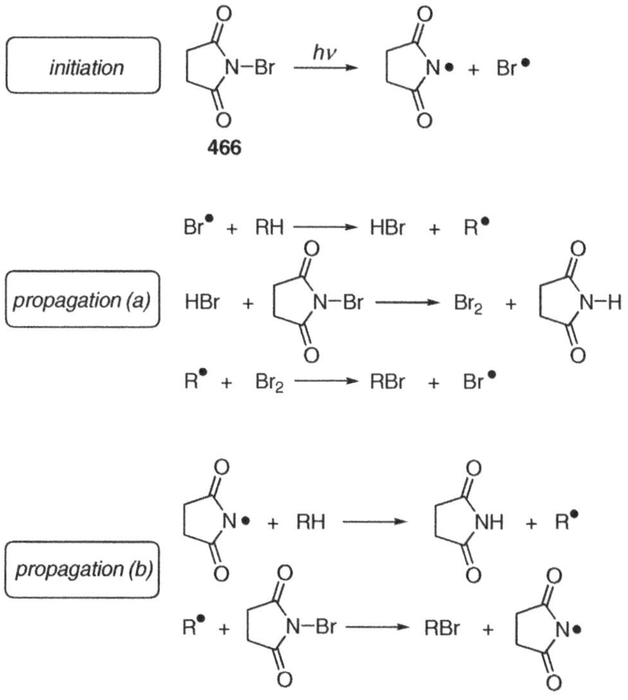

**Scheme 6.224**

## Special Topic 6.20: Organic photochemistry in industry

Although organic photochemistry has experienced remarkable growth, applications of photochemical synthetic methods in the chemical industry have been mostly limited to radical reactions, such as photohalogenation (this section), photopolymerization (Section 6.8.1) and to some extent photosulfochlorination, photooxidation (Section 6.7) and photonitrosylation, although some other reactions are also being used.[155]

Photochlorination of aromatic compounds, for example, can lead efficiently to fully chlorinated products. Benzene is converted to 1,2,3,4,5,6-hexachlorocyclohexane with $\Phi = 2500$ (Scheme 6.225).[1308] One of the stereoisomers, lindane, is a well-known insecticide (banned today), which was produced and used globally in agriculture in annual amounts of $\sim 10^6$ tonnes.[1309]

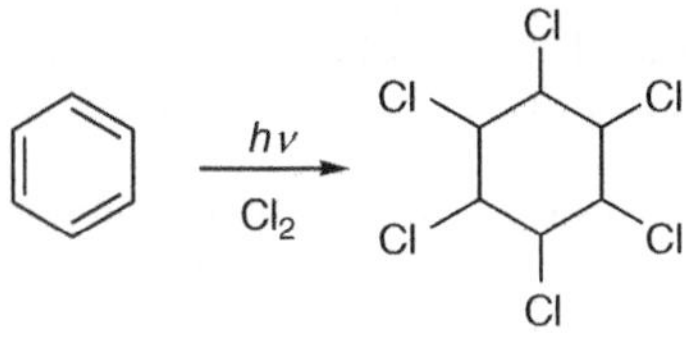

**Scheme 6.225**

In another example of an industrial process, cyclohexane photonitrosylation, leads to a nitrosocyclohexane, which is converted to caprolactam, a precursor to nylon-6 polymerization (Scheme 6.226).[155]

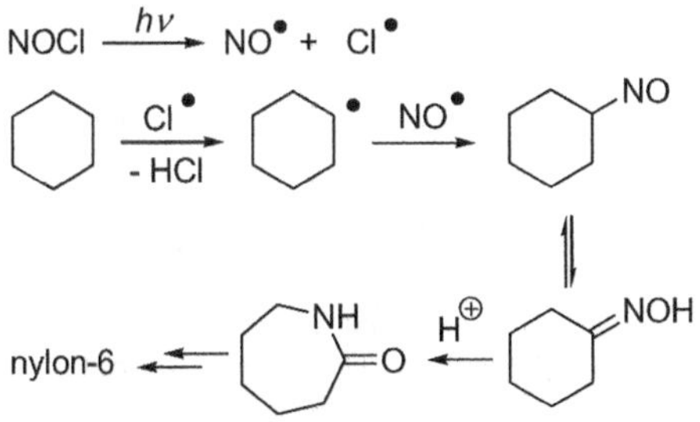

**Scheme 6.226**

Cholesterol can be converted to vitamin D photochemically from 7-dehydrocholesterol (provitamin D) (Special Topic 6.4) and this procedure is still used in industry.[616] Vitamin D can also be made by irradiating yeasts rich in ergosterol. In addition, vitamin A (retinol) (see Special Topic 6.1) is synthesized by *E–Z* photoisomerization (Section 6.1.1), sensitized by chlorophyll or other chromophores (Section 6.8) of its 11-*cis* isomer, which is produced industrially by conventional synthetic steps.[1310]

Rose oxide (**467**), a valuable perfume additive, is manufactured by an ene reaction (Section 6.7.3) of citronellol (**468**) (derived from lemon grass) with oxygen by irradiation in the presence of Methylene Blue as a triplet photosensitizer (Scheme 6.227).[1311,1312]

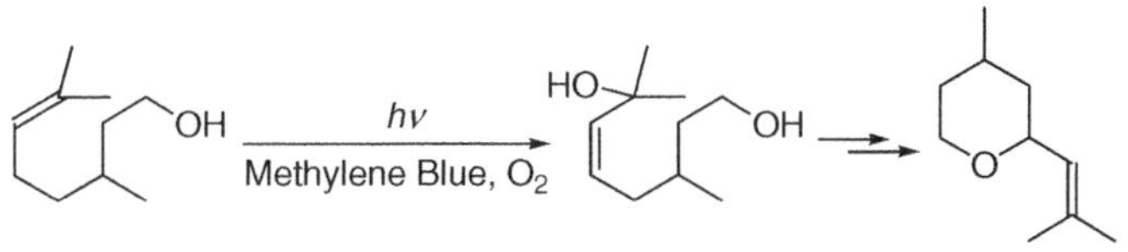

**468**                                    **467**

*Scheme 6.227*

Photopolymerization, that is, a polymerization process initiated by photochemical production of the reactive species – *initiators* (Section 6.8.1), is largely used in industry. Polymers can be synthesized by photoinitiation, but can also be modified by photochemical cross-linking (photochemical hardening or UV curing; photoimaging; photolithography; see Special Topic 6.27).

The most important and widespread applications of photochemistry are, of course, photoimaging techniques, such as photography and xerography (Special Topic 6.32).[1313] Some other large-scale applications, such as environmental remediation of anthropogenic pollutants by photocatalytic processes (Special Topic 6.28) and potential applications such as photoelectrochemical cells (Special Topic 6.31) and artificial photosynthesis (Special Topic 6.26), are also discussed in the following text.

### 6.6.2  Organic Halogen Compounds: Photofragmentation, Photoreduction and Nucleophilic Photosubstitution

*Recommended review articles.*[836,838–842,849,852,1314–1321]
*Selected theoretical and computational photochemistry references.*[1322–1333]

*Alkyl Halides*

Simple alkyl chlorides and bromides do not absorb above 250 nm, whereas the absorption band of an n,σ* transition in methyl iodide, for example, has a maximum at $\lambda_{max} = 258$ nm.[1334] Direct irradiation of alkyl halides results in homolytic cleavage of the C−X bond (X = halogen). In general, the excitation energy ($E_S$) in alkyl iodides or bromides (380–475 kJ mol$^{-1}$) is higher than the C−X bond dissociation energies (220–300 kJ mol$^{-1}$).[1314] The quantum yield of the fragmentation does not have to be unity even in the gas phase because of possible concurrent non-radiative or radiative energy dissipation processes.[1335] The photoreaction usually involves the excited singlet state and reaction is initiated by homolytic cleavage, possibly followed by electron transfer within the caged radical pair (Scheme 6.228). The radical species undergo radical reactions, such as hydrogen abstraction or rearrangement, while the resulting carbocation can be attacked by nucleophiles (nucleophilic photosubstitution; S$_N$1*).

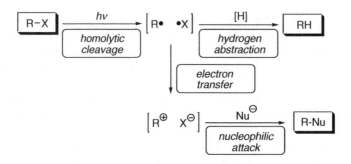

*Scheme 6.228*

Irradiation of primary alkyl bromides or iodides affords typically a mixture of reduction and elimination products. For example, a radical pair produced by photolysis of the halooctanes **469** undergoes in the initial stage either competing diffusion and hydrogen abstraction from the solvent molecules (to give **470**) or electron transfer giving rise to elimination product (**471**) – a favoured process in case of alkyl iodides (Scheme 6.229).[1336] Formation of the nucleophilic substitution product **472** is not observed, because solvent does not participate in carbocation generation.

$$n\text{-}C_7H_{15}CH_2X \xrightarrow[\text{MeOH}]{h\nu} n\text{-}C_7H_{15}CH_3 \ + \ n\text{-}C_6H_{13}CH=CH_2 \ + \ n\text{-}C_7H_{15}CH_2OMe$$

| **469** | **470** | **471** | **472** |
|---|---|---|---|
| X = Br | 66% | 31% | not detected |
| X = I | 38% | 56% | not detected |

*Scheme 6.229*

Photolysis of norbornan-1-yl (tertiary) halides (**473**) in methanol affords both reduction and the nucleophilic photosubstitution products **474** and **475**, respectively (Scheme 6.230).[1336]

Since no detectable incorporation of deuterium was observed to occur in CH$_3$OD, **475** cannot be produced via an elimination process to give a bridgehead alkene (bicyclo[2.2.1]hept-1-ene, **476**), followed by acid-catalysed addition of an alcohol. The ionic (electron transfer) mechanism is apparently preferred again in the photolysis of iodides, although the interplay between ionic versus radical mechanism is also dependent on the ionization potentials of the alkyl radicals (that subsequently form carbocation intermediates). For example, an electron-withdrawing hydroxy group in position 3 of **473** stimulates formation of the corresponding reduction product because the ionization potential of the 3-hydroxynorbornan-1-yl radical is higher than that of norbornan-1-yl radical.[1337]

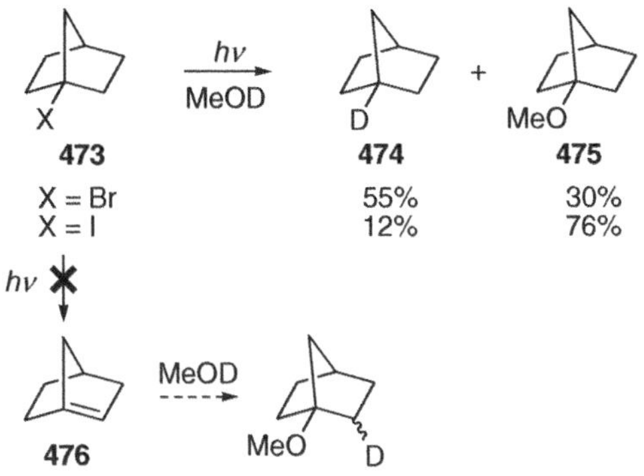

*Scheme 6.230*

Photochemistry of polyhaloalkanes results in a C−X bond fission of the weakest bond in the initial step. For example, irradiation of bromotrichloromethane affords trichloromethyl and bromine radicals, which add to styrene to obtain 1-bromo-3,3,3-trichloropropylbenzene (**477**) in 78% chemical yield (Scheme 6.231).[1338]

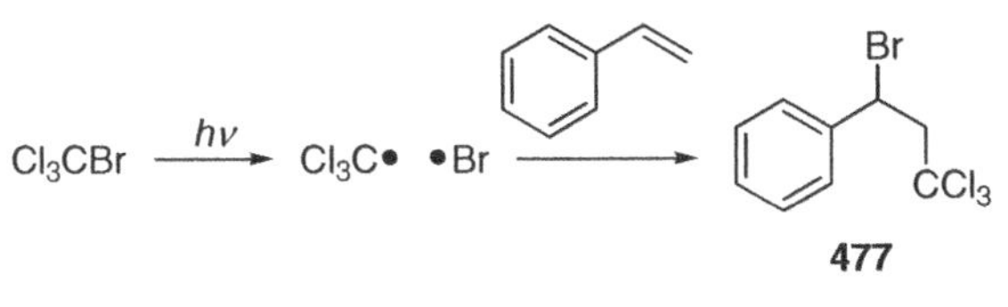

*Scheme 6.231*

## Vinyl Halides

Photolysis of vinyl iodides is a convenient method for the generation of vinyl cations, highly reactive intermediates that are difficult to generate by thermal processes. Irradiation of 1-iodocyclohexene (**478**, X = I) in methanol in the presence of zinc as an iodine scavenger affords the major nucleophilic substitution product **479** and small amounts of the radical-derived reduction product cyclohexene (**480**) (Scheme 6.232).[1339] In the absence of zinc, the ketal **481** is apparently obtained via an acid-catalysed addition of methanol to **479**. In contrast, vinyl bromide (**478**, X = Br) mainly provides the photoreduction product **480**, which parallels closely the photobehaviour of the saturated analogues.[1314]

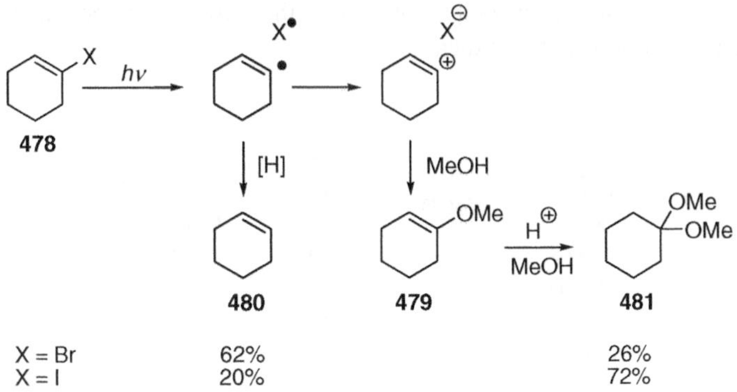

**Scheme 6.232**

*Aryl Halides*

Simple aromatic halides display two principal superimposing absorption bands, $\pi,\pi^*$ and $\sigma,\sigma^*$, tailing over 250 nm. Intersystem crossing (ISC) to $T_1$ can be fast and efficient due to increased spin–orbit coupling attributed to the 'heavy-atom effect' (Section 4.8) of halogen atoms (Br, I).[842,849,1316] The subsequent C—X bond homolysis may afford aryl radicals and halogen atoms as the primary products (Scheme 6.233), which requires that the energy of the reactive excited state be greater than the corresponding C—X dissociation energy. Fission is, therefore, thermodynamically unfavourable for aryl fluorides, for example fluorobenzene ($E_T = 353\,\text{kJ mol}^{-1}$; $D_{C-F} = 526\,\text{kJ mol}^{-1}$), and for most aryl chlorides, for example 1-chloronaphthalene ($E_T = 248\,\text{kJ mol}^{-1}$; $D_{C-Cl} = 403\,\text{kJ mol}^{-1}$).[849] In the latter

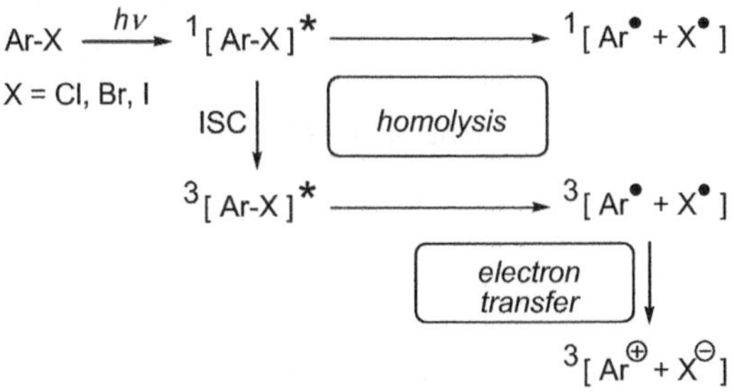

**Scheme 6.233**

case, inefficient photodegradation may take place from the singlet state.[1340] Triplet state homolysis can be exothermic in iodo- and bromoarenes. The resulting triplet radical pair can be involved in subsequent radical reactions or undergo in-cage *electron transfer* to form an aryl cation, which is readily attacked by nucleophiles such as water.

Photodehalogenation (photoreduction) – replacement of the halogen atom by hydrogen – in the presence of hydrogen-atom donors is an interesting alternative to Raney nickel-catalysed reduction. Irradiation of 2-bromophenol (**482**) in propan-2-ol, for example, leads to phenol in >90% chemical yield via a radical mechanism (Scheme 6.234).[1341]

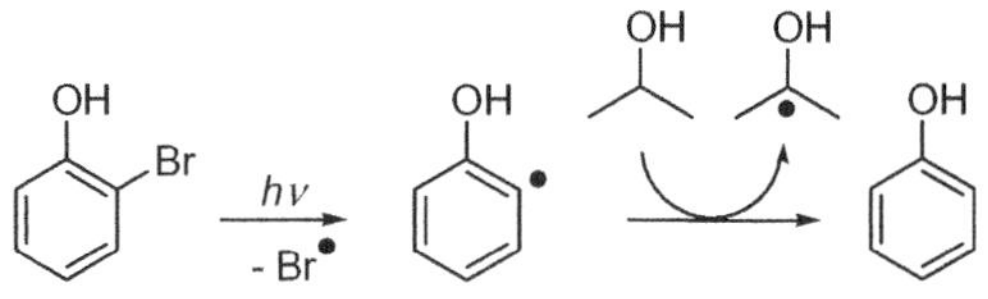

**Scheme 6.234**

There is much evidence that photodehalogenation reactions of triplet excited halogenated biphenyls (**483**), naphthalenes and other aromatic compounds (which exhibit efficient intersystem crossing) in the presence of good electron donors, such as amines, is accompanied by full or partial electron transfer to form exciplex intermediates (Scheme 6.235).[667,849] Escape of a free radical ion **484** may give the corresponding hydrocarbon in the presence of a proton donor such as water or a good hydrogen-atom donor ([H]).

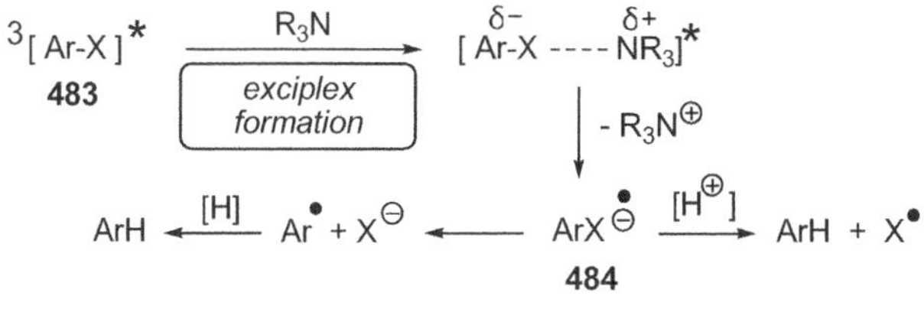

**Scheme 6.235**

The synthetically useful ring closure reaction (intramolecular cyclization) of 1-(*o*-chlorophenyl)naphthalene (**485**) gives fluoranthene (**486**) in 72% chemical yield (Scheme 6.236).[1342] This reaction proceeds via initial homolysis of the C−Cl bond, followed by radical aromatic substitution.

Some important *persistent environmental pollutants*, such as 1,1,1-trichloro-2,2-bis (4-chlorophenyl)ethane (DDT), polychlorinated biphenyls (PCBs), dibenzo-*p*-dioxins (PCDDs) and polybrominated diphenyl ethers (PBDEs), are aryl halides; therefore, studies

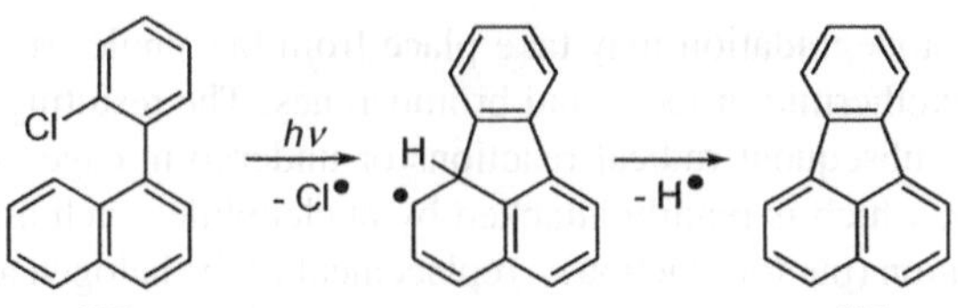

**485**                                                                    **486**

*Scheme 6.236*

of the photochemistry of halogenated aromatic compounds have also been stimulated by environmental concerns.[849] Photolysis of chlorinated biphenyls in non-nucleophilic media, for example, exhibits a preferential dehalogenation of the *o*-chlorine atoms, because their loss relieves the steric strain (Scheme 6.237).[849,1343]

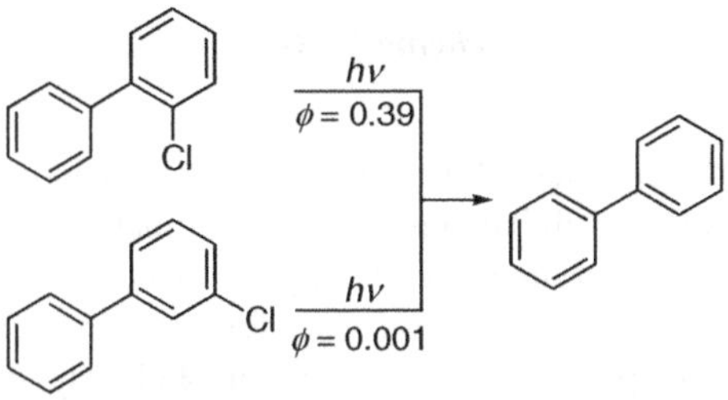

*Scheme 6.237*

In aqueous solutions, 2-halophenols, such as 2-bromophenol, are known to undergo photo-Wolff rearrangement (see also Scheme 6.171). In this reaction, the cyclopentadienecarboxylic acid derivatives **487** are formed via the singlet α-ketocarbene (**488**) and ketene (**489**) intermediates, followed by nucleophilic addition of water (Scheme 6.238).[1344–1346]

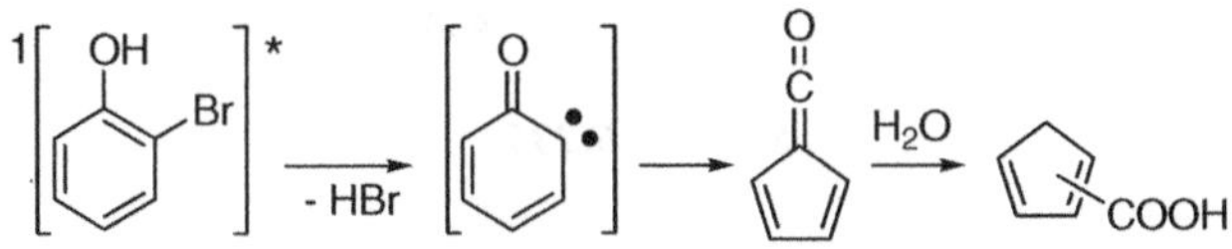

**488**        **489**        **487**

*Scheme 6.238*

Irradiation of 3,4-difluoroacetophenone (**490**) in acidic aqueous solutions leads to an unusual product of solvolysis, 4-fluoro-3-hydroxyacetophenone (**491**), in nearly quantitative chemical yield (Scheme 6.239).[332] Nucleophilic addition of water to the *meta*-position proceeds via O-protonated triplet acetophenone.

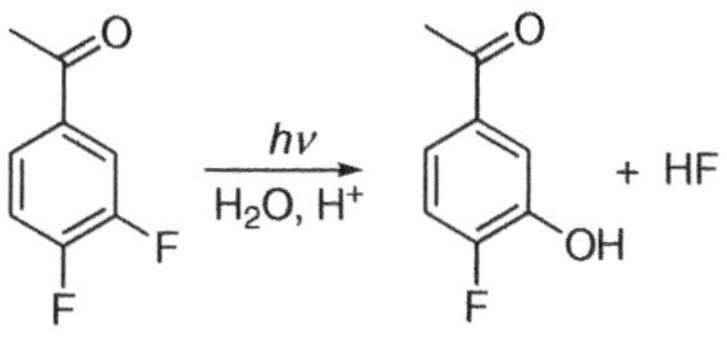

**Scheme 6.239**

Nucleophilic photosubstitutions are typical reactions of aromatic halides. The variability of the process in terms of mechanism is presented in detail, along with many examples, in Section 6.2.3; therefore only one example of the reaction is mentioned here. Multiple photocyanation of hexachlorobenzene (**492**) takes place in aqueous acetonitrile containing a cyanide ion to give pentacyanophenolate (**493**) (Scheme 6.240).[1347] The reaction involves the triplet state apparently via an $S_N2Ar^*$ (Scheme 6.93) pathway.

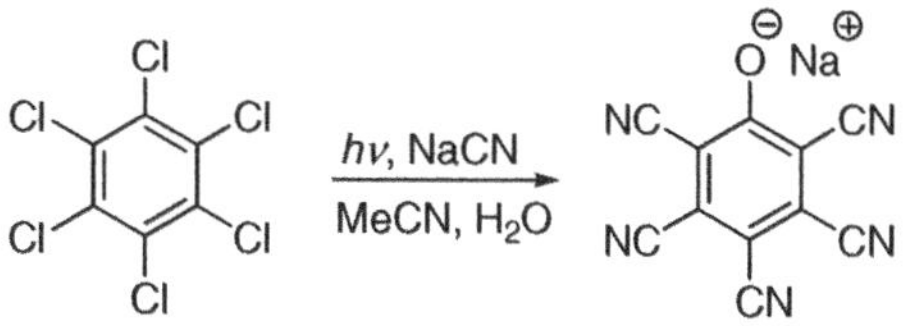

**Scheme 6.240**

## Case Study 6.35: Synthesis of cage compounds – cubane

Photolysis of 1,4-diiodocubane (**494**) in nucleophilic solvents, such as methanol, was found to provide effectively the substitution product **495** (Scheme 6.241).[1348] It was argued that the reaction proceeds via the formation of a caged radical pair and electron transfer to give iodide ion and a carbocation, which is trapped by the solvent ($S_N1^*$ pathway). Hyperconjugative stabilization in cubyl cation cannot assist much from an extremely strained olefin cubene. However, the authors expected that photochemical fission of the C–I bond would still provide sufficient energy for production of the cation intermediate.

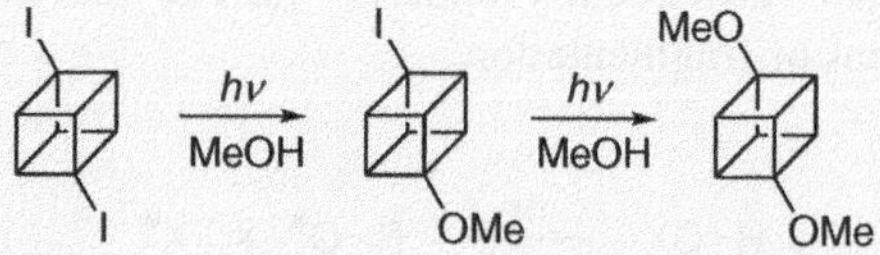

**Scheme 6.241**

*Experimental details.*[1348] 1,4-Diiodocubane (**494**; 0.2 mmol) suspended in dry methanol (6.0 ml) in a quartz tube equipped with a magnetic stirrer was irradiated in a Rayonet reactor equipped with 16 low-pressure mercury lamps ($\lambda_{irr} = 254$ nm) (Figure 3.10) under a nitrogen atmosphere at 40 °C. About 32% monosubstituted 1-iodo-4-methoxycubane was detected by GC after 6 h. Further irradiation gave a significant production of 1,4-dimethoxycubane (**495**; ~50%).

### Hypervalent Iodine Compounds

Hypervalent iodine compounds, especially organoiodine(III) species, are commonly used in synthetic organic chemistry, but photochemical applications are still rare.[1349] Direct photolysis of diphenyliodonium triflate (**496**), for example, affords a singlet excited species that undergoes a heterolytic cleavage leading to phenyl cation and iodobenzene. In addition, intersystem crossing leads to the triplet state, which is homolytically fragmented to give a caged pair of iodobenzene radical cation and phenyl radical. The primary isolated products in H-atom donating solvents are 2-, 3- and 4-iodobiphenyl, iodobenzene, benzene and trifluoromethanesulfonic acid (Scheme 6.242).[1350] Such a photochemically generated strong acid has been successfully used as an initiator of cationic polymerizations[1351] (Section 6.8.1).

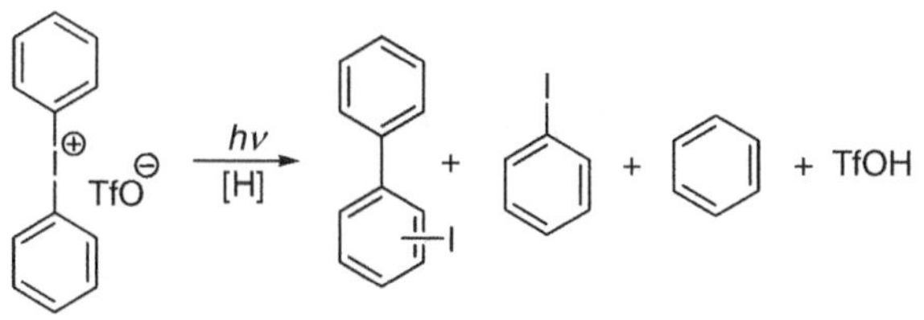

*Scheme 6.242*

### Alkyl Hypohalites

Hypohalites represent another class of photolabile organic halogen compounds, which exhibit two absorption peaks above 250 nm; alkyl hypochlorites have $\lambda_{max}$ at ~260 and ~320 nm, while the absorption of alkyl hypobromites and hypoiodites is bathochromically shifted.[1319] Irradiation of alkyl hypohalites leads to a homolytic cleavage to give alkoxyl and halogen radicals with a high photochemical efficiency (Scheme 6.243).[1352] These primary intermediates can subsequently undergo various reactions, such as hydrogen abstraction, rearrangement or fragmentation.

$$R\text{--}OX \xrightarrow{\ h\nu\ } R\text{--}O^{\bullet} + X^{\bullet}$$

*Scheme 6.243*

Alkyl hypochlorites and hypobromites are relatively stable molecules; alkyl hypoiodites can only be prepared *in situ*, usually by the reaction of alcohols with metal acetates or oxides and iodine or by the reaction of alcohols with a hypervalent iodine compound and iodine.[1319] Alkyl hypochlorites and hypoiodites can be utilized in reactions that parallel those of the organic nitrites (Barton reaction; Section 6.4.2). For example, the photochemistry of the steroidal hypoiodite **497**, prepared by the reaction of the corresponding alcohol with iodine oxide (I$_2$O) generated from mercury(II) oxide and molecular iodine *in situ*, affords a new five-membered ring via an O–I bond fission, 1,5-hydrogen abstraction and substitution (Scheme 6.244).[1353]

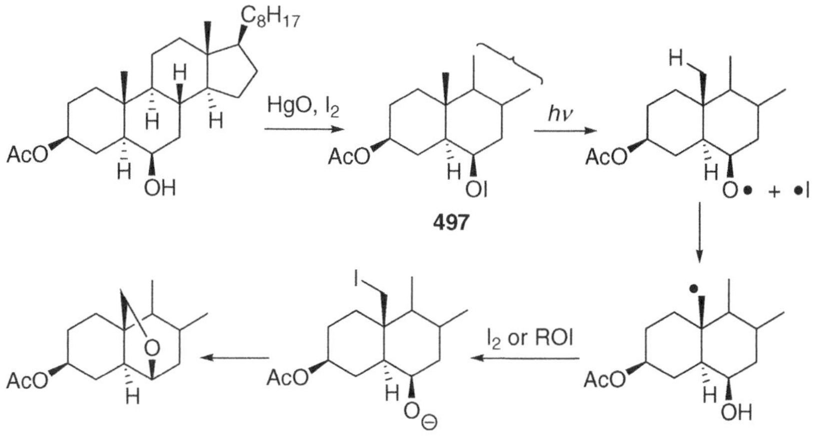

**Scheme 6.244**

## 6.6.3 Problems

1. Explain the following concepts and keywords: photoinduced radical chain mechanism; photonitrosylation; nucleophilic photosubstitution; photodehalogenation.
2. Suggest the mechanisms for the following reactions:

(a)

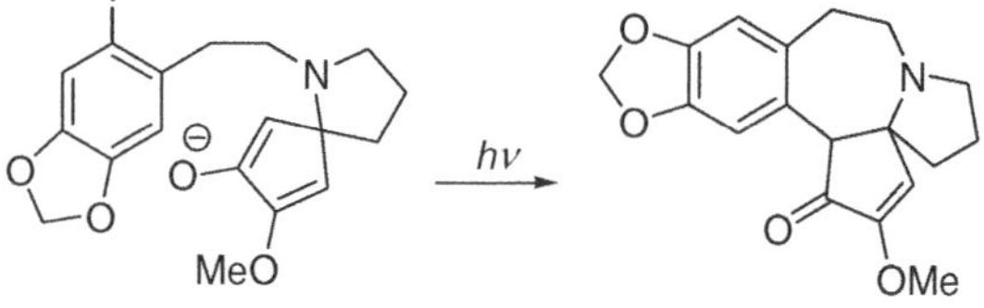

[ref. 1354]

(b)

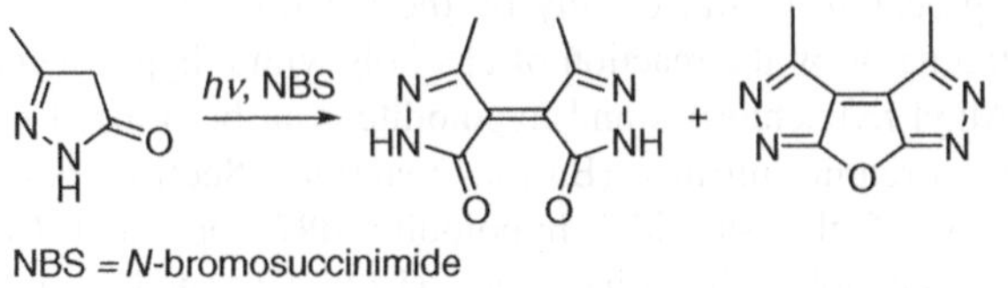

NBS = *N*-bromosuccinimide

[ref. 1355]

3. Predict the major photoproduct(s):

(a)

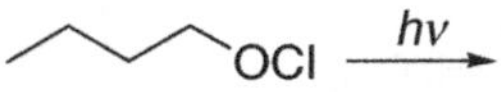

[ref. 1356]

(b)

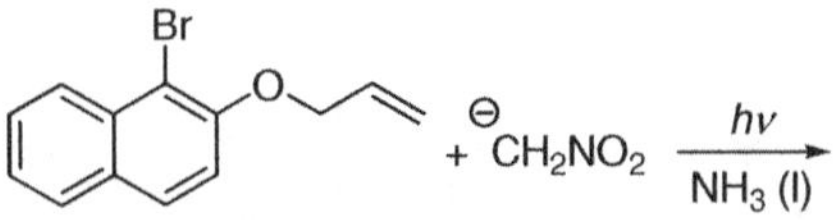

[ref. 1357]

## 6.7  Molecular Oxygen

The photophysical properties of oxygen were discussed in Section 2.2.5. Molecular oxygen can be involved in photoreactions as either the ground-state (triplet; $^3O_2$) or an excited (singlet; $^1O_2$) species. Section 6.7.1 discusses the oxidation reactions of excited organic compounds with ground-state oxygen, called oxygenations (Table 6.20, entry 1). When a triplet sensitizer, such as Methylene Blue, Rose Bengal or porphyrin, is introduced, its excitation is readily transferred to ground-state oxygen to form $^1O_2$ (entry 2). Singlet oxygen can also be formed by thermal decomposition of unstable molecules, such as endoperoxides (entry 3).

$^1O_2$ is a reactive, short-lived molecule, which can undergo several specific reactions with organic unsaturated hydrocarbons. Oxygenation of a carbon–carbon double bond to give 1,2-endoperoxides (entry 4) can be formally interpreted as [2+2] cycloaddition, whereas addition of singlet oxygen to a 1,3-diene moiety, such as a buta-1,3-diene derivative or benzene ring, to form 1,4-endoperoxides (entry 5) correspond to [2 + 4] cycloaddition. Endoperoxides may be further photolysed to rearrange to diepoxides (entry 6) or to revert back to the corresponding diene and oxygen molecule. When an alkene reactant contains at least one allylic hydrogen, the ene reaction is another competing process (entry 7).

**Table 6.20** *Photoprocesses involving molecular oxygen*

| Entry | Starting material | Product(s) | Mechanism | Section |
|---|---|---|---|---|
| 1 | $R* + {}^3O_2$ | Oxidation products | Oxygenation with ground state oxygen | 6.7.1 |
| 2 | ${}^3O_2$ + sensitizer* | ${}^1O_2$ + sensitizer | Singlet oxygen production | 6.7.1 |
| 3 |  |  + ${}^1O_2$ | Singlet oxygen production | 6.7.1 |
| 4 |  + ${}^1O_2$ |  | [2 + 2] photooxygenation | 6.7.2 |
| 5 |  + ${}^1O_2$ |  | [4 + 2] photooxygenation | 6.7.2 |
| 6 |  |  | Endoperoxide photorearrangement | 6.7.2 |
| 7 |  + ${}^1O_2$ |  | Ene reaction | 6.7.3 |

## 6.7.1 Molecular Oxygen: Ground State and Excited State

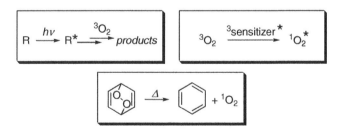

*Recommended review articles.*[135,136,1358–1363]
*Selected theoretical and computational photochemistry references.*[135,136,1364]

*Ground-state Molecular Oxygen*

Ground-state molecular oxygen (dioxygen), an open-shell triplet (biradical) denoted by spectroscopists ${}^3\Sigma_g^-$, is one of the most abundant and important life-supporting species on Earth. It can participate in some photochemically initiated reactions, sometimes called *type I oxygenations.*[1365,1366] In such a case, a radical **498**, typically formed via hydrogen atom transfer between an excited species (sensitizer) and an organic hydrogen donor, is trapped by oxygen to form a peroxy radical intermediate **499**, which undergoes subsequent reactions (Scheme 6.245).[1365] In contrast, photoinduced electron transfer from an electron donor, such as an radical anion **500**, to ground-state oxygen can give a reactive superoxide anion ($O_2{}^{\bullet-}$; **501**), which readily reacts with the radical cation **502** formed in the first step (Scheme 6.246).[1366]

$$\text{sensitizer}^* + \text{substrate-H} \longrightarrow [\text{sensitizer-H}]^\bullet + \text{substrate}^\bullet$$
$$\textbf{498}$$

$$\text{substrate}^\bullet + {}^3O_2 \longrightarrow \text{substrate-O-O}^\bullet \longrightarrow \text{oxidation products}$$
$$\textbf{499}$$

*Scheme 6.245*

$$\text{sensitizer}^* + \text{substrate} \longrightarrow \text{sensitizer}^{\bullet\ominus} + \text{substrate}^{\bullet\oplus}$$
$$\textbf{500} \qquad\qquad \textbf{502}$$

$$\text{sensitizer}^{\bullet\ominus} + {}^3O_2 \longrightarrow \text{sensitizer} + O_2{}^{\bullet\ominus}$$
$$\textbf{501}$$

$$\text{substrate}^{\bullet\oplus} + O_2{}^{\bullet\ominus} \longrightarrow \text{oxidation products}$$

*Scheme 6.246*

---

### Special Topic 6.21: Atmospheric photochemistry

Photochemical reactions have played a decisive role in the evolution of the atmosphere and of life on Earth. Such processes generally involve simple species, many of which are otherwise considered to be stable and unreactive. Special Topic 6.17 has already discussed photochemistry in a prebiotic atmosphere. Since geochemical inorganic processes cannot be responsible for the current level of oxygen in the atmosphere (21%), it must be almost exclusively the product of biological activity (Special Topic 6.25).

There are two principle photoreactions connected to atmospheric oxygen.[1367] When stratospheric (approximately 15–35 km above the Earth's surface) molecular oxygen is exposed to UVC ($\lambda < 240$ nm), oxygen atoms are produced and react with $O_2$ to form ozone. This important compound strongly absorbs UVB ($\lambda < 300$ nm) radiation (Figure 1.1),[1368] so that only a small part of the life-threatening UV radiation from the Sun reaches the Earth's surface:

$$O_2 + h\nu \rightarrow 2O$$

$$O + O_2 \rightarrow O_3$$

$$O_3 + h\nu \rightarrow O_2 + O$$

Ozone in the stratosphere is depleted by reactions with halogen atoms. Depletion of stratospheric ozone, commonly referred to as the *ozone hole*, usually occurs over the Earth's cold regions. The main source of chlorine atoms in the stratosphere is photodissociation of chlorofluorocarbon (CFC) compounds,[1369] commonly called *Freons*, e.g.:

$$CFCl_3 + h\nu \rightarrow {}^\bullet CFCl_2 + Cl^\bullet$$

$$O_3 + Cl^\bullet \rightarrow O_2 + ClO^\bullet$$

Nitrogen oxide, formed by photofragmentation of $NO_2$ or chlorine nitrate ($ClONO_2$), hydroxyl radicals and some other reactive species are also responsible for stratospheric ozone depletion. These compounds may have both natural and anthropogenic (combustion, etc.) origin. Atmospheric chemistry also takes place in aerosol particles, cloud droplets[1370] and ice crystals.[1371,1372]

Ozone depletion in the stratosphere (causing enhanced levels of the penetrating UV radiation) leads to increased tropospheric ozone levels. Ground-level ozone can pose a health risk, because it is a strong oxidant. Furthermore, other airborne tropospheric pollutants, such as $SO_2$, $NO_2$ and $HNO_2$, are also photolabile and may be responsible for the formation of secondary pollutants.[1367] Hundreds of individual volatile organic compounds (VOCs), such as alkanes, alkenes, aromatic hydrocarbons, oxygen and nitrogen-containing compounds, are important trace constituents of the atmosphere. Photochemical reactions also participate in their production.[1373] For example, nitrous acid undergoes rapid photolysis after sunrise (in so-called photochemical smog[1374]), which leads to early-morning production of hydroxyl radicals. They may abstract hydrogen from VOCs,[1375] such as methane, to give a radical, which reacts with oxygen and subsequently with NO to form formaldehyde:

$$HNO_2 + h\nu \rightarrow HO^\bullet + NO^\bullet$$

$$CH_4 + HO^\bullet \rightarrow {}^\bullet CH_3 + H_2O$$

$${}^\bullet CH_3 + O_2 \rightarrow H_3COO^\bullet$$

$$H_3COO^\bullet + NO \rightarrow H_3CO^\bullet + NO_2$$

$$2H_3CO^\bullet + {}^1\!/_2\, O_2 \rightarrow 2H_2CO + H_2O$$

Other harmful compounds, such as peroxyacyl nitrates (PANs), respiratory and eye irritants, are also produced in photochemical smog:

$$hydrocarbons + O_2 + NO_2 + h\nu \rightarrow RC(O)OONO_2$$

*Excited-State Molecular Oxygen*

Two excited singlet states of oxygen, $^1\Delta_g$ and $^1\Sigma_g$, have energies only 95 and 158 kJ mol$^{-1}$ above that of the ground-state, respectively (see Section 2.2.5).[135,136] In $^1\Delta_g$, both electrons are paired in one of the $\pi^*$ orbitals and this species is rather expected to be involved in polar reactions; in the latter case ($^1\Sigma_g$), electrons are located in two degenerate $\pi^*$ orbitals. Such a molecule has a (bi)radical-like character, similarly to that of $^3\Sigma_g$. Since direct excitation of oxygen by irradiation is not easily accomplished in the photochemical laboratory (only liquid or gaseous oxygen under high pressure exhibits sufficient absorption in the visible or near-UV region),[135,1364] sensitization is often utilized for *in situ* generation of $^1\Delta_g$. This species is called *singlet oxygen* and is involved in most photochemically induced oxygenation reactions (Sections 6.7.2 and 6.7.3). Thanks to its unusually low excitation energy, ground-state oxygen ($^3\Sigma_g$) can efficiently quench most triplet excited species. In addition, energy transfer between a singlet excited compound and the ground-state oxygen can also be feasible.[139]

Scheme 6.247 illustrates a general sensitization process, in which energy from a triplet excited sensitizer is transferred to ground-state triplet oxygen. The quantum yield of singlet oxygen production depends on many parameters, such as the quantum yield of singlet excited sensitizer intersystem crossing (ISC), the efficiency of triplet excited sensitizer quenching or simply local oxygen and sensitizer concentrations. Sensitizers (dyes) with low triplet energies, such as Rose Bengal (**503**) ($E_T = 164$–177 kJ mol$^{-1}$)[1376] and Methylene Blue (**504**) ($E_T = 138$ kJ mol$^{-1}$),[157] usually used in polar (aqueous or alcoholic) solutions, and tetraphenylporphyrin (**505**) ($E_T = 140$ kJ mol$^{-1}$), applied in nonpolar solutions, are typical examples.[135,136] Since the quantum yield of bimolecular singlet oxygen formation is usually very high and dyes have very high absorption coefficients, concentrations of a dye in the reaction mixture can be low.[1377] Singlet oxygen has a relatively long lifetime in common solvents (microseconds in protic solvents, such as water and alcohols; milliseconds in nonprotic solvents), even in the presence of reactive targets.[1358,1363] It has been demonstrated that it can also be relatively long-lived in a cell ($\tau \sim 3$ µs; Special Topic 6.23) and is capable of diffusing over substantial distances and crossing the cell membranes into the extracellular environment.[1378] Excited oxygen can deactivate by radiative and radiationless processes,[1361,1379] act as an electron acceptor (to form superoxide anion $O_2^{\bullet-}$) in the presence of good electron donors[1380] such as amines or be involved in subsequent reactions, sometimes called type II oxygenation[1366] reactions (Sections 6.7.2 and 6.7.3).

$$\text{sensitizer} \xrightarrow{h\nu} {}^1\text{sensitizer}^* \xrightarrow{\text{ISC}} {}^3\text{sensitizer}^*$$

$$^3\text{sensitizer}^* + {}^3O_2\,(^3\Sigma_g) \longrightarrow {}^1\text{sensitizer} + {}^1O_2\,(^1\Delta_g)$$

Sensitizers:

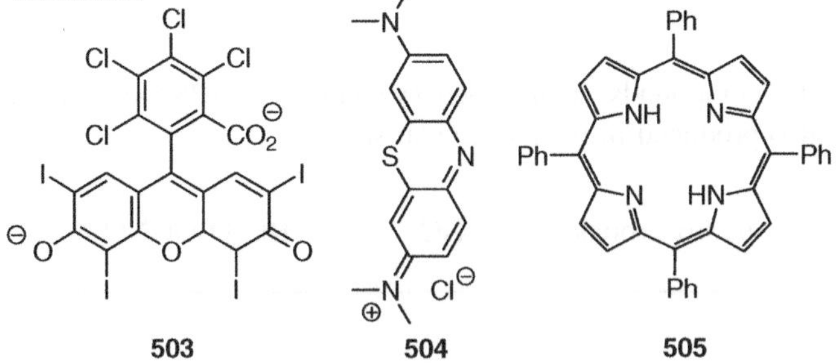

**503**          **504**          **505**

**Scheme 6.247**

---

## Special Topic 6.22: Phototoxicity and photoallergy

Photosensitivity reactions are adverse skin responses to the combined action of a new chemical (such as a drug) and light.[1381,1382] The mechanisms are usually complex but we can distinguish two types of reactions: phototoxic and photoallergic. More common phototoxic reactions result from direct tissue damage caused by *in situ* formation of singlet oxygen. In contrast, a photoallergy is an immunologically mediated hypersensitivity developed to a small amount of photochemically produced compound. The list of chemicals associated with photosensitivity includes several common antibiotics, such as sulfonamides and tetracyclines, diuretics, tranquillizers, cancer medicines and psoralen (Special Topic 6.8), fragrances, and so on. Most phototoxic agents absorb radiation in the UVA range.

For example, tetracycline (**506**), a broad-spectrum antibiotic, and its derivatives are known to induce phototoxic or photoallergic reactions that involve photosensitization of biomolecules by the drug or the formation of one or more photoproducts and their subsequent photoreactions.[1383] Singlet oxygen is probably involved.

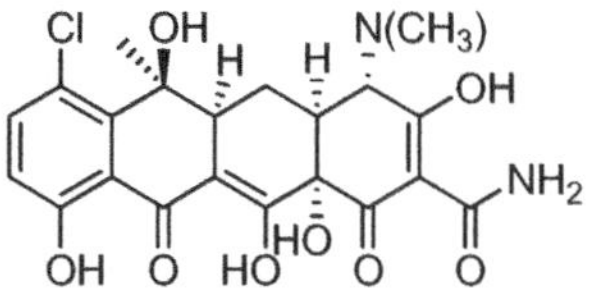

**506**

---

Singlet oxygen can also be obtained by thermal decomposition of unstable molecules, such as arene endoperoxides (e.g. **507**) (Section 6.7.2) or ozonides (e.g. **508**),[135,1384,1385] by photolysis of ozone[1386] or arene endoperoxides (such as **507**),[1384,1387] or by microwave discharge in the presence of ground-state oxygen (Scheme 6.248).[1388]

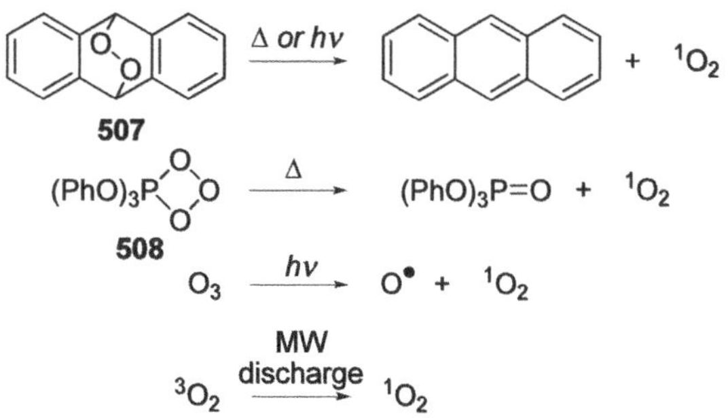

*Scheme 6.248*

Singlet oxygen is involved in many important chemical processes and photochemical applications, including photodynamic therapy (Special Topic 6.23), photocarcinogeneity (Special Topic 6.7) and phototoxicity (Special Topic 6.22), chemiluminescence (Section 5.6), atmospheric photochemistry (Special Topic 6.21), polymer degradation (Special Topic 6.13), photosynthesis[1389] (Special Topic 6.25) or industrial organic synthesis (Special Topic 6.20).

## Special Topic 6.23: Photodynamic therapy

Photodynamic therapy (PDT) is a well-known and commonly used method for treatment of cancer (e.g. lung, oesophagus) and other hyperproliferative diseases.[1390–1398] It involves a photosensitizer, light and tissue oxygen to produce *in vivo* singlet *oxygen* as a cytotoxic agent (see also Case Study 6.36). In general, red and near-infrared light penetrate well into most human tissues (3–5 and 6–10 mm at $\lambda = 600$ and 800 nm, respectively), because it is not significantly absorbed by the chromophores present in tissues, except for melanin.[1391,1399] Therefore, dyes (photosensitizers), such as porphyrin derivatives, are most frequently utilized in PDT since they possess significant absorption bands in the 600–800 nm region and show no detectable cytotoxicity at the doses applied. The first sensitizer used in clinical PDT was an ill-defined mixture of haematoporphyrin oligomeric derivatives (**509**, Figure 6.17) known as Photofrin.[1392,1400] Its illumination with 630 nm radiation is usually carried out 48 h after systemic administration by injection. Today, new generations of PDT chromophores have been developed in order to improve their biological, photophysical, pharmacokinetic and phototherapeutic properties. For example, *m*-tetrahydroxyphenyl chlorin (**510**) gives the same photodynamic effect as **509** at concentrations about two orders of magnitude lower, which is related to the difference in their molar absorption coefficients at 630 nm. An alternative to the administration of exogenous photosensitizers is to stimulate cellular synthesis of endogenous photosensitizers. In one type of treatment, aminolevulinic acid (ALA) (**511**) is applied and a natural photosensitizer, protoporphyrin IX (**512**), also associated with some types of porphyria, a disorder of the haem biosynthetic pathway, is subsequently slowly (~24 h) produced by the enzyme ALA synthase from glycine and succinyl-CoA (Scheme 6.249). Both conventional and coherent sources of light, such as wavelength-tunable dye lasers, solid-state lasers and red-emitting diode lasers coupled with optical fibres, are used today. The exposure typically lasts from a few minutes to a few tens of minutes and is then repeated.

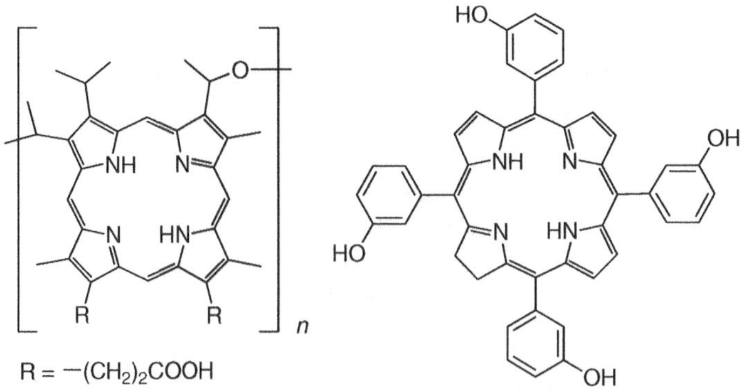

R = —(CH$_2$)$_2$COOH

**509**                 **510**

**Figure 6.17**    Haematoporphyrin and *m*-tetrahydroxyphenyl chlorin

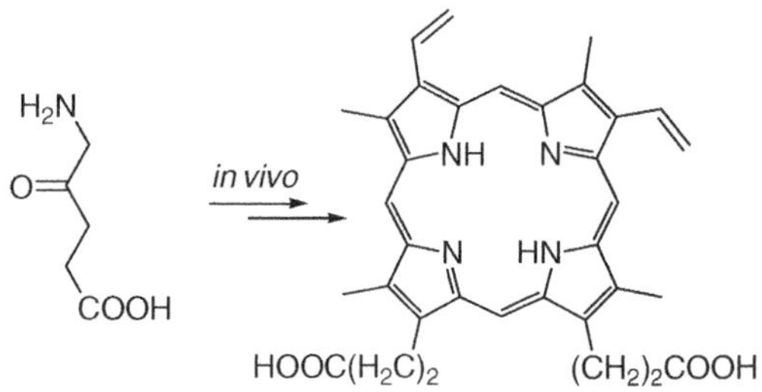

**Scheme 6.249**

Singlet oxygen, generated in an essentially heterogeneous biological cell, can be quenched with the biomolecules in the vicinity, resulting in cell killing through controlled cell death (apoptosis)[1401] or necrosis, associated with loss of plasma membrane integrity.[1400] There are many molecular targets of photodynamic effect in biological systems, including aromatic and sulfur-containing amino acids, unsaturated lipids, steroids and guanosine nucleotides, that undergo various oxidation reactions (type II photochemistry; see also the following sections).[1391] The type I mechanism, according to which the excited triplet sensitizer participates in electron transfer reactions with molecules in the vicinity to produce free radical intermediates that react with ground-state oxygen to produce peroxy radicals and other reactive oxygen species (Scheme 6.246), is believed to be far less important.[1400]

The extent of the photodynamic effect is connected particularly to the lifetime of singlet oxygen and its concentration in a cell. Contrary to earlier reports, it has been found to diffuse over appreciable distances and across cell membranes into the extracellular environment.[1378] Using a direct detection method called singlet oxygen microscopy, which has been developed to generate images of singlet oxygen (time-resolved) phosphorescence in a range of materials including single cells, the singlet oxygen lifetime in a cell has been determined to be about 3 μs, allowing for its diffusion over approximately 130 nm.[1402]

## Special Topic 6.24: Environmental aquatic and snow photochemistry

Sunshine can initiate photochemical reactions in atmospheric water droplets, in surface and marine waters and even in snow. Solar short-wavelength light can reach ecologically significant depths in freshwater and marine ecosystems.[1367,1403] In coastal waters with high turbidity and larger concentrations of 'yellow matter', a naturally occurring complex mixture of coloured organic compounds collectively called *humic acids*, UVB (280–315 nm) penetrates only tens of centimetres, whereas penetration to depths of tens of metres can occur in clear oceanic waters.[1404] Solar UV radiation at current or potentially enhanced levels due to stratospheric ozone depletion (Special Topic 6.21) has been found to affect, either positively or adversely, most forms of life found in shallow waters. A large proportion of dissolved organic material (DOM), including compounds of anthropogenic origin, absorb in the wavelength range

300–400 nm. Thus light drives many important chemical transformations of DOM within the oceanic boundary layer, freshwaters and rainwater.[1405–1407] Such transformations affect the fluctuations of important atmospheric trace gases (including $CO_2$, CO and dimethyl sulfide) and the processes regulating the marine food web.[1408] Humic acids, aromatic macromolecules with various hydroxy, amino, carbonyl and carboxylic functional groups, play a special role in aquatic photochemistry. They have relatively strong UV absorption and are also thought to be involved in singlet oxygen production via sensitization[1409] and photooxidation of phenols and carbonate ions.[1410,1411] Natural photochemical processes, including transition metal photocatalysis (Section 6.8.2), may also involve other reactive oxygen species, such as peroxides, superoxide ions and hydroxyl radicals.[1412,1413] In contrast, short-wavelength radiation may impair photosynthesis (Special Topic 6.25) of phytoplankton (photoinhibition).[1414]

Photochemical production of a variety of chemicals has recently been reported to occur in snow and ice. Subsequent release of these chemicals may significantly impact the chemistry of the overlying atmosphere in the cryosphere (a seasonal maximum of 40% of land is covered by snow or ice and a significant percentage of the world's oceans are covered by sea ice).[1372] Apart from some simple reactive gases, such as NO and $NO_2$, produced by photolysis of inorganic nitrates, some anthropogenic organic pollutants may also be subject to photochemical degradation in the snowpack.[1415–1418]

### 6.7.2  Singlet Oxygen: [2 + 2] and [4 + 2] Photooxygenation and Related Photoreactions

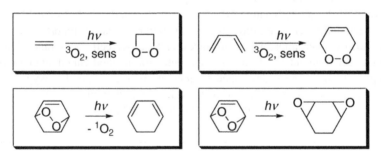

*Recommended review articles.*[135,136,1358,1359,1384,1419–1426]
*Selected theoretical and computational photochemistry references.*[1427–1433]

Singlet oxygen ($^1\Delta_g$) generated by triplet sensitization (Section 6.7.1) can readily oxygenate alkenes, dienes and aromatic compounds; the reactions are formally interpreted as [2π + 2π] ([2 + 2]) or [4π + 2π] ([4 + 2]) cycloaddition reactions, producing 1,2- or 1,4-endoperoxides, respectively. Scheme 6.250 shows examples of the formation of 1,2-dioxetane (**513**), 3,6-dihydro-1,2-dioxine (**514**) and 2,3-dioxabicyclo[2.2.2]octa-5,7-diene (**515**).

*[2 + 2] Photooxygenation*

Several photooxygenation mechanisms, such as concerted supra-antara [2 + 2] cycloaddition and stepwise additions involving charge-transfer complex (**516**), perepoxide

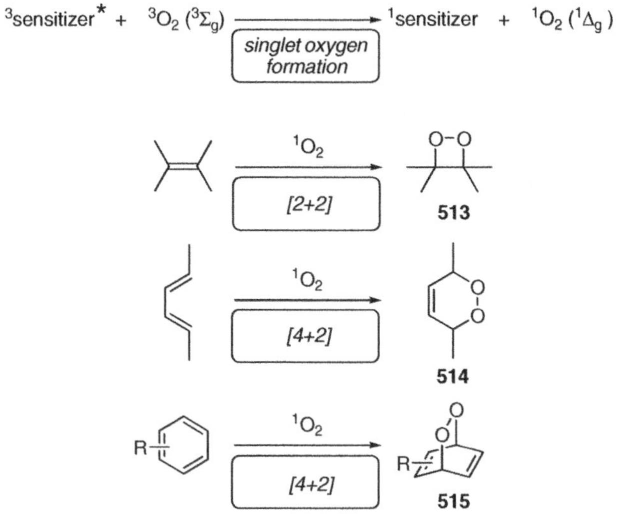

Scheme 6.250

(**517**), 1,4-biradical (**518**) and zwitterion (**519**) intermediates, have been proposed (Scheme 6.251);[1421] however, the formation of polar species seems to be the most probable.[1434] The photooxygenation reaction rates are almost completely determined by entropy changes.[1435]

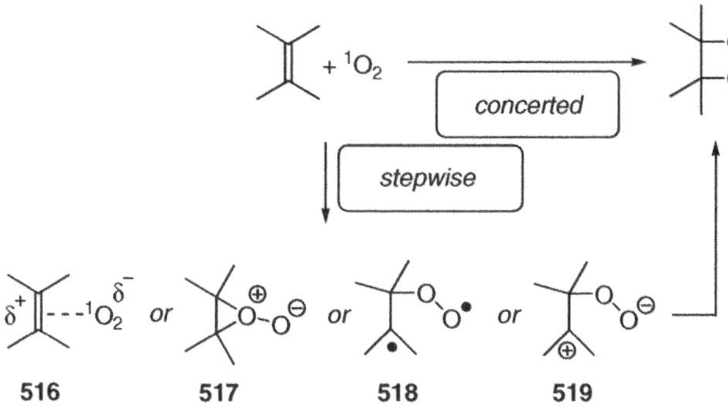

Scheme 6.251

Photooxygenation [2 + 2] is a unique method of preparing 1,2-endoperoxides, some of which can be relatively stable. For example, adamantylideneadamantane-1,2-dioxetane (**520**) was prepared by irradiation of biadamantylidene (**521**) in the presence of Methylene Blue and oxygen at $\lambda_{irr} > 350$ nm in 66% chemical yield (Scheme 6.252).[1436,1437]

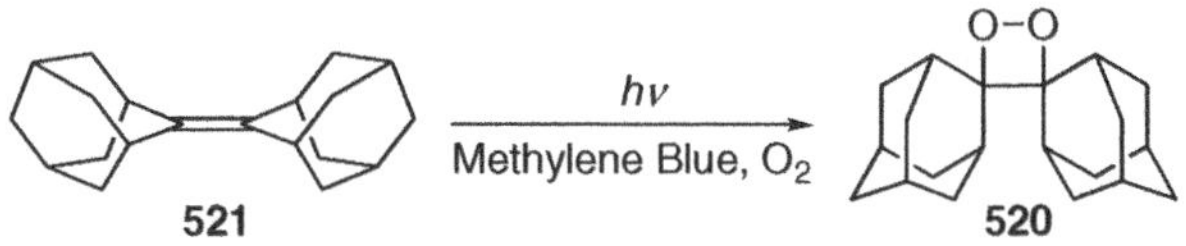

Scheme 6.252

Possible side-reactions, such as the formation of allyl hydroperoxides from compounds having an allylic hydrogen (a competitive ene reaction; Section 6.7.3) or 1,4-endoperoxides from 1,3-dienes, may be suppressed by changing the experimental conditions.[1359,1438] For example, the extent of [2 + 2] and [4 + 2] photooxygenation in styrenes, such as the propenylanisole **522** (Scheme 6.253), is controlled by solvent polarity and pH, possibly due to protonation of a perepoxide/zwitterion intermediate.[1421,1439] The [4 + 2] product **523** is preferentially produced in non-polar benzene or chloroform, whereas the 1,2-dioxetane **524** is almost exclusively formed in methanol or acidified non-polar solvents.

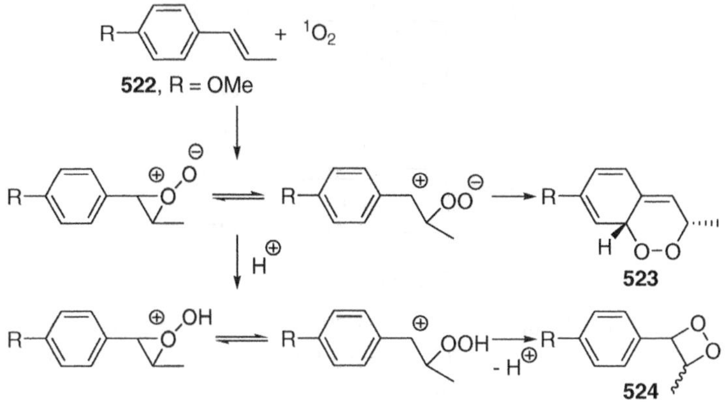

**522**, R = OMe

*Scheme 6.253*

1,2-Dioxetanes such as tetramethyl-1,2-dioxetane (**525**) are known to undergo thermal decomposition to form two carbonyl compounds via a concerted or stepwise (radical) mechanism, accompanied by chemiluminescence (Section 5.6) (Scheme 6.254).[135,511,1440] The degradation of **525** results principally in acetone phosphorescence ($\lambda_{max} = 430$ nm) and the reaction is very sensitive to quenching by oxygen.

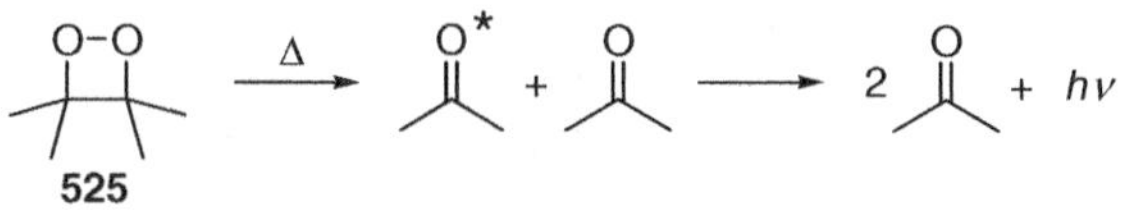

**525**

*Scheme 6.254*

Dioxetanes are useful precursors to various oxidation products, such as carbonyl and carboxyl compounds. Photooxidative double bond cleavage, an alternative to ozonolysis, can successfully be adopted for the preparation of large heterocyclic rings or for ring-opening procedures. For example, photooxygenation of enamine **526** leads to the 10-membered product **527** in 84% chemical yield via a dioxetane intermediate **528** (Scheme 6.255).[1441]

In another example, the dioxetane **529**, obtained from the oxathiin **530** by irradiation in the presence of oxygen and tetraphenylporphyrin (TPP), is converted to **531** in nearly quantitative chemical yield (Scheme 6.256).[1442]

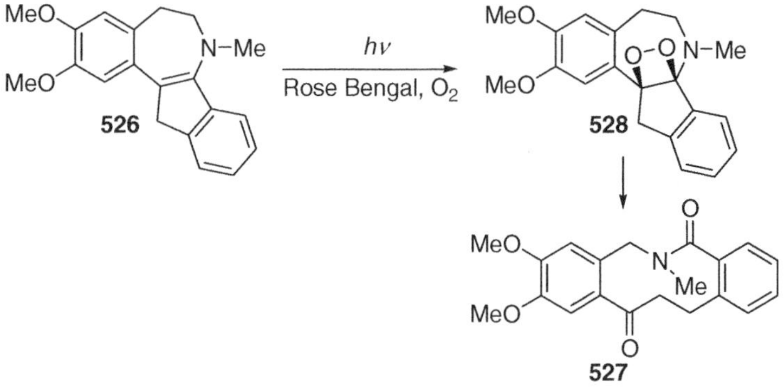

*Scheme 6.255*

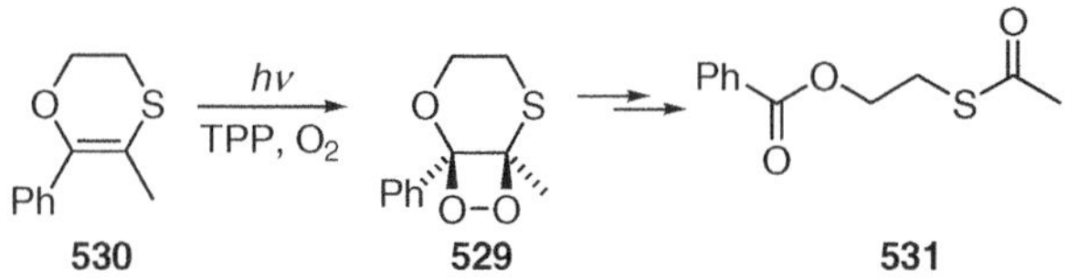

*Scheme 6.256*

1,2-Dioxetanes can also be transformed into 1,2-diols by reduction with LiAlH$_4$ or into epoxides by treatment with phosphines in the dark.[1421]

## [4 + 2] Photooxygenation

This process, formally related to the Diels–Alder reaction, may also proceed by various mechanisms (Scheme 6.257)[1421,1443] similar to those of [2 + 2] cycloaddition (Scheme 6.251), such as a concerted process or formation of charge-transfer (exciplex, **532**), biradical (**533**), zwitterion (**534**) or perepoxide (**535**) intermediates. A concerted pathway[1444] and exciplex[1445] intermediacy was proposed to be involved in most cases. The [4 + 2] photooxygenation may be accompanied by other related processes (e.g. [2 + 2]).

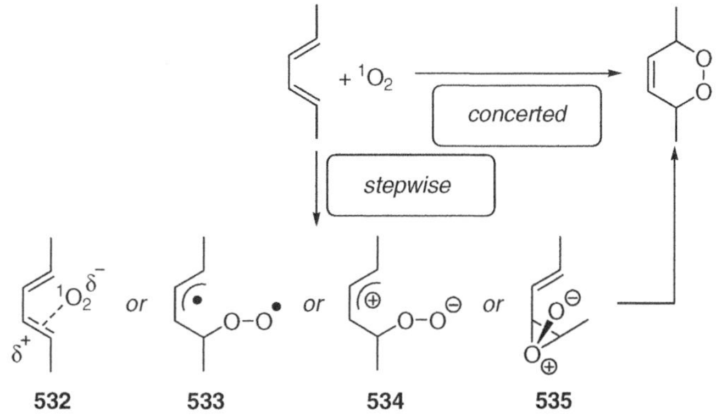

*Scheme 6.257*

Since oxygen is a highly symmetric molecule, the potential stereoselectivity of photooxygenation reaction should be largely controlled by the structure of the substrate, especially when a concerted or highly synchronous pathway is expected.[1421] For example, formation of an endoperoxide stereoisomer **536** is selectively enhanced when a bulky silyl group prevents the attack of oxygen from the same face (Scheme 6.258).[1446] Tetraphenylporphyrin (TPP) was utilized here to generate the singlet oxygen.

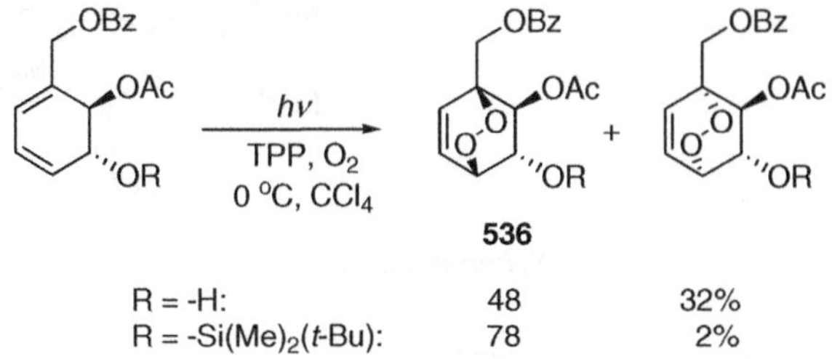

|  | **536** |  |
| --- | --- | --- |
| R = -H: | 48 | 32% |
| R = -Si(Me)$_2$(*t*-Bu): | 78 | 2% |

*Scheme 6.258*

Only activated aromatic compounds, such as benzenes with good electron-donating substituents or polycyclic aromatic hydrocarbons, can undergo [4 + 2] photooxygenation with singlet oxygen.[1421,1425] For example, $^1O_2$ addition to 1,4-dimethylnaphthalene, anthracene and pentacene proceeds with reaction rate constants on the order of $10^4$, $10^5$ and $10^9 \, mol^{-1} \, s^{-1}$, respectively, evidently increasing with increasing electron-donating abilities of the systems.[1425] The mechanism is believed to be a concerted addition through a symmetrical transition state involving charge transfer from the aromatic compound to oxygen.[1385] This addition is reversible; endoperoxides easily decompose either *thermally* to the starting aromatic compounds and singlet or ground state oxygen, or *photochemically* (Scheme 6.262).[1387] Scheme 6.259 shows the photooxidation of heterocoerdianthrone (**537**); the resulting endoperoxide **538** can be deoxygenated by irradiation at 313 nm ($\Phi = 0.27$).[218] This compound can be used as an actinometer.[214]

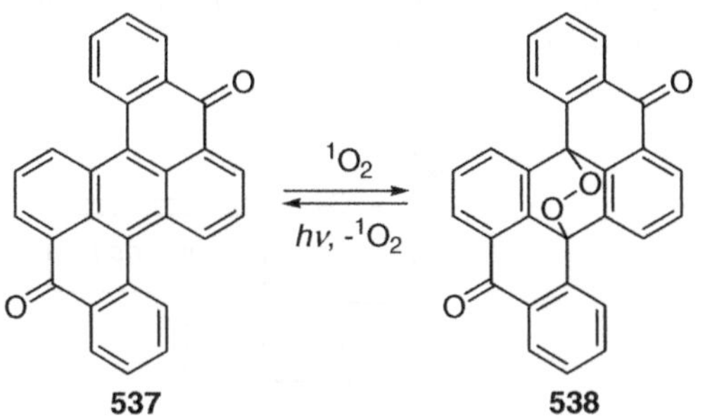

*Scheme 6.259*

Scheme 6.260 shows an example of the application of photooxygenation in the total synthesis of dysidiolide (**539**), a cdc25A protein phosphatase inhibitor, involving regioselective oxidation of the furan moiety of **540** in nearly quantitative chemical yield in the last step of the synthetic procedure.[1447]

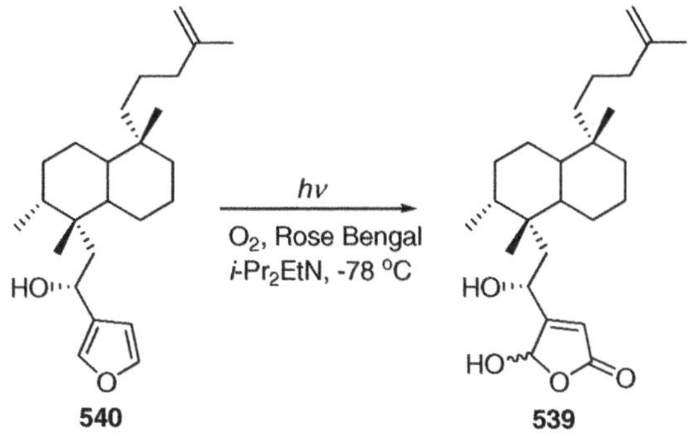

**Scheme 6.260**

---

## Case Study 6.36: Photomedicine – photooxygenation of DNA

Singlet oxygen is a well known cytotoxic agent (Special Topic 6.23) and may react with guanine or other parts of cellular DNA. Since the determination of the UVA-mediated damage to cellular DNA is still a difficult analytical problem, laboratory experiments with DNA models have become the best means to reveal the mechanistic details of photooxygenation reactions.

For example, the guanosine derivative **541** is oxidized by singlet oxygen, generated using tetraphenylporphyrin (TPP) as a sensitizer, to 7,8-dihydro-8-oxoguanosine (**542**) (Scheme 6.261).[1448] The proposed mechanism involves initial formation of an endo-peroxide **543** and its rearrangement to a very reactive 8-hydroperoxy derivative **544**, which reacts further to **542**. The product was obtained in preparative irradiation experiments, and the singlet oxygen quenching rates were determined by time-resolved measurements (Section 3.7) of singlet oxygen phosphorescence decay at 1270 nm.

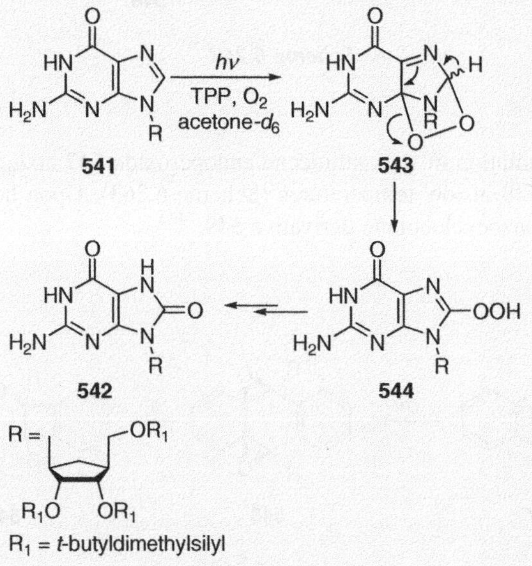

**Scheme 6.261**

*Experimental details.*[1448] A solution of **541** (~0.1 M) and TPP ($6 \times 10^{-5}$ M) in acetone-$d_6$, was continuously purged with oxygen in an NMR tube immersed in a bath to keep the sample temperature constant. This was followed by irradiation of the sample with a xenon lamp (300 W) through a 1% potassium dichromate filter solution that cut off the wavelengths below 500 nm (Figure 3.9). The reaction progress was followed by NMR measurements.

## Photochemistry of Endoperoxides

Endoperoxides, usually obtained by photooxygenation, are photolabile compounds, which can either lose *singlet oxygen* to regenerate the diene precursor (e.g. **545**) when irradiated at short wavelengths (cycloreversion; see also the preceding text) or to undergo a rearrangement to form diepoxides (e.g. **546**) when irradiated at longer wavelengths (Scheme 6.262).[1384,1385,1387] The mechanism of the photochemical cycloreversion remains a matter of controversy.

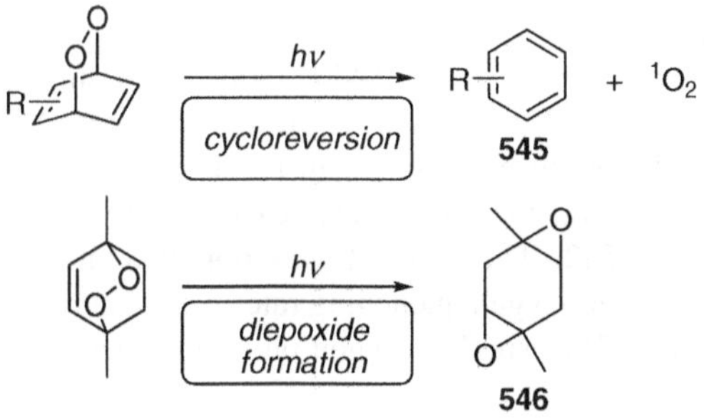

Scheme 6.262

For example, irradiation of the anthracene endoperoxide **547** at $\lambda_{irr} = 400$ nm leads to the *syn*-diepoxide **548** at low temperatures (Scheme 6.263). Upon heating, the product isomerizes to the benzocyclobutane derivative **549**.[1449]

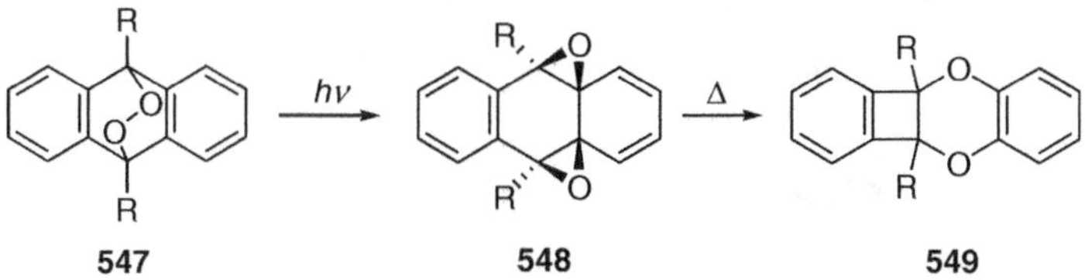

Scheme 6.263

### 6.7.3 Singlet Oxygen: ene Reaction

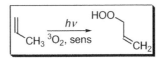

*Recommended review articles.*[135,136,1358,1359,1376,1377,1422,1423,1450]
*Selected theoretical and computational photochemistry references.*[1428,1429,1451–1453]

The *ene* or *Schenck*[1454] reaction is another type of photooxygenation reaction that involves singlet oxygen (Section 6.7.1) and an alkene containing at least one allylic hydrogen.[1376,1377,1422,1423,1450] The reaction intermediates proposed for the [2 + 2] and [4 + 2] photooxygenation (Section 6.7.2; Scheme 6.250) can also be considered for the ene reactions (Scheme 6.264); nevertheless, highly negative activation entropies may indicate a concerted mechanism[1455] or an exciplex **550** involvement.[1456] There is also strong evidence that a perepoxide intermediate **551** is formed in certain cases.[1450,1456–1459]

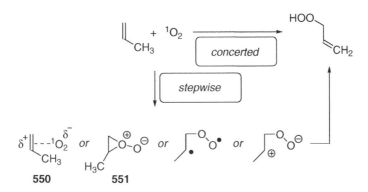

**Scheme 6.264**

Competitive [4 + 2] and ene reactions of 2,4-dimethylpenta-1,3-diene (**552**) were suggested to occur by either a concerted or a perepoxide intermediate pathway (Scheme 6.265).[1460] The major [4 + 2] product **553** is formed in nonpolar solvents nearly exclusively, whereas formation of the ene product (**554**) via a polar perepoxide intermediate **555** is enhanced in polar solvents, such as methanol. The less stable *s-cis*-conformation of **552** is required for the concerted process. Another competing process, physical quenching of singlet oxygen, was found to be at least as efficient as the chemical processes.

A concerted pathway for ene reaction was ruled out by a stereochemical investigation of photooxygenation of an optically pure, deuterated 3,4-dimethyl-1-phenylpent-2-ene (**556**) (Scheme 6.266).[1461] The reaction was suggested to proceed suprafacially and

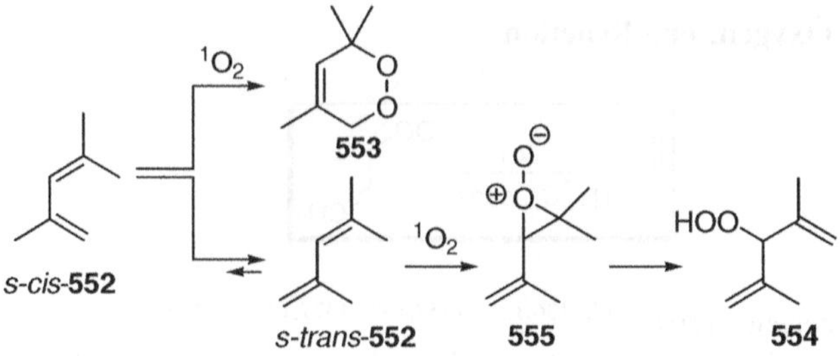

**Scheme 6.265**

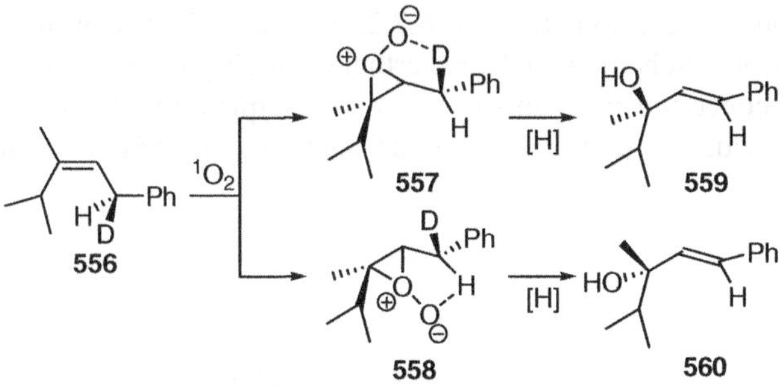

**Scheme 6.266**

stereospecifically via perepoxide intermediates **557** and **558** (or exciplex species), in which the bulky phenyl group is kept away from the reaction centre and either a hydrogen or deuterium atom is abstracted. Because no stereochemical (the product concentration ratio [**559**]:[**560**] was equal to 1) and no isotopic ($k_H/k_D = 1$) discrimination was observed, the rate-determining step must be the attack of singlet oxygen on the double bond.[1377] A reduction step ([H]) is necessary to give the corresponding alcohols instead of primary hydroperoxides.

Both steric and electronic effects can influence the regioselectivity of ene reactions. For example, a strong preference for hydrogen abstraction from the more substituted side of the double bond of the perepoxide intermediates, generated from trisubstituted alkenes, is observed.[1376,1377] This effect has been called the *cis* effect and explained on the basis of orbital interactions[1423] and activation entropy differences.[1457] Scheme 6.267 shows the product distribution after photooxygenation of three alkenes (**561–563**).[1461] There is an apparent steric effect of the methyl versus 2-propyl substituents in the first two cases. The cyclopropyl moiety in the last example remains unreacted, which rules out the formation of a biradical intermediate (see also Special Topic 6.10).

Scheme 6.268 shows the synthesis of the antimalarial drug artemisinin (**564**) from dihydroartemisinic acid (**565**): photooxygenation of **565** in the presence of Methylene Blue gives **566**, which is oxidized by ground-state oxygen in ~30% chemical yield.[1462]

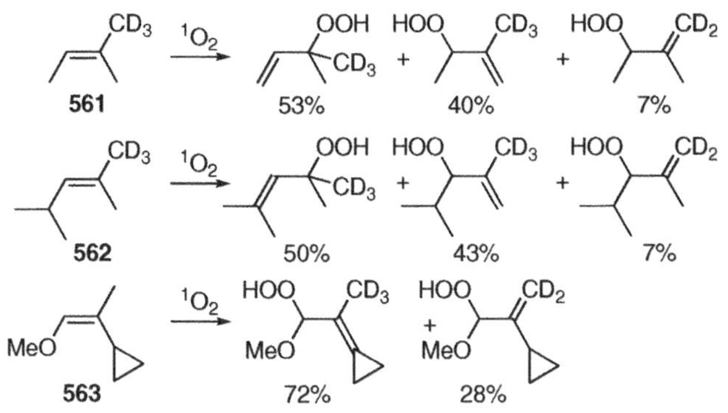

**Scheme 6.267**

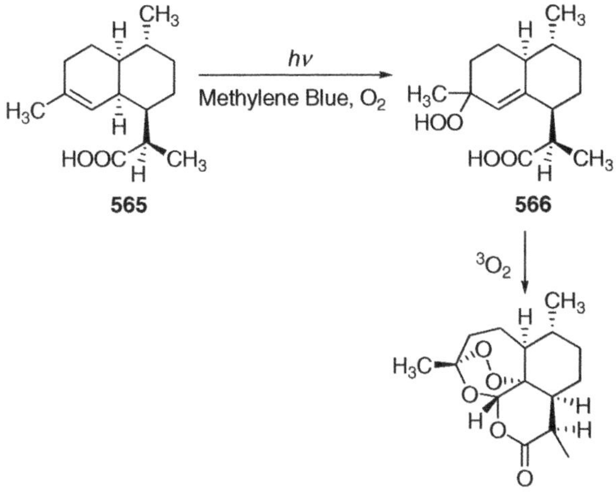

**Scheme 6.268**

---

## Case Study 6.37: Solid support synthesis – ene reaction in zeolites

Contrary to reactions in solution (see the preceding text), the less substituted side of some trisubstituted alkenes becomes more vulnerable to ene photooxygenation if it takes place in thionin-supported zeolite Na-Y (Scheme 6.269).[1463] Synchronous stabilization (less hindered) interaction of the alkali metal cation (Na$^+$; in zeolite) with the former C=C bond of the alkene and oxygen during perepoxide formation (**567**) is apparently responsible for this anomalous behaviour. Thionin sensitizer used to generate singlet oxygen in the solid support was placed within the zeolite via an aqueous-based cation-exchange process.[1464]

*Experimental details.*[1463] Thionin-exchanged zeolite (1 g; prepared according to the literature;[1464] the loading level was lower than one dye molecule per 100 supercages) and an alkene (10 mg) were added to dry hexane (15 ml). The slurry was irradiated with visible light (>450 nm) (Figure 3.9). The resulting hydroperoxides were extracted with tetrahydrofuran and analysed by NMR spectroscopy.

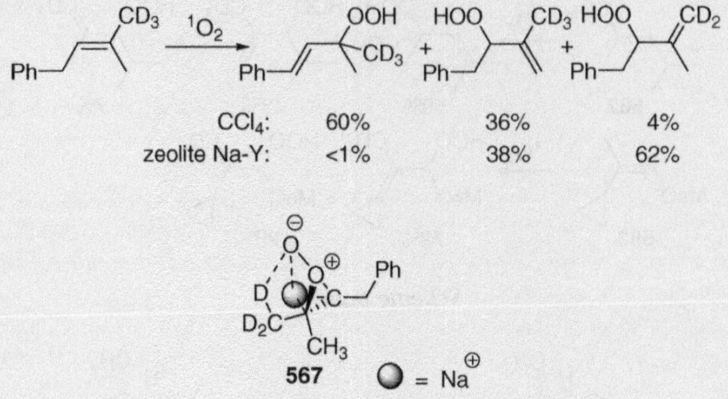

|  | CCl$_4$: | 60% | 36% | 4% |
| zeolite Na-Y: | <1% | 38% | 62% |

**Scheme 6.269**

## 6.7.4 Problems

1. Explain the following concepts and keywords: photooxygenation; stratospheric ozone depletion; singlet oxygen; photochemical smog; photosensitivity; phototoxicity; photoallergy; photodynamic therapy; photoinhibition; chemiluminescence; perepoxide.

2. Suggest the mechanisms for the following reactions:

(a)

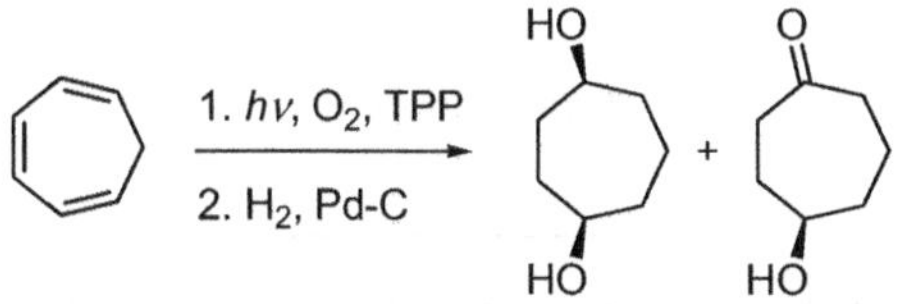

(TPP = tetraphenylporphyrin)

[ref. 1465]

(b)

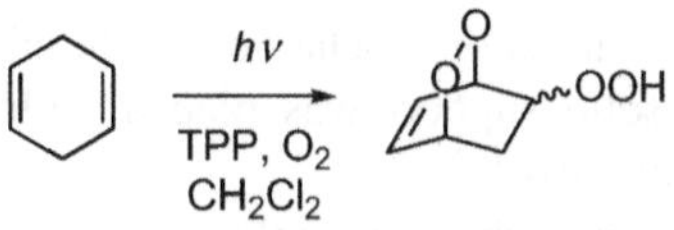

[ref. 1466]

3. Predict the major photoproduct(s):

(a)

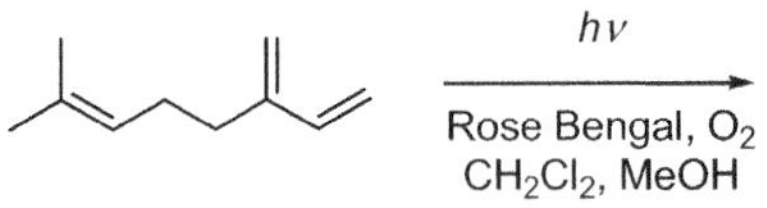

$$\xrightarrow[\substack{\text{Rose Bengal, O}_2 \\ \text{CH}_2\text{Cl}_2, \text{ MeOH}}]{h\nu}$$

[ref. 1467]

(b)

$$\xrightarrow[\text{Rose Bengal, O}_2]{h\nu}$$ (light emission in the final step)

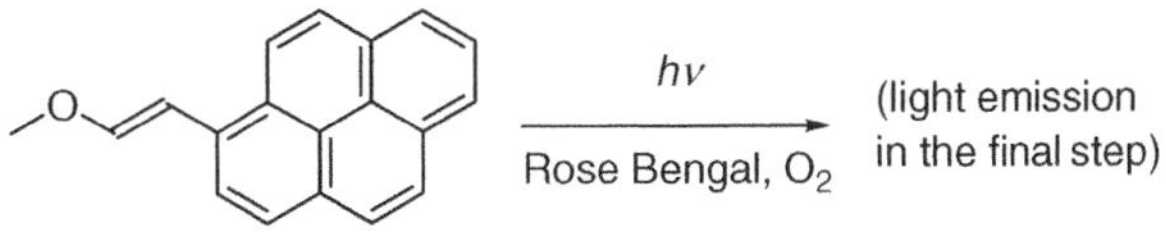

[ref. 1468]

## 6.8 Photosensitizers, Photoinitiators and Photocatalysts

We have encountered examples of photochemically activated, indirect initiation of chemical reactions via energy or electron transfer many times throughout the text. It has been shown that they require the utilization of specific reaction conditions and experimental arrangements. They are extremely useful in many laboratory experiments and also for practical photochemical applications, especially when direct excitation by light is difficult to accomplish or unwanted side-reactions take place. Also, such reactions can be very effective, so they can be successfully implemented when a high degree of reactant degradation is needed. In Section 6.8.1, various applications of sensitization (Table 6.21, entry 1) and also photoinduced electron transfer techniques (entry 2) are presented, involving organic photosensitizers, photocatalysts and photoinitiators for organic synthesis, biology or material/surface sciences. Both excited photosensitizer and photocatalyst are typically regenerated in processes such as deactivation or electron transfer, whereas photoinitiators are generally consumed in the reaction.

Electron transfer processes involving excited transition metal photocatalysts are examined in Section 6.8.2. Reactive radical ions are usually generated from neutral

*Table 6.21* *Photoprocesses involving photosensitizers, photocatalysts and photoinitiators*

| Entry | Starting material[a] | Product(s)[a] | Process | Section |
|---|---|---|---|---|
| 1 | R + sensitizer* | R* + sensitizer | Sensitization | 6.8.1 |
| 2 | D + A + $h\nu$ | $D^{\bullet +}$ + $A^{\bullet -}$ | Photoinduced electron transfer | 6.8.1 |
| 3 | $Fe^{3+}$ + $H_2O$ + $h\nu$ | $Fe^{2+}$ + $HO^{\bullet}$ + $H^+$ | Homogeneous photocatalysis | 6.8.2 |
| 4 | D + A + $TiO_2$ + $h\nu$ | $D^{\bullet +}$ + $A^{\bullet -}$ + $TiO_2$ | Heterogeneous photocatalysis | 6.8.2 |

[a]D = electron donor; A = electron acceptor.

organic molecules, which then undergo subsequent secondary transformations. Photoredox photo-Fenton's process (entry 3) is an example of the catalysis in a homogeneous solution. A photocatalytic process, utilizing insoluble excited semiconductor such as titanium dioxide which promote interfacial electron transfer between the excited semiconductor surface and adsorbed molecules, is portrayed in entry 4.

### 6.8.1  Organic Photosensitizers, Photocatalysts and Photoinitiators

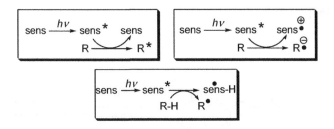

*Recommended review articles.*[602,1134,1469–1471]

*Organic Photosensitizers: Energy Transfer*

Most of the chemical processes described in the previous sections occur directly from an excited state precursor (R), which can be obtained either directly by absorption of light (Scheme 6.270a) or indirectly by photosensitization (via energy transfer; Scheme 6.270b). In the latter case, the sensitizer (sens) is regenerated as a ground-state molecule and therefore it must be electronically excited again in order to be involved in a new sensitization cycle.

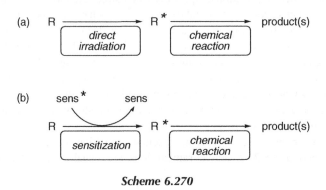

*Scheme 6.270*

Photosensitization is one of the most common and practical ways to generate excited states, particularly triplet states, especially when excitation at a desired wavelength cannot be achieved or when it does not lead to the desired excited state (Section 2.2). Good photosensitizers should absorb strongly in the region of interest and have sufficient excited-state lifetimes. For efficient energy transfer, the process should be exergonic. Table 6.22 lists some of the common photosensitizers and their typical applications.

**Table 6.22** *Examples of photosensitizers*

| Photosensitizer | Characteristics[157] | Utilized in | Section |
|---|---|---|---|
| $(COOR)_x$ | Singlet photosensitizer $(E_S = 420\,kJ\,mol^{-1})$ | Synthesis | 6.1.1; 6.1.4 |
| $C_{20}H_{38}OOC$ COOMe (bacteriochlorophyll a) | Singlet photosensitizer $(E_S = 177\,kJ\,mol^{-1})$ | Photosynthesis | This section |
| | Triplet photosensitizer $(E_T = 337\,kJ\,mol^{-1})$ | Synthesis | 6.1.4 |
| | Triplet photosensitizer $(E_T = 332\,kJ\,mol^{-1})$ | Synthesis | 6.1.1; 6.3.5 |
| | Triplet photosensitizer $(E_T = 310\,kJ\,mol^{-1})$ | Synthesis | 6.3; 6.1.5 |
| Ph Ph | Triplet photosensitizer $(E_T = 288\,kJ\,mol^{-1})$ | Synthesis | 6.1.5 |
| $C_5H_{11}O_4$ (riboflavin) | Triplet photosensitizer $(E_T = 210\,kJ\,mol^{-1})^{1472}$ | Natural photosensitizer | 6.3.3 |
| (Rose Bengal) | Triplet photosensitizer $(E_T = {\sim}170\,kJ\,mol^{-1})$ | Singlet oxygen production | 6.7.1 |
| (tetraphenylporphyrin) | Triplet photosensitizer $(E_T = 140\,kJ\,mol^{-1})$ | Singlet oxygen production | 6.7.1 |

*Organic Photocatalysts: Electron Transfer*

The term photosensitization can also be employed for a process initiated by electron transfer (Section 6.4.4), in which the photosensitizer serves as an electron donor or acceptor.[1469] Such a compound will be called a photocatalyst (PC) if it is regenerated in an associated (auxiliary) electron transfer reaction elsewhere in the reaction cycle (Scheme 6.271). Lower than stoichiometric amounts of PC can logically be used in this case.

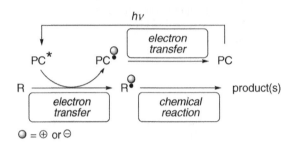

*Scheme 6.271*

Section 6.4.4 already presented several examples of photocatalysts such as metalloporphyrins and other organic metal complexes. In another example, a singlet excited photocatalyst, 1,4-dicyanonaphthalene (DCN), initiates the photochemical electron transfer-induced ring opening of an azirine **568** (Scheme 6.272).[1473] Subsequent addition of an intermediate **569** to acrylonitrile leads to **570** in two steps, while DCN is regenerated.[883]

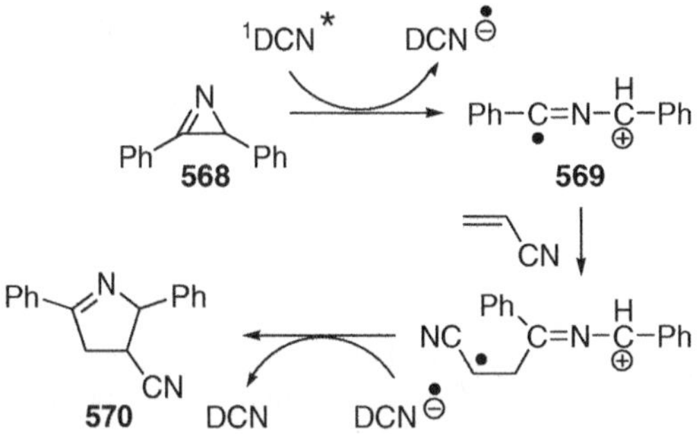

*Scheme 6.272*

Scheme 6.273 illustrates the photocatalytic addition of a tertiary amine (**571**) to the furanone **572**,[1474] which is initiated by electron transfer from the amine to a photoexcited benzophenone to form a radical ion pair.[1475] The ketyl radical anion abstracts a proton from an amine radical cation and subsequently donates a hydrogen atom to the previously coupled radical intermediate **573** to regenerate the benzophenone molecule and to give the final addition product **574** (<94% chemical yield).

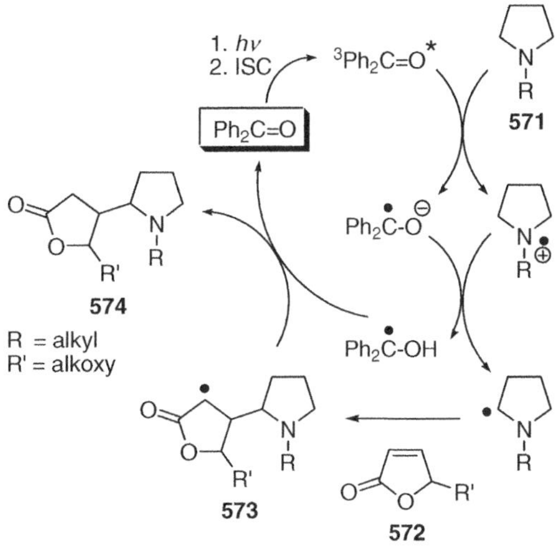

**Scheme 6.273**

## Special Topic 6.25: Photosynthesis in bacteria and plants

Photosynthesis is the conversion of light energy into chemical potential in living organisms. It involves direct excitation, rapid and efficient energy and electron transfer processes and subsequent dark chemical reactions.

The simplest photosynthetic system is that found in *purple bacteria*.[1476,1477] The initial event is absorption of a photon by the *light-harvesting antenna* system LH2, consisting of a ring of 18 (or nine) bacteriochlorophyll (BC) molecules (see the absorption spectrum in Figure 6.9) and associated with the organic pigments carotenoids within the walls of two helical protein molecules (Figure 6.18).[1478,1479] BC molecules are largely *exciton-coupled* (see also Special Topic 6.29), i.e., the exciton state is delocalized over several chromophores. The excitation energy is then rapidly transferred to a similar but larger antenna complex (LH1) consisting of 32 BC molecules, surrounding the *reaction centre* (RC; see below). The excitation energy migration among the rings occurs on a time scale of picoseconds. The singlet–singlet resonance energy transfer (section 2.2.2) takes place possibly through quantum coherence to locate the most efficient transfer path.[1480] There are several LH2 complexes surrounding LH1. The energy transfer from LH1 to RC is then slower by an order of magnitude.

The primary photosynthetic process is carried out by a pigment–protein complex – the reaction centre (RC) – embedded in a lipid bilayer membrane (Figure 6.19) and surrounded by light-harvesting complexes.[1477,1481,1482] Thus energy is transferred from LH1 to a bacteriochlorophyll 'special pair' (P) and then through a bacteriochlorophyll molecule (BC; monomer) to bacteriopheophytin (BP; a chlorophyll molecule lacking the central $Mg^{2+}$ ion), followed by electron transfer to a quinone $Q_A$ in hundreds of ps. The neutral P is then restored by electron transfer from the nearest intermembrane space protein cytochrome $c$ (Cyt $c$) in hundreds of ns. The rate constants of the

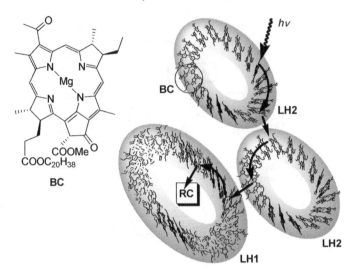

**Figure 6.18**   Energy transfer in light-harvesting complexes

electron transfer steps are shown in Figure 6.20; these charge separation processes over large distances (across the membrane) are highly efficient and essentially non-reversible because charge recombination rates lie in the Marcus inverted region (Section 5.2).[1477] The whole charge separation process converts light energy in about 10 ms and with a quantum yield of near unity, but only a fraction of the excited-state energy is conserved in the final charge-separated state. The final electron acceptor $Q_A$ is reduced in a two-electron–two-proton process to a hydroquinone ($Q_AH_2$). This compound diffuses to cytochrome *bc*, also called a proton pump, where it oxidizes back to $Q_A$ (Figure 6.21). The energy released is used to transfer a proton through the membrane to ATP synthase, the enzyme that synthesizes ATP from ADP and inorganic phosphate ($P_i$). ATP is then used as a high-energy compound to drive various endergonic biochemical reactions.

The biophysics of anaerobic photosynthesis in purple bacteria described above is now well understood. Photosynthesis in *green plants* and *cyanobacteria* is similar but more complex and differs mainly by the presence of two associated light-harvesting centres and *oxygen production*. In the case of plants, photosynthesis occurs in organelles called chloroplasts utilizing two protein photosystems. Photosystem II, absorbing light at $\lambda_{max} = 680$ nm, provides an oxidation potential that is sufficient to remove an electron from water in a manganese-containing metalloenzyme (called an oxygen-evolving complex) to produce oxygen.[1483] The electron is transferred via several carriers to an electron-deficient photosystem I where chlorophyll *a* serves as the reaction centre. Upon absorption of light at $\lambda_{max} \leq 700$ nm, an electron is transported to the iron–sulfur protein ferredoxin, where it reduces coenzyme $NADP^+$ to NADPH or provides the energy for the phosphorylation of ADP to ATP.

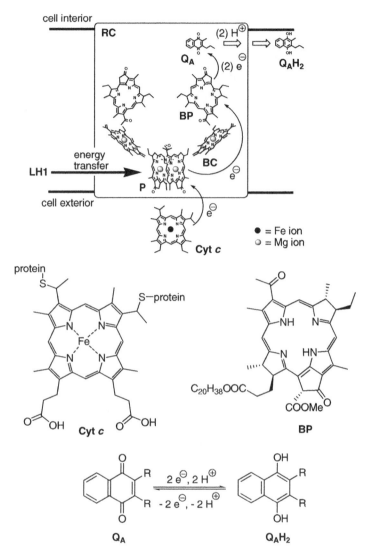

**Figure 6.19** Photosynthetic reaction centre and charge separation processes[1477]

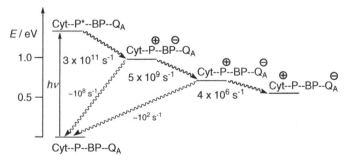

**Figure 6.20** Energy level diagram of charge separation processes[1477]

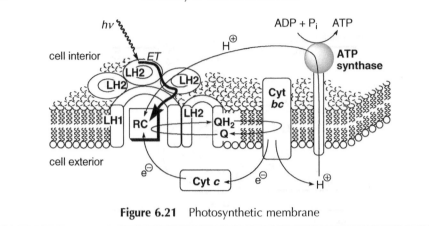

**Figure 6.21**   Photosynthetic membrane

---

## Special Topic 6.26: Artificial photosynthesis

There is increasing demand for inexpensive, clean energy sources. Although this demand could be met with fossil (limited) fuels, $CO_2$ emissions in the atmosphere have already risen and may be an important contributor to global warming. Novel chemical systems capable of solar energy conversion, that is, artificial nanoscale devices and rudimentary semibiological hybrids (artificial photosynthetic systems) that can perform many of the functions of the natural photosynthetic process (Special Topic 6.25), are being developed.[5,7,1248,1477,1484–1487] The conversion of solar energy into the form of fuels is challenging and attractive at the same time. Molecular hydrogen, as the most promising and environmentally friendly fuel accessible by a photoinduced water splitting process, is expected to be widely used in fuel cell-powered vehicles in the future. This process, by which water can be converted to hydrogen and oxygen using light:

$$2H_2O \xrightarrow[\text{cat.}]{h\nu} 2H_2 + O_2$$

$$2H_2O \longrightarrow O_2 + 4H^{\oplus} + 4e^{\ominus}$$

$$4H^{\oplus} + 4e^{\ominus} \longrightarrow 2H_2$$

can be photocatalysed by many inorganic semiconductors.[1484,1488] As yet, no catalytic process is capable of making the conversion with a quantum yield that is sufficient for commercially profitable applications (i.e. >10%) using visible light.[1489] Hydrogen molecule can, however, be successfully produced also from various organic molecules and materials.[1486] Alkanes or alcohols are, for example, dehydrogenated photo-chemically in the presence of various rhodium phosphine complexes, such as $Rh^I(PR_3)_2(CO)Cl$ or $Rh^I(PR_3)_3Cl$.

   Mimicking natural photosynthesis requires that all synthesized chemical building blocks linked covalently by self-assembly into a working molecular device, which basically consists of a light-harvesting antenna to capture light, reaction centres converting the excitation energy to chemical potential in the form of long-lived charge-separated systems, a catalyst converting the redox intermediates hereby formed to storable fuel and systems performing the physical separation of the products.

A light-harvesting antenna is a complex multi-chromophore system in which each component can absorb (visible) light and is able to transfer the excitation energy to the nearest neighbour chromophore before undergoing deactivation. Its efficiency depends on the distance and relative orientation of the chromophores. In recent years, the development of supramolecular chemistry has allowed the construction of very complex and efficient light-harvesting antennae, based for example on dendritic structures.[1248,1477] Porphyrin, metalloporphyrins or phthalocyanine arrays are the first logical choice as the synthetic analogues of natural systems. Energy transfer occurs, for example, in 10–100 ps between Mg or Zn porphyrin (MP) and free-base porphyrin (P) assemblies. Figure 6.22 shows an example of MP and P assembly.

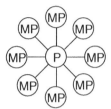

**Figure 6.22**   Light-harvesting antenna

The simplest reaction centre is a dyad, consisting of weakly electronically coupled electron donor (or acceptor) and electron acceptor (or donor) components to produce a charge-separated ion pair upon photoinitiated electron transfer ($k_{et}$). Figure 6.23 implies that this system may undergo excited-state deactivation ($k_d$) or charge recombination ($k_{cr}$); the former process is overcome in nature with an array of efficient short-range electron-transfer steps, thereby creating a charge separation that extends over a much greater distance (Special Topic 6.25).[1477,1487]

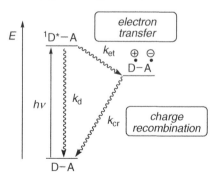

**Figure 6.23**   Dyad[1477]

The scheme for a three-component system (triad),[1485] which produces the charge separation over two bonds, is shown in Figure 6.24. After the initial excitation of the primary chromophore C (step 1), an electron is exergonically transferred to an acceptor (step 2), followed by a second thermal electron transfer step from a donor moiety to the oxidized chromophore C. Again, deactivation and charge recombination processes compete.

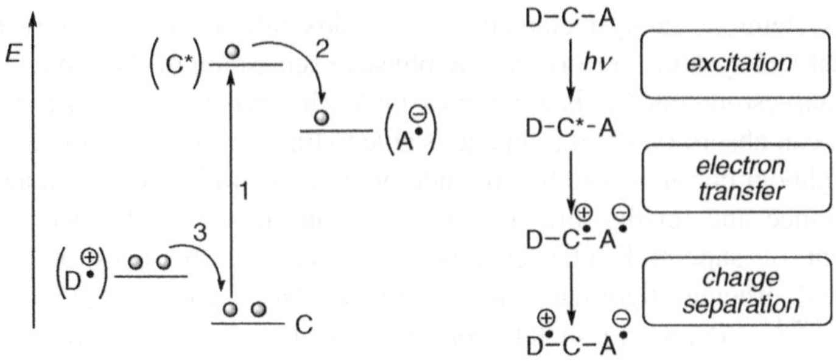

**Figure 6.24** Triad

Figure 6.25 shows an example of a six-component array (hexad) composed of four zinc porphyrin units (3 ZnP and 1 ZnP$_C$) connected to a reaction centre (fullerene as an electron acceptor; F) through a free-base porphyrin (P) via acetylene bridges, which ensure weak electronic coupling.[1490] Thus, excitation of any one of the peripheral zinc porphyrin moieties (ZnP*–ZnP$_C$–P–F) is followed by singlet–singlet energy transfer to the central zinc porphyrin (ZnP$_C$) in 50 ps to form ZnP–ZnP$_C$*–P–F. The excitation is then passed on to the free base porphyrin in 240 ps (ZnP–ZnP$_C$–P*–F), which decays by electron transfer to the fullerene with a lifetime of 3 ps to produce a charge-separated pair (ZnP–ZnP$_C$–P$^{\bullet+}$–F$^{\bullet-}$) having a lifetime of ~1 ns.

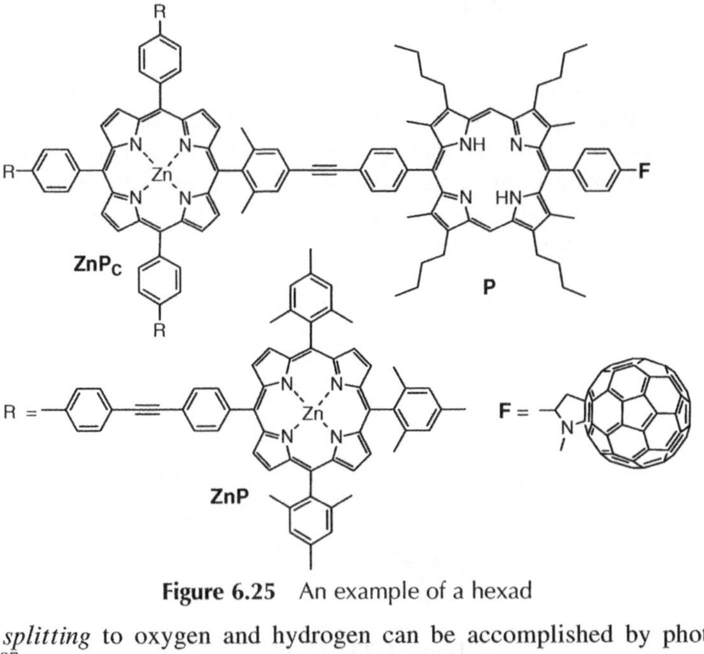

**Figure 6.25** An example of a hexad

*Water splitting* to oxygen and hydrogen can be accomplished by photocatalysis.[1484,1487] Some progress has been made in the field of artificial photosynthesis. However, full integration of all components mimicking the natural process has not yet been achieved.[1477] Figure 6.26 provides a schematic representation of water splitting using artificial photosynthesis. A triad D–C–A produces a charge-separated pair D$^{\bullet+}$—C—A$^{\bullet-}$ upon excitation through a light-harvesting antenna (C′). A membrane physically separates the two reaction sites, the hydrogen- and oxygen-evolving centres.

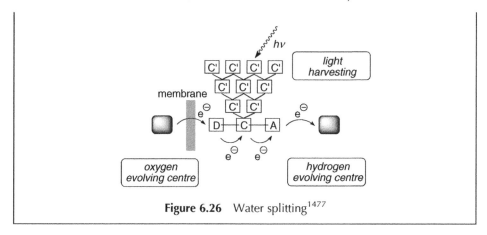

**Figure 6.26** Water splitting[1477]

*Organic Photocatalysts: Radical Reaction*

In a few less common procedures, the photocatalyst serves as a carrier of the hydrogen atom, subsequently forming a radical intermediate that must be regenerated via another hydrogen transfer step. For instance, a lower than stoichiometric amount of benzophenone is sufficient to trigger a photocatalysed addition of alkynes substituted with electron-withdrawing groups to cycloalkanes (**575** is produced in less than 70% chemical yield; Scheme 6.274).[1491] In addition, photosensitized isomerization via the Schenck mechanism (Scheme 6.15 in Section 6.1.1) may also represent a photocatalytic radical process.

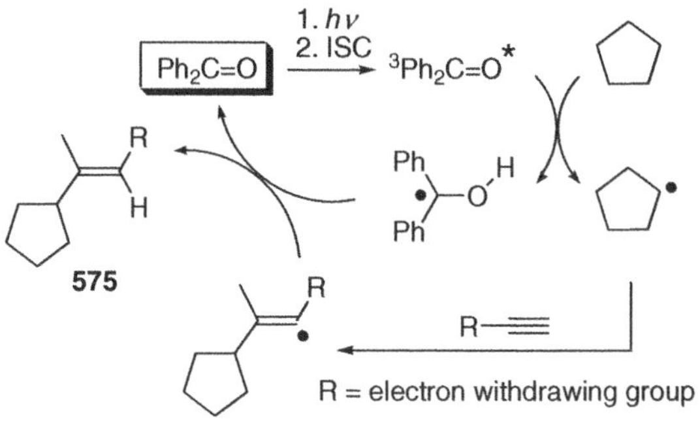

*Scheme 6.274*

*Organic Photoinitiators*

In the event that the photoactive auxiliary molecule irreversibly degrades, that is, it is not regenerated and cannot be involved in another reaction cycle, the initial process should be more generically termed *photoinitiation* rather than photosensitization and the absorbing molecule a *photoinitiator* (Scheme 6.275; PI). If no chain propagation steps follow, equimolar amounts of both photoinitiator and substrate are required (see, for example, Scheme 6.200 in Section 6.4.4).

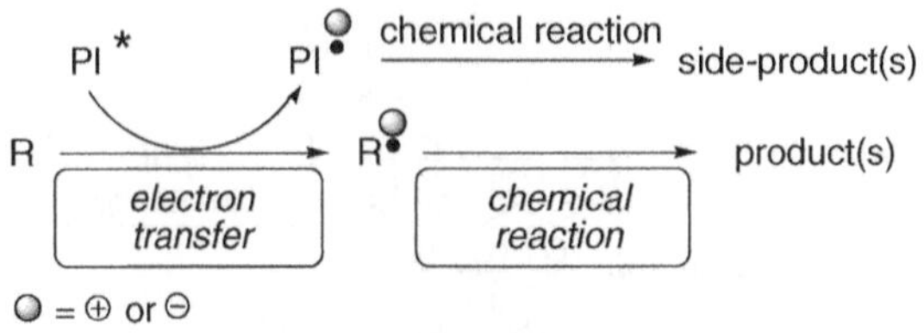

*Scheme 6.275*

Not only radical ions, generated via electron transfer from a photoinitiator (Scheme 6.275), but also photoproduced ions or radicals can be the key intermediates in the photoinitiation processes. Today, the term photoinitiator is mostly connected with photoinitiated (chain) polymerization reactions,[1134,1470,1471] in which the reactive intermediates are generated from relatively small amounts of an excited initiator (initiation step) to start a chain reaction (propagation steps). Initiation of polymerization by light, rather than by heat, has several advantages, such as high reaction rates at ambient temperature and spatial control of the process.[1492]

The literature describes two general classes of polymerization photoinitiators. Type I photoinitiators undergo a unimolecular fragmentation upon irradiation to yield radicals or other reactive species, whereas type II photoinitiators undergo a bimolecular reaction, in which the excited state of a photoinitiator reacts with an auxiliary molecule to generate the reactive species.[1493–1495] Both kinds of photoinitiators can generally be excited by UV irradiation; however, the type II species are often designed to absorb also in the visible region.[1496,1497] The following schemes show examples of type I and II photoinitiated reactions: type I radical formation of poly(vinyl pivalate) initiated by AIBN photofragmentation (Scheme 6.276; see also Section 6.4.2),[1498,1499] type II cationic polymerization catalysed by a photochemically generated acid (HX) from a hypervalent iodine compound **576** (Scheme 6.277; see also Section 6.6.2),[1349,1351,1500] type II initiated radical polymerization of an alkene via photoreduction (Section 6.3.1) of thioxanthone (**577**) by benzoxazine as a hydrogen donor (Scheme 6.278)[1501] and type II photoinduced electron transfer between excited 1-chloro-4-propoxythioxanthone (**578**) and an amine (see also Section 6.4.4) to produce a reactive radical cation[1502] (Scheme 6.279).

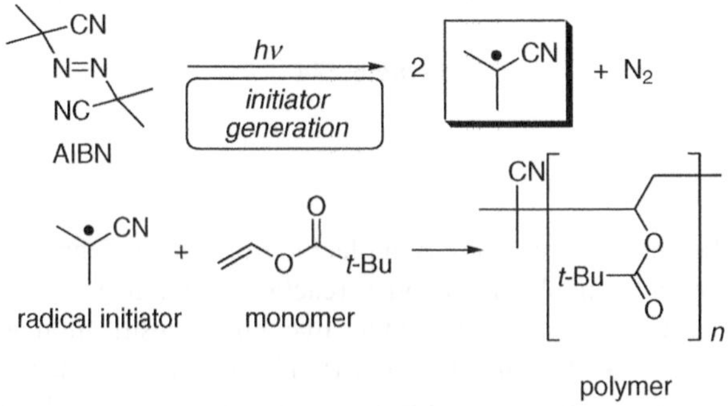

*Scheme 6.276*

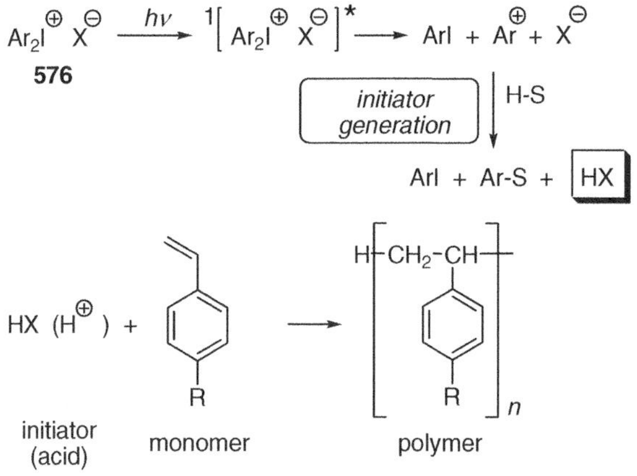

Ar$_2$I$^{\oplus}$ X$^{\ominus}$ $\xrightarrow{h\nu}$ $^1\left[ \text{Ar}_2\text{I}^{\oplus} \text{ X}^{\ominus} \right]^*$ $\longrightarrow$ ArI + Ar$^{\oplus}$ + X$^{\ominus}$

**576**

initiator generation | H-S

ArI + Ar-S + HX

HX (H$^{\oplus}$) + [monomer: 4-substituted styrene] $\longrightarrow$ [polymer: poly(4-substituted styrene)]

initiator (acid)  monomer  polymer

*Scheme 6.277*

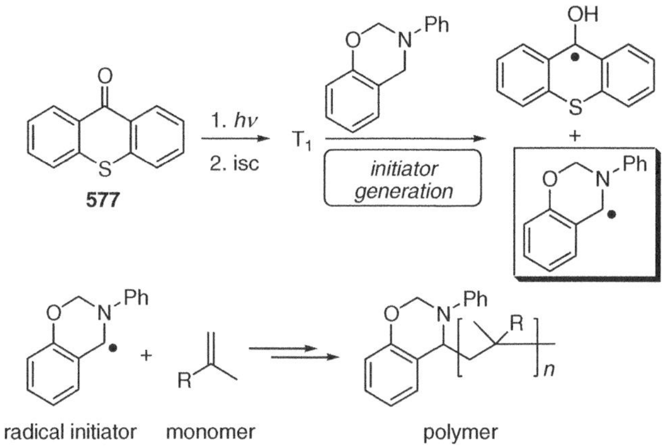

**577** $\xrightarrow[\text{2. isc}]{\text{1. } h\nu}$ T$_1$ | initiator generation

radical initiator + monomer $\Longrightarrow$ polymer

*Scheme 6.278*

**578** $\xrightarrow[\text{2. isc}]{\text{1. } h\nu}$ T$_1$ | N(CH$_2$CH$_3$)$_3$ / $\bullet^{\oplus}$ − N(CH$_2$CH$_3$)$_3$

products $\longleftarrow$ [radical intermediate] $\xleftarrow{- \text{Cl}^{\ominus}}$ [radical intermediate]

$\bullet^{\oplus}$N(CH$_2$CH$_3$)$_3$ $\xrightarrow[\text{initiator generation}]{\text{base}}$ $\bullet$CH$_3$CHN(CH$_2$CH$_3$)$_2$

radical initiator

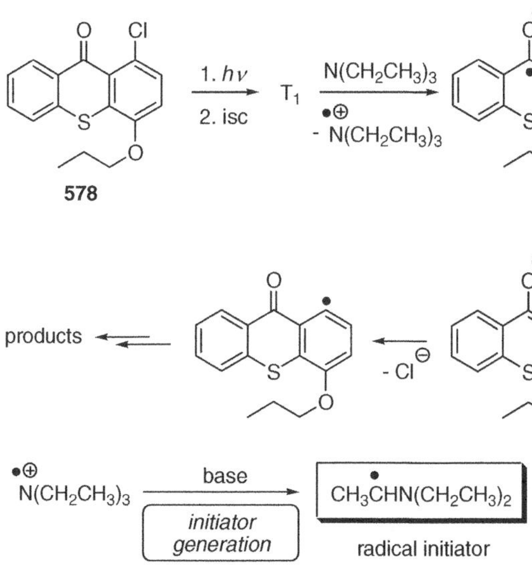

*Scheme 6.279*

**Case Study 6.38: Macromolecular chemistry – photoinitiated polymerization of methacrylate**

The extent of dimethylaminoethyl methacrylate (**579**) photopolymerization (Scheme 6.280) was found to be dependent on the type of photoinitiator.[1503] The type I photoinitiator bis(2,6-dimethoxybenzoyl)-2,4,4-trimethylpentylphosphine oxide (**580**), producing radicals by homolytic fragmentation (Scheme 6.281), was found to be more effective in promoting polymerization than the type II photoinitiator consisting of a mixture of 1-hydroxycyclohexyl phenyl ketone (**581**) and benzophenone. The ketone **581** apparently undergoes α-cleavage (Section 6.3.3) to initiate radical polymerization,[1504] while electron followed by proton transfer (see Scheme 6.100 in Section 6.3.1), involving benzophenone or **581** and an amino group of the monomer (**579**) or polymer (Scheme 6.282), then promotes radical branching and cross-linking processes among macromolecules.

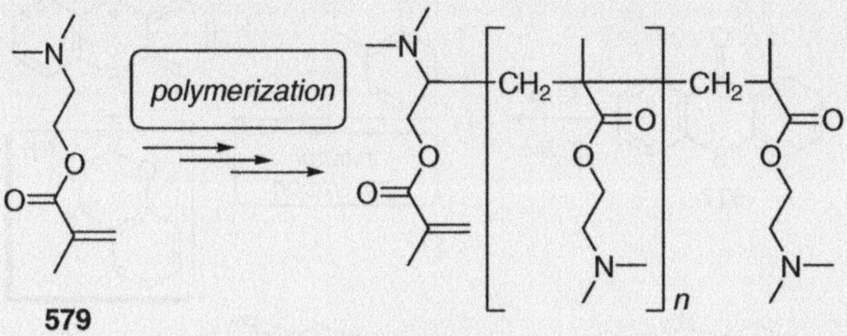

**579**

*Scheme 6.280*

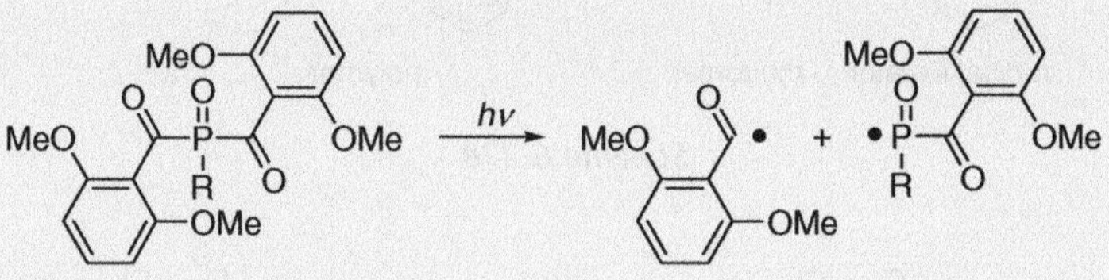

**580**, R = 2,4,4-trimethylpentyl

*Scheme 6.281*

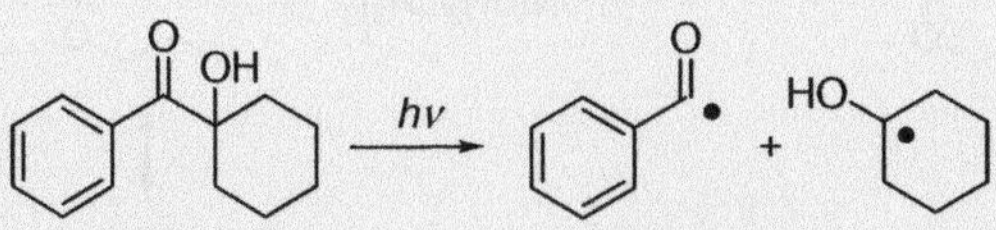

**581**

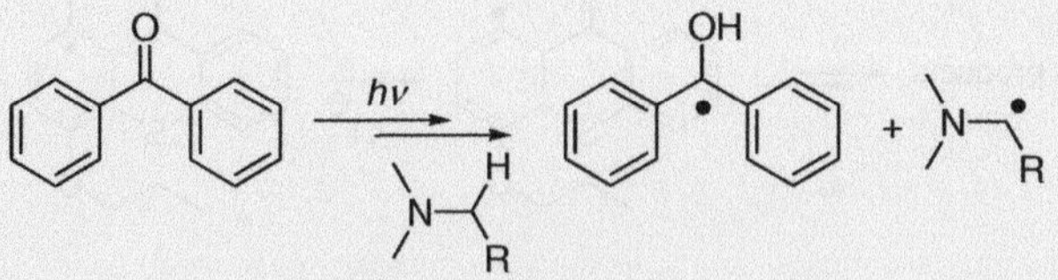

monomer or polymer

*Scheme 6.282*

*Experimental details.*[1503] A mixture of the monomer **579** (1–3 mg) and a photoinitiator (~3 wt%) in a sample (open-air) cell was irradiated with a high-pressure mercury discharge lamp (200 W) (Figure 3.11). In order to prevent evaporation of the monomer, the cell was covered with a thin poly(ethylene terephthalate) film. The extent of polymerization was evaluated by differential photocalorimetry.

Application of photoinitiators is one of the most efficient methods for achieving fast polymerization or controlled chemical modification of polymers. The use of photoinitiators is common in many valuable industrial applications,[174,1134,1505] such as photolithography and UV curing (Special Topic 6.27).

## Special Topic 6.27: Photolithography and UV curing

Photolithography (or optical lithography) is a photoimaging process (see also Special Topic 6.32) designed to form microscale and nanoscale patterns (in so-called micro- and nanofabrication, respectively) on a solid surface.[1505,1506] Together with particle beam lithography, photolithography is the most common lithographic technique utilized in the microelectronics industry. The process employs a photoresist, an organic material that, upon irradiation through a mask, either degrades to become more soluble (positive image generation) or cross-links to become insoluble (negative image generation) under specified conditions for development (Figure 6.27).[1507,1508] The irradiated photoresist layer prior to the development is called a latent image. In subsequent chemical treatment (etching), the upper layer of the substrate in the areas that are not protected by photoresist is removed, followed by etching the unprotected layers and stripping of the photoresist layer. The current highest reproducible resolution of a feature on the surface that can be achieved by photolithography is about 90 nm.[1506]

In positive-tone imaging, a high-sensitivity photoresist is usually composed of a photoinitiator and a polymer, the degradation of which is typically based on chemical

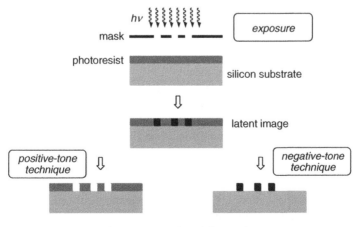

**Figure 6.27** Photolithography

amplification, that is, a chain reaction initiated by photoproduction of a catalyst.[1508] For example, a mixture of poly[4-(*tert*-butoxycarbonyloxy)styrene] (*t*-BOC resist) (**582**), a polymer with acid-labile *tert*-butyl carbonate groups, and triphenylsulfonium triflate (**583**), a photoinitiator that generates a strong acid ($CF_3SO_3H$) upon photolysis[1509] (see also the photolysis of hypervalent iodine compounds in Scheme 6.242), creates a latent image upon irradiation at 250 nm through a mask. The exposed areas contain an acid, which catalyses the decomposition of the *t*-BOC groups to form poly(4-hydroxystyrene) (**584**), $CO_2$ and isobutene with concurrent recovery of a proton upon heating to 100 °C (Scheme 6.283). Thanks to the relatively low temperatures used, the unexposed surface of the resist remains unaffected, whereas that irradiated (containing phenol moieties) can be washed away with a solution of a base. This photoresist system can also be used for negative-tone imaging, when a developer – a less polar organic solvent, such as anisole – dissolves the unmodified polymer but not the poly(4-hydroxystyrene).

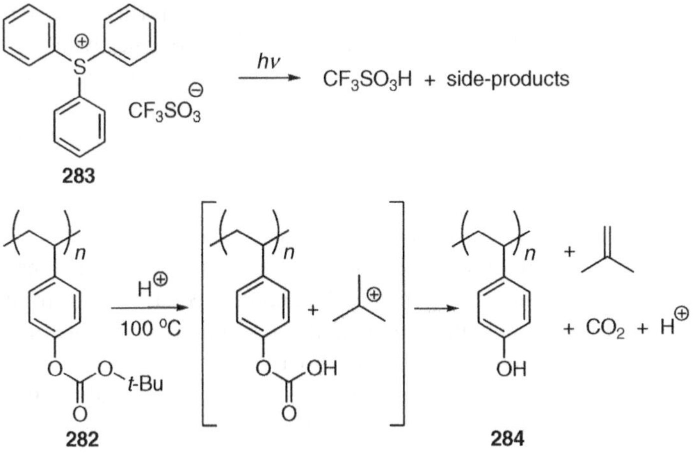

**Scheme 6.283**

Novolak – diazonaphthoquinone positive-tone resists, the most important imaging system of semiconductor production today[1510,1511] – is an archetypal example of the industrial applications of photochemistry. Novolak is a phenol–formaldehyde polymer (Bakelite) that dissolves in aqueous hydroxide, but the addition of a small amount of the diazonaphthoquinone **585** dramatically decreases the solubility. When irradiated, **585** undergoes the photo-Wolff rearrangement (see also Scheme 6.171), leading to ring contraction and subsequently to carboxylic acid formation (Scheme 6.284). Such a photochemically altered site is readily soluble and can be removed with a basic developer solution.

Another industrial application, *UV curing*, is a technologically advanced process by which monomers/oligomers/polymers, usually in the presence of photoinitiators, undergo polymerization and cross-linking upon UV irradiation. The process is utilized

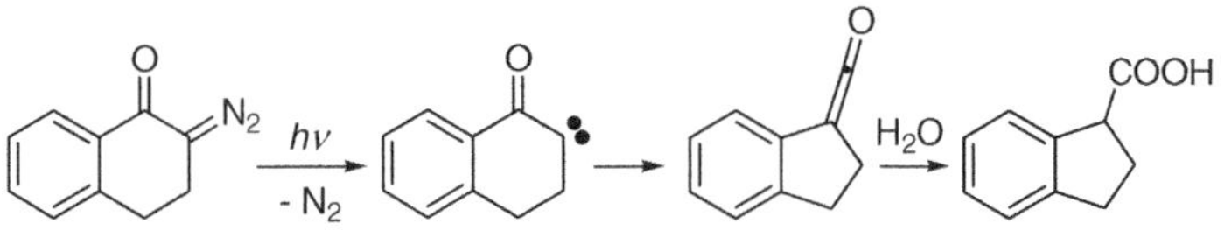

**585**

*Scheme 6.284*

to cure (harden, dry) inks, paints,[1512] coatings,[1513] adhesives,[1513] and dental glues,[1514] and also resists.[174] This method has also recently been introduced in coating and printing processes for CD and DVD manufacture.[1515]

The most common dental polymers, used for prosthetic purposes and restorative dentistry (filling material), are polymethacrylates.[1514] The polymerization process performed directly in the dental cavity has to meet strict demands; the reaction must be fast at a temperature below $50\,°C$ and it must avoid the formation of a toxic product. These requirements can be fulfilled by UV curing. For example, a mixture of camphorquinone (**586**), a chromophore (photoinitiator) with an absorption maximum at 468 nm and an amine **587** as a co-initiator (see also Scheme 6.100), initiates a radical polymerization reaction of the acrylate monomer **588** upon photolysis using a conventional blue lamp or laser (Scheme 6.285).

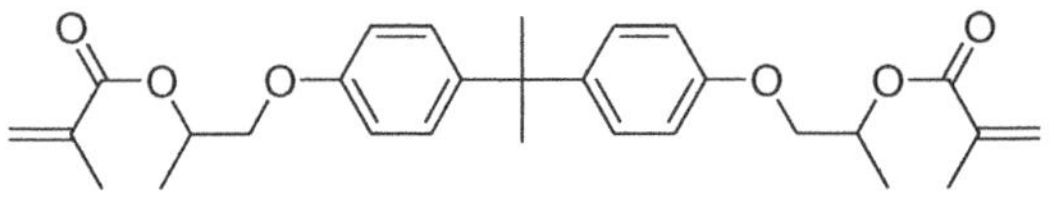

**588**

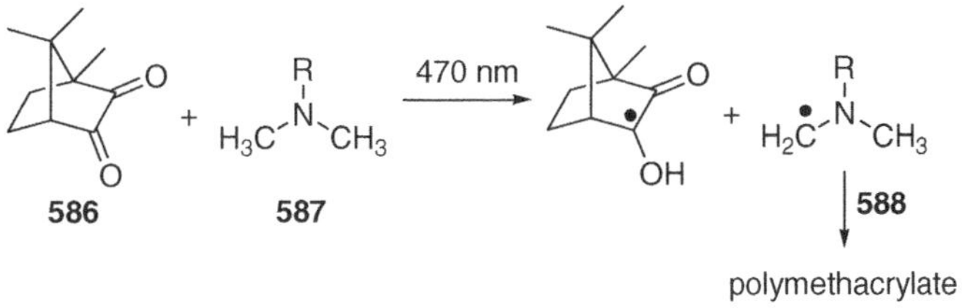

**586**     **587**

polymethacrylate

*Scheme 6.285*

In contrast, UV curing of poly(vinyl cinnamate) produces a cross-linked polymer (Scheme 6.286) via $[2+2]$ photocycloaddition (Section 6.1.5) in the absence of an photoinitiator. This reaction can be used in negative-tone photolithography because the complex photoproduct is poorly soluble.[1516] UV curing can also be used to produce the high-definition images by UV *nanoimprint lithography*, in which a topographic pattern from a rigid mould (such as a silicon matrix) is transferred into a low-viscosity monomer that is polymerized by UV irradiation to form solid structures on surfaces.[1506,1517]

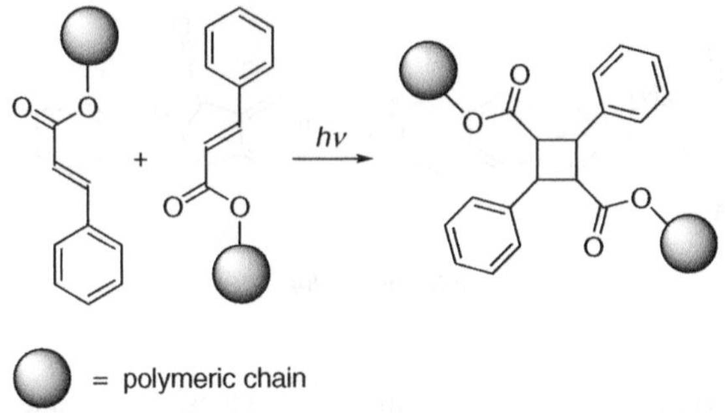

= polymeric chain

*Scheme 6.286*

## 6.8.2   Transition Metal Photocatalysts

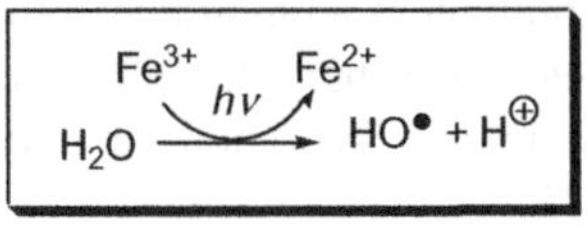

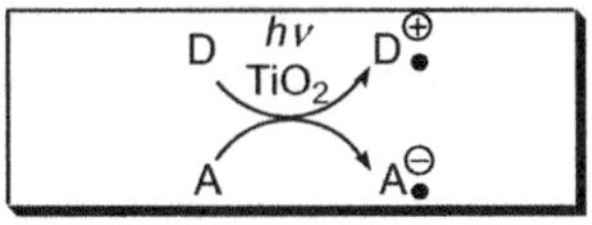

*Recommended review articles.*[170,1248,1518–1526]

*Homogeneous Transition Metal Photocatalysis*

Photoreactions that involve transition metal ions, complexes or compounds can usually be classified as (photo)redox (simultaneous oxidation and reduction) processes. A representative non-photoassisted catalytic system is Fenton's reagent that produces HO· radicals on reaction of ferrous ions ($Fe^{2+}$) and hydrogen peroxide (Scheme 6.287a). Its photochemical counterpart is the photo-Fenton process,[1527] in which ferric ($Fe^{3+}$) complexes in aqueous solutions (absorbing over 300 nm) are reduced to ferrous ions (Scheme 6.287b).

(a)     $Fe^{2+}$  +  $H_2O_2$  $\longrightarrow$  $Fe^{3+}$ + $HO^{\ominus}$ + $HO^{\bullet}$

(b)     $Fe^{3+}$  +  $H_2O$  $\xrightarrow{h\nu}$  $Fe^{2+}$ + $HO^{\bullet}$ + $H^{\oplus}$

*Scheme 6.287*

## Special Topic 6.28: Environmental remediation

Homogeneous transition metal photocatalysis reactions generating HO· radicals, often referred to as advanced oxidation processes (AOP),[1518] are powerful methods to remediate (i.e. to remove or to destroy) organic pollutants from aqueous solutions.[1519]

Nonselective and efficient consecutive oxidation reactions ultimately lead to nontoxic mineralization products, such as $CO_2$ and $H_2O$.[1519,1520] For example, an improved version of the photo-Fenton system, utilizing ferrioxalate ion, very efficiently oxidizes organic compounds present in the aqueous solution.[1528] This process affords a reactive $C_2O_4^{\cdot-}$ intermediate, which generates a superoxide radical anion ($O_2^{\cdot-}$) from dissolved oxygen or directly attacks relatively inert molecules such as $CCl_4$ (Scheme 6.288).[1529]

$$[Fe^{III}(C_2O_4)_3]^{3-} \xrightarrow{h\nu} [Fe^{II}(C_2O_4)_2]^{2-} + C_2O_4^{\cdot \ominus}$$

$$C_2O_4^{\cdot \ominus} + O_2 \longrightarrow 2\,CO_2 + O_2^{\cdot \ominus}$$

$$C_2O_4^{\cdot \ominus} + CCl_4 \longrightarrow 2\,CO_2 + {}^{\cdot}CCl_3 + Cl^{\ominus}$$

*Scheme 6.288*

Transition metal catalysts can also be used in photochemical organic synthesis. For example, the photo-Bergman reaction (cycloaromatization; see also Scheme 6.62) of (Z)-hex-3-en-1,5-diyne [(Z)-**589**], which is obtained by photostationary E–Z isomerization (Section 6.1.1) from (E)-**589**, occurs via transition metal-catalysed cycloaromatization reaction followed by photochemical dissociation of the arene ligand from the complex **590** (Scheme 6.289).[1530] The product, 1,2-di(n-propyl)benzene (**591**), is obtained in 91% chemical yield.

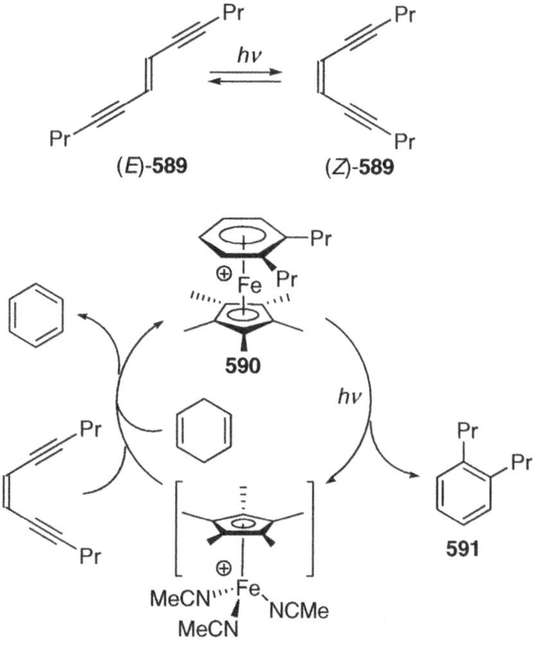

*Scheme 6.289*

In another example, iron pentacarbonyl photochemically decarbonylates (see Section 6.3.9) to form the π-allyl complex **592** with an allyl alcohol **593**, which then undergoes various dark reactions to give **594** (Scheme 6.290).[1531]

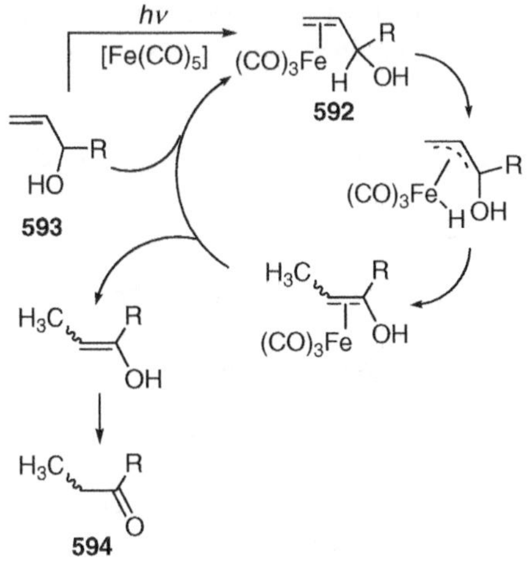

**Scheme 6.290**

*Heterogeneous Transition Metal Photocatalysis*

The most common photocatalytic processes, in terms of both mechanistic analysis and practical use, involve insoluble semiconductor metal oxides or sulfides, which upon irradiation undergo dual interfacial electron transfer between the excited semiconductor surface and adsorbed donor (D) and/or acceptor (A) molecules (Scheme 6.291). Titanium dioxide ($TiO_2$) is a particularly popular photocatalyst due to its good redox properties (see also Special Topic 6.29), high stability, low toxicity and low price.

$$TiO_2 + D + A \xrightarrow{h\nu} TiO_2 + D^{\oplus \bullet} + A^{\ominus \bullet}$$

**Scheme 6.291**

---

**Special Topic 6.29: Excitons and redox reactions on a semiconductor**

Irradiation of a semiconductor with photons having an energy exceeding its band-gap (the energy difference between the valence and conduction bands, usually considered to be lower than 3 eV in $TiO_2$) results in electron transfer from the valence to the conduction band.[170,1248,1521–1524] The generated negative charge (electron, $e^-$) in the conduction band associated with $Ti^{III}$ centres of $TiO_2$, and positive charge (a quasi-particle, electron hole, $h^+$) in the valence band associated with $Ti^{IV}$ centres of $TiO_2$, represent a bound excited state of an electron called an *exciton* that is formed within

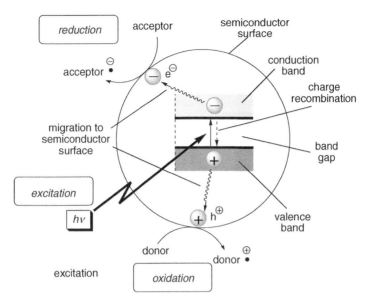

**Figure 6.28** Semiconductor photochemistry

femtoseconds.[1532] The exciton energy is slightly lower than that of the unbound electrons and holes because it gains from the binding interaction of the electron with its hole. The charges recombine (in 10–100 ns) to generate heat (radiationless pathway) or photon emission or they can migrate to the semiconductor surface and be trapped by ambient molecules via electron transfer processes (Figure 6.28). The charge transfer probability for electrons and holes is related to the redox potentials of the adsorbed species and the positions of the band edges. For interfacial charge transfer, oxidation of a donor is faster (on the order of 100 ns) than reduction of an acceptor (on the order of milliseconds).

Both donor and acceptor molecules are indispensable for the accomplishment of the photoredox reaction and the electrochemical potentials of the donor (D/D$^{\bullet +}$) and acceptor (A/A$^{\bullet -}$) couples should lie within the semiconductor band gap. Oxidation reactions, photocatalysed by TiO$_2$, are usually performed in the presence of easily reducible molecular oxygen as an electron acceptor, thereby generating a superoxide radical ion (O$_2^{\bullet -}$) and subsequently hydroxyl radicals. The resulting holes on the semiconductor surface can oxidize many compounds (Scheme 6.292), including alcohols, hydroxyl anion and even water.[1522,1523]

Photocatalytic reductions are less common because the reducing ability of an electron in the conduction band is considerably lower than the oxidizing ability of a positively charged valence band. Furthermore, reduction of most organic compounds usually cannot compete kinetically with that of oxygen.[170] Photoreductions in aqueous solutions are often accompanied by the production of molecular hydrogen.

Semiconductors, such as TiO$_2$, can be prepared in the laboratory by annealing (pyrolysis) of various Ti$^{IV}$ compounds, such as titanium isopropoxide, Ti[OCH(CH$_3$)$_2$]$_4$.[170] The resulting semiconductor properties, such as surface area, morphology, particle size and

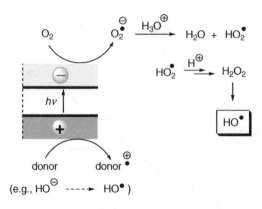

**Scheme 6.292**

surface character, then strongly influence its photophysical and photochemical behaviour. Semiconductors having high surface area can adsorb large amounts of organic molecules. Titanium dioxide in the anatase form is the most photoactive and most used in applications, such as water or air purification and water disinfection.[1523] In photochemical experiments, the semiconductor can be irradiated in the form of powder suspended in a stirred solvent or deposited on a solid support, such as a zeolite. The irradiation wavelength below 400 nm is sufficient to photoactivate neat $TiO_2$ (it is a white solid).

Titanium dioxide can also be modified by metal(0) deposition, which may significantly enhance its photoreaction efficiency. For example, $Pt^0$ or $Au^0$ deposited on $TiO_2$ increases the charge separation lifetime. Further, the absorption properties of $TiO_2$ can be improved by organic dye sensitization (see also Section 6.8.1). A sensitizer (S) then absorbs visible light to form an excited state ($S^*$), which transfers an electron to the conducting band of the semiconductor (Scheme 6.293). The radical cation thereby generated ($S^{\cdot +}$) is reduced by an auxiliary electron donor (D) and, as a result, the sensitizer is regenerated. An example of such a sensitizer is tris(2,2'-bipyridine)ruthenium(II) (see also Section 6.4.4), which can be used in solar (photovoltaic) cells (Special Topic 6.31).[1524,1533]

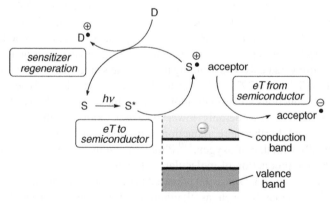

**Scheme 6.293**

TiO$_2$-based photocatalysts are undoubtedly the most commonly used, but some other heterogeneous catalysts also display promising photoactivity. Zinc and cadmium sulfides and iron, zinc and vanadium oxides are examples.[170] Metal chalcogenides, cadmium sulfide (CdS) and zinc sulfide (ZnS), have smaller band gaps, hence they can be irradiated at longer wavelengths (visible light).[1534,1535] Compared with TiO$_2$, which has a strong photooxidizing power, metal chalcogenides have strong reductive properties. The electrons in the conduction band of CdS have sufficient negative potential to reduce water to form molecular hydrogen (see also Special Topic 6.26).[1534] Unfortunately, metal chalcogenides are known to suffer *photocorrosion*, a process in which charge carriers in the illuminated semiconductor are not transferred to molecular donors or acceptors but oxidize or reduce the semiconductor itself, leading to its dissolution (i.e. releasing the metal ions to the solution). Some metal oxides, such as haematite ($\alpha$-Fe$_2$O$_3$), absorb in the visible region. They generally display considerably lower photocatalytic activity than TiO$_2$, although ZnO was shown to produce hydrogen peroxide more efficiently than TiO$_2$.[1536]

---

### Special Topic 6.30: Quantum dots

The absorption properties of semiconductors are dependent on the size of the particles. Small semiconductor particles (usually 2–10 nm in diameter) are called quantum dots,[1537–1539] when they confine the migration of a small finite number of conduction band electrons and valence band holes (i.e. excitons) due to electrostatic potentials and semiconductor surface/interface properties. Discrete absorption (but also emission) spectra are thus directly related to band-gap energies, which are larger in smaller particles (the colour of colloid solutions is changed with the size of the particles). For example, Figure 6.29 compares the optical properties of CdS sols formed in water; this mixture is yellow–green and the average particle size is ~3 nm,

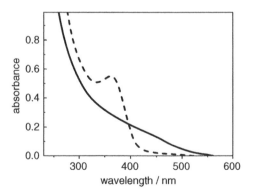

**Figure 6.29** CdS in water at 298 K (—) and in propan-2-ol at 213 K (- - -). Adapted from ref. 1534, copyright 1984. Reproduced with permission from IUPAC

corresponding to an agglomeration number of 300 CdS units. When generated in propan-2-ol, the mixture is colourless because the average size of the particles is less than 3 nm.

---

## Special Topic 6.31: Photovoltaic and photoelectrochemical cells

Photovoltaic and photoelectrochemical cells convert solar energy to electric energy.[5,1477,1533,1540,1541] In a *photovoltaic cell*,[1542] light excites electrons across the band gap of an inorganic semiconductor to form electron ($e^-$)–hole ($h^+$) pairs, which are subsequently separated by a p–n junction, that is, by an interface between doped n-type and p-type semiconductor layers (Figure 6.30). This drives the electrons in one direction and holes in the other, generating a potential difference at the external electrodes (with a load). Today, the best commercially available solar cells based on crystalline or amorphous silicon have 15–20% efficiency and lifetimes of about 30 years.[1543,1544] The cell production is a trade-off between price (mostly dictated by the purity of the material used) and overall operational efficiency. Even for cells that operate close to the theoretical efficiency limit (the maximum power divided by the input light irradiance and the surface area of the solar cell), cost-effective cell manufacturing and electric power storage are still the greatest challenge for both research and technological development. Recent scientific efforts have been focused on utilizing nanocrystalline and conducting polymer devices, which might be relatively cheap to fabricate.

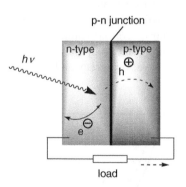

**Figure 6.30**  Photovoltaic cell

*Photoelectrochemical cells* that benefit from the sensitization of wide-gap semiconductors (Scheme 6.293) by organic dyes are referred to as *Grätzel cells*.[1533,1540,1545,1546] Figure 6.31 shows a cell, consisting of (1) a photoanode, a mesoporous semiconductor [typically $TiO_2$ (anatase) film] deposited on a conductive glass (current collector) by screen printing, with a photosensitizer adsorbed on a semiconductor surface, (2) a metallic cathode and (3) an electrolyte solution containing a redox mediator. In such an arrangement, light is absorbed by a dye, which upon excitation injects an electron to the conduction band of a semiconductor and is subsequently reduced by a mediator (usually the $I_3^-/I^-$ couple). The oxidized mediator travels to the cathode and is discharged to promote electric current when the external circuit has a load. The semiconductor structure is typically 10 mm thick and with a high porosity, which provides a large surface area for efficient chemisorption of a dye. Many different photosensitizers have been studied. An exceptional photovoltaic performance has so far been accomplished with polypyridyl complexes of ruthenium and osmium (e.g. **595**).[1545] They exhibit metal-to-ligand charge transfer and often bear carboxylic groups attached to the semiconductor surface. Overall solar conversion efficiencies by photoelectrochemical cells of over 10% have already been reached in the laboratory.[1545]

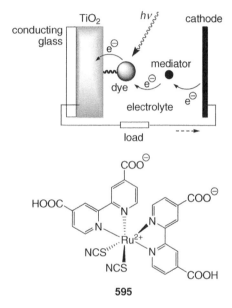

**595**

**Figure 6.31** Photoelectrochemical cell and a sensitizer

## Special Topic 6.32: Photography, xerography and photoconductivity

The *photographic process*, another photoimaging technique (see also Special Topic 6.27), is still the most widespread application of photochemistry.[1547] IR, visible, UV and X-ray irradiation of photosensitive silver halide ($Ag^+X^-$) crystals in a gelatine emulsion, coated on a film base, glass or paper substrate, causes reduction of interstitial silver ions to form atomic silver ($Ag^0$). Silver halides are known to exhibit an increase in electrical conductivity due to the absorption of light (*photoconductivity*), in which electrons are excited from the valence band to the conduction band; this excited state (exciton) either decays or forms an electron–hole pair. The process then involves the migration of the electrons and the interstitial silver ions to preferential sites. Their combination then produces silver clusters, which further efficiently trap photoelectrons produced in the vicinity. As a result, specks of silver are gradually formed:

$$X^- + h\nu \rightarrow X^* + e^-$$

$$Ag^+ + e^- \rightarrow Ag^0$$

$$2\,Ag^+ + e^- \rightarrow Ag_2^+$$

$$Ag_2^+ + e^- \rightarrow Ag_2, \text{ etc.}$$

The electron holes are also mobile; they 'diffuse' towards the surface of the grain, where oxidation of halide ions can take place. A short exposure to light creates a low concentration of metallic silver atoms on silver halide grains, producing the so-called latent image, which becomes enhanced and stabilized in a developer solution containing an organic reducing agent, such as hydroquinone or *N*-methyl-4-aminophenol. Prolonged exposure causes darkening of the surface due to extensive metallic silver formation (print-out effect). Silver halides absorb considerably in the blue, violet and ultraviolet regions; however, addition of specific dyes (sensitizers; Section 6.8.1) to a photographic emulsion increases the sensitivity of AgX to other wavelengths. Sensitization specifically to the three primary colours (blue, green and red) is utilized in conventional colour photography, in which the dyes are located on the surfaces of three superimposed layers.

*Xerography* (from Greek, 'dry writing') is a photocopying technique that combines electrostatic printing with photography.[1548] In this technique, the surface of a cylindrical drum, coated with a photoconductive material, such as selenium or an organic dye,[1549] is electrostatically negatively charged at a high voltage (Figure 6.32). In the next step, an illuminated original document reflects the light on to a drum surface. The sites that are exposed to light (corresponding to bright areas in the document) become conductive and discharge to ground. Positively charged toner attaches to the non-illuminated areas that are still negatively charged and is subsequently transferred to a paper with a higher electrostatic negative charge than that of the drum. *Laser printers* utilize the same xerographic printing process; however, the primary image is obtained by sequential irradiation of the photosensitive charged surface with the laser beam.

The semiconductor selenium is an excellent photoconductor; however, its photoresponsiveness decreases sharply at wavelengths over 550 nm, which has motivated research on organic photoconductive materials.[1550,1551] Many different organic compounds display photoconductive behaviour, such as phthalocyanines (**596**; M = metal) or squaraines (**597**) (Figure 6.33); today more than 90% of xerographic photoreceptors are composed of organic photoconductors.

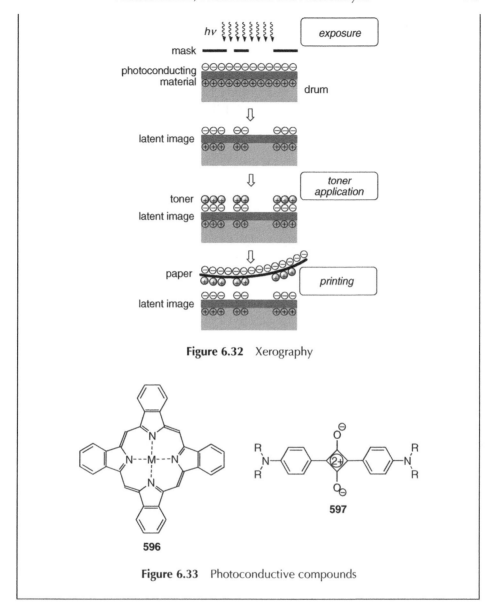

**Figure 6.32** Xerography

**Figure 6.33** Photoconductive compounds

Semiconductor-photocatalysed transformations of functional groups often provide high chemical yields, although the reaction quantum yields are usually not very high because a semiconductor surface may reflect radiation.[170,1521,1525,1526] In general, an organic functionality that contains a π-bond or a lone electron pair can be oxidized, which on the other hand considerably reduces the synthetic potential for applications involving substrates that have multiple reactive functional groups.

For example, alkenes such as propenylbenzene (**598**) and 1,1-diphenylethene (**599**) are selectively oxidized to the corresponding ketones in a very high chemical yield in air-saturated acetonitrile solutions (Scheme 6.294).[1552] Only a small amount of other side-products, such as epoxides and aldehydes, are formed in the former case. Irradiation of the $TiO_2$ band gap in acetonitrile promotes an electron into the conduction band (the band edge is $-0.8$ V vs SCE), forming a hole in the valence band ($+2.4$ V vs SCE). As a result, molecules with standard potentials of oxidation less positive than 2.4 V may react, provided that the reduction couple has a redox potential less negative than $-0.8$ V, for instance, because the standard potential of reduction of oxygen in acetonitrile is $-0.78$ V.[1525]

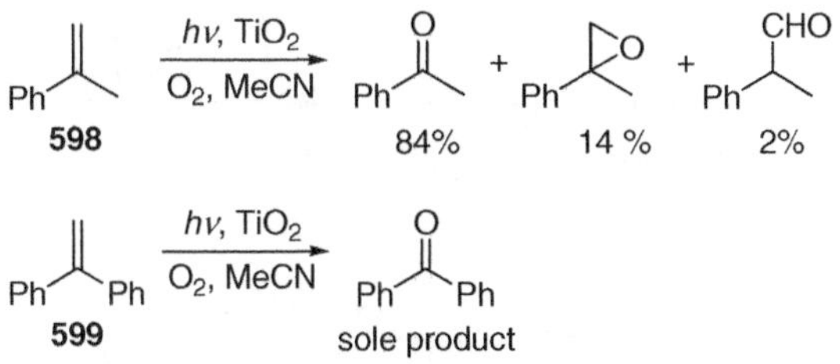

Scheme 6.294

Photocatalytic oxidation of lactic acid (**600**) on platinized $TiO_2$ or CdS is strongly regioselective (Scheme 6.295).[1553] Irradiation in the presence of $Pt/TiO_2$ at 360–420 nm leads to the cleavage oxidation products acetaldehyde and carbon dioxide, whereas pyruvic acid is exclusively obtained by irradiation of the Pt–CdS mixture at 440–520 nm. Since the standard potential of oxidation of acetic acid is known to be more positive than the valence band edge of CdS (CdS has a less positive valence band edge than does $TiO_2$), both catalysts readily oxidize aliphatic alcohols and the CdS photocatalyst is apparently capable of specific oxidation.

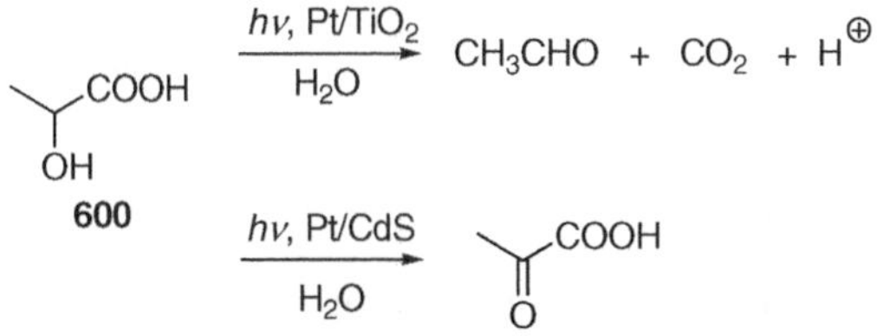

Scheme 6.295

The mechanism of photoinduced oxidation of aromatic compounds mediated by $TiO_2$ in aqueous media is demonstrated by the reaction of 4-chlorophenol (**601**). Its degradation is principally based on oxidation by photocatalytically produced hydroxyl radicals, most likely adsorbed on the surface of a semiconductor catalyst.[1554,1555] The initial reaction affords a 4-chlorodihydroxycyclodienyl radical **602**, which releases the chlorine atom to form hydroquinone in a radical substitution reaction or loses the hydrogen atom via hydrogen abstraction by a neighbouring radical species (Scheme 6.296). Prolonged irradiation leads ultimately to mineralization products, such as $CO_2$ and HCl. Total photocatalytic destruction (remediation) of organic compounds, such as anthropogenic pollutants, can be beneficially adapted in large-scale water or air purification processes (see also Special Topic 6.28).[1523]

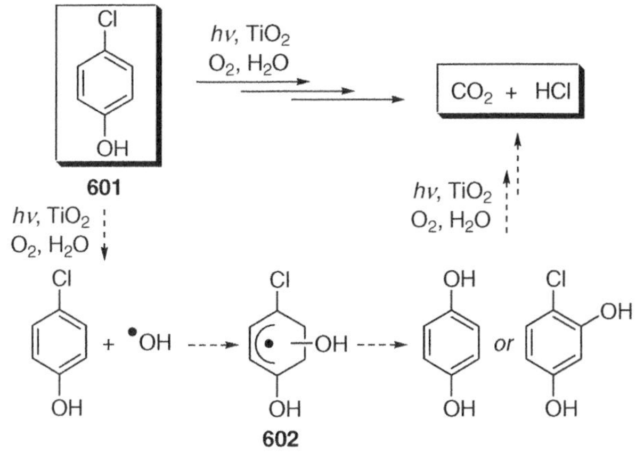

*Scheme 6.296*

As was mentioned above, the reducing ability of a conduction band electron in $TiO_2$ is appreciably lower than the oxidizing ability of a valence band hole, and reduction of most organic compounds generally cannot compete kinetically with that of molecular oxygen. Therefore, these reactions are not very common, although they may be possible when good electron trapping agents (acceptors), such as methyl viologen dication (603; Scheme 6.297) or protons of aqueous mineral acid, are used.[170,1521,1525] Co-catalysts such as platinum or palladium metal are often indispensable to the reaction. Scheme 6.298 shows an example of a selective reduction of the double bond in cyclohexenedicarboxylic acid (604) in the presence of platinized $TiO_2$ and aqueous nitric acid as an electron acceptor, under a nitrogen atmosphere (i.e. in the absence of oxygen).[1525]

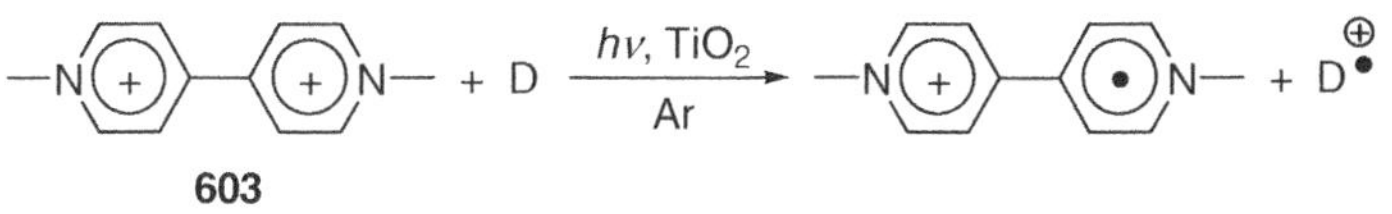

*Scheme 6.297*

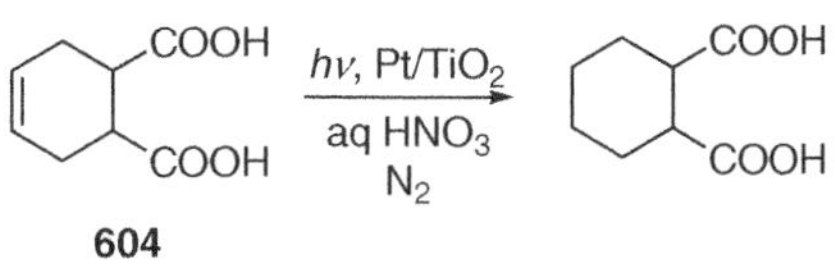

*Scheme 6.298*

## Case Study 6.39: Photocatalysis – oxidation of lactams and *N*-acylamines

The photocatalytic (TiO$_2$) oxidation of the amides **605** and **606** shows a high degree of selectivity at the carbon α to the amide nitrogen (Scheme 6.299).[1556] The reactants were found to be inert to photolysis in the absence of a suitable oxidant, such as O$_2$. The authors initially argued that photoproduced hydrogen peroxide and superoxide radical anion (discussed earlier) could serve to oxidize the organic reactant. However, the experiments established that water scavenges the valence band holes to generate HO$^\bullet$ radicals, which are consequently responsible for the amide oxidation.

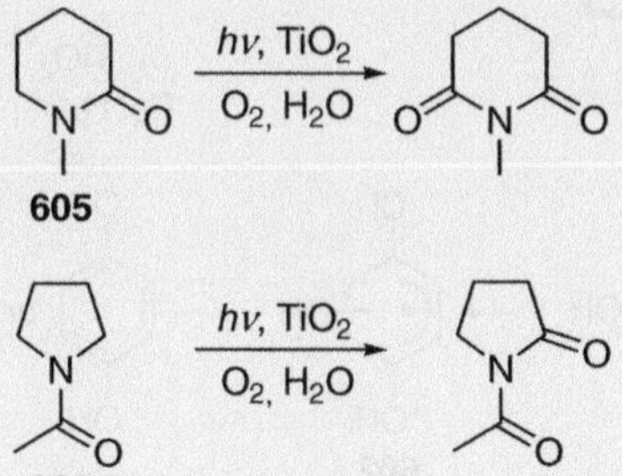

**605**

**606**

*Scheme 6.299*

*Experimental details.*[1556] An aqueous solution of an amide (0.2 M, 10 ml) was irradiated with Pyrex filtered light from a high pressure mercury lamp (200 W) (Figure 3.9) in the presence of suspended unreduced anatase (TiO$_2$; 100 mg) under a constant stream of oxygen. Products were isolated using preparative GC.

### 6.8.3  Problems

1. Explain the following concepts and keywords: photosensitizer; photocatalyst; photoinitiator; light-harvesting antenna; photosynthesis; photoinduced water splitting; dyad; photoimaging; UV curing; photoresist; photo-Fenton process; advanced oxidation processes; conduction band of semiconductor; quantum dot; photovoltaic cell; photoconductivity; photocopying.

2. Suggest the mechanisms for the following reactions:

(a)

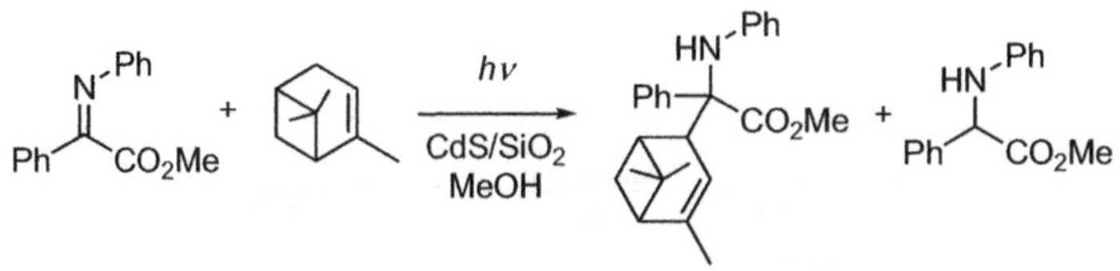

[ref. 1557]

(b)

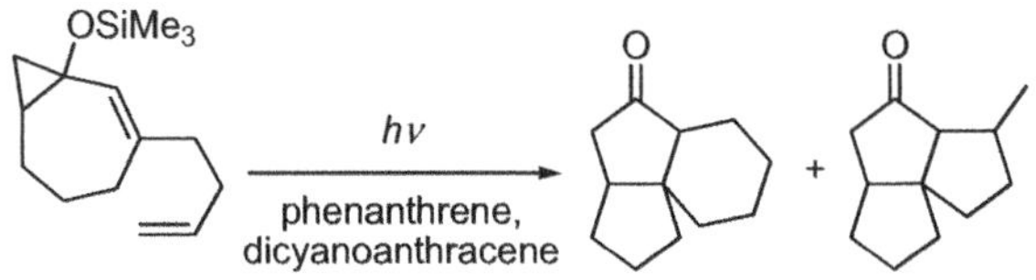

[ref. 1558]

3. Predict the major photoproduct(s):

(a)

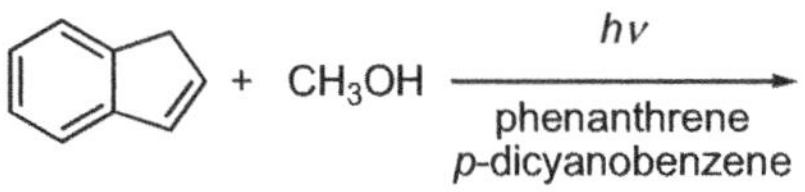

[ref. 1231]

(b)

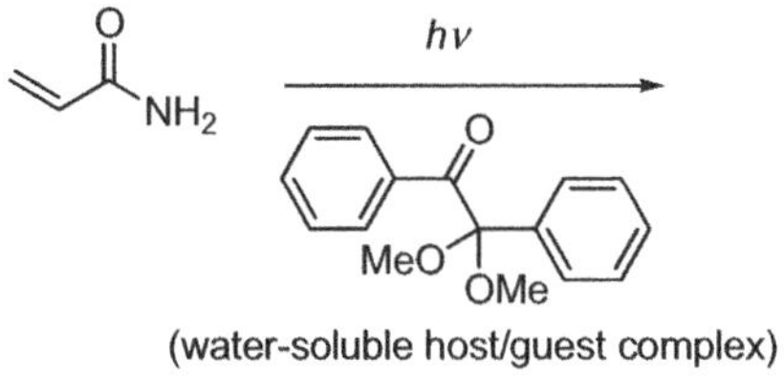

(water-soluble host/guest complex)

[ref. 1559]

# 7

# Retrosynthetic Photochemistry

Although photochemistry is widely used in many chemical and biological applications (Chapter 6), photochemical synthesis unfortunately remains far from being adopted by the general chemical community, despite the fact that a great number of procedures have been developed that afford products in high chemical yields and structures that can be made only with difficulty by conventional methods. Once a photochemical reaction is selected for an organic synthesis, it typically represents only one step in a multistep construction process of other 'dark' reactions. Such a light-induced transformation, then, represents a challenge for the synthetic chemist, who should be relatively knowledgeable about certain theoretical and experimental aspects.

Photochemistry is a very promising tool in many applications indeed:

- Photochemical reactions can be *green* to the extent that no other reagents but light are used.
- Unlike thermal processes, photochemical reactions usually do not require heating; quite the opposite, they can be performed at very low temperatures and even in the solid phase.
- Photochemistry offers many unique means of reaction control, such as selective irradiation wavelength, multi-photon absorption, indirect (sensitizer-mediated) excitation, sensitizer-mediated electron transfer initiation, temperature and phase variations, timing and spatial control.
- Reactions proceeding after the excitation can be very fast and efficient, because reactive (unstable) intermediates, such as radicals, biradicals, radical ions, carbenes and strained rings, are often involved.

This short chapter may serve as a problem-solving list of *target structures*, which can be synthesized by photochemical means. The information is sorted according to structural types of products and is limited to the reactions described in Chapter 6. A double arrow ($\Rightarrow$) indicates the retrosynthetic direction,[1560] starting from the product and showing the most important photochemical step in the course of the reaction. This does not necessarily mean that we can choose the corresponding procedure under all circumstances. Potential

*Photochemistry of Organic Compounds: From Concepts to Practice*   Petr Klán and Jakob Wirz
Copyright © 2009 P. Klán and J. Wirz

obstacles and limitations must be considered carefully, because no known (photo)chemical procedure is truly universally applicable.

The following list of potential obstacles may prompt the reader to go back over the preceding chapters, which provide sufficient details to handle most of the photochemical procedures successfully. The obstacles are listed here not to discourage, but, on the contrary, to accentuate that careful preparation for the experiment and consideration of possible difficulties will help in achieving fast and efficient photoprocesses, which selectively produce the desired compounds in very high chemical yields. Thus we should be aware that:

- The excited chromophore may undergo other competing chemical processes, which are dependent on the substrate structure and also on the presence of other compounds in the reaction environment.
- The excited chromophore may undergo physical deactivation, which lowers the quantum yield of the reaction.
- Other chromophores present in the substrate or the reaction system may impede the reaction and/or undergo unwanted reactions themselves.
- Solvent may interfere with the reaction as a reagent or as an optical filter.
- The presence of acids, bases, electrophilic and nucleophilic species or good electron donors and acceptors may strongly affect many photoreactions.
- The reaction intermediates may attack the reacting molecules.
- Oxygen and many other species can quench the excited state.
- Singlet oxygen produced *in situ* by sensitization can react with most molecules.
- The products or long-lived intermediates may absorb light and thereby become optical filters.
- The target product may undergo subsequent unwanted thermal or photochemical transformations.

Tables 7.1–7.13 are self-explanatory and the reactions represent only simple examples. New target product skeletons or bonds are shown in bold. Detailed mechanistic descriptions and reaction limitations are discussed in Chapter 6.

**Table 7.1** *Three-membered rings*

| Key retrosynthetic step(s) | Reaction | Section |
|---|---|---|
| | Di-π-methane rearrangement | 6.1.3 |
| | Oxa-di-π-methane rearrangement | 6.1.3 |
| | Sigmatropic photorearrangement | 6.1.6 |
| | α-Cleavage (Norrish type I reaction) | 6.3.3 |
| | Intramolecular H-abstraction | 6.3.5 |
| | Intramolecular H-abstraction | 6.5.1 |
| | Photodecarboxylation | 6.3.8 |
| | Carbene photoformation; carbene addition | 6.4.2 |
| | Photoelimination of $N_2$; nitrene addition | 6.4.2 |

**Table 7.2**   *Four-membered rings*

| Key retrosynthetic step(s) | Reaction | Section |
|---|---|---|
| | Photocycloaddition | 6.1.5<br>6.1.6<br>6.3.7 |
| | Photocycloaddition | 6.2.2 |
| | Photocycloaddition | 6.1.5 |
| | Electrocyclic photorearrangement | 6.1.2 |
| | Yang cyclization | 6.3.5<br>6.3.6 |
| | Oxetane formation<br>(Paternò–Büchi reaction) | 6.3.2<br>6.3.7 |
| | Thietane formation | 6.5.1 |
| | Photooxygenation | 6.7.2 |

**Table 7.3**  *Five-membered rings*

| Key retrosynthetic step(s) | Reaction | Section |
|---|---|---|
| <br>X = O, CR$_2$ | Intramolecular<br>H-abstraction | 6.3.5 |
| | Intramolecular<br>H-abstraction | 6.5.1 |
| <br>LG = leaving group | Photoenolization | 6.3.6 |
| <br>Hal = Cl, Br, I | Photosubstitution | 6.6.2 |

**Table 7.4**   *Six-membered and larger rings*

| Key retrosynthetic step(s) | Reaction | Section |
|---|---|---|
| | Electrocyclic photorearrangement | 6.1.2 |
| | Photocycloaddition | 6.2.2 6.1.5 |
| | Photoenolization; cycloaddition | 6.3.6 |
| X = 1 or more atoms | Intramolecular H-abstraction (various ring sizes) | 6.3.5 6.4.4 |
| | Electrocyclic photorearrangement | 6.1.2 |
| | Photoamination (various ring sizes) | 6.4.4 |
| LG = leaving group | de Mayo reaction (various ring sizes) | 6.1.6 |
| | Beckmann photoisomerization | 6.4.1 |
| Nu = nucleophile | Nitrene formation; ring expansion | 6.4.2 |
| | Photooxygenation | 6.7.2 |

**Table 7.5** *Condensed aromatic rings*

| Key retrosynthetic step(s) | Reaction | Section |
|---|---|---|
| | Photoelectrocyclization; oxidation | 6.1.2 |
| | Cycloaromatization (photo-Bergman reaction) | 6.1.5 |

**Table 7.6** *Cage compounds*

| Key retrosynthetic step(s) | Reaction | Section |
|---|---|---|
| | Photocycloaddition | 6.1.5 |
| | Photorearrangement | 6.2.1 |
| | Photocycloaddition | 6.2.2 |
| | Photocycloaddition | 6.2.2 |
| | Oxetane formation | 6.3.2 |

**Table 7.7**    *Spiro compounds*

| Key retrosynthetic step(s) | Reaction | Section |
|---|---|---|
| | Photoelectrocyclization | 6.1.2<br>6.4.1 |
| | Oxetane formation | 6.3.2<br>6.3.7 |

**Table 7.8**    *E-configuration of a double bond*

| Key retrosynthetic step(s) | Reaction | Section |
|---|---|---|
| | $E-Z$ photoisomerization | 6.1.1 |
| <br>X = C, N | $E-Z$ photoisomerization | 6.4.1 |

**Table 7.9** *C–C bond*

| Key retrosynthetic step(s) | Reaction | Section |
|---|---|---|
| | Photocatalysed radical addition | 6.8.1 |
| | Photochemical addition via e$^-$ and H$^+$ transfer | 6.1.4<br>6.4.4 |
| | Pinacol formation | 6.3.1 |
| | α-Cleavage (Norrish type I reaction) | 6.3.3 |
| | Photoelimination of N$_2$ | 6.4.1<br>6.4.2 |
| | Photoelimination of SO$_2$ | 6.5.2 |
| | Phototransposition | 6.2.1 |
| <br>LG = leaving group (halogen, CN ...) | Photosubstitution | 6.2.3 |
| <br>Hal = Cl, Br, I | Photosubstitution | 6.6.2 |
| | Photo-Fries rearrangement | 6.3.8 |
| | Photoelimination of N$_2$ | 6.4.2 |
| | Norrish type II elimination | 6.3.4 |

**Table 7.10**   C–O bonds

| Key retrosynthetic step(s) | Reaction | Section |
|---|---|---|
| $\xrightarrow[\text{ROH (H}^+\text{)}]{hv}$    R = H, alkyl | Photoinduced nucleophile (often acid-catalysed) addition | 6.1.4 |
| $\xrightarrow[\substack{\text{ea; [H];}\\ \text{ROH}}]{hv}$    ea = electron acceptor | Photoinduced (co-sensitized) nucleophilic addition | 6.1.4 |
| R—OR′ $\xrightarrow[\text{R′OH}]{hv}$ R—Hal    R = alkyl, aryl   R′ = H, alkyl; Hal = Cl, Br, I | Photosubstitution | 6.2.3 6.6.2 |
| (phenol) –OH $\xrightarrow[\text{H}_2\text{O}]{hv}$ (aryl) –N$_2^\oplus$ | Photosubstitution | 6.4.2 |
| $\xrightarrow{\text{ROH}}$ C=O $\xrightarrow[-\text{N}_2]{hv}$ =N$_2$ | Ketene formation | 6.4.2 |
| R—OR′ $\xrightarrow[-\text{SO}_2]{hv}$ R–S(O$_2$)–O–R′ | Photoelimination of SO$_2$ | 6.5.2 |
| $\xrightarrow{hv}$ LG; NO / NO$_2$   LG = leaving group | Intramolecular H-abstraction | 6.4.3 |
| R—ONO $\xrightarrow{hv}$ R–NO$_2$ | Photorearrangement | 6.4.3 |
| O–O $\xrightarrow[\text{sens}]{hv}$ $^1\text{O}_2$ $\Rightarrow$ $^3\text{O}_2$ | Singlet oxygen formation and addition | 6.7.2 |
| (endoperoxide) $\xrightarrow[\text{sens}]{hv}$ $^1\text{O}_2$ $\Rightarrow$ $^3\text{O}_2$ | Singlet oxygen formation and endoperoxide formation | 6.7.2 |
| HOO–...=CR$_2$ $\Rightarrow$ CHR$_2$ $\xrightarrow[\text{sens}]{hv}$ $^1\text{O}_2$ $\Rightarrow$ $^3\text{O}_2$ | Singlet oxygen formation and ene reaction | 6.7.3 |

**Table 7.11** C–N bonds

| Key retrosynthetic step(s) | Reaction | Section |
|---|---|---|
| | Nitrene formation and insertion | 6.4.2 |
| | Barton reaction | 6.4.2 |
| | Beckmann photoisomerization | 6.4.1 |
| | Photoamination | 6.4.4 |

**Table 7.12** C–halogen atom bonds

| Key retrosynthetic step(s) | Reaction | Section |
|---|---|---|
| | Photohalogenation | 6.6.1 |
| | Photohalogenation | 6.6.1 |

**Table 7.13**   C–H, N–H bonds

| Key retrosynthetic step(s) | Reaction | Section |
|---|---|---|
| | Photoreduction | 6.3.1 |
| | Reduction of aryl diazonium salts | 6.4.2 |
| | Photodecarboxylation | 6.3.8 |
| | Ester photofragmentation | 6.3.8 |
| | Photoreduction | 6.4.4 |
| | Photoreduction | 6.4.3 |
| | Photoreduction | 6.6.2 |
| | Photoreduction | 6.6.2 |
| | Photoreduction via electron transfer | 6.6.2 |

# 8

# Information Sources, Tables

*Scientific photochemical journals*

- Journal of Photochemistry and Photobiology A: Chemistry (www.sciencedirect.com/science/journal/10106030)
- Journal of Photochemistry and Photobiology B: Biology (www.sciencedirect.com/science/journal/10111344)
- Journal of Photochemistry and Photobiology C: Photochemistry Reviews (www.sciencedirect.com/science/journal/13895567)
- Photochemistry and Photobiology (www3.interscience.wiley.com/journal/118493575/home)
- Photochemical & Photobiological Sciences (www.rsc.org/publishing/journals/pp/index.asp)
- The Spectrum (journal free of charge) (www.bgsu.edu/departments/photochem/spectrum/index.html)

*Homepages of the photochemical societies and organizations*

- American Society for Photobiology (www.pol-us.net/ASP_Home)
- Asian and Oceanian Photochemistry Association (www.asianphotochem.com)
- European Photochemistry Association (www.photochemistry.eu)
- European Society for Photobiology (www.esp-photobiology.it)
- Inter-American Photochemical Society (www.i-aps.org)
- International Commission on Illumination (www.cie.co.at/index_ie.html)
- Intergovernmental Panel on Climate Change (www.ipcc.ch)
- International Commission on Illumination (www.cie.co.at/index_ie.html)
- Japanese Photochemistry Association (photochemistry.jp/ENGLISH/index.htm)
- Photobiology Association of Japan (www.cherry.bio.titech.ac.jp/photobio/en/photon.html)
- U.S. Department of Energy, Basic research needs for solar energy utilization (www.sc.doe.gov/bes/BES.html)

*Other information sources*

- Absorption spectra.[280,1561]
- Photophysical data, optical materials, light sources, etc.[109,137,157,159]
- Photochemistry and Photobiology Database (w3.chemres.hu/pchem)
- Simple HMO programs available on the web (www.chem.ucalgary.ca/SHMO; neon. chem.swin.edu.au/modules/mod3/huckel.html)

This text uses SI units.[10] In quantum mechanics, the use of atomic units (a.u.) is more convenient because in these units the numerical value of many fundamental physical constants is unity (Table 8.1). An authoritative compilation of definitions and concepts is available in the *Glossary of Terms Used in Photochemistry*, 3rd edition,[22] which can be downloaded from the IUPAC web pages (www.iupac.org). See also the *Glossary of Terms Used in Theoretical Organic Chemistry*[1562] and the *Glossary of Terms Used in Physical Organic Chemistry*,[1563] available also from the IUPAC web page.

**Table 8.1**  *Fundamental physical constants[a]*

| Physical quantity | Symbol | Value | |
|---|---|---|---|
| | | In a.u. | In SI units |
| Planck's constant | $h$ | $2\pi$ | $6.626 \times 10^{-34}$ J s |
| Angular momentum | $\hbar = h/2\pi$ | 1 | $1.055 \times 10^{-34}$ J s |
| Boltzmann's constant | $k_B$ | | $1.381 \times 10^{-23}$ J K$^{-1}$ |
| Electron mass | $m_e$ | 1 | $9.109 \times 10^{-31}$ kg |
| Unified atomic mass unit | $u = m_u = \mathrm{Da}$ | | $1.661 \times 10^{-27}$ kg |
| Elementary charge | $e$ | 1 | $1.602 \times 10^{-19}$ C |
| Speed of light in vacuum | $c$ | | $2.998 \times 10^{8}$ m s$^{-1}$ |
| Electric constant[b] | $\varepsilon_0$ | | $8.854 \times 10^{-12}$ C$^2$ J$^{-1}$ m$^{-1}$ |
| Bohr radius | $a_0 = 4\pi\varepsilon_0\hbar^2/m_e e^2$ | 1 | $5.292 \times 10^{-11}$ m |
| Bohr magneton | $\mu_B = e\hbar/2m_e$ | 0.5 | $9.274 \times 10^{-24}$ J T$^{-1}$ |
| Nuclear magneton | $\mu_N$ | | $5.051 \times 10^{-27}$ J T$^{-1}$ |
| Energy (hartree) | $E_h = \hbar^2/m_e a_0^2$ | 1 | $4.360 \times 10^{-18}$ J |
| Avogadro's constant | $N_A$ | | $6.022 \times 10^{23}$ mol$^{-1}$ |
| Faraday constant | $F = N_A e$ | | $9.649 \times 10^{4}$ C mol$^{-1}$ |

[a]High precision values and additional data are available in CODATA[1564] and in the IUPAC Green Book.[10]
[b]Permittivity of vacuum.
[c]1 T = 1 kg s$^{-2}$ A$^{-1}$.

**Table 8.2**  *Numerical energy conversion factors[a]*

| | J | kJ mol$^{-1}$ | eV | $E_h$ | $\mu$m$^{-1}$ |
|---|---|---|---|---|---|
| J = W s | 1 | $6.0221 \times 10^{20}$ | $6.2415 \times 10^{18}$ | $2.2937 \times 10^{17}$ | $5.0341 \times 10^{18}$ |
| kJ mol$^{-1}$ | $1.6605 \times 10^{-21}$ | 1 | $1.0364 \times 10^{-2}$ | $3.8088 \times 10^{-4}$ | $8.3593 \times 10^{-3}$ |
| eV | $1.6022 \times 10^{-19}$ | 96.485 | 1 | $3.6749 \times 10^{-2}$ | 0.80655 |
| $E_h$ | $4.3597 \times 10^{-18}$ | $2.6255 \times 10^{3}$ | 27.211 | 1 | 21.947 |
| $\mu$m$^{-1}$ | $1.9864 \times 10^{-19}$ | 119.63 | 1.2398 | $4.5563 \times 10^{-2}$ | 1 |

[a]Derived from the relations $E_p = h\nu = hc/\lambda = hc\tilde{\nu}$; $E_m = N_A E$; 1 cal ≡ 4.184 J.

**Table 8.3** Viscosities $\eta/mP^a$ of pure solvents and diffusion-controlled rate constants $k_d$ (calculated using Equation 2.29) at $T = 298\,K$, $p = 0.1\,MPa^b$

| Solvent | $\eta/mP$ | $k_d/(\text{M}^{-1}\,\text{s}^{-1})^b$ | Ref. |
|---|---|---|---|
| Water | 8.90 | $7.4 \times 10^9$ | 1565 |
| Methanol | 5.54 | $1.2 \times 10^{10}$ | 1566 |
| Ethanol | 10.8 | $6.1 \times 10^9$ | 1566 |
| Propan-1-ol | 19.7 | $3.4 \times 10^9$ | 1566 |
| Propan-2-ol | 20.5 | $3.2 \times 10^9$ | 1566 |
| Butan-1-ol | 25.7 | $2.6 \times 10^9$ | 1566 |
| n-Hexane | 2.98 | $2.2 \times 10^{10}$ | 1567 |
| n-Heptane | 3.88 | $1.7 \times 10^{10}$ | 1567 |
| n-Octane | 5.05 | $1.3 \times 10^{10}$ | 1567 |
| n-Nonane | 6.52 | $1.0 \times 10^{10}$ | 1567 |
| n-Decane | 8.44 | $7.8 \times 10^9$ | 1567 |
| Cyclohexane | 8.83 | $7.5 \times 10^9$ | 1567 |
| Methylcyclohexane | 6.92 | $9.5 \times 10^9$ | 1565 |
| Carbon tetrachloride | 9.06 | $7.3 \times 10^9$ | 1568 |
| Benzene | 6.00 | $1.1 \times 10^{10}$ | 1569 |
| Toluene | 5.54 | $1.2 \times 10^{10}$ | 1570 |
| Acetonitrile | 3.42 | $1.9 \times 10^{10}$ | 1566 |
| Liquid paraffins | 300–2000 | $(3–20) \times 10^7$ | 1566 |
| Glycerol | 10000 | $6.6 \times 10^6$ | 1571 |

$^a 1\,P = 0.1\,\text{kg}\,\text{m}^{-1}\,\text{s}^{-1} = 0.1\,\text{Pa}\,\text{s}.$
$^b$For further data, see refs 1565 and 1572.

**Table 8.4** Properties of some lasers that are commonly used in photochemistry

| Laser | Medium | $\lambda/nm$ (fundamental) | Optical power or pulse energy | Pulse duration |
|---|---|---|---|---|
| He—Ne | Gas | 632.8 (1152, 3392) | 1–100 mW | cw |
| Argon ion | Gas | Many lines, ~500 | 0.01–100 W | cw |
| Excimer | Gas | 193 (ArF), 248 (KrF), 308 (XeCl), 351 (XeF) | 100 mJ | 5–30 ns |
| Nd:YAG$^a$ | Solid | 1064 | 1 J | 10 ns or 30 ps (mode locked) |
| Ti:Sa | Solid | ~800 | 5 mJ | 200 fs |

$^a$Yttrium–aluminium garnet.

**Table 8.5** List of suggested HMO parameters (Section 4.3) for heteroatoms.[271,1573]

| Element | Bond type | Example | $\delta\alpha_\mu/\beta$ | $\delta\beta_{\mu\nu}/\beta$ |
|---|---|---|---|---|
| C | C----C | Benzene | 0.0 | 0.0 |
| Me (inductive) | C—CH$_3{}^a$ | Propene | −0.333 | 0.0 |
| N$^\bullet$ | C----N | Pyridine | 0.5 | 0.0 |
| N$^{\bullet+}$ | C----NH | Pyridinium | 2.0 | 0.0 |
| N: | C—N | Aniline | 1.5 | −0.2 |
| O$^\bullet$ | C=O | Ketone | 1.0 | 0.0 |
| O: | C—O | Phenol | 2.0 | −0.2 |
| O$^{\bullet+}$ | C=O$^+$H | Ketone–H$^+$ | 2.5 | −0.2 |
| F | C—F | Fluorobenzene | 3.0 | −0.3 |
| Cl | C—Cl | Chlorobenzene | 2.0 | −0.6 |

$^a$Inductive effect of methyl substitution at atom $C_\mu$.

**Table 8.6**  *Photophysical properties of some organic chromophores*[109,137]

| Compound | $\tau_f$/ns | $\Phi_f$ | $E_S$/kJ mol$^{-1}$ | $\Phi_T$ | $E_T$/kJ mol$^{-1}$ |
|---|---|---|---|---|---|
| Naphthalene | 262 | 0.2 | 385 | 0.8 | 254 |
| Anthracene | 5 | 0.3 | 319 | 0.7 | 176 |
| Pyrene | 500 | 0.3 | 322 | 0.7 | 202 |
| Azulene[a] | 1.7 (S$_2$) | 0.031 (S$_2$) | 171 | 0.0 | 165 |
| Acetone | 1.7 | 0.001 | 386 | 1.0 | 332 |
| Acetophenone | <0.01 | 0.00 | 334 | 1.0 | 310 |
| Xanthone[b] | 0.06 | $5 \times 10^{-4}$ | 324 | 1.0 | 310 |
| Benzophenone | 0.006[c] | 0.00 | 313 | 1.0 | 288 |
| Tetraphenylporphyrin | 12 | 0.1 | 180 | 0.8 | 139 |

[a]Ref. 293; $\Phi_f(S_1) = 7 \times 10^{-5}$, $\Phi_{IC}(S_2) = 0.97$.[1574]
[b]Ref. 127.
[c]Ref. 332.

# References

1. Mattay, J., Griesbeck, A., Photochemical Key Steps in Organic Synthesis, VCH, Weinheim, 1994.
2. Haag, W., Hoigné, J., Singlet Oxygen in Surface Waters. 3. Photochemical Formation and Steady-state Concentrations in Various Types of Waters, *Environ. Sci. Technol.* 1986, **20**, 341–348.
3. BP., *Statistical Review of World Energy*, http://www.bp.com, 2008.
4. Lewis, N. S., *Basic Research Needs for Solar Energy Utilization*, http://www.sc.doe.gov/bes/reports/files/SEU_rpt.pdf, US Department of Energy, 2005.
5. Lewis, N. S., Nocera, D. G., Powering the Planet: Chemical Challenges in Solar Energy Utilization, *Proc. Natl. Acad. Sci. USA* 2006, **103**, 15729–15735.
6. Solomon, S., Qin, D., Manning, M., Chen, Z., Marquis, M., Averyt, K. B., Tignor, M., Miller, H. L. (eds), *Climate Change 2007: the Physical Science Basis*, Cambridge University Press, Cambridge, 2007.
7. Chu, C.-C., Bassani, D. M., Challenges and Opportunities for Photochemists on the Verge of Solar Energy Conversion, *Photochem. Photobiol. Sci.* 2008, **7**, 521–530.
8. Ciamician, G., The Photochemistry of the Future, *Science* 1912, **36**, 385–394.
9. Albini, A., Fagnoni, M., 1908: Giacomo Ciamician and the Concept of Green Chemistry, *ChemSusChem* 2008, **1**, 63–66.
10. Cohen, E. R., Cvitaš, T., Frey, J. G., Holmström, B., Kuchitsu, K., Marquardt, R., Mills, I., Pavese, F., Quack, M., Stohner, J., Strauss, H. L., Takami, M., Thor, A. J., *Quantities Units and Symbols in Physical Chemistry*, 3rd edn, Royal Society of Chemistry, Cambridge, 2007.
11. Lee, J., Malpractices in Chemical Calculations, *J. Chem. Educ.* 2003, **7**, 27–32.
12. Millikan, R. A., A Direct Photoelectric Determination of Planck's "*h*", *Phys. Rev.* 1916, **7**, 355–388.
13. Smithson, H. E., Sensory, Computational and Cognitive Components of Human Colour Constancy, *Philos. Trans. R. Soc. Lond. B Biol. Sci.* 2005, **360**, 1329–1346.
14. Cramer, C. J., *Essentials of Computational Chemistry*, John Wiley & Sons, Inc., Hoboken, NJ, 2002.
15. Klessinger, M., Michl, J., *Excited States and Photochemistry of Organic Molecules*, VCH, New York, 1995.
16. Michl, J., Bonacic-Koutecky, V., *Electronic Aspects of Organic Photochemistry*, John Wiley & Sons, Inc., New York, 1990.

17. Olivucci, M. (ed), *Computational Photochemistry*, Theoretical and Computational Chemistry, Vol. **16**, Elsevier, Amsterdam, 2005.

18. Atkins, P. W., Friedman, R. S., *Molecular Quantum Mechanics*, 3rd edn, Oxford University Press, New York, 1999.

19. Bally, T., Borden, W. T., Calculations on Open-shell Molecules: a Beginner's Guide, *Rev. Comput. Chem.* 1999, **13**, 1–97.

20. Shaik, S., Hiberty, P. C. (eds), *A Chemist's Guide to Valence Bond Theory*, John Wiley & Sons, Inc, Hoboken, NJ, 2008.

21. Kiang, N. Y., The Color of Plants on Other Worlds, *Sci. Am.* 2008, **288**, 28–35.

22. Braslavsky, S. E., Glossary of Terms used in Photochemistry, 3rd edn, *Pure Appl. Chem.* 2007, **79**, 293–465.

23. Lewis, G. N., Kasha, M., Phosphorescence and the Triplet State, *J. Am. Chem. Soc.* 1944, **66**, 2100–2116.

24. Kasha, M., Fifty Years of the Jabłonski Diagram, *Acta Phys. Pol.* 1987, **A71**, 661–670.

25. Jablonski, A., Über den Mechanismus der Photolumineszenz von Farbstoffphosphoren, *Z. Phys.* 1935, **94**, 38–46.

26. Delorme, R., Perrin, F., Durées de Fluorescence des Sels d'Uranyle Solides et de leurs Solutions, *J. Phys., Sér. VI* 1929, **10**, 177–186.

27. Lewis, G. N., Lipkin, D., Magel, T. T., Reversible Photochemical Processes in Rigid Media. A Study of the Phosphorescent State *J. Am. Chem. Soc.* 1941, **63**, 3005–3018.

28. Terenin, A. N., Photochemical Processes in Aromatic Compounds, *Acta Physicochim. URSS* 1943, **18**, 210–241.

29. Hutchison, C. A. Jr., Mangum, B. W., Paramagnetic Resonance Absorption in Naphthalene in Its Phosphorescent State, *J. Chem. Phys.* 1958, **29**, 952–953.

30. Valeur, B., Historical Aspects of Fluorescence. In Valeur, B., Brochon, J.-C. (eds), *New Trends in Fluorescence*, Springer, Berlin, 2001, pp. 3–6.

31. Strickler, S. J., Berg, R. A., Relationship between Absorption Intensity and Fluorescence Lifetime of Molecules, *J. Chem. Phys.* 1962, **37**, 814–822.

32. Parker, C. A., *Photoluminescence of Solutions*, Elsevier, Amsterdam, 1968.

33. Dirac, P. A. M., The Quantum Theory of Emission and Absorption of Radiation, *Proc. R. Soc. London, Ser. A* 1927, **114**, 243–265.

34. Robinson, G. W., Frosch, R. P., Theory of Electronic Relaxation in the Solid Phase, *J. Chem. Phys.* 1962, **37**, 1962–1973.

35. Englman, R., Jortner, J., The Energy Gap Law for Radiationless Transitions in Large Molecules, *Mol. Phys.* 1979, **18**, 145–164.

36. Bixon, M., Jortner, J., Intramolecular Radiationless Transitions, *J. Chem. Phys.* 1968, **48**, 715–726.

37. Freed, K. F., Radiationless Transitions in Molecules, *Acc. Chem. Res.* 1978, **11**, 74–80.

38. Avouris, P., Gelbart, W. M., El Sayed, M. A., Nonradiative Electronic Relaxation under Collision-free Conditions, *Chem. Rev.* 1977, **77**, 793–833.

39. Siebrand, W., Radiationless Transitions in Polyaromatic Molecules. II. Triplet–Ground-state Transitions in Aromatic Hydrocarbons, *J. Chem. Phys.* 1967, **47**, 2411–2422.

40. Doubleday, C. Jr., Turro, N. J., Wang, J.-F., Dynamics of Flexible Triplet Biradicals, *Acc. Chem. Res.* 1989, **22**, 199–205.

41. El Sayed, M. A., Vanishing First- and Second-order Intramolecular Heavy-atom Effects on the $(\pi^* \rightarrow n)$ Phosphorescence in Carbonyls, *J. Chem. Phys.* 1964. **41**, 2462–2467 and references therein.

42. Aloïse, S., Ruckenbusch, C., Blanchet, L., Réhault, J., Buntinx, G., Huvenne, J.-P., The Benzophenone $S1(n,\pi^*) \rightarrow T1(n,\pi^*)$ States Intersystem Crossing Reinvestigated by Ultrafast Absorption Spectroscopy and Multivariate Curve Resolution, *J. Phys. Chem. A* 2008, **112**, 224–231.

43. Beer, M., Longuet-Higgins, H. C., Anomalous Light Emission of Azulene, *J. Chem. Phys.* 1955, **23**, 1390–1391.

44. Leupin, W., Magde, D., Persy, G., Wirz, J., 1,4,7-Triazacycl[3.3.3]azine: Basicity, Photoelectron Spectrum, Photophysical Properties, *J. Am. Chem. Soc.* 1986, **108**, 17–22.

45. Leupin, W., Wirz, J., Low-lying Electronically Excited States of Cycl[3.3.3]azine, a Bridged 12$\pi$-Perimeter, *J. Am. Chem. Soc.* 1980, **102**, 6068–6075.

46. Kropp, J. L., Stanley, C. C., The Temperature Dependence of Ovalene Fluorescence, *Chem. Phys. Lett.* 1971, **9**, 534–538.

47. Amirav, A., Even, U., Jortner, J., Intermediate Level Structure in the $S_2$ State of the Isolated Ultracold Ovalene Molecule, *Chem. Phys. Lett.* 1980, **69**, 14–17.

48. Nickel, B., Delayed Fluorescence from Upper Excited Singlet States $S_n$ ($n > 1$) of the Aromatic Hydrocarbons 1,2-Benzanthracene, Fluoranthene, Pyrene, and Chrysene in Methylcyclohexane, *Helv. Chim. Acta* 1978, **61**, 198–222.

49. Nickel, B., Karbach, H.-J., Complete Spectra of the Delayed Luminescence from Aromatic Compounds in Liquid Solutions: on the Observability of Direct Radiative Triplet–Triplet Annihilation, *Chem. Phys.* 1990, **148**, 155–182.

50. Longfellow, R. J., Moss, D. B., Parmenter, C. S., Rovibrational Level Mixing Below and Within the Channel Three Region of $S_1$ Benzene, *J. Phys. Chem.* 1988, **92**, 5438–5449.

51. Siegrist, A. E., Eckhardt, C., Kaschig, J., Schmidt, E., Optical Brighteners. In *Ullmann's Encyclopedia of Industrial Chemistry*, Wiley-VCH Verlag, GmbH, Weinheim, 2003.

52. Wong-Wah-Chung, P., Mailhot, G., Aguer, J.-P., Bolte, M., Fate of a Stilbene-type Fluorescent Whitening Agent (DSBP) in the Presence of Fe(III) Aquacomplexes: from the Redox Process to the Photodegradation, *Chemosphere* 2006, **65**, 2185–2192.

53. Saltiel, J., Atwater, B. W., *Spin-statistical Factors in Diffusion-controlled Reactions*, Advances in Photochemistry, Vol. 14, John Wiley & Sons, Inc., New York, 1988, pp. 1–90.

54. Kincaid, J. F., Eyring, H., Stearn, A. E., The Theory of Absolute Reaction Rates and Its Application to Viscosity and Diffusion in the Liquid State, *Chem. Rev.* 1941, **41**, 301–365.

55. Kierstead, A. H., Turkevich, J., Viscosity and Structure of Pure Hydrocarbons, *J. Chem. Phys.* 1944, **12**, 24–27.

56. Alwattar, A. H., Lumb, M. D., Birks, J. B., Diffusion-controlled Rate Processes. In Birks, J. B., (ed), *Organic Molecular Photophysics*, John Wiley & Sons, Ltd, London, 1973, Vol. **1**, pp. 403–456.

57. Schuh, H.-H., Fischer, H., The Kinetics of the Bimolecular Self-termination of *t*-Butyl Radicals in Solution, *Helv. Chim. Acta* 1978, **61**, 2130–2164.

58. Fischer, H., Paul, H., Rate Constants for Some Prototype Radical Reactions in Liquids by Kinetic Electron Spin Resonance, *Acc. Chem. Res.* 1987, **20**, 200–206.

59. Noyes, R. M., Kinetics of Competitive Processes when Reactive Fragments are Produced in Pairs, *J. Am. Chem. Soc.* 1955, **77**, 2042–2045.

60. Porter, G., Wright, M. R., Intramolecular and Intermolecular Energy Conversion Involving Change of Multiplicity, *Discuss. Faraday Soc.* 1959, No. 27, 18–27.

61. Gijzeman, O. L. J., Kaufman, F., Porter, G., Oxygen Quenching of Aromatic Triplet States in Solution. 1, *J. Chem. Soc., Faraday Trans. 2* 1973, **69**, 708–720.

62. Charlton, J. L., Dabestani, R., Saltiel, J., Role of Triplet–Triplet Annihilation in Anthracene Dimerization *J. Am. Chem. Soc.* 1983, **105**, 3473–3476.

63. Van Der Meer, B. W., Coker, G., Chen, S.Y., *Resonance Energy Transfer*, VCH, Weinheim, 1994.

64. Andrews, D. L., Demidov, A. A., *Resonance Energy Transfer*, Wiley-VCH Verlag GmbH, Weinheim, 1999.

65. Barber, J., Photosystem II: an Enzyme of Global Significance, *Biochem. Soc. Trans.* 2006, **34**, 615–631.

66. Saltiel, J., Townsend, D. E., Sykes, A., Role of Higher Triplet States in the Anthracene-sensitized Photoisomerization of Stilbene and 2,4-Hexadiene, *J. Am. Chem. Soc.* 1983, **105**, 2530–2538.

67. Förster, T., *Fluoreszenz Organischer Verbindungen*, Vandenhoeck & Ruprecht, Göttingen, 1951.

68. Förster, T., Transfer Mechanisms of Electronic Excitation, *Discuss. Faraday Soc.* 1959, **27**, 7–17.

69. Förster, T., In Sinanoglu, O. (ed), *Modern Quantum Chemistry*, Istanbul Lectures, Academic Press, New York, 1965.

70. Braslavsky, S. E., Fron, E., Rodríguez, H. B., San Román, E., Scholes, G. D., Schweitzer, G., Valeur, B., Wirz, J., Pitfalls and Limitations in the Practical Use of Förster's Theory of Resonance Energy Transfer, *Photochem. Photobiol. Sci.* 2008, **7**, 1444–1448.

71. Yang, J., Winnik, M. A., The Orientation Parameter for Energy Transfer in Restricted Geometries Including Block Copolymer Interfaces: a Monte Carlo Study, *J. Phys. Chem. B* 2005, **109**, 18408–18417.

72. Torgenson, P. M., Morales, M. F., Application of the Dale Eisinger Analysis to Proximity Mapping in the Contractile System, *Proc. Natl. Acad. Sci. USA* 1984, **81**, 3723–3727.

73. dos Remedios, C. G., Moens, P. D. J., Fluorescence Resonance Energy Transfer Spectroscopy Is a Reliable "Ruler" for Measuring Structural Changes in Proteins: Dispelling the Problem of the Unknown Orientation Factor, *J. Struct. Biol.* 1995, **115**, 175–185.

74. Stryer, L., Haugland, R. P., Energ, Transfer: a Spectroscopic Ruler, *Proc. Natl. Acad. Sci. USA* 1967, **58**, 719–726.

75. Klostermeier, D., Millar, D. P., Time-resolved Fluorescence Resonance Energy Transfer: a Versatile Tool for the Analysis of Nucleic Acids, *Biopolymers (Nucleic Acid Sci.)* 2002, **61**, 159–179.

76. Wang, D., Geva, E., Protein Structur and Dynamics from Single-molecule Fluorescence Resonance Energy Transfer, *J. Chem. Phys. B* 2005, **109**, 1626–1634.

77. Buchner, J., Kiefhaber, T. (eds), *Protein Folding Handbook*, Vol.5, Wiley-VCH Verlag GmbH, Weinheim, 2005, p. 445

78. Michalet, X., Weiss, S., Jäger, M., Single-molecule Fluorescence Studies of Protein Folding and Conformational Dynamics, *Chem. Rev.* 2006, **106**, 1785–1813.

79. Schuler, B., Lipman, E. A., Eaton, W. A., Probing the Free-energy Surface for Protein Folding with Single-molecule Fluorescence Spectroscopy, *Nature* 2002, **419**, 743–747.

80. Hoffmann, A., Kane, A., Nettls, D., Hertzog, D. E., Baumgartel, P., Lengefeld, J., Reichardt, G., Horsley, D. A., Seckler, R., Bakajin, O., Schuler, B., Mapping Protein Collapse with Single-molecule Fluorescence and Kinetic Synchrotron Radiation Circular Dichroism Spectroscopy, *Proc. Natl. Acad. Sci. USA* 2007, **104**, 105–110.

81. Nienhaus, G. U., Exploring Protein Structure and Dynamics under Denaturing Conditions by Single-molecule FRET Analysis, *Macromol. Biosci.* 2006, **6**, 907–922.

82. Greenleaf, W. J., Woodside, M. T., Block, S. M., High-Resolution, Single-molecule Measurements of Biomolecular Motion, *Annu. Rev. Biophys. Biomol. Struct.* 2007, **36**, 171–190.

83. Roeffaers, M. B. J., De Cremer, G., Uji-i, H., Muls, B., Sels, B. F., Jacobs, P. A., De Schryver, F. C., De Vos, D. E., Hofkens, J., Single-molecule Fluorescence Spectroscopy in (Bio)catalysis, *Proc. Natl. Acad. Sci. USA* 2007, **104**, 12603–12609.

84. Wagner, P. J., Klán, P., Intramolecular Triplet Energy Transfer in Flexible Molecules: Electronic, Dynamic, and Structural Aspects, *J. Am. Chem. Soc.* 1999, **212**, 9626–9635.

85. Vrbka, L., Klán, P., Kriz, Z., Koca, J., Wagner, P. J., Computer Modeling and Simulations on Flexible Bifunctional Systems: Intramolecular Energy Transfer Implications, *J. Phys. Chem. A* 2003, **107**, 3404–3413.

86. Bieri, O., Wirz, J., Hellrung, B., Schutkowski, M., Drewello, M., Kiefhaber, T., The Speed Limit for Protein Folding Measured by Triplet–Triplet Energy Transfer, *Proc. Natl. Acad. Sci. USA* 1999, **96**, 9597–9601.

87. Fierz, B., Kiefhaber, T., End-to-End vs Interior Loop Formation Kinetics in Unfolded Polypeptide Chains, *J. Am. Chem. Soc.* 2007, **129**, 672–679.

88. Fierz, B., Satzger, H., Root, C., Gilch, P., Zinth, W., Kiefhaber, T., Loop Formation in Unfolded Polypeptide Chains on the Picoseconds to Microseconds Time Scale, *Proc. Natl. Acad. Sci. USA* 2007, **104**, 2163–2168.

89. Roccatano, D., Sahoo, H., Zacharias, M., Nau, W. M., Temperature Dependence of Looping Rates in a Short Peptide, *J. Chem. Phys. B* 2007, **111**, 2639–2646.

90. Hennig, A., Florea, M., Roth, D., Enderle, T., Nau, W. M., Design of Peptide Substrates for Nanosecond Time-resolved Fluorescence Assays of Proteases: 2,3-Diazabicyclo[2.2.2]oct-2-ene as a Noninvasive Fluorophore, *Anal. Biochem.* 2007, **360**, 255–265.

91. Wang, X., Nau, W. M., Kinetics of End-to-End Collision in Short Single-stranded Nucleic Acids, *J. Am. Chem. Soc.* 2004, **126**, 808–813.

92. Minkowski, C., Calzaferri, G., Förster-type Energy Transfer along a Specified Axis, *Angew. Chem. Int. Ed.* 2005, **44**, 5325–5329.

93. Kuhn, H., Information, Electron Energy Transfer in Surface Layers, *Pure Appl. Chem.* 1981, **53**, 2105–2122.

94. Ajayaghosh, A., Praveen, V.K., π-Organogels of Self-assembled *p*-Phenylenevinylenes: Soft Materials with Distinct Size, Shape, and Functions, *Acc. Chem. Res.* 2007, **40**, 644–656.

95. Markovitsi, D., Marguet, S., Gallos, L. K., Sigal, H., Millié, P., Argyrakis, P., Ringsdorf, H., Kumar, S., Electronic Coupling Responsible for Energy Transfer in Columnar Liquid Crystals, *Chem. Phys. Lett.* 1999, **306**, 163–167.

96. Feng, X., Pisula, W., Zhi, L., Takase, M., Müllen, K., Controlling the Columnar Orientation of $C_3$-symmetric "Superbenzenes" through Alternating Polar/Apolar Substituents, *Angew. Chem. Int. Ed.* 2008, **47**, 1703–1706.

97. Cotlet, M., Gronheid, R., Habuchi, S., Stefan, A., Barbafina, A., Müllen, K., Hofkens, J., De Schryver, F. C., Intramolecular Directional Förster Resonance Energy Transfer at the Single-molecule Level in a Dendritic System, *J. Am. Chem. Soc.* 2003, **125**, 13609–13617.

98. Cotlet, M., Vosch, T., Habuchi, S., Weil, T., Müllen, K., Hofkens, J., De Schryver, F. C., Probing Intramolecular Förster Resonance Energy Transfer in a Naphthaleneimide–Perileneimide–Terrilenediimide-based Dendrimer by Ensemble and Single-molecule Fluorescence Spectroscopy, *J. Am. Chem. Soc.* 2005, **127**, 9760–9768.

99. Andreasson, J., Straight, S. D., Bandyopadhyay, S., Mitchell, R. H., Moore, T. A., Moore, A. L., Gust, D., Molecular 2:1 Digital Multiplexer, *Angew. Chem. Int. Ed.* 2007, **46**, 958–961.

100. Andreasson, J., Straight, S. D., Bandyopadhyay, S., Mitchell, R. H., Moore, T. A., Moore, A. L., Gust, D., A Molecule-based 1:2 Digital Demultiplexer, *J. Phys. Chem. C* 2007, **111**, 14274–14278.

101. Nakamura, Y., Aratani, N., Osuka, A., Cyclic Porphyrin Arrays as Artificial Antenna: Synthesis and Excitation Transfer, *Chem. Soc. Rev.* 2007, **36**, 831–845.

102. Berera, R., van Stokkum, I. H. M., Kodis, G., Keirstead, A. E., Pillai, S., Herrero, C., Palacios, R. E., Vengris, M., van Grondelle, R., Gust, D., Moore, T. A., Moore, A. L., Kennis, J. T. M., Energy Transfer, Excited-state Deactivation and Exciplex Formation in Artificial Carotene–Phthalocyanine Light-harvesting Antennas, *J. Chem. Phys. B* 2007, **111**, 6868–6877.

103. Scholes, G. D., Long-range Resonance Energy Transfer in Molecular Systems, *Annu. Rev. Phys. Chem.* 2003, **54**, 57–87.

104. Ermolaev, V. L., Sveshnikova, E. V., Inductive-resonance Energy Transfer from Aromatic Molecules in the Triplet State, *Dokl. Akad. Nauk SSSR* 1963, **149**, 1295–1298.

105. Bennett, R. G., Schwenker, R. P., Kellog, R. E., Radiationless Intermolecular Energy Transfer. II. Triplet–Singlet Transfer, *J. Chem. Phys.* 1964, **41**, 3040–3041.

106. Roller, R. S., Winnik, M. A., The Determination of the Förster Distance ($R_0$) for Phenanthrene and Anthracene Derivatives in Poly(methyl methacrylate) Films, *J. Phys. Chem. B* 2005, **109**, 12261–12269.

107. Kinka, G. W., Faulkner, L. R., Wurster's Blue as a Fluorescence Quencher for Anthracene, Perylene, and Fluoranthene, *J. Am. Chem. Soc.* 1976, **98**, 3897–3901.

108. Tyson, D. S., Gryczynski, I., Castellano, F. N., Long-range Resonance Enegry Transfer to $[Ru(bpy)_3]^{2+}$, *J. Phys. Chem. A* 2000, **104**, 2919–2924.

109. Birks, J. B., *Photophysics of Aromatic Molecules*, Wiley-Interscience, London, 1970.

110. Klein, U. K. A., Frey, R., Hauser, M., Gösele, U., Theoretica, Experimental Investigations of Combined Diffusion and Long-range Energy Transfer, *Chem. Phys. Lett.* 1976, **41**, 139–142.

111. Sandros, K., Transfer of Triplet State Energy in Fluid Solutions. III. Reversible Energy Transfer, *Acta Chem. Scand.* 1964, **18**, 2355–2374.

112. Hammond, G. S., Saltiel, J., Lamola, A. A., Turro, N. J., Bradshaw, J. S., Cowan, D. O., Counsell, R. C., Vogt, V., Dalton, C., Mechanisms of Photochemical Reactions in Solution. XXII. Photochemical *cis–trans* Isomerization, *J. Am. Chem. Soc.* 1964, **86**, 3197–3217.

113. Balzani, V., Bolletta, F., Scandola, F., Vertica, "Nonvertical" Energy Transfer Processes. A General Classical Treatment, *J. Am. Chem. Soc.* 1980, **102**, 2152–2163.

114. Saltiel, J., Marchand, G. R., Kirkor-Kaminska, E., Smothers, W. K., Mueller, W. B., Charlton, J. L., Nonvertical Triplet Excitation Transfer to *cis*- and *trans*-Stilbene, *J. Am. Chem. Soc.* 1984, **106**, 3144–3151.

115. Saltiel, J., Mace, J. E., Watkins, L. P., Gormin, D. A., Clark, R. J., Dmitrenko, O., Biindanylidenes: Role of Central Bond Torsion in Nonvertical Triplet Excitation Transfer to the Stilbenes, *J. Am. Chem. Soc.* 2003, **125**, 16158–16159.

116. Frutos, L. M., Castano, O., Andres, J. L., Merchan, M., Acuna, A. U., A Theory of Nonvertical Triplet Energy Transfer in Terms of Accurate Potential Energy Surfaces: the Transfer Reaction from $\pi,\pi^*$ Triplet Donors to 1,3,5,7-Cyclooctatetraene, *J. Chem. Phys.* 2004, **120**, 1208–1216.

117. Wirz, J., Krebs, A., Schmalstieg, H., Angliker, H., Electron Structur. The Electronic Triplet State of a Peralkylated Cyclobutadiene, *Angew. Chem. Int. Ed. Engl.* 1981, **20**, 192–193.

118. Picot, A., D'Aléo, A., Patrice, L., Baldeck, P. L., Grichine, A., Duperray, A., Andraud, C., Maury, O., Long-lived Two-photon Excited Luminescence of Water-soluble Europium Complex: Applications in Biological Imaging Using Two-photon Scanning Microscopy, *J. Am. Chem. Soc.* 2008, **130**, 1532–1533.

119. Selvin, P. R., Principle, Biophysical Applications of Lanthanide-based Probes, *Annu. Rev. Biophys. Biomol. Struct.* 2002, **31**, 275–302.

120. Förster, T., Kasper, K., Concentration Reversal of the Fluorescence of Pyrene, *Z. Elektrochem.* 1956, **59**, 976–980.

121. Andriessen, R., Boens, N., Ameloot, M., De Schryver, F. C., Non-*a-priori* Analysis of Fluorescence Decay Surfaces of Excited-state Processes. 2. Intermolecular Excimer Formation of Pyrene *J. Phys. Chem.* 1991, **95**, 2047–2058.

122. Zachariasse, K. A., Kuehnle, W., Leinhos, U., Reynders, P., Striker, G., Time-resolved Monomer and Excimer Fluorescence of 1,3-Di(1-pyrenyl)propane at Different Temperatures: No Evidence for Distributions from Picosecond Laser Experiments with Nanosecond Time Resolution, *J. Phys. Chem.* 1991, **95**, 5476–5488.

123. Zachariasse, K. A., Macanita, A. L., Kuehnle, W., Chain Length Dependence of Intramolecular Excimer Formation with 1,*n*-Bis(1-pyrenylcarboxy)alkanes for $n = 1$–16, 22, and 32, *J. Chem. Phys. B* 1999, **103**, 9356–9365.

124. Winnik, F. M., Photophysics of Preassociated Pyrenes in Aqueous Polymer Solutions and in Other Organized Media, *Chem. Rev.* 1993, **93**, 587–614.

125. Nickel, B., Rodriguez Prieto, M. F., On the Alleged Triplet-Excimer Phosphorescence from Liquid Solutions of Naphthalene and Di-α-naphthylalkanes in Isooctane, *Chem. Phys. Lett.* 1988, **146**, 125–132.

126. Kang, H. K., Kang, D. E., Boo, B. H., Yoo, S. J., Lee, J. K., Lim, E. C., Existence of Intramolecular Triplet Excimer of Bis(9-fluorenyl)methane, *J. Phys. Chem. A* 2005, **109**, 6799–6804.

127. Heinz, B., Schmidt, B., Root, C., Satzger, H., Milota, F., Fierz, B., Kiefhaber, T., Zinth, W., Gilch, P., On the Unusual Fluorescence Properties of Xanthone in Water, *Phys. Chem. Chem. Phys.* 2006, **8**, 3432–3439.

128. Parker, C. A., Hatchard, C. G., Delayed Fluorescence of Pyrene in Ethanol, *Trans. Faraday Soc.* 1963, **59**, 284–295.

129. Adam, W., The Singlet Oxygen Story, *Chem. Unserer Zeit* 1981, **15**, 190–196.

130. Greer, A., Christopher Foote's Discovery of the Role of Singlet Oxygen [$^1O_2$ ($^1\Delta_g$)] in Photosensitized Oxidation Reactions, *Acc. Chem. Res.* 2006, **39**, 797–804.

131. Kautsky, H., de Bruijn, H., Neuwirth, R., Baumeister, W., Energy Transfers at Surfaces. VII. Photosensitized Oxidation as the Action of an Active, Metastable State of the Oxygen Molecule, *Ber. Bunsen-Ges. Phys. Chem.* 1933, **66B**, 1588–1600.

132. Khan, A., Kasha, M., Red Chemiluminescence of Oxygen in Aqueous Solution, *J. Chem. Phys.* 1963, **39**, 2105–2106.

133. Foote, C. S., Wexler, S., Singlet Oxygen. A Probable Intermediate in Photosensitized Autoxidations, *J. Am. Chem. Soc.* 1964, **86**, 3880–3881.

134. Herzberg, G., *Molecular Spectra and Molecular Structure*, Vols I–III Van Nostrand, Toronto, 1966.

135. Kearns, D. R., Physical and Chemical Properties of Singlet Molecular Oxygen, *Chem. Rev.* 1971, **71**, 395–427.

136. Wasserman, H. H., Murray, R. W., *Singlet Oxygen*, Academic Press, New York, 1979.

137. Murov, S. L., Carmichael, I., Hug, G. L., *Handbook of Photochemistry*, 2nd edn, Marcel Dekker, New York, 1993.

138. Ayman, A., Abdel-Shafi, A. A., Ward, M. D., Schmidt, R., Mechanism of Quenching by Oxygen of the Excited States of Ruthenium(II) Complexes in Aqueous Media. Solvent Isotope Effect and Photosensitized Generation of Singlet Oxygen, $O_2(^1\Delta_g)$, by [Ru(diimine)(CN)$_4$]$^{2-}$ Complex Ions, *Dalton Trans.* 2007, 2517–2527.

139. Abdel-Shafi, A. A., Wilkinson, F., Charge Transfer Effects on the Efficiency of Singlet Oxygen Production Following Oxygen Quenching of Excited Singlet and Triplet States of Aromatic Hydrocarbons in Acetonitrile, *J. Phys. Chem. A* 2000, **104**, 5747–5757.

140. Hatz, S., Lambert, J. D. C., Ogilby, P. R., Measuring the Lifetime of Singlet Oxygen in a Single Cell: Addressing the Issue of Cell Viability, *Photochem. Photobiol. Sci.* 2007, **6**, 1106–1116.

141. Weldon, D., Ogilby, P. R., Time-resolved Absorption Spectrum of Singlet Oxygen in Solution, *J. Am. Chem. Soc.* 1998, **120**, 12978–12979.

142. Schmidt, R., Quantitative Determination of $^1\Sigma_{g+}$ and $1\Delta_g$ Singlet Oxygen in Solvents of Very Different Polarity. General Energy Gap Law for Rate Constants of Electronic Energy Transfer to and from $O_2$ in the Absence of Charge Transfer Interactions, *J. Phys. Chem. A* 2006, **110**, 2622–2628.

143. Ogilby, P. R., Foote, C. S., The Effect of Solvent, Solvent Isotopic Substitution, and Temperature on the Lifetime of Singlet Molecular Oxygen, *J. Am. Chem. Soc.* 1983, **105**, 3423–3430.

144. Rodgers, M. A. J., Snowden, P. T., Lifetime of $O_2(^1\Delta_g)$ in Liquid Water as Determined by Time-resolved Infrared Luminescence Measurements, *J. Am. Chem. Soc.* 1982, **104**, 5541–5543.

145. Schmidt, R., Afshari, E., Collisional Deactivation of $O_2(^1\Delta_g)$ by Solvent Molecules, *Ber. Bunsen-Ges. Phys. Chem.* 1992, **96**, 788–794.

146. Solomon, M., Sivaguru, J., Jockusch, S., Adam, W., Turro, N. J., Vibrational Deactivation of Singlet Oxygen: Does it Play a Role in Stereoselectivity During Photooxygenation?, *Photochem. Photobiol. Sci.* 2008, **7**, 531–533.

147. Gouterman, M., Oxygen Quenching of Luminescence of Pressure Sensitive Paint for Wind Tunnel Research, *J. Chem. Educ.* 1997, **74**, 697–702.

148. Khalil, G. E., Chang, A., Gouterman, M., Callis, J. B., Dalton, L. R., Turro, N. J., Jockusch, S., Oxygen Pressure Measurement Using Singlet Oxygen Emission, *Rev. Sci. Instrum.* 2005, **76**, art. no. 054101.

149. Born, R., Fischer, W., Heger, D., Tokarczyk, B., Wirz, J., Photochromism of Phenoxy-naphthacenequinones: Diabatic or Adiabatic Phenyl Group Transfer?, *Photochem. Photobiol. Sci.* 2007, **6**, 552–559.

150. Longuet-Higgins, H. C., Intersection of Potential Energy Surfaces in Polyatomic Molecules, *Proc. R. Soc. London, Ser. A* 1975, **344**, 147–156.

151. Teller, E., Crossing of Potential Surfaces, *J. Phys. Chem.* 1937, **41**, 109–116.

152. Wigner, E., Witmer, E. E., Über die Struktur der Zweiatomigen Molekelspektren nach der Quantenmechanik, *Z. Phys.* 1928, **51**, 859–886.

153. Jasper, A. W., Nangia, S., Zhu, C., Truhlar, D. G., Non-Born–Oppenheimer Molecular Dynamics, *Acc. Chem. Res.* 2006, **39**, 101–108.

154. Wittig, C., The Landau–Zener Formula, *J. Phys. Chem. B* 2005, **109**, 8428–8430.

155. Braun, A. M., Maurette, A.-T., Oliveros, E., *Photochemical Technology*, John Wiley & Sons, Ltd, Chichester, 1991, pp. 202–396.

156. Calvert, J. G., Pitts, J. N., Experimental Methods in Photochemistry. In: *Photochemistry*, John Wiley & Sons, Inc, New York, 1966, Chapter 7, pp. 686–798.

157. Montalti, M., Credi, A., Prodi, L., Gandolfi, M. T., *Handbook of Photochemistry*, 3rd edn, CRC Press, Boca Raton, FL, 2006.

158. Rabek, J. F., *Experimental Methods in Photochemistry and Photophysics*, John Wiley and Sons, Inc., New York, 1982.

159. Scaiano, J. C., *CRC Handbook of Organic Photochemistry* (2 Volumes), CRC Press, Boca Raton, FL, 1987.

160. Maiman, T. H., Stimulated Optical Radiation in Ruby, *Nature* 1960, **187**, 493–494.

161. Milloni, P. W., Eberly, J. H., *Lasers*, John Wiley & Sons, Inc, New York, 1988.

162. Klimov, V.I., Ivanov, S. A., Nanda, J., Achermann, M., Bezel, I., McGuire, J. A., Piryatinski, A., Single-exciton Optical Gain in Semiconductor Nanocrystals, *Nature* 2007, **447**, 441–446.

163. Nanda, J., Ivanov, S. A., Achermann, M., Bezel, I., Piryatinski, A., Klimov, V. I., Light Amplification in the Single-exciton Regime Using Exciton–Exciton Repulsion in Type-II Nanocrystal Quantum Dots, *J. Phys. Chem.* 2007, **111**, 15382–15390.

164. Hänsch, T. W., Passion for Precision, *Ann. Phys. (Leipzig)* 2006, **15**, 627–652.

165. Yersin, H., *Highly Efficient OLEDs with Phosphorescent Materials*, Solid State Chemistry, Wiley-VCH, Verlag GmbH, Weinheim, 2007.

166. Jähnisch, K., Hessel, V., Löwe, H., Baerns, M., Chemistry in Microstructured Reactors, *Angew. Chem. Int. Ed.* 2004, **43**, 406–446.

167. Klán, P., Cirkva, V., Microwaves in Photochemistry. In Loupy, A. (ed), *Microwaves in Organic Synthesis*, 2nd edn, Wiley-VCH Verlag GmbH, Weinheim, 2006, pp. 860–896.

168. Klán, P. Hájek, M., Cirkva, V., The Electrodeless Discharge Lamp: a Prospective Tool for Photochemistry. Part 3. The Microwave Photochemistry Reactor, *J. Photochem. Photobiol. A* 2001, **140**, 185–189.

169. Klán, P. Literák, J., Hájek, M., The Electrodeless Discharge Lamp: a Prospective Tool for Photochemistry, *J. Photochem. Photobiol. A* 1999, **128**, 145–149.

170. Fox, M. A., Dulay, M. T., Heterogeneous Photocatalysis, *Chem. Rev.* 1993, **93**, 341–357.

171. Tung, C. H., Song, K., Wu, L.-Z., Li, H.-R., Zhang, L.-P., Microreactor-controlled Product Selectivity in Organic Photochemical Reactions. In Ramamurthy, V., Schanze, K. (eds), *Understanding and Manipulating Excited-state Processes*, Marcel Dekker, New York, 2001, pp. 317–383.

172. Ramamurthy, V., Organic Photochemistry in Organized Media, *Tetrahedron* 1986, **42**, 5753–5839.

173. Weiss, R. G., Ramamurthy, V., Hammond, G. S., Photochemistry in Organized and Confining Media – A Model, *Acc. Chem. Res.* 1993, **26**, 530–536.

174. Decker, C., Photoinitiated Crosslinking Polymerisation, *Prog. Polym. Sci.* 1996, **21**, 593–650.

175. Valeur, B., *Molecular Fluorescence, Principles and Applications*, Wiley-VCH Verlag GmbH, Weinheim, 2002.

176. Eaton, D. F., Reference Materials for Fluorescence Measurements, *Pure Appl. Chem.* 1988, **60**, 1107–1114.

177. Brower, F., San Roman, E., *Reference Methods, Standards and Applications of Photoluminescence*, IUPAC. 2008, in preparation.

178. Evans, D. F., Magnetic Perturbation of Singlet–Triplet Transitions. Part IV. Unsaturated Compounds, *J. Chem. Soc.* 1960, 1735–1745.

179. Demas, J. N., Crosby, G. A., The Measurement of Photoluminescence Quantum Yields. A Review, *J. Phys. Chem.* 1971, **75**, 991–1024.

180. Eaton, D. F., Recommended Methods for Fluorescence Decay Analysis, *Pure Appl. Chem.* 1990, **62**, 1631–1648.

181. Ernsting, N. P., Fluorescence Upconversion, *Pure Appl. Chem.* in preparation.

182. Michl, J., Thulstrup, E. W., *Spectroscopy with Polarized Light*, VCH, New York, 1986.

183. Thulstrup, E. W., Michl, J., Eggers, J. H., Polarization Spectra in Stretched Polymer Sheets. Physical Significance of the Orientation Factors and Determination of $\pi,\pi^*$ Transition Moment Directions in Molecules of Low Symmetry, *J. Phys. Chem.* 1970, **74**, 3878–3884.

184. Wilkinson, F., Editorial: Special Issue in Commemoration of Lord George Porter FRSC FRS OM, *Photochem. Photobiol. Sci.* 2003, **2**, ix–x.

185. Carmichael, I., Hug, G. L., Spectroscopy, Intramolecular Photophysics of Triplet States. In Scaiano, J. C. (ed), *CRC Handbook of Organic Photochemistry*, Vol. I, CRC Press, Boca Raton, FL, 1987, pp. 369–403.

186. Lobastov, V. A., Weissenrieder, J., Tang, J., Zewail, A. H., Ultrafast Electron Microscopy (UEM): Four-dimensional Imaging and Diffraction of Nanostructures During Phase Transitions, *Nano Lett.* 2007, **7**, 2552–2558.

187. Diau, E. W. G., Kotting, C., Solling, T. I., Zewail, A. H., Femtochemistry of Norrish Type-I Reactions: III. Highly Excited Ketones – Theoretical, *ChemPhysChem* 2002, **3**, 57–78.

188. Scaiano, J. C., Nanosecond Laser Flash Photolysis: a Tool for Physical Organic Chemistry. In Moss, R. A., Platz, M. S., Jones, M., Jr. (eds), *Reactive Intermediate Chemistry*, John Wiley, & Sons, Inc., Hoboken, NJ, 2004, pp. 847–871.

189. Wilkinson, F., Kelly, G., Diffuse Reflectance Flash Photolysis. In Scaiano, J. C. (ed), *Handbook of Organic Photochemistry*, Vol. I, CRC Press, Boca Raton, FL, 1989, pp. 293–314.

190. Carmichael, I., Hug, G. L., A Unified Analysis of Noncomparative Methods for Measuring the Molar Absorptivity of Triplet–Triplet Transitions, *Appl. Spectrosc.* 1987, **41**, 1033.

191. Bonneau, R., Wirz, J., Zuberbühler, A. D., Methods for the Analysis of Transient Absorbance Data (Technical Report), *Pure Appl. Chem.* 1997, **69**, 979–992.

192. Baum, P., Zewail, A. H., Breaking Resolution Limits in Ultrafast Electron Diffraction and Microscopy, *Proc. Natl. Acad. Sci. USA* 2006, **103**, 16105–16110.

193. Yang, D.-S., Gedik, N., Zewail, A. H., Ultrafast Electron Crystallography. 1. Nonequilibrium Dynamics of Nanometer-scale Structures, *J. Phys. Chem. C* 2007, **111**, 4889–4919.

194. Kovalenko, S. A., Dobryakov, A. L., Ruthmann, J., Ernsting, N. P., Femtosecond Spectroscopy of Condensed Phases with Chirped Supercontinuum Probing, *Phys. Rev. A* 1999, **59**, 2369–2384.

195. Hilinski, E. F., The Picosecond Realm. In Moss, R. A., Platz, M. S., Jones, M. Jr, (eds), *Reactive Intermediate Chemistry*, John Wiley & Sons, Inc, Hoboken, NJ, 2004, pp. 873–897.

196. Bazin, M., Ebbesen, T. W., Distortions in Laser Flash Photolysis Absorption Measurements. The Overlap Problem., *Photochem. Photobiol.* 1983, **37**, 675–678.

197. Fron, E., Pilot, R., Schweitzer, G., Qu, J., Herrmann, A., Muellen, K., Hofkens, J., Van der Auweraer, M., De Schryver, F. C., Photoinduced Electron Transfer in Perylenediimide Triphenylamine-based Dendrimers: Single Photon Timing and Femtosecond Transient Absorption Spectroscopy *Photochem. Photobiol. Sci.* 2008, **7**, 597–604.

198. Mauser, H., Gauglitz, G., *Photokinetics, Chemical Kinetics*, Vol. 36, Elsevier, Amsterdam, 1998.

199. Andraos, J., Lathioor, E. C., Leigh, W. J., Simultaneous pH–Rate Profiles Applied to the Two-step Consecutive Sequence A → B → C: a Theoretical Analysis and Experimental Verification, *J. Chem. Soc., Perkin Trans. 2* 2000, 365–373.

200. Grellmann, K. H., Scholz, H.-G., Determination of Decay Constants with a Sampling Flash Apparatus. The Triplet State Lifetimes of Anthracene and Pyrene in Fluid Solutions *Chem. Phys. Lett.* 1979, **62**, 64–71.

201. Malinowski, E. R., *Factor Analysis in Chemistry*, 3rd edn, John Wiley & Sons, Inc, New York, 2002.

202. Maeder, M., Neuhold, Y.-M., *Practical Data Analysis in Chemistry*, Data Handling in Science and Technology, Vol. 26, Elsevier, New York, 2007.

203. Maeder, M., Evolving Factor Analysis for the Resolution of Overlapping Chromatographic Peaks, *Anal. Chem.* 1987, **59**, 527–530.

204. von Frese, J., Kovalenko, S. A., Ernsting, N. P., Interactive Curve Resolution by Using Latent Projections in Polar Coordinates, *J. Chemom.* 2007, **21**, 2–9.

205. Maeder, M., Neuhold, Y.-M., Kinetic Modeling of Multivariate Measurements with Nonlinear Regression. In Gemperline, P. (ed), *Practical Guide to Chemometrics*, 2nd edn, CRC Press, Boca Raton, FL, 2006, pp. 217–261.

206. Maeder, M., Zuberbühler, A. D., Nonlinear Least-squares Fitting of Multivariate Absorption Data, *Anal. Chem.* 1990, **62**, 2220–2224.

207. Ma, C., Kwok, W. M., Chan, W. S., Du, Y., Kan, J. T. W., Phillips, D. L., Ultrafast Time-resolved Transient Absorption and Resonance Raman Spectroscopy Study of the Photodeprotection and Rearrangement Reactions of *p*-Hydroxyphenacyl Caged Phosphates, *J. Am. Chem. Soc.* 2006, **128**, 2558–2570.

208. Kukura, P., Yoon, S., Mathies, R. A., Femtosecond Stimulated Raman Spectroscopy, *Anal. Chem.* 2006, **78**, 5952–5959.

209. Shim, S., Mathies, R. A., Development of a Tunable Femtosecond Stimulated Raman Apparatus and Its Application to β-Carotene, *J. Phys. Chem. B* 2008, **112**, 4826–4832.

210. Kukura, P., McCamant, D. W., Yoon, S., Wandschneider, D. B., Mathies, R. A., Structural Observation of the Primary Isomerization in Vision With Femtosecond-Stimulated Raman, *Science* 2005, **310**, 1006–1009.

211. Rödig, C., Siebert, F., Error and Artifacts in Time-resolved Step-scan FT-IR Spectroscopy, *Appl. Spectrosc.* 1999, **53**, 893–901.

212. Wang, Y., Yuzawa, T., Hamaguchi, H., Toscano, J. P., Time-resolved IR Sudies of 2-Naphthyl (carbomethoxy)carbene: Reactivity and Direct Experimental Estimate of the Singlet/Triplet Energy Gap, *J. Am. Chem. Soc.* 1999, **212**, 2875–2882.

213. Barth, A., Time-resolved IR Spectroscopy with Caged Compounds: an Introduction. In Goeldner, M., Givens, R. S. (eds), *Dynamic Studies in Biology*, Wiley-VCH Verlag GmbH, Weinheim, 2005, pp. 369–399.

214. Kuhn, H. J., Braslavsky, S. E., Schmidt, R., Chemical Actinometry, *Pure Appl. Chem.* 2004, **76**, 2105–2146.

215. Hatchard, C. G., Parker, C. A., A New Sensitive Chemical Actinometer. II. Potassium Ferrioxalate as a Standard Chemical Actinometer, *Proc. R. Soc. London, Ser. A* 1956, **235**, 518–536.

216. Goldstein, S., Rabani, J., The Ferrioxalate and Iodide–Iodate Actinometers in the UV Region, *J. Photochem. Photobiol. A* 2008, **193**, 50–55.

217. Uhlmann, E., Gauglitz, G., New Aspects in the Photokinetics of Aberchrome 540, *J. Photochem. Photobiol. A* 1996, **98**, 45–49.

218. Schmidt, R., Brauer, H. D., Self-sensitized Photooxidation of Aromatic Compounds and Photocycloreversion of Endoperoxides – Applications in Chemical Actinometry, *J. Photochem.* 1984, **25**, 489–499.

219. Bowman, W. D., Demas, J. N., Ferrioxalate Actinometry. A Warning on Its Correct Use, *J. Phys. Chem.* 1976, **80**, 2434–2435.

220. Mauser, H., Zur Spektroskopischen Untersuchung der Kinetik Chemischer Reaktionen, II. Extinktionsdifferenzendiagramme, *Z. Naturforsch., Teil B* 1968, **23**. 1025,

221. Mauser, H., *Formale Kinetik*, Experimentelle Methoden der Physik und Chemie, Vol. I, Bertelsmann Universitätsverlag, Düsseldorf, 1974.

222. Gauglitz, G., Hubig, S., Photokinetische Grundlagen Moderner Chemischer Aktinometer, *Z. Phys. Chem., N. F.* 1984, **139**, 237–246.

223. Gauglitz, G., Hubig, S., Chemical Actinometry in the UV by Azobenzene Actinometry in Concentrated Solution, *J. Photochem.* 1985, **30**, 121–125.

224. Serpone, N., Salinaro, A., Terminology, Relative Photonic Efficiencies and Quantum Yields in Heterogeneous Photocatalysis. Part I, *Pure Appl. Chem.* 1999, **71**, 303–320.

225. Salinaro, A., Emeline, A. V., Zhao, J., Hidaka, H., Ryabchuk, V. A., Serpone, N., Terminology, Relative Photonic Efficiencies and Quantum Yields in Heterogeneous Photocatalysis. Part II: Experimental Determination of Quantum Yields, *Pure Appl. Chem.* 1999, **71**, 321–335.

226. Dulin, D., Mill, T., Development and Evaluation of Sunlight Actinometers, *Environ. Sci. Technol.* 1982, **16**, 815–820.

227. Oakes, J., Photofading of Textile Dyes, *Rev. Prog. Color. Relat. Top.* 2001, **31**, 21–28.

228. Pugh, S. L., Guthrie, J. T., The Development of Light Fastness Testing and Light Fastness Standards, *Rev. Prog. Color. Relat. Top.* 2001, **31**, 42–56.

229. Linschitz, H., Pekkarinen, L., Studies on Metastable States of Porphyrins. 2., *J. Am. Chem. Soc.* 1960, **82**, 2407–2411.

230. Wagner, P. J., Steady-state Kinetics. In Scaiano, J. C. (ed), *Handbook of Organic Photochemistry*, Vol. II, CRC Press, Boca Raton, FL, 1989, pp. 251–269.

231. Keizer, J., Diffusion Effects on Rapid Bimolecular Chemical Reactions, *Chem. Rev.* 1987, **87**, 167–180.

232. Mac, M., Wirz, J., Deriving Intrinsic Electron-transfer Rates from Nonlinear Stern–Volmer Dependencies for Fluorescence Quenching of Aromatic Molecules by Inorganic Anions in Acetonitrile, *Chem. Phys. Lett.* 1993, **211**, 20–26.

233. Allonas, X., Ley, C., Bibaut, C., Jacques, P., Fouassier, J. P., Investigation of the Triplet Quantum Yield of Thioxanthone by Time-resolved Thermal Lens Spectroscopy: Solvent and Population Lens Effects, *Chem. Phys. Lett.* 2000, **322**, 483–490.

234. Bonneau, R., Carmichael, I., Hug, G. L., Molar Absorption Coefficients of Transient Species in Solution, *Pure Appl. Chem.* 1991, **63**, 289–299.

235. Bensasson, R., Goldschmidt, C. R., Land, E. J., Truscott, T. G., Triplet Excited State of Furocoumarins: Reaction with Nucleic Acid Bases and Amino Acids, *Photochem. Photobiol.* 1978, **28**, 277–281.

236. Adam, W., Fragale, G., Klapstein, D., Nau, W. M., Wirz, J., Phosphorescenc and Transient Absorption of Azoalkane Triplet States, *J. Am. Chem. Soc.* 1995, **117**, 12578–12592.

237. Murasecco-Suardi, P., Gassmann, E., Braun, A. M., Oliveros, E., Determination of the Quantum Yield of Intersystem Crossing of Rose Bengal, *Helv. Chim. Acta* 1987, **70**, 1760–1773.

238. Horrocks, A. R., Medinger, T., Wilkinson, F., New Accurate Method for Determining the Quantum Yields of Triplet State Production of Aromatic Molecules in Solution, *Chem. Commun.* 1965, 452.

239. Bachilo, S. M., Weisman, R. B., Determination of Triplet Quantum Yields from Triplet–Triplet Annihilation Fluorescence, *J. Phys. Chem. A* 2000, **104**, 7711–7714.

240. Bally, T., Matrix Isolation. In Moss, R. A., Platz, M. S., Jones, M. (eds), *Reactive Intermediate Chemistry*, John Wiley & Sons, Inc., Hoboken, NJ, 2004, pp. 797–845.

241. Maier, G., Reisenauer, H. P., Preiss, T., Pacl, H., Juergen, D., Tross, R., Senger, S., Highly Reactive Molecules: Examples for the Interplay Between Theory and Experiment, *Pure Appl. Chem.* 1997, **69**, 113–118.

242. Bondybey, V. E., Smith, A. M., Agreiter, J., New Developments in Matrix Isolation Spectroscopy, *Chem. Rev.* 1996, **96**, 2113–2134.

243. Dunkin, I. R., Matrix Photochemistry. In Horspool, W. M., Lenci, F. (eds), *CRC Handbook of Organic Photochemistry and Photobiology*, 2nd edn, CRC Press, Boca Raton, FL, 2004, Chapter 14, pp. 1–27.

244. Jacox, M. E., Vibrational and Electronic Spectra of Neutral and Ionic Combustion Reaction Intermediates Trapped in Rare-gas Matrixes, *Acc. Chem. Res.* 2004, **37**, 727–734.

245. Whittle, E., Dows, D. A., Pimentel, G. C., Matrix Isolation Method for the Experimental Study of Unstable Species, *J. Chem. Phys.* 1954, **22**, 1943–1943.

246. Jonkman, H. T., Michl, J., Secondary Ion Mass Spectrometry: a Tool for Identification of Matrix-isolated Species, *J. Chem. Soc., Chem. Commun.* 1978, 751–752.

247. Maier, J. P., Electronic Spectroscopy of Carbon Chains, *Chem. Soc. Rev.* 1997, **26**, 21–28.

248. Harshbarger, W. R., Robin, M. B., The Opto-acoustic Effect: Revival of an Old Technique for Molecular Spectroscopy, *Acc. Chem. Res.* 1973, **6**, 329–334.

249. Braslavsky, S., Heibel, G. E., Time-resolved Photothermal and Photoacoustic Methods Applied to Photoinduced Processes in Solution, *Chem. Rev.* 1992, **92**, 1381–1410.

250. Gensch, T., Viappiani, C., Time-resolved Photothermal Methods: Accessing Time-resolved Thermodynamics of Photoinduced Processes in Chemistry and Biology, *Photochem. Photobiol. Sci.* 2003, **2**, 699–721.

251. Ni, T., Caldwell, R. A., Melton, L. A., The Relaxed and Spectroscopic Energies of Olefin Triplets, *J. Am. Chem. Soc.* 1989, **111**, 457–464.

252. Arnaut, L. G., Caldwell, R. A., Elbert, J. E., Melton, L. A., Recent Advances in Photoacoustic Calorimetry: theoretical Basis and Improvements in Experimental Design, *Rev. Sci. Instrum.* 1992, **63**, 5381–5389.

253. Andres, G. O., Cabrerizo, F. M., Martinez-Junza, V., Braslavsky, S. E., A Large Entropic Term Due to Water Rearrangement is Concomitant with the Photoproduction of Anionic Free-base Porphyrin Triplet States in Aqueous Solutions, *Photochem. Photobiol.* 2007, **83**, 503–510.

254. Andres, G. O., Martinez-Junza, V., Crovetto, L., Braslavsky, S. E., Photoinduced Electron Transfer from Tetrasulfonated Porphyrin to Benzoquinone Revisited. The Structural Volume-normalized Entropy Change Correlates with Marcus Reorganization Energy, *J. Phys. Chem. A* 2006, **110**, 10185–10190.

255. Hou, H. J. M., Mauzerall, D., The A($-$)F(x) to F–A/B Step in Synechocystis 6803 Photosystem I is Entropy Driven, *J. Am. Chem. Soc.* 2006, **128**, 1580–1586.

256. Herbrich, R. P., Schmidt, R., Investigation of the Pyrene/*N*,*N'*-Diethylaniline Exciplex by Photoacoustic Calorimetry and Fluorescence Spectroscopy, *J. Photochem. Photobiol. A* 2000, **133**, 149–158.

257. Helmchen, F., Denk, W., Deep Tissue Two-photon Microscopy, *Nature Methods* 2005, **2**, 932–940.

258. Barbara, P. F., Single-molecule Spectroscopy, *Acc. Chem. Res.* 2005, **38**, 503.

259. Silbey, R. J., Single-molecule Chemistry and Biology. Special Feature, *Proc. Natl. Acad. Sci.* 2007, **104**, 12596–12602.

260. Cornish, P. V., Ha, T., A Survey of Single-molecule Techniques in Chemical Biology, *ACS Chem. Biol.* 2007, **2**, 53–61.

261. Ambrose, W. P., Goodwin, P. M., Jett, J. H., Van Orden, A., Werner, J. H., Keller, J. H., Single Molecule Fluorescence Spectroscopy at Ambient Temperature, *Chem. Rev.* 1999, **99**, 2929–2956.

262. Widengren, J., Kudryavtsev, V., Antonik, M., Berger, S., Gerken, M., Seidel, C. A. M., Single-molecule Detection, Identification of Multiple Species by Multiparameter Fluorescence Detection, *Anal. Chem.* 2006, **78**, 2039–2050.

263. Al-Soufi, W., Reija, B., Novo, M., Felekyan, S., Kuehnemuth, R., Seidel, C. A. M., Fluorescence Correlation Spectroscopy, a Tool to Investigate Supramolecular Dynamics: Inclusion Complexes of Pyronines with Cyclodextrin, *J. Am. Chem. Soc.* 2005, **127**, 8775–8784.

264. Lampe, M., Briggs, J. A. G., Endress, T., Glass, B., Riegelsberger, S., Kraeusslich, H.-G., Lamb, D. C., Braeuchle, C., Mueller, B., Double-labelled HIV-1 Particles for Study of Virus–Cell Interaction, *Virology* 2007, **360**, 92–104.

265. Betzig, E., Patterson, G. H., Sougrat, R., Lindwasser, O. W., Olenych, S., Bonifacio, J. S., Davidson, M. W., Lippincott-Schwartz, J., Hess, H. F., Imaging Intracellular Fluorescent Proteins at Nanometer Resolution, *Science* 2006, **313**, 1642–1645.

266. Koopmans, W. J. A., Brehm, A., Logie, C., Schmidt, T., van Noort, J., Single-pair FRET Microscopy Reveals Mononucleosome Dynamics, *J. Fluoresc.*, 2007, **17**, 785–795.

267. Zimmer, M., Green Fluorescent Protein (GFP): Applications, Structure, and Related Photophysical Behavior, *Chem. Rev.* 2002, **102**, 759–781.

268. Tsien, R. Y., The Green Fluorescent Protein, *Annu. Rev. Biochem.* 1998, **67**, 509–544.

269. Orte, A., Craggs, T. D., White, S. S., Jackson, S. E., Klenerman, D., Evidence of an Intermediate and Parallel Pathways in Protein Unfolding from Single-molecule Fluorescence, *J. Am. Chem. Soc.* 2008, **130**, 7898–7907.

270. Inoue, H., Ezaki, A., Hide, M., Mechanism of the Photocycloaddition of 1-Aminoanthraquinones to Olefins by Visible Light Irradiation, *J. Chem. Soc., Perkin Trans. 2* 1982, 833–839.

271. Heilbronner, E., Bock, H., *Hückel Molecular Orbital Model and Its Application*, John Wiley & Sons, Inc., New York, 1976.

272. Borden, W. T., *Modern Molecular Orbital Theory for Chemists*, Prentice-Hall, London, 1975.

273. Bishop, D. M., *Group Theory and Chemistry*, Dover, Mineola, NY, 1993.

274. Cotton, A. F., *Chemical Applications of Group Theory*, 3rd edn, Wiley-Interscience, New York, 1990.

275. Walton, P. H., *Beginning Group Theory for Chemistry*, Oxford University Press, Oxford, 1998.

276. Sondheimer, F., Ben-Efraim, D. A., Wolosovsky, R., Unsaturated Macrocyclic Compounds. XVII. The Prototropic Rearrangement of Linear 1,5-Enynes to Conjugated Polyenes. The Synthesis of a Series of Vinylogs of Butadiene, *J. Am. Chem. Soc.* 1961, **83**, 1675–1681.

277. Pino, T., Ding, H., Güthe, F., Maier, J. P., Electronic Spectra of the Chains $HC_{2n}H$ ($n = 8$–13) in the Gas Phase, *J. Chem. Phys.* 2001, **114**, 2208–2212.

278. Malhotra, S. S., Whiting, M. C., Researches on Polyenes. Part VIL. The Preparation and Electronic Absorption Spectra of Homologous Series of Simple Cyanines, Merocyanines, and Oxonols, *J. Chem. Soc.* 1960, 3812–3821.

279. Clar, E., *Aromatische Kohlenwasserstoffe, Polycyclische Systeme.* 2nd edn, Organische Chemie in Einzeldarstellungen, Vol. 2, Springer, Berlin, 1952, p. 481

280. Perkampus, H.-H., *UV–VIS Atlas of Organic Compounds*, 2nd edn, Wiley-VCH Verlag GmbH, Weinheim, 1992.

281. Angliker, H., Rommel, E., Wirz, J., Electronic Spectra of Hexacene in Solution (Ground State, Triplet State, Dication and Dianion), *Chem. Phys. Lett.* 1982, **87**, 208–212.

282. Heilbronner, E., Murrell, J. N., The Prediction of the Spectra of Aromatic Hydrocarbons, *J. Chem. Soc.* 1962, 2611–2615.

283. Clar, E., Robertson, J. M., Schlögl, R., Schmidt, W., Photoelectron Spectra of Polynuclear Aromatics. 6. Applications to Structural Elucidation: "Circumanthracene", *J. Am. Chem. Soc.* 1981, **103**, 1320–1328.

284. Dewar, M. J. S., Dougherty, R. C., *The PMO Theory of Organic Chemistry*, Plenum Press, New York, 1975.

285. Dewar, M. J. S., *The Molecular Orbital Theory of Organic Chemistry*, McGraw-Hill, New York, 1969.

286. Coulson, C. A., Rushbrooke, G. S., Note on the Method of Molecular Orbitals, *Proc. Camb. Philos. Soc.* 1940, **36**, 193–200.

287. Longuet-Higgins, H. C., Studies in MO Theory. II: Ionisation Constants of Heteroatomic Amines, *J. Chem. Phys.* 1950, **18**, 275–282.

288. Dewar, M. J. S., A Molecular-orbital Theory of Organic Chemistry. I. General Principles, *J. Am. Chem. Soc.* 1952, **74**, 3341–3345.

289. Dougherty, R. C., Perturbation Molecular Orbital Treatment of Photochemical Reactivity. Nonconservation of Orbital Symmetry in Photochemical Pericyclic Reactions, *J. Am. Chem. Soc.* 1971, **93**, 7187–7201.

290. Heilbronner, E., Murrell, J., The Effect of Alkyl Groups on the Electronic Spectra of Benzenoid Hydrocarbons, *Theor. Chim. Acta* 1963, **1**, 235–244.

291. Heilbronner, E., Hoshi, T., Von Rosenberg, J., Hafner, K., Alkyl-induced, Natural Hyposochromic Shifts of the $2A \leftarrow 2X$ and $2B \leftarrow 2X$ Transitions of Azulene and Naphthalene Radical Cations, *Nouv. J. Chim.* 1977, **1**, 105–112.

292. Michl, J., Thulstrup, E. W., Why is Azulene Blue and Anthracene White? A Simple MO Picture, *Tetrahedron* 1976, **32**, 205–209.

293. Nickel, B., Klemp, D., The Lowest Triplet State of Azulene-$h_8$, and Azulene-$d_8$ in Liquid Solution. II. Phosphorescence and E-Type Delayed Fluorescence, *Chem. Phys.* 1993, **174**, 319–330.

294. Pariser, R., Parr, R. G., A Semi-empirical Theory of the Electronic Spectra and Electronic Structure of Complex Unsaturated Molecules. I, *J. Chem. Phys.* 1953, **21**, 466–471.

295. Pople, J. A., Electron Interaction in Unsaturated Hydrocarbons, *Trans. Faraday Soc.* 1953, **49**, 1375–1385.

296. Suzuki, H., *Electronic Absorption Spectra and Geometry of Organic Molecules*, Academic Press, New York, 1967.

297. Klevens, H. B., Platt, J. R., Spectral Resemblances of Cata-condensed Hydrocarbons, *J. Chem. Phys.* 1949, **17**, 470–481.

298. Platt, J. R., Classification of Spectra of Cata-condensed Hydrocarbons, *J. Chem. Phys.* 1959, **17**, 484–495.

299. Whipple, M. R., Vasak, M., Michl, J., Magnetic Circular Dichroism of Cyclic π-Electron Systems. 8. Derivatives of Naphthalene, *J. Am. Chem. Soc.* 1978, **100**, 6844–6852.

300. Blattmann, H.-R., Böll, W. A., Heilbronner, E., Hohlneicher, G., Vogel, E., Weber, J.-P., Die Elektronenzustände von Perimeter-π-Systemen: I. Die Elektronenspektren 1,6-überbückter [10]Annulene, *Helv. Chim. Acta* 1966, **49**, 2017–2038.

301. Baumann, H., Oth, J. F. M., The Low-temperature UV/VIS Absorption Spectrum of [14] Annulene, *Helv. Chim. Acta* 1995, **78**, 679–692.

302. Blattmann, H.-R., Heilbronner, E., Wagnière, G., Electronic States of Perimeter π Systems. IV. The Electronic Spectrum of [18]Annulene, *J. Am. Chem. Soc.* 1968, **90**, 4786–4789.

303. Boguslavskiy, A. E., Ding, H., Maier, J. P., Gas-phase Electronic Spectra of $C_{18}$ and $C_{22}$ Rings, *J. Chem. Phys.* 2005, **123**, 0343051–0343057.

304. Boguslavskiy, A. E., Maier, J. P., Gas-phase Electronic Spectrum of the $C_{14}$ Ring, *Phys. Chem. Chem. Phys.* 2007, **9**, 127–130.

305. Sassara, A., Zerza, G., Chergui, M., Negri, F., Orlandi, G., The Visible Emission and Absorption Spectrum of $C_{60}$, *J. Chem. Phys.* 1997, **107**, 8731–8741.

306. Leach, S., Vervloet, M., Deprès, A., Bréheret, E., Hare, J. P., Dennis, T. J., Kroto, H. W., Taylor, R., Walton, D. R. M., Electronic Spectra and Transitions of the Fullerene $C_{60}$, *Chem. Phys.* 1992, **160**, 451–466.

307. Wirz, J., Electronic Structure and Photophysical Properties of Planar Conjugated Hydrocarbons with a 4n-Membered Ring, *Jerusalem Symp. Quantum Chem. Biochem.* 1977, **10**, 283–294.

308. Höweler, U., Downing, J. W., Fleischhauer, J., Michl, J., MCD of Non-aromatic Cyclic π-Electron Systems. Part 1. The Perimeter Model for Antiaromatic 4N-electron [n]annulene biradicals [erratum to document], *J. Chem. Soc., Perkin Trans. 2* 1998, 2323.

309. Fleischhauer, J., Michl, J., MCD of Nonaromatic Cyclic π-Electron Systems. 4. Explicit Relations between Molecular Structure and Spectra *J. Phys. Chem. A* 2000, **104**, 7776–7784.

310. Shida, T., *Electronic Absorption Spectra of Radical Ions*, Elsevier, Amsterdam, 1988.

311. Hudson, B. S., Kohler, B. E., A Low-lying Weak Transition in the Polyene α,ω-Diphenyloctatetraene, *Chem. Phys. Lett.* 1972, **14**, 299–304.

312. Schulten, K., Karplus, M., On the Origin of a Low-lying Forbidden Transition in Polyenes and Related Molecules, *Chem. Phys. Lett.* 1972, **14**, 305–309.

313. Zechmeister, L., LeRosen, A. L., Schroeder, W. A., Polgar, A., Pauling, L., Spectral Characteristic, Configuration of Some Stereoisomeric Carotenoids Including Prolycopene and Pro-γ-carotene, *J. Am. Chem. Soc.* 1943, **65**, 1940–1951.

314. Zechmeister, L., Polgar, A., *Cis–trans* Isomerization and *cis*-Peak Effect in the α-Carotene Set and in Some Other Stereoisomeric Sets, *J. Am. Chem. Soc.* 1944, **66**, 137–144.

315. Lunde, K., Zechmeister, L., *Cis–trans* Isomeric 1,6-Diphenylhexatrienes, *J. Am. Chem. Soc.* 1954, **76**, 2308–2313.

316. Zhao, Y., Truhlar, D.G., Density Functionals with Broad Applicability in Chemistry, *Acc. Chem. Res.* 2008, **41**, 157–167.

317. Borden, W. T., The Partnership between Electronic Structure Calculations and Experiments in the Study of Reactive Intermediates. In Moss, R. A., Platz, M. S., Jones, M. Jr. (eds), *Reactive Intermediate Chemistry*, John Wiley & Sons, Inc, Hoboken, NJ, 2004, pp. 961–1004.

318. Salem, L., Rowland, C., The Electronic Properties of Diradicals, *Angew. Chem. Int. Ed. Engl.* 1972, **11**, 92–111.

319. Michl, J., Havlas, Z., Spin–Orbit Coupling in Biradicals: Structural Aspects, *Pure Appl. Chem.* 1997, **69**, 785–790.

320. Michl, J., Spin–Orbit Coupling in Biradicals. 1. The 2-Electrons-in-2-Orbitals Model Revisited, *J. Am. Chem. Soc.* 1996, **118**, 3568–3579.

321. Havlas, Z., Michl, J., Prediction of an Inverse Heavy-atom Effect in $H-C-CH_2Br$: Bromine Substituent as a π Acceptor, *J. Am. Chem. Soc.* 2002, **124**, 5606–5607.

322. Wan, P., Budac, D., Photodecarboxylation of Acids and Lactones. In: Horspool, W. M., Song, P.-S., (eds), *CRC Handbook of Organic Photochemistry and Photobiology*, CRC Press, Boca Raton, FL, 1995, pp. 384–392.

323. Budac, D., Wan, P., Photodecarboxylation – Mechanism and Synthetic Utility, *J. Photochem. Photobiol. A* 1992, **67**, 135–166.

324. Zimmerman, H. E., Alabugin, I. V., Energy Distribution and Redistribution and Chemical Reactivity. The Generalized Delta Overlap-density Method for Ground State and Electron

Transfer Reactions: a New Quantitative Counterpart of Electron Pushing, *J. Am. Chem. Soc.* 2001, **123**, 2265–2270.

325. Van Riel, H. C. H. A., Lodder, G., Havinga, E., Photochemical Methoxide Exchange in Some Nitromethoxybenzenes – the Role of the Nitro Group in $S_N 2Ar^*$ Reactions, *J. Am. Chem. Soc.* 1981, **103**, 7257–7262.

326. Havinga, E., de Jongh, R. O., Dorst, W., Photochemical Acceleration of the Hydrolysis of Nitrophenyl Phosphates and Nitrophenyl Sulfates, *Recl. Trav. Chim. Pays-Bas Belg.* 1956, **75**, 378–383.

327. Zimmerman, H. E., Sandel, V. R., Mechanistic Organic Photochemistry. II. Solvolytic Photochemical Reactions, *J. Am. Chem. Soc.* 1963, **85**, 915–922.

328. Seiler, P., Wirz, J., Structure and Photochemical Reactivity, Photohydrolysis of Trifluoromethyl-substituted Phenols and Naphthols, *Helv. Chim. Acta* 1972, **55**, 2693–2712.

329. Seiler, P., Wirz, J., Photohydrolysis of Eight Isomeric Trifluoromethylnaphthols, *Tetrahedron Lett.* 1971, 1683–1686.

330. Pincock, J. A., Photochemistry of Arylmethyl Esters in Nucleophilic Solvents: Radical Pair and Ion Pair Intermediates, *Acc. Chem. Res.* 1997, **30**, 43–49.

331. Dauben, W. G., Salem, L., Turro, N. J., Classification of Photochemical Reactions, *Acc. Chem. Res.* 1975, **8**, 41–54.

332. Ramseier, M., Senn, P., Wirz, J., Photohydration of Benzophenone in Aqueous Acid, *J. Phys. Chem. A* 2003, **107**, 3305–3315.

333. Hou, Y., Wan, P., Formal Intramolecular Photoredox Chemistry of Anthraquinones in Aqueous Solution: Photodeprotection for Alcohols, Aldehydes and Ketones, *Photochem. Photobiol. Sci.* 2008, **7**, 588–596.

334. Hou, Y., Wan, P., A Pentacene Intermediate Via Formal Intramolecular Photoredox of a 6,13-Pentacenequinone in Aqueous Solution, *Can. J. Chem.* 2007, **85**, 1023–1032.

335. Basaric, N., Mitchell, D., Wan, P., Substituent Effects in the Intramolecular Photoredox Reactions of Benzophenones in Aqueous Solution, *Can. J. Chem.* 2007, **85**, 561–571.

336. Woodward, R. B., Hoffmann, R., The Conservation of Orbital Symmetry, *Angew. Chem. Int. Ed. Engl.* 1969, **8**, 781–853.

337. Wirz, J., Persy, G., Rommel, E., Murata, I., Nakasuji, K., Photoisomerization Pathways of 8,16-Methano[2.2]metacyclophane-1,9-diene. A Model Case for Adiabatic Electrocyclic Ring Closure in the Excited Singlet State, *Helv. Chim. Acta* 1984, **67**, 305–317.

338. Shaik, S., Is My Chemical Universe Localized or Delocalized? Is There a Future for Chemical Concepts? *New J. Chem.* 2007, **31**, 2015–2028.

339. Steiner, U. E., Magnetic Field Effects in Chemical Kinetics and Related Phenomena, *Chem. Rev.* 1989, **89**, 51–147.

340. Gould, I. R., Ege, D., Moser, J. E., Farid, S., Efficiencies of Photoinduced Electron-transfer Reactions: Role of the Marcus Inverted Region in Return Electron Transfer within Geminate Radical-ion Pairs, *J. Am. Chem. Soc.* 1990, **112**, 4290–4301.

341. Roth, H. D., Return Electron Transfer in Radical Ion Pairs of Triplet Multiplicity, *Pure Appl. Chem.* 2005, **77**, 1075–1085.

342. Knibbe, H., Rehm, D., Weller, A., Thermodynamics of the Formation of Excited EDA (Electron Donor–Acceptor) Complexes, *Ber. Bunsen-Ges. Phys. Chem.* 1969, **73**, 839–845.

343. Rossi, M., Buser, U., Haselbach, E., Multiple Charge-transfer Transitions in Alkylbenzene–TCNE Complexes, *Helv. Chim. Acta*, 1976, **59**, 1039–1053.

344. Rehm, D., Weller, A., Kinetic and Mechanism of Electron Transfer in Fluorescence Quenching in Acetonitrile *Ber. Bunsen-Ges. Phys. Chem.* 1969, **73**, 834–839.

345. Knibbe, H., Rehm, D., Weller, A., Intermediate, Kinetics of Fluorescence Quenching by Electron Transfer, *Ber. Bunsen-Ges. Phys. Chem.* 1968, **72**, 257–263.

346. Marcus, R. A., Exchange Reaction, Electron Transfer Reactions Including Isotopic Exchange, *Discuss. Faraday Soc.* 1960, **29**, 21–31.

347. Rehm, D., Weller, A., Kinetics of Fluorescence Quenching by Electron and H-Atom Transfer, *Isr. J. Chem.* 1970, **8**, 259–271.

348. Zachariasse, K. A., Electron Transfer in the Weller group at the Free University of Amsterdam (1964–1971), *Spectrum* 2006, **19**, 22–28.

349. Creutz, C., Taube, H., A Direct Approach to Measuring the Franck–Condon Barrier to Electron Transfer between Metal Ions, *J. Am. Chem. Soc.* 1969, **91**, 3988–3989.

350. Kaim, W., Klein, A., Glöckle, M., Exploration of Mixed-valence Chemistry: Inventing New Analogues of the Creutz–Taube Ion, *Acc. Chem. Res.* 2000, **33**, 755–763.

351. Day, P., Hush, N. S., Clark, R. J. H., Mixed Valence: Origins and Developments, *Philos. Trans. R. Soc. London, Ser. A* 2007, **366**, 5–14.

352. Miller, J. R., Calcaterra, L. T., Closs, G. L., Intramolecular Long-distance Electron Transfer in Radical Anions. The Effect of Free Energy and Solvent on the Reaction Rates, *J. Am. Chem. Soc.*, 1984, **106**, 3047–3049.

353. Liang, N., Miller, J. R., Closs, G. L., Temperature-independent Long-range Electron Transfer Reactions in the Marcus Inverted Region *J. Am. Chem. Soc.* 1990, **112**, 5353–5354.

354. Siders, P., Marcus, R. A., Quantum Effects for Electron-transfer Reaction in the "Inverted Region" *J. Am. Chem. Soc.* 1981, **103**, 748–752.

355. Bixon, M., Jortner, J., Charge Separation and Recombination in Isolated Supermolecules, *J. Phys. Chem.* 1993, **97**, 13061–13066.

356. Gould, I. R., Farid, S., Dynamics of Bimolecular Photoinduced Electron-transfer Reactions, *Acc. Chem. Res.* 1996, **29**, 522–528.

357. Serpa, C., Gomes, P. J. S., Arnaut, L. G., de Melo, J. S., Formosinho, S. J., Temperature Dependence of Ultra-exothermic Charge Recombinations, *ChemPhysChem* 2006, 2533–2539.

358. Scholes, G. D., Jones, M., Kumar, S., Energetics of Photoinduced Electron-transfer Reactions Decided by Quantum Confinement, *J. Phys. Chem. C* 2007, **111**, 13777–13785.

359. Khundkar, L. R., Perry, J. W., Hanson, J. E., Dervan, P. B., Weak Temperature Dependence of Electron Transfer Rates in Fixed-Distance Porphyrin–Quinone Model Systems, *J. Am. Chem. Soc.* 1994, **116**, 9700–9709.

360. Closs, G. L., Johnson, M. D., Miller, J. R., Piotrowiak, P., A Connection between Intramolecular Long-range Electron, Hole, and Triplet Energy Transfers, *J. Am. Chem. Soc.* 1989, **111**, 3751–3753.

361. Lippert, E., Lueder, W., Moll, F., Naegele, W., Boos, H., Prigge, H., Seibold-Blankenstein, I., Transformation of Electron Excitation Energy, *Angew. Chem. Int. Ed. Engl.* 1961, **73**, 695–706.

362. Druzhinin, S. I., Ernsting, N. P., Kovalenko, S. A., Lustres, L. P., Senyushkina, T. A., Zachariasse, K. A., Dynamics of Ultrafast Intramolecular Charge Transfer with 4-(Dimethylamino)benzonitrile in Acetonitrile, *J. Phys. Chem. A* 2006, **110**, 2955–2969.

363. Grabowski, Z. R., Rotkiewicz, K., Rettig, W., Structural Changes Accompanying Intramolecular Electron Transfer: Focus on Twisted Intramolecular Charge-transfer States and Structures, *Chem. Rev.* 2003, **103**, 3899–4031.

364. Kang, T. J., Kahlow, M. A., Giser, D., Swallen, S., Nagarajan, V., Jarzeba, W., Barbara, P. F., Dynamic Solvent Effects in the Electron-transfer Kinetics of $S_1$ Bianthryls, *J. Phys. Chem.* 1988, **92**, 6800–6807.

365. Roth, H. D., Biradicals by Triplet Recombination of Radical Ion Pairs, *Photochem. Photobiol. Sci.* 2008, **7**, 540–546.

366. Stubbe, J., Nocera, D. G., Yee, C. S., Chang, M. C. Y., Radical Initiation in the Class I Ribonucleotide Reductase: Long-range Proton-coupled Electron Transfer?, *Chem. Rev.* 2003, **103**, 2167–2201.

367. Lewis, F. D., Zhu, H., Daublain, P., Fiebig, T., Raytchev, M., Wang, Q., Shafirovich, V., Crossover from Superexchange to Hopping as the Mechanism for Photoinduced Charge Transfer in DNA Hairpin Conjugates, *J. Am. Chem. Soc.* 2005, **128**, 791–800.

368. Lewis, F. D., Zhu, H., Daublain, P., Sigmund, K., Fiebig, T., Raytchev, M., Wang, Q., Shafirovich, V., Getting to Guanine: Mechanism And Dynamics of Charge Separation and Charge Recombination in DNA Revisited, *Photochem. Photobiol. Sci.* 2008, **7**, 534–539.

369. Lewis, F. D., Daublain, P., Zhang, L., Cohen, B., Vura-Weis, J., Wasielewski, M. R., Shafirovich, V., Wang, Q., Raytchev, M., Fiebig, T., Reversible Bridge-mediated Excited-state Symmetry Breaking in Stilbene-linked DNA Dumbbells, *J. Chem. Phys. B* 2008, **112**, 3838–3843.

370. Lewis, F. D., DNA Photonics, *Pure Appl. Chem.* 2006, **78**, 2287–2295.

371. Schuster, G. B., Long-range Charge Transfer in DNA: Transient Structural Distortions Control the Distance Dependence, *Acc. Chem. Res.* 2000, **33**, 253–260.

372. Gosh, A., Joy, A., Schuster, G. B., Douki, T., Cadet, J., Selective One-electron Oxidation of Duplex DNA Oligomers: Reaction at Thymines, *Org. Biomol. Chem.* 2008, **6**, 916–928.

373. Giese, B., Napp, M., Jacques, O., Boudebous, H., Taylor, A. M., Wirz, J., Multistep Electron Transfer in Oligopeptides: Direct Observation of Radical Cation Intermediates, *Angew. Chem. Int. Ed.* 2005, **44**, 4073–4075.

374. Cordes, M., Jacques, O., Kottgen, A., Jasper, C., Boudebous, H., Giese, B., Development of a Model System for the Study of Long Distance Electron Transfer in Peptides, *Adv. Synth. Catal.* 2008, **350**, 1053–1062.

375. Giese, B., Electron Transfer through DNA and Peptides, *Bioorg. Med. Chem.* 2006, **14**, 6139–6143.

376. Giese, B., Long-distance Charge Transport in DNA: the Hopping Mechanism, *Acc. Chem. Res.* 2000, **33**, 631–636.

377. Turro, N. J., Barton, J. K., Paradigm, Supermolecules Electron Transfer, Chemistry at a Distance. What's the Problem? The Science or the Paradigm? *J. Biol. Inorg. Chem.* 1998, **3**, 201–209.

378. Jortner, J., Bixon, M., Langenbacher, T., Michel-Beyerle, M. E., Charge Transfer and Transport in DNA, *Proc. Natl. Acad. Sci. USA* 1998, **95**, 12759–12765.

379. Fiebig, T., Wagenknecht, H.-A., DNA Photonics– Photoinduced Electron Transfer in Synthetic DNA-Donor–Acceptor Systems, *Chimia* 2007, **61**, 133–139.

380. Takada, T., Kawai, K., Fujitsuka, M., Majima, T., Contributions of the Distance-dependent Reorganization Energy and Proton-transfer to the Hole-transfer Process in DNA, *Chem. Eur. J.* 2005, **11**, 3835–3842.

381. Fink, H.-W., Schönenberger, C., Electrical Conduction through DNA Molecules, *Nature* 1999, **398**, 407–410.

382. Jackson, G., Porter, G., Acidity Constants in the Triplet State, *Proc. R. Soc. London, Ser. A* 1961, **260**, 13–30.

383. Conrad, P. G., Givens, R. S., Hellrung, B., Rajesh, C. S., Ramseier, M., Wirz, J, *p*-Hydroxyphenacyl Phototriggers: the Reactive Excited State of Phosphate Photorelease, *J. Am. Chem. Soc.* 2000, **122**, 9346–9347.

384. Bell, R. P., *The Tunnel Effect in Chemistry*, Chapman and Hall, London, 1980.

385. Formosinho, S. J., Radiationless Transitions and Photochemical Reactivity, *Pure Appl. Chem.* 1986, **58**, 1173–1178.

386. Al-Soufi, W., Eychmüller, A., Grellmann, K. H., Kinetics of the Photoenolization of 5,8-Dimethyl-1-tetralone. Hydrogen-transfer Tunnel Effects in the Excited Triplet State, *J. Phys. Chem.* 1991, **95**, 2022–2026.

387. Campos, L. M., Warrier, M. V., Peterfy, K., Houk, K. N., Garcia-Garibay, M. A., Secondary $\alpha$ Isotope Effects on Deuterium Tunneling in Triplet *o*-Methylanthrones: Extraordinary Sensitivity to Barrier Width, *J. Am. Chem. Soc.* 2005, **127**, 10178–10179.

388. Johnson, B. A., Kleinman, M. H., Turro, N. J., Garcia-Garibay, M. A., Hydrogen Atom Tunneling in Triplet *o*-Methylbenzocycloalkanones: Effects of Structure on Reaction Geometry and Excited State Configuration, *J. Org. Chem.* 2002, **67**, 6944–6953.

389. Grellmann, K. H., Weller, H., Tauer, E., Tunnel Effect on the Kinetics of Hydrogen Shifts. The Enol–Ketone Transformation of 2-Methylacetophenone, *Chem. Phys. Lett.* 1983, **95**, 195–199.

390. Moss, R. A., Platz, M. S., Jones, M. (eds), *Reactive Intermediate Chemistry*, Wiley-Interscience, New York, 2004.

391. Platz, M. S., Moss, R. A., Jones, M. (eds), *Reviews of Reactive Intermediate Chemistry*, John Wiley & Sons, Inc., Hoboken, NJ, 2007.

392. Shavitt, I., Geometry and Singlet–Triplet Energy Gap in Methylene: a Critical Review of Experimental and Theoretical Determinations, *Tetrahedron* 1985, **41**, 1531–1542.

393. Leopold, D. G., Murray, K. K., Miller, A. E. S., Lineberger, W. C., Methylene: a Study of the $X^3B_1$ and $a^1A_1$ States by Photoelectron Spectroscopy of $CH_2^-$ and $CD_2$, *J. Chem. Phys.* 1985, **83**, 4849–4865.

394. Irikura, K. K., Goddard, W. A. III, Beauchamp, J. L., Singlet–Triplet Gaps in Substituted Carbenes CXY (X, Y = H, Fluoro, Chloro, Bromo, Iodo, Silyl), *J. Am. Chem. Soc.* 1992, **114**, 48–51.

395. Worthington, S. E., Cramer, C. J., Density Functional Calculations of the Influence of Substitution on Singlet–Triplet Gaps in Carbenes and Vinylidenes, *J. Phys. Org. Chem.* 1997, **10**, 755–767.

396. Griller, D., Hadel, L., Nazran, A. S., Platz, M. S., Wong, P. C., Scaiano, J. C., Fluorenylidene: Kinetic, Mechanisms *J. Am. Chem. Soc.* 1984, **106**, 2227–2235.

397. Bonneau, R., Hellrung, B., Liu, M. T. H., Wirz, J., Adamantylidene Revisited: Flash Photolysis of Adamantanediazirine, *J. Photochem. Photobiol. A* 1998, **116**, 9–19.

398. Bally, T., Matzinger, S., Truttmann, L., Platz, M. S., Morgan, S., Matrix Spectroscopy of 2-Adamantylidene, a Dialkylcarbene with Singlet Ground State, *Angew. Chem. Int. Ed. Engl.* 1994, 1994, 1964–1966.

399. Kawano, M., Hirai, K., Tomioka, H., Ohashi, Y., Structure Determination of Triplet Diphenylcarbenes by *In Situ* X-ray Crystallographic Analysis, *J. Am. Chem. Soc.* 2007, **129**, 2383–2391.

400. Itoh, T., Nakata, Y., Hirai, K., Tomioka, H., Triplet Diphenylcarbenes Protected by Trifluoromethyl and Bromine Groups. A Triplet Carbene Surviving a Day in Solution at Room Temperature, *J. Am. Chem. Soc.* 2006, **128**, 957–967.

401. Bertrand, G., Stable Singlet Carbenes. In Moss, R. A., Platz, M. S., Jones, M. (eds), *Reactive Intermediate Chemistry*, John Wiley & Sons, Inc., Hoboken, NJ, 2004, pp. 329–373.

402. Zhu, Z., Bally, T., Stracener, L. L., McMahon, R. J., Reversible Interconversion between Singlet and Triplet 2-Naphthyl(carbomethoxy)carbene, *J. Am. Chem. Soc.* 1999, **121**, 2863–2874.

403. Wang, Y., Hadad, C. M., Toscano, J. P., Solvent Dependence of the 2-Naphthyl(carbomethoxy) carbene Singlet–Triplet Energy Gap, *J. Am. Chem. Soc.* 2002, **124**, 1761–1767.

404. Hess, G. C., Kohler, B., Likhotvorik, I., Peon, J., Platz, M. S., Ultrafast Carbonylcarbene Formation and Spin Equilibration, *J. Am. Chem. Soc.* 2000, **122**, 8087–8088.

405. Wang, J.-L., Likhotvorik, I., Platz, M. S., A Laser Flash Photolysis Study of 2-Naphthyl (carbomethoxy)carbene, *J. Am. Chem. Soc.* 1999, **121**, 2883–2890.

406. Gritsan, N. P., Platz, M. S., Kinetic, Spectroscopy, and Computational Chemistry of Arylnitrenes, *Chem. Rev.* 2006, **106**, 3844–3867.

407. Borden, W. T., Gritsan, N. P., Hadad, C. M., Karney, W. L., Kemnitz, C. R., Platz, M. S., The Interplay of Theory and Experiment in the Study of Phenylnitrene, *Acc. Chem. Res.* 2000, **33**, 765–771.

408. Gritsan, N. P., Likhotvorik, I., Zhu, Z., Platz, M. S., Observation of Perfluoromethylnitrene in Cryogenic Matrixes, *J. Phys. Chem. A* 2001, **105**, 3039–3041.

409. Huisgen, R., Appl, M., The Mechanism of the Ring Enlargement in the Decomposition of Phenyl Azide in Aniline, *Chem. Ber.* 1958, **91**, 12–21.

410. Doering, W. v.E., Odum, R. A., Ring Enlargement in the Photolysis of Phenyl Azide, *Tetrahedron*, 1966, **22**, 81–93.

411. Leyva, E., Platz, M. S., Persy, G., Wirz, J., Photochemistry of Phenyl Azide – the Role of Singlet and Triplet Phenylnitrene as Transient Intermediates, *J. Am. Chem. Soc.* 1986, **108**, 3783–3790.

412. McClelland, R. A., Kahley, M. J., Davidse, P. A., Hadzialic, G., Acid–Base Properties of Arylnitrenium Ions *J. Am. Chem. Soc.* 1996, **118**, 4794–4803.

413. Winter, A. H., Thomas, S. I., Kung, A. C., Falvey, D. E., Photochemical Generation of Nitrenium Ions from Protonated 1,1-Diarylhydrazines, *Org. Lett.* 2004, **6**, 4671–4674.

414. Born, R., Burda, C., Senn, P., Wirz, J., Transient Absorption Spectra and Reaction Kinetics of Singlet Phenyl Nitrene and Its 2,4,6-Tribromo Derivative in Solution, *J. Am. Chem. Soc.* 1997, **119**, 5061–5062.

415. Fishbein, J. C., McClelland, R. A., Halide Ion Trapping of Nitrenium Ions Formed in the Bamberger Rearrangement of *N*-Arylhydroxylamines. Lifetime of the Parent Phenylnitrenium Ion in Water, *Can. J. Chem.* 1996, **74**, 1321–1328.

416. Wang, J., Kubicki, J., Platz, M. S., An Ultrafast Study of Phenyl Azide: the Direct Observation of Phenylnitrenium Ion, *Org. Lett.* 2007, **9**, 3973–3976.

417. McClelland, R. A., Ahmad, A., Dicks, A. P., Licence, V. E., Spectroscopic Characterization of the Initial $C_8$ Intermediate in the Reaction of the 2-Fluorenylnitrenium Ion with 2′-Deoxyguanosine, *J. Am. Chem. Soc.* 1999, **121**, 3303–3310.

418. McClelland, R. A., Postigo, A., Solvent Effects on the Reactivity of Fluorenyl Nitrenium Ion with DNA-like Probes, *Biophys. Chem.* 2006, **119**, 213–218.

419. Winter, A. H., Gibson, H. H., Falvey, D. E., Carbazolyl Nitrenium Ion: Electron Configuration and Antiaromaticity Assessed by Laser Flash Photolysis, Trapping Rate Constants, Product Analysis, and Computational Studies, *J. Org. Chem.* 2007, **72**, 8186–8195.

420. Cline, M. R., Mandel, S. M., Platz, M. S., Identification of the Reactive Intermediates Produced upon Photolysis of *p*-Azidoacetophenone and Its Tetrafluoro Analogue in Aqueous and Organic Solvents: Implications for Photoaffinity Labeling, *Biochemistry* 2007, **46**, 1981–1987.

421. McCulla, R. D., Gohar, G. A., Hadad, C. M., Platz, M. S., Computational Study of the Curtius-like Rearrangements of Phosphoryl, Phosphinyl, and Phosphinoyl Azides and Their Corresponding Nitrenes, *J. Org. Chem.* 2007, **72**, 9426–9438.

422. Rizk, M. S., Shi, X., Platz, M. S., Lifetimes and Reactivities of Some 1,2-Didehydroazepines Commonly Used in Photoaffinity Labeling Experiments in Aqueous Solutions, *Biochemistry* 2006, **45**, 543–551.

423. Dietlin, C., Allonas, X., Defoin, A., Fouassier, J.-P., Theoretical and Experimental Study of the Norrish I Photodissociation of Aromatic Ketones, *Photochem. Photobiol. Sci.* 2008, **7**, 558–565.

424. Fischer, H., Knuhl, B., Marque, S. R. A., Absolute Rate Constants for the Addition of the 1-(*t*-Butoxy)Carbonylethyl Radical onto Cyclic Alkenes in Solution, *Helv. Chim. Acta* 2006, **89**, 2327–2329.

425. Henry, D. J., Coote, M. L., Gomez-Balderas, R., Radom, L., Comparison of the Kinetics and Thermodynamics for Methyl Radical Addition to C=C, C=O, and C=S Double Bonds, *J. Am. Chem. Soc.* 2004, **126**, 1732–1740.

426. Maillard, B., Ingold, K. U., Scaiano, J. C., Rate Constants for the Reactions of Free Radicals with Oxygen in Solution, *J. Am. Chem. Soc.* 1983, **105**, 5095–5099.

427. Font-Sanchis, E., Aliaga, C., Bejan, E. V., Cornejo, R., Scaiano, J.C., Generation and Reactivity Toward Oxygen of Carbon-centered Radicals Containing Indane, Indene, and Fluorenyl Moieties, *J. Org. Chem.* 2003, **68**, 3199–3204.

428. Bargon, J., Fischer, H., Johnsen, U., Nuclear Magnetic Resonance Emission Lines During Fast Radical Reactions. I. Recording Methods and Examples, *Z. Naturforsch., Teil A* 1967, **22**, 1551–1555.

429. Ward, H. R., Lawler, R. G., Nuclear Magnetic Resonance Emission and Enhanced Absorption in Rapid Organometallic Reactions, *J. Am. Chem. Soc.* 1967, **89**, 5518–5519.

430. Bargon, J., The Discovery of Chemically Induced Dynamic Polarization (CIDNP), *Helv. Chim. Acta* 2006, **89**, 2082–2102.

431. Goez, M., An Introduction to Chemically Induced Dynamic Nuclear Polarization, *Concepts Magn. Reson.* 1995, **7**, 69–86.

432. Pine, S. H., Chemically-induced Dynamic Nuclear Polarization, *J. Chem. Educ.* 1972, **49**, 664–668.

433. Kaptein, R., Chemically Induced Dynamic Nuclear Polarization. VIII. Spin Dynamics and Diffusion of Radical Pairs, *J. Am. Chem. Soc.* 1972, **94**, 6251–6262.

434. Turro, N. J., Magnetic Field and Magnetic Isotope Effects in Organic Photochemical Reactions. A Novel Probe of Reaction Mechanisms and a Method for Enrichment of Magnetic Isotopes, *Acc. Chem. Res.* 1980, **13**, 369–377.

435. Turro, N. J., Photochemistry of Organic Molecules in Microscopic Reactors, *Pure Appl. Chem.* 1986, **58**, 1219–1228.

436. Roth, H. D., Organic Radical Ions. In Moss, R. A., Platz, M. S., Jones, M. Jr. (eds), *Reactive Intermediate Chemistry*, John Wiley & Sons, Inc, Hoboken, NJ, 2004, pp. 205–272.

437. Weller, A., Staerk, H., Treichel, R., Magnetic-Field Effects on Geminate Radical-pair Recombination, *Faraday Discuss. Chem. Soc.* 1984, **78**, 271–278.

438. Schulten, K., Weller, A., Exploring Fast Electron Transfer Processes by Magnetic Fields, *Biophys. J.* 1978, **24**, 295–305.

439. Steenken, S., Warren, C. J., Gilbert, B. C., Generation of Radical Cations from Naphthalene and Some Derivatives, Both by Photoionization and Reaction with $SO_4^-$: Formation and Reactions Studied by Laser Flash Photolysis, *J. Chem. Soc., Perkin Trans.* 1990, **2**, 335–342.

440. Brugman, C. J. M., Rettschnick, R. P. H., Hoytink, G. J., Quartet–Doublet Phosphorescence from an Aromatic Radical. Decacyclene Mononegative Ion, *Chem. Phys. Lett.* 1971, **8**, 263–264.

441. Kothe, G., Kim, S. S., Weissman, S. I., Transient Magnetic Resonance of a Photoexcited Quartet State, *Chem. Phys. Lett.* 1980, **71**, 445–447.

442. Hochlaf, M., Taylor, S., Eland, J. H. D., Quartet States of the Acetylene Cation: Electronic Structure Calculations and Spin–Orbit Coupling Terms, *J. Chem. Phys.* 2006, **125**, 214301/214301–214301/214308.

443. Komiha, N., Rosmus, P., Maier, J. P., Low Lying Quartet States in Diacetylene, Triacetylene and Benzene Radical Cations, *Mol. Phys.* 2007, **105**, 893–897.

444. Borden, W. T. (ed), *Diradicals*, John Wiley & Sons, Inc., New York, 1982.

445. Wirz, J., Spectroscopic and Kinetic Investigations of Conjugated Biradical Intermediates, *Pure Appl. Chem.* 1984, **56**, 1289–1300.

446. Borden, W. T., Davidson, E. R., Effects of Electron Repulsion in Conjugated Hydrocarbon Diradicals, *J. Am. Chem. Soc.* 1976, **99**, 4587–4594.

447. Wenthold, P. G., Hu, J., Squires, R. R., Lineberger, W. C., Photoelectron Spectroscopy of the Trimethylenemethane Negative Ion, *J. Am. Soc. Mass Spectrom.* 1999, **10**, 800–809.

448. Brabec, J., Pittner, J., The Singlet–Triplet Gap in Trimethylenmethane and the Ring-opening of Methylenecyclopropane, *J. Phys. Chem. A* 2006, **110**, 11765–11769.

449. Hangarter, M.-A., Hoermann, A., Kamdzhilov, Y., Wirz, J., Primary Photoreactions of Phylloquinone (Vitamin $K_1$) and Plastoquinone-1 in Solution, *Photochem. Photobiol. Sci.* 2003, **2**, 524–535.

450. Kita, F., Adam, W., Jordan, P., Nau, W. M., Wirz, J., 1,3-Cyclopentanediyl Diradicals: Substituent and Temperature Dependence of Triplet–Singlet Intersystem Crossing, *J. Am. Chem. Soc.* 1999, **121**, 9265–9275.

451. Abe, M., Adam, W., Borden, W. T., Hattori, M., Hrovat, D. A., Nojima, M., Nozaki, K., Wirz, J., Effects of Spiroconjugation on the Calculated Singlet–Triplet Energy Gap in 2,2-Dialkoxycyclopentane-1,3-diyls and on the Experimental Electronic Absorption Spectra of Singlet 1,3-Diphenyl Derivatives. Assignment of the Lowest-energy Electronic Transition of Singlet Cyclopentane-1,3-diyls, *J. Am. Chem. Soc.* 2004, **126**, 574–582.

452. McMasters, D. R., Wirz, J., Spectroscopy, Reactivity of Kekulé Hydrocarbons with Very Small Singlet–Triplet Gaps, *J. Am. Chem. Soc.* 2001, **123**, 238–246.

453. Burnett, M. N., Boothe, R., Clark, E., Gisin, M., Hassaneen, H. M., Pagni, R. M., Persy, G., Smith, R. J., Wirz, J., 1,4-Perinaphthadiyl. Singlet- and Triplet-state Reactivity of a Conjugated Hydrocarbon Biradical, *J. Am. Chem. Soc.* 1988, **110**, 2527–2538.

454. Adam, W., Platsch, H., Wirz, J., Oxygen Trapping and Thermochemistry of a Hydrocarbon Singlet Biradical: 1,3-Diphenylcyclopentane-1,3-diyl, *J. Am. Chem. Soc.* 1989, **111**, 6896–6898.

455. Coms, F. D., Dougherty, D. A., Diphenylbicyclo[2.1.0]pentane. A Persistent Hydrocarbon with a Very Weak Carbon–Carbon Bond, *J. Am. Chem. Soc.* 1989, **111**, 6894–6896.

456. Adam, W., Platsch, H., Sendelbach, J., Wirz, J., Determination of the Triplet Lifetimes of 1,3-Cyclopentanediyl Biradicals Derived from the Photodenitrogenation of Azoalkanes with Time-resolved Photoacoustic Calorimetry, *J. Org. Chem.* 1993, **58**, 1477–1482.

457. Adam, W., Reinhard, G., Platsch, H., Wirz, J., Effect of 2,2-Dimethyl Substitution on the Lifetimes of Cyclic Hydrocarbon Triplet 1,3-Biradicals, *J. Am. Chem. Soc.* 1990, **112**, 4570–4571.

458. Hasler, E., Gassmann, E., Wirz, J., Conjugated Biradical Intermediates: Spectroscopic, Kinetic, and Trapping Studies of 2,2-Dimethyl-1,3-Perinaphthadiyl, *Helv. Chim. Acta* 1985, **68**, 777–788.

459. Engel, P. S., Lowe, K. L., The Lifetime of the 2,2-Dimethylcyclopentane-1,3-diyl Biradical by the Cyclopropylcarbinyl Radical Clock Method, *Tetrahedron Lett.* 1994, **35**, 2267–2270.

460. De Feyter, S., Diau, E. W. G., Scala, A. A., Zewail, A. H., Femtosecond Dynamics of Diradicals: Transition States, Entropic Configurations and Stereochemistry, *Chem. Phys. Lett.* 1999, **303**, 249–260.

461. De Feyter, S., Diau, E. W. G., Zewail, A. H., Femtosecond Dynamics of Norrish Type-II Reactions: Nonconcerted Hydrogen-transfer and Diradical Intermediacy, *Angew. Chem. Int. Ed.* 2000, **39**, 260–263.

462. Givens, R. S., Heger, D., Hellrung, B., Kamdzhilov, Y., Mac, M., Conrad, P. G., Cope, E., Lee, J. I., Mata-Segreda, J. F., Schowen, R. L., Wirz, J., The Photo-Favorskii Reaction of *p*-Hydroxyphenacyl Compounds is Initiated by Water-assisted, Adiabatic Extrusion of a Triplet Biradical, *J. Am. Chem. Soc.* 2008, **130**, 3307–3309.

463. McClelland, R. A., Carbocations. In Moss, R. A., Platz, M. S., Jones, M. (eds), *Reactive Intermediate Chemistry*, John Wiley & Sons, Inc., Hoboken, NJ, 2004, pp. 3–40.

464. van Dorp, J. W. J., Lodder, G., Substituent Effects on the Photogeneration and Selectivity of Triaryl Vinyl Cations, *J. Org. Chem.* 2008, **73**, 5416–5428.

465. Steenken, S., Ashokkumar, M., Maruthamuthu, P., McClelland, R. A., Making Photochemically Generated Phenyl Cations Visible by Addition to Aromatics: Production of Phenylcyclohexadienyl Cations and Their Reactions with Bases/Nucleophiles, *J. Am. Chem. Soc.* 1998, **120**, 11925–11931.

466. Mecklenburg, S. L., Hilinski, E. F., Picosecond Spectroscopic Characterization of the 9-Fluorenyl Cation in Solution, *J. Am. Chem. Soc.* 1989, **111**, 5471–5472.

467. McClelland, R. A., Mathivanan, N., Steenken, S., Laser Flash Photolysis of 9-Fluorenol. Production and Reactivities of the 9-Fluorenol Radical Cation and the 9-Fluorenyl Cation, *J. Am. Chem. Soc.* 1990, **112**, 4857–4861.

468. Gurzadyan, G. G., Steenken, S., Solvent-dependent C−OH Homolysis and Heterolysis in Electronically Excited 9-Fluorenol: the Life and Solvation Time of the 9-Fluorenyl Cation in Water, *Chem. Eur. J.* 2001, **7**, 1808–1815.

469. O'Neill, M. A., Cozens, F. L., Schepp, N. P., Generation and Direct Observation of the 9-Fluorenyl Cation in Non-acidic Zeolites, *Tetrahedron* 2000, **56**, 6969–6977.

470. Courtney, M. C., MacCormack, A. C., More O'Ferrall, R. A., Comparison of p$K_R$ Values of Fluorenyl and Anthracenyl Cations, *J. Phys. Org. Chem.* 2002, **15**, 529–539.

471. Wirz, J., Kinetics of Proton Transfer Reactions Involving Carbon, *Pure Appl. Chem.* 1998, **70**, 2221–2232.

472. Kresge, A. J., Keto–Enol Tautomerism of Phenols in Aqueous Solution, *Chemtracts* 2002, **15**, 212–215.

473. Rappoport, Z., *The Chemistry of Enols*, John Wiley & Sons, Ltd, Chichester, 1990.

474. Haspra, P., Sutter, A., Wirz, J., Acidity of Acetophenone Enol in Aqueous Solution, *Angew. Chem. Int. Ed. Engl.* 1978, **18**, 617–619.

475. Capponi, M., Gut, I. G., Hellrung, B., Persy, G., Wirz, J., Ketonization Equilibria of Phenol in Aqueous Solution, *Can. J. Chem.* 1999, **77**, 605–613.

476. Chiang, Y., Kresge, A. J., Hochstrasser, R., Wirz, J., Phenyl- and Mesitylnol. The First Generation and Direct Observation of Hydroxyacetylenes in Solution, *J. Am. Chem. Soc.* 1989, **111**, 2355–2357.

477. Chiang, Y., Grant, A. S., Kresge, A. J., Pruszynski, P., Schepp, N. P., Wirz, J., Generation and Study of Ynamines by Laser Photolysis in Aqueous Solution, *Angew. Chem. Int. Ed. Engl.* 1991, **103**, 1407–1408.

478. Chiang, Y., Grant, A. S., Guo, H. X., Kresge, A. J., Paine, S. W., Flash Photolytic De-carbonylation and Ring-opening of 2-(*N*-(Pentafluorophenyl)amino)-3-phenylcyclopropenone. Isomerization of the Resulting Ynamine to a Ketenimine, Hydration of the Ketenimine, and Hydrolysis of the Enamine Produced by Ring Opening, *J. Org. Chem.* 1997, **62**, 5363–5370.

479. Saltiel, J., Waller, A. S., Sears, D. F. Jr., The Temperature and Medium Dependencies of *cis*-Stilbene Fluorescence. The Energetics of Twisting in the Lowest Excited Singlet State, *J. Am. Chem. Soc.* 1993, **115**, 2453–2465.

480. Bernardi, F., Olivucci, M., Robb, M. A., The Role of Conical Intersections and Excited State Reaction Paths in Photochemical Pericyclic Reactions, *J. Photochem. Photobiol. A* 1997, **105**, 365–371.

481. Wismonski-Knittel, T., Fischer, G., Fischer, E., Temperature Dependence of Photoisomerization. VIII. Excited-state Behavior of 1-Naphthyl-2-phenyl- and 1,2-Dinaphthylethylenes and Their Photocyclization Products, and Properties of the Latter, *J. Chem. Soc., Perkin Trans. 2* 1974, 1930–1940.

482. Sension, R. J., Repinec, S. T., Szarka, A. Z., Hochstrasser, R. M., Femtosecond Laser Studies of the *cis*-Stilbene Photoisomerization Reactions, *J. Chem. Phys.* 1993, **98**, 6291–6315.

483. Ishii, K., Takeuchi, S., Tahara, T., Pronounced Non-Condon Effect as the Origin of the Quantum Beat Observed in the Time-resolved Absorption Signal from Excited-state *cis*-Stilbene, *J. Phys. Chem. A* 2008, **112**, 2219–2227.

484. Saltiel, J., Waller, A. S., Sears, D. F. Jr., Garrett, C.Z., Fluorescence Quantum Yields of *trans*-stilbene-$d_0$ and -$d_2$ in *n*-Hexane and *n*-Tetradecane: Medium and Deuterium Isotope Effects on Decay Processes, *J. Phys. Chem.* 1993, **97**, 2516–2522.

485. Fischer, G., Seger, G., Muszkat, K. A., Fischer, E., Emissions of Sterically Hindered Stilbene Derivatives and Related Compounds. IV. Large Conformational Differences Between Ground and Excited States of Sterically Hindered Stilbenes. Implications Regarding Stokes Shifts and Viscosity or Temperature Dependence of Fluorescence Yields, *J. Chem. Soc., Perkin Trans. 2* 1975, 1569–1576.

486. Yoshihara, K., Namiki, A., Sumitani, M., Nakashima, N., Picosecond Flash Photolysis of *cis*- and *trans*-Stilbene. Observation of an Intense Intramolecular Charge-Resonance Transition, *J. Chem. Phys.* 1979, **71**, 2892–2895.

487. Gilbert, A., Cyclization of Stilbene and Its Derivatives. In Horspool, W. M., Lenci, F. (eds), *CRC Handbook of Organic Photochemistry and Photobiology*, 2nd edn, CRC Press LLC, Boca Raton, FL, 2004, Chapter 33. pp. 1–11.

488. Saltiel, J., Sun, Y. P., Intrinsic Potential Energy Barrier for Twisting in the *trans*-Stilbene $S_1$ State in Hydrocarbon Solvents, *J. Phys. Chem.* 1989, **93**, 6246–6250.

489. Kim, S. K., Fleming, G. R., Reorientation and Isomerization of *trans*-Stilbene in Alkane Solutions, *J. Phys. Chem.* 1988, **92**, 2168–2172.

490. Hohlneicher, G., Wrzal, R., Lenoir, D., Frank, R., Two-photon Spectra of Stiff Stilbenes: a Contribution to the Assignment of the Low Lying Electronically Excited States of the Stilbene System, *J. Phys. Chem. A* 1999, **103**, 8969–8975.

491. Goerner, H., Schulte-Frohlinde, D., Observation of the Triplet State of Stilbene in Fluid Solution. Determination of the Equilibrium Constant ($^3t^* - {}^3p^*$) and of the Rate Constant for Intersystem Crossing ($^3p^* - {}^1p$), *J. Phys. Chem.* 1981, **85**, 1835–1841.

492. Saltiel, J., Ganapathy, S., Werking, C., The $\Delta H$ for Thermal *trans/cis*-Stilbene Isomerization: Do $S_0$ and $T_1$ Potential Energy Curves Cross?, *J. Phys. Chem.* 1987, **91**, 2755–2758.

493. Bearpark, M. J., Bernardi, F., Clifford, S., Olivucci, M., Robb, M. A., Vreven, T., Cooperating Rings in *cis*-Stilbene Lead to an $S_0/S_1$ Conical Intersection, *J. Phys. Chem. A* 1997, **101**, 3841–3847.

494. Gagliardi, L., Orlandi, G., Molina, V., Malmqvist, P.-A., Roos, B., Theoretical Study of the Lowest $^1B_u$ States of *trans*-Stilbene, *J. Phys. Chem. A* 2002, **106**, 7355–7361.

495. Improta, R., Santoro, F., Excited-state Behavior of *trans* and *cis* Isomers of Stilbene and Stiff Stilbene: A TD-DFT Study, *J. Phys. Chem. A* 2005, **109**, 10058–10067.

496. Quenneville, J., Martinez, T. J., *Ab Initio* Study of *cis–trans* Photoisomerization in Stilbene and Ethylene, *J. Phys. Chem. A* 2003, **107**, 829–837.

497. Orlandi, G., Siebrand, W., Model for the Direct Photoisomerization of Stilbene, *Chem. Phys. Lett.* 1975, **30**, 352–354.

498. Saltiel, J., Dmitrenko, O., Pillai, Z. S., Klima, R., Wang, S., Wharton, T., Huang, Z., N., van de Burgt, L. J., Arranz, J., Triplet and Ground State Potential Energy Surfaces of 1,4-Diphenyl-1,3-Butadiene: Theory and Experiment, *Photochem. Photobiol. Sci.* 2008, **7**, 566–577.

499. Saltiel, J., Crowder, J. M., Wang, S., Mapping the Potential Energy Surfaces of the 1,6-Diphenyl-1,3,5-hexatriene Ground and Triplet States, *J. Am. Chem. Soc.* 1999, **121**, 895–902.

500. Arai, T., Tokumaru, K., Photochemical One-way Adiabatic Isomerization of Aromatic Olefins, *Chem. Rev.* 1993, **93**, 23–39.

501. Saltiel, J., Tarkalanov, N., Sears, D. F. Jr., Conformer-specific Adiabatic *cis* → *trans* Photo-isomerization of *cis*-1-(2-Naphthyl)-2-phenylethene. A Striking Application of the NEER Principle, *J. Am. Chem. Soc.* 1995, **117**, 5586–5587.

502. Adam, W., Kazakov, D. V., Kazakov, V. P., Singlet-oxygen Chemiluminescence in Peroxide Reactions, *Chem. Rev.* 2005, **105**, 3371–3387.

503. Carpenter, B. K., Electronically Nonadiabatic Thermal Reactions of Organic Molecules, *Chem. Soc. Rev.* 2006, **35**, 736–747.

504. Fraga, H., Firefly Luminescence: a Historical Perspective and Recent Developments, *Photochem. Photobiol. Sci.* 2008, **7**, 146–158.

505. Matsumoto, M., Advanced Chemistry of Dioxetane-based Chemiluminescent Substrates Originating from Bioluminescence, *J. Photochem. Photobiol. C* 2004, **5**, 27–53.

506. McCapra, F., Chemical Generation of Excited States: the Basis of Chemiluminescence and Bioluminescence, *Methods Enzymol.* 2000, **305**, 3–47.

507. Turro, N. J., Lechtken, P., Schore, N. E., Schuster, G., Steinmetzer, H.-C., Yekta, A., Tetramethyl-1,2-Dioxetane – Experiments in Chemiexcitation, Chemiluminescence, Photochemistry, Chemical Dynamics, and Spectroscopy, *Acc. Chem. Res.* 1974, **7**, 97–105.

508. Richter, M. M., Electrochemiluminescence (ECL), *Chem. Rev.* 2004, **104**, 3003–3036.

509. Givens, R. S., Schowen, R. L., The Peroxyoxalate Chemiluminescence Reaction. In Birks, J. B. (ed), *Chemiluminescence and Photochemical Reaction Detection in Chromatography*, VCH, New York, 1989, pp. 125–147.

510. Chandross, E. A., A New Chemiluminescent System, *Tetrahedron Lett.* 1963, 761–765.

511. Rauhut, M. M., Chemiluminescence from Concerted Peroxide Decomposition Reactions, *Acc. Chem. Res.* 1969, **2**, 80–87.

512. Tonkin, S. A., Bos, R., Dyson, G. A., Lim, K. F., Russell, R. A., Watson, S. P., Hindson, C. M., Barnett, N. W., Studies on the Mechanism of the Peroxyoxalate Chemiluminescence Reaction, *Anal. Chim. Acta* 2008, **614**, 173–181.

513. Orlovic, M., Schowen, R. L., Givens, R. S., Alvarez, F., Matuszewski, B., Parekh, N., A Simplified Model for the Dynamics of Chemiluminescence in the Oxalate–Hydrogen Peroxide System: Toward a Reaction Mechanism, *J. Org. Chem.* 1989, **54**, 3606–3610.

514. Zhang, Y., Zeng, X.-R., You, X.-Z., A New Complete Basis Set Model (CBS-QB3) Study on the Possible Intermediates in Chemiluminescence, *J. Chem. Phys.* 2000, **113**, 7731–7734.

515. Silva, S. M., Wagner, K., Weiss, D., Beckert, R., Stevani, C. V., Baader, W. J., Studies on the Chemiexcitation Step in Peroxyoxalate Chemiluminescence Using Steroid-substituted Activators, *Luminescence* 2002, **17**, 362–369.

516. Schuster, G. B., Chemiluminescence of Organic Peroxides. Conversion of Ground-state Reactants to Excited-state Products by the Chemically Initiated Electron-exchange Luminescence Mechanism, *Acc. Chem. Res.* 1979, **12**, 366–373.

517. De Vico, L., Liu, Y.-J., Krogh, J. W., Lindh, R., Chemiluminescence of 1,2-Dioxetane. Reaction Mechanism Uncovered, *J. Phys. Chem. A* 2007, **111**, 8013–8019.

518. Viviani, V. R., Arnoldi, F. G. C., Neto, A. J. S., Oehlmeyer, T. L., Bechara, E. J. H., Ohmiya, Y., The Structural Origin and Biological Function of pH Sensitivity in Firefly Luciferases, *Photochem. Photobiol. Sci.* 2008, **7**, 159–169.

519. Berson, J. A., Non-Kekulé Molecules as Reactive Intermediates. In Moss, R. A., Platz, M. S., Jones, M. (eds), *Reactive Intermediate Chemistry*, John Wiley & Sons, Inc., Hoboken, NJ, 2004, pp. 165–203.

520. Linstrom, J., Mallard, W. G. (eds), *NIST Chemistry WebBook*. NIST Standard Reference Database No. 69, National Institute of Standards and Technology, Gaithersburg, MD, 2005, http://webbook.nist.gov.

521. Saltiel, J., Waller, A. S., Sears, D. F., Dynamics of *cis*-Stilbene Photoisomerization – the Adiabatic Pathway to Excited *trans*-Stilbene, *J. Photochem. Photobiol. A* 1992, **65**, 29–40.

522. Friedrich, S., Griebel, R., Hohlneicher, G., Metz, F., Schneider, S., Description of the Radiative Properties of Unsaturated Molecules Containing a Triple Bond Based on a Quasi $\pi$-Model, *Chem. Phys.* 1973, **1**, 319–329.

523. Kropp, P. J., Photorearrangement and Fragmentation of Alkenes. In Horspool, W. M., Lenci, F. (eds), *CRC Handbook of Organic Photochemistry and Photobiology*, 2nd edn, CRC Press LLC, Boca Raton, 2004, Chapter 13, pp. 1–15.

524. Mori, T., Inoue, Y., C=C Photoinduced Isomerization Reactions. In Griesbeck, A. G., Mattay, J. (eds), *Synthetic Organic Photochemistry*, Marcel Dekker, New York, 2005, pp. 417–452.

525. Saltiel, J., Sears, D. F. Jr., Ko, D.-H., Park, K.-M., *Cis–trans* Isomerization of Alkenes. In Horspool, W. M., Song, P.-S. (eds), *CRC Handbook of Organic Photochemistry and Photobiology*, CRC Press, Boca Raton, FL, 1995, pp. 3–15.

526. Arai, T., Photochemical *cis–trans* Isomerization in the Triplet State. In Ramamurthy, V., Schanze, K. (eds), *Organic Molecular Photochemistry*, Vol. 3, Marcel Dekker, New York, 1999, pp. 131–167.

527. Rao, V. J., Photochemical *cis–trans* Isomerization from the Singlet Excited State. In Ramamurthy, V., Schanze, K. (eds), *Organic Molecular Photochemistry*, Vol. 3, Marcel Dekker, New York, 1999, pp. 169–209.

528. Kropp, P. J., Photorearrangement, Fragmentation of Alkenes. In Horspool, W. M., Song, P.-S. (eds), *CRC Handbook of Organic Photochemistry and Photobiology*, CRC Press, Boca Raton, FL, 1995, pp. 16–28.

529. Waldeck, D. H., Photoisomerization Dynamics of Stilbenes, *Chem. Rev.* 1991, **91**, 415–436.

530. Mori, T., Inoue, Y., Photochemical Isomerization of Cycloalkenes. In Horspool, W. M., Lenci, F. (eds), *CRC Handbook of Organic Photochemistry and Photobiology*, 2nd edn, CRC Press LLC, Boca Raton, FL, 2004, Chapter 16, pp. 1–16.

531. Liu, R. S. H., Hammond, G. S., Hula-Twist: a Photochemical Reaction Mechanism Involving Simultaneous Configurational and Conformational Isomerization. In Horspool, W. M., Lenci, F. (eds), *CRC Handbook of Organic Photochemistry and Photobiology*, 2nd edn, CRC Press LLC, Boca Raton, FL, 2004, Chapter 26, pp. 1–11.

532. Mattay, J., Griesbeck, A. G. (eds), Alkenes, Arylalkenes and Cycloalkenes. In *Photochemical Key Steps in Organic Synthesis*, VCH, Weinheim, 1994, pp. 201–267.

533. Mazzucato, U., Momicchioli, F., Rotational Isomerism in *trans*-1,2-Diarylethylenes, *Chem. Rev.* 1991, **91**, 1679–1719.

534. Olivucci, M. (ed), *Computational Photochemistry*, Elsevier, Amsterdam, 2005.

535. Kutateladze, A. G. (ed), *Computational Methods in Photochemistry*, Vol. 13, CRC Press LLC, Boca Raton, FL, 2005.

536. Ben-Nun, M., Molnar, F., Schulten, K., Martinez, T. J., The Role of Intersection Topography in Bond Selectivity of *cis–trans* Photoisomerization, *Proc. Natl. Acad. Sci. USA* 2002, **99**, 1769–1773.

537. Ben-Nun, M., Martinez, T. J., Photodynamics of Ethylene: *Ab Initio* Studies of Conical Intersections, *Chem. Phys.* 2000, **259**, 237–248.

538. Ben-Nun, M., Quenneville, J., Martinez, T. J., *Ab Initio* Multiple Spawning: Photochemistry from first Principles Quantum Molecular Dynamics, *J. Phys. Chem. A* 2000, **104**, 5161–5175.

539. Olivucci, M., Ragazos, I. N., Bernardi, F., Robb, M. A., A Conical Intersection Mechanism for the Photochemistry of Butadiene – A MC-SCF Study, *J. Am. Chem. Soc.* 1993, **115**, 3710–3721.

540. Ben-Nun, M., Martinez, T. J., *Ab Initio* Molecular Dynamics Study of *cis–trans* Photo-isomerization in Ethylene, *Chem. Phys. Lett.* 1998, **298**, 57–65.

541. Buenker, R. J., Bonacic-Koutecky, V., Pogliani, L., Potential-energy and Dipole-moment Surfaces for Simultaneous Torsion and Pyramidalization of Ethylene in Its Lowest-lying Singlet Excited States – A CI Study of the Sudden Polarization Effect, *J. Chem. Phys.* 1980, **73**, 1836–1849.

542. Bonacic-Koutecky, V., Persico, M., Dohnert, D., Sevin, A., CI Study of Geometrical Relaxation in the Excited States of Butadiene – Energy Surfaces and Properties for Simultaneous Torsion and Elongation of One Double Bond, *J. Am. Chem. Soc.* 1982, **104**, 6900–6907.

543. Ohmine, I., Morokuma, K., Photoisomerization of Polyenes – Reaction Coordinate and Trajectory in Triplet Mechanism, *J. Chem. Phys.* 1981, **74**, 564–569.

544. Bonacic-Koutecky, V., Koutecky, J., Michl, J., Neutral and Charged Biradicals, Zwitterions, Funnels in S₁, and Proton Translocation – Their Role in Photochemistry, Photophysics, and Vision, *Angew. Chem. Int. Ed. Engl.* 1987, **26**, 170–189.

545. Migani, A., Robb, M. A., Olivucci, M., Relationship between Photoisomerization Path and Intersection Space in a Retinal Chromophore Model, *J. Am. Chem. Soc.* 2003, **125**, 2804–2808.

546. Fantacci, S., Migani, A., Olivucci, M., CASPT2/CASSCF and TDDFT/CASSCF Mapping of the Excited State Isomerization Path of a Minimal Model of the Retinal Chromophore, *J. Phys. Chem. A* 2004, **108**, 1208–1213.

547. Ferre, N., Olivucci, M., Probing the Rhodopsin Cavity with Reduced Retinal Models at the CASPT2/CASSCF/AMBER Level of Theory, *J. Am. Chem. Soc.* 2003, **125**, 6868–6869.

548. Orlandi, G., Palmieri, P., Poggi, G., *Ab Initio* Study of the *cis–trans* Photoisomerization of Stilbene, *J. Am. Chem. Soc.* 1979, **101**, 3492–3497.

549. Molina, V., Merchan, M., Roos, B. O., Theoretical Study of the Electronic Spectrum of *trans*-Stilbene, *J. Phys. Chem. A* 1997, **101**, 3478–3487.

550. Garavelli, M., Smith, B. R., Bearpark, M. J., Bernardi, F., Olivucci, M., Robb, M. A., Relaxation Paths and Dynamics of Photoexcited Polyene Chains: Evidence for Creation and Annihilation of Neutral Soliton Pairs, *J. Am. Chem. Soc.* 2000, **122**, 5568–5581.

551. Ruiz, D. S., Cembran, A., Garavelli, M., Olivucci, M., Fuss, W., Structure of the Conical Intersections Driving the *cis–trans* Photoisomerization of Conjugated Molecules, *Photochem. Photobiol.* 2002, **76**, 622–633.

552. Strambi, A., Coto, P. B., Frutos, L. M., Ferre, N., Olivucci, M., Relationship between the Excited State Relaxation Paths of Rhodopsin and Isorhodopsin, *J. Am. Chem. Soc.* 2008, **130**, 3382–3388.

553. Saltiel, J., D'Agostino, J., Megarity, E. D., Metts, D., Neuberger, K. R., Wrighton, M., Zafiriou, O.C., The *cis–trans* Photoisomerization of Olefins, *Org. Photochem.* 1973, **3**, 1–113.

554. Merer, A. J., Mulliken, R. S., Ultraviolet Spectra and Excited States of Ethylene and Its Alkyl Derivatives, *Chem. Rev.* 1969, **69**, 639–656.

555. Inoue, Y., Daino, Y., Tai, A., Hakushi, T., Okada, T., Synchrotron-radiation Study of Weak Fluorescence from Neat Liquids of Simple Alkenes – Anomalous Excitation-Spectra as Evidence for Wavelength-dependent Photochemistry, *J. Am. Chem. Soc.* 1989, **111**, 5584–5586.

556. Inoue, Y., Mukai, T., Hakushi, T., Wavelength-dependent Photochemistry of 2,3-Dimethyl-2-butene and 2-Octene in Solution, *Chem. Lett.* 1983, 1665–1668.

557. Kropp, P. J., Snyder, J. J., Rawlings, P. C., Fravel, H. G., Photochemistry of Cycloalkenes. 9. Photodimerization of Cyclohexene, *J. Org. Chem.* 1980, **45**, 4471–4474.

558. Ziegler, K., Wilms, H., Über Vielgliedrige Ringsysteme XIII: Ungesättigte Kohlenwasserstoff-8-Ringe, *Liebigs Ann. Chem.* 1950, **567**, 1–43.

559. Dyck, R. H., McClure, D. S., Ultraviolet Spectra of Stilbene, *p*-Monohalogen Stilbenes, and Azobenzene and *trans* to *cis* Photoisomerization Process, *J. Chem. Phys.* 1962, **36**, 2326–2345.

560. Sension, R. J., Repinec, S. T., Szarka, A. Z., Hochstrasser, R. M., Femtosecond Laser Studies of the *cis*-Stilbene Photoisomerization Reactions, *J. Chem. Phys.* 1993, **98**, 6291–6315.

561. Mallory, F. B., Wood, C. S., Gordon, J. T., Photochemistry of Stilbenes. 3. Some Aspects of Mechanism of Photocyclization to Phenanthrenes, *J. Am. Chem. Soc.* 1964, **86**, 3094–3102.

562. Hayakawa, J., Momotake, A., Arai, T., Water-soluble Stilbene Dendrimers, *Chem. Commun.* 2003, 94–95.

563. Lewis, F. D., Oxman, J. D., Gibson, L. L., Hampsch, H. L., Quillen, S. L., Lewis Acid Catalysis of Photochemical Reactions. 4. Selective Isomerization of Cinnamic Esters, *J. Am. Chem. Soc.* 1986, **108**, 3005–3015.

564. Jacobs, H. J. C., Havinga, E., Photochemistry of Vitamin D and Its Isomers and of Simple Trienes, *Adv. Photochem.* 1979, **11**, 305–373.

565. Saltiel, J., Krishna, T. S. R., Turek, A. M., Photoisomerization of *cis*-1-(2-Naphthyl)-2-phenylethene in Methylcyclohexane at 77 K: No Hula-Twist, *J. Am. Chem. Soc.* 2005, **127**, 6938–6939.

566. Warshel, A., Bicycle-pedal Model for 1st Step in Vision Process, *Nature* 1976, **260**, 679–683.

567. Saltiel, J., Krishna, T. S. R., Turek, A. M., Clark, R. J., Photoisomerization of *cis,cis*-1,4-Diphenyl-1,3-Butadiene in Glassy Media at 77 K: the Bicycle-pedal Mechanism, *Chem. Commun.* 2006, 1506–1508.

568. Saltiel, J., Krishna, T. S. R., Laohhasurayotin, S., Fort, K., Clark, R. J., Photoisomerization of *cis,cis*- to *trans,trans*-1,4-Diaryl-1,3-butadienes in the Solid State: the Bicycle-pedal Mechanism, *J. Phys. Chem. A* 2008, **112**, 199–209.

569. Liu, R. S. H., Photoisomerization by Hula Twist: a Fundamental Supramolecular Photochemical Reaction, *Acc. Chem. Res.* 2001, **34**, 555–562.

570. Yang, L. Y., Liu, R. S. H., Boarman, K. J., Wendt, N. L., Liu, J., New Aspects of Diphenylbutadiene Photochemistry. Regiospecific Hula-twist Photoisomerization, *J. Am. Chem. Soc.* 2005, **127**, 2404–2405.

571. Saltiel, J., Bremer, M. A., Laohhasurayotin, S., Krishna, T. S. R., Photoisomerization of *cis,cis*- and *cis,trans*-1,4-di-*o*-tolyl-1,3-butadiene in Glassy Media at 77 K: One-bond Twist and Bicycle-pedal Mechanisms, *Angew. Chem.* 2008, **47**, 1237–1240.

572. Maessen, P. A., Jacobs, H. J. C., Cornelisse, J., Havinga, E., Studies of Vitamin D and Related Compounds. 30. Photochemistry of Previtamin $D_3$ at 92 K – Formation of an Unstable Tachysterol Rotamer, *Angew. Chem. Int. Ed. Engl.* 1983, **22**, 718–719.

573. Muller, A. M., Lochbrunner, S., Schmid, W. E., Fuss, W., Low-temperature Photochemistry of Previtamin D: a Hula-twist Isomerization of a Triene, *Angew. Chem. Int. Ed.* 1998, **37**, 505–507.

574. Shichida, Y., Yoshizawa, T., Photochemical Aspect of Rhodopsin. In Horspool, W. M., Lenci, F. (eds), *CRC Handbook of Organic Photochemistry and Photobiology*, 2nd edn, CRC Press LLC, Boca Raton, FL, 2004, Chapter 125, pp. 1–13.

575. Frutos, L. M., Andruniow, T., Santoro, F., Ferre, N., Olivucci, M., Tracking the Excited-state Time Evolution of the Visual Pigment with Multiconfigurational Quantum Chemistry, *Proc. Natl. Acad. Sci. USA* 2007, **104**, 7764–7769.

576. Wang, Q., Schoenlein, R. W., Peteanu, L. A., Mathies, R. A., Shank, C. V., Vibrationally Coherent Photochemistry in the Femtosecond Primary Event of Vision, *Science* 1994, **266**, 422–424.

577. Song, L., El-Sayed, M. A., Lanyi, J. K., Protein Catalysis of the Retinal Subpicosecond Photoisomerization in the Primary Process of Bacteriorhodopsin Photosynthesis, *Science* 1993, **261**, 891–894.

578. McDonagh, A. F., Palma, L. A., Lightner, D. A., Blue Light and Bilirubin Excretion, *Science* 1980, **208**, 145–151.

579. McDonagh, A. F., Palma, L. A., Trull, F. R., Lightner, D. A., Phototherapy for Neonatal Jaundice – Configurational Isomers of Bilirubin, *J. Am. Chem. Soc.* 1982, **104**, 6865–6867.

580. Zietz, B., Gillbro, T., Initial Photochemistry of Bilirubin Probed by Femtosecond Spectroscopy, *J. Chem. Phys. B* 2007, **111**, 11997–12003.

581. Snyder, J. J., Tise, F. P., Davis, R. D., Kropp, P. J., Photochemistry of Alkenes. 7. *E–Z* Isomerization of Alkenes Sensitized with Benzene and Derivatives, *J. Org. Chem.* 1981, **46**, 3609–3611.

582. Swenton, J. S., Photoisomerization of *cis*-Cyclooctene to *trans*-Cyclooctene, *J. Org. Chem.* 1969, **34**, 3217–3218.

583. Inoue, Y., Yamasaki, N., Yokoyama, T., Tai, A., Enantiodifferentiating *Z–E* Photoisomerization of Cyclooctene Sensitized by Chiral Polyalkyl Benzenepolycarboxylates, *J. Org. Chem.* 1992, **57**, 1332–1345.

584. Inoue, Y., Asymmetric Photochemical Reactions in Solution, *Chem. Rev.* 1992, **92**, 741–770.

585. Rau, H., Direct Asymmetric Photochemistry with Circularly Polarized Light. In Inoue, Y., Ramamurthy, V. (eds), *Chiral Photochemistry*, Vol. 11, Marcel Dekker, New York, 2004, pp. 1–44.

586. Feringa, B. L., van Delden, R. A., Koumura, N., Geertsema, E. M., Chiroptical Molecular Switches, *Chem. Rev.* 2000, **100**, 1789–1816.

587. Bernstein, W. J., Calvin, M., Buchardt, O., Absolute Asymmetric Synthesis. 1. On Mechanism of Photochemical Synthesis of Nonracemic Helicenes with Circularly Polarized Light – Wavelength Dependence of Optical Yield of Octahelicene, *J. Am. Chem. Soc.* 1972, **94**, 494–498.

588. Sonoda, Y., [2 + 2] Photocycloadditions in the Solid State. In Horspool, W. M., Lenci, F. (eds), *CRC Handbook of Organic Photochemistry and Photobiology*, 2nd edn, CRC Press LLC, Boca Raton, FL, 2004, Chapter 73, pp. 1–15.

589. Koshima, H., Chiral Solid-state Photochemistry Including Supramolecular Approaches. In Inoue, Y., Ramamurthy, V. (eds), *Chiral Photochemistry*, Vol. 11, Marcel Dekker, New York, 2004, pp. 485–532.

590. Scheffer, J. R., The Solid-state Ionic Chiral Auxiliary Approach to Asymmetric Induction in Photochemical Reactions. In Inoue, Y., Ramamurthy, V. (eds), *Chiral Photochemistry*, Vol. 11, Marcel Dekker, New York, 2004, pp. 463–484.

591. Grosch, B., Bach, T., Template-induced Enantioselective Photochemical Reactions in Solution. In Inoue, Y., Ramamurthy, V. (eds), *Chiral Photochemistry*, Vol. 11, Marcel Dekker, New York, 2004, pp. 315–340.

592. Ramamurthy, V., Natarajan, A., Lakshmi, S. K., Karthikeyan, S., Shailaja, J., Sivaguru, J., Chiral Photochemistry Within Zeolites. Inoue, Y., Ramamurthy, V. (eds), *Chiral Photochemistry*, Vol. 11, Marcel Dekker, New York, 2004, pp. 563–632.

593. Inoue, Y., Yamasaki, N., Yokoyama, T., Tai, A., Highly Enantiodifferentiating Photoisomerization of Cyclooctene by Congested and or Triplex-forming Chiral Sensitizers, *J. Org. Chem.* 1993, **58**, 1011–1018.

594. Schenck, G. O., Steinmetz, R., Neuartige durch Benzophenon Photosensibilisierte Additionen von Maleinsäureanhydrid an Benzol und andere Aromaten, *Tetrahedron Lett.* 1960, **1**, 1–8.

595. Saltiel, J., Neuberger, K. R., Wrighton, M., The Nature of the Intermediates in the Sensitized *cis–trans* Photoisomerization of Alkenes, *J. Am. Chem. Soc.* 1969, **91**, 3658–3659.

596. Leigh, W. J., Diene/Cyclobutene Photochemistry. In Horspool, W. M., Song, P.-S. (eds), *CRC Handbook of Organic Photochemistry and Photobiology*, CRC Press, Boca Raton, FL, 1995, pp. 123–142.

597. Laarhoven, W. H., Jacobs, H. J. C., Photochemistry of Acyclic 1,3, 5-Trienes and Related Compounds. In Horspool, W. M., Song, P.-S. (eds), *CRC Handbook of Organic Photochemistry and Photobiology*, CRC Press, Boca Raton, FL, 1995, pp. 143–154.

598. Jacobs, H. J. C., Photochemistry of Vitamin D and Related Compounds. In Horspool, W. M., Song, P.-S. (eds), *CRC Handbook of Organic Photochemistry and Photobiology*, CRC Press, Boca Raton, FL, 1995, pp. 155–164.

599. Gilbert, A., Cyclization of Stilbene and Its Derivatives. In Horspool, W. M., Song, P.-S. (eds), *CRC Handbook of Organic Photochemistry and Photobiology*, CRC Press, Boca Raton, FL, 1995, pp. 291–300.

600. Dunkin, I. R., Low Temperature Matrix Photochemistry of Alkenes. In Horspool, W. M., Lenci, F. (eds), *CRC Handbook of Organic Photochemistry and Photobiology*, 2nd edn, CRC Press LLC, Boca Raton, FL, 2004, Chapter 12, pp. 1–17.

601. Beaudry, C. M., Malerich, J. P., Trauner, D., Biosynthetic and Biomimetic Electrocyclizations, *Chem. Rev.* 2005, **105**, 4757–4778.

602. Hoffmann, N., Photochemical Reactions as Key Steps in Organic Synthesis, *Chem. Rev.* 2008, **108**, 1052–1103.

603. Li, X. Y., He, F. C., Tian, A. M., Yan, G. S., Study on Photochemistry of Concerted [1,$j$]-Sigmatropic and [$i$,$j$]-Sigmatropic Rearrangements by the Classical Path Method, *THEOCHEM* 1995, **342**, 181–186.

604. Morihashi, K., Kikuchi, O., Suzuki, K., Non-adiabatic Coupling Constants for the Disrotatory and Conrotatory Isomerization Paths Between Butadiene and Cyclobutene, *Chem. Phys. Lett.* 1982, **90**, 346–350.

605. Share, P. E., Kompa, K. L., Peyerimhoff, S. D., Vanhemert, M. C., An MRD-CI Investigation of the Photochemical Isomerization of Cyclohexadiene to Hexatriene, *Chem. Phys.* 1988, **120**, 411–419.

606. Fuss, W., Hering, P., Kompa, K. L., Lochbrunner, S., Schikarski, T., Schmid, W. E., Trushin, S. A., Ultrafast Photochemical Pericyclic Reactions and Isomerizations of Small Polyenes, *Ber. Bunsen-Ges. Phys. Chem.* 1997, **101**, 500–509.

607. Bernardi, F., Olivucci, M., Robb, M. A., Tonachini, G., Can a Photochemical Reaction Be Concerted? – A Theoretical Study of the Photochemical Sigmatropic Rearrangement of But-1-ene, *J. Am. Chem. Soc.* 1992, **114**, 5805–5812.

608. Garavelli, M., Bernardi, F., Moliner, V., Olivucci, M., Intrinsically Competitive Photoinduced Polycyclization and Double-bond Shift through a Boatlike Conical Intersection, *Angew. Chem. Int. Ed.* 2001, **40**, 1466–1468.

609. Garavelli, M., Frabboni, B., Fato, M., Celani, P., Bernardi, F., Robb, M. A., Olivucci, M., Photochemistry of Highly Alkylated Dienes: Computational Evidence for a Concerted Formation of Bicyclobutane, *J. Am. Chem. Soc.* 1999, **121**, 1537–1545.

610. Garavelli, M., Page, C. S., Celani, P., Olivucci, M., Schmid, W. E., Trushin, S. A., Fuss, W., Reaction Path of a sub-200 fs Photochemical Electrocyclic Reaction, *J. Phys. Chem. A* 2001, **105**, 4458–4469.

611. Vanderlinden, P., Boue, S., Direct Photolysis of Penta-1,3-Dienes –Recognition of Wavelength-dependent Behavior in Solution, *J. Chem. Soc., Chem. Commun.* 1975, 932–933.

612. Dauben, W. G., Kellogg, M. S., Photochemistry of *cis*-Fused Bicyclo[4.*n*.0]-2,4-Dienes. Ground-state Conformational Control, *J. Am. Chem. Soc.* 1980, **102**, 4456–4463.

613. Gielen, J. W. J., Jacobs, H. J. C., Havinga, E., Influence of Wavelength on Photochemistry of *E*-Hexa-1,3,5-Trienes and *Z*-Hexa-1,3,5-Trienes, *Tetrahedron Lett.* 1976, 3751–3754.

614. Bois, F., Gardette, D., Gramain, J. C., A New Asymmetric Synthesis of (*S*)-(+)-Pipecoline and (*S*)-(+)- and (*R*)-(−)-Coniine by Reductive Photocyclization of Dienamides, *Tetrahedron Lett.* 2000, **41**, 8769–8772.

615. Holick, M. F., Vitamin D Deficiency, *N. Engl. J. Med.* 2007, **357**, 266–281.

616. Zhu, G. D., Okamura, W. H., Synthesis of Vitamin D (Calciferol), *Chem. Rev.* 1995, **95**, 1877–1952.

617. Saltiel, J., Cires, L., Turek, A. M., Conformer-specific Photochemistry in the Vitamin D Field. In Horspool, W. M., Lenci, F. (eds), *CRC Handbook of Organic Photochemistry and Photobiology*, 2nd edn, CRC Press LLC, Boca Raton, FL, 2004.

618. Jacobs, H. J. C., Photochemistry of Conjugated Trienes – Vitamin D Revisited, *Pure Appl. Chem.* 1995, **67**, 63–70.

619. Havinga, E., Schlatmann, J., Remarks on Specificities of Photochemical and Thermal Transformations in Vitamin D Field, *Tetrahedron* 1961, **16**, 146–152.

620. Dauben, W. G., Disanayaka, B., Funhoff, D. J. H., Kohler, B. E., Schilke, D. E., Zhou, B. L., Polyene $2^1A_g$ and $1^1B_u$ States and the Photochemistry of Previtamin $D_3$, *J. Am. Chem. Soc.* 1991, **113**, 8367–8374.

621. Enas, J. D., Palenzuela, J. A., Okamura, W. H., Studies on Vitamin-D (Calciferol) and Its Analogs. 37. *A*-Homo-11-hydroxy-3-deoxyvitamin D – Ring Size and π-Facial Selectivity Effects on the 1,7-Sigmatropic Hydrogen Shift of Previtamin D to Vitamin D, *J. Am. Chem. Soc.* 1991, **113**, 1355–1363.

622. Laarhoven, W. H., Photochemical Cyclizations and Intramolecular Cycloadditions of Conjugated Arylolefins.1. Photocyclization with Dehydrogenation, *Recl. Trav. Chim. Pays-Bas* 1983, **102**, 185–204.

623. Martin, R. H., Marchant, M. J., Baes, M., Syntheses in Field of Polycyclic Aromatic Compounds. 31. Rapid Syntheses of Hexa and Heptahelicene, *Helv. Chim. Acta* 1971, **54**, 358–360.

624. Bouas-Laurent, H., Durr, H., Organic Photochromism, *Pure Appl. Chem.* 2001, **73**, 639–665.

625. Yamaguchi, T., Uchida, K., Irie, M., Asymmetric Photocyclization of Diarylethene Derivatives, *J. Am. Chem. Soc.* 1997, **119**, 6066–6071.

626. Cookson, R. C., Hudec, J., Sharma, M., Retention of Configuration at C-1 of Allyl Group in a Photochemical 1,3-Allylic Shift of a Benzyl Group, *J. Chem. Soc. D* 1971, 108.

627. Kiefer, E. F., Tanna, C. H., Alternative Electrocyclic Pathways. Photolysis and Thermolysis of Dimethylallene Dimers, *J. Am. Chem. Soc.* 1969, **91**, 4478–4480.

628. Singh, V., Alam, S. Q., Intramolecular Diels–Alder Reaction in 1-Oxaspiro[2.5]octa-5,7-dien-4-one and Sigmatropic Shifts in Excited States: Novel Route to Sterpuranes and Linear Triquinanes: Formal Total Synthesis of (±)-Coriolin, *Chem. Commun.* 1999, 2519–2520.

629. Armesto, D., Ortiz, M. J., Agarrabeitia, A. R., Recent Advances in Di-π-methane Photochemistry – A New Look at a Classical Reaction. In Ramamurthy, V., Schanze, K. (eds), *Molecular and Supramolecular Photochemistry*, Vol. 9, Marcel Dekker, New York, 2003, pp. 1–42.

630. Armesto, D., Ortiz, M. J., Agarrabeitia, A. R., Di-π-methane Rearrangement. In Griesbeck, A. G., Mattay, J. (eds), *Synthetic Organic Photochemistry*, Vol. 12, Marcel Dekker, New York, 2005, pp. 161–187.

631. Rao, V. J., Griesbeck, A. G., Oxa-di-π-methane Rearrangement. In Griesbeck, A.G., Mattay, J. (eds), *Synthetic Organic Photochemistry*, Vol. 12, Marcel Dekker, New York, 2005, pp. 189–210.

632. Zimmerman, H. E., Armesto, D., Synthetic Aspects of the Di-π-methane Rearrangement, *Chem. Rev.* 1996, **96**, 3065–3112.

633. Zimmerman, H. E., The Di-π-methane Rearrangement. In Horspool, W. M., Song, P.-S. (eds), *CRC Handbook of Organic Photochemistry and Photobiology*, CRC Press, Boca Raton, FL, 1995, pp. 184–193.

634. Tsuno, T., Sugiyama, K., The Photochemical Reactivity of the Allenyl–Vinyl Methane System. Horspool, W.M., Lenci, F. (eds), *CRC Handbook of Organic Photochemistry and Photobiology*, 2nd edn, CRC Press LLC, Boca Raton, FL, 2004, Chapter 30, pp. 1–15.

635. Singh, V., Photochemical Rearrangements in β, γ-Unsaturated Enones: the Oxa-di-π-methane Rearrangement. In Horspool, W. M., Lenci, F. (eds), *CRC Handbook of Organic Photochemistry and Photobiology*, 2nd edn, CRC Press LLC, Boca Raton, FL, 2004, pp. 1–34.

636. Armesto, D., Ortiz, M. J., Agarrabeitia, A. R., Novel Di-π-methane Rearrangements Promoted by Photoelectron Transfer and Triplet Sensitization. In Horspool, W. M., Lenci, F. (eds), *CRC Handbook of Organic Photochemistry and Photobiology*, 2nd edn, CRC Press LLC, Boca Raton, FL, 2004, Chapter 95, pp. 1–16.

637. Liao, C.-C., Yang, P.-H., Photorearrangements of Benzobarrelenes and Related Analogues. In Horspool, W. M., Song, P.-S. (eds), *CRC Handbook of Organic Photochemistry and Photobiology*, CRC Press, Boca Raton, FL, 1995, pp. 194–203.

638. Scheffer, J. R., Yang, J., The Photochemistry of Dibenzobarrelene (9,10-Ethenoanthracene) and Its Derivatives. In Horspool, W. M., Song, P.-S. (eds), *CRC Handbook of Organic Photochemistry and Photobiology*, CRC Press, Boca Raton, FL, 1995, pp. 204–221.

639. Liao, C.-C., Peddinti, R. K., Photochemistry of Heteroarene-fused Barrelenes (Chapter 32). In Horspool, W. M., Lenci, F. (eds), *CRC Handbook of Organic Photochemistry and Photobiology*, 2nd edn, CRC Press LLC, Boca Raton, FL, 2004, pp. 1–17.

640. Margaretha, P., Photorearrangement Reactions of Cyclohex-2-enones. In Horspool, W. M., Lenci, F. (eds), *CRC Handbook of Organic Photochemistry and Photobiology*, 2nd edn, CRC Press LLC, Boca Raton, FL, 2004, Chapter 76, pp. 1–12.

641. Schuster, D. I., Photorearrangement Reactions of Cyclohexenones. In Horspool, W. M., Song, P.-S. (eds), *CRC Handbook of Organic Photochemistry and Photobiology*, CRC Press, Boca Raton, FL, 1995, pp. 579–592.

642. Armesto, D., The Aza-di-π-methane Rearrangement. In Horspool, W. M., Song, P.-S. (eds), *CRC Handbook of Organic Photochemistry and Photobiology*, CRC Press, Boca Raton, FL, 1995, pp. 915–936.

643. Hixson, S. S., Mariano, P. S., Zimmerman, H. E., Di-π-methane and Oxa-di-π-methane Rearrangements, *Chem. Rev.* 1973, **73**, 531–551.

644. Zimmerman, H. E., Sulzbach, H. M., Tollefson, M. B., Experimental and Theoretical Exploration of the Detailed Mechanism of the Rearrangement of Barrelenes to Semibullvalenes – Diradical Intermediates and Transition-states, *J. Am. Chem. Soc.* 1993, **115**, 6548–6556.

645. Zimmerman, H. E., Kutateladze, A. G., Maekawa, Y., Mangette, J. E., Excited-state Reactivity as a Function of Diradical Structure – Evidence for 2 Triplet Cyclopropyldicarbinyl Diradical Intermediates with Differing Reactivity, *J. Am. Chem. Soc.* 1994, **116**, 9795–9796.

646. Zimmerman, H. E., Kutateladze, A. G., Novel Dissection-analysis of Spin Orbit Coupling in the Type-B Cyclohexenone Photorearrangement – What Controls Photoreactivity? – Mechanistic and Exploratory Organic Photochemistry, *J. Org. Chem.* 1995, **60**, 6008–6009.

647. Garavelli, M., Bernardi, F., Cembran, A., Castano, O., Frutos, L. M., Merchan, M., Olivucci, M., Cyclooctatetraene Computational Photo- and Thermal Chemistry: a Reactivity Model for Conjugated Hydrocarbons, *J. Am. Chem. Soc.* 2002, **124**, 13770–13789.

648. Zimmerman, H. E., Grunewald, G. L., Chemistry of Barrelene. III. A Unique Photo-isomerization to Semibullvalene, *J. Am. Chem. Soc.* 1966, **88**, 183–184.

649. Zimmerman, H. E., Mariano, P. S., Di-π-methane Rearrangement. Interaction of Electronically Excited Vinyl Chromophores. Mechanistic and Exploratory Organic Photochemistry 41, *J. Am. Chem. Soc.* 1969, **91**, 1718–1727.

650. Zimmerman, H. E., Binkley, R. W., Givens, R. S., Sherwin, M. A., Mechanistic Organic Photochemistry. XXIV. Mechanism of Conversion of Barrelene to Semibullvalene. A General Photochemical Process, *J. Am. Chem. Soc.* 1967, **89**, 3932–3933.

651. Quenemoen, K., Borden, W. T., Davidson, E. R., Feller, D., Some Aspects of the Triplet Di-π-methane Rearrangement – Comparison of the Ring-opening of Cyclopropyldicarbinyl and Cyclopropylcarbinyl, *J. Am. Chem. Soc.* 1985, **107**, 5054–5059.

652. Demuth, M., Lemmer, D., Schaffner, K., Electron Spin-resonance and IR Evidence for Intermediates in the Di-π-methane Photorearrangement of a Naphthobarrelene-like Compound after Low-temperature UV Irradiation, *J. Am. Chem. Soc.* 1980, **102**, 5407–5409.

653. Reguero, M., Bernardi, F., Jones, H., Olivucci, M., Ragazos, I. N., Robb, M. A., A Concerted Nonadiabatic Reaction Path for the Singlet Di-π-methane Rearrangement, *J. Am. Chem. Soc.* 1993, **115**, 2073–2074.

654. Zimmerman, H. E., Pratt, A. C., Unsymmetrical Substitution and Direction of Di-π-methane Rearrangement – Mechanistic and Exploratory Organic Photochemistry, 56, *J. Am. Chem. Soc.* 1970, **92**, 6259–6267.

655. Zimmerman, H. E., Klun, R. T., Mechanistic and Exploratory Organic Photochemistry. 112. Di-π-methane Rearrangement of Systems with Simple Vinyl Moieties, *Tetrahedron* 1978, **34**, 1775–1803.

656. Look, S. A., Fenical, W., Vanengen, D., Clardy, J., Erythrolides – Unique Marine Diterpenoids Interrelated by a Naturally-occurring Di-π-methane Rearrangement, *J. Am. Chem. Soc.* 1984, **106**, 5026–5027.

657. Tenney, L. P., Boykin, D. W., Lutz, R. E., Novel Photocyclization of a Highly Phenylated β,γ-Unsaturated Ketone to a Cyclopropyl Ketone Involving Benzoyl Group Migration, *J. Am. Chem. Soc.* 1966, **88**, 1835–1836.

658. Dauben, W. G., Kellogg, M. S., Seeman, J. I., Spitzer, W. A., Photochemical Rearrangement of an Acyclic β,γ-Unsaturated Ketone to a Conjugated Cyclopropyl Ketone – An Oxa-di-π-methane Rearrangement, *J. Am. Chem. Soc.* 1970, **92**, 1786–1787.

659. Houk, K. N., Photochemistry and Spectroscopy of β, γ-Unsaturated Carbonyl Compounds, *Chem. Rev.* 1976, **76**, 1–74.

660. Zimmerman, H. E., Cassel, J. M., Unusual Rearrangements in Di-π-methane Systems – Mechanistic and Exploratory Organic Photochemistry, *J. Org. Chem.* 1989, **54**, 3800–3816.

661. Armesto, D., Martin, J. A. F., Perezossorio, R., Horspool, W. M., A Novel Aza-di-π-methane Rearrangement: the Photoreaction of 4,4-Dimethyl-1,6,6-triphenyl-2-aza-hexa-2,5-diene, *Tetrahedron Lett.* 1982, **23**, 2149–2152.

662. Kropp, P. J., Photoreactions of Alkenes in Protic Media. In Horspool, W. M., Lenci, F. (eds), *CRC Handbook of Organic Photochemistry and Photobiology*, 2nd edn, CRC Press LLC, Boca Raton, FL, 2004, Chapter 9, pp. 1–11.

663. Kropp, P. J., Photoreactions of Alkenes in Protic Media. In Horspool, W. M., Song, P.-S. (eds), *CRC Handbook of Organic Photochemistry and Photobiology*, CRC Press, Boca Raton, FL, 1995, pp. 105–114.

664. Kropp, P. J., Photochemistry of Alkenes in Solution, *Org. Photochem.* 1979, **4**, 1–142.

665. Lewis, F. D., Crompton, E. M., SET Addition of Amines to Alkenes. In Horspool, W. M., Lenci, F. (eds), *CRC Handbook of Organic Photochemistry and Photobiology*, 2nd edn, CRC Press LLC, Boca Raton, FL, 2004, Chapter 7, pp. 1–18.

666. Lewis, F. D., Bassani, D.M., Reddy, G.D., Styrene–Amine and Stilbene–Amine Intramolecular Addition Reactions, *Pure Appl. Chem.* 1992, **64**, 1271–1277.

667. Bunce, N. J., Photochemical Reactions of Arenes with Amines. In Horspool, W. M., Song, P.-S. (eds), *CRC Handbook of Organic Photochemistry and Photobiology*, CRC Press, Boca Raton, FL, 1995, pp. 266–279.

668. Trinquier, G., Paillous, N., Lattes, A., Malrieu, J. P., Ground and Excited-state Conformations of an *ortho*-Allyl Aniline and Their Photochemical Implications a PCILO Study, *New J. Chem.* 1977, **1**, 403–411.

669. Kavarnos, G. J., Turro, N. J., Photosensitization by Reversible Electron Transfer – Theories, Experimental Evidence, and Examples, *Chem. Rev.* 1986, **86**, 401–449.

670. Kavarnos, G. J., *Fundamentals of Photoinduced Electron Transfer*, VCH, New York, 1993.

671. Kropp, P. J., Reardon, E. J., Gaibel, Z. L. F., Williard, K. F., Hattaway, J. H., Photochemistry of Alkenes. Direct Irradiation in Hydroxylic Media, *J. Am. Chem. Soc.* 1973, **95**, 7058–7067.

672. Leigh, W. J., Cook, B. H. O., Stereospecific (Conrotatory) Photochemical Ring Opening of Alkylcyclobutenes in the Gas Phase and in Solution. Ring Opening from the Rydberg Excited State or by Hot Ground State Reaction? *J. Org. Chem.* 1999, **64**, 5256–5263.

673. McEwen, J., Yates, K., Photohydration of Styrenes and Phenylacetylenes – General Acid Catalysis and Brønsted Relationships, *J. Am. Chem. Soc.* 1987, **109**, 5800–5808.

674. Wan, P., Culshaw, S., Yates, K., Photohydration of Aromatic Alkenes and Alkynes, *J. Am. Chem. Soc.* 1982, **104**, 2509–2515.

675. Chiang, Y., Kresge, A. J., Capponi, M., Wirz, J., Direct Observation of Acetophenone Enol Formed by Photohydration of Phenylacetylene, *Helv. Chim. Acta* 1986, **69**, 1331–1332.

676. Wan, P., Davis, M. J., Teo, M. A., Photoaddition of Water and Alcohols to 3-Nitrostyrenes – Structure Reactivity and Solvent Effects, *J. Org. Chem.* 1989, **54**, 1354–1359.

677. Shim, S. C., Kim, D. S., Yoo, D. J., Wada, T., Inoue, Y., Diastereoselectivity Control in Photosensitized Addition of Methanol to (R)-(+)-Limonene, *J. Org. Chem.* 2002, **67**, 5718–5726.

678. Lewis, F. D., Ho, T. I., *trans*-Stilbene–Amine Exciplexes – Photochemical Addition of Secondary and Tertiary Amines to Stilbene, *J. Am. Chem. Soc.* 1977, **99**, 7991–7996.

679. Lewis, F. D., Reddy, G. D., Bassani, D. M., Schneider, S., Gahr, M., Chain-length-dependent and Solvent-dependent Intramolecular Proton Transfer in Styrene Amine Exciplexes, *J. Am. Chem. Soc.* 1994, **116**, 597–605.

680. Fleming, S. A., Photocycloaddition of Alkenes to Excited Alkenes. In Griesbeck, A. G., Mattay, J. (eds), *Synthetic Organic Photochemistry*, Marcel Dekker, New York, 2005, pp. 141–160.

681. Schreiber, S. L., [2 + 2] Photocycloadditions in the Synthesis of Chiral Molecules, *Science* 1985, **227**, 857–863.

682. Oppolzer, W., Intramolecular [2 + 2] Photoaddition Cyclobutane Fragmentation Sequence in Organic Synthesis, *Acc. Chem. Res.* 1982, **15**, 135–141.

683. Grosch, B., Bach, T., Enantioselective Photocycloaddition Reactions in Solution. In Horspool, W.M., Lenci, F. (eds), *CRC Handbook of Organic Photochemistry and Photobiology*, 2nd edn, CRC Press LLC, Boca Raton, FL, 2004, Chapter 61, pp. 1–14.

684. Ghosh, S., Copper(I)-catalyzed Inter- and Intramolecular [2 + 2] Photocycloaddition Reactions of Alkenes. In Horspool, W. M., Lenci, F. (eds), *CRC Handbook of Organic Photochemistry and Photobiology*, 2nd edn, CRC Press LLC, Boca Raton, FL, 2004, Chapter 18, pp. 1–24.

685. Kaupp, G., [2 + 2] Cyclobutane Synthesis (Liquid Phase). Horspool, W. M., Song, P.-S. (eds), *CRC Handbook of Organic Photochemistry and Photobiology*, CRC Press, Boca Raton, FL, 1995, pp. 29–49.

686. Kaupp, G., Cyclobutane Synthesis in the Solid Phase. In Horspool, W. M., Song, P.-S. (eds), *CRC Handbook of Organic Photochemistry and Photobiology*, CRC Press, Boca Raton, FL, 1995, pp. 50–63.

687. Gleiter, R., Treptow, B., Photochemical Synthesis of Cage Compounds: Propellaprismanes and Their Precursors. In Horspool, W. M., Song, P.-S. (eds), *CRC Handbook of Organic Photochemistry and Photobiology*, CRC Press, Boca Raton, FL, 1995, pp. 64–83.

688. Langer, K., Mattay, J., Copper(I)-catalyzed Intra- and Intermolecular Photocycloaddition Reactions of Alkenes. In Horspool, W. M., Song, P.-S. (eds), *CRC Handbook of Organic Photochemistry and Photobiology*, CRC Press, Boca Raton, FL, 1995, pp. 84–104.

689. Bach, T., Stereoselective Intermolecular [2 + 2]-Photocycloaddition Reactions and their Application in Synthesis, *Synthesis* 1998, 683–703.

690. Sieburth, S. M., Photocycloadditions of Alkenes (Dienes) to Dienes ([4 + 2]/[4 + 4]). In Griesbeck, A. G., Mattay, J. (eds), *Synthetic Organic Photochemistry*, Marcel Dekker, New York, 2005, pp. 239–268.

691. Nishimura, J., Nakamura, Y., Yamazaki, T., Inokuma, S., Photochemical Synthesis of Cyclophanes. In Horspool, W. M., Lenci, F. (eds), *CRC Handbook of Organic Photochemistry and Photobiology*, 2nd edn, CRC Press LLC, Boca Raton, FL, 2004, Chapter 19, pp. 1–15.

692. Winkler, J. D., Bowen, C. M., Liotta, F., [2 + 2]-Photocycloaddition/Fragmentation Strategies for the Synthesis of Natural and Unnatural Products, *Chem. Rev.* 1995, **95**, 2003–2020.

693. Lee-Ruff, E., Mladenova, G., Enantiomerically Pure Cyclobutane Derivatives and their Use in Organic Synthesis, *Chem. Rev.* 2003, **103**, 1449–1483.

694. Namyslo, J. C., Kaufmann, D. E., The Application of Cyclobutane Derivatives in Organic Synthesis, *Chem. Rev.* 2003, **103**, 1485–1537.

695. Bernardi, F., Olivucci, M., Robb, M. A., Predicting Forbidden and Allowed Cycloaddition Reactions – Potential Surface-Topology and Its Rationalization, *Acc. Chem. Res.* 1990, **23**, 405–412.

696. Bernardi, F., De, S., Olivucci, M., Robb, M. A., Mechanism of Ground-state Forbidden Photochemical Pericyclic Reactions – Evidence for Real Conical Intersections, *J. Am. Chem. Soc.* 1990, **112**, 1737–1744.

697. Wittekindt, C., Klessinger, M., Intramolecular Photocycloaddition of Nonconjugated Dienes, *J. Inform. Rec.* 1998, **24**, 229–233.

698. Allen, F. H., Mahon, M. F., Raithby, P. R., Shields, G. P., Sparkes, H. A., New Light on the Mechanism of the Solid State [2 + 2] Cycloaddition of Alkenes: a Database Analysis, *New J. Chem.* 2005, **29**, 182–187.

699. Bernardi, F., Bottoni, A., Olivucci, M., Venturini, A., Robb, M. A., *Ab Initio* MC-SCF Study of Thermal and Photochemical [2 + 2] Cycloadditions, *J. Chem. Soc., Faraday Trans.* 1994, **90**, 1617–1630.

700. Bentzien, J., Klessinger, M.,Theoretical Investigations on the Regiochemistry and Stereochemistry of the Photochemical [2 + 2] Cycloaddition of Propene *J. Org. Chem.* 1994, **59**, 4887–4894.

701. Caldwell, R. A., Diaz, J. F., Hrncir, D. C., Unett, D. J., Alken, Triplets as 1,2-Biradicals – The Photoaddition of *p*-Acetylstyrene to Styrene, *J. Am. Chem. Soc.* 1994, **116**, 8138–8145.

702. Unett, D. J., Caldwell, R. A., Hrncir, D. C., Photodimerization of 1-Phenylcyclohexene. A Novel Transient–Transient Component, *J. Am. Chem. Soc.* 1996, **118**, 1682–1689.

703. Takahashi, Y., Okitsu, O., Ando, M., Miyashi, T., Electron Transfer-induced Intramolecular [2 + 2] Cycloaddition of 2,6-Diarylhepta-1,6-Dienes, *Tetrahedron Lett.* 1994, **35**, 3953–3956.

704. Asaoka, S., Ooi, M., Jiang, P. Y., Wada, T., Inoue, Y., Enantiodifferentiating Photocyclodimerization of Cyclohexa-1,3-diene Sensitized by Chiral Arenecarboxylates, *J. Chem. Soc., Perkin Trans. 2* 2000, 77–84.

705. Salomon, R. G.,Homogeneous Metal Catalysis in Organic Photochemistry, *Tetrahedron* 1983, **39**, 485–575.

706. Salomon, R. G., Folting, K., Streib, W. E., Kochi, J. K., Copper(I) Catalysis in Photocycloadditions. 2. Cyclopentene, Cyclohexene, and Cycloheptene, *J. Am. Chem. Soc.* 1974, **96**, 1145–1152.

707. Trecker, D. J., Henry, J. P., McKeon, J. E., Photodimerization of Metal-complexed Olefins, *J. Am. Chem. Soc.* 1965, **87**, 3261–3263.

708. Zimmerman, H. E., Kamm, K. S., Werthemann, D. P., Mechanisms of Electron Demotion – Direct Measurement of Internal Conversion and Intersystem Crossing Rates – Mechanistic Organic Photochemistry, *J. Am. Chem. Soc.* 1975, **97**, 3718–3725.

709. Yamazaki, H., Cvetanovic, R. J., Irwin, R. S., Kinetics of Stereospecific Photochemical Cyclodimerization of 2-Butene in Liquid Phase, *J. Am. Chem. Soc.* 1976, **98**, 2198–2205.

710. Avasthi, K., Raychaudhuri, S. R., Salomon, R. G., Copper(I) Catalysis of Olefin Photoreactions. 13. Synthesis of Bicyclic Vinylcyclobutanes via Copper(I)-catalyzed Intramolecular $2\pi + 2\pi$ Photocycloadditions of Conjugated Dienes to Alkenes, *J. Org. Chem.* 1984, **49**, 4322–4324.

711. Panda, J., Ghosh, S., A New Stereoselective Route to the Carbocyclic Nucleoside Cyclobut-A, *J. Chem. Soc., Perkin Trans. 1* 2001, 3013–3016.

712. Syamala, M. S., Ramamurthy, V., Consequences of Hydrophobic Association in Photoreactions – Photodimerization of Stilbenes in Water *J. Org. Chem.* 1986, **51**, 3712–3715.

713. Ramamurthy, V., *Photochemistry in Organized and Constrained Media*, John Wiley & Sons, Inc., New York, 1991.

714. Schmidt, G. M. J., Topochemistry. 3. Crystal Chemistry of Some *trans*-Cinnamic Acids, *J. Chem. Soc.* 1964, 2014–2021.

715. Addadi, L., Vanmil, J., Lahav, M., Photo-polymerization in Chiral Crystals. 4. Engineering of Chiral Crystals for Asymmetric [$2\pi + 2\pi$] Photo-polymerization – Execution of an Absolute Asymmetric Synthesis with Quantitative Enantiomeric Yield, *J. Am. Chem. Soc.* 1982, **104**, 3422–3429.

716. Novak, K., Enkelmann, V., Wegner, G., Wagener, K. B., Crystallographic Study of a Single-crystal to Single-crystal Photodimerization and Its Thermal Reverse Reaction, *Angew. Chem. Int. Ed. Engl.* 1993, **32**, 1614–1616.

717. Lee, T. S., Lee, S. J., Shim, S. C., [2 + 2] Photocycloaddition Reaction of Aryl-1,3-Butadiynes with Some Olefins, *J. Org. Chem.* 1990, **55**, 4544–4549.

718. Ward, S. C., Fleming, S. A., [2 + 2] Photocycloaddition of Cinnamyloxy Silanes, *J. Org. Chem.* 1994, **59**, 6476–6479.

719. Bradford, C .L., Fleming, S. A., Ward, S. C., Regio-controlled Ene–Yne Photochemical [2 + 2] Cycloaddition Using Silicon as a Tether, *Tetrahedron Lett.* 1995, **36**, 4189–4192.

720. Nishimura, J., Horikoshi, Y., Wada, Y., Takahashi, H., Sato, M., Intramolecular [2 + 2] Photocycloaddition. 10. Conformationally Stable *syn*-[2.2]Metacyclophanes, *J. Am. Chem. Soc.* 1991, **113**, 3485–3489.

721. Arnold, D. R., Trecker, D. J., Whipple, E. B., Stereochemistry of Pentacyclo[8.2.1.1$^{4,7}$.O$^{2,9}$. O$^{3,8}$]tetradecanes and Dienes. Norbornene and Norbornadiene Dimers, *J. Am. Chem. Soc.* 1965, **87**, 2596–2602.

722. Arnold, D. R., Hinman, R. L., Glick, A. H., Chemical Properties of the Carbonyl Normal π* State – The Photochemical Preparation of Oxetanes, *Tetrahedron Lett.* 1964, 1425–1430.

723. Liu, R. S. H., Turro, N. J., Hammond, G. S., Mechanisms of Photochemical Reactions in Solution. 31. Activation and Deactivation of Conjugated Dienes by Energy Transfer, *J. Am. Chem. Soc.* 1965, **87**, 3406–3412.

724. Olah, G. A., *Cage Hydrocarbons*, Wiley-VCH Verlag GmbH, Weinheim, 1990.

725. Shinmyozu, T., Nogita, R., Akita, M., Lim, C., Photochemical Synthesis of Cage Compounds. In Horspool, W. M., Lenci, F. (eds), *CRC Handbook of Organic Photochemistry and Photobiology*, 2nd edn, CRC Press LLC, Boca Raton, FL, 2004, Chapter 22, pp. 1–21.

726. Eaton, P. E., Cole, T. W., Cubane, *J. Am. Chem. Soc.* 1964, **86**, 3157–3157.

727. Gleiter, R., Brand, S., Generation of Octamethylcuneane and Octamethylcubane from *syn*-Octamethyltricyclo[4.2.0.0$^{2,5}$]octa-3,7-diene, *Tetrahedron Lett.* 1994, **35**, 4969–4972.

728. Friedel, M. G., Cichon, M. K., Carell, T., DNA Damage and Repair: Photochemistry. In Horspool, W. M., Lenci, F. (eds), *CRC Handbook of Organic Photochemistry and Photobiology*, 2nd edn, CRC Press LLC, Boca Raton, FL, 2004, Chapter 141, pp. 1–22.

729. Mitchell, D. L., DNA Damage and Repair. In Horspool, W. M., Lenci, F. (eds), *CRC Handbook of Organic Photochemistry and Photobiology*, 2nd edn, CRC Press LLC, Boca Raton, FL, 2004, Chapter 140, pp. 1–8.

730. Broo, A., A Theoretical Investigation of the Physical Reason for the Very Different Luminescence Properties of the Two Isomers Adenine and 2-Aminopurine, *J. Phys. Chem. A* 1998, **102**, 526–531.

731. Crespo-Hernandez, C. E., Cohen, B., Hare, P. M., Kohler, B., Ultrafast Excited-state Dynamics in Nucleic Acids, *Chem. Rev.* 2004, **104**, 1977–2019.

732. Abo-Riziq, A., Grace, L., Nir, E., Kabelac, M., Hobza, P., de Vries, M. S., Photochemical Selectivity in Guanine–Cytosine Base-pair Structures, *Proc. Natl. Acad. Sci. USA* 2005, **102**, 20–23.

733. Cadet, J., Berger, M., Douki, T., Morin, B., Raoul, S., Ravanat, J. L., Spinelli, S., Effects of UV and Visible Radiation on DNA – Final Base Damage, *Biol. Chem.* 1997, **378**, 1275–1286.

734. Ravanat, J. L., Douki, T., Cadet, J., Direct and Indirect Effects of UV Radiation on DNA and Its Components, *J. Photochem. Photobiol. B* 2001, **63**, 88–102.

735. Zhang, R. B., Eriksson, L. A., A Triplet Mechanism for the Formation of Cyclobutane Pyrimidine Dimers in UV-irradiated DNA, *J. Chem. Phys. B* 2006, **110**, 7556–7562.

736. Stern, R. S., Psorale, Ultraviolet a Light Therapy for Psoriasis, *N. Engl. J. Med.* 2007, **357**, 682–690.

737. Dall'Acqua, F., Viola, G., Vedaldi, D., Molecular Basis of Psoralen Photochemotherapy. In Horspool, W. M., Lenci, F. (eds), *CRC Handbook of Organic Photochemistry and Photobiology*, 2nd edn, CRC Press LLC, Boca Raton, FL, 2004, Chapter 142, pp. 1–17.

738. Bergman, R. G., Reactive 1,4-Dehydroaromatics, *Acc. Chem. Res.* 1973, **6**, 25–31.

739. Jones, G. B., Russell, K. C., The Photo-Bergman Cycloaromatization of Enediynes. In Horspool, W. M., Lenci, F. (eds), *CRC Handbook of Organic Photochemistry and Photobiology*, 2nd edn, CRC Press LLC, Boca Raton, FL, 2004, pp. 1–21.

740. Evenzahav, A., Turro, N. J., Photochemical Rearrangement of Enediynes: Is a "Photo-Bergman" Cyclization a Possibility? *J. Am. Chem. Soc.* 1998, **120**, 1835–1841.

741. Schuster, D. I., Lem, G., Kaprinidis, N. A., New Insights into an Old Mechanism – [2 + 2]-Photocycloaddition of Enones to Alkenes, *Chem. Rev.* 1993, **93**, 3–22.

742. Crimmins, M. T., Synthetic Applications of Intramolecular Enone Olefin Photocycloadditions, *Chem. Rev.* 1988, **88**, 1453–1473.

743. Schuster, D. I., Mechanistic Issues in [2 + 2]-Photocycloadditions of Cyclic Enones to Alkenes. In Horspool, W. M., Lenci, F. (eds), *CRC Handbook of Organic Photochemistry and Photobiology*, 2nd edn, CRC Press LLC, Boca Raton, FL, 2004, Chapter 72, pp. 1–24.

744. Pete, J. P., [2 + 2]-Photocycloaddition Reactions of Cyclopentenones with Alkenes. In Horspool, W. M., Lenci, F. (eds), *CRC Handbook of Organic Photochemistry and Photobiology*, 2nd edn, CRC Press LLC, Boca Raton, FL, 2004, Chapter 71, pp. 1–14.

745. Margaretha, P., Photocycloaddition of Cycloalk-2-enones to Alkenes. In Griesbeck, A.G., Mattay, J. (eds), *Synthetic Organic Photochemistry*, Marcel Dekker, New York, 2005, pp. 211–237.

746. Mattay, J., Intramolecular Photocycloadditions to Enones: Influence of the Chain Length. In Horspool, W. M., Song, P.-S. (eds), *CRC Handbook of Organic Photochemistry and Photobiology*, CRC Press, Boca Raton, FL, 1995, pp. 618–633.

747. Weedon, A. C., Photocycloaddition Reactions of Cyclopentenones with Alkenes. In Horspool, W. M., Song, P.-S. (eds), *CRC Handbook of Organic Photochemistry and Photobiology*, CRC Press, Boca Raton, FL, 1995, pp. 634–651.

748. Schuster, D. I., [2 + 2] Photocycloaddition of Cyclohexenones to Alkenes. In Horspool, W. M., Song, P.-S. (eds), *CRC Handbook of Organic Photochemistry and Photobiology*, CRC Press, Boca Raton, FL, 1995, pp. 652–669.

749. Weedon, A. C., [2 + 2] Photocycloaddition Reactions of Enolized 1,3-Diketones with Alkenes: the de Mayo Reaction. In Horspool, W. M., Song, P.-S. (eds), *CRC Handbook of Organic Photochemistry and Photobiology*, CRC Press, Boca Raton, FL, 1995, pp. 670–684.

750. Eaton, P. E., Photochemical Reactions of Simple Alicyclic Enones, *Acc. Chem. Res.* 1968, **1**, 50–57.

751. Mattay, J., Griesbeck, A. G. (eds), Carbonyl Compounds. In *Photochemical Key Steps in Organic Synthesis*, VCH, Weinheim, 1994, pp. 11–118.

752. Garcia-Exposito, E., Bearpark, M. J., Ortuno, R. M., Robb, M. A., Branchadell, V., Theoretical Study of the Photochemical [2 + 2]-Cycloadditions of Cyclic and Acyclic α,β-Unsaturated Carbonyl Compounds to Ethylene, *J. Org. Chem.* 2002, **67**, 6070–6077.

753. Wilsey, S., Gonzalez, L., Robb, M. A., Houk, K. N., Ground- and Excited-state Surfaces for the [2 + 2]-Photocycloaddition of α,β-Enones to Alkenes, *J. Am. Chem. Soc.* 2000, **122**, 5866–5876.

754. Zimmerman, H. E., Sebek, P., Photochemistry in a Crystalline Cage. Control of the Type-B Bicyclic Reaction Course: Mechanistic and Exploratory Organic Photochemistry, *J. Am. Chem. Soc.* 1997, **119**, 3677–3690.

755. Zimmerman, H. E., Zhu, Z. N., General Theoretical Treatments of Solid-state Photochemical Rearrangements and a Variety of Contrasting Crystal Versus Solution Photochemistry, *J. Am. Chem. Soc.* 1995, **117**, 5245–5262.

756. Zimmerman, H. E., Mitkin, O. D., Conical Intersection Control of Heterocyclic Photochemical Bond Scission, *J. Am. Chem. Soc.* 2006, **128**, 12743–12749.

757. Zimmerman, H. E., Suryanarayan, V., Organic Photochemical Rearrangements of Triplets and Zwitterions, Mechanistic and Exploratory Organic Photochemistry, *Eur. J. Org. Chem.* 2007, 4091–4102.

758. Grob, C. A., Baumann, W., Die 1,4-Eliminierung Unter Fragmentierung, *Helv. Chim. Acta* 1955, **38**, 594–610.

759. Corey, E. J., Lemahieu, R., Mitra, R. B., Bass, J. D., Study of Photochemical Reactions of 2-Cyclohexenones with Substituted Olefins, *J. Am. Chem. Soc.* 1964, **86**, 5570–5583.

760. de Mayo, P., Enone Photoannelation, *Acc. Chem. Res.* 1971, **4**, 41–47.

761. Kearns, D. R., Marsh, G., Schaffne, K., Excited Singlet and Triplet States of a Cyclic Conjugated Enone, *J. Chem. Phys.* 1968, **49**, 3316–3317.

762. Hastings, D. J., Weedon, A. C., Origin of the Regioselectivity in the Photochemical Cycloaddition Reactions of Cyclic Enones with Alkenes – Chemical Trapping Evidence for the Structures, Mechanism of Formation, and Fates of 1,4-Biradical Intermediates, *J. Am. Chem. Soc.* 1991, **113**, 8525–8527.

763. Shimada, Y., Nakamura, M., Suzuka, T., Matsui, J., Tatsumi, R., Tsutsumi, K., Morimoto, T., Kurosawa, H., Kakiuchi, K., A New Route for the Construction of the AB-ring Core of Taxol, *Tetrahedron Lett.* 2003, **44**, 1401–1403.

764. Becker, D., Nagler, M., Sahali, Y., Haddad, N., Regiochemistry and Stereochemistry of Intramolecular [2 + 2] Photocycloaddition of Carbon Carbon Double Bonds to Cyclohexenones, *J. Org. Chem.* 1991, **56**, 4537–4543.

765. Srinivas, R., Carlough, K. H., Mercury ($^3P_1$) Photosensitized Internal Cycloaddition Reactions in 1,4-, 1,5-, and 1,6-Dienes, *J. Am. Chem. Soc.* 1967, **89**, 4932–4936.

766. Schroder, C., Wolff, S., Agosta, W. C., Biradical Reversion in the Intramolecular Photochemistry of Carbonyl-substituted 1,5-Hexadienes, *J. Am. Chem. Soc.* 1987, **109**, 5491–5497.

767. de Mayo, P., Takeshit, H., Photochemical Syntheses. 6. Formation of Heptandiones from Acetylacetone and Alkenes, *Can. J. Chem.*, 1963, **41**, 440–449.

768. Begley, M. J., Mellor, M., Pattenden, G., New Approach to Fused Carbocycles – Intramolecular Photocyclizations of 1,3-Dione Enol Acetates, *J. Chem. Soc., Chem. Commun.* 1979, 235–236.

769. Minter, D. E., Winslow, C. D., A Photochemical Approach to the Galanthan Ring System, *J. Org. Chem.* 2004, **69**, 1603–1606.

770. Cruciani, G., Margaretha, P., Photorearrangement of 4,4-Dimethylcyclohex-2-Enones with Alkyl or Fluoro Substituents at C(5) and C(6) – in Search of the Mechanism, *Helv. Chim. Acta* 1990, **73**, 890–895.

771. Zimmerman, H. E., Lewis, R. G., McCullough, J. J., Padwa, A., Staley, S. W., Semmelha, M., Mechanistic, Exploratory Photochemistry, X. V., The relation of Cyclohexenone to Cylohexadienone Rearrangements, *J. Am. Chem. Soc.* 1966, **88**, 1965–1973.

772. Dauben, W. G., Spitzer, W. A., Kellogg, M. S., Photochemistry of 4-Methyl-4-Phenyl-2-Cyclohexenone – Effect of Solvent on Excited State, *J. Am. Chem. Soc.* 1971, **93**, 3674–3677.

773. Zimmerman, H. E., Nesterov, E. E., Development of Experimental and Theoretical Crystal Lattice Organic Photochemistry: the Quantitative Cavity. Mechanistic and Exploratory Organic Photochemistry, *Acc. Chem. Res.* 2002, **35**, 77–85.

774. Zimmerman, H. E., Schuster, D. I., A New Approach to Mechanistic Organic Photochemistry. 4. Photochemical Rearrangements of 4,4-Diphenylcyclohexadienone, *J. Am. Chem. Soc.* 1962, **84**, 4527–4540.

775. Schuster, D. I., Mechanisms of Photochemical Transformations of Cross-conjugated Cyclohexadienones, *Acc. Chem. Res.* 1978, **11**, 65–73.

776. Matlin, A. R., Photocycloaddition/Trapping Reactions of Cross-conjugated Cyclic Dienones: Capture of Oxyallyl Intermediates. In Horspool, W. M., Lenci, F. (eds), *CRC Handbook of Organic Photochemistry and Photobiology*, 2nd edn, CRC Press LLC, Boca Raton, FL, 2004, Chapter 81, pp. 1–12.

777. Blay, G., Photochemical Rearrangements of 6/6- and 6/5-Fused Cross-conjugated Cyclohexadienones. In Horspool, W. M., Lenci, F. (eds), *CRC Handbook of Organic*

*Photochemistry and Photobiology*, 2nd edn, CRC Press LLC, Boca Raton, FL, 2004, Chapter 80, pp. 1–21.

778. Arigoni, D., Bosshard, H., Bruderer, H., Büchi, G., Jeger, O., Krebaum, L. J., Photochemische Reaktionen. 2. Uber Gegenseitige Beziehungen und Umwandlungen bei Bestrahlungsprodukten des Santonins, *Helv. Chim. Acta* 1957, **40**, 1732–1749.

779. Schaffner-Sabba, K., α-Santonin – Photochemistry in Protic Solvents and Pyrolysis, *Helv. Chim. Acta* 1969, **52**, 1237–1249.

780. Dauben, W. G., Van Riel, H. C. H. A., Robbins, J. D., Wagner, G. J., Photochemistry of *cis*-1-Phenylcyclohexene. Proof of Involvement of *trans* Isomer in Reaction Processes, *J. Am. Chem. Soc.* 1979, **101**, 6383–6389.

781. Nickon, A., Ilao, M. C., Stern, A. G., Summers, M. F., Hydrogen Trajectories in Alkene to Carbene Rearrangements. Unequal Deuterium Isotope Effects for the Axial and Equatorial Paths, *J. Am. Chem. Soc.* 1992, **114**, 9230–9232.

782. McMurry, J. E., Choy, W., Total Synthesis of α-Panasinsene and β-Panasinsene, *Tetrahedron Lett.* 1980, **21**, 2477–2480.

783. Lo, P. C. K., Snapper, M. L., Intramolecula, [2 + 2] Photocycloaddition/Thermal Fragmentation Approach toward 5–8–5 Ring Systems, *Org. Lett.* 2001, **3**, 2819–2821.

784. Pincock, J. A., The Photochemistry of Substituted Benzenes: Phototranspositions and the Photoadditions of Alcohols. In Horspool, W. M., Lenci, F. (eds), *CRC Handbook of Organic Photochemistry and Photobiology*, 2nd edn, CRC Press LLC, Boca Raton, FL, 2004, Chapter 46, pp. 1–19.

785. Gilbert, A., Ring Isomerization of Benzene and Naphthalene Derivatives. In Horspool, W. M., Song, P.-S. (eds), *CRC Handbook of Organic Photochemistry and Photobiology*, CRC Press, Boca Raton, FL, 1995, pp. 229–236.

786. Mariano, P. S., A New Look at Pyridinium Salt Photochemistry. Horspool, W. M., Lenci, F. (eds), *CRC Handbook of Organic Photochemistry and Photobiology*, 2nd edn, CRC Press LLC, Boca Raton, FL, 2004, Chapter 100, pp. 1–10.

787. Pavlik, J. W., Phototransposition, Photo-ring Contraction Reactions of 4-Pyrones and 4-Hydroxypyrylium Cations. In Horspool, W. M., Song, P.-S. (eds), *CRC Handbook of Organic Photochemistry and Photobiology*, CRC Press, Boca Raton, FL, 1995, pp. 237–249.

788. Bryce-Smith, D., Gilbert, A., Organic Photochemistry of Benzene. Part 1, *Tetrahedron* 1976, **32**, 1309–1326.

789. Johnson, R. P., Daoust, K. J., Electrocyclic Ring Opening Modes of Dewar Benzenes: *Ab Initio* Predictions for Mobius Benzene and *trans*-Dewar Benzene as New $C_6H_6$ Isomers, *J. Am. Chem. Soc.* 1996, **118**, 7381–7385.

790. Dreyer, J., Klessinger, M., The Photochemical Formation of Fulvene from Benzene via Prefulvene – A Theoretical Study, *Chem. Eur. J.* 1996, **2**, 335–341.

791. Frank, I., Grimme, S., Peyerimhoff, S. D., *Ab Initio* Study of the Isomerization of Substituted Benzenes and [6]Paracyclophanes to the Dewar Benzene Isomers, *J. Am. Chem. Soc.* 1994, **116**, 5949–5953.

792. Palmer, M. H., A Reinterpretation of the UV-photoelectron Spectra of Dewar Benzene, Norbornadiene and Barrelene by *Ab Initio* Configuration Interaction Calculations, *J. Mol. Struct.* 1987, **161**, 333–345.

793. Sobolewski, A. L., Domcke, W., Photophysically Relevant Potential Energy Functions of Low-lying Singlet States of Benzene, Pyridine and Pyrazine – An *Ab Initio* Study, *Chem. Phys. Lett.* 1991, **180**, 381–386.

794. Xu, X. F., Cao, Z. X., Zhang, Q. E., What Definitively Controls the Photochemical Activity of Methylbenzonitriles and Methylanisoles? Insights from Theory, *J. Phys. Chem. A* 2007, **111**, 5775–5783.

795. Wilzbach, K. E., Ritscher, J. S., Kaplan, L., Benzvalene Tricyclic Valence Isomer of Benzene, *J. Am. Chem. Soc.* 1967, **89**, 1031–1032.

796. Bryce-Smith, D., Gilbert, A., Robinson, D. A., Direct Transformation of Second Excited Singlet State of Benzene into Dewar Benzene, *Angew. Chem. Int. Ed. Engl.* 1971, **10**, 745–746.

797. Wilzbach, K. E., Kaplan, L., Photoisomerization of Dialkylbenzenes, *J. Am. Chem. Soc.* 1964, **86**, 2307–2308.

798. Den Besten, I. E., Kaplan, L., Wilzbach, K. E., Photoisomerization of tri-*t*-Butylbenzenes. Photochemical Interconversion of Benzvalenes, *J. Am. Chem. Soc.* 1968, **90**, 5868–5872.

799. Pavlik, J. W., Laohhasurayotin, S., The Vapor-phase Phototransposition Chemistry of Pyridine: Deuterium Labeling Studies, *J. Org. Chem.* 2008, **73**, 2746–2752.

800. Barltrop, J. A., Summers, A. J. H., Dawes, K., Day, A. C., Photohydrolysis of 2,4,6-Trimethylpyrylium Perchlorate – Evidence for Isomerization to an Oxoniabenzvalene Intermediate, *J. Chem. Soc., Chem. Commun.* 1972, 1240–1241.

801. Pavlik, J. W., Spada, A. P., Snead, T. E., Photochemistry of 4-Hydroxypyrylium Cations in Aqueous Sulfuric Acid *J. Org. Chem.* 1985, **50**, 3046–3050.

802. Wender, P. A., Dore, T. M., Intra- and Intermolecular Cycloadditions of Benzene Derivatives. In Horspool, W. M., Song, P.-S. (eds), *CRC Handbook of Organic Photochemistry and Photobiology*, CRC Press, Boca Raton, FL, 1995, pp. 280–290.

803. Sieburth, S. M., Photochemical Reactivity of Pyridones. In Horspool, W. M., Lenci, F. (eds), *CRC Handbook of Organic Photochemistry and Photobiology*, 2nd edn, CRC Press LLC, Boca Raton, FL, 2004, Chapter 103, pp. 1–18.

804. Corneliesse, J., de Haan, R., *Ortho* Photocycloaddition Reactions of Aromatic Compounds. In Ramamurthy, V., Schanze, K. (eds), *Understanding and Manipulating Excited-state Processes*, Marcel Dekker, New York, 2001, pp. 1–126.

805. Mizuno, K., Maeda, H., Sugimoto, A., Chiyonobu, K., Photocycloaddition and Photoaddition Reactions of Aromatic Compounds. In Ramamurthy, V., Schanze, K. (eds), *Understanding and Manipulating Excited-state Processes*, Marcel Dekker, New York, 2001, pp. 127–241.

806. Wender, P. A., Ternansky, R., Delong, M., Singh, S., Olivero, A., Rice, K., Arene Alkene Cycloadditions and Organic Synthesis, *Pure Appl. Chem.* 1990, **62**, 1597–1602.

807. Cornelisse, J., The *meta* Photocycloaddition of Arenes to Alkenes, *Chem. Rev.* 1993, **93**, 615–669.

808. Dekeukeleire, D., He, S. L., Photochemical Strategies for the Construction of Polycyclic Molecules, *Chem. Rev.* 1993, **93**, 359–380.

809. Hoffmann, N., *Ortho-*, *meta-*, and *para*-Photocycloaddition of Arenes. In Griesbeck, A. G., Mattay, J. (eds), *Synthetic Organic Photochemistry*, Marcel Dekker, New York, 2005, pp. 529–552.

810. Hoffmann, N., Photochemical Cycloaddition between Benzene Derivatives and Alkenes, *Synthesis* 2004, 481–495.

811. Chappell, D., Russell, A. T., From α-Cedrene to Crinipellin B and Onward: 25 Years of the Alkene–Arene *m*-Photocycloaddition Reaction in Natural Product Synthesis, *Org. Biomol. Chem.* 2006, **4**, 4409–4430.

812. McCullough, J. J., Photoadditions of Aromatic Compounds, *Chem. Rev.* 1987, **87**, 811–860.

813. Sotzmann, A., Mattay, J., Photochemical Reaction of Fullerenes and Fullerene Derivatives. In Horspool, W. M., Lenci, F. (eds), *CRC Handbook of Organic Photochemistry and Photobiology*, 2nd edn, CRC Press LLC, Boca Raton, FL, 2004, Chapter 28, pp. 1–42.

814. Stehouwer, A. M., van der Hart, J. A., Mulder, J. J. C., Cornelisse, J., The *meta* Photocycloaddition of Benzene to Ethylene – *Ab Initio* Calculation of a Symmetrical Pathway, *THEOCHEM* 1992, **92**, 333–338.

815. van der Hart, J. A., Mulder, J. J. C., Cornelisse, J., The *meta* Photocycloaddition of Benzene to Ethylene: Semi-empirical Calculations, *THEOCHEM* 1987, **36**, 1–10.

816. Gilbert, A., Intra- and Intermolecular Cycloadditions of Benzene Derivatives. In Horspool, W. M., Lenci, F. (eds), *CRC Handbook of Organic Photochemistry and Photobiology*, 2nd edn, CRC Press LLC, Boca Raton, FL, 2004, Chapter 41, pp. 1–11.

817. Bryce-Smith, D., Deshpande, R. R., Gilbert, A., Mechanism of Photoaddition of Maleic Anhydride to Benzene, *Tetrahedron Lett.* 1975, 1627–1630.

818. Ohashi, M., Tanaka, Y., Yamada, S., [2 + 2] Cycloaddition versus Substitution in Photochemical Reactions of Methoxybenzene–Acrylonitrile Systems, *Tetrahedron Lett.* 1977, **18**, 3629–3632.

819. Sket, B., Zupan, M., [2 + 2] Photoaddition of Acetylenes to Hexafluorobenzene – Isolation of Bicyclo[4.2.0]octatriene Derivatives, *J. Am. Chem. Soc.* 1977, **99**, 3504–3505.

820. Aoyama, H., Arata, Y., Omote, Y., Intramolecular Photocycloaddition of *N*-Benzylstyrylacetamides – [2 + 2] Addition of Styrenes to Benzenes, *J. Chem. Soc., Chem. Commun.* 1990, 736–737.

821. Maeda, H., Waseda, S., Mizuno, K., Stereoselective, [$2\pi + 2\pi$] Photocycloaddition of Arylalkenes to Chrysene, *Chem. Lett.* 2000, 1238–1239.

822. Sakamoto, M., Sano, T., Takahashi, M., Yamaguchi, K., Fujita, T., Watanabe, S., Photochemistry of Heteroaromatics – A Novel Photocycloaddition of 2-Alkoxy-3-cyanopyridines with Methacrylonitrile, *Chem. Commun.* 1996, 1349–1350.

823. Guldi, D. M., Prato, M., Excited-state Properties of C-60 Fullerene Derivatives, *Acc. Chem. Res.* 2000, **33**, 695–703.

824. Foote, C. S., Photophysical and Photochemical Properties of Fullerenes, *Top. Curr. Chem.* 1994, **169**, 347–363.

825. Vassilikogiannakis, G., Orfanopoulos, M., [2 + 2] Photocycloadditions of *cis/trans*-4-Propenylanisole to $C_{60}$. A Step-wise Mechanism, *Tetrahedron Lett.* 1997, **38**, 4323–4326.

826. Guo, L. W., Gao, X., Zhang, D. W., Wu, S. H., Wu, H. M., Li, Y. J., Wilson, S. R., Richardson, C. F., Schuster, D. I., Alkaloid–Fullerene Systems Through Photocycloaddition Reactions, *J. Org. Chem.* 2000, **65**, 3804–3810.

827. Cheng, K. L., Wagner, P. J., Biradical Rearrangements During Intramolecular Cycloaddition of Double Bonds to Triplet Benzenes, *J. Am. Chem. Soc.* 1994, **116**, 7945–7946.

828. Bowry, V. W., Lusztyk, J., Ingold, K. U., Calibration of a New Horology of Fast Radical Clocks – Ring-opening Rates for Ring Alkyl-substituted and α-Alkyl-substituted Cyclopropylcarbinyl Radicals and for the Bicyclo[2.1.0]pent-2-yl Radical, *J. Am. Chem. Soc.* 1991, **113**, 5687–5698.

829. Griller, D., Ingold, K. U., Free-radical Clocks, *Acc. Chem. Res.* 1980, **13**, 317–323.

830. Mattay, J., Runsink, J., Piccirilli, J. A., Jans, A. W. H., Cornelisse, J., Selectivity and Charge Transfer in Photoreactions of Donor–Acceptor Systems. 8. Photochemical Cycloadditions of 1,3-Dioxoles to Anisole, *J. Chem. Soc., Perkin Trans. 1* 1987, 15–20.

831. Timmermans, J. L., Wamelink, M. P., Lodder, G., Cornelisse, J., Diastereoselective Intramolecular *meta*-Photocycloaddition of Side-chain-substituted 5-(2-Methoxyphenyl) pent-1-enes, *Eur. J. Org. Chem.* 1999, 463–470.

832. Okada, K., Samizo, F., Oda, M., Photochemical Reactions of (9-Anthryl)methyl Methyl Fumarate and Maleate – Application to Asymmetric [4 + 2]-Photocycloaddition Reaction, *Tetrahedron Lett.* 1987, **28**, 3819–3822.

833. Albini, A., Fasani, E., Faiardi, D., Charge-transfer and Exciplex Pathway in the Photocycloaddition of 9-Anthracenecarbonitrile with Anthracene and Naphthalenes, *J. Org. Chem.* 1987, **52**, 155–157.

834. Lim, C., Yasutake, M., Shinmyozu, T., Formation of a Novel Cage Compound with a Pentacyclo[6.3.0$^{14,11}$.0$^{2,6}$.0$^{5,10}$]dodecane Skeleton by Photolysis of [3$_4$](1,2,4,5)Cyclophane, *Angew. Chem. Int. Ed.* 2000, **39**, 578–580.

835. Okamoto, H., Satake, K., Ishida, H., Kimura, M., Photoreaction of a 2,11-Diaza[3.3] paracyclophane Derivative: Formation of Octahedrane by Photochemical Dimerization of Benzene, *J. Am. Chem. Soc.* 2006, **128**, 16508–16509.

836. Karapire, C., Icli, S., Photochemical Aromatic Substitution. In Horspool, W. M., Lenci, F. (eds), *CRC Handbook of Organic Photochemistry and Photobiology*, 2nd edn, CRC Press LLC, Boca Raton, FL, 2004, Chapter 37, pp. 1–14.

837. Mangion, D., Arnold, D. R., The Photochemical Nucleophile–Olefin Combination, Aromatic Substitution (Photo-NOCAS) Reaction. In Horspool, W. M., Lenci, F. (eds), *CRC Handbook of Organic Photochemistry and Photobiology*, 2nd edn, CRC Press LLC, Boca Raton, FL, 2004, Chapter 40, pp. 1–40.

838. Fagnoni, M., Albini, A., Photonucleophilic Substitution Reactions. In Ramamurthy, V., Schanze, K. S. (eds), *Organic Photochemistry and Photophysics*, Vol. 14, CRC Press, Boca Raton, FL, 2006, pp. 131–177.

839. Corneliesse, J., Photochemical Aromatic Substitution. In Horspool, W. M., Song, P.-S. (eds), *CRC Handbook of Organic Photochemistry and Photobiology*, CRC Press, Boca Raton, FL, 1995, pp. 250–265.

840. Rossi, R. A., Photoinduced Aromatic Nucleophilic Substitution Reactions. In Griesbeck, A. G., Mattay, J. (eds), *Synthetic Organic Photochemistry*, Marcel Dekker, New York, 2005, pp. 495–527.

841. Beugelmans, R., The Photostimulated S$_{RN}$1 Process: Reactions of Haloarenes with Enolates. In Horspool, W. M., Song, P.-S. (eds), *CRC Handbook of Organic Photochemistry and Photobiology*, Boca Raton, FL, 1995, pp. 1200–1217.

842. Bunce, N. J., Photochemical C–X Bond Cleavage in Arenes. In Horspool, W. M., Song, P.-S. (eds), *CRC Handbook of Organic Photochemistry and Photobiology*, CRC Press, Boca Raton, FL, 1995, pp. 1181–1192.

843. Arnold, D. R., Chan, M. S. W., McManus, K. A., Photochemical Nucleophile–Olefin Combination, Aromatic Substitution (Photo-NOCAS) Reaction. 12. Factors Controlling the Regiochemistry of the Reaction with Alcohol as the Nucleophile, *Can. J. Chem.* 1996, **74**, 2143–2166.

844. Galli, C., Gentili, P., Guarnieri, A., Calculation of the LUMO Energy of the Substitution Product of the Aromatic S$_{RN}$1 Reaction, *Gazz. Chim. Ital.* 1997, **127**, 159–164.

845. Freccero, M., Fagnoni, M., Albini, A., Homolytic vs. Heterolytic Paths in the Photochemistry of Haloanilines, *J. Am. Chem. Soc.* 2003, **125**, 13182–13190.

846. Pinter, B., De Proft, F., Veszpremi, T., Geerlings, P., Theoretical Study of the Orientation Rules in Photonucleophilic Aromatic Substitutions, *J. Org. Chem.* 2008, **73**, 1243–1252.

847. Mella, M., Coppo, P., Guizzardi, B., Fagnoni, M., Freccero, M., Albini, A., Photoinduced, Ionic Meerwein Arylation of Olefins, *J. Org. Chem.* 2001, **66**, 6344–6352.

848. Choudhry, G. G., Webster, G. R. B., Hutzinger, O., Environmentally Significant Photochemistry of Chlorinated Benzenes and Their Derivatives in Aquatic Systems, *Toxicol. Environ. Chem.* 1986, **13**, 27–83.

849. Schutt, L., Bunce, N. J., Photodehalogenation of Aryl Halides. In Horspool, W. M., Lenci, F. (eds), *CRC Handbook of Organic Photochemistry and Photobiology*, 2nd edn, CRC Press LLC, Boca Raton, FL, 2004, Chapter 38, pp. 1–18.

850. Mangion, D., Arnold, D. R., Photochemical Nucleophile–Olefin Combination, Aromatic Substitution Reaction. Its Synthetic Development and Mechanistic Exploration, *Acc. Chem. Res.* 2002, **35**, 297–304.

851. Torriani, R., Mella, M., Fasani, E., Albini, A., On the Mechanism of the Photochemical Reaction between 1,4-Dicyanobenzene and 2,3-Dimethylbutene in the Presence of Nucleophiles, *Tetrahedron* 1997, **53**, 2573–2580.

852. Rossi, R. A., Penénori, A. B., The Photostimulated $S_{RN}1$ Process: Reaction of Haloarenes with Carbanions. In Horspool, W. M., Lenci, F. (eds), *CRC Handbook of Organic Photochemistry and Photobiology*, 2nd edn, CRC Press LLC, Boca Raton, FL, 2004, Chapter 47, pp. 1–24.

853. Borosky, G. L., Pierini, A. B., Rossi, R. A., Differences in Reactivity of Stabilized Carbanions with Haloarenes in the Initiation and Propagation Steps of the $S_{RN}1$ Mechanism in DMSO, *J. Org. Chem.* 1992, **57**, 247–252.

854. Chowdhry, V., Westheimer, F. H., Photoaffinity Labeling of Biological Systems, *Annu. Rev. Biochem.* 1979, **48**, 293–325.

855. Cantos, A., Marquet, J., Morenomanas, M., Gonzalezlafont, A., Lluch, J. M., Bertran, J., On the Regioselectivity of 4-Nitroanisole Photosubstitution with Primary Amines – A Mechanistic and Theoretical Study, *J. Org. Chem.* 1990, **55**, 3303–3310.

856. Howell, N., Pincock, J. A., Stefanova, R., The Phototransposition in Acetonitrile and the Photoaddition of 2,2,2-Trifluoroethanol to the Six Isomers of Dimethylbenzonitrile, *J. Org. Chem.* 2000, **65**, 6173–6178.

857. Nuss, J. M., Chinn, J. P., Murphy, M. M., Substituent Control of Excited-state Reactivity. The Intramolecular *ortho* Arene–Olefin Photocycloaddition, *J. Am. Chem. Soc.* 1995, **117**, 6801–6802.

858. deLijser, H. J. P., Arnold, D. R., Radical Ions in Photochemistry. 44. The Photo-NOCAS Reaction with Acetonitrile as the Nucleophile, *J. Org. Chem.* 1997, **62**, 8432–8438.

859. Fessner, W. D., Sedelmeier, G., Spurr, P. R., Rihs, G., Prinzbach, H., "Pagodane": the Efficient Synthesis of a Novel, Versatile Molecular Framework, *J. Am. Chem. Soc.* 1987, **109**, 4626–4642.

860. Al-Jalal, N., Gilbert, A., Heath, P., Photocycloaddition of Ethenes to Cyanoanisoles, *Tetrahedron* 1988, **44**, 1449–1459.

861. Vanossi, M., Mella, M., Albini, A., 2 + 2 + 2 Cycloaddition vs. Radical Ion Chemistry in the Photoreactions of 1,2,4,5-Benzenetetracarbonitrile with Alkenes in Acetonitrile, *J. Am. Chem. Soc.* 1994, **116**, 10070–10075.

862. Rubin, M. B., Photoinduced Intermolecular Hydrogen Abstraction Reactions of Ketones. In Horspool, W. M., Song, P.-S. (eds), *CRC Handbook of Organic Photochemistry and Photobiology*, CRC Press, Boca Raton, FL, 1995, pp. 430–436.

863. Wagner, P. J., Park, B.-S., Photoinduced Hydrogen Atom Abstraction by Carbonyl Compounds. In Padwa, A. (ed), *Organic Photochemistry*, Vol. 11, Marcel Dekker, New York, 1991, pp. 227–366.

864. Nau, W. M., Pischel, U., Photoreactivity of n,π* Excited Azoalkanes and Ketones. In Ramamurthy, V., Schanze, K. S. (eds), *Organic Photochemistry and Photophysics*, Vol. 14, CRC Press, Boca Raton, FL, 2006, pp. 75–129.

865. Scaiano, J. C., Intermolecular Photoreductions of Ketones, *J. Photochem.* 1973/74, **2**, 81–118.

866. Nau, W. M., Greiner, G., Wall, J., Rau, H., Olivucci, M., Robb, M. A., The Mechanism for Hydrogen Abstraction by n,π* Excited Singlet States: Evidence for Thermal Activation and Deactivation through a Conical Intersection, *Angew. Chem. Int. Ed.* 1998, **37**, 98–101.

867. Nau, W. M., Greiner, G., Rau, H., Olivucci, M., Robb, M. A., Discrimination Between Hydrogen Atom and Proton Abstraction in the Quenching of n,π* Singlet-Excited States by Protic Solvents, *Ber. Bunsen-Ges. Phys. Chem.* 1998, **102**, 486–492.

868. Sinicropi, A., Pogni, R., Basosi, R., Robb, M. A., Gramlich, G., Nau, W. M., Olivucci, M., Fluorescence Quenching by Sequential Hydrogen, Electron, and Proton Transfer in the Proximity of a Conical Intersection, *Angew. Chem. Int. Ed.* 2001, **40**, 4185–4189.

869. Jacques, P., [3]CTC as the Main Step in the Quenching of [3]BP by Electron and Hydrogen Donors, *J. Photochem. Photobiol. A* 1991, **56**, 159–163.

870. Sinicropi, A., Pischel, U., Basosi, R., Nau, W. M., Olivucci, M., Conical Intersections in Charge-transfer Induced Quenching, *Angew. Chem. Int. Ed.* 2000, **39**, 4582–4586.

871. Ciamician, G., Silber, P., Chemische Lichtwirkungen, *Chem. Ber.* 1900, **33**, 2913–2914.

872. Rubin, M. B., Hydrogen Abstraction Reactions of α-Diketones. In Horspool, W. M., Song, P.-S. (eds), *CRC Handbook of Organic Photochemistry and Photobiology*, CRC Press, Boca Raton, FL, 1995, pp. 437–448.

873. Paul, H., Small, R. D., Scaiano, J. C., Hydrogen Abstraction by *t*-Butoxy Radicals – Laser Photolysis and Electron Spin Resonance Study, *J. Am. Chem. Soc.* 1978, **100**, 4520–4527.

874. Luo, Y.-R., *Handbook of Bond Dissociation Energies in Organic Compounds*, CRC Press, Boca Raton, FL, 2003.

875. Cohen, S. G., Parola, A., Parsons, G. H. Jr., Photoreduction by Amines, *Chem. Rev.* 1973, **73**, 141–161.

876. Cossy, J., Belotti, D., Generation of Ketyl Radical Anions by Photoinduced Electron Transfer (PET) between Ketones and Amines. Synthetic Applications, *Tetrahedron* 2006, **62**, 6459–6470.

877. Reynolds, J. L., Erdner, K. R., Jones, P. B., Photoreduction of Benzophenones by Amines in Room-Temperature Ionic Liquids, *Org. Lett.* 2002, **4**, 917–919.

878. Chiappe, C., Pieraccini, D., Ionic Liquids: Solvent Properties and Organic Reactivity, *J. Phys. Org. Chem.* 2005, **18**, 275–297.

879. Griesbeck, A. G., Bondock, S., Oxetane Formation: Stereocontrol. In Horspool, W. M., Lenci, F. (eds), *CRC Handbook of Organic Photochemistry and Photobiology*, 2nd edn, CRC Press LLC, Boca Raton, FL, 2004, Chapter 59, pp. 1–19.

880. Griesbeck, A. G., Bondock, S., Oxetane Formation: Intermolecular Additions. In Horspool, W. M., Lenci, F. (eds), *CRC Handbook of Organic Photochemistry and Photobiology*, 2nd edn, CRC Press LLC, Boca Raton, FL, 2004, Chapter 60, pp. 1–21.

881. Griesbeck, A.G., Photocycloadditions of Alkenes to Excited Carbonyls. In Griesbeck, A.G., Mattay, J. (eds), *Synthetic Organic Photochemistry*, Marcel Dekker, New York, 2005, pp. 89–139.

882. Griesbeck, A. G., Abe, M., Bondock, S., Selectivity Control in Electron Spin Inversion Processes: Regio- and Stereochemistry of Paternò–Büchi Photocycloadditions as a Powerful Tool for Mapping Intersystem Crossing Processes, *Acc. Chem. Res.* 2004, **37**, 919–928.

883. Muller, F., Mattay, J., Photocycloadditions – Control by Energy and Electron Transfer, *Chem. Rev.* 1993, **93**, 99–117.

884. Minaev, B. F., Agren, H., Spin–Orbit Coupling in Oxygen Containing Diradicals, *THEOCHEM* 1998, **434**, 193–206.

885. Kutateladze, A. G., Conformational Analysis of Singlet–Triplet State Mixing in Paternò–Büchi Diradicals, *J. Am. Chem. Soc.* 2001, **123**, 9279–9282.

886. Griesbeck, A. G., Spin Selectivity in Photochemistry: a Tool for Organic Synthesis, *Synlett* 2003, 451–472.

887. Paternò, E., Chieffi, G., Sintesi in Chimica Organica per Mezzo della Luce. Nota II. Composti degli Idrocarburi non Saturi con Aldeidi e Chetoni, *Gazz. Chim. Ital.* 1909, **39**, 341–361.

888. Büchi, G., Inman, C. G., Lipinsky, E. S., Light-catalyzed Organic Reactions. I. The Reaction of Carbonyl Compounds with 2-Methyl-2-butene in the Presence of Ultraviolet Light, *J. Am. Chem. Soc.* 1954, **76**, 4327–4331.

889. Turro, N. J., Wriede, P., The Photocycloaddition of Acetone to 1-Methoxy-1-butene. A Comparison of Singlet and Triplet Mechanism and Singlet and Triplet Biradical Intermediates, *J. Am. Chem. Soc.* 1970, **92**, 320–329.

890. Freilich, S. C., Peters, K. S., Observation of the 1,4-Biradical in the Paternò–Büchi Reaction, *J. Am. Chem. Soc.* 1981, **103**, 6255–6257.

891. Freilich, S. C., Peters, K. S., Picosecond Dynamics of the Paternò–Büchi Reaction, *J. Am. Chem. Soc.* 1985, **107**, 3819–3822.

892. Turro, N. J., Farrington, G. L., Quenching of the Fluorescence of 2-Norbornanone and Derivatives by Electron-rich and Electron-poor Ethylenes, *J. Am. Chem. Soc.* 1980, **102**, 6051–6055.

893. Turro, N. J., Dalton, C. J., Dawes, K., Farrington, G., Hautala, R., Morton, D., Niemczyk, M., Schore, N., Molecular Photochemistry of Alkanones in Solution. α-Cleavage, Hydrogen Abstraction, Cycloaddition, and Sensitization Reactions, *Acc. Chem. Res.* 1972, **5**, 92–101.

894. Ashby, E. C., Boone, J. R., Mechanism of Lithium Aluminum Hydride Reduction of Ketones. Kinetics of Reduction of Mesityl Phenyl Ketone, *J. Am. Chem. Soc.* 1976, **98**, 5524–5531.

895. Griesbeck, A. G., Mauder, H., Stadtmuller, S., Intersystem Crossing in Triplet 1,4-Biradicals – Conformational Memory Effects on the Stereoselectivity of Photocycloaddition Reactions, *Acc. Chem. Res.* 1994, **27**, 70–75.

896. Palmer, I. J., Ragazos, I. N., Bernardi, F., Olivucci, M., Robb, M. A., An MC-SCF Study of the (Photochemical) Paternò–Büchi Reaction, *J. Am. Chem. Soc.* 1994, **116**, 2121–2132.

897. Yang, N. C., Kimura, M., Eisenhardt, W., Paternò–Büchi Reactions of Aromatic Aldehydes with 2-Butenes and their Implication on the Rate of Intersystem Crossing of Aromatic Aldehydes, *J. Am. Chem. Soc.* 1973, **95**, 5058–5060.

898. Griesbeck, A. G., Mauder, H., Peters, K., Peters, E. M., Vonschnering, H. G., Photocycloadditions with α-Naphthaldehyde and β-Naphthaldehyde – Complete Inversion of Diastereoselectivity as a Consequence of Differently Configurated Electronic States, *Chem. Ber.* 1991, **124**, 407–410.

899. Griesbeck, A. G., Buhr, S., Fiege, M., Schmickler, H., Lex, J., Stereoselectivity of Triplet Photocycloadditions: Diene–Carbonyl Reactions and Solvent Effects, *J. Org. Chem.* 1998, **63**, 3847–3854.

900. Bach, T., Bergmann, H., Harms, K., High Facial Diastereoselectivity in the Photocycloaddition of a Chiral Aromatic Aldehyde and an Enamide Induced by Intermolecular Hydrogen Bonding, *J. Am. Chem. Soc.* 1999, **121**, 10650–10651.

901. Bach, T., Bergmann, H., Brummerhop, H., Lewis, W., Harms, K., The [2 + 2]-Photocycloaddition of Aromatic Aldehydes and Ketones to 3,4-Dihydro-2-pyridones: Regioselectivity, Diastereoselectivity, and Reductive Ring Opening of the Product Oxetanes, *Chem. Eur. J.* 2001, **7**, 4512–4521.

902. Iriondo-Alberdi, J., Perea-Buceta, J. E., Greaney, M. F., A Paternò–Büchi Approach to the Synthesis of Merrilactone A, *Org. Lett.* 2005, **7**, 3969–3971.

903. Garcia-Garibay, M. A., Campos, L. M., Photochemical Decarbonylation of Ketones: Recent Advances and Reactions in Crystalline Solids. In Horspool, W. M., Lenci, F. (eds), *CRC Handbook of Organic Photochemistry and Photobiology*, 2nd edn, CRC Press LLC, Boca Raton, FL, 2004, Chapter 48, pp. 1–41.

904. Bohne, C., Norrish Type I Processes of Ketones: Selected Examples and Synthetic Applications. In Horspool, W. M., Song, P.-S. (eds), *CRC Handbook of Organic Photochemistry and Photobiology*, CRC Press, Boca Raton, FL, 1995, pp. 423–429.

905. Bohne, C., Norrish Type I Processes of Ketones: Basic Concepts. In Horspool, W. M., Song, P.-S. (eds), *CRC Handbook of Organic Photochemistry and Photobiology*, CRC Press, Boca Raton, FL, 1995, pp. 416–422.

906. Roberts, S. M., Carbene Formation in the Photochemistry of Cyclic Ketones. In Horspool, W. M., Lenci, F. (eds), *CRC Handbook of Organic Photochemistry and Photobiology*, 2nd edn, CRC Press LLC, Boca Raton, FL, 2004, Chapter 49, pp. 1–6.

907. Rubin, M. B., Photochemistry of Vicinal Polycarbonyl Compounds. In Horspool, W. M., Lenci, F. (eds), *CRC Handbook of Organic Photochemistry and Photobiology*, 2nd edn, CRC Press LLC, Boca Raton, FL, 2004, Chapter 50, pp. 1–14.

908. Shinmyozu, T., Nogita, R., Akita, M., Lim, C., Photochemical Routes to Cyclophanes Involving Decarbonylation Reactions and Related Process. In Horspool, W. M., Lenci, F. (eds), *CRC Handbook of Organic Photochemistry and Photobiology*, 2nd edn, CRC Press LLC, Boca Raton, FL, 2004, Chapter 51, pp. 1–6.

909. Chatgilialoglu, C., Crich, D., Komatsu, M., Ryu, I., Chemistry of Acyl Radicals, *Chem. Rev.* 1999, **99**, 1991–2069.

910. Jackson, W. M., Okabe, H., Photodissociation Dynamics of Small Molecules, *Adv. Photochem.* 1986, **13**, 1–94.

911. Diau, E. W. G., Kotting, C., Zewail, A. H., Femtochemistry of Norrish Type I Reactions: I. Experimental and Theoretical Studies of Acetone and Related Ketones on the $S_1$ Surface, *ChemPhysChem* 2001, **2**, 273–293.

912. Diau, E. W. G., Kotting, C., Zewail, A. H., Femtochemistry of Norrish Type I Reactions: II. The Anomalous Predissociation Dynamics of Cyclobutanone on the $S_1$ Surface, *ChemPhysChem* 2001, **2**, 294–309.

913. Sakurai, H., Kato, S., A Theoretical Study of the Norrish Type I Reaction of Acetone, *THEOCHEM* 1999, **462**, 145–152.

914. He, H. Y., Fang, W.H., Phillips, D. L., Photochemistry of Butyrophenone: Combined Complete-active-space Self-consistent Field and Density Functional Theory Study of Norrish Type I and II Reactions, *J. Phys. Chem. A* 2004, **108**, 5386–5392.

915. Campos, L. M., Mortko, C. J., Garcia-Garibay, M. A., Norrish Type I vs. Norrish–Yang Type II in the Solid State Photochemistry of *cis*-2,6-Di(1-cyclohexenyl)cyclohexanone: a Computational Study, *Mol. Cryst. Liq. Cryst.* 2006, **456**, 15–24.

916. Dietlin, C., Allonas, X., Defoin, A., Fouassier, J.-P., Theoretical and Experimental Study of the Norrish I Photodissociation of Aromatic Ketones, *Photochem. Photobiol. Sci.* 2008, **7**, 558–565.

917. Norrish, R. G. W., Appleyard, M. E. S., Primary Photochemical Reactions. Part IV. Decomposition of Methyl Ethyl Ketone and Methyl Butyl Ketone, *J. Chem. Soc.* 1934, 874–880.

918. Lee, E. K. C., Hemminger, J. C., Rusbult, C. F., Unusual Photochemistry of Cyclobutanone Near Its Predissociation Threshold, *J. Am. Chem. Soc.* 1971, **93**, 1867–1871.

919. Lewis, F. D., Magyar, J. G., Photoreduction and α-Cleavage of Aryl Alkyl Ketones, *J. Org. Chem.* 1972, **37**, 2102–2107.

920. Weiss, R. G., The Norrish Type I Reaction in Cycloalkanone Photochemistry. In Padwa, A. (ed), *Organic Photochemistry*, Vol. 5, Marcel Dekker, New York, 1981, pp. 347–420.

921. Lewis, F. D., Magyar, J. G., Cage Effects in the Photochemistry of (*S*)-(+)-2-Phenylpropiophenone, *J. Am. Chem. Soc.* 1973, **95**, 5973–5976.

922. Yang, N.-C., Feit, E. D., Hui, M. H., Turro, N. J., Dalton, J. C., Photochemistry of di-*tert*-Butyl Ketone and Structural Effects on the Rate and Efficiency of Intersystem Crossing of Aliphatic Ketones, *J. Am. Chem. Soc.* 1970, **92**, 6974–6976.

923. Kraus, J. W., Calvert, J. G., The Disproportionation and Combination Reactions of Butyl Free Radicals, *J. Am. Chem. Soc.* 1957, **79**, 5921–5926.

924. Ramamurthy, V., Eaton, D. F., Caspar, J. V., Photochemical and Photophysical Studies of Organic Molecules Included Within Zeolites, *Acc. Chem. Res.* 1992, **25**, 299–307.

925. Engel, P. S., Photochemistry of Dibenzyl Ketone, *J. Am. Chem. Soc.* 1970, **92**, 6074–6076.

926. Gould, I. R., Baretz, B. H., Turro, N. J., Primary Processes in the Type I Photocleavage of Dibenzyl Ketones – A Pulsed Laser and Photochemically Induced Dynamic Nuclear-Polarization Study, *J. Phys. Chem.* 1987, **91**, 925–929.

927. Turro, N. J., Cherry, W. R., Photoreactions in Detergent Solutions. Enhancement of Regioselectivity Resulting from the Reduced Dimensionality of Substrates Sequestered in a Micelle, *J. Am. Chem. Soc.* 1978, **100**, 7431–7432.

928. Keating, A. E., Garcia-Garibay, M. A., Photochemical Solid-to-Solid Reactions. In Ramamurthy, V., Schanze, K. S. (eds), *Molecular and Supramolecular Photochemistry*, Vol. 2, Marcel Dekker, New York, 1998, pp. 195–248.

929. Choi, T., Cizmeciyan, D., Khan, S. I., Garcia-Garibay, M. A., An Efficient Solid-to-Solid Reaction via a Steady-state Phase Separation Mechanism, *J. Am. Chem. Soc.* 1995, **117**, 12893–12894.

930. Morton, D. R., Lee-Ruff, E., Southam, R. M., Turro, N. J., Molecular Photochemistry. XXVII. Photochemical Ring Expansion of Cyclobutanone, Substituted Cyclobutanones, and Related Cyclic Ketones, *J. Am. Chem. Soc.* 1970, **92**, 4349–4357.

931. Yates, P., Loutfy, R. O., Photochemical Ring Expansion of Cyclic Ketones via Cyclic Oxacarbenes, *Acc. Chem. Res.* 1975, **8**, 209–216.

932. Burns, C. S., Heyerick, A., De Keukeleire, D., Forbes, M. D. E., Mechanism for Formation of the Lightstruck Flavor in Beer Revealed by Time-resolved Electron Paramagnetic Resonance, *Chem. Eur. J.* 2001, **7**, 4553–4561.

933. Wagner, P. J., Klán, P., Norrish Type II Photoelimination of Ketones: Cleavage of 1,4-Biradicals Formed by α-Hydrogen Abstraction. In Horspool, W. M., Lenci, F. (eds), *CRC Handbook of Organic Photochemistry and Photobiology*, 2nd edn, CRC Press LLC, Boca Raton, FL, 2004, Chapter 52, pp. 1–31.

934. Wagner, P. J., 1,5-Biradicals and 5-Membered Rings Generated by δ-Hydrogen Abstraction in Photoexcited Ketones, *Acc. Chem. Res.* 1989, **22**, 83–91.

935. Wagner, P. J., Type II Photoelimination and Photocyclization of Ketones, *Acc. Chem. Res.* 1971, **4**, 168–177.

936. Weiss, R. G., Norrish Type II Processes of Ketones: Influence of Environment. In Horspool, W. M., Song, P.-S. (eds), *CRC Handbook of Organic Photochemistry and Photobiology*, CRC Press, Boca Raton, FL, 1995, pp. 471–483.

937. Hasegawa, T., Norrish Type II Processes of Ketones: Influence of Environment. In Horspool, W. M., Lenci, F. (eds), *CRC Handbook of Organic Photochemistry and Photobiology*, 2nd edn, CRC Press LLC, Boca Raton, FL, 2004, Chapter 55, pp. 1–14.

938. Wagner, P. J., Abstraction of γ-Hydrogens by Excited Carbonyls. In Griesbeck, A. G., Mattay, J. (eds), *Synthetic Organic Photochemistry*, Marcel Dekker, New York, 2005, pp. 11–39.

939. Scaiano, J. C., Laser Flash-photolysis Studies of the Reactions of Some 1,4-Biradicals, *Acc. Chem. Res.* 1982, **15**, 252–258.

940. Chandra, A. K., Rao, V. S., Two-dimensional Tunneling of Hydrogen in a Norrish Type II Process of a Ketone, *Chem. Phys. Lett.* 1997, **270**, 87–92.

941. Rao, V. S., Chandra, A. K., An *Ab Initio* Treatment of the Norrish Type II Process in Pentane-2-one and the Role of Tunneling of Hydrogen, *Chem. Phys.* 1997, **214**, 103–112.

942. Sauers, R. R., Edberg, L. A., Modeling of Norrish Type II Reactions by Semiempirical and *Ab Initio* Methodology, *J. Org. Chem.* 1994, **59**, 7061–7066.

943. Morita, A., Kato, S., Theoretical Study on the Intersystem Crossing Mechanism of a Diradical in Norrish Type II Reactions in Solution, *J. Phys. Chem.* 1993, **97**, 3298–3313.

944. Dewar, M. J. S., Doubleday, C., MINDO-3 Study of Norrish Type II Reaction of Butanal, *J. Am. Chem. Soc.* 1978, **100**, 4935–4941.

945. Ihmels, H., Scheffer, J. R., The Norrish Type II Reaction in the Crystalline State: Toward a Better Understanding of the Geometric Requirements for γ-Hydrogen Atom Abstraction, *Tetrahedron* 1999, **55**, 885–907.

946. Sengupta, D., Chandra, A. K., Studies on the Norrish Type II Reactions of Aliphatic α-Diketones and the Accompanying Cyclization and Disproportionation of 1,4-Biradicals, *J. Photochem. Photobiol. A* 1993, **75**, 151–162.

947. Yang, N. C., Yang, D.-H., Cyclobutanol Formation from Irradiation of Ketones, *J. Am. Chem. Soc.* 1958, **80**, 2913–2914.

948. Wagner, P. J., Differences between Singlet and Triplet State Type II Photoelimination of Aliphatic Ketones, *Tetrahedron Lett.* 1968, **9**, 5385–5388.

949. Wagner, P. J., Kemppainen, A. E., Schott, H. N., Effects of Ring Substituents on the Type II Photoreactions of Phenyl Ketones. How Interactions Between Nearby Excited Triplets Affect Chemical Reactivity, *J. Am. Chem. Soc.* 1973, **95**, 5604–5614.

950. Small, R. D., Scaiano, J. C., Photochemistry of Phenyl Alkyl Ketones – Lifetime of Intermediate Biradicals, *J. Phys. Chem.* 1977, **81**, 2126–2131.

951. Wagner, P. J., Kelso, P. A., Kemppainen, A. E., McGrath, J. M., Schott, H. N., Zepp, R. G., Type II Photoprocesses of Phenyl Ketones. A Glimpse at the Behavior of 1,4-Biradicals, *J. Am. Chem. Soc.* 1972, **94**, 7506–7512.

952. Wagner, P. J., Kemppainen, A. E., Type II Photoprocesses of Phenyl Ketones. Triplet State Reactivity as a Function of $\gamma$-and $\delta$-Substitution, *J. Am. Chem. Soc.* 1972, **94**, 7495–7499.

953. Zepp, R. G., Gumz, M. M., Miller, W. L., Gao, H., Photoreaction of Valerophenone in Aqueous Solution, *J. Phys. Chem. A* 1998, **102**, 5716–5723.

954. Rabeck, J. F., *Polymer Photodegradation: Mechanisms and Experimental Methods*, Chapman and Hall, London, 1995.

955. Golemba, F. J., Guillet, J. E., Photochemistry of Ketone Polymers. VII. Polymers and Copolymers of Phenyl Vinyl Ketone, *Macromolecules* 1972, **5**, 212–216.

956. Zweig, A., Henderson, W. A., Singlet Oxygen and Polymer Photooxidations. 1. Sensitizers, Quenchers, and Reactants, *J. Polym. Sci., Part A – Polym. Chem.* 1975, **13**, 717–736.

957. Chen, S., Patrick, B. O., Scheffer, J. R., Enantioselective Photochemical Synthesis of a Simple Alkene via the Solid State Ionic Chiral Auxiliary Approach, *J. Org. Chem.* 2004, **69**, 2711–2718.

958. Chong, K. C. W., Scheffer, J. R., Thermal and Photochemical Transformation of Conformational Chirality into Configurational Chirality in the Crystalline State, *J. Am. Chem. Soc.* 2003, **125**, 4040–4041.

959. Wagner, P. J., Yang Photocyclization: Coupling of Biradicals Formed by Intramolecular Hydrogen Abstraction of Ketones. In Horspool, W. M., Lenci, F. (eds), *CRC Handbook of Organic Photochemistry and Photobiology*, 2nd edn, CRC Press LLC, Boca Raton, FL, 2004, Chapter 58, pp. 1–70.

960. Wessig, P., Mühling, O., Abstraction of ($\gamma \pm n$)-Hydrogen by Excited Carbonyls. In Griesbeck, A. G., Mattay, J. (eds), *Synthetic Organic Photochemistry*, Vol. 12, Marcel Dekker, New York, 2005, pp. 41–87.

961. Wessig, P., Regioselective Photochemical Synthesis of Carbo- and Heterocyclic Compounds: the Norrish/Yang Reaction. In Horspool, W. M., Lenci, F. (eds), *CRC Handbook of Organic Photochemistry and Photobiology*, 2nd edn, CRC Press LLC, Boca Raton, FL, 2004, Chapter 57, pp. 1–20.

962. Oelgemöller, M., Griesbeck, A. G., Photoinduced Electron Transfer Processes of Phthalimides. In Horspool, W. M., Lenci, F. (eds), *CRC Handbook of Organic Photochemistry and Photobiology*, 2nd edn, CRC Press LLC, Boca Raton, FL, 2004, Chapter 84, pp. 1–19.

963. Yoshioka, M., Miyazoe, S., Hasegawa, T., Photochemical Reaction of 3-Hydroxy-1-(*o*-methylaryl)alkan-1-ones. Formation of Cyclopropane-1,2-diols and Benzocyclobutenols through $\beta$-Hydrogen and $\gamma$-Hydrogen Abstractions, *J. Chem. Soc., Perkin Trans. 1* 1993, 2781–2786.

964. Lewis, F. D., Hilliard, T. A., Photochemistry of Methyl-substituted Butyrophenones, *J. Am. Chem. Soc.* 1972, **94**, 3852–3858.

965. Lewis, F. D., Johnson, R. W., Ruden, R. A., Photochemistry of Bicycloalkyl Phenyl Ketones, *J. Am. Chem. Soc.* 1972, **94**, 4292–4297.

966. Wagner, P. J., Chiu, C., Preferential 1,4- vs. 1,6-Hydrogen Transfer in a 1,5-Biradical. Photochemistry of $\beta$-Ethoxypropiophenone, *J. Am. Chem. Soc.* 1979, **101**, 7134–7135.

967. Giese, B., Wettstein, P., Stahelin, C., Barbosa, F., Neuburger, M., Zehnder, M., Wessig, P., Memory of Chirality in Photochemistry, *Angew. Chem. Int. Ed.* 1999, **38**, 2586–2587.

968. Zhou, B., Wagner, P. J., Long-range Triplet Hydrogen Abstraction – Photochemical Formation of 2-Tetralols from β-Arylpropiophenones, *J. Am. Chem. Soc.* 1989, **111**, 6796–6799.

969. Kraus, G. A., Wu, Y. S., 1,5-Hydrogen and 1,9-Hydrogen Atom Abstractions – Photochemical Strategies for Radical Cyclizations, *J. Am. Chem. Soc.* 1992, **114**, 8705–8707.

970. Hu, S. K., Neckers, D. C., Photochemical Reactions of Alkoxy-containing Alkyl Phenylglyoxylates: Remote Hydrogen Abstraction, *J. Chem. Soc., Perkin Trans. 2* 1997, 1751–1754.

971. Griesbeck, A. G., Oelgemöller, M., Lex, J., Photochemistry of MTM- and MTE-esters of ω-Phthalimido Carboxylic Acids: Macrocyclization versus Deprotection, *J. Org. Chem.* 2000, **65**, 9028–9032.

972. Yoon, U. C., Jin, Y. X., Oh, S. W., Park, C. H., Park, J. H., Campana, C. F., Cai, X. L., Duesler, E. N., Mariano, P. S., A Synthetic Strategy for the Preparation of Cyclic Peptide Mimetics Based on SET-promoted Photocyclization Processes, *J. Am. Chem. Soc.* 2003, **125**, 10664–10671.

973. Cossy, J., Belotti, D., Leblanc, C., Total Synthesis of (±)-Actinidine and of (±)-Isooxyskytanthine, *J. Org. Chem.* 1993, **58**, 2351–2354.

974. Sammes, P. G., Photoenolisation, *Tetrahedron* 1976, **32**, 405

975. Leonenko, Z. V., Gritsan, N. P., Quantum Chemical Study of the Intramolecular Transfer of Hydrogen in *o*-Methylacetophenone and 1-Alkylanthraquinones, *J. Struct. Chem.* 1997, **38**, 536–543.

976. Garcia-Garibay, M. A., Gamarnik, A., Bise, R., Pang, L., Jenks, W. S., Primary Isotope Effects on Excited-state Hydrogen-atom Transfer Reactions – Activated and Tunneling Mechanisms in an *ortho*-Methylanthrone, *J. Am. Chem. Soc.* 1995, **117**, 10264–10275.

977. Sengupta, D., Chandra, A. K., Role of Tunneling of Hydrogen in Photoenolization of a Ketone, *Int. J. Quantum Chem.* 1994, **52**, 1317–1328.

978. Yang, N. C., Rivas, C., A New Photochemical Primary Process. The Photochemical Enolization of *o*-Substituted Benzophenones, *J. Am. Chem. Soc.* 1961, **83**, 2213–2213.

979. Haag, R., Wirz, J., Wagner, P. J., The Photoenolization of 2-Methylacetophenone and Related Compounds, *Helv. Chim. Acta* 1977, **60**, 2595–2607.

980. Das, P. K., Encinas, M. V., Small, R. D., Scaiano, J. C., Photoenolization of *o*-Alkyl-substituted Carbonyl Compounds – Use of Electron Transfer Processes to Characterize Transient Intermediates, *J. Am. Chem. Soc.* 1979, **101**, 6965–6970.

981. Wagner, P. J., Sobczak, M., Park, B. S., Stereoselectivity in *o*-Alkylphenyl Ketone Photochemistry: How Many *o*-Xylylenes Can One Ketone Form? *J. Am. Chem. Soc.* 1998, **120**, 2488–2489.

982. Connolly, T. J., Durst, T., Photochemically Generated Bicyclic *o*-Quinodimethanes: Photoenolization of Bicyclic Aldehydes and Ketones, *Tetrahedron* 1997, **53**, 15969–15982.

983. Nicolaou, K. C., Gray, D. L. F., Tae, J. S., Total Synthesis of Hamigerans and Analogues thereof. Photochemical Generation and Diels–Alder Trapping of Hydroxy-*o*-Quinodimethanes, *J. Am. Chem. Soc.* 2004, **126**, 613–627.

984. Bergmark, W. R., Barnes, C., Clark, J., Paparian, S., Marynowski, S., Photoenolization with α-Chloro Substituents, *J. Org. Chem.* 1985, **50**, 5612–5615.

985. Pelliccioli, A. P., Klán, P., Zabadal, M., Wirz, J., Photorelease of HCl from *o*-Methylphenacyl Chloride Proceeds through the Z-Xylylenol, *J. Am. Chem. Soc.* 2001, **123**, 7931–7932.

986. Zabadal, M., Pelliccioli, A. P., Klán, P., Wirz, J., 2,5-Dimethylphenacyl Esters: a Photoremovable Protecting Group for Carboxylic Acids, *J. Phys. Chem. A* 2001, **105**, 10329–10333.

987. Plistil, L., Solomek, T., Wirz, J., Heger, D., Klán, P., Photochemistry of 2-Alkoxymethyl-5-methylphenacyl Chloride and Benzoate, *J. Org. Chem.* 2006, **71**, 8050–8058.

988. Pospisil, T., Veetil, A. T., Lovely Angel, P. A., Klán, P., Photochemical Synthesis of Substituted Indan-1-ones Related to Donepezil, *Photochem. Photobiol. Sci.* 2008, **7**, 625–632.

989. Literák, J., Wirz, J., Klán, P., 2,5-Dimethylphenacyl Carbonates: A Photoremovable Protecting Group for Alcohols and Phenols, *Photochem. Photobiol. Sci.* 2005, **4**, 43–46.

990. Gilbert, A., 1,4-Quinone Cycloaddition Reactions with Alkenes, Alkynes, and Related Compounds. In Horspool, W. M., Lenci, F. (eds), *CRC Handbook of Organic Photochemistry and Photobiology*, 2nd edn, CRC Press LLC, Boca Raton, FL, 2004, Chapter 87, pp. 1–16.

991. Kokubo, K., Oshima, T., Photochemistry of Homoquinones. In Horspool, W. M., Lenci, F. (eds), *CRC Handbook of Organic Photochemistry and Photobiology*, 2nd edn, CRC Press LLC, Boca Raton, FL, 2004, Chapter 74, pp. 1–16.

992. Oelgemöller, M., Mattay, J., The "Photochemical Friedel–Crafts Acylation" of Quinones: from the Beginnings of Organic Photochemistry to Modern Solar Chemical Applications. In Horspool, W. M., Lenci, F. (eds), *CRC Handbook of Organic Photochemistry and Photobiology*, 2nd edn, CRC Press LLC, Boca Raton, FL, 2004, Chapter 88, pp. 1–16.

993. Creed, D., 1,4-Quinone Cycloaddition Reactions with Alkenes, Alkynes, and Related Compounds. Horspool, W. M., Song, P.-S. (eds), *CRC Handbook of Organic Photochemistry and Photobiology*, Boca Raton, FL, 1995, pp. 737–747.

994. Maruyama, K., Kubo, Y., Photochemical Hydrogen Abstraction Reactions of Quinones. In Horspool, W. M., Song, P.-S. (eds), *CRC Handbook of Organic Photochemistry and Photobiology*, Boca Raton, FL, 1995, pp. 748–756.

995. Bruce, J. M., Light-induced Reactions of Quinones, *Q. Rev.* 1967, **21**, 405–428.

996. Honda, Y., Hada, M., Ehara, M., Nakatsuji, H., Ground and Excited States of Singlet, Cation Doublet, and Anion Doublet States of *o*-Benzoquinone: a Theoretical Study, *J. Phys. Chem. A* 2007, **111**, 2634–2639.

997. Honda, Y., Hada, M., Ehara, M., Nakatsuji, H., Excited and Ionized States of *p*-Benzoquinone and Its Anion Radical: SAC-CI Theoretical Study, *J. Phys. Chem. A* 2002, **106**, 3838–3849.

998. Pan, Y., Fu, Y., Liu, S. X., Yu, H. Z., Gao, Y. H., Guo, Q. X., Yu, S. Q., Studies on Photoinduced H-atom and Electron Transfer Reactions of *o*-Naphthoquinones by Laser Flash Photolysis, *J. Phys. Chem. A* 2006, **110**, 7316–7322.

999. Chiang, Y., Kresge, A.J., Hellrung, B., Schunemann, P., Wirz, J., Flash Photolysis of 5-Methyl-1,4-Naphthoquinone in Aqueous Solution: Kinetics and Mechanism of Photoenolization and of Enol Trapping, *Helv. Chim. Acta* 1997, **80**, 1106–1121.

1000. Schnapp, K.A., Wilson, R.M., Ho, D.M., Creed, D., Caldwell, R.A., Benzoquinone–Olefin Exciplexes – the Observation and Chemistry of the *p*-Benzoquinone Tetraphenylallene Exciplex, *J. Am. Chem. Soc.* 1990, **112**, 3700–3702.

1001. Goez, M., Frisch, I., Photocycloadditions of Quinones with Quadricyclane and Norbornadiene – a Mechanistic Study, *J. Am. Chem. Soc.* 1995, **117**, 10486–10502.

1002. Kobayashi, K., Shimizu, H., Sasaki, A., Suginome, H., Photoinduced Molecular Transformations. 140. New One-step General Synthesis of Naphtho[2,3-*b*]furan-4,9-diones and their 2,3-Dihydro Derivatives by the Regioselective [3 + 2] Photoaddition of 2-Hydroxy-1,4-naphthoquinones with Various Alkynes and Alkenes – Application of the Photoaddition to a Two-step Synthesis of Maturinone, *J. Org. Chem.* 1993, **58**, 4614–4618.

1003. Maruyama, K., Otsuki, T., Naruta, Y., Photochemical Reaction of 9,10-Phenanthrenequinone with Hydrogen Donors – Behavior of Radicals in Solution as Studied by CIDNP, *Bull. Chem. Soc. Jpn.* 1976, **49**, 791–795.

1004. Oelgemöller, M., Schiel, C., Frohlich, R., Mattay, J., The "Photo-Friedel–Crafts Acylation" of 1,4-Naphthoquinones, *Eur. J. Org. Chem.* 2002, 2465–2474.

1005. Kraus, G. A., Kirihara, M., Quinone Photochemistry – a General Synthesis of Acylhydroquinones, *J. Org. Chem.* 1992, **57**, 3256–3257.

1006. Schiel, C., Oelgemöller, M., Ortner, J., Mattay, J., Green Photochemistry: the Solar-Chemical Photo-Friedel–Crafts Acylation of Quinones, *Green Chem.* 2001, **3**, 224–228.

1007. Oelgemöller, M., Jung, C., Mattay, J., Green Photochemistry: Production of Fine Chemicals with Sunlight, *Pure Appl. Chem.* 2007, **79**, 1939–1947.

1008. Oelgemöller, M., Healy, N., de Oliveira, L., Jung, C., Mattay, J., Green Photochemistry: Solar–Chemical Synthesis of Juglone with Medium Concentrated Sunlight, *Green Chem.* 2006, **8**, 831–834.

1009. Esser, P., Pohlmann, B., Scharf, H. D., The Photochemical Synthesis of Fine Chemicals with Sunlight, *Angew. Chem. Int. Ed.* 1994, **33**, 2009–2023.

1010. Jung, C., Funken, K. H., Ortner, J., PROPHIS: Parabolic Trough-facility for Organic Photochemical Syntheses in Sunlight, *Photochem. Photobiol. Sci.* 2005, **4**, 409–411.

1011. Pitchumani, K., Madhavan, D., Induced Diastereoselectivity in Photodecarboxylation Reactions. In Pitchumani, K., Madhavan, D. (eds), *CRC Handbook of Organic Photochemistry and Photobiology*, 2nd edn, CRC Press LLC, Boca Raton, FL, 2004, Chapter 65, pp. 1–14.

1012. Miranda, M. G., Galindo, F., Photo-Fries Reaction and Related Processes. In Pitchumani, K., Madhavan, D. (eds), *CRC Handbook of Organic Photochemistry and Photobiology*, 2nd edn, CRC Press LLC, Boca Raton, FL, 2004, Chapter 42, pp. 1–11.

1013. Pincock, J. A., The Photochemistry of Esters of Carboxylic Acids. In Horspool, W. M., Lenci, F. (eds), *CRC Handbook of Organic Photochemistry and Photobiology*, 2nd edn, CRC Press LLC, Boca Raton, FL, 2004, Chapter 66, pp. 1–17.

1014. Dalko, P. I., The Photochemistry of Barton Esters. In Horspool, W. M., Lenci, F. (eds), *CRC Handbook of Organic Photochemistry and Photobiology*, 2nd edn, CRC Press LLC, Boca Raton, FL, 2004, Chapter 67, pp. 1–23.

1015. Givens, R. S., Conrad, I. P. G., Yousef, A. L., Lee, J.-I., Photoremovable Protecting Groups. In Horspool, W. M., Lenci, F. (eds), *CRC Handbook of Organic Photochemistry and Photobiology*, 2nd edn, CRC Press LLC, Boca Raton, FL, 2004, Chapter 69, pp. 1–46.

1016. Piva, O., Photodeconjugation of Enones and Carboxylic Acid Derivatives. In Horspool, W. M., Lenci, F. (eds), *CRC Handbook of Organic Photochemistry and Photobiology*, 2nd edn, CRC Press LLC, Boca Raton, FL, 2004, Chapter 70, pp. 1–18.

1017. Pincock, J. A., The Photochemistry of Esters of Carboxylic Acids. In Horspool, W. M., Song, P.-S. (eds), *CRC Handbook of Organic Photochemistry and Photobiology*, CRC Press, Boca Raton, FL, 1995, pp. 393–407.

1018. Coyle, J. D., Photochemistry of Carboxylic Acid Derivatives, *Chem. Rev.* 1978, **78**, 97–123.

1019. Fang, W. H., Liu, R. Z., Photodissociation of Acrylic Acid in the Gas Phase: an *Ab Initio* Study, *J. Am. Chem. Soc.* 2000, **122**, 10886–10894.

1020. Steuhl, H. M., Klessinger, M., Excited States of Cyclic Conjugated Anions: a Theoretical Study of the Photodecarboxylation of Cycloheptatriene and Cyclopentadiene Carboxylate Anions, *J. Chem. Soc., Perkin Trans. 2* 1998, 2035–2038.

1021. Li, J., Zhang, F., Fang, W. H., Probing Photophysical and Photochemical Processes of Benzoic Acid from *Ab Initio* Calculations, *J. Phys. Chem. A* 2005, **109**, 7718–7724.

1022. Staikova, M., Oh, M., Donaldson, D. J., Overtone-induced Decarboxylation: a Potential Sink for Atmospheric Diacids, *J. Phys. Chem. A* 2005, **109**, 597–602.

1023. Ding, W. J., Fang, W. H., Liu, R. Z., Mechanisms of Unimolecular Reactions for Ground-state Pyruvic Acid, *Acta Phys. Chim. Sin.* 2004, **20**, 911–916.

1024. Pritchina, E. A., Gritsan, N. P., Burdzinski, G. T., Platz, M. S., Study of Acyl Group Migration by Femtosecond Transient Absorption Spectroscopy and Computational Chemistry, *J. Phys. Chem. A* 2007, **111**, 10483–10489.

1025. Boscá, F., Marín, M. L., Miranda, M. A., Photodecarboxylation of Acids and Lactones: Antiinflammatory Drugs. In Horspool, W. M., Lenci, F. (eds), *CRC Handbook of Organic Photochemistry and Photobiology*, 2nd edn, CRC Press LLC, Boca Raton, FL, 2004, Chapter 64, pp. 1–10.

1026. Epling, G. A., Lopes, A., Fragmentation Pathways in Photolysis of Phenylacetic Acid, *J. Am. Chem. Soc.* 1977, **99**, 2700–2704.

1027. Meiggs, T. O., Grosswei, Li, Miller, S. I., Extinction Coefficient and Recombination Rate of Benzyl Radicals.1. Photolysis of Sodium Phenylacetate, *J. Am. Chem. Soc.* 1972, **94**, 7981–7986.

1028. Bosca, F., Miranda, M. A., Photosensitizing Drugs Containing the Benzophenone Chromophore, *J. Photochem. Photobiol. B* 1998, **43**, 1–26.

1029. Monti, S., Sortino, S., DeGuidi, G., Marconi, G., Photochemistry of 2-(3-Benzoylphenyl) propionic Acid (Ketoprofen). 1. A Picosecond and Nanosecond Time-resolved Study in Aqueous Solution, *J. Chem. Soc., Faraday Trans.* 1997, **93**, 2269–2275.

1030. Decosta, D. P., Pincock, J. A., Photochemistry of Substituted 1-Naphthylmethyl Esters of Phenylacetic and 3-Phenylpropanoic Acid – Radical Pairs, Ion Pairs, and Marcus Electron-transfer, *J. Am. Chem. Soc.* 1993, **115**, 2180–2190.

1031. Havinga, E., Dejongh, R. O., Dorst, W., Photochemical Acceleration of the Hydrolysis of Nitrophenyl Phosphates and Nitrophenyl Sulphates, *Recl. Trav. Chim. Pays-Bas* 1956, **75**, 378–383.

1032. Zimmerman, H. E., The *meta* Effect in Organic Photochemistry – Mechanistic and Exploratory Organic Photochemistry, *J. Am. Chem. Soc.* 1995, **117**, 8988–8991.

1033. Zimmerman, H. E., *Meta–ortho* Effect in Organic Photochemistry: Mechanistic and Exploratory Organic Photochemistry, *J. Phys. Chem. A* 1998, **102**, 5616–5621.

1034. Goeldner, M., Givens, R. S., *Dynamic Studies in Biology*, Wiley-VCH Verlag GmbH, Weinheim, 2005.

1035. Pelliccioli, A. P., Wirz, J., Photoremovable Protecting Groups: Reaction Mechanisms and Applications, *Photochem. Photobiol. Sci.* 2002, **1**, 441–458.

1036. Conrad, P. G., Givens, R. S., Weber, J. F. W., Kandler, K., New Phototriggers: Extending the *p*-Hydroxyphenacyl $\pi$–$\pi^*$ Absorption Range, *Org. Lett.* 2000, **2**, 1545–1547.

1037. Kandler, K., Givens, R. S., Katz, L. C., Photostimulation with Caged Glutamate. In Yuste, R., Lanni, R. F., Konnerth, A. (eds), *Imaging Neurons: a Laboratory Manual*, Cold Spring Harbor Laboratory Press, Cold Spring Harbor, NY, 1997, Chapter 27, pp. 1–9.

1038. Murai, A., Abiko, A., Ono, M., Masamune, T., Studies on the Phytoalexins. 31. Synthetic Studies of Rishitin and Related-Compounds. 11. Synthesis of Aubergenone, a Sesquiterpenoid Phytoalexin from Diseased Eggplants, *Bull. Chem. Soc. Jpn.* 1982, **55**, 1191–1194.

1039. Sato, T., Niino, H., Yabe, A., Consecutive Photolyses of Naphthalenedicarboxylic Anhydrides in Low Temperature Matrixes: Experimental and Computational Studies on Naphthynes and Benzocyclopentadienylideneketenes, *J. Phys. Chem. A* 2001, **105**, 7790–7798.

1040. Wenk, H. H., Winkler, M., Sander, W., One Century of Aryne Chemistry, *Angew. Chem. Int. Ed.* 2003, **42**, 502–528.

1041. Cai, X., Sakamoto, M., Yamaji, M., Fujitsuka, M., Majima, T., C–O Bond Cleavage of Esters with a Naphthyl Group in the Higher Triplet Excited State during Two-color Two-laser Flash Photolysis, *Chem. Eur. J.* 2007, **13**, 3143–3149.

1042. Andrew, D., Islet, B. T. D., Margaritis, A., Weedon, A. C., Photo-Fries Rearrangement of Naphthyl Acetate in Supercritical Carbon Dioxide – Chemical Evidence for Solvent–Solute Clustering, *J. Am. Chem. Soc.* 1995, **117**, 6132–6133.

1043. Bitterwolf, T. E., Organometallic Photochemistry at the End of Its First Century, *J. Organomet. Chem.* 2004, **689**, 3939–3952.

1044. Lees, A. J., Quantitative Photochemistry of Organometallic Complexes: Insight to Their Photophysical and Photoreactivity Mechanisms, *Coord. Chem. Rev.* 2001, **211**, 255–278.

1045. Sun, S. S., Lees, A. J., Transition Metal Based Supramolecular Systems: Synthesis, Photophysics, Photochemistry and their Potential Applications as Luminescent Anion Chemosensors, *Coord. Chem. Rev.* 2002, **230**, 171–192.

1046. Vlcek, A., The Life and Times of Excited States of Organometallic and Coordination Compounds, *Coord. Chem. Rev.* 2000, **200**, 933–977.

1047. Vogler, A., Kunkely, H., Charge Transfer Excitation of Organometallic Compounds – Spectroscopy and Photochemistry, *Coord. Chem. Rev.* 2004, **248**, 273–278.

1048. Klassen, J. K., Selke, M., Sorensen, A. A., Yang, G. K., Metal–Ligand Bond-dissociation Energies in $CpMn(Co)_2L$ Complexes, *J. Am. Chem. Soc.* 1990, **112**, 1267–1268.

1049. Pannell, K. H., Rozell, J. M., Hernandez, C., Organometalloidal Derivatives of the Transition Metals. 21. Synthesis and Photochemical Deoligomerizations of a Series of Isomeric Disilyliron Complexes – $(\eta^5\text{-}C_5H_5)Fe(Co)_2Si_2Ph_{3-n}Me_{2+n}$, *J. Am. Chem. Soc.* 1989, **111**, 4482–4485.

1050. Attig, T. G., Wojcicki, A., Stereospecificity at Iron Center of Decarbonylation of an Iron–Acetyl Complex, *J. Organomet. Chem.* 1974, **82**, 397–415.

1051. Lian, T., Bromberg, S. E., Yang, H., Proulx, G., Bergman, R. G., Harris, C. B., Femtosecond IR Studies of Alkane C−H Bond Activation by Organometallic Compounds: Direct Observation of Reactive Intermediates in Room Temperature Solutions, *J. Am. Chem. Soc.* 1996, **118**, 3769–3770.

1052. Friedrich, L. E., Bower, J. D., Detection of an Oxetene Intermediate in Photoreaction of Benzaldehyde with 2-Butyne, *J. Am. Chem. Soc.* 1973, **95**, 6869–6870.

1053. Wagner, P. J., Liu, K. C., Noguchi, Y., Monoradical Rearrangements of the 1,4-Biradicals Involved in Norrish Type-II Photoreactions, *J. Am. Chem. Soc.* 1981, **103**, 3837–3841.

1054. Garcia, H., Iborra, S., Primo, J., Miranda, M. A., 6-Endo-dig vs. 5-Exo-dig Ring Closure in *o*-Hydroxyaryl Phenylethynyl Ketones. A New Approach to the Synthesis of Flavones and Aurones, *J. Org. Chem.* 1986, **51**, 4432–4436.

1055. Kurabaya, K., Mukai, T., Organic Photochemistry 21. Photochemical and Thermal Behavior of Bicyclo[4.2.1]nona-2,4,7-trien-9-one, *Tetrahedron Lett.* 1972, 1049–1052.

1056. Yoshioka, M., Suzuki, T., Oka, M., Photochemical-Reaction of Phenyl-substituted 1,3-Diketones, *Bull. Chem. Soc. Jpn.* 1984, **57**, 1604–1607.

1057. Rau, H., Spectroscopic Properties of Organic Azo-Compounds, *Angew. Chem. Int. Ed. Engl.* 1973, **12**, 224–235.

1058. Brinton, R. K., Volman, D. H., The Ultraviolet Absorption Spectra of Gaseous Diazomethane and Diazoethane – Evidence for the Existence of Ethylidine Radicals in Diazoethane Photolysis, *J. Chem. Phys.* 1951, **19**, 1394–1395.

1059. Nau, W. M., Zhang, X. Y., An Exceedingly Long-lived Fluorescent State as a Distinct Structural and Dynamic Probe for Supramolecular Association: an Exploratory Study of Host–Guest Complexation by Cyclodextrins, *J. Am. Chem. Soc.* 1999, **121**, 8022–8032.

1060. Spellane, P. J., Gouterman, M., Antipas, A., Kim, S., Liu, Y .C., Porphyrins. 40. Electronic Spectra and 4-Orbital Energies of Free-base, Zinc, Copper, and Palladium Tetrakis (perfluorophenyl)porphyrins, *Inorg. Chem.* 1980, **19**, 386–391.

1061. Suginome, H., *E,Z*-Isomerization and Accompanying Photoreactions of Oximes, Oxime Ethers, Nitrones, Hydrazones, Imines, Azo and Azoxy Compounds, and Various Applications. In Horspool, W. M., Lenci, F. (eds), *CRC Handbook of Organic Photochemistry and Photobiology*, 2nd edn, CRC Press LLC, Boca Raton, FL, 2004, Chapter 94, pp. 1–55.

1062. Knoll, K., Photoisomerism of Azobenzenes. In Horspool, W. M., Lenci, F. (eds), *CRC Handbook of Organic Photochemistry and Photobiology*, 2nd edn, CRC Press LLC, Boca Raton, FL, 2004, Chapter 89, pp. 1–89.

1063. Engel, P. S., Mechanism of the Thermal and Photochemical Decomposition of Azoalkanes, *Chem. Rev.* 1980, **80**, 99–150.

1064. Engel, P. S., Steel, C., Photochemistry of Aliphatic Azo-Compounds in Solution, *Acc. Chem. Res.* 1973, **6**, 275–281.

1065. Pratt, A. C., Photochemistry of Imines, *Chem. Soc. Rev.* 1977, **6**, 63–81.

1066. Padwa, A., Photochemistry of Carbon–Nitrogen Double Bond, *Chem. Rev.* 1977, **77**, 37–68.

1067. Spence, G. G., Taylor, E. C., Buchardt, O., Photochemical Reactions of Azoxy Compounds, Nitrones, and Aromatic Amine *N*-Oxides, *Chem. Rev.* 1970, **70**, 231–265.

1068. Kumar, G. S., Neckers, D. C., Photochemistry of Azobenzene-containing Polymers, *Chem. Rev.* 1989, **89**, 1915–1925.

1069. Rau, H., Photoisomerization of Azobenzenes. In Rabeck, J. (ed), *Photochemistry and Photophysics*, Vol. 1, CRC Press, Boca Raton, FL, 1990, pp. 119–141.

1070. Griffith, J., Selected Aspects of Photochemistry. 2. Photochemistry of Azobenzene and Its Derivatives, *Chem. Soc. Rev.* 1972, **1**, 481–493.

1071. Ishikawa, T., Noro, T., Shoda, T., Theoretical Study on the Photoisomerization of Azobenzene, *J. Chem. Phys.* 2001, **115**, 7503–7512.

1072. Cattaneo, P., Persico, M., An *Ab Initio* Study of the Photochemistry of Azobenzene, *Phys. Chem. Chem. Phys.* 1999, **1**, 4739–4743.

1073. Cisnetti, F., Ballardini, R., Credi, A., Gandolfi, M. T., Masiero, S., Negri, F., Pieraccini, S., Spada, G. P., Photochemical and Electronic Properties of Conjugated Bis(azo) Compounds: an Experimental and Computational Study, *Chem. Eur. J.* 2004, **10**, 2011–2021.

1074. Gagliardi, L., Orlandi, G., Bernardi, F., Cembran, A., Garavelli, M., A Theoretical Study of the Lowest Electronic States of Azobenzene: the Role of Torsion Coordinate in the *cis–trans* Photoisomerization, *Theor. Chem. Acc.* 2004, **111**, 363–372.

1075. Cembran, A., Bernardi, F., Garavelli, M., Gagliardi, L., Orlandi, G., On the Mechanism of the *cis–trans* Isomerization in the Lowest Electronic States of Azobenzene: $S_0$, $S_1$, and $T_1$, *J. Am. Chem. Soc.* 2004, **126**, 3234–3243.

1076. Norikane, Y., Tamaoki, N., Photochemical and Thermal *cis/trans* Isomerization of Cyclic and Noncyclic Azobenzene Dimers: Effect of a Cyclic Structure on Isomerization, *Eur. J. Org. Chem.* 2006, 1296–1302.

1077. Tiago, M. L., Ismail-Beigi, S., Louie, S. G., Photoisomerization of Azobenzene from First Principles Constrained Density-Functional Calculations, *J. Chem. Phys.* 2005, **122**, 094311.

1078. Granucci, G., Persico, M., Excited State Dynamics with the Direct Trajectory Surface Hopping Method: Azobenzene and Its Derivatives as a Case Study, *Theor. Chem. Acc.* 2007, **117**, 1131–1143.

1079. Conti, I., Garavelli, M., Orlandi, G., The Different Photoisomerization Efficiency of Azobenzene in the Lowest n,π* and π,π* Singlets: the Role of a Phantom State, *J. Am. Chem. Soc.* 2008, **130**, 5216–5230.

1080. Monti, S., Orlandi, G., Palmieri, P., Features of the Photochemically Active State Surfaces of Azobenzene, *Chem. Phys.* 1982, **71**, 87–99.

1081. Rau, H., Luddecke, E., On the Rotation–Inversion Controversy on Photoisomerization of Azobenzenes – Experimental Proof of Inversion, *J. Am. Chem. Soc.* 1982, **104**, 1616–1620.

1082. Bonacic-Koutecky, V., Michl, J., Photochemical *syn–anti* Isomerization of a Schiff Base – a Two-dimensional Description of a Conical Intersection in Formaldimine, *Theor. Chim. Acta* 1985, **68**, 45–55.

1083. Engel, P. S., Melaugh, R. A., Page, M. A., Szilagyi, S., Timberlake, J. W., Stable *cis*-Dialkyldiazenes (Azoalkanes) – *cis*-Di-1-adamantyldiazene and *cis*-Di-1-norbornyldiazene, *J. Am. Chem. Soc.* 1976, **98**, 1971–1972.

1084. Gisin, M., Wirz, J., Photolysis of Azo Precursors of 2,3-Naphthoquinodimethane and 1,8-Naphthoquinodimethane, *Helv. Chim. Acta* 1976, **59**, 2273–2277.

1085. Adam, W., Fragale, G., Klapstein, D., Nau, W. M., Wirz, J., Phosphorescence and Transient Absorption of Azoalkane Triplet States, *J. Am. Chem. Soc.* 1995, **117**, 12578–12592.

1086. Engel, P. S., Horsey, D. W., Scholz, J. N., Karatsu, T., Kitamura, A., Intramolecular Triplet Energy-transfer in Ester-linked Bichromophoric Azoalkanes and Naphthalenes, *J. Phys. Chem.* 1992, **96**, 7524–7535.

1087. Caldwell, R. A., Helms, A. M., Engel, P. S., Wu, A. Y., Triplet Energy and Lifetime of 2,3-Diazabicyclo[2.2.2]oct-1-ene, *J. Phys. Chem.* 1996, **100**, 17716–17717.

1088. Clark, W.D. K., Steel, C., Photochemistry of 2,3-Diazabicyclo[2.2.2]Oct-2-Ene, *J. Am. Chem. Soc.* 1971, **93**, 6347–6355.

1089. Nagele, T., Hoche, R., Zinth, W., Wachtveitl, J., Femtosecond Photoisomerization of *cis*-Azobenzene, *Chem. Phys. Lett.* 1997, **272**, 489–495.

1090. Bortolus, P., Monti, S., *Cis–trans* Photoisomerization of Azobenzene – Solvent and Triplet Donor Effects, *J. Phys. Chem.* 1979, **83**, 648–652.

1091. Forber, C. L., Kelusky, E. C., Bunce, N. J., Zerner, M. C., Electronic Spectra of *cis*-Azobenzenes and *trans*-Azobenzenes – Consequences of *ortho*-Substitution, *J. Am. Chem. Soc.* 1985, **107**, 5884–5890.

1092. Hasler, E., Hormann, A., Persy, G., Platsch, H., Wirz, J., Singlet and Triplet Biradical Intermediates in the Valence Isomerization of 2,7-Dihydro-2,2,7,7-Tetramethylpyrene, *J. Am. Chem. Soc.* 1993, **115**, 5400–5409.

1093. Crano, J. C., Guglielmetti, R. J., *Organic Photochromic and Thermochromic Compounds. Vol. 1: Main Photochromic Families*, Plenum Press, New York, 1999.

1094. Crano, J. C., Guglielmetti, R. J., *Organic Photochromic and Thermochromic Compounds. Vol. 2: Physicochemical Studies, Biological Applications, and Thermochromism*, Plenum Press, New York, 1999.

1095. Dürr, H., Bouas-Laurent, H., *Photochromism: Molecules and Systems*, Elsevier Science, Amsterdam, 2003.

1096. Bechinger, C., Ferrer, S., Zaban, A., Sprague, J., Gregg, B. A., Photoelectrochromic Windows and Displays, *Nature* 1996, **383**, 608–610.

1097. Irie, M., Diarylethenes for Memories and Switches, *Chem. Rev.* 2000, **100**, 1685–1716.

1098. Yokoyama, Y., Fulgides for Memories and Switches, *Chem. Rev.* 2000, **100**, 1717–1739.

1099. Berkovic, G., Krongauz, V., Weiss, V., Spiropyrans and Spirooxazines for Memories and Switches, *Chem. Rev.* 2000, **100**, 1741–1753.

1100. Kawata, S., Kawata, Y., Three-dimensional Optical Data Storage Using Photochromic Materials, *Chem. Rev.* 2000, **100**, 1777–1788.

1101. Ahmed, S. A., Abdel-Wahab, A.-M.A., Dürr, H., Photochromic Nitrogen-containing Compounds. In Horspool, W. M., Lenci, F. (eds), *CRC Handbook of Organic Photochemistry and Photobiology*, 2nd edn, CRC Press LLC, Boca Raton, FL, 2004, Chapter 96, pp. 1–25.

1102. Crano, J. C., Flood, T., Knowles, D., Kumar, A., VanGemert, B., Photochromic Compounds: Chemistry and Application in Ophthalmic Lenses, *Pure Appl. Chem.* 1996, **68**, 1395–1398.

1103. Feringa, B. L., Jager, W. F., Delange, B., Organic Materials for Reversible Optical Data Storage, *Tetrahedron* 1993, **49**, 8267–8310.

1104. Balzani, V., Venturi, M., Credi, A., *Molecular Devices and Machines*, Wiley-VCH Verlag GmbH, Weinheim, 2003.

1105. Ichimura, K., Photoalignment of Liquid-crystal Systems, *Chem. Rev.* 2000, **100**, 1847–1873.

1106. Mustroph, H., Stollenwerk, M., Bressau, V., Current Developments in Optical Data Storage with Organic Dyes, *Angew. Chem. Int. Ed.* 2006, **45**, 2016–2035.

1107. Cook, M. J., Nygard, A. M., Wang, Z. X., Russell, D. A., An Evanescent Field Driven Monomolecular Layer Photoswitch: Coordination and Release of Metallated Macrocycles, *Chem. Commun.* 2002, 1056–1057.

1108. Shinkai, S., Nakaji, T., Ogawa, T., Shigematsu, K., Manabe, O., Photoresponsive Crown Ethers. 2. Photocontrol of Ion Extraction and Ion Transport by a Bis(crown Ether) with a Butterfly-like Motion, *J. Am. Chem. Soc.* 1981, **103**, 111–115.

1109. Suginome, H., Takahashi, H., Stereochemical Aspects of Photo-Beckmann Rearrangement – Stereochemical Integrity of Terminus of Migrating Carbon in Photo-Beckmann Rearrangements of 5-α-Cholestan-6-one and 5-β-Cholestan-6-one Oximes, *Bull. Chem. Soc. Jpn.* 1975, **48**, 576–582.

1110. Celius, T. C., Wang, Y., Toscano, J. P., Photochemical Reactivity of α-Diazocarbonyl Compounds. In Horspool, W. M., Lenci, F. (eds), *CRC Handbook of Organic Photochemistry and Photobiology*, 2nd edn, CRC Press LLC, Boca Raton, FL, 2004, Chapter 90, pp. 1–16.

1111. Grimshaw, J., Photochemistry of Aryl Diazonium Salts, Triazoles and Tetrazoles. In Horspool, W. M., Lenci, F. (eds), *CRC Handbook of Organic Photochemistry and Photobiology*, 2nd edn, CRC Press LLC, Boca Raton, FL, 2004, Chapter 43, pp. 1–17.

1112. Bucher, G., Photochemical Reactivity of Azides. In Horspool, W. M., Lenci, F. (eds), *CRC Handbook of Organic Photochemistry and Photobiology*, 2nd edn, CRC Press LLC, Boca Raton, FL, 2004, Chapter 44, pp. 1–31.

1113. Abraham, H.-W., Photogenerated Nitrene Addition to π-Bonds. In Griesbeck, A. G., Mattay, J. (eds), *Synthetic Organic Photochemistry*, Marcel Dekker, New York, 2005, pp. 391–416.

1114. Dürr, H., Abdel-Wahab, A.-M.A., Carbene Formation by Extrusion of Nitrogen. In Horspool, W. M., Song, P.-S. (eds), *CRC Handbook of Organic Photochemistry and Photobiology*, CRC Press, Boca Raton, FL, 1995, pp. 954–983.

1115. Abdel-Wahab, A.-M.A., Ahmed, S. A., Dürr, H., Carbene Formation by Extrusion of Nitrogen. In Horspool, W. M., Lenci, F. (eds), *CRC Handbook of Organic Photochemistry and Photobiology*, 2nd edn, CRC Press LLC, Boca Raton, FL, 2004, Chapter 91, pp. 1–37.

1116. Padwa, A., Azirine Photochemistry, *Acc. Chem. Res.* 1976, **9**, 371–378.

1117. Pavlik, J. W., Photoisomerization of Some Nitrogen-containing Hetero Aromatic Compounds. In Horspool, W. M., Lenci, F. (eds), *CRC Handbook of Organic Photochemistry and Photobiology*, 2nd edn, CRC Press LLC, Boca Raton, FL, 2004, Chapter 97, pp. 1–22.

1118. Cattaneo, P., Persico, N., Semiclassical Simulations of Azomethane Photochemistry in the Gas Phase and in Solution, *J. Am. Chem. Soc.* 2001, **123**, 7638–7645.

1119. Cattaneo, P., Persico, M., Diabatic and Adiabatic Potential Energy Surfaces for Azomethane Photochemistry, *Theor. Chem. Acc.* 2000, **103**, 390–398.

1120. Cattaneo, P., Persico, M., Semi-classical Treatment of the Photofragmentation of Azomethane, *Chem. Phys. Lett.* 1998, **289**, 160–166.

1121. Liu, R. F., Cui, Q., Dunn, K. M., Morokuma, K., *Ab Initio* Molecular Orbital Study of the Mechanism of Photodissociation of *trans*-Azomethane, *J. Chem. Phys.* 1996, **105**, 2333–2345.

1122. Maier, G., Eckwert, J., Bothur, A., Reisenauer, H. P., Schmidt, C., Photochemical Fragmentation of Unsubstituted Tetrazole, 1,2,3-Triazole, and 1,2,4-Triazole: First Matrix-spectroscopic Identification of Nitrilimine HCNNH, *Liebigs Ann. Chem.* 1996, 1041–1053.

1123. Bigot, B., Sevin, A., Devaquet, A., Theoretical *Ab Initio* SCF Investigation of Photochemical Behavior of 3-Membered Rings. 2. Azirine, *J. Am. Chem. Soc.* 1978, **100**, 6924–6929.

1124. Shustov, G. V., Liu, M. T. H., Houk, K. N., Facile Formation of Azines from Reactions of Singlet Methylene and Dimethylcarbene with Precursor Diazirines: Theoretical Explorations, *Can. J. Chem.* 1999, **77**, 540–549.

1125. Samartzis, P. C., Wodtke, A. M., Casting a New Light on Azide Photochemistry: Photolytic Production of Cyclic-$N_3$, *Phys. Chem. Chem. Phys.* 2007, **9**, 3054–3066.

1126. Abu-Eittah, R. H., Mohamed, A. A., Al-Omar, A. M., Theoretical Investigation of the Decomposition of Acyl Azides: Molecular Orbital Treatment, *Int. J. Quantum Chem.* 2006, **106**, 863–875.

1127. Kemnitz, C. R., Karney, W. L., Borden, W. T., Why Are Nitrenes More Stable Than Carbenes? An *Ab Initio* Study, *J. Am. Chem. Soc.* 1998, **120**, 3499–3503.

1128. O'Brien, J. M., Czuba, E., Hastie, D. R., Francisco, J. S., Shepson, P. B., Determination of the Hydroxy Nitrate Yields from the Reaction of $C_2$–$C_6$ Alkenes with OH in the Presence of NO, *J. Phys. Chem. A* 1998, **102**, 8903–8908.

1129. Klessinger, M., Bornemann, C., Theoretical Study of the Ring Opening Reactions of 2*H*-Azirines – A Classification of Substituent Effects, *J. Phys. Org. Chem.* 2002, **15**, 514–518.

1130. Arenas, J. F., Lopez-Tocon, I., Otero, J. C., Soto, J., Carbene Formation in Its Lower Singlet State from Photoexcited 3*H*-Diazirine or Diazomethane. A Combined CASPT2 and *Ab Initio* Direct Dynamics Trajectory Study, *J. Am. Chem. Soc.* 2002, **124**, 1728–1735.

1131. Smith, P., Rosenberg, A. M., The Kinetics of the Photolysis of 2,2'-Azo-bis-isobutyronitrile, *J. Am. Chem. Soc.* 1959, **81**, 2037–2043.

1132. Jaffe, A. B., Skinner, K. J., McBride, J. M., Solvent Steric Effects. II. The Free-radical Chemistry of Azobisisobutyronitrile and Azobis-3-cyano-3-pentane in Viscous and Crystalline Media, *J. Am. Chem. Soc.* 1972, **94**, 8510–8515.

1133. Schmittel, M., Ruchardt, C., Aliphatic Azo Compounds. 16. Stereoisomerization and Homolytic Decomposition of *cis*- and *trans*-Bridgehead Diazenes, *J. Am. Chem. Soc.* 1987, **109**, 2750–2759.

1134. Decker, C., The Use of UV Irradiation in Polymerization, *Polym. Int.* 1998, **45**, 133–141.

1135. Padwa, A., Rosenthal, R. J., Dent, W., Filho, P., Turro, N. J., Hrovat, D. A., Gould, I. R., Steady-state and Laser Photolysis Studies of Substituted 2*H*-Azirines – Spectroscopy, Absolute Rates, and Arrhenius Behavior for the Reaction of Nitrile Ylides with Electron Deficient Olefins, *J. Org. Chem.* 1984, **49**, 3174–3180.

1136. Orton, E., Collins, S. T., Pimentel, G. C., Molecular Structure of the Nitrile Ylide Derived from 3-Phenyl-2*H*-azirine in a Nitrogen Matrix, *J. Phys. Chem.* 1986, **90**, 6139–6143.

1137. Inui, H., Murata, S., Control of C–C and C–N Bond Cleavage of 2*H*-Azirine by Means of the Excitation Wavelength: Studies in Matrices and in Solutions, *Chem. Lett.* 2001, 832–833.

1138. Gilgen, P., Heimgartner, H., Schmid, H., Review on Photochemistry of 2*H*-Azirines, *Heterocycles* 1977, **6**, 143–212.

1139. Inui, H., Murata, S., Mechanism of Photochemical Rearrangement of 2*H*-Azirines in Low-temperature Matrices: Chemical Evidences for the Participation of Vibrationally Hot Molecules, *Chem. Phys. Lett.* 2002, **359**, 267–272.

1140. Padwa, A., Dharan, M., Smolanoff, J., Wetmore, S. I., Observations on Scope of Photoinduced 1,3-Dipolar Addition Reactions of Arylazirines, *J. Am. Chem. Soc.* 1973, **95**, 1945–1954.

1141. Padwa, A., Smolanoff, J., Wetmore, S. I., Photochemical Transformations of Small Ring Heterocyclic Compounds. 48. Further Studies on Photocycloaddition and Photodimerization Reactions of Arylazirines, *J. Org. Chem.* 1973, **38**, 1333–1340.

1142. Liu, M. T. H., The Thermolysis and Photolysis of Diazirines, *Chem. Soc. Rev.* 1982, **11**, 127–140.

1143. Smith, R. A. G., Knowles, J. R., Preparation and Photolysis of 3-Aryl-3*H*-diazirines, *J. Chem. Soc., Perkin Trans. 2* 1975, 686–694.

1144. Platz, M. S., Huang, H. Y., Ford, F., Toscano, J., Photochemical Rearrangements of Diazirines and Thermal Rearrangements of Carbenes *Pure Appl. Chem.* 1997, **69**, 803–807.

1145. Celius, T. C., Toscano, J. P., The Photochemistry of Diazirines. In Horspool, W. M., Lenci, F. (eds), *CRC Handbook of Organic Photochemistry and Photobiology*, 2nd edn, CRC Press LLC, Boca Raton, FL, 2004, Chapter 92, pp. 1–10.

1146. Branchadell, V., Muray, E., Oliva, A., Ortuno, R. M., Rodriguez-Garcia, C., Theoretical Study of the Mechanism of the Addition of Diazomethane to Ethylene and Formaldehyde. Comparison of Conventional *Ab Initio* and Density Functional Methods, *J. Phys. Chem. A* 1998, **102**, 10106–10112.

1147. Dormán, G., Photoaffinity Labeling in Biological Signal Transduction, *Top. Curr. Chem.* 2001, **211**, 169–225.

1148. Ye, T., McKervey, M. A., Organic Synthesis with α-Diazocarbonyl Compounds, *Chem. Rev.* 1994, **94**, 1091–1160.

1149. Fenwick, J., Frater, G., Ogi, K., Strausz, O. P., Mechanism of Wolff Rearrangement. 4. Role of Oxirene in Photolysis of α-Diazo Ketones and Ketenes, *J. Am. Chem. Soc.* 1973, **95**, 124–132.

1150. Tomioka, H., Okuno, H., Izawa, Y., Mechanism of the Photochemical Wolff Rearrangement – The Role of Conformation in the Photolysis of α-Diazo Carbonyl Compounds, *J. Org. Chem.* 1980, **45**, 5278–5283.

1151. Kirmse, W., 100 Years of the Wolff Rearrangement, *Eur. J. Org. Chem.* 2002, 2193–2256.

1152. Barra, M., Fisher, T. A., Cernigliaro, G. J., Sinta, R., Scaiano, J. C., On the Photo-decomposition Mechanism of *ortho*-Diazonaphthoquinones, *J. Am. Chem. Soc.* 1992, **114**, 2630–2634.

1153. Almstead, J. I. K., Urwyler, B., Wirz, J., Flash Photolysis of α-Diazonaphthoquinones in Aqueous Solution – Determination of Rates and Equilibria for Keto–Enol Tautomerization of 1-Indene-3-carboxylic Acid, *J. Am. Chem. Soc.* 1994, **116**, 954–960.

1154. Cava, M. P., Spangler, R. J., 2-Carbomethoxybenzocyclobutenone. Synthesis of a Photochemically Sensitive Small-ring System by a Pyrolytic Wolff Rearrangement, *J. Am. Chem. Soc.* 1967, **89**, 4550–4551.

1155. Banks, J. T., Scaiano, J. C., Laser Drop and Low Intensity Photolysis of 2-Diazo-1,3-Indandione: Evidence for a Propadienone Intermediate, *J. Photochem. Photobiol. A* 1996, **96**, 31–33.

1156. Trozzolo, A. M., Electronic Spectroscopy of Arylmethylenes, *Acc. Chem. Res.* 1968, **1**, 329–335.

1157. Eisenthal, K. B., Turro, N. J., Aikawa, M., Butcher, J. A., Dupuy, C., Hefferon, G., Hetherington, W., Korenowski, G. M., McAuliffe, M. J., Dynamic and Energetics of the Singlet–Triplet Interconversion of Diphenylcarbene, *J. Am. Chem. Soc.* 1980, **102**, 6563–6565.

1158. Eisenthal, K. B., Turro, N. J., Sitzmann, E. V., Gould, I. R., Hefferon, G., Langan, J., Cha, Y., Singlet–Triplet Interconversion of Diphenylmethylene – Energetics, Dynamics and Reactivities of Different Spin States, *Tetrahedron* 1985, **41**, 1543–1554.

1159. Guella, G., Ascenzi, D., Franceschi, P., Tosi, P., Gas-phase Synthesis and Detection of the Benzenediazonium Ion, $C_6H_5N_2{}^+$. A Joint Atmospheric Pressure Chemical Ionization and Guided Ion Beam Experiment, *Rapid Commun. Mass Spectrom.* 2005, **19**, 1951–1955.

1160. Winkler, M., Sander, W., Generation and Reactivity of the Phenyl Cation in Cryogenic Argon Matrices: Monitoring the Reactions with Nitrogen and Carbon Monoxide Directly by IR Spectroscopy, *J. Org. Chem.* 2006, **71**, 6357–6367.

1161. Slegt, M., Overkleeft, H.S., Lodder, G., Fingerprints of Singlet and Triplet Phenyl Cations, *Eur. J. Org. Chem.* 2007, 5364–5375.

1162. Labbe, G., Decomposition and Addition Reactions of Organic Azides, *Chem. Rev.* 1969, **69**, 345–363.

1163. Lwowski, W., Demauriac, R. A., Thompson, M., Wilde, R. E., Chen, S. Y., Curtius and Lossen Rearrangements. 3. Photolysis of Certain Carbamoyl Azides, *J. Org. Chem.* 1975, **40**, 2608–2612.

1164. Tsuchia, T., Nitrene Formation by Photoextrusion of Nitrogen from Azides. In Horspool, W. M., Song, P.-S. (eds), *CRC Handbook of Organic Photochemistry and Photobiology*, CRC Press, Boca Raton, FL, 1995, pp. 984–991.

1165. Schrock, A. K., Schuster, G. B., Photochemistry of Phenyl Azide – Chemical Properties of the Transient Intermediates, *J. Am. Chem. Soc.* 1984, **106**, 5228–5234.

1166. Kotzyba-Hibert, F., Kapfer, I., Goeldner, M., Recent Trends in Photoaffinity Labeling, *Angew. Chem. Int. Ed. Engl.* 1995, **34**, 1296–1312.

1167. Vodovozova, E. L., Photoaffinity Labeling and Its Application in Structural Biology, *Biochemistry (Moscow)* 2007, **72**, 1–20.

1168. Tomohiro, T., Hashimoto, M., Hatanaka, Y., Cross-linking Chemistry and Biology: Development of Multifunctional Photoaffinity Probes, *Chem. Record* 2005, **5**, 385–395.

1169. Dormán, G., Prestwich, G. D., Benzophenone Photophores in Biochemistry, *Biochemistry* 1994, **33**, 5661–5673.

1170. Fleming, S. A., Chemical Reagents in Photoaffinity Labeling, *Tetrahedron* 1995, **51**, 12479–12520.

1171. Miller, D. J., Moody, C. J., Synthetic Applications of the O–H Insertion Reactions of Carbenes and Carbenoids Derived from Diazocarbonyl and Related Diazocompounds, *Tetrahedron* 1995, **51**, 10811–10843.

1172. Admasu, A., Gudmundsdottir, A. D., Platz, M. S., Watt, D. S., Kwiatkowski, S., Crocker, P. J., A Laser Flash Photolysis Study of *p*-Tolyl(trifluoromethyl)carbene, *J. Chem. Soc., Perkin Trans. 2* 1998, 1093–1099.

1173. Husi, H., Luyten, M. A., Zurini, M. G. M., Mapping of the Immunophilin-immunosuppressant Site of Interaction on Calcineurin, *J. Biol. Chem.* 1994, **269**, 14199–14204.

1174. Rasenick, M. M., Talluri, M., Dunn, W. J., Photoaffinity Guanosine 5′-Triphosphate Analogs as a Tool for the Study of GTP-binding Proteins, *Methods Mol. Biol.* 1994, **237**, 100–110.

1175. Albini, A., Bettinetti, G. F., Minoli, G., 1,3-Benzoxazepine and 3,1-Benzoxazepine, *Tetrahedron Lett.* 1979, 3761–3764.

1176. Suginome, H., Remote Functionalization by Alkoxyl Radicals Generated by the Photolysis of Nitrite Esters: the Barton Reaction and Related Reactions of Nitrite Esters. In Horspool, W. M., Lenci, F. (eds), *CRC Handbook of Organic Photochemistry and Photobiology*, 2nd edn, CRC Press LLC, Boca Raton, FL, 2004, Chapter 102, pp. 1–16.

1177. Barton, D. H. R., Beaton, J. M., Geller, L. E., Pechet, M. M., A New Photochemical Reaction, *J. Am. Chem. Soc.* 1960, **82**, 2640–2641.

1178. Akhtar, M., Pechet, M. M., Mechanism of Barton Reaction, *J. Am. Chem. Soc.* 1964, **86**, 265–268.

1179. Wang, H., Burda, C., Persy, G., Wirz, J., Photochemistry of 1*H*-Benzotriazole in Aqueous Solution: a Photolatent Base, *J. Am. Chem. Soc.* 2000, **122**, 5849–5855.

1180. Barltrop, J. A., Colinday, A., Ring Permutations – Novel Approach to Aromatic Phototransposition Reactions, *J. Chem. Soc., Chem. Commun.* 1975, 177–179.

1181. Chyba, C. F., Atmospheric Science – Rethinking Earth's Early Atmosphere, *Science* 2005, **308**, 962–963.

1182. Lazcano, A., Miller, S. L., The Origin and Early Evolution of Life: Prebiotic Chemistry, the Pre-RNA World, and Time, *Cell* 1996, **85**, 793–798.

1183. Dondi, D., Merli, D., Pretali, L., Fagnoni, M., Albini, A., Serpone, N., Prebiotic Chemistry: Chemical Evolution of Organics on the Primitive Earth under Simulated Prebiotic Conditions, *Photochem. Photobiol. Sci.* 2007, **6**, 1210–1217.

1184. Allamandola, L. J., Bernstein, M. P., Sandford, S. A., Walker, R. L., Evolution of Interstellar Ices, *Space Sci. Rev.* 1999, **90**, 219–232.

1185. Bernstein, M. P., Allamandola, L. J., Sandford, S. A., Complex Organics in Laboratory Simulations of Interstellar/Cometary Ices, *Adv. Space Res.* 1997, **19**, 991–998.

1186. Meierhenrich, U. J., Thiemann, W. H. P., Photochemical Concepts on the Origin of Biomolecular Asymmetry, *Orig. Life Evol. Biosph.* 2004, **34**, 111–121.

1187. Döpp, D., Photochemical Reactivity of the Nitro Group. In Horspool, W. M., Song, P.-S. (eds), *CRC Handbook of Organic Photochemistry and Photobiology*, CRC Press, Boca Raton, FL, 1995, pp. 1019–1062.

1188. Li, Y. M., Sun, J. L., Yin, H. M., Han, K. L., He, G. Z., Photodissociation of Nitrobenzene at 266 nm: Experimental and Theoretical Approach, *J. Chem. Phys.* 2003, **118**, 6244–6249.

1189. Li, Y. M., Sun, J. L., Han, K. L., He, G. Z., Li, Z. J., The Dynamics of NO Radical Formation in the UV 266 nm Photodissociation of Nitroethane, *Chem. Phys. Lett.* 2006, **421**, 232–236.

1190. Dunkin, I. R., Gebicki, J., Kiszka, M., Sanin-Leira, D., Phototautomerism of *o*-Nitrobenzyl Compounds: *o*-Quinonoid *aci*-Nitro Species Studied by Matrix Isolation and DFT Calculations, *J. Chem. Soc., Perkin Trans.* 2001, **2**, 1414–1425.

1191. Glenewinkel-Meyer, T., Crim, F. F., The Isomerization of Nitrobenzene to Phenylnitrite, *THEOCHEM* 1995, **337**, 209–224.

1192. Hashimot, S., Kana, K., Photochemical Reduction of Nitrobenzene and Its Reduction Intermediates. 10. Photochemical Reduction of Monosubstituted Nitrobenzenes in 2-Propanol, *Bull. Chem. Soc. Jpn.* 1972, **45**, 549–553.

1193. Takezaki, M., Hirota, N., Terazima, M., Excited State Dynamics of Nitrobenzene Studied by the Time-resolved Transient Grating Method *Prog. Nat. Sci.* 1996, **6**, S453–S456.

1194. Wubbels, G. G., Jordan, J. W., Mills, N. S., Hydrochloric Acid Catalyzed Photoreduction of Nitrobenzene by 2-Propanol – Question of Protonation in Excited-state, *J. Am. Chem. Soc.* 1973, **95**, 1281–1285.

1195. Wubbels, G. G., Letsinger, R. L., Photoreactions of Nitrobenzene and Monosubstituted Nitrobenzenes with Hydrochloric Acid – Evidence Concerning Reaction Mechanism, *J. Am. Chem. Soc.* 1974, **96**, 6698–6706.

1196. Schworer, M., Wirz, J., Photochemical Reaction Mechanisms of 2-Nitrobenzyl Compounds in Solution I. 2-Nitrotoluene: Thermodynamic and Kinetic Parameters of the *aci*-Nitro Tautomer, *Helv. Chim. Acta* 2001, **84**, 1441–1458.

1197. Morrison, H., *Biological Applications of Photochemical Switches* Vol., 2, John Wiley and Sons, Inc., New York, 1993.

1198. Mayer, G., Heckel, A., Biologically Active Molecules with a "Light Switch", *Angew. Chem. Int. Ed.* 2006, **45**, 4900–4921.

1199. Pillai, V. N. R., Photoremovable Protecting Groups in Organic Synthesis, *Synthesis* 1980, 1–26.

1200. Bochet, C. G., Photolabile Protecting Groups and Linkers, *J. Chem. Soc., Perkin Trans. 1* 2002, 125–142.

1201. Derrer, S., Flachsmann, F., Plessis, C., Stang, M., Applied Photochemistry Light Controlled Perfume Release *Chimia* 2007, **61**, 665–669.

1202. Denk, W., Strickler, J. H., Webb, W. W., 2-Photon Laser Scanning Fluorescence Microscopy, *Science* 1990, **248**, 73–76.

1203. Kaplan, J. H., Forbush, B., Hoffman, J. F., Rapid Photolytic Release of Adenosine 5′-Triphosphate from a Protected Analog – Utilization by Na–K Pump of Human Red Blood-cell Ghosts, *Biochemistry* 1978, **17**, 1929–1935.

1204. Walker, J. W., Reid, G. P., McCray, J. A., Trentham, D. R., Photolabile 1-(2-nitrophenyl)ethyl Phosphate Esters of Adenine Nucleotide Analogs – Synthesis and Mechanism of Photolysis, *J. Am. Chem. Soc.* 1988, **110**, 7170–7177.

1205. Il'ichev, Y. V., Schworer, M. A., Wirz, J., Photochemical Reaction Mechanisms of 2-Nitrobenzyl Compounds: Methyl Ethers and Caged ATP, *J. Am. Chem. Soc.* 2004, **126**, 4581–4595.

1206. Hellrung, B., Kamdzhilov, Y., Schworer, M., Wirz, J., Photorelease of Alcohols from 2-Nitrobenzyl Ethers Proceeds via Hemiacetals and May Be Further Retarded by Buffers Intercepting the Primary *aci*-Nitro Intermediates, *J. Am. Chem. Soc.* 2005, **127**, 8934–8935.

1207. Gaplovsky, M., Il'ichev, Y.V., Kamdzhilov, Y., Kombarova, S.V., Mac, M., Schworer, M.A., Wirz, J., Photochemical Reaction Mechanisms of 2-Nitrobenzyl Compounds: 2-Nitrobenzyl Alcohols form 2-Nitroso Hydrates by Dual Proton Transfer, *Photochem. Photobiol. Sci.* 2005, **4**, 33–42.

1208. Nicolaou, K. C., Winssinger, N., Pastor, J., DeRoose, F., A General and Highly Efficient Solid Phase Synthesis of Oligosaccharides. Total Synthesis of a Heptasaccharide Phytoalexin Elicitor (HPE), *J. Am. Chem. Soc.* 1997, **119**, 449–450.

1209. Guerrero, L., Smart, O. S., Weston, C. J., Burns, D. C., Woolley, G. A., Allemann, R. K., Photochemical Regulation of DNA-binding Specificity of MyoD, *Angew. Chem. Int. Ed.* 2005, **44**, 7778–7782.

1210. Bennett, I. M., Farfano, H. M. V., Bogani, F., Primak, A., Liddell, P. A., Otero, L., Sereno, L., Silber, J. J., Moore, A. L., Moore, T. A., Gust, D., Active Transport of $Ca^{2+}$ by an Artificial Photosynthetic Membrane, *Nature* 2002, **420**, 398–401.

1211. Blanc, A., Bochet, C. G., Wavelength-controlled Orthogonal Photolysis of Protecting Groups, *J. Org. Chem.* 2002, **67**, 5567–5577.

1212. Yasuda, M., Shiragami, T., Matsumoto, J., Yamashita, T., Shima, K., Photoamination with Ammonia and Amines. In Ramamurthy, V., Schanze, K. S. (eds), *Organic Photochemistry and Photophysics*, Vol. 14, CRC Press, Boca Raton, FL, 2006, pp. 207–253.

1213. Burdzinski, G., Kubicki, J., Maciejewski, A., Steer, R. P., Velate, S., Yeow, E. K. L., Photochemistry and Photophysics of Highly Excited Valence States of Polyatomic Molecules: Thioketones, and Metalloporphyrins. In Ramamurthy, V., Schanze, K. S. (eds), *Organic Photochemistry and Photophysics*, Vol. 14, CRC Press, Boca Raton, FL, 2006, pp. 75–129.

1214. Meunier, B., Metalloporphyrins as Versatile Catalysts for Oxidation Reactions and Oxidative DNA Cleavage, *Chem. Rev.* 1992, **92**, 1411–1456.

1215. Oelgemöller, M., Bunte, J. O., Mattay, J., Photoinduced Electron Transfer Cyclizations via Radical Ions. In Griesbeck, A. G., Mattay, J. (eds), *Synthetic Organic Photochemistry*, Marcel Dekker, New York, 2005, pp. 269–297.

1216. Whitten, D. G., Photoinduced Electron Transfer Reactions of Metal Complexes in Solution, *Acc. Chem. Res.* 1980, **13**, 83–90.

1217. Gould, I. R., Young, R. H., Moody, R. E., Farid, S., Contact and Solvent-separated Geminate Radical Ion Pairs in Electron Transfer Photochemistry, *J. Phys. Chem.* 1991, **95**, 2068–2080.

1218. Weller, A., Photoinduced Electron Transfer in Solution – Exciplex and Radical Ion-pair Formation Free Enthalpies and Their Solvent Dependence, *Z. Phys. Chem.* 1982, **133**, 93–98.

1219. Mattay, J., Vondenhof, M., Contact and Solvent-separated Radical Ion Pairs in Organic Photochemistry, *Top. Curr. Chem.* 1991, **159**, 219–255.

1220. Veillard, A., *Ab Initio* Calculations of Transition Metal Organometallics – Structure and Molecular Properties, *Chem. Rev.* 1991, **91**, 743–766.

1221. Newton, M. D., Quantum Chemical Probes of Electron Transfer Kinetics – The Nature of Donor–Acceptor Interactions, *Chem. Rev.* 1991, **91**, 767–792.

1222. Lewis, F. D., Reddy, G. D., Schneider, S., Gahr, M., Photophysical and Photochemical Behavior of Intramolecular Styrene–Amine Exciplexes, *J. Am. Chem. Soc.* 1991, **113**, 3498–3506.

1223. Sobolewski, A. L., Domcke, W., Photoinduced Electron and Proton Transfer in Phenol and Its Clusters with Water and Ammonia, *J. Phys. Chem. A* 2001, **105**, 9275–9283.

1224. Jacques, P., Burget, D., Allonas, X., On the Quantitative Appraisal of the Free Energy Change $\Delta G_{ET}$ in Photoinduced Electron Transfer, *New J. Chem.* 1996, **20**, 933–937.

1225. Lewis, F. D., Kultgen, S. G., Photochemical Synthesis of Medium-ring Azalactams from *N*-(Aminoalkyl)-2-stilbene Carboxamides, *J. Photochem. Photobiol. A* 1998, **112**, 159–164.

1226. Wubbels, G. G., Sevetson, B. R., Sanders, H., Competitive Catalysis and Quenching by Amines of Photo-Smiles Rearrangement as Evidence for a Zwitterionic Triplet as the Proton-donating Intermediate, *J. Am. Chem. Soc.* 1989, **111**, 1018–1022.

1227. Falvey, D. E., Sundararajan, C., Photoremovable Protecting Groups Based on Electron Transfer Chemistry, *Photochem. Photobiol. Sci.* 2004, **3**, 831–838.

1228. Banerjee, A., Falvey, D. E., Direct Photolysis of Phenacyl Protecting Groups Studied by Laser Flash Photolysis: an Excited State Hydrogen Atom Abstraction Pathway Leads to Formation of Carboxylic Acids and Acetophenone, *J. Am. Chem. Soc.* 1998, **120**, 2965–2966.

1229. Literák, J., Dostalová, A., Klán, P., Chain Mechanism in the Photocleavage of Phenacyl and Pyridacyl Esters in the Presence of Hydrogen Donors, *J. Org. Chem.* 2006, **71**, 713–723.

1230. Banerjee, A., Falvey, D. E., Protecting Groups that Can Be Removed Through Photochemical Electron Transfer: Mechanistic and Product Studies on Photosensitized Release of Carboxylates from Phenacyl Esters, *J. Org. Chem.* 1997, **62**, 6245–6251.

1231. Majima, T., Pac, C., Nakasone, A., Sakurai, H., Redox-photosensitized Reactions. 7. Aromatic Hydrocarbon-photosensitized Electron-transfer Reactions of Furan, Methylated Furans,1,1-Diphenylethylene, and Indene with *p*-Dicyanobenzene *J. Am. Chem. Soc.* 1981, **103**, 4499–4508.

1232. Zhang, X. M., Mariano, P. S., Mechanistic Details for SET-promoted Photoadditions of Amines to Conjugated Enones Arising from Studies of Aniline–Cyclohexenone Photoreactions, *J. Org. Chem.* 1991, **56**, 1655–1660.

1233. Jeon, Y. T., Lee, C. P., Mariano, P. S., Radical Cyclization Reactions of α-Silyl Amine α,β-Unsaturated Ketone and Ester Systems Promoted by Single Electron Transfer Photosensitization, *J. Am. Chem. Soc.* 1991, **113**, 8847–8863.

1234. Goeller, F., Heinemann, C., Demuth, M., Investigations of Cascade Cyclizations of Terpenoid Polyalkenes via Radical Cations. A Biomimetic-type Synthesis of (±)-3-Hydroxy-spongian-16-one, *Synthesis* 2001, 1114–1116.

1235. Durham, B., Caspar, J. V., Nagle, J. K., Meyer, T. J., Photochemistry of Ru(Bpy)$_3^{2+}$, *J. Am. Chem. Soc.* 1982, **104**, 4803–4810.

1236. Segawa, H., Shimidzu, T., Honda, K., A Novel Photosensitized Polymerization of Pyrrole, *J. Chem. Soc., Chem. Commun.* 1989, 132–133.

1237. Kato, H., Asakura, K., Kudo, A., Highly Efficient Water Splitting into H$_2$ and O$_2$ Over Lanthanum-doped NaTaO$_3$ Photocatalysts with High Crystallinity and Surface Nanostructure, *J. Am. Chem. Soc.* 2003, **125**, 3082–3089.

1238. Elvington, M., Brown, J., Arachchige, S. M., Brewer, K. J., Photocatalytic Hydrogen Production from Water Employing a Ru, Rh, Ru Molecular Device for Photoinitiated Electron Collection, *J. Am. Chem. Soc.* 2007, **129**, 10644–10645.

1239. Ballardini, R., Balzani, V., Credi, A., Gandolfi, M. T., Venturi, M., Artificial Molecular-level Machines: Which Energy to Make them Work? *Acc. Chem. Res.* 2001, **34**, 445–455.

1240. Browne, W. R., Feringa, B. L., Making Molecular Machines Work, *Nat. Nanotech.* 2006, **1**, 25–35.

1241. Balzani, V., Credi, A., Ferrer, B., Silvi, S., Venturi, M., Artificial Molecular Motors and Machines: Design Principles and Prototype Systems, *Top. Curr. Chem.* 2005, **262**, 1–28.

1242. Balzani, V., Credi, A., Raymo, F. M., Stoddart, J. F., Artificial Molecular Machines, *Angew. Chem. Int. Ed.* 2000, **39**, 3349–3391.

1243. Vacek, J., Michl, J., Artificial Surface-mounted Molecular Rotors: Molecular Dynamics Simulations, *Adv. Funct. Mater.* 2007, **17**, 730–739.

1244. Sauvage, J. P., Transition Metal-containing Rotaxanes and Catenanes in Motion: Toward Molecular Machines and Motors, *Acc. Chem. Res.* 1998, **31**, 611–619.

1245. van Delden, R. A., ter Wiel, M. K. J., Pollard, M. M., Vicario, J., Koumura, N., Feringa, B. L., Unidirectional Molecular Motor on a Gold Surface, *Nature* 2005, **437**, 1337–1340.

1246. Balzani, V., Clemente-Leon, M., Credi, A., Ferrer, B., Venturi, M., Flood, A. H., Stoddart, J. F., Autonomous Artificial Nanomotor Powered by Sunlight, *Proc. Natl. Acad. Sci. USA* 2006, **103**, 1178–1183.

1247. Stemp, E. D. A., Arkin, M. R., Barton, J. K., Oxidation of Guanine in DNA by Ru(phen)(II) (dppz)$^{3+}$ Using the Flash-quench Technique, *J. Am. Chem. Soc.* 1997, **119**, 2921–2925.

1248. Szacilowski, K., Macyk, W., Drzewiecka-Matuszek, A., Brindell, M., Stochel, G., Bioinorganic Photochemistry: Frontiers and Mechanisms, *Chem. Rev.* 2005, **105**, 2647–2694.

1249. Boyle, R. W., Dolphin, D., Structure and Biodistribution Relationships of Photodynamic Sensitizers, *Photochem. Photobiol.* 1996, **64**, 469–485.

1250. Tsuchiya, S., Intramolecular Electron Transfer of Diporphyrins Comprised of Electron-deficient Porphyrin and Electron-rich Porphyrin with Photocontrolled Isomerization, *J. Am. Chem. Soc.* 1999, **121**, 48–53.

1251. Barton, D. H., Beaton, J. M., A Synthesis of Aldosterone Acetate, *J. Am. Chem. Soc.* 1961, **83**, 4083–4089.

1252. Horspool, W. M., Kershaw, J. R., Murray, A. W., Stevenson, G. M., Photolysis of 4-substituted 1,2,3-Benzotriazine-3-*N*-oxides, *J. Am. Chem. Soc.* 1973, **95**, 2390–2391.

1253. Xu, W., Jeon, Y. T., Hasegawa, E., Yoon, U. C., Mariano, P. S., Novel Electron-transfer Photocyclization Reactions of α-Silyl Amine α,β-Unsaturated Ketone and Ester Systems, *J. Am. Chem. Soc.* 1989, **111**, 406–408.

1254. Moriconi, E. J., Murray, J. J., Pyrolysis and Photolysis of 1-Methyl-3-Diazooxindole. Base Decomposition of Isatin 2-Tosylhydrazone, *J. Org. Chem.* 1964, **29**, 3577–3584.

1255. Gee, K. R., Niu, L., Schaper, K., Jayaraman, V., Hess, G. P., Synthesis and Photochemistry of a Photolabile Precursor of *N*-Methyl-D-aspartate (NMDA) that is Photolyzed in the Microsecond Time Region and is Suitable for Chemical Kinetic Investigations of the NMDA Receptor, *Biochemistry* 1999, **38**, 3140–3147.

1256. Suginome, H., Kaji, M., and Yamada, S., Photo-Induced Molecular Transformations. 87. Regiospecific Photo-Beckmann Rearrangement of Steroidal α,β-Unsaturated Ketone Oximes: Synthesis of Some Steroidal Enamino Lactams, *J. Chem. Soc., Perkin Trans. 1*, 1988, 321–326.

1257. Falk, K. J., Steer, R. P., Photophysics and Intramolecular Photochemistry of Adamantanethione, Thiocamphor, and Thiofenchone Excited to Their 2nd Excited Singlet States – Evidence for Subpicosecond Photoprocesses, *J. Am. Chem. Soc.* 1989, **111**, 6518–6524.

1258. Petiau, M., Fabian, J., The Thiocarbonyl Chromophore. A Time-dependent Density-functional Study, *THEOCHEM* 2001, **538**, 253–260.

1259. Pushkara Rao, V., Nageshwer Rao, B., Ramamurthy, V., Thiocarbonyl: Photochemical Hydrogen Abstraction. In Horspool, W. M., Song, P.-S. (eds), *CRC Handbook of Organic Photochemistry and Photobiology*, CRC Press, Boca Raton, FL, 1995, pp. 793–802.

1260. Ramamurthy, V., Nageshwer Rao, B., Pushkara Rao, V., Solution Photochemistry of Thioketones. In Horspool, W. M., Song, P.-S. (eds), *CRC Handbook of Organic Photochemistry and Photobiology*, CRC Press, Boca Raton, FL, 1995, pp. 775–792.

1261. Maciejewski, A., Steer, R. P., The Photophysics, Physical Photochemistry, and Related Spectroscopy of Thiocarbonyls, *Chem. Rev.* 1993, **93**, 67–98.

1262. Steer, R. P., Ramamurthy, V., Photophysics and Intramolecular Photochemistry of Thiones in Solution, *Acc. Chem. Res.* 1988, **21**, 380–386.

1263. Sakamoto, M., Photochemical Aspects of Thiocarbonyl Compounds in the Solid-state, *Org. Solid State React.* 2005, **254**, 207–232.

1264. Coyle, J. D., The Photochemistry of Thiocarbonyl Compounds, *Tetrahedron* 1985, **41**, 5393–5425.

1265. Lin, L., Ding, W. J., Fang, W. H., Liu, R. Z., Theoretical Prediction of the Structures and Reactivity of Thiocarbonyl Compounds in Excited States, *Acta Chim. Sin.* 2003, **61**, 1–7.

1266. Sikorski, M., Khmelinskii, I. V., Augustyniak, W., Wilkinson, F., Triplet State Decay of Some Thioketones in Solution, *J. Chem. Soc., Faraday Trans.* 1996, **92**, 3487–3490.

1267. Sustmann, R., Sicking, W., Huisgen, R., Regiochemistry in Cycloadditions of Diazomethane to Thioformaldehyde and Thioketones *J. Org. Chem.* 1993, **58**, 82–89.

1268. Rao, V. P., Chandrasekhar, J., Ramamurthy, V., Thermal and Photochemical Cycloaddition Reactions of Thiocarbonyls – A Qualitative Molecular-Orbital Analysis, *J. Chem. Soc., Perkin Trans.* 2 1988, 647–659.

1269. Sumathi, K., Chandra, A. K., A Comparative Study of the Photochemical Hydrogen-Abstraction Reactions of a Ketone and a Thione, *J. Photochem. Photobiol. A* 1988, **43**, 313–327.

1270. Bigot, B., Theoretical Study on the Origin of the Behavior of Ketones and Thiones in Hydrogen Photoabstraction Reactions, *Isr. J. Chem.* 1983, **23**, 116–123.

1271. Fu, T. Y., Scheffer, J. R., Trotter, J., Crystal Structure–Reactivity Relationships in the Solid State Photochemistry of 2,4,6-Triisopropylthiobenzophenone: C=O···H versus C=S···H Abstraction Geometry, *Tetrahedron Lett.* 1996, **37**, 2125–2128.

1272. Ho, K. W., de Mayo, P., Thione Photochemistry. On the Mechanism of Photocyclization of Aralkyl Thiones, *J. Am. Chem. Soc.* 1979, **101**, 5725–5732.

1273. Okazaki, R., Ishii, A., Fukuda, N., Oyama, H., Inamoto, N., Thermal and Photochemical Reactions of a Stable Thiobenzaldehyde, 2,4,6-Tri-*t*-butylthiobenzaldehyde, *Tetrahedron Lett.* 1984, **25**, 849–852.

1274. Couture, A., Gomez, J., de Mayo, P., Thione Photochemistry. 31. Abstraction and Cyclization at the β-Position of Aralkyl Thiones from 2 Excited States, *J. Org. Chem.* 1981, **46**, 2010–2016.

1275. Kumar, C. V., Qin, L., Das, P. K., Aromatic Thioketone Triplets and Their Quenching Behavior Towards Oxygen and Di-*t*-butylnitroxy Radical – A Laser-flash Photolysis Study, *J. Chem. Soc, Faraday Trans.* 2, 1984, **80**, 783–793.

1276. Ohno, A., Ohnishi, Y., Tsuchihashi, G., Photocycloaddition of Thiocarbonyl Compounds to Olefins. Reaction of Thiobenzophenone with Various Types of Olefins. V, *J. Am. Chem. Soc.* 1969, **91**, 5038–5045.

1277. Still, I. W. J., Photochemistry of Organic Sulfur Compounds. In Bernardi, F., Csizmadia, I. G., Mangini, A. (eds), *Studies in Organic Chemistry*, Vol. 19, Elsevier, Amsterdam, 1985, pp. 596–658.

1278. Coyle, J. D., Photochemistry of Organic Sulfur Compounds, *Chem. Soc. Rev.* 1975, **4**, 523–532.

1279. Pillai, V. N. R., Photochemical Methods for Protection and Deprotection of Sulfur-containing Compounds. In Horspool, W. M., Song, P.-S. (eds), *CRC Handbook of Organic Photochemistry and Photobiology*, CRC Press, Boca Raton, FL, 1995, pp. 766–774.

1280. Cubbage, J. W., Jenks, W. S., Computational Studies of the Ground and Excited State Potentials of DMSO and H$_2$SO: Relevance to Photostereomutation, *J. Phys. Chem. A* 2001, **105**, 10588–10595.

1281. Gerwens, H., Jug, K., SINDO1 Study of the Photoreaction of Tetramethylene Sulfone, *J. Comput. Chem.* 1995, **16**, 405–413.

1282. Langler, R. F., Marini, Z. A., Pincock, J. A., Photochemistry of Benzylic Sulfonyl Compounds – Preparation of Sulfones and Sulfinic Acids, *Can. J. Chem.* 1978, **56**, 903–907.

1283. Izawa, Y., Kuromiya, N., Photolysis of Methyl Benzenesulfonate in Methanol, *Bull. Chem. Soc. Jpn.* 1975, **48**, 3197–3199.

1284. Givens, R. S., Matuszewski, B., Photoextrusion Reactions – Comparative Mechanistic Study of SO$_2$ Photoextrusion, *Tetrahedron Lett.* 1978, 861–864.

1285. Givens, R. S., Olsen, R. J., Wylie, P. L., Mechanistic Studies in Photochemistry. 21. Photoextrusion of Sulfur Dioxide – General Route to [2.2]Cyclophanes, *J. Org. Chem.* 1979, **44**, 1608–1613.

1286. Pete, J. P., Portella, C., Photolysis of Alkyl Arenesulfonates – Influence of Bases and Reaction Mechanism, *Bull. Soc. Chim. Fr.* 1980, 275–282.

1287. Zen, S., Tashima, S., Koto, S., Photolysis of Sugar Tosylate. A New Procedure for the De-*O*-tosylation, *Bull. Chem. Soc. Jpn.* 1968, **41**, 3025–3025.

1288. Masnovi, J., Koholic, D. J., Berki, R. J., Binkley, R. W., Reductive Cleavage of Sulfonates – Deprotection of Carbohydrate Tosylates by Photoinduced Electron Transfer, *J. Am. Chem. Soc.* 1987, **109**, 2851–2853.

1289. Binkley, R. W., Koholic, D. J., Photoremovable Hydroxyl Group Protection – Use of the *p*-Tolysulfonyl Protecting Group in β-Disaccharide Synthesis, *J. Org. Chem.* 1989, **54**, 3577–3581.

1290. Schultz, A. G., Schlessinger, R. H., The Role of Sulphenate Esters in Sulphoxide Photoracemization, *J. Chem. Soc. D* 1970, 1294–1295.

1291. Sakamoto, M., Aoyama, H., Omote, Y., Photochemical Reactions of Acyclic Monothioimides – A Novel Photorearrangement Involving 1,2-Thiobenzoyl Shift, *J. Org. Chem.* 1984, **49**, 1837–1838.

1292. Harmata, M., Herron, B. F., Photochemical Activation of Trimethylsilylmethyl Allylic Sulfones for Intramolecular 4 + 3 Cycloaddition, *Tetrahedron Lett.* 1993, **34**, 5381–5384.

1293. Bolton, J. R., Chen, K. S., Lawrence, A. H., de Mayo, P., Thione Photochemistry. Photoreduction of Adamantanethione, *J. Am. Chem. Soc.* 1975, **97**, 1832–1837.

1294. Givens, R. S., Hrinczenko, B., Liu, J. H. S., Matuszewski, B., Tholencollison, J., Photoextrusion of $SO_2$ from Arylmethyl Sulfones. Exploration of the Mechanism by Chemical Trapping, Chiral, and CIDNP Probes, *J. Am. Chem. Soc.* 1984, **106**, 1779–1789.

1295. Ingold, K. U., Lusztyk, J., Raner, K. D., The Unusual and the Unexpected in an Old Reaction – The Photochlorination of Alkanes with Molecular Chlorine in Solution, *Acc. Chem. Res.* 1990, **23**, 219–225.

1296. Dneprovskii, A. S., Kuznetsov, D. V., Eliseenkov, E. V., Fletcher, B., Tanko, J. M., Free Radical Chlorinations in Halogenated Solvents: Are There Any Solvents Which Are Truly Noncomplexing? *J. Org. Chem.* 1998, **63**, 8860–8864.

1297. Alexander, A. J., Kim, Z. H., Kandel, S. A., Zare, R. N., Rakitzis, T. P., Asano, Y., Yabushita, S., Oriented Chlorine Atoms as a Probe of the Nonadiabatic Photodissociation Dynamics of Molecular Chlorine, *J. Chem. Phys.* 2000, **113**, 9022–9031.

1298. Kim, Z. H., Alexander, A. J., Kandel, S. A., Rakitzis, T. P., Zare, R. N., Orientation as a Probe of Photodissociation Dynamics, *Faraday Discuss.* 1999, 27–36.

1299. Rangel, C., Navarrete, M., Corchado, J. C., Espinosa-Garcia, J., Potential Energy Surface, Kinetics, and Dynamics Study of the $Cl + CH_4 \rightarrow HCl + CH_3$ Reaction, *J. Chem. Phys.* 2006, 124.

1300. Murray, C., Orr-Ewing, A. J., The Dynamics of Chlorine Atom Reactions with Polyatomic Organic Molecules, *Int. Rev. Phys. Chem.* 2004, **23**, 435–482.

1301. Rudic, S., Murray, C., Harvey, J. N., Orr-Ewing, A. J., On-the-fly *Ab Initio* Trajectory Calculations of the Dynamics of Cl Atom Reactions with Methane, Ethane and Methanol, *J. Chem. Phys.* 2004, **120**, 186–198.

1302. Smith, G. W., Williams, H. D., Some Reactions of Adamantane and Adamantane Derivatives, *J. Org. Chem.* 1961, **26**, 2207–2212.

1303. Poutsma, M. L., Chlorination Studies of Unsaturated Materials in Nonpolar Media. 3. Competition between Ionic and Free-radical Reactions During Chlorination of Isomeric Butenes and Allyl Chloride, *J. Am. Chem. Soc.* 1965, **87**, 2172–2183.

1304. Russell, G. A., Deboer, C., Substitutions at Saturated Carbon–Hydrogen Bonds Utilizing Molecular Bromine or Bromotrichloromethane, *J. Am. Chem. Soc.* 1963, **85**, 3136–3139.

1305. Calo, V., Lopez, L., Pesce, G., Regio-selective and Stereo-selective Free-radical Bromination of Steroidal α,β-Unsaturated Ketones, *J. Chem. Soc., Perkin Trans. 2* 1976, 247–248.

1306. Chow, Y. L., Zhao, D. C., The Presence of 2 Reactive Intermediates in the Photolysis of *N*-Bromosuccinimide – Kinetic Proofs, *J. Org. Chem.* 1989, **54**, 530–534.

1307. Tanner, D. D., Ruo, T. C.S., Takiguchi, H., Guillaume, A., Reed, D. W., Setiloane, B. P., Tan, S. L., Meintzer, C. P., On the Mechanism of *N*-Bromosuccinimide Brominations – Bromination of Cyclohexane and Cyclopentane with *N*-Bromosuccinimide, *J. Org. Chem.* 1983, **48**, 2743–2747.

1308. Fischer, M., Industrial Applications of Photochemical Syntheses, *Angew. Chem. Int. Ed. Engl.* 1978, **17**, 16–26.

1309. Heinisch, E., Kettrup, A., Bergheim, W., Martens, D., Wenzel, S., Persistent Chlorinated Hydrocarbons (PCHC), Source Oriented Monitoring in Aquatic Media – 3. The Isomers of Hexachlorocyclohexane, *Fresenius Environ. Bull.* 2005, **14**, 444–462.

1310. Liu, R. S.H., Asato, A. E., Photochemistry and Synthesis of Stereoisomers of Vitamin A, *Tetrahedron* 1984, **40**, 1931–1969.

1311. Pape, M., Industrial Applications of Photochemistry, *Pure Appl. Chem.* 1975, **41**, 535–558.

1312. Monnerie, N., Ortner, J., Economic Evaluation of the Industrial Photosynthesis of Rose Oxide via Lamp or Solar Operated Photooxidation of Citronellol, *J. Solar Energy Eng.* 2001, **123**, 171–174.

1313. Alt, L., The Photochemical Industry: Historical Essays in Business Strategy and Internationalization, *Bus. Econ. Hist.* 1987, **16**, 183–190.

1314. Kropp, P. J., Photobehavior of Alkyl Halides. In Horspool, W. M., Lenci, F. (eds), *CRC Handbook of Organic Photochemistry and Photobiology*, 2nd edn, CRC Press LLC, Boca Raton, FL, 2004, Chapter 1, pp. 1–32.

1315. Kitamura, T., C−X Bond Fission in Alkene Systems. In Horspool, W. M., Song, P.-S. (eds), *CRC Handbook of Organic Photochemistry and Photobiology*, CRC Press, Boca Raton, FL, 1995, pp. 1171–1180.

1316. Grimshaw, J., Desilva, A. P., Photochemistry and Photocyclization of Aryl Halides, *Chem. Soc. Rev.* 1981, **10**, 181–203.

1317. Kropp, P. J., Photobehavior of Alkylhalides in Solution – Radical, Carbocation, and Carbene Intermediates, *Acc. Chem. Res.* 1984, **17**, 131–137.

1318. Kessar, S. V., Mankotia, A. K.S., Photocyclization of Haloarenes. In Horspool, W. M., Song, P.-S. (eds), *CRC Handbook of Organic Photochemistry and Photobiology*, CRC Press, Boca Raton, FL, 1995, pp. 1218–1228.

1319. Suginome, H., Reaction and Synthetic Application of Oxygen-centered Radicals Photochemically Generated from Alkyl Hypohalites. In Horspool, W. M., Lenci, F. (eds), *CRC Handbook of Organic Photochemistry and Photobiology*, 2nd edn, CRC Press LLC, Boca Raton, FL, 2004, Chapter 109, pp. 1–16.

1320. Albini, A., Fagnoni, M., Photoinduced CX Cleavage of Benzylic Substrates. In Griesbeck, A. G., Mattay, J. (eds), *Synthetic Organic Photochemistry*, Marcel Dekker, New York, 2005, pp. 453–494.

1321. Fleming, S. A., Pincock, J. A., Photochemical Cleavage reactions of Benzyl–Heteroatom Sigma Bonds. In Ramamurthy, V., Schanze, K. (eds), *Organic Molecular Photochemistry*, Vol. 3, Marcel Dekker, New York, 1999, pp. 211–281.

1322. Rozgonyi, T., Gonzalez, L., Photochemistry of $CH_2BrCl$: an *Ab Initio* and Dynamical Study, *J. Phys. Chem. A* 2002, **106**, 11150–11161.

1323. Liu, K., Zhao, H. M., Wang, C. X., Zhang, A. H., Ma, S. Y., Li, Z. H., A Theoretical Study of Bond Selective Photochemistry in $CH_2BrI$, *J. Chem. Phys.* 2005, **122**, 44310.

1324. Tian, Y. C., Liu, Y. J., Fang, W. H., Theoretical Investigation on *o*-, *m*-, and *p*-Chlorotoluene Photodissociations at 193 and 266 nm, *J. Chem. Phys.* 2007, **127**, 044309.

1325. Ji, L., Tang, Y., Zhu, R. S., Wei, Z. R., Zhang, B., Photodissociation Dynamics of Allyl Bromide at 234, 265, and 267 nm *J. Chem. Phys.* 2006, **125**, 164307.

1326. Parsons, B. F., Butler, L. J., Ruscic, B., Theoretical Investigation of the Transition States Leading to HCl Elimination in 2-Chloropropene, *Mol. Phys.* 2002, **100**, 865–874.

1327. Zhu, R. S., Zhang, H., Wang, G. J., Gu, X. B., Han, K. L., He, G. Z., Lou, N. Q., Photodissociation of *o*-Dichlorobenzene at 266 nm, *Chem. Phys.* 1999, **248**, 285–292.

1328. Katayanagi, H., Yonekuira, N., Suzuki, T., C–Br Bond Rupture in 193 nm Photodissociation of Vinyl Bromide, *Chem. Phys.* 1998, **231**, 345–353.

1329. Wang, G. J., Zhu, R. S., Zhang, H., Han, K. L., He, G. Z., Lou, N. Q., Photodissociation of Chlorobenzene at 266 nm, *Chem. Phys. Lett.* 1998, **288**, 429–432.

1330. Liu, Y. J., Tian, Y. C., Fang, W. H., Spin–Orbit *Ab Initio* Investigation of the Photolysis of *o*-, *m*-, and *p*-Bromotoluene, *J. Chem. Phys.* 2008, **128**, 064307.

1331. Liu, Y. J., Ajitha, D., Krogh, J. W., Tarnovsky, A. N., Lindh, R., Spin–Orbit *Ab Initio* Investigation of the Photolysis of Bromoiodomethane, *ChemPhysChem* 2006, **7**, 955–963.

1332. Zhu, R. S., Tang, B. F., Ji, L., Tang, Y., Zhang, B., Measurement of Photolysis Branching Ratios of 2-Bromopropane at 234 and 267 nm, *Chem. Phys. Lett.* 2005, **405**, 58–62.

1333. Tang, B. F., Zhu, R. S., Tang, Y., Ji, L., Zhang, B., Photodissociation of Bromobenzene at 267 and 234 nm: Experimental and Theoretical Investigation of the Photodissociation Mechanism, *Chem. Phys. Lett.* 2003, **381**, 617–622.

1334. Balasubramanian, A., Substituent and Solvent Effects on n,σ* Transition of Alkyl Iodides, *Indian J. Chem.* 1963, **1**, 329–330.

1335. Shepson, P. B., Heicklen, J., Photooxidation of Ethyl Iodide at 22 °C, *J. Phys. Chem.* 1981, **85**, 2691–2694.

1336. Kropp, P. J., Poindexter, G. S., Pienta, N. J., Hamilton, D. C., Photochemistry of Alkyl Halides. 4. 1-Norbornyl, 1-Norbornylmethyl, 1-Adamantyl and 2-Adamantyl, and 1-Octyl Bromides and Iodides, *J. Am. Chem. Soc.* 1976, **98**, 8135–8144.

1337. Luef, W., Keese, R., Synthesis of 1-Hydroxynorbornene from Norcamphor, *Chimia* 1982, **36**, 81–82.

1338. Kharasch, M. S., Reinmuth, O., Urry, W. H., Reactions of Atoms and Free Radicals in Solution. 11. The Addition of Bromotrichloromethane to Olefins, *J. Am. Chem. Soc.* 1947, **69**, 1105–1110.

1339. McNeely, S. A., Kropp, P. J., Photochemistry of Alkyl Halides. 3. Generation of Vinyl Cations, *J. Am. Chem. Soc.* 1976, **98**, 4319–4320.

1340. Bunce, N. J., Bergsma, J. P., Bergsma, M. D., Degraaf, W., Kumar, Y., Ravanal, L., Structure and Mechanism in the Photoreduction of Aryl Chlorides in Alkane Solvents, *J. Org. Chem.* 1980, **45**, 3708–3713.

1341. Pinhey, J. T., Rigby, R. D.G., Photoreduction of Chloro- and Bromo-aromatic Compounds, *Tetrahedron Lett.* 1969, 1267–1270.

1342. Henderson, W. A., Lopresti, R., Zweig, A., Photolytic Rearrangement and Halogen-dependent Photocylization of Halophenylnaphthalenes. II., *J. Am. Chem. Soc.* 1969, **91**, 6049–6057.

1343. Bunce, N. J., Kumar, Y., Ravanal, L., Safe, S., Photochemistry of Chlorinated Biphenyls in *iso*-Octane Solution, *J. Chem. Soc., Perkin Trans. 2* 1978, 880–884.

1344. Bonnichon, F., Richard, C., Grabner, G., Formation of an α-Ketocarbene by Photolysis of Aqueous 2-Bromophenol *Chem. Commun.* 2001, 73–74.

1345. Urwyler, B., Wirz, J., The Tautomeric Equilibrium between Cyclopentadienyl-1-carboxylic Acid and Fulvene-6,6-diol in Aqueous Solution, *Angew. Chem. Int. Ed. Engl.* 1990, **29**, 790–792.

1346. Guyon, C., Boule, P., Lemaire, J., Photochemistry and the Environment. 3. Formation of Cyclopentadienic Acid by Irradiation of 2-Chlorophenol in Basic Aqueous Solution, *Tetrahedron Lett.* 1982, **23**, 1581–1584.

1347. Konstantinov, A., Kingsmill, C. A., Ferguson, G., Bunce, N. J., Successive Photosubstitution of Hexachlorobenzene with Cyanide Ion, *J. Am. Chem. Soc.* 1998, **120**, 5464–5468.

1348. Reddy, D. S., Sollott, G. P., Eaton, P. E., Photolysis of Cubyl Iodides – Access to the Cubyl Cation, *J. Org. Chem.* 1989, **54**, 722–723.

1349. Kitamura, T., Photochemistry of Hypervalent Iodine Compounds. In Horspool, W. M., Lenci, F. (eds), *CRC Handbook of Organic Photochemistry and Photobiology*, 2nd edn, CRC Press LLC, Boca Raton, FL, 2004, Chapter 110, pp. 1–16.

1350. Dektar, J. L., Hacker, N. P., Photochemistry of Diaryliodonium Salts, *J. Org. Chem.* 1990, **55**, 639–647.

1351. Crivello, J. V., Lam, J. H.W., Diaryliodonium Salts – New Class of Photoinitiators for Cationic Polymerization, *Macromolecules* 1977, **10**, 1307–1315.

1352. Anbar, M., Ginsburg, D., Organic Hypohalites, *Chem. Rev.* 1954, **54**, 925–958.

1353. Akhtar, M., Barton, D. H.R., Reactions at Position 19 in Steroid Nucleus. Convenient Synthesis of 19-Norsteroids, *J. Am. Chem. Soc.* 1964, **86**, 1528–1536.

1354. Semmelhack, M. F., Bargar, T., Photostimulated Nucleophilic Aromatic Substitution for Halides with Carbon Nucleophiles. Preparative and Mechanistic Aspects, *J. Am. Chem. Soc.* 1980, **102**, 7765–7774.

1355. Ahmed, S. A., Awad, I. M.A., Abdel-Wahab, A. M.A., A Highly Efficient Photochemical Bromination as a New Method for Preparation of Mono, Bis and Fused Pyrazole Derivatives, *Photochem. Photobiol. Sci.* 2002, **1**, 84–86.

1356. Jenner, E. L., Intramolecular Hydrogen Abstraction in a Primary Alkoxy Radical, *J. Org. Chem* **1962**, 1031, **27**.

1357. Vaillard, S. E., Postigo, A., Rossi, R. A., Syntheses of 3-Substituted 2,3-Dihydrobenzo-furanes,1,2-Dihydronaphtho(2,1-*b*)furanes, and 2,3-Dihydro-1*H*-indoles by Tandem Ring Closure – $S_{RN}1$ Reactions *J. Org. Chem.* 2002, **67**, 8500–8506.

1358. Lissi, E. A., Lemp, E., Zanocco, A. L., Singlet-oxygen Reaction: Solvent and Compartmentalization Effects. In Ramamurthy, V., Schanze, K. (eds), *Understanding and Manipulating Excited-state Processes*, Marcel Dekker, New York, 2001, pp. 287–316.

1359. Foote, C. S., Photosensitized Oxygenations and Role of Singlet Oxygen, *Acc. Chem. Res.* 1968, **1**, 104–110.

1360. Schweitzer, C., Schmidt, R., Physical Mechanisms of Generation and Deactivation of Singlet Oxygen, *Chem. Rev.* 2003, **103**, 1685–1757.

1361. Ogilby, P. R., Solvent Effects on the Radiative Transitions of Singlet Oxygen, *Acc. Chem. Res.* 1999, **32**, 512–519.

1362. Slanger, T. G., Copeland, R. A., Energetic Oxygen in the Upper Atmosphere and the Laboratory, *Chem. Rev.* 2003, **103**, 4731–4765.

1363. Lissi, E. A., Encinas, M. V., Lemp, E., Rubio, M. A., Singlet Oxygen $O_2$ ($^1\Delta_g$) Bimolecular Processes – Solvent and Compartmentalization Effects, *Chem. Rev.* 1993, **93**, 699–723.

1364. Keilin, D., Hartree, E. F., Absorption Spectrum of Oxygen, *Nature* 1950, **165**, 543–544.

1365. Gollnick, K., Photooxygenation Reactions in Solution, *Adv. Photochem.* 1968, **6**, 1–114.

1366. Foote, C. S., Definition of Type I and Type II Photosensitized Oxidation, *Photochem. Photobiol.* 1991, **54**, 659–659.

1367. Hutzinger, O., *The Handbook of Environmental Chemistry*, Springer, Berlin, 1982.

1368. Molina, L. T., Molina, M. J., Absolute Absorption Cross-sections of Ozone in the 185-nm to 350-nm Wavelength Range, *J. Geophys. Res.* 1986, **91**, 14501–14508.

1369. Molina, M. J., Rowland, F. S., Stratospheric Sink for Chlorofluoromethanes – Chlorine Atomic-catalysed Destruction of Ozone, *Nature* 1974, **249**, 810–812.

1370. Molina, M. J., Molina, L. T., Kolb, C. E., Gas-phase and Heterogeneous Chemical Kinetics of the Troposphere and Stratosphere, *Annu. Rev. Phys. Chem.* 1996, **47**, 327–367.

1371. Domine, F., Shepson, P. B., Air–Snow Interactions and Atmospheric Chemistry, *Science* 2002, **297**, 1506–1510.

1372. Grannas, A. M., Jones, A. E., Dibb, J., Ammann, M., Anastasio, C., Beine, H. J., Bergin, M., Bottenheim, J., Boxe, C. S., Carver, G., Chen, G., Crawford, J. H., Domine, F., Frey, M. M., Guzman, M. I., Heard, D. E., Helmig, D., Hoffmann, M. R., Honrath, R. E., Huey, L. G.,

Hutterli, M., Jacobi, H. W., Klán, P., Lefer, B., McConnell, J., Plane, J., Sander, R., Savarino, J., Shepson, P. B., Simpson, W. R., Sodeau, J. R., von Glasow, R., Weller, R., Wolff, E. W., Zhu, T., An Overview of Snow Photochemistry: Evidence, Mechanisms and Impacts, *Atmos. Chem. Phys.* 2007, **7**, 4329–4373.

1373. Atkinson, R., Atmospheric Chemistry of VOCs and $NO_x$, *Atmos. Environ.* 2000, **34**, 2063–2101.

1374. Chameides, W. L., Lindsay, R. W., Richardson, J., Kiang, C. S., The Role of Biogenic Hydrocarbons in Urban Photochemical Smog – Atlanta as a Case Study, *Science* 1988, **241**, 1473–1475.

1375. Pitts, J. N., Khan, A. U., Smith, E. B., Wayne, R. P., Singlet Oxygen in Environmental Sciences Singlet Molecular Oxygen and Photochemical Air Pollution, *Environ. Sci. Technol.* 1969, **3**, 241–247.

1376. Griesbeck, A. G., El-Idreesy, T. T., Adam, W., Krebs, O., Ene-reactions with Singlet Oxygen. In Horspool, W. M., Lenci, F. (eds), *CRC Handbook of Organic Photochemistry and Photobiology*, 2nd edn, CRC Press LLC, Boca Raton, FL, 2004, Chapter 8, pp. 1–20.

1377. Orfanopoulos, M., Singlet-oxygen Ene-sensitized Photooxygenations: Stereochemistry and Mechanisms. In Ramamurthy, V., Schanze, K. (eds), *Understanding and Manipulating Excited-state Processes*, Marcel Dekker, New York, 2001, pp. 243–285.

1378. Skovsen, E., Snyder, J. W., Lambert, J. D.C., Ogilby, P. R., Lifetime and Diffusion of Singlet Oxygen in a Cell, *J. Chem. Phys. B* 2005, **109**, 8570–8573.

1379. Schmidt, R., Brauer, H. D., Radiationless Deactivation of Singlet Oxygen ($^1\Delta_g$) by Solvent Molecules, *J. Am. Chem. Soc.* 1987, **109**, 6976–6981.

1380. Garner, A., Wilkinson, F., Quenching of Triplet States by Molecular Oxygen and Role of Charge-transfer Interactions, *Chem. Phys. Lett.* 1977, **45**, 432–435.

1381. Harber, L. C., Bickers, D. R., Drug Induced Photosensitivity (Phototoxic and Photoallergic Drug Reactions). In Harber, L. C., Bickers, D. R. (eds), *Photosensitivity Diseases. Principles of Diagnosis and Treatment*, 2nd edn, Marcel Decker, Toronto, 1986, pp. 160–202.

1382. Moore, D. E., Drug-induced Cutaneous Photosensitivity – Incidence, Mechanism, Prevention and Management, *Drug Safety* 2002, **25**, 345–372.

1383. Hasan, T., Kochevar, I. E., McAuliffe, D. J., Cooperman, B. S., Abdulah, D., Mechanism of Tetracycline Phototoxicity, *J. Invest. Dermatol.* 1984, **83**, 179–183.

1384. Griesbeck, A. G., Gudipati, M. S., Endoperoxides: Thermal and Photochemical Reactions and Spectroscopy. In Horspool, W. M., Lenci, F. (eds), *CRC Handbook of Organic Photochemistry and Photobiology*, 2nd edn, CRC Press LLC, Boca Raton, FL, 2004, Chapter 108, pp. 1–15.

1385. Aubry, J. M., Pierlot, C., Rigaudy, J., Schmidt, R., Reversible Binding of Oxygen to Aromatic Compounds, *Acc. Chem. Res.* 2003, **36**, 668–675.

1386. Jones, I. T.N., Wayne, R. P., Photolysis of Ozone by 254-, 313-, and 334-nm Radiation, *J. Chem. Phys.* 1969, **51**, 3617–3618.

1387. Rigaudy, J., Photorearrangements of Endoperoxides. In Horspool, W. M., Song, P.-S. (eds), *CRC Handbook of Organic Photochemistry and Photobiology*, CRC Press, Boca Raton, FL, 1995, pp. 325–334.

1388. Getoff, N., Generation of $^1O_2$ by Microwave Discharge and Some Characteristic Reactions – A Short Review, *Radiat. Phys. Chem.* 1995, **45**, 609–614.

1389. Foyer, C. H., Descourvieres, P., Kunert, K. J., Protection against Oxygen Radicals – An Important Defense Mechanism Studied in Transgenic Plants, *Plant Cell Environ.* 1994, **17**, 507–523.

1390. Dougherty, T. J., Levy, J. G., Clinical Applications of Photodynamic Therapy. In Horspool, W. M., Lenci, F. (eds), *CRC Handbook of Organic Photochemistry and Photobiology*, 2nd edn, CRC Press LLC, Boca Raton, FL, 2004, Chapter 147, pp. 1–17.

1391. Jori, G., Photodynamic Therapy: Basic and Preclinical Aspects. In Horspool, W. M., Lenci, F. (eds), *CRC Handbook of Organic Photochemistry and Photobiology*, 2nd edn, CRC Press LLC, Boca Raton, FL, 2004, Chapter 146, pp. 1–10.

1392. Henderson, B. W., Gollnick, S. O., Mechanistic Principles of Photodynamic Therapy. In Horspool, W. M., Lenci, F. (eds), *CRC Handbook of Organic Photochemistry and Photobiology*, 2nd edn, CRC Press LLC, Boca Raton, FL, 2004, Chapter 145, pp. 1–25.

1393. Bonnett, R., *Chemical Aspects of Photodynamic Therapy*, CRC Press, Boca Raton, FL, 2000.

1394. Patrice, T., *Photodynamic Therapy*, Royal Society of Chemistry, Cambridge, 2004.

1395. Dougherty, T. J., Gomer, C. J., Henderson, B. W., Jori, G., Kessel, D., Korbelik, M., Moan, J., Peng, Q., Photodynamic Therapy, *J. Nat. Cancer Inst.* 1998, **90**, 889–905.

1396. Bonnett, R., Photosensitizers of the Porphyrin and Phthalocyanine Series for Photodynamic Therapy, *Chem. Soc. Rev.* 1995, **24**, 19–33.

1397. Phillips, D., The Photochemistry of Sensitizers for Photodynamic Therapy, *Pure Appl. Chem.* 1995, **67**, 117–126.

1398. Henderson, B. W., Dougherty, T. J., How Does Photodynamic Therapy Work, *Photochem. Photobiol.* 1992, **55**, 145–157.

1399. Karu, T., Primary and Secondary Mechanisms of Action of Visible to near-IR Radiation on Cells, *J. Photochem. Photobiol. B* 1999, **49**, 1–17.

1400. Oleinick, N. L., Morris, R. L., Belichenko, T., The Role of Apoptosis in Response to Photodynamic Therapy: What, Where, Why, and How, *Photochem. Photobiol. Sci.* 2002, **1**, 1–21.

1401. Kochevar, I. E., Lynch, M. C., Zhuang, S. G., Lambert, C. R., Singlet Oxygen, But Not Oxidizing Radicals, Induces Apoptosis in HL-60 Cells, *Photochem. Photobiol.* 2000, **72**, 548–553.

1402. Snyder, J. W., Zebger, I., Gao, Z., Poulsen, L., Frederiksen, P. K., Skovsen, E., McIlroy, S. P., Klinger, M., Andersen, L. K., Ogilby, P. R., Singlet Oxygen Microscope: from Phase-separated Polymers to Single Biological Cells, *Acc. Chem. Res.* 2004, **37**, 894–901.

1403. Häder, D.-P., Photoecolog and Environmental Photobiology. In Horspool, W. M., Song, P.-S. (eds), *CRC Handbook of Organic Photochemistry and Photobiology*, CRC Press, Boca Raton, FL, 1995, pp. 1392–1401.

1404. Zepp, R. G., Callaghan, T. V., Erickson, D. J., Effects of Enhanced Solar Ultraviolet Radiation on Biogeochemical Cycles, *J. Photochem. Photobiol. B* 1998, **46**, 69–82.

1405. Moran, M. A., Zepp, R. G., Role of Photoreactions in the Formation of Biologically Labile Compounds from Dissolved Organic Matter, *Limnol. Oceanogr.* 1997, **42**, 1307–1316.

1406. Kieber, R. J., Willey, J. D., Whitehead, R. F., Reid, S. N., Photobleaching of Chromophoric Dissolved Organic Matter (CDOM) in Rainwater, *J. Atmos. Chem.* 2007, **58**, 219–235.

1407. Canonica, S., Laubscher, H.-U., Inhibitory Effect of Dissolved Organic Matter on Triplet-induced Oxidation of Aquatic Contaminants, *Photochem. Photobiol. Sci.* 2008, **7**, 547–551.

1408. Mopper, K., Zhou, X. L., Kieber, R. J., Kieber, D. J., Sikorski, R. J., Jones, R. D., Photochemical Degradation of Dissolved Organic Carbon and Its Impact on the Oceanic Carbon Cycle, *Nature* 1991, **353**, 60–62.

1409. Frimmel, F. H., Photochemical Aspects Related to Humic Substances, *Environ. Int.* 1994, **20**, 373–385.

1410. Canonica, S., Kohn, T., Mac, M., Real, F. J., Wirz, J., Von Gunten, U., Photosensitizer Method to Determine Rate Constants for the Reaction of Carbonate Radical with Organic Compounds, *Environ. Sci. Technol.* 2005, **39**, 9182–9188.

1411. Canonica, S., Hellrung, B., Wirz, J., Oxidation of Phenols by Triplet Aromatic Ketones in Aqueous Solution, *J. Phys. Chem. A* 2000, **104**, 1226–1232.

1412. Lesser, M. P., Oxidative Stress in Marine Environments: Biochemistry and Physiological Ecology, *Annu. Rev. Physiol.* 2006, **68**, 253–278.

1413. Miles, C. J., Brezonik, P. L., Oxygen Consumption in Humic Colored Waters by a Photochemical Ferrous–Ferric Catalytic Cycle, *Environ. Sci. Technol.* 1981, **15**, 1089–1095.

1414. Haberlein, A., Häder, D. P., UV Effects on Photosynthetic Oxygen Production and Chromoprotein Composition in a Fresh-water Flagellate Cryptomonas, *Acta Protozool.* 1992, **31**, 85–92.

1415. Klán, P., Klánová, J., Holoubek, I., Cupr, P., Photochemical Activity of Organic Compounds in Ice Induced by Sunlight Irradiation: the Svalbard Project, *Geophys. Res. Lett.* 2003, **30**, art. no. 1313.

1416. Klán, P., Holoubek, I., Ice (Photo)chemistry. Ice as a Medium for Long-term (Photo)chemical Transformations – Environmental Implications, *Chemosphere* 2002, **46**, 1201–1210.

1417. Matykiewiczová, N., Klánová, J., Klán, P., Photochemical Degradation of PCBs in Snow, *Environ. Sci. Technol.* 2007, **41**, 8308–8314.

1418. Dolinová, J., Ruzicka, R., Kurková, R., Klánová, J., Klán, P., Oxidation of Aromatic and Aliphatic Hydrocarbons by OH Radicals Photochemically Generated from $H_2O_2$ in Ice, *Environ. Sci. Technol.* 2006, **40**, 7668–7674.

1419. Adam, W., Griesbeck, A. G., Photooxygenation of 1, 3-Dienes. In Horspool, W. M., Song, P.-S. (eds), *CRC Handbook of Organic Photochemistry and Photobiology*, CRC Press, Boca Raton, FL, 1995, pp. 311–324.

1420. Griesbeck, A. G., Ene Reactions with Singlet Oxygen. In Horspool, W. M., Song, P.-S. (eds), *CRC Handbook of Organic Photochemistry and Photobiology*, CRC Press, Boca Raton, FL, 1995, pp. 301–310.

1421. Iesce, M. R., Photooxygenation of the [4 + 2] and [2 + 2] Type. In Griesbeck, A. G., Mattay, J. (eds), *Synthetic Organic Photochemistry*, Marcel Dekker, New York, 2005, pp. 299–363.

1422. Frimer, A. A., Reaction of Singlet Oxygen with Olefins – Question of Mechanism, *Chem. Rev.* 1979, **79**, 359–387.

1423. Stephenson, L. M., Grdina, M. J., Orfanopoulos, M., Mechanism of the Ene Reaction Between Singlet Oxygen and Olefins, *Acc. Chem. Res.* 1980, **13**, 419–425.

1424. Adam, W., Saha-Moller, C. R., Schambony, S. B., Schmid, K. S., Wirth, T., Stereocontrolled Photooxygenations – A Valuable Synthetic Tool, *Photochem. Photobiol.* 1999, **70**, 476–483.

1425. Albini, A., Fagnoni, M., Oxidation of Aromatics. In Horspool, W. M., Lenci, F. (eds), *CRC Handbook of Organic Photochemistry and Photobiology*, 2nd edn, CRC Press LLC, Boca Raton, FL, 2004, Chapter 45, pp. 1–19.

1426. Solomon, M., Sivaguru, J., Jockusch, S., Adam, W., Turro, N. J., Vibrational Deactivation of Singlet Oxygen: Does It Play a Role in Stereoselectivity During Photooxygenation? *Photochem. Photobiol. Sci.* 2008, **7**, 531–533.

1427. Maranzana, A., Ghigo, G., Tonachini, G., Diradical and Peroxirane Pathways in the $[\pi 2 + \pi 2]$ Cycloaddition Reactions of $^1\Delta_g$ Dioxygen with Ethene, Methyl Vinyl Ether, and Butadiene: a Density Functional and Multireference Perturbation Theory Study, *J. Am. Chem. Soc.* 2000, **122**, 1414–1423.

1428. Singleton, D. A., Hang, C., Szymanski, M. J., Meyer, M. P., Leach, A. G., Kuwata, K. T., Chen, J. S., Greer, A., Foote, C. S., Houk, K. N., Mechanism of Ene Reactions of Singlet Oxygen. A Two-step No-intermediate Mechanism, *J. Am. Chem. Soc.* 2003, **125**, 1319–1328.

1429. Ghigo, G., Maranzana, A., Causa, M., Tonachini, G., Theoretical Mechanistic Studies on Oxidation Reactions of Some Saturated and Unsaturated Organic Molecules, *Theor. Chem. Acc.* 2007, **117**, 699–707.

1430. Bobrowski, M., Liwo, A., Oldziej, S., Jeziorek, D., Ossowski, T., CAS MCSCF/CAS MCQDPT2 Study of the Mechanism of Singlet Oxygen Addition to 1,3-Butadiene and Benzene, *J. Am. Chem. Soc.* 2000, **122**, 8112–8119.

1431. McCarrick, M. A., Wu, Y. D., Houk, K. N., Hetero-Diels–Alder Reaction Transition Structures – Reactivity, Stereoselectivity, Catalysis, Solvent Effects, and the *exo*-Lone Pair Effect, *J. Org. Chem.* 1993, **58**, 3330–3343.

1432. Jursic, B. S., Zdravkovski, Z., Reaction of Imidazoles with Ethylene and Singlet Oxygen – An *Ab Initio* Theoretical Study, *J. Org. Chem.* 1995, **60**, 2865–2869.

1433. Corral, I., Gonzalez, L., The Electronic Excited States of a Model Organic Endoperoxide: a Comparison of TD-DFT and *Ab Initio* Methods, *Chem. Phys. Lett.* 2007, **446**, 262–267.

1434. Jefford, C. W., The Photooxygenation of Olefins and the Role of Zwitterionic Peroxides, *Chem. Soc. Rev.* 1993, **22**, 59–66.

1435. Gorman, A. A., Gould, I. R., Hamblett, I., Time-resolved Study of the Solvent and Temperature Dependence of Singlet Oxygen ($^1\Delta_g$) Reactivity toward Enol Ethers – Reactivity Parameters Typical of Rapid Reversible Exciplex Formation, *J. Am. Chem. Soc.* 1982, **104**, 7098–7104.

1436. Wieringa, J. H., Wynberg, H., Strating, J., Adam, W., Adamantylideneadamantane Peroxide, a Stable 1,2-Dioxetane, *Tetrahedron Lett.* 1972, 169–172.

1437. Schuster, G. B., Turro, N. J., Steinmetzer, H. C., Schaap, A. P., Faler, G., Adam, W., Liu, J. C., Adamantylideneadamantane-1,2-Dioxetane – Investigation of Chemiluminescence and Decomposition Kinetics of an Unusually Stable 1,2-Dioxetane, *J. Am. Chem. Soc.* 1975, **97**, 7110–7118.

1438. Wilkinson, F., Helman, W. P., Ross, A. B., Rate Constants for the Decay and Reactions of the Lowest Electronically Excited Singlet State of Molecular Oxygen in Solution – An Expanded and Revised Compilation, *J. Phys. Chem. Ref. Data* 1995, **24**, 663–1021.

1439. Greer, A., Vassilikogiannakis, G., Lee, K. C., Koffas, T. S., Nahm, K., Foote, C. S., Reaction of Singlet Oxygen with *trans*-4-Propenylanisole. Formation of [2 + 2] Products with Added Acid, *J. Org. Chem.* 2000, **65**, 6876–6878.

1440. Turro, N. J., Lechtken, P., Schore, N. E., Schuster, G., Steinmetzer, H. C., Yekta, A., Tetramethyl-1,2-Dioxetane – Experiments in Chemiexcitation, Chemiluminescence, Photochemistry, Chemical Dynamics, and Spectroscopy, *Acc. Chem. Res.* 1974, **7**, 97–105.

1441. Orito, K., Kurokawa, Y., Itoh, M., On the Synthetic Approach to the Protopine Alkaloids, *Tetrahedron* 1980, **36**, 617–621.

1442. Cermola, F., Iesce, M. R., Substituent and Solvent Effects on the Photosensitized Oxygenation of 5,6-Dihydro-1,4-Oxathiins. Intramolecular Oxygen Transfer vs. Normal Cleavage of the Dioxetane Intermediates, *J. Org. Chem.* 2002, **67**, 4937–4944.

1443. Adam, W., Prein, M., π-Facial Diastereoselectivity in the [4 + 2] Cycloaddition of Singlet Oxygen as a Mechanistic Probe, *Acc. Chem. Res.* 1996, **29**, 275–283.

1444. Sevin, F., McKee, M. L., Reactions of 1,3-Cyclohexadiene with Singlet Oxygen. A Theoretical Study, *J. Am. Chem. Soc.* 2001, **123**, 4591–4600.

1445. Aubry, J. M., Mandardcazin, B., Rougee, M., Bensasson, R. V., Kinetic Studies of Singlet Oxygen [4 + 2] -Cycloadditions with Cyclic 1,3-Dienes in 28 Solvents, *J. Am. Chem. Soc.* 1995, **117**, 9159–9164.

1446. Shing, T. K. M., Tam, E. K. W., (−)-Quinic Acid in Organic Synthesis. 8. Enantiospecific Syntheses of (+)-Crotepoxide, (+)-Boesenoxide, (+)-β-Senepoxide, (+)-Pipoxide acetate, (−)-*iso*-Crotepoxide, (−)-Senepoxide, and (−)-Tingtanoxide from (−)-Quinic Acid, *J. Org. Chem.* 1998, **63**, 1547–1554.

1447. Corey, E. J., Roberts, B. E., Total Synthesis of Dysidiolide, *J. Am. Chem. Soc.* 1997, **119**, 12425–12431.

1448. Sheu, C., Foote, C. S., Reactivity toward Singlet Oxygen of a 7,8-Dihydro-8-Oxoguanosine (8-Hydroxyguanosine) Formed by Photooxidation of a Guanosine Derivative, *J. Am. Chem. Soc.* 1995, **117**, 6439–6442.

1449. Rigaudy, J., Scribe, P., Breliere, C., Photochemical Transformations of Endoperoxides Derived from Polycyclic Aromatic Hydrocarbons. 1. Case of the Photooxide of 9,10-Diphenyl Anthracene – Synthesis and Properties of the Diepoxide Isomer, *Tetrahedron* 1981, **37**, 2585–2593.

1450. Clennan, E. L., Photooxygenation of the Ene Type. In Griesbeck, A. G., Mattay, J. (eds), *Synthetic Organic Photochemistry*, Marcel Dekker, New York, 2005, pp. 365–390.

1451. Maranzana, A., Ghigo, G., Tonachini, G., The $^1\Delta_g$ Dioxygen Ene Reaction with Propene: a Density Functional and Multireference Perturbation Theory Mechanistic Study, *Chem. Eur. J.* 2003, **9**, 2616–2626.

1452. Adam, W., Bottke, N., Engels, B., Krebs, O., An Experimental and Computational Study on the Reactivity and Regioselectivity for the Nitrosoarene Ene Reaction: Comparison with Triazolinedione and Singlet Oxygen, *J. Am. Chem. Soc.* 2001, **123**, 5542–5548.

1453. Yoshioka, Y., Tsunesada, T., Yamaguchi, K., Saito, I., CASSCF, MP2, and CASMP2 Studies on Addition Reaction of Singlet Molecular Oxygen to Ethylene Molecule, *Int. J. Quantum Chem.* 1997, **65**, 787–801.

1454. Schenck, G. O., Eggert, H., Denk, W., Photochemische Reaktionen. 3. Über die Bildung von Hydroperoxyden bei Photosensibilisierten Reaktionen von $O_2$ mit Geeigneten Akzeptoren, Insbesondere mit α-Pinen und β-Pinen, *Justus Liebigs Ann. Chem.* 1953, **584**, 177–198.

1455. Yamaguchi, K., Fueno, T., Saito, I., Matsuura, T., Houk, K. N., On the Concerted Mechanism of the Ene Reaction of Singlet Molecular Oxygen with Olefins – an *Ab Initio* MO Study, *Tetrahedron Lett.* 1981, **22**, 749–752.

1456. Poon, T. H.W., Pringle, K., Foote, C. S., Reaction of Cyclooctenes with Singlet Oxygen – Trapping of a Perepoxide Intermediate, *J. Am. Chem. Soc.* 1995, **117**, 7611–7618.

1457. Hurst, J. R., Wilson, S. L., Schuster, G. B., The Ene Reaction of Singlet Oxygen – Kinetic and Product Evidence in Support of a Perepoxide Intermediate, *Tetrahedron* 1985, **41**, 2191–2197.

1458. Gorman, A. A., Hamblett, I., Lambert, C., Spencer, B., Standen, M. C., Identification of Both Preequilibrium and Diffusion Limits for Reaction of Singlet Oxygen, $O_2(^1\Delta_g)$, with Both Physical and Chemical Quenchers – Variable-temperature, Time-resolved Infrared Luminescence Studies, *J. Am. Chem. Soc.* 1988, **110**, 8053–8059.

1459. Orfanopoulos, M., Stephenson, L. M., Stereochemistry of the Singlet Oxygen Olefin Ene Reaction, *J. Am. Chem. Soc.* 1980, **102**, 1417–1418.

1460. Griesbeck, A. G., Fiege, M., Gudipati, M. S., Wagner, R., Photooxygenation of 2,4-Dimethyl-1,3-Pentadiene: Solvent Dependence of the Chemical (Ene Reaction and [4 + 2] Cycloaddition) and Physical Quenching of Singlet Oxygen, *Eur. J. Org. Chem* 1998, 2833–2838.

1461. Orfanopoulos, M., Grdina, S. M.B., Stephenson, L. M., Site Specificity in the Singlet Oxygen–Trisubstituted Olefin Reaction, *J. Am. Chem. Soc.* 1979, **101**, 275–276.

1462. Acton, N., Roth, R. J., On the Conversion of Dihydroartemisinic Acid into Artemisinin, *J. Org. Chem.* 1992, **57**, 3610–3614.

1463. Stratakis, M., Froudakis, G., Site Specificity in the Photooxidation of Some Trisubstituted Alkenes in Thionin-supported Zeolite Na-Y. On the Role of the Alkali Metal Cation, *Org. Lett.* 2000, **2**, 1369–1372.

1464. Ramamurthy, V., Sanderson, D. R., Eaton, D. F., Control of Dye Assembly Within Zeolites – Role of Water, *J. Am. Chem. Soc.* 1993, **115**, 10438–10439.

1465. Adam, W., Balci, M., Photooxygenation of 1,3,5-Cycloheptatriene. Isolation and Characterization of Endoperoxides, *J. Am. Chem. Soc.* 1979, **101**, 7537–7541.

1466. Salamci, E., Secen, H., Sutbeyaz, Y., Balci, M., A Concise and Convenient Synthesis of DL-*proto*-Quercitol and DL-*gala*-Quercitol via Ene Reaction of Singlet Oxygen Combined with [2 + 4] Cycloaddition to Cyclohexadiene, *J. Org. Chem.* 1997, **62**, 2453–2457.

1467. Matsumoto, M., Kondo, K., Sensitized Photooxygenation of Linear Monoterpenes Bearing Conjugated Double Bonds, *J. Org. Chem.* 1975, **40**, 2259–2260.

1468. Thompson, A., Lever, J. R., Canella, K. A., Miura, K., Posner, G. H., Seliger, H. H., Chemiluminescence Mechanism and Quantum Yield of Synthetic Vinylpyrene Analogs of Benzo[*a*]pyrene-7,8-Dihydrodiol, *J. Am. Chem. Soc.* 1986, **108**, 4498–4504.

1469. Fagnoni, M., Dondi, D., Ravelli, D., Albini, A., Photocatalysis for the Formation of the C—C Bond, *Chem. Rev.* 2007, **107**, 2725–2756.

1470. Gruber, H. F., Photoinitiators for Free-radical Polymerization, *Prog. Polym. Sci.* 1992, **17**, 953–1044.

1471. Monroe, B. M., Weed, G. C., Photoinitiators for Free-radical-initiated Photoimaging Systems, *Chem. Rev.* 1993, **93**, 435–448.

1472. Heelis, P. F., The Photophysical and Photochemical Properties of Flavins (Isoalloxazines), *Chem. Soc. Rev.* 1982, **11**, 15–39.

1473. Muller, F., Mattay, J., [3 + 2] Cycloadditions with Azirine Radical Cations – A New Synthesis of *N*-Substituted Imidazoles, *Angew. Chem. Int. Ed. Engl.* 1991, **30**, 1336–1337.

1474. Bertrand, S., Glapski, C., Hoffmann, N., Pete, J. P., Highly Efficient Photochemical Addition of Tertiary Amines to Electron Deficient Alkenes. Diastereoselective Addition to (5*R*)-5-Menthyloxy-2-5*H*-furanone, *Tetrahedron Lett.* 1999, **40**, 3169–3172.

1475. de Alvarenga, E. S., Cardin, C. J., Mann, J., An Unexpected Photoadduct of *N*-Carbomethoxymethylpyrrolidine with a Chiral Butenolide and Benzophenone: X-ray Structure and Synthetic Utility, *Tetrahedron* 1997, **53**, 1457–1466.

1476. Blankenship, R. E., *Molecular Mechanisms of Photosynthesis*, Blackwell Science, Oxford, 2002.

1477. Balzani, V., Credi, A., Venturi, M., Photochemical Conversion of Solar Energy, *ChemSusChem* 2008, **1**, 2–35.

1478. McDermott, G., Prince, S. M., Freer, A. A., Hawthornthwaitelawless, A. M., Papiz, M. Z., Cogdell, R. J., Isaacs, N. W., Crystal Structure of an Integral Membrane Light-harvesting Complex from Photosynthetic Bacteria, *Nature* 1995, **374**, 517–521.

1479. Pullerits, T., Sundstrom, V., Photosynthetic Light-harvesting Pigment Protein Complexes: Toward Understanding How and Why, *Acc. Chem. Res.* 1996, **29**, 381–389.

1480. Engel, G. S., Calhoun, T. R., Read, E. L., Ahn, T. K., Mancal, T., Cheng, Y. C., Blankenship, R. E., Fleming, G. R., Evidence for Wavelike Energy Transfer Through Quantum Coherence in Photosynthetic Systems, *Nature* 2007, **446**, 782–786.

1481. Mathis, P., Photosynthetic Reaction Centers. In Horspool, W. M., Lenci, F. (eds), *CRC Handbook of Organic Photochemistry and Photobiology*, 2nd edn, CRC Press LLC, Boca Raton, FL, 2004, Chapter 118, pp. 1–12.

1482. Deisenhofer, J., Epp, O., Miki, K., Huber, R., Michel, H., Structure of the Protein Subunits in the Photosynthetic Reaction Center of *Rhodopseudomonas viridis* at 3 Å Resolution, *Nature* 1985, **318**, 618–624.

1483. Barber, J., Photosystem II: An Enzyme of Global Significance, *Biochem. Soc. Trans.* 2006, **34**, 619–631.

1484. Osterloh, F. E., Inorganic Materials as Catalysts for Photochemical Splitting of Water, *Chem. Mater.* 2008, **20**, 35–54.

1485. Gust, D., Moore, T. A., Moore, A. L., Mimicking Photosynthetic Solar Energy Transduction, *Acc. Chem. Res.* 2001, **34**, 40–48.

1486. Esswein, M. J., Nocera, D. G., Hydrogen Production by Molecular Photocatalysis, *Chem. Rev.* 2007, **107**, 4022–4047.

1487. Fukuzumi, S., Development of Bioinspired Artificial Photosynthetic Systems, *Phys. Chem. Chem. Phys.* 2008, **10**, 2283–2297.

1488. Ritterskamp, P., Kuklya, A., Wustkamp, M. A., Kerpen, K., Weidenthaler, C., Demuth, M., A Titanium Disilicide Derived Semiconducting Catalyst for Water Splitting Under Solar Radiation – Reversible Storage of Oxygen and Hydrogen, *Angew. Chem. Int. Ed.* 2007, **46**, 7770–7774.

1489. Nowotny, J., Sorrell, C. C., Bak, T., Sheppard, L. R., Solar Hydrogen: Unresolved Problems in Solid-state Science, *Solar Energy* 2005, **78**, 593–602.

1490. Kuciauskas, D., Liddell, P. A., Lin, S., Johnson, T. E., Weghorn, S. J., Lindsey, J. S., Moore, A. L., Moore, T. A., Gust, D., An Artificial Photosynthetic Antenna Reaction Center Complex, *J. Am. Chem. Soc.* 1999, **121**, 8604–8614.

1491. Geraghty, N. W.A., Hannan, J. J., Functionalisation of Cycloalkanes: the Photomediated Reaction of Cycloalkanes with Alkynes, *Tetrahedron Lett.* 2001, **42**, 3211–3213.

1492. Fouassier, J.-P., *Photoinitiation, Photopolymerization, and Photocuring: Fundamentals and Applications*, Carl Hanser Verlag, Munich, 1995.

1493. Kaur, M., Srivastava, A. K., Photopolymerization: a Review, *J. Macromol. Sci., Polym. Rev.* 2002, **C42**, 481–512.

1494. Allen, N. S., Marin, M. C., Edge, M., Davies, D. W., Garrett, J., Jones, F., Navaratnam, S., Parsons, B. J., Photochemistry and Photoinduced Chemical Crosslinking Activity of Type I & II Co-reactive Photoinitiators in Acrylated Prepolymers *J. Photochem. Photobiol. A* 1999, **126**, 135–149.

1495. Rutsch, W., Dietliker, K., Leppard, D., Kohler, M., Misev, L., Kolczak, U., Rist, G., Recent Developments in Photoinitiators, *Prog. Org. Catal.* 1996, **27**, 227–239.

1496. Valdes-Aguilera, O., Pathak, C. P., Shi, J., Watson, D., Neckers, D. C., Photopolymerization Studies Using Visible-light Photoinitiators, *Macromolecules* 1992, **25**, 541–547.

1497. Chatterjee, S., Davis, P. D., Gottschalk, P., Kurz, M. E., Sauerwein, B., Yang, X. Q., Schuster, G. B., Photochemistry of Carbocyanine Alkyltriphenylborate Salts: Intra-ion-pair Electron-transfer and the Chemistry of Boranyl Radicals, *J. Am. Chem. Soc.* 1990, **112**, 6329–6338.

1498. Lyoo, W. S., Ha, W. S., Preparation of Syndiotacticity-rich High Molecular Weight Poly(vinyl alcohol) Microfibrillar Fiber by Photoinitiated Bulk Polymerization and Saponification, *J. Polym. Sci., Part A – Polym. Chem.* 1997, **35**, 55–67.

1499. Terazima, M., Nogami, Y., Tominaga, T., Diffusion of a Radical from an Initiator of a Free Radical Polymerization: a Radical from AIBN *Chem. Phys. Lett.* 2000, **332**, 503–507.

1500. Crivello, J. V., Cationic Polymerization – Iodonium and Sulfonium Salt Photoinitiators, *Adv. Polym. Sci.* 1984, **62**, 1–48.

1501. Tasdelen, M. A., Kiskan, B., Yagci, Y., Photoinitiated Free Radical Polymerization using Benzoxazines as Hydrogen Donors, *Macromol. Rapid Commun.* 2006, **27**, 1539–1544.

1502. Allen, N. S., Mallon, D., Timms, A. W., Green, W. A., Catalina, F., Corrales, T., Navaratnam, S., Parsons, B. J., Photochemistry and Photocuring Activity of Novel 1-Halogeno-4-Propoxythioxanthones, *J. Chem. Soc., Faraday Trans.* 1994, **90**, 83–92.

1503. Pamedytyte, V., Abadie, M. J. M., Makuska, R., Photopolymerization of *N,N*-Dimethylaminoethylmethacrylate Studied by Photocalorimetry *J. Appl. Polym. Sci.* 2002, **86**, 579–588.

1504. Jockusch, S., Landis, M. S., Freiermuth, B., Turro, N. J., Photochemistry and Photophysics of α-Hydroxy Ketones, *Macromolecules* 2001, **34**, 1619–1626.

1505. Endruweit, A., Johnson, M. S., Long, A. C., Curing of Composite Components by Ultraviolet Radiation: a Review, *Polym. Compos.* 2006, **27**, 119–128.

1506. Gates, B. D., Xu, Q. B., Stewart, M., Ryan, D., Willson, C. G., Whitesides, G. M., New Approaches to Nanofabrication: Molding, Printing, and Other Techniques, *Chem. Rev.* 2005, **105**, 1171–1196.

1507. Willson, C. G., Trinque, B. C., The Evolution of Materials for the Photolithographic Process, *J. Photopolym. Sci. Technol.* 2003, **16**, 621–627.

1508. MacDonald, S. A., Willson, C. G., Frechet, J. M.J., Chemical Amplification in High-resolution Imaging Systems, *Acc. Chem. Res.* 1994, **27**, 151–158.

1509. Dektar, J. L., Hacker, N. P., Triphenylsulfonium Salt Photochemistry – New Evidence for Triplet Excited-state Reactions, *J. Org. Chem.* 1988, **53**, 1833–1835.

1510. Reiser, A., Huang, J. P., He, X., Yeh, T. F., Jha, S., Shih, H. Y., Kim, M. S., Han, Y. K., Yan, K., The Molecular Mechanism of Novolak–Diazonaphthoquinone Resists, *Eur. Polym. J.* 2002, **38**, 619–629.

1511. Reiser, A., Shih, H. Y., Yeh, T. F., Huang, J. P., Novolak–Diazoquinone Resists: the Imaging Systems of the Computer Chip, *Angew. Chem. Int. Ed. Engl.* 1996, **35**, 2428–2440.

1512. Maag, K., Lenhard, W., Loffles, H., New UV Curing systems for Automotive Applications, *Prog. Org. Catal.* 2000, **40**, 93–97.

1513. Allen, N. S., Photoinitiators for UV and Visible Curing of Coatings: Mechanisms and Properties, *J. Photochem. Photobiol. A* 1996, **100**, 101–107.

1514. Lindén, L. A., Applied Photochemistry in Dental Science, *Proc. Indian Acad. Sci., Chem. Sci.* 1993, **105**, 405–419.

1515. Moon, J. H., Han, H. S., Shul, Y. G., Jang, D. H., Ro, M. D., Yun, D. S., A Study on UV-curable Coatings for HD-DVD: Primer and Top Coats *Prog. Org. Catal.* 2007, **59**, 106–114.

1516. Subramanian, K., Krishnasamy, V., Nanjundan, S., Reddy, A. V.R., Photosensitive Polymer: Synthesis, Characterization and Properties of a Polymer Having Pendant Photocrosslinkable Group, *Eur. Polym. J.* 2000, **36**, 2343–2350.

1517. Schmitt, H., Frey, L., Ryssel, H., Rommel, M., Lehrer, C., UV Nanoimprint Materials: Surface Energies, Residual Layers, and Imprint Quality, *J. Vac. Sci. Technol. B* 2007, **25**, 785–790.

1518. Andreozzi, R., Caprio, V., Insola, A., Marotta, R., Advanced Oxidation Processes (AOP) for Water Purification and Recovery, *Catal. Today* 1999, **53**, 51–59.

1519. Legrini, O., Oliveros, E., Braun, A. M., Photochemical Processes for Water Treatment, *Chem. Rev.* 1993, **93**, 671–698.

1520. Burrows, H. D., Canle, M., Santaballa, J. A., Steenken, S., Reaction Pathways and Mechanisms of Photodegradation of Pesticides, *J. Photochem. Photobiol. B* 2002, **67**, 71–108.

1521. Fox, M. A., Organic Heterogeneous Photocatalysis – Chemical Conversions Sensitized by Irradiated Semiconductors, *Acc. Chem. Res.* 1983, **16**, 314–321.

1522. Linsebigler, A. L., Lu, G. Q., Yates, J. T., Photocatalysis on $TiO_2$ Surfaces – Principles, Mechanisms, and Selected Results, *Chem. Rev.* 1995, **95**, 735–758.

1523. Hoffmann, M. R., Martin, S. T., Choi, W. Y., Bahnemann, D. W., Environmental Applications of Semiconductor Photocatalysis, *Chem. Rev.* 1995, **95**, 69–96.

1524. Kalyanasundaram, K., Grätzel, M., Applications of Functionalized Transition Metal Complexes in Photonic and Optoelectronic Devices, *Coord. Chem. Rev.* 1998, **177**, 347–414.

1525. Fox, M. A., Selective Formation of Organic Compounds by Photoelectrosynthesis at Semiconductor Particles, *Top. Curr. Chem.* 1987, **142**, 71–99.

1526. Koval, C. A., Howard, J. N., Electron-transfer at Semiconductor Electrode Liquid Electrolyte Interfaces, *Chem. Rev.* 1992, **92**, 411–433.

1527. Zepp, R. G., Faust, B. C., Hoigne, J., Hydroxyl Radical Formation in Aqueous Reactions (pH 3–8) of Iron(II) with Hydrogen Peroxide – The Photo-Fenton Reaction, *Environ. Sci. Technol.* 1992, **26**, 313–319.

1528. Safarzadeh-Amiri, A., Bolton, J. R., Cater, S. R., Ferrioxalate-mediated Photodegradation of Organic Pollutants in Contaminated Water, *Water Res.* 1997, **31**, 787–798.

1529. Fallmann, H., Krutzler, T., Bauer, R., Malato, S., Blanco, J., Applicability of the Photo-Fenton Method for Treating Water Containing Pesticides, *Catal. Today* 1999, **54**, 309–319.

1530. O'Connor, J. M., Friese, S. J., Rodgers, B. L., A Transition-metal-catalyzed Enediyne Cycloaromatization, *J. Am. Chem. Soc.* 2005, **127**, 16342–16343.

1531. Branchadell, V., Crevisy, C., Gree, R., From Allylic Alcohols to Aldols by Using Iron Carbonyls as Catalysts: Computational Study on a Novel Tandem Isomerization-Aldolization Reaction, *Chem. Eur. J.* 2004, **10**, 5795–5803.

1532. Bastard, G., Brum, J. A., Ferreira, R., Electronic States in Semiconductor Heterostructures, *Solid State Phys., Adv. Res. Appl.* 1991, **44**, 229–415.

1533. Hagfeldt, A., Grätzel, M., Molecular Photovoltaic, *Acc. Chem. Res.* 2000, **33**, 269–277.

1534. Henglein, A.,Catalysis of Photochemical Reactions by Colloidal Semiconductors, *Pure Appl. Chem.* 1984, **56**, 1215–1224.

1535. Peng, X. G., Manna, L., Yang, W. D., Wickham, J., Scher, E., Kadavanich, A., Alivisatos, A. P., Shape Control of CdSe Nanocrystals, *Nature* 2000, **404**, 59–61.

1536. Hoffman, A. J., Carraway, E. R., Hoffmann, M. R., Photocatalytic Production of $H_2O_2$ and Organic Peroxides on Quantum-sized Semiconductor Colloids, *Environ. Sci. Technol.* 1994, **28**, 776–785.

1537. Alivisatos, A. P., Semiconductor Clusters, Nanocrystals, and Quantum Dots, *Science* 1996, **271**, 933–937.

1538. Dayal, S., Burda, C., One- and Two-photon Induced QD-based Energy Transfer and the Influence of Multiple QD Excitations, *Photochem. Photobiol. Sci.* 2008, **7**, 605–613.

1539. Burda, C., Chen, X. B., Narayanan, R., El-Sayed, M. A., Chemistry and Properties of Nanocrystals of Different Shapes, *Chem. Rev.* 2005, **105**, 1025–1102.

1540. Grätzel, M., Photoelectrochemical Cells, *Nature* 2001, **414**, 338–344.

1541. Brabec, C. J., Organic Photovoltaics: Technology and Market, *Solar Energy Mater.* 2004, **83**, 273–292.

1542. Green, M. A., *Third Generation Photovoltaics: Advanced Solar Energy Conversion*, Springer, New York, 2003.

1543. Lewis, N. S., Toward Cost-Effective Solar Energy Use, *Science* 2007, **315**, 798–801.

1544. Crabtree, G. W., Lewis, N. S., Solar Energy Conversion, *Phys. Today* 2007, **60**, 37–42.

1545. Grätzel, M., Dye-sensitized Solar Cells, *J. Photochem. Photobiol. C* 2003, **4**, 145–153.

1546. Paci, I., Johnson, J. C., Chen, X. D., Rana, G., Popovic, D., David, D. E., Nozik, A. J., Ratner, M. A., Michl, J., Singlet Fission for Dye-sensitized Solar Cells: Can a Suitable Sensitizer be Found? *J. Am. Chem. Soc.* 2006, **128**, 16546–16553.

1547. Tadaaki, T., *Photographic Sensitivity: Theory and Mechanisms*, Oxford University Press, Oxford, 1995.

1548. Borsenberger, P. M., Weiss, D. S., *Organic Photoreceptors for Xerography*, Marcel Dekker, New York, 1998.

1549. Petritz, R. L., Theory of Photoconductivity in Semiconductor Films, *Phys. Rev.* 1956, **104**, 1508–1516.

1550. Law, K. Y., Organic Photoconductive Materials – Recent Trends and Developments, *Chem. Rev.* 1993, **93**, 449–486.

1551. Loutfy, R. O., Hor, A. M., Hsiao, C. K., Baranyi, G., Kazmaier, P., Organic Photoconductive Materials, *Pure Appl. Chem.* 1988, **60**, 1047–1054.

1552. Fox, M. A., Chen, C. C., Mechanistic Features of the Semiconductor Photocatalyzed Olefin-to-Carbonyl Oxidative Cleavage, *J. Am. Chem. Soc.* 1981, **103**, 6757–6759.

1553. Harada, H., Sakata, T., Ueda, T., Effect of Semiconductor on Photocatalytic Decomposition of Lactic Acid, *J. Am. Chem. Soc.* 1985, **107**, 1773–1774.

1554. Mills, A., Morris, S., Davies, R., Photomineralisation of 4-Chlorophenol Sensitized by Titanium Dioxide – A Study of the Intermediates, *J. Photochem. Photobiol. A* 1993, **70**, 183–191.

1555. Mills, A., Morris, S., Photomineralization of 4-Chlorophenol Sensitized by Titanium Dioxide – A Study of the Initial Kinetics of Carbon Dioxide Photogeneration, *J. Photochem. Photobiol. A* 1993, **71**, 75–83.

1556. Pavlik, J. W., Tantayanon, S., Photocatalytic Oxidations of Lactams and *N*-Acylamines, *J. Am. Chem. Soc.* 1981, **103**, 6755–6757.

1557. Hopfner, M., Weiss, H., Meissner, D., Heinemann, F. W., Kisch, H., Semiconductor Photocatalysis Type B: Synthesis of Unsaturated α-Amino Esters from Imines and Olefins Photocatalyzed by Silica-supported Cadmium Sulfide, *Photochem. Photobiol. Sci.* 2002, 696–703.

1558. Rinderhagen, H., Mattay, J., Synthetic Applications in Radical/Radical Cationic Cascade Reactions, *Chem. Eur. J.* 2004, **10**, 851–874.

1559. Li, S. J., Wu, F. P., Li, M. Z., Wang, E. J., Host/Guest Complex of Me-β-CD/2,2-Dimethoxy-2-phenyl Acetophenone for Initiation of Aqueous Photopolymerization: Kinetics and Mechanism, Polymer 2005, **46**, 11934–11939.

1560. Corey, E. J., Chang, X.-M., *The Logic of Chemical Synthesis*, John Wiley and Sons, Inc., New York, 1995.

1561. *NIST Standard Reference Database No. 69*, National Institute of Standards and Technology, Gaithersburg, MD, 2005, http://webbook.nist.gov.

1562. Mirkin, V. I., Glossary of Terms Used in Theoretical Organic Chemistry, *Pure Appl. Chem.* 1999, **71**, 1919–1981.

1563. Muller, P., Glossary of Terms Used in Physical Organic Chemistry, *Pure Appl. Chem.* 1994, **66**, 1077–1184.

1564. Mohr, P. J., Taylor, B. N., Newell, D. B., CODATA Recommended Values of the Fundamental Physical Constants: 2006, *Rev. Mod. Phys.* 2008, **80**, 633–730.

1565. *Landolt-Börnstein, Zahlenwerte und Funktionen*, Springer, Berlin, 1969, Transportphänomene, Vol. 5a.

1566. Nikam, P. S., Shirsat, L. N., Hasan, M., Density and Viscosity Studies, *J. Chem. Eng. Data* 1998, **43**, 732–737.

1567. Aminabhavi, T. M., Patil, V. B., Aralaguppi, M. I., Phayde, H. T.S., Density, Viscosity, and Refractive Index of the Binary Mixtures of Cyclohexane with Hexane, Heptane, Octane, Nonane, and Decane, *J. Chem. Eng. Data* 1996, **41**, 521–525.

1568. Hien, N. Q., Ponter, A. P., Peier, W., Density and Viscosity of Carbon Tetrachloride Solutions, *J. Chem. Eng. Data* 1978, **23**, 54–55.

1569. Joshi, S. S., Aminabhavi, T. M., Shukla, S. S., Densities and Viscosities of Binary Liquid Mixtures of Anisole with Methanol and Benzene, *J. Chem. Eng. Data* 1990, **35**, 187–189.

1570. Harris, K. R., Malhotra, R., Woolf, L. A., Temperature and Density Dependence of the Viscosity of Octane and Toluene, *J. Chem. Eng. Data* 1997, **42**, 1254–1260.

1571. Magazu, S., Migliardo, F., Malomuzh, N. P., Blazhnov, I. V., Theoretical and Experimental Models on Viscosity: I. Glycerol, *J. Phys. Chem. B* 2007, **111**, 9563–9570.

1572. Viswanath, D. S., Natarajan, G., *Data Book on the Viscosity of Liquids*, Hemisphere, New York, 1989.

1573. Streitwieser, A., *Molecular Orbital Theory for Organic Chemists*, John Wiley & Sons, Inc., New York, 1961.

1574. Klemp, D., Nickel, B., Relative Quantum Yield of the $S_2 \rightarrow S_1$ Fluorescence from Azulene, *Chem. Phys. Lett.* 1986, **130**, 493–497.

# Index

Printed and bound by CPI Group (UK) Ltd, Croydon, CR0 4YY

27/10/2024

14580201-0003